Earth Accretionary Systems in Space and Time

The Geological Society of London
Books Editorial Committee

Chief Editor
BOB PANKHURST (UK)

Society Books Editors
JOHN GREGORY (UK)
JIM GRIFFITHS (UK)
JOHN HOWE (UK)
PHIL LEAT (UK)
NICK ROBINS (UK)
JONATHAN TURNER (UK)

Society Books Advisors
MIKE BROWN (USA)
ERIC BUFFETAUT (FRANCE)
JONATHAN CRAIG (ITALY)
RETO GIERÉ (GERMANY)
TOM MCCANN (GERMANY)
DOUG STEAD (CANADA)
RANDELL STEPHENSON (NETHERLANDS)

Geological Society books refereeing procedures

The Society makes every effort to ensure that the scientific and production quality of its books matches that of its journals. Since 1997, all book proposals have been refereed by specialist reviewers as well as by the Society's Books Editorial Committee. If the referees identify weaknesses in the proposal, these must be addressed before the proposal is accepted.

Once the book is accepted, the Society Book Editors ensure that the volume editors follow strict guidelines on refereeing and quality control. We insist that individual papers can only be accepted after satisfactory review by two independent referees. The questions on the review forms are similar to those for *Journal of the Geological Society*. The referees' forms and comments must be available to the Society's Book Editors on request.

Although many of the books result from meetings, the editors are expected to commission papers that were not presented at the meeting to ensure that the book provides a balanced coverage of the subject. Being accepted for presentation at the meeting does not guarantee inclusion in the book.

More information about submitting a proposal and producing a book for the Society can be found on its web site: www.geolsoc.org.uk.

It is recommended that reference to all or part of this book should be made in one of the following ways:

CAWOOD, P. A. & KRÖNER, A. (eds) 2009. *Earth Accretionary Systems in Space and Time*. Geological Society, London, Special Publications, **318**.

CLIFT, P. D., SCHOUTEN, H. & VANNUCCHI, P. 2009. Arc–continent collisions, sediment recycling and the maintenance of the continental crust. *In*: CAWOOD, P. A. & KRÖNER, A. (eds) 2009. *Earth Accretionary Systems in Space and Time*. Geological Society, London, Special Publications, **318**, 75–103.

GEOLOGICAL SOCIETY SPECIAL PUBLICATION NO. 318

Earth Accretionary Systems in Space and Time

EDITED BY

P. A. CAWOOD
University of Western Australia, Australia

and

A. KRÖNER
Institut für Geowissenschaften, Germany

2009
Published by
The Geological Society
London

THE GEOLOGICAL SOCIETY

The Geological Society of London (GSL) was founded in 1807. It is the oldest national geological society in the world and the largest in Europe. It was incorporated under Royal Charter in 1825 and is Registered Charity 210161.

The Society is the UK national learned and professional society for geology with a worldwide Fellowship (FGS) of over 9000. The Society has the power to confer Chartered status on suitably qualified Fellows, and about 2000 of the Fellowship carry the title (CGeol). Chartered Geologists may also obtain the equivalent European title, European Geologist (EurGeol). One fifth of the Society's fellowship resides outside the UK. To find out more about the Society, log on to www.geolsoc.org.uk.

The Geological Society Publishing House (Bath, UK) produces the Society's international journals and books, and acts as European distributor for selected publications of the American Association of Petroleum Geologists (AAPG), the Indonesian Petroleum Association (IPA), the Geological Society of America (GSA), the Society for Sedimentary Geology (SEPM) and the Geologists' Association (GA). Joint marketing agreements ensure that GSL Fellows may purchase these societies' publications at a discount. The Society's online bookshop (accessible from www.geolsoc.org.uk) offers secure book purchasing with your credit or debit card.

To find out about joining the Society and benefiting from substantial discounts on publications of GSL and other societies worldwide, consult www.geolsoc.org.uk, or contact the Fellowship Department at: The Geological Society, Burlington House, Piccadilly, London W1J 0BG: Tel. +44 (0)20 7434 9944; Fax +44 (0)20 7439 8975; E-mail: enquiries@geolsoc.org.uk.

For information about the Society's meetings, consult *Events* on www.geolsoc.org.uk. To find out more about the Society's Corporate Affiliates Scheme, write to enquiries@geolsoc.org.uk.

Published by The Geological Society from:
The Geological Society Publishing House, Unit 7, Brassmill Enterprise Centre, Brassmill Lane, Bath BA1 3JN, UK

(*Orders*: Tel. +44 (0)1225 445046, Fax +44 (0)1225 442836)
Online bookshop: www.geolsoc.org.uk/bookshop

The publishers make no representation, express or implied, with regard to the accuracy of the information contained in this book and cannot accept any legal responsibility for any errors or omissions that may be made.

© The Geological Society of London 2009. All rights reserved. No reproduction, copy or transmission of this publication may be made without written permission. No paragraph of this publication may be reproduced, copied or transmitted save with the provisions of the Copyright Licensing Agency, 90 Tottenham Court Road, London W1P 9HE. Users registered with the Copyright Clearance Center, 27 Congress Street, Salem, MA 01970, USA: the item-fee code for this publication is 0305-8719/09/$15.00.

British Library Cataloguing in Publication Data

A catalogue record for this book is available from the British Library.
ISBN 978-1-86239-278-6

Typeset by Techset Composition Ltd., Salisbury, UK
Printed by MPG Books Ltd., Bodmin, UK

Distributors

North America
For trade and institutional orders:
The Geological Society, c/o AIDC, 82 Winter Sport Lane, Williston, VT 05495, USA
Orders: Tel. +1 800-972-9892
            Fax +1 802-864-7626
            E-mail: gsl.orders@aidcvt.com

For individual and corporate orders:
AAPG Bookstore, PO Box 979, Tulsa, OK 74101-0979, USA
Orders: Tel. +1 918-584-2555
            Fax +1 918-560-2652
            E-mail: bookstore@aapg.org
            Website: http://bookstore.aapg.org

India
Affiliated East-West Press Private Ltd, Marketing Division, G-1/16 Ansari Road, Darya Ganj, New Delhi 110 002, India
Orders: Tel. +91 11 2327-9113/2326-4180
            Fax +91 11 2326-0538
            E-mail: affiliat@vsnl.com

Contents

Preface	vii
CAWOOD, P. A., KRÖNER, A., COLLINS, W. J., KUSKY, T. M., MOONEY, W. D. & WINDLEY, B. F. Accretionary orogens through Earth history	1
BROWN, M. Metamorphic patterns in orogenic systems and the geological record	37
CLIFT, P. D., SCHOUTEN, H. & VANNUCCHI, P. Arc–continent collisions, sediment recycling and the maintenance of the continental crust	75
SCHOLL, D. W. & VON HUENE, R. Implications of estimated magmatic additions and recycling losses at the subduction zones of accretionary (non-collisional) and collisional (suturing) orogens	105
NUTMAN, A. P., BENNETT, V. C., FRIEND, C. R. L., JENNER, F., YUSHENG, W. & DUNYI, L. Eoarchaen crustal growth in West Greenland (Itsaq Gneiss Complex) and in northeastern China (Anshan area): review and synthesis	127
POLAT, A., KERRICH, R. & WINDLEY, B. Archean crustal growth processes in southern West Greenland and the southern Superior Province: geodynamic and magmatic constraints	155
ST-ONGE, M. R., VAN GOOL, J. A. M., GARDE, A. A. & SCOTT, D. J. Correlation of Archaean and Palaeoproterozoic units between northeastern Canada and western Greenland: constraining the pre-collisional upper plate accretionary history of the Trans-Hudson orogen	193
LAHTINEN, R., KORJA, A., NIRONEN, M. & HEIKKINEN, P. Palaeoproterozoic accretionary processes in Fennoscandia	237
SNYDER, D. B., PILKINGTON, M., CLOWES, R. M. & COOK, F. A. The underestimated Proterozoic component of the Canadian Cordillera accretionary margin	257
COLPRON, M. & NELSON, J. L. A Palaeozoic Northwest Passage: incursion of Caledonian, Baltican and Siberian terranes into eastern Panthalassa, and the early evolution of the North American Cordillera	273
TIZZARD, A. M., JOHNSTON, S. T. & HEAMAN, L. M. Arc imbrication during thick-skinned collision within the northern Cordilleran accretionary orogen, Yukon, Canada	309
FOSTER, D. A., GRAY, D. R., SPAGGIARI, C., KAMENOV, G. & BIERLEIN, F. P. Palaeozoic Lachlan orogen, Australia; accretion and construction of continental crust in a marginal ocean setting: isotopic evidence from Cambrian metavolcanic rocks	329
HALL, R. The Eurasian SE Asian Margin as a modern example of an accretionary orogen	351
MORLEY, C. K. Evolution from an oblique subduction back-arc mobile belt to a highly oblique collisional margin: the Cenozoic tectonic development of Thailand and eastern Myanmar	373
Index	405

Preface

Accretionary systems are the result of plate tectonics and form at sites of subduction of oceanic lithosphere. They consist of magmatic arc systems along with material accreted from the downgoing plate and eroded from the upper plate. These long-lived systems have contributed significantly to crustal growth through Earth history and are the most important 'factories' for generating, recycling and maturing continental crust (e.g. Condie 2007; Foster *et al.*). These systems received less attention in the past than shorter-lived collisional systems resulting from continental amalgamation.

In 2003 a group of international scientists interested in accretionary systems established a forum for discussion entitled ERAS (EaRth Accretionary Systems in space and time), following the philosophy that an integrated, multi-disciplinary and comprehensive programme of research in selected accretionary systems of all ages will provide a common framework to better understand their development. Recognition of the importance of accretionary systems has been hindered by the lack of a unifying model, with different possible evolutionary paths, to explain their evolution, or recognition of a common suite of processes that operate in many of these systems. The first field workshop of ERAS was held in Taipei, Taiwan, in May 2004, organized by Bor-ming Jahn, followed by a field excursion to the Coastal Range on the west and SW coast of Taiwan to view a modern accretionary orogen as it actively undergoes arc–continent collision. ERAS was formally established in 2005 as a 5 year Research Project under the International Lithosphere Program (ILP) as Task Force 1 and has since organized thematic sessions at several international conferences. A second field workshop was held in Kochi, Japan, in September 2006, organized by Kimura Gaku, Yukio Isoztaki and M. Santosh, and sponsored by the Japan Society of Promoting Sciences, followed by a field trip to central Shikoku and Inuyama to study aspects of ocean-floor stratigraphy and accretion tectonics.

The first four papers of the volume discuss general aspects of accretionary systems, and the following 10 contributions deal with specific terranes or orogenic belts, beginning with the early Archaean in West Greenland and ending with the Cenozoic in SE Asia. **Cawood *et al.*** provide an overview of accretionary systems through Earth history, defining types of accretionary orogens, discussing driving mechanisms, and emphasizing that these systems have contributed significantly to the growth of continental lithosphere. **Brown** reviews regional metamorphic processes in orogenic systems and emphasizes the complexity of metamorphic assemblages that may originally have evolved in accretionary systems but were subsequently overprinted during collisional orogeny, the final fate of many accretionary belts. **Clift *et al.*** discuss sediment recycling and crustal loss in subduction zones and argue that net crustal growth predominantly occurs through accretion of oceanic arcs to passive continental margins. **Scholl & von Huene** argue that volumetrically small or missing accretionary masses along modern subduction zones are due to subduction erosion and sediment subduction, and that recycling losses of lower plate crust during plate convergence may lead to disappearance of geological evidence for accretionary processes.

The regional contributions begin with an account by **Nutman *et al.*** on what is probably the oldest preserved accretionary system in West Greenland, the 3.87–3.60 Ga Itsaq Gneiss Complex. These authors relate the episodic formation of a voluminous tonalite–trondhjemite–granodiorite gneissic suite to an Eoarchaean subduction zone with short-lived episodes of mantle wedge melting and subsequent melting of subducted crust. **Polat *et al.*** compare Archaean crustal growth processes in southern West Greenland and the Superior Province of Canada, and use petrological and geochemical arguments to suggest that these terranes constitute large subduction–accretion complexes formed during Phanerozoic-style plate convergence. **St. Onge *et al.*** using a different approach, also compare Archaean and Palaeoproterozoic tectonic processes in West Greenland and northeastern Canada, and develop a generalized evolutionary scenario for the period 2.7–1.8 Ga on the basis of tectonostratigraphic, structural and age data. Their model of crustal accretion during the growth of northeastern Laurentia in the Palaeoproterozoic may be comparable with the growth of the upper plate Asian continent prior to collision with India. **Lahtinen *et al.*** describe one of the best documented examples of later crustal growth and accretion in the Palaeoproterozoic from the Svecofennian orogen in Scandinavia. They use published geological and geophysical data to document magmatic growth episodes between 2.1 and 1.8 Ga with additions of *c.* 2.1–1.8 Ga microcontinents and juvenile arcs and major Andean-type vertical magmatic additions at 1.9–1.8 Ga.

This was followed by tectonic accretion and reworking of the older rocks at 1.7 Ga. **Snyder et al.** use several geophysical methods to argue that much of the continental crust underlying the Canadian Cordillera consists of a thick Proterozoic sedimentary package shed off the Canadian shield along a passive margin between 1.84 and 0.54 Ga and forming a prograding wedge. This implies that the tectonically overlying Phanerozoic accreted terranes are only a few kilometres thick.

Moving to the Phanerozoic, **Colpron & Nelson** discuss the early evolution of the North American Cordillera and argue for a Caribbean- or Scotia-style subduction system between northern Laurentia and Siberia in the mid-Palaeozoic. They postulate that upper mantle flow out of the shrinking Iapetus–Rheic oceans opened a mid-Palaeozoic 'gateway' between Laurentia and Siberia, which progressively developed and led to propagation of subduction along western Laurentia. **Tizzard et al.** discuss arc imbrication and thick-skinned collision of the oceanic Stikinia arc terrane in Yukon, Canada, as it is accreted into the Cordilleran orogen during development of a Jurassic crustal-scale shear zone. **Foster et al.** review the tectonic history of the Australian Lachlan orogen and provide geochemical data for a Cambrian marginal oceanic basin that formed the basement for a thick Palaeozoic turbidite fan.

The last two papers of the volume provide examples of accretionary processes along the southeastern margin of Eurasia. **Hall** summarizes the evolution of the Indonesian Archipelago and the Philippines, which consist of a continental core of blocks rifted off Gondwana and surrounded by subduction zones for much of the Mesozoic and Cenozoic. This is a mountain belt in the process of formation and serves as a modern example of a complex accretionary orogen exemplifying episodic crustal growth.

Finally, **Morley** portrays the Tertiary tectonic evolution of Thailand and eastern Myanmar from an oblique subduction back-arc orogen to a highly oblique collisional margin resulting from collision of India with the Burma block.

The editors are grateful to the following, who kindly reviewed the manuscripts in this volume: J. Ali, C. Barnes, S. Barr, B. Bingen, D. Bradley, D. Brown, P. Charusiri, S. Daly, G. Ernst, C. Fergusson, I. Fitzsimons, A. Garde, G. Gehrels, J. Hall, S. Johnson, R. Korsch, I. Metcalfe, B. Murphy, D. Nelson, O. Oncken, G. Ross, T. Rivers, H. Smithies, D. Tappin, M. van Kranendonk, C. van Staal and G. Zhao.

Reference

CONDIE, K. C. 2007. Accretionary orogens in space and time. *In*: HATCHER, R. D., JR, CARLSON, M. P., MCBRIDE, J. H. & CATALÁN, J. R. (eds) *4-D Framework of Continental Crust*. Geological Society of America, Memoirs, **200**, 1–14.

A. KRÖNER & P. A. CAWOOD

Accretionary orogens through Earth history

PETER A. CAWOOD[1]*, ALFRED KRÖNER[2], WILLIAM J. COLLINS[3], TIMOTHY M. KUSKY[4], WALTER D. MOONEY[5] & BRIAN F. WINDLEY[6]

[1]*School of Earth and Environment, University of Western Australia, 35 Stirling Highway, Crawley, WA 6009, Australia*

[2]*Institut für Geowissenschaften, Universität Mainz, 55099 Mainz, Germany*

[3]*School of Earth Sciences, James Cook University, Townsville, Qld 4811, Australia*

[4]*Department of Earth and Atmospheric Sciences, St. Louis University, St. Louis, MO 63103, USA*

[5]*US Geological Survey, 345 Middlefield Road, Menlo Park, CA 94025, USA*

[6]*Department of Geology, University of Leicester, Leicester LE1 7RH, UK*

*Corresponding author (e-mail: Peter.Cawood@uwa.edu.au)

Abstract: Accretionary orogens form at intraoceanic and continental margin convergent plate boundaries. They include the supra-subduction zone forearc, magmatic arc and back-arc components. Accretionary orogens can be grouped into retreating and advancing types, based on their kinematic framework and resulting geological character. Retreating orogens (e.g. modern western Pacific) are undergoing long-term extension in response to the site of subduction of the lower plate retreating with respect to the overriding plate and are characterized by back-arc basins. Advancing orogens (e.g. Andes) develop in an environment in which the overriding plate is advancing towards the downgoing plate, resulting in the development of foreland fold and thrust belts and crustal thickening. Cratonization of accretionary orogens occurs during continuing plate convergence and requires transient coupling across the plate boundary with strain concentrated in zones of mechanical and thermal weakening such as the magmatic arc and back-arc region. Potential driving mechanisms for coupling include accretion of buoyant lithosphere (terrane accretion), flat-slab subduction, and rapid absolute upper plate motion overriding the downgoing plate. Accretionary orogens have been active throughout Earth history, extending back until at least 3.2 Ga, and potentially earlier, and provide an important constraint on the initiation of horizontal motion of lithospheric plates on Earth. They have been responsible for major growth of the continental lithosphere through the addition of juvenile magmatic products but are also major sites of consumption and reworking of continental crust through time, through sediment subduction and subduction erosion. It is probable that the rates of crustal growth and destruction are roughly equal, implying that net growth since the Archaean is effectively zero.

Classic models of orogens involve a Wilson cycle of ocean opening and closing with orogenesis related to continent–continent collision. These imply that mountain building occurs at the end of a cycle of ocean opening and closing, and marks the termination of subduction, and that the mountain belt should occupy an internal location within an assembled continent (supercontinent). The modern Alpine–Himalayan chain exemplifies the features of this model, lying between the Eurasian and colliding African and Indian plates (Fig. 1). The Palaeozoic Appalachian–Caledonian orogen (Wilson 1966; Dewey 1969), the Mesoproterozoic Grenville orogen (Gower *et al.* 1990; Hoffman 1991; Gower 1996), and the Palaeoproterozoic Trans-Hudson (Ansdell 2005), Ketilidian (Garde *et al.* 2002), Capricorn (Cawood & Tyler 2004) and Limpopo (Kröner *et al.* 1999) orogens are inferred ancient examples. Such models, however, do not explain the geological history of a significant number of orogenic belts throughout the world. Such belts lie at plate margins in which deformation, metamorphism and crustal growth took place in an environment of continuing subduction and accretion. These belts are termed accretionary orogens but have also been referred to as non-collisional or exterior orogens, Cordilleran-, Pacific-, Andean-, Miyashiro- and Altaid-type orogens, or zones of type-B subduction (Matsuda & Uyeda 1971; Crook 1974; Bally 1981; Murphy & Nance 1991; Windley 1992; Şengör 1993; Şengör & Natal'in 1996; Maruyama 1997; Ernst 2005). Accretionary

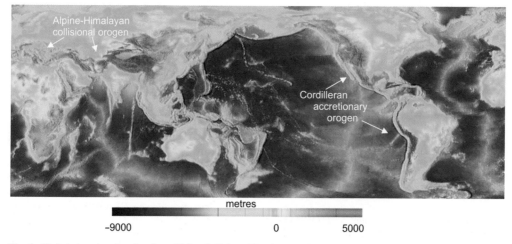

Fig. 1. Global elevation showing the collisional Alpine–Himalayan orogen (which occupies an internal location between the Eurasian and colliding African and Indian plates) and the accretionary North and South American Cordilleran orogens (which lie at plate margins involving continuing subduction). Image modified from NOAA, National Geophysical Data Center.

orogens appear to have been active throughout much of Earth history and constitute major sites of continental growth (Cawood et al. 2006). The accretionary orogens of the western and northern Pacific extending from Indonesia via the Philippines and Japan to Alaska, and the North and South American Cordillera are archetypical modern examples, with ancient examples represented by the Phanerozoic Terra Australis and Central Asian orogens, the Proterozoic orogens of the Avalon–Cadomian belt of the North Atlantic borderlands, Birimian of West Africa, Svecofennian of Finland and Sweden, Cadomian of western Europe, Mazatzal–Yavapai in southwestern USA, and the Arabian–Nubian Shield, and Archaean greenstone terranes (Windley 1992; Kusky & Polat 1999; Karlstrom et al. 2001; Johnson & Woldehaimanot 2003; Cawood 2005; Kröner et al. 2008; Murphy & Nance 1991).

Accretionary orogens form at sites of subduction of oceanic lithosphere. They include accretionary wedges, containing material accreted from the downgoing plate and eroded from the upper plate, island arcs, back-arcs, dismembered ophiolites, oceanic plateaux, old continental blocks, post-accretion granitic rocks and metamorphic products up to the granulite facies, exhumed high- or ultrahigh-pressure metamorphic rocks, and clastic sedimentary basins. Accretionary orogens contain significant mineral deposits (Groves & Bierlein 2007), and thus provide the mineralization potential of many countries such as Australia, Canada, Chile, Ghana, Zimbabwe, Saudi Arabia, China, Kazakhstan, Mongolia and Indonesia. All accretionary orogens are ultimately involved in a collisional phase when oceans close and plate subduction ceases, and this may lead to significant structural modification of the accreted material and to partial or complete obliteration as a result of thrusting and extensive crustal shortening (e.g. Central Asian Orogenic Belt; India–Asia collision).

Our understanding of the processes for the initiation and development of accretionary orogens is moderately well established in modern orogens such as the Andes, Japan, Indonesia and Alaska, the broad structure and evolution of which are constrained by plate kinematics, seismic profiles, tomography, field mapping, palaeontology, and isotope geochemistry and geochronology (e.g. Oncken et al. 2006a; Fuis et al. 2008). However, the processes responsible for cratonization and incorporation of accretionary orogens into continental nuclei and the mechanisms of formation of most pre-Mesozoic accretionary orogens are less well understood. In a uniformitarian sense many of the features and processes of formation of modern accretionary orogens have been rarely applied to pre-Mesozoic orogens, and hence to elucidating Earth evolution.

Our aim is to outline the broad features of accretionary orogens and discuss their implications for understanding models of crustal growth. We believe that future research into accretionary orogens will increase our understanding of tectonic processes and crustal evolution just as work on geosynclines, plate tectonics and mountain belts, terranes, and supercontinents provided a stimulus to orogenic and geological research in past decades (Kay 1951; Aubouin 1965; Wilson 1966; Dewey 1969; Coney et al. 1980; Dalziel 1991; Hoffman 1991; Moores 1991).

Classification of orogens in space and time

Orogens are major linear zones of the Earth's crust that contain variably deformed rocks that accumulated over a long period, dominantly in a marine environment, and that show distinctive distribution of sedimentary facies, deformational styles and metamorphic patterns often aligned approximately parallel to the belt. They are preserved as mountain belts through crustal thickening, magmatism and metamorphism during one or more tectonothermal events (orogenies), which are generally of short duration with respect to the overall age range of the orogen, and which ultimately stabilize and cratonize the orogen.

Codifying orogens is fraught with difficulty as each has unique characteristics. However, we believe they can be grouped within a spectrum of three end-member types: collisional, accretionary and intracratonic (Figs 2 and 3). Collisional orogens form through collision of continental lithospheric fragments, accretionary orogens form at sites of continuing oceanic plate subduction, and intracratonic orogens lie within a continent, away from an active plate margin.

Plate-tectonic models of orogenic cycles have been dominated by work on collisional orogens involving a Wilson cycle (Dewey & Spall 1975) of opening and closing of an ocean basin with deformation and metamorphism related to subduction followed by collision of continental blocks to generate mountain belts (Wilson 1966; Dewey 1969; Brown 2009). This in part reflects the historical focus, and resultant development of geological ideas, on the classic orogens of Europe and eastern North American, which all formed in a collisional setting; the Appalachian–Caledonian, Hercynian and Alpine–Himalayan systems (see Miyashiro et al. 1982). However, it has also long been recognized that the Palaeozoic to Recent history of the Circum-Pacific region, where orogeny is in progress, does not readily fit such a model and that alternative mechanisms for this type of orogenesis are required (Matsuda & Uyeda 1971; Coney 1973; Crook 1974; Packham & Leitch 1974).

Accretionary and collisional orogens (excluding aulacogens) form at sites of subduction of oceanic lithosphere and are end-members of a spectrum of orogen types (Figs 2 and 3). An early stage is represented by island arc accretion in, for example, Japan (Isozaki 1996; Maruyama 1997) and Alaska (Sisson et al. 2003). Such offshore arcs may accrete to one another and to an active continental margin, where they are incorporated into an Andean-type batholith and orogen; for example, the Coastal batholith of Peru, which engulfed the Casma volcanic arc (Petford & Atherton 1995), and the Peninsular batholith of Southern California, and elsewhere along the Cordillera of North and South America (Lee et al. 2007). Such arc-generated orogens as old as Neoarchaean have been recognized (Windley & Smith 1976; Windley & Garde 2009).

Final continental collision and termination of subduction within collisional orogens is generally preceded by an accretionary phase of subduction-related activity linked to ocean closure. Examples include a series of magmatic arcs developed within and along the margins of the Iapetus ocean of the Appalachian–Caledonian orogen (Cawood & Suhr 1992; van Staal et al. 1998) and the accreted Kohistan island arc in the western Himalaya orogen that was intruded by the Andean-style Kangdese batholith before final collision between India and Asia and closure of the Tethys ocean (Bignold & Treloar 2003); also, in the Palaeoproterozoic the Trans-Hudson orogen similarly formed during ocean closure and arc accretion events prior to collision of cratons (Lucas et al. 1999; St-Onge et al. 2009). The Indonesian island arc is currently in transition from a simple system involving underthrusting of oceanic lithosphere in the west to collision with Australian continental lithosphere in the east (Hamilton 1979; Snyder et al. 1996b). Conversely, accretionary orogens may contain accreted continental fragments such as in the Central Asian Orogenic Belt (Badarch et al. 2002; Kröner et al. 2007) and the Abas, Afif and Al-Mahfid terranes in the Arabian–Nubian Shield (Windley et al. 1996; Johnson & Woldehaimanot

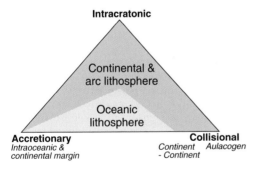

Fig. 2. Classification of orogen types into three interrelated end-members: collisional, accretionary and intracratonic. Accretionary orogens include those formed at continental margins and in intraoceanic settings, and collisional orogens include those formed from failed rifts (aulacogens) and through continent–continent collision. Accretionary and continent–continent collisional orogens lie at plate margins and form through the subduction of oceanic lithosphere, with the former forming at sites of continuing subduction and the latter at the termination of subduction.

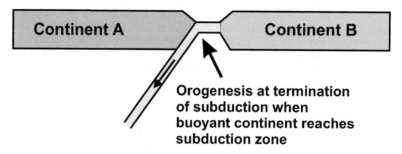

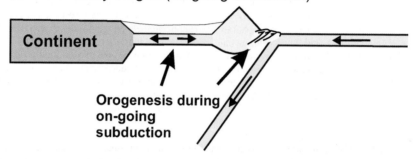

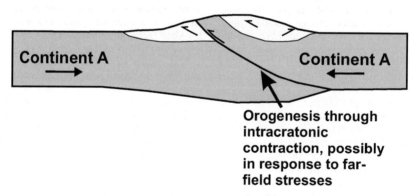

Fig. 3. Schematic cross-sections through (**a**) collisional, (**b**) accretionary and (**c**) intracratonic orogens.

2003; Stern *et al.* 2004). The long-lived accretionary Central Asian Orogenic Belt completed its history with a Himalayan-style collisional orogen in northern China (Xiao *et al.* 2003, 2004; Windley *et al.* 2007). Nevertheless, accretionary orogens stand out as an integral, well-defined group of orogens that are further characterized by significant crustal growth (Samson & Patchett 1991; Şengör & Natal'in 1996; Jahn *et al.* 2000b; Wu *et al.* 2000; Jahn 2004).

Continental extension that fails to lead to ocean opening and subsequently undergoes compression can occur at failed arms of ocean basins (aulacogens) and at intracontinental settings isolated from plate margins (Figs 2 and 3). The former represents a specific subset of collisional-type orogens that lack any evidence for the production and subsequent subduction of oceanic lithosphere, and the resultant converging continental fragments are the same as those that underwent initial extension (Hoffman

1973; Hoffman *et al.* 1974; Burke 1977; Şengör *et al.* 1978; Şengör 1995). The location of aulacogens adjacent to sites of successful ocean opening means that they are linked to subsequent sites of collisional or accretionary orogens; for example, the Oklahoma aulacogen, SE Laurentia, lies marginal to the Appalachian–Ouachita orogen. The degree of deformation and metamorphism during compressional reactivation of aulacogens is generally minimal (Hoffman *et al.* 1974). Aulacogens or rifts may also form at a high angle to the orogenic trend during collision (Şengör 1976; Şengör *et al.* 1978).

Sites of intracontinental subsidence are sites of thermal and/or rheological weakening that can be reactivated during compression, often in response to far-field stresses (see Sandiford *et al.* 2001). Examples include the late Mesoproterozoic to early Neoproterozoic successions in the North Atlantic (Cawood *et al.* 2004, 2007), and the Neoproterozoic Centralian Supergroup of Central Australia, which was deformed during the late Neoproterozoic Peterman and Palaeozoic Alice Springs orogenies (Collins & Teyssier 1989; Walter *et al.* 1995; Sandiford & Hand 1998; Hand & Sandiford 1999; Cawood & Korsch 2008). These examples are associated with the transformation of Rodinia into Gondwana, and the deformation and metamorphism of the sedimentary successions overlaps with events at the plate margin edges of these supercontinents (e.g. Cawood *et al.* 2004).

Some sites of intracratonic orogenesis are ultimately related to continental margin subduction. For example, intracratonic orogenic activity up to 1300 km inboard of the inferred plate margin occurred in the mid-Palaeozoic and Permo-Triassic in the Terra Australis orogen in eastern Australian and South African segments, respectively (Trouw & De Wit 1999; Collins 2002*a*), in China in the Mesozoic (Kusky *et al.* 2007*b*; Li & Li 2007) and along South America in the Tertiary (e.g. Kay & Mpodozis 2002). This activity parallels the plate margin and is related to strain localization in rheologically weak back-arc lithosphere (see Hyndman *et al.* 2005), possibly associated with flat-slab subduction and, hence, is part of the accretionary orogen deformation cycle. Other sites of intracratonic deformation occur inboard of zones of continent–continent collision (Fig. 1, Tianshan) and relate to stress transmission through weak quartz-dominated continental rheologies (see Dewey *et al.* 1986).

Characteristics of accretionary orogens

Accretionary orogens comprise a range of mafic to silicic igneous rocks and their sedimentary derivatives that develop on oceanic (e.g. West Pacific) or continental (e.g. Andes and Japan) lithospheric substrates during continuing plate convergence. Magmatic arc activity is characteristically calc-alkaline in composition but ranges from low-K tholeiite to shoshonitic, depending in part on the nature of the interaction of the magma with the arc substrate (e.g. Tatsumi & Eggins 1995). Magmatic activity occurs when the subducting, dehydrating slab of oceanic lithosphere interacts with mantle in the wedge of the overriding plate, and generally initiates at a depth of around 100 km or more above the downgoing slab (Tatsumi 2005). Importantly, and in contrast to magmatic activity at mid-ocean ridges or within-plate settings, arc magmas contain up to a few per cent water and other volatile phases and, as a consequence, are highly explosive (e.g. Krakatoa). Thus the extrusive products of arc volcanism are often reworked as pyroclastic and volcaniclastic deposits into intra-arc and arc flanking basins of the adjoining forearc and back-arc. Water also influenced the crystallization of the plutonic sections of arcs producing amphibole-bearing gabbros and ultramafic rocks, as demonstrated experimentally by Müntener *et al.* (2001) and geologically by Claeson & Meurer (2004) in the Proterozoic Trans-Scandinavian arc belt of Sweden.

Accretionary orogens are variably deformed and metamorphosed by tectonothermal events, commonly in dual, parallel, high-T and high-P regimes up to granulite and eclogite facies (Miyashiro 1973*a*; Ernst 2005; Brown 2006, 2009). Deformational features include structures formed in extensional and compressive environments during steady-state convergence (arc or back-arc v. accretionary prism) that are overprinted by short regional compressive orogenic events (Kusky & Bradley 1999; Collins 2002*a*).

Still-evolving accretionary orogens, such as those around the Pacific, have long, narrow aspect ratios, but completed orogens may be as broad as long (e.g. Central Asian Orogenic Belt, Arabian–Nubian Shield, and Superior and Yilgarn provinces). However, at least with the Superior and Yilgarn provinces, this appears to reflect the subsequent tectonic events such that the original linear extent of these bodies is unknown.

Lithotectonic elements of convergent plate margin systems include an accretionary prism incorporating accreted and tectonically dismembered ocean plate strata, forearc basin and its substrate, magmatic arc and back-arc basin. Some may also incorporate accreted arcs and oceanic plateaux and slices of convergent margin assemblages that have moved along the margin through strike-slip activity.

Crustal structure in accretionary orogens

Geophysical studies of accretionary orogens, including seismic reflection, refraction, seismic tomography and teleseismic data, have provided fundamental insights into structures of these orogens and their formation. Because of pronounced lateral variations across oceanic trenches and subduction zones, studies that combine multiple seismic and non-seismic data have been the most successful at determining the deep structure (e.g. Wannamaker *et al.* 1989).

Tonanki and Nankai subduction zones (Japan)

Japanese subduction zones are among the best-studied circum-Pacific accretionary complexes, with a lithospheric structure that is considered typical of many convergent plate margins. Two crustal models derived from seismic refraction–wide-angle reflection and gravity data collected across the Nankai Trough show the geometry of the subducting oceanic crust as it descends beneath the Japan volcanic arc (Fig. 4; Sagiya & Thatcher 1999; Kodaira *et al.* 2000; Nakanishi *et al.* 2002; Wells *et al.* 2003). These models define the geometry of the thick sedimentary basins that are located between the Nankai Trough and continental Japan. The subducting lower plate is divisible into three crustal layers separated from the upper mantle by a marked velocity jump. The wedge-shaped upper plate consists of a low-seismic velocity sedimentary package and the higher velocity igneous crust of the

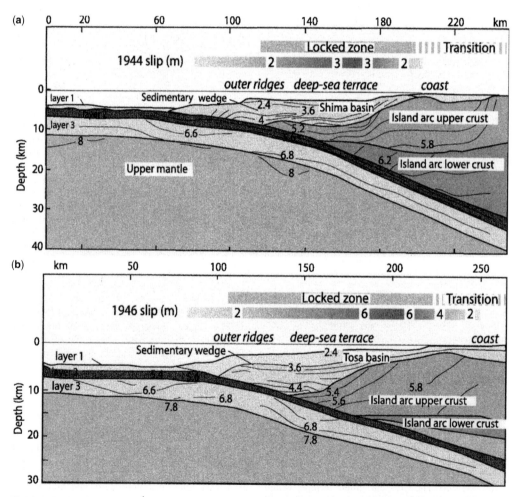

Fig. 4. Seismic velocity (km s^{-1}) structure of (**a**) the Tonanki subduction zone, site of the 1944 earthquake, and (**b**) the Nankai subduction zone, site of the 1946 earthquake. The relationship between the crustal structure and the locked zone is shown (Hyndman *et al.* 1995; Wang *et al.* 1995; Sagiya & Thatcher 1999; Kodaira *et al.* 2000; Nakanishi *et al.* 2002). The crustal structure is typical of many subduction zone complexes and includes a prominent low-seismic velocity sedimentary wedge and the higher-velocity igneous crust of the island arc (after Wells *et al.* 2003).

island arc, which is internally divisible into upper and lower crust on the basis of differences in seismic velocity.

Western margin of North America

Western North America is composed of a series of accreted oceanic domains (Coney et al. 1980; Samson et al. 1989; Fuis & Mooney 1990; Fuis 1998; Fuis et al. 2008). A detailed seismic transect from the active plate boundary at the Aleutian Trench in the Gulf of Alaska to the orogenic foreland fold-and-thrust belt on the margin of the Arctic Ocean shows a history of continental growth through magmatism, accretion and underplating (Fuis & Plafker 1991; Fuis et al. 2008; Fig. 5a). The edge of the Pacific plate (labelled 'A' in Fig. 5a and b) has velocities of 6.9 km s^{-1} and is covered by a thin upper layer of the lower oceanic crust of the Yakutat terrane with velocities of 6.1–6.4 km s^{-1}. This difference results in a structurally induced crustal doubling and contrasts with the situation inboard where the subducting oceanic crust has a normal 5–10 km thickness (Fig. 5b). Previously accreted oceanic lithosphere (B1, B2; Fig. 5b) is Mesozoic to early Cenozoic in age and contains magnetic, intermediate-velocity rocks of the Peninsular terrane, as well as interpreted regions of the Kula plate, which have velocities of 5.6–7.7 km s^{-1} at depth. The Cenozoic accretionary prism (C; Fig. 5a and b) is the Prince William terrane and the Mesozoic accretionary prism (D'; Fig. 5b) is a tectonic wedge, and includes the Chugach terrane and Border Ranges ultramafic–mafic assemblage (BRUMA; Kusky et al. 2007a). The backstop to the Mesozoic prism (E; Fig. 5b) is composed of the Peninsular and Wrangellia terranes. Near the Arctic margin the Brooks Range reveals crustal thickening attributed to the development of a foreland fold-and-thrust belt that overlies a tectonic wedge of North Slope lithosphere (Fuis et al. 1997). Crustal underplating in southern Alaska and crustal thrusting in northern Alaska overlapped in the Palaeogene and can be related to an orogenic float model in which a décollement extended northward from the subduction zone in the south to the Brooks Range in the north (Oldow et al. 1990; Fuis et al. 2008).

Southern Vancouver Island in British Columbia (Fig. 6a) contains material accreted from the subducting Juan de Fuca plate (A) (Clowes et al. 1987, 1995, 1997; Hyndman et al. 1990; Fuis & Clowes 1993). The Crescent–Siletz terrane, which is similar to the Yakutat terrane in Alaska, as well as unidentified tectonically underplated rocks and possible fragments were accreted in the Cenozoic era (B2, B3; Fig. 6a), together with a possible remnant of an oceanic plateau from the Palaeocene–Eocene. A thick wedge of Cenozoic accretionary prism lies near the toe of the overriding plate (C; Fig. 6a), whereas the Pacific Rim terrane is the Mesozoic accretionary prism (D; Fig. 6a). The Wrangellia terrane, the West Coast Plutonic Complex, and Nanaimo sediments in Georgia Strait form the backstop to the Mesozoic prism (E; Fig. 6a). An undivided lower crust is not defined, and other Cenozoic rocks include sedimentary strata on the continental shelf and Pacific Ocean basin (G; Fig. 6a). In contrast, the seismic transects in northern British Columbia and the Yukon show that the accreted Mesozoic and younger terranes are thin-skinned and constrained to the upper crust, and are underlain by a wedge of Proterozoic sedimentary rocks derived from cratonic Laurentia (Snyder et al. 2002, 2009). The deeper portions of the accreted blocks must have detached or underthrust the wedge during accretion. MacKenzie et al. (2005) reported evidence for ultrahigh-pressure garnet peridotite in the Canadian Cordillera that indicated that the thickness of the Proterozoic lithosphere was >100 km (and possibly up to 150 km).

A transect from Santa Cruz to Modesto in the San Francisco Bay region (Fig. 6b) is representative of accreted regions in northern and central California (Fuis & Mooney 1990; Page & Brocher 1993). There is no actively subducting oceanic crust, but oceanic lithosphere that accreted in the Cenozoic (B3; Fig. 6b) is preserved. The accreted material is interpreted as a 5–18 km layer at the base of the crust that can be traced seaward to the Pacific oceanic crust, containing intermediate- to high-velocity rocks. Other transects from California contain lithosphere accreted in the Mesozoic, which includes ophiolite complexes from the Great Valley, characterized as magnetic, dense, intermediate- to high-velocity rocks. The Cenozoic accretionary prism (Fig. 6b) is present as unidentified Cenozoic sedimentary rocks interfaulted with the Franciscan assemblage. The Mesozoic accretionary orogen (D', E and G; Fig. 6b) comprises the Franciscan terranes, Coast Range ophiolite, the Great Valley sequence and Sierran foothills.

Songpan–Ganzi terrane (China)

The triangular-shaped Songpan–Ganzi terrane in the central Tibetan plateau lies between the Qinling–Qilian orogen to the north and the Qiangtang terrane to the south (Fig. 7). The crust consists of a vast tract of highly deformed and locally metamorphosed Triassic deep marine sedimentary rocks interpreted as the fill of a diachronously closing remnant ocean basin (Nie et al. 1994; Ingersoll et al. 1995; Zhou & Graham 1996). This terrane formed during Jurassic deformation and

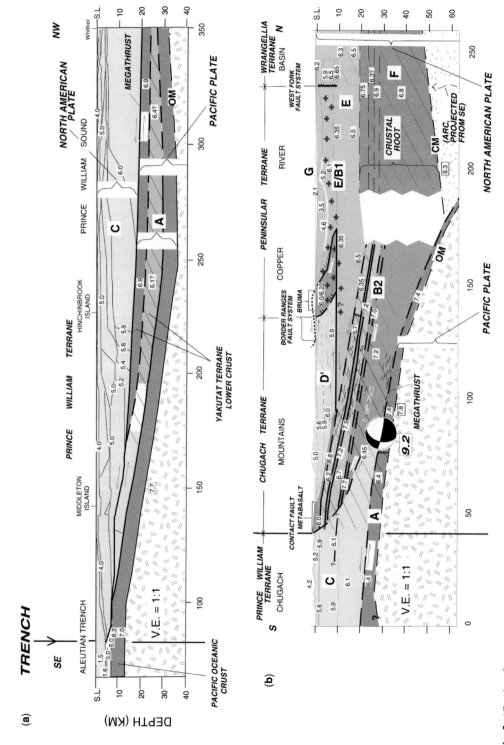

Fig. 5. (Continued).

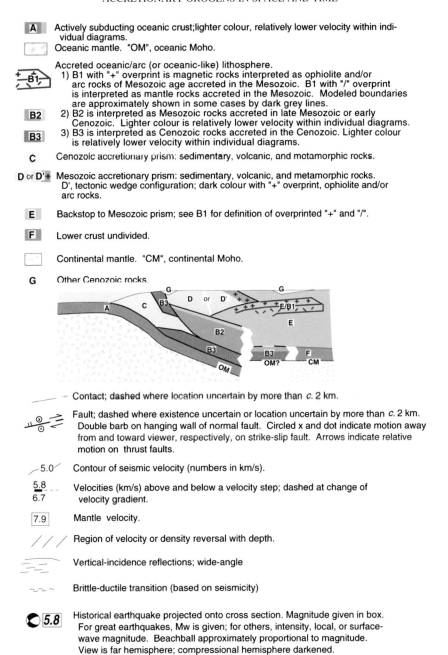

Fig. 5. Deep structure of the accretionary margin of southern Alaska, North American Cordillera (Fuis 1998). (**a**) The Pacific plate subducts at a shallow angle beneath metasedimentary rocks of the Prince William terrane (C) and underplated oceanic crust (Yakutat terrane lower crust). This active subduction zone maintains a low angle beneath the 250 km wide accretionary terrane. (**b**) The transition from the Prince William terrane (C) to the Chugach terrane (D) is marked by a pronounced transition from metasedimentary rocks to thick imbricated sheets of igneous oceanic crust. North of the Chugach terrane, the crust thickens to at least 50 km beneath the arc-related Peninsular terrane. This composite accretionary margin is highly diverse in terms of the crustal lithology. Section (b) is offset 150 km across strike from section (a).

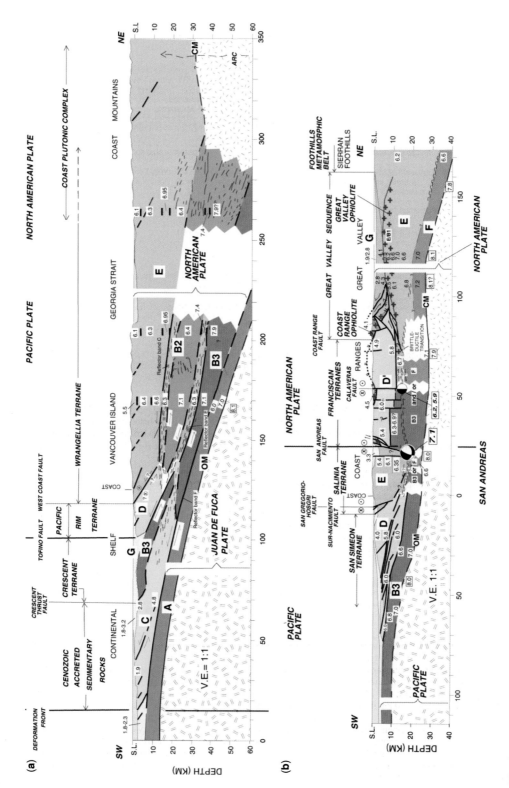

Fig. 6. (*Continued*).

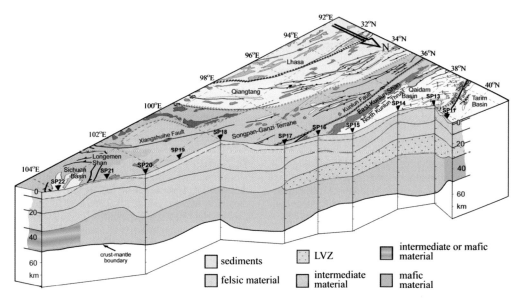

Fig. 7. Three-dimensional perspective view of the crustal structure beneath the Songpan–Ganzi accretionary orogen within a recent deep seismic transect from the southern Tarim basin to the Sichuan basin (Wang *et al.* in press). Colouring within the different layers indicates composition as derived from Poisson's ratio and P-wave velocity. Major faults and thrusts transected by the profile are qualitatively extrapolated at depth. The accreted upper and middle crust reaches a maximum thickness of 40 km beneath the Songpan–Ganzi terrane. The lower crust is 20–30 km thick and may consist of underthrust crystalline crust of continental affinity.

greenschist-facies metamorphism (Ratschbacher *et al.* 1996; Xiao *et al.* 1998) and was elevated above sea level at *c.* 20 Ma (e.g. Tapponnier *et al.* 2001). The terrane is inferred to be underlain by continental crust of the South China Block (Luo 1991). Recent seismic measurements show that the felsic upper crust (flysch? $V_p = 5.95$ km s^{-1}) is at least 10–20 km thick, and seismic velocities remain low ($V_p = 6.25$ km s^{-1}) to 40 km depth (Fig. 7; Wang *et al.* 2009). This implies that accreted material may attain a thickness of 40 km. Furthermore, the total crustal thickness is *c.* 70 km beneath the northern Songpan–Ganzi terrane (Fig. 7). The origin of the 20–30 km thick lower crust beneath this terrane is enigmatic, but the lack of surficial volcanic rocks indicates that an igneous origin is unlikely. A more plausible model would involve underthrusting of crystalline continental crust from the north and east. We note that the thickness of the lower crust is three to five times greater than the 7 km thickness of typical oceanic crust. Crustal thickness decreases towards the eastern border of the Tibetan plateau and reaches *c.* 48 km beneath the Sichuan basin (SP22, Fig. 7). Despite more than 14 km of crustal thinning, the topography remains constant across the Songpan–Ganzi terrane at an elevation of *c.* 4 km and then abruptly drops by 3.4 km from the elevated Longmen Shan into the low-lying Sichuan basin (elevation 0.6 km). Thinning of the crust along this portion of the profile is therefore mainly caused by thinning of the upper crust ($V_p = 5.95$ km s^{-1}).

Fig. 6. Cross-sections of North American Cordillera (Fuis 1998). (**a**) Accreted margin across Vancouver Island, British Columbia, showing some features similar to southern Alaska (Fig. 5b), particularly with respect to the diversity of crustal lithologies, including accreted sedimentary (C) and volcanic rocks (B3 and D) and tectonically imbricated sheets of igneous oceanic crust (B2). The total volume of igneous rocks at this margin exceeds that of sedimentary rocks. The crust thickens to about 50 km some 200 km east of the deformation front. (**b**) Central California, southern San Francisco Bay region: Santa Cruz to Sierra Nevada foothills. This accretionary margin has undergone hundreds of kilometres of lateral displacement along several faults, most prominently the San Andreas Fault. Some elements in common with the margins of southern Alaska (Fig. 5) and British Columbia (a) are the presence of (1) accreted sedimentary rocks (San Simeon and Franciscan terranes), (2) an arc terrane (Salina) and (3) underplated oceanic crust (B3). Deep geophysical data across these and other accretionary systems provide strong evidence for pronounced lithological diversity within the crust. Legend as for Figure 5.

Seismic data from Precambrian accretionary orogens

Seismic traverses across the Palaeoproterozoic accretionary orogens of the Svecofennian domain of NW Europe and the Hottah terrane of NW Canada have delineated mantle reflections dipping at about 30° from the Moho to about 100 km depths that are interpreted as fossil subduction zones (BABEL Working Group 1993b; Cook et al. 1999). Seismic reflection, refraction and geoelectric data imply the Svecofennian to be a collage of microcontinental blocks with intervening basins (Korja et al. 1993; Korja & Heikkinen 2005; Lahtinen et al. 2005, 2009). The reflection seismic data (BABEL, FIRE) revealed well-preserved pre-, syn- to post-collisional structures (e.g. a fossilized arc margin with an attached accretionary prism; BABEL Working Group, 1993a, b), whereas geochemical and petrogenetic studies suggest that the juxtaposed pieces were of Palaeoproterozoic origin. Comparative seismic reflection studies integrated with geological data for the Palaeoproterozoic Svecofennian, Scottish and Trans-Hudson orogens demonstrated in each case that 1.9–1.8 Ga lithosphere was wedged into crustal flakes that overrode Archaean margins (Snyder et al. 1996a). The Svecofennian accretionary orogen could serve as an analogue of the future accretionary-turned-collisional orogen that will be preserved when the Indonesian archipelago, with its variable size and age, is squeezed between Eurasia and Australia.

Crustal sections (Figs 4–7) cover a spectrum from active convergent plate margins to cratonized equivalents preserved in an accretionary orogen and reveal a range of processes in the development of continental crust. Actively subducting margins (Figs 4, 5 and 6a) reveal a coherent downgoing plate and an overlying forearc sedimentary wedge developed on an igneous arc basement, which at accreting margins form a backstop to offscraped sedimentary slivers. Stabilizing of the arc system occurs through underplating and accretion of oceanic material (Fig. 5, Yakutat terrane) and the progressive oceanward progression of the plate margin through accretion of trench sediments and of older arc systems (Figs 5 and 6). Termination of subduction as a result of changing plate kinematics (Fig. 6b) or continental collision and extensive crustal thickening (Fig. 7) results in final cratonization of the arc system.

Accretionary orogen types

Accretionary orogens can be grouped into two end-member types (Fig. 3), namely retreating and advancing (see Royden 1993b), based on their contrasting geological character, and modern examples from the eastern and western Pacific reflect a gross long-term kinematic framework with respect to an asthenospheric reference frame (Uyeda & Kanamori 1979; Dewey 1980; Lallemand et al. 2008). Retreating orogens are undergoing long-term extension in response to lower plate retreat (trench rollback), with respect to the overriding plate (Royden 1993a), resulting in upper plate extension, including back-arc basin opening as exemplified by the Tertiary history of the western Pacific (Taylor & Karner 1983; Leitch 1984; Schellart et al. 2006). Advancing orogens develop in an environment in which the overriding plate is advancing towards the downgoing plate at a rate equal to, or greater than, the rate of lower plate slab retreat, and this results in overall upper plate compression (Lallemand et al. 2005, 2008). For the eastern Pacific this corresponds to westward motion of the North and South American plates (Russo & Silver 1996a; Silver et al. 1998; Oncken et al. 2006a). This resulted in accretion (and strike-slip motion) of previously rifted arc and microcontinental ribbons, and the development of extensive retro-arc fold-and-thrust belts (e.g. Johnston 2001). Husson et al. (2008) argued that trench advance in South America is driven by high Andean topography and that this westward push is strong enough to shear the entire Pacific upper mantle with a surface velocity of 30 mm a^{-1}.

Advancing and retreating settings of accretionary orogens are simplified 2D representations of what is likely to be a more complex response to an overall environment of oblique convergence. Oblique accretion has played an important role in the assembly of the Cordillera in western North America (Johnston 2001; Colpron & Nelson 2006, 2009; Colpron et al. 2007), and probably also in many other orogens.

Retreating orogens

Retreating plate margins develop where the rate of rollback of the downgoing plate exceeds the rate of advance of the overriding plate, resulting in crustal extension in the latter. Rollback is driven by the negative buoyancy of the downgoing slab with respect to the underlying mantle, which in turn induces a backward sinking of the slab and retreat of the slab hinge, causing the overriding plate to extend (Elsasser 1971; Schellart & Lister 2004).

Upper plate extension leads to the development of intra-arc and arc-flanking basins culminating in rifting of the arc and development of back-arc basins (Dickinson 1995; Marsaglia 1995; Smith & Landis 1995). Many retreating orogens have multiple back-arc basins that generally, but not always, young outboard, towards the retreating plate margin. The preservation and incorporation of such basin fills within a retreating accretionary orogen is dependent on features or processes active during continuing subduction (e.g. thickness of sediment cover on the downgoing plate, rate of rollback) and the character of tectonothermal events that deform and stabilize the orogen in the rock record.

The process of rollback of the downgoing plate and the consequent development of back-arc basins is well developed in the SW Pacific. Between 82 and 52 Ma east- and NE-directed rollback of the Pacific plate by some 750 km was accommodated by opening of the New Caledonia, South Loyalty, Coral Sea and Pocklington back-arc basins (Schellart et al. 2006). Change in the relative motion of Pacific–Australia at 50 Ma resulted in subduction of the South Loyalty and Pocklington basins. This subduction was followed by two additional phases of rollback of the Pacific slab of some 650 and 400 km during opening of the South Fiji and Norfolk basins between 25 and 15 Ma and the Lau Basin from 5 to 0 Ma, respectively (Schellart et al. 2006). Slab rollback and back-arc basin extension is also argued to have played a fundamental role in the development and subsequent cratonization of the arc systems in the Terra Australis orogen in eastern Australia (Collins 2002a; Foster et al. 2009). The eastern third of Australia is composed of arc systems that developed along, and were accreted to, the rifted margin of East Gondwana following the initiation of subduction in the late Neoproterozoic (Cawood 2005). Subduction is inferred to have commenced at or near the continent–ocean boundary of East Gondwana. The width of eastern Australia and New Zealand prior to opening of the Tasman Sea was some 2000 km. Since that time, this region has undergone overall orogenic foreshortening of the order of 50%, but local foreshortening may have been considerably higher; for example, the western Lachlan segment of the Terra Australis orogen has a current width of 330 km and restored original width of between 800 and 1200 km (Gray & Foster 2006; Foster & Gray 2007). Thus, overall rollback of the proto-Pacific plate since the start of subduction towards the end of the Neoproterozoic until now is of the order of 6000 km, comprising some 4000 km during the Palaeozoic and Mesozoic that is preserved in the geological record of Australia and New Zealand and another 1800 km in the Cenozoic as documented by Schellart et al. (2006) in the SW Pacific. Rollback has not been continuous throughout this time frame and was undoubtedly interspersed with periods when rollback was either stationary or, with respect to the overriding plate, was advancing and driving periods of orogenesis (Collins 2002a). Extension was accommodated by both back-arc opening (Coney et al. 1990; Coney 1992; Fergusson & Coney 1992) and offscraping and accretion of material from the downgoing plate (Cawood 1982; Fergusson 1985).

Advancing orogens

Advancing orogens are characterized by widespread crustal shortening and uplift (Fig. 8), including the development of retroarc fold-and-thrust belts. The modern South American Cordillera is an example where deformation patterns can be placed into a plate-tectonic framework. Oncken et al. (2006b) provided a quantitative analysis of the spatial and temporal distribution of deformation in the Central Andes, and concluded that the amount and rate of shortening of the upper plate as well as lateral variability at the leading edge of the plate is primarily controlled by the difference between the upper plate velocity and the oceanic plate slab rollback velocity (see Russo & Silver 1996a; Silver et al. 1998; Schellart 2008). This allows the upper plate to override the downgoing plate, resulting in coupling and deformation across the plate boundary, including subduction erosion (Scholl & von Huene

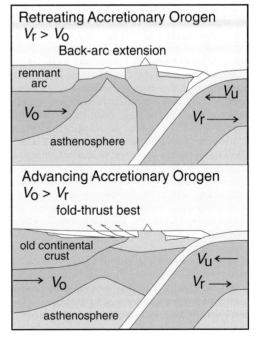

Fig. 8. Accretionary orogen types (after Cawood & Buchan 2007). For the retreating type the velocity of slab retreat (V_r) for the underriding plate (V_u) is greater than that of the overriding plate (V_o), whereas for the advancing orogen the velocity of the overriding plate is greater.

2009). Furthermore, Oncken et al. showed that the influence of differential trench–upper plate velocity is modulated by factors such as a phase of high trench sediment accumulation that reduced coupling across the plate boundary between about 45 and 33 Ma and a stage of reduced slab dip between 33 and 20 Ma that accelerated shortening. The magnitudes of post-Eocene crustal shortening across the Andes reach a maximum value of some 250–275 km in the central Andes and decrease to the north and south (Oncken et al. 2006b). The analysis of Schellart et al. (2007; see also Schellart 2008) suggests that this lateral variation in shortening may be related to along-strike variations in the rate of trench rollback, relative to the westward motion of the overriding South American plate. Schellart et al. (2007) used a global analysis of subduction zone width to show that trench migration rate is inversely related to slab width and depends on proximity to a lateral slab edge. Thus, for the 7000 km long South American trench, slab retreat is greatest at the northern and southern ends of the trench and least in the Central Andes in the vicinity of the Bolivian orocline.

The Central Andes has remained essentially stationary for the last 25 Ma and is the site of greatest crustal thickening. However, based on palaeomagnetic data, formation of the Bolivian orocline occurred in the last 10 Ma (Rousse et al. 2003), and not over the 25–50 Ma timeframe required by the model of Schellart et al. (2007). Russo & Silver (1996b) proposed that lateral variations in rollback are probably accommodated by trench-parallel flow in the mantle of the subducting slab. Seismic anisotropy beneath the Nazca Plate suggests trench-parallel mantle flow in the north and south but negligible flow in the central region, below the Bolivian orocline. Alternatively, Iaffaldano & Bunge (2008) argued that topographic load of mountain belts leads to increased frictional forces between the downgoing and overriding plates and that, specifically, uplift of the Andes over the last 10 Ma was linked to the slowdown in convergence between the Nazca and South American plates over this timeframe (see Iaffaldano et al. 2006).

Foreland fold-and-thrust belts, located inboard of the magmatic arc (Jordan 1995), are well developed in advancing orogens as a result of horizontal shortening, crustal thickening and resultant loading. They are well developed along the continental interior of the North and South American cordilleras.

Tectonic switching

Accretionary plate margins and orogenic systems can switch between phases of advance and retreat. Collins (2002a) proposed that episodes of orogenesis in the Lachlan segment of the Terra Australis orogen was driven by periodic advance of the downgoing plate through flat-slab subduction of an ocean plateau, which was otherwise undergoing long-term retreat with respect to the overriding plate. Lister et al. (2001) suggested that an accretionary orogen can undergo multiple cycles of tectonic mode switching (see also Beltrando et al. 2007). They proposed that accretion of a continental ribbon at a convergent plate margin results in shortening and burial of the overriding plate followed by a stepping out of the subduction zone beyond the accreted terrane and a new phase of rollback of the downgoing plate causing extension and exhumation in the overriding plate. This concept was extended by Lister & Forster (2006) into two types of tectonic mode switches: pull–push cycle, in which a retreating margin changes to an advancing margin, and the opposing push–pull cycle, in which the margin changes from advancing to retreating.

Sedimentary successions and accretionary orogens

Accretionary prisms

Structures formed during steady-state subduction are focused at the interface between the overriding and downgoing plates and are associated with the offscraping and underplating of material from the downgoing plate to form a subduction complex–accretionary prism (Fig. 9). Material accreted to the overriding plate can be subsequently removed and carried into the mantle through subduction erosion along the subduction channel (Scholl et al. 1980; Scholl & von Huene 2007). The subduction channel is the boundary zone between the upper and lower plates (Shreve & Cloos 1986; Beaumont et al. 1999). Variations in the strength and width of the subduction channel, which in part reflect the strength and thickness of material on the downgoing plate, can affect the behaviour of the overriding plate (De Franco et al. 2008). The interplay of advancing v. retreating accretionary plate margin with either the offscraping of material from the downgoing plate and its incorporation into an accretionary prism or the subduction (or erosion) of this material leads to the recognition of four types of plate margins (De Franco et al. 2008): accretionary prism with back-arc compression (e.g. Alaska, Sumatra, Nankai margin of Japan); erosive margin with back-arc extension (e.g. Central America, Marianas, Tonga); accretionary prism with back-arc extension (e.g. Lesser Antilles, Aegean, Makran); erosive margin with back-arc compression (e.g. Peru, Honshu margin of Japan, Kurile).

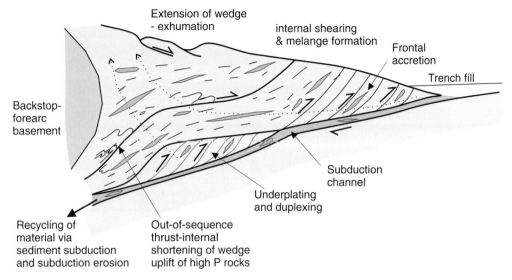

Fig. 9. Schematic section of accretionary prism showing sediment accretion through frontal accretion and basal accretion, and internal deformation of the wedge through contraction and extension (adapted from Kusky *et al.* 1997*b*). Particle paths are simplified net vectors and are drawn relative to backstop (see van Gool & Cawood 1994). Backstop is arc basement, and is composed of magmatic arc igneous rock or ophiolitic material.

Accretion of material results in the retreat of the trench axis seaward, away from the margin, and the active widening of the margin with time through progressive accretion from the downgoing plate (e.g. Seeley *et al.* 1974). According to Scholl & von Huene (2007), of the *c*. 42 000 km of active convergent margin subduction zones only some 30% have well-developed subduction complexes, and these are commonly characterized by a trench floor with a well-developed turbidite sequence and an orthogonal convergence rate of less than 40–50 km Ma^{-1}. Examples include Nankai, East Alaska, Cascadia and Sumatra–Andaman margins. Some high-latitude convergent margins (southern Chile and Alaska), contain small, young prisms developed through enhanced input of glacial age turbidites to trenches (Scholl & von Huene 2009). In southern Chile, the long-term processes on the margin are non-accreting or erosive (Bangs & Cande 1997).

A range of physical conditions extends from the trench floor to deep within the subduction complex. High fluid pressure and shearing lead to a spectrum of structures within the prism ranging from discrete thrust imbrication of relatively coherent sedimentary packages to chaotic mélange formation. Where least disrupted, the sequence displays a distinctive ocean plate stratigraphy consisting, from bottom to top, of a succession of mid-ocean ridge basalt (MORB), chert, hemipelagic mudstone, turbidite or sandstone and conglomerate (Fig. 10). This sequence records the history of sedimentation on the ocean floor as it travels from a mid-ocean ridge spreading centre to a trench. The biostratigraphy, structure, and geochemistry of this offscraped sequence has been studied in, for example, Phanerozoic circum-Pacific orogens in Japan (Isozaki *et al.* 1990; Matsuda & Isozaki 1991; Kimura & Hori 1993; Kato *et al.* 2002), California (Cowan & Page 1975; Sedlock & Isozaki 1990; Isozaki & Blake 1994), Alaska (Kusky & Bradley 1999; Kusky & Young 1999), Eastern Australia (Cawood 1982, 1984; Fergusson 1985) and New Zealand (Coombs *et al.* 1976; Mortimer 2004). Imbricated ocean plate stratigraphy is also increasingly recognized in Precambrian orogens; for example, in the 600 Ma Mona Complex of Anglesey, North Wales (Kawai *et al.* 2006, 2007; Maruyama *et al.* in press), the 2.7 Ga Point Lake greenstone belt (Kusky 1991), and possibly the 3.5 Ga chert–clastic sequence in the Archaean Pilbara craton (Kato *et al.* 1998; Kato & Nakamura 2003) and the 3.8 Ga Isua greenstone belt, West Greenland (Komiya *et al.* 1999). However, the validity of the Pilbara successions as an imbricated ocean-plate has been questioned by Van Kranendonk *et al.* (2007), who favoured a plume-related intracontinental setting. These rocks are interlayered with felsic volcanic rocks, and Williams & Collins (1990) have pointed out that they are commonly intruded by granites of the same age.

Systematic disruption of the ocean plate and trench sequence results in the production of broken

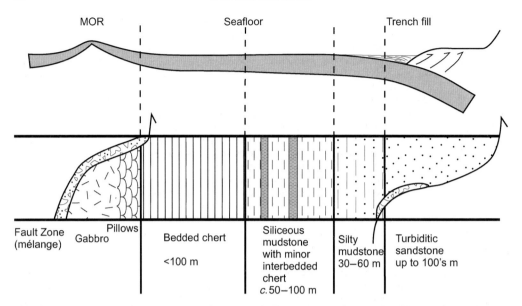

Fig. 10. Schematic section through a coherent thrust slice of offscraped ocean plate showing lithotectonic associations that develop on the plate as it moves from a mid-ocean ridge spreading centre to a trench.

formation and mélange (Greenly 1919; Hsü 1971). Results from southern Alaska and eastern Australia have shown that controls on whether chaotic mélange or coherent flysch are accreted at convergent margins may lie in the thickness of the sedimentary pile being subducted on the downgoing plate (Fergusson 1984; Kusky et al. 1997b; Kusky & Bradley 1999; Sisson et al. 2003). Downgoing plates with a thin sediment cover tend to concentrate shear strain in a thin zone that includes oceanic basement structural highs, whereas plates with thick sediment cover tend to disperse shear strains through a thick stratigraphic section with resulting less obvious deformation.

Bathymetric highs on the downgoing oceanic plate, such as guyots, may be offscraped into, and disrupt, the accretionary prism. Modern examples of seamount subduction appear to be associated with both sedimentary and tectonic disruption of the accretionary wedge (Ballance et al. 1989; Cawood 1990). Inferred seamount material, including alkali basalt and oceanic reef limestones, occurs in late Palaeozoic to Mesozoic accretionary prisms in Japan (Isozaki et al. 1990; Tatsumi et al. 1990; Isozaki 1997), including the huge late Jurassic Sorachi oceanic plateau, which was accreted in the early Cretaceous (Kimura et al. 1994). The eclogitic slab on the Sanbagawa mountains was derived from part of an oceanic plateau that was accreted, subducted and exhumed (Maruyama, pers. comm.). The Izu–Bonin arc collided with the Honshu arc in the late Cenozoic to give rise to spectacular indentation and curvature of the whole of central Japan (Soh et al. 1998).

Differentiating sedimentary successions within the accretionary orogen

Continuing convergent plate margin processes (subduction, magmatism, accretion, tectonic erosion), as well as subsequent processes involved in the incorporation and stabilization of convergent plate margin elements within continental lithosphere, destroy the original geometry so that accretionary orogens rarely contain a continuous and idealized distribution of lithotectonic elements. This structural complication may lead to uncertainty in ascertaining original settings and affinities for these elements, particularly in relation to their previous tectonothermal evolution, location of the original arc, apparent lack of accretionary prisms, occasional large dimension, and the composition and tectonic setting of the ophiolite slivers.

Many accretionary orogens, particularly the larger, less well-understood varieties, contain vast accumulations of deep-water turbidites that are tectonically intercalated with arc terranes and intruded by post-tectonic granites. The critical question here is whether these turbidites are accretionary wedge material, with ophiolitic slivers interpreted as ocean-floor lithosphere derived from the subducted plate, or whether they are back-arc basin fills, interlayered with the ophiolitic slivers representing remnants of the back-arc basin (see Foster & Gray 2000; Collins 2002a; Foster et al. 2009).

One of the major uncertainties has been the origin and timing of the heat budget within such turbidite assemblages, as many display a high-temperature (T), low-pressure (P) secondary mineral assemblage. If the turbidites are interpreted as offscraped subduction-related sequences then the mineral assemblage requires the migration of the magmatic arc, the inferred source of the heat, into the subduction complex (see Matsuda & Uyeda 1971) during oceanward propagation of the plate boundary during slab retreat. In this situation, the models that envisage accretionary orogens simply as subduction–accretion complexes require that the high-T regimes associated with arc magmatism should be superimposed upon low-T high-P regimes, including blueschist-facies terranes related to subduction accretion. Such overprints have been documented in the Chugach complex in Alaska and the New England segment of the Terra Australis orogen (Dirks et al. 1992, 1993), but many high-T, low-P metamorphic terranes bear no record of such overprinting. It is possible that the high-T metamorphism was sufficient to destroy all high-P evidence, but equally, many extensive turbidite complexes preserve original bedding and stratigraphic continuity over large areas, so it can be demonstrated that they never experienced high-P metamorphism associated with an evolving prism. In such instances, such turbidite piles could have filled back-arc basins, with the first metamorphism being associated with emplacement of igneous rocks, within either an arc (intra-arc basin) or a back-arc setting.

Another determinant of original setting of accretionary orogen turbidite assemblages is sedimentary lithotype. Accretionary prisms typically receive detritus from the adjacent arc and hence, commonly, are lithic-rich, whereas back-arc basins are likely to include detritus shed from the adjacent continental interior, and are likely to be more quartzose (Dickinson & Suczek 1979; Dickinson & Valloni 1980; Cawood 1983, 1990, 1991a, b; Dickinson 1985). Accordingly, arc–trench sandstones (forearc basins–accretionary prisms) also should contain a much higher proportion of young arc-derived zircons than old cratonic grains (Cawood et al. 1999; Cawood & Nemchin 2001). None the less, this is not always definitive, as far-travelled continental detritus can be deposited in accretionary prisms, such as the Barbados Ridge in the Lesser Antilles, which largely consists of Andean detritus shed via the Amazon (Parra et al. 1997).

The composition and structural relations of ophiolites are also probably discriminants between accretionary prisms and back-arc basins. Greenstones of accretionary prisms are typically fault-bound slivers ranging from those at the base of a relatively coherent thrust sheet (Fig. 10) to dismembered tectonic lozenges in a mélange. They can show either MORB or ocean island basalt (OIB) geochemical signatures (e.g. Cawood 1984), with the former rocks being detached from their base and occurring at the bottom of an ocean-plate sedimentary sequence (Fig. 10), where stratigraphic relations are preserved. Ocean island basalts can occur interstratified within the sedimentary sequence and may, if originating as guyots, be overlain or associated with limestone lenses. In contrast, back-arc basins are likely to contain sills and flows of basalts, which typically preserve original contact relations. Moreover, these basalts are MORB-like, but with a subtle subducted slab flux component, evident as elevated large ion lithophile element (LILE) abundances, which form the typical spiked pattern of subduction-related arc basalts on spidergrams, although this pattern is more subdued than that of arc basalts (Jenner et al. 1987; Hawkins 1994; Collins 2002b). In the 1970s and 1980s, such subtle LILE additions were commonly perceived as metasomatic effects from metamorphism, or as the products of melting mantle lithosphere enriched during a previous subduction event. However, through the Ocean Drilling Program, it became evident that these basalts were the dominant type of oceanic back-arc basins (e.g. Smellie 1994). These basalts also happen to be the most common type in retreating accretionary orogens.

The presence or absence of silicic tuff horizons is another possible discriminant of sedimentary successions in supra-subduction zone settings. Back-arc basins reflect the transitional tectonic stage between extending arc and formation of oceanic back-arc basins. They commonly receive the volcanic products associated with crustal melting during the initial stages of back-arc extension, and the explosive nature of arc magmas means that pyroclastic and volcaniclastic detritus is easily redistributed into the back-arc, and directly overlies the oceanic lithosphere and may be interstratified with, and dispersed within, any hemi-pelagic successions. In contrast, the ocean plate sequence incorporated into accretionary prisms, commonly located at least several hundred kilometres outboard from the prism, is less likely to contain volcanic layers, particularly ash flow ignimbrites. Accordingly, the presence of silicic volcanic rocks (with or without mafic counterparts), particularly in the deeper part of the stratigraphic succession, and intermittent silicic tuff horizons higher in the turbidite pile, are indicators of back-arc basin environments. It should be noted, however, that tuffs occur interbedded with cherts as part of an ocean plate stratigraphy in the Ordovician Ballantrae ophiolite in Scotland, which were then imbricated into an interpreted forearc accretionary environment (Sawaki et al. in press).

Metamorphic patterns in accretionary orogens

Miyashiro (1961, 1972, 1973b) highlighted that convergent plate margins are characterized by two regional, paired metamorphic belts of inferred similar age but contrasting mineral assemblages representing discrete $P-T$ regimes. These are a high-P belt formed under a low geothermal gradient, which lies on the oceanward side of a low-P belt formed under a high geothermal gradient (Oxburgh & Turcotte 1970, 1971). The high-P belt was equated with the zone of the accretionary prism and the low-P belt with the magmatic arc and back-arc (Miyashiro 1972, 1973b). This duality of metamorphic belts is a characteristic feature of accretionary plate margins (Ernst 2005). The age equivalence and primary across-strike setting of paired belts was, however, critically investigated by Brown (1998b) for part of the Japanese arc. He proposed that the inboard, low-P–high-T Ryoke Belt and the outboard high-P–low-T Sambagawa Belt originally lay along strike and were juxtaposed by sinistral strike-slip motion along the Median Tectonic Line. The high-T metamorphism in belts such as the Ryoke and Abukuma in Japan are, according to Brown (1998a), the result of ridge subduction and slab window formation just inboard of the trench and downgoing plate (Bradley et al. 2003). This represents an alternative mechanism to the magmatic arc and back-arc for producing high-T metamorphism. The distribution of low-P, high-T metamorphic assemblages and any associated magmatism will be limited to the site of ridge subduction, and age relationships will be diachronous, reflecting ridge–trench migration.

'Back-arc basin' orogeny

One of the more enigmatic features of accretionary orogens is the presence of peak (high-T) metamorphic assemblages during contraction (Thompson et al. 2001; Collins 2002a), which is impossible to reproduce during the structural evolution of an accretionary prism (Jamieson et al. 1998). Hyndman et al. (2005) realized that back-arcs were always regions of high heat flow, irrespective of whether the orogen was advancing or retreating. They showed that around the entire Pacific Rim, elevated heat flow and thin crust are normal in back-arcs. Even in the North American Cordillera, where high mountains, fold-and-thrust belts, and foreland basins attest to long-term crustal shortening associated with an advancing orogen, the crust of the Cordillera is only 30–35 km thick and heat flow ($c.$ 75 mW m^{-2}) is almost twice that of the adjacent craton ($c.$ 40 mW m^{-2}). The high heat flow exists because the entire lithosphere is only 50–60 km thick (see also Currie & Hyndman 2006; Currie et al. 2008). This hot, thin zone of lithospheric weakness becomes the focus of shortening during periods of increased compressive stress, and the heat is a natural consequence of shallow convection in the hydrous mantle wedge above the subducting plate (Hyndman et al. 2005). As a result, compressional features in accretionary orogens, which form above the subducting plate, develop up to 1000 km away from the accretionary prism.

Furthermore, the high heat flow and corresponding rheological weakness of the back-arc region make it the likely site for the focusing of deformation within the accretionary orogen system. Deformation in the eastern Myanmar–western Thailand region of SE Asia (Shan-Thai block) is focused within a pre-existing back-arc basin subjected to oblique strain related to the India–Asia collision (Morley 2009). Strain partitioning is characteristic of this region and is heterogeneous, with adjoining regions of cold lithosphere, corresponding to a forearc basin setting (e.g. Central Basin in Myanmar), remaining undeformed and the site of continuing sedimentation.

Cratonization and driving mechanisms of orogenesis in accretionary orogens

Conversion of convergent plate margins into stable continental crust typically involves deformation and crustal thickening during one or more tectonothermal events. This conversion occurs in an environment of continuing plate convergence such that orogenesis involves transfer and concentration of stress in the upper plate through transitory coupling across the plate boundary (Cawood & Buchan 2007). Potential mechanisms of coupling include: (1) subduction of buoyant oceanic lithosphere (flat-slab subduction); (2) accretion of buoyant lithosphere (terrane accretion); (3) plate reorganization causing an increase in convergence across the boundary (Fig. 11). The effects of flat-slab subduction (e.g. Ramos et al. 2002) and suspect terrane accretion (e.g. Maxson & Tikoff 1996) should be spatially limited to the region of either the flat slab or the accretion zone, which in turn should result in short-lived orogenesis and/or diachronous events that migrate along the convergent margin in harmony with the subducted plate movement vector. These are local mechanisms in which the effect (e.g. orogenesis) is directly linked to the cause (buoyant slab causing coupling). These mechanisms can be observed in modern orogens (Kay & Mpodozis 2001; Mann & Taira 2004) and are commonly invoked as a mechanism in the geological record (e.g. Holm et al. 2005; St-Onge et al.

(a) Flat slab subduction

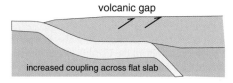

(b) Terrane accretion

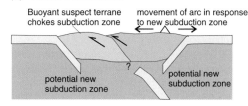

(c) Global plate kinematic reorganization

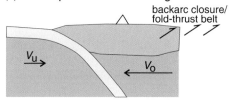

Fig. 11. Three possible modes of plate coupling that can drive orogenesis in an accretionary orogen: (**a**) flat-slab subduction; (**b**) terrane accretion; (**c**) plate reorganization leading to increased convergence.

2009). Accretion of arc and continental fragments has played a major role in the growth of SE Asia and the North American Cordillera (Coney et al. 1980; Metcalfe 1996a, b, 2002; Hall 2002, 2009). In contrast, the effects of plate reorganization are broader in scale and should extend synchronously along an orogen or plate boundary, albeit with variable effects, and reflect widespread and possibly long-term changes in orogenic character. Plate reorganization may traverse plate boundaries and be inter-orogen in extent. This reorganization is a regional mechanism in which the cause (plate reorganization) is not directly preserved at the site of its effect (orogenesis).

Crucial in establishing the potential contribution of these different coupling mechanisms to orogenesis is a detailed understanding of the spatial and temporal distribution of the tectonothermal effects of orogenic events in accretionary orogens. Synchroneity and cyclicity of accretionary orogenesis on an intra- and inter-orogen scale would suggest plate reorganization as the possible driver for orogenesis. Diachroneity, and/or orogenic events or belts restricted within an orogen would favour local events associated with terrane accretion or flat-slab subduction, or ridge subduction.

Subduction of buoyant oceanic lithosphere (flat-slab subduction)

Subduction of buoyant oceanic lithosphere will induce a flattening of the slab and can result from either the migration of young lithosphere associated with a spreading ridge or the thickened lithosphere of a hotspot (Fig. 11a; Gutscher et al. 2000; Gutscher 2002). Flat-slab subduction is currently occurring in a number of regions around the world, notably in southern Japan and South America (Ramos et al. 2002; van Hunen et al. 2002). Orogenesis driven by flat-slab subduction should be spatially limited to the region above the buoyant subducting lithosphere and will be relatively short-lived and diachronous, moving in concert with the subducting plate movement vector. Kay & Mpodozis (2001, 2002) argued that the thermal consequences of changing slab dip, combined with subduction of the Juan Fernandez Ridge hotspot track, have left a predictable magmatic and mineralization record in the Andes. Murphy et al. (1998) suggested that plume subduction led to flattening of the downgoing slab, generating 'plume-modified orogeny' (see Murphy et al. 1999, 2003; Dalziel et al. 2000). Flat-slab subduction and the resultant transitory plate coupling has been invoked as an important mechanism of orogenesis in the accretionary Lachlan orogen (Collins 2002a), in the North American Cordillera (Dickinson & Snyder 1978; Saleeby 2003), and in development of the Japanese accretionary orogen (Osozawa 1988; Underwood 1993; Isozaki 1996; Maeda & Kagami 1996; Brown 1998a). The mechanism for increased buoyancy with flat-slab subduction depends on the nature and rate of input of the thermal anomaly (Bradley et al. 2003; Kusky et al. 2003). Ridge subduction will be associated with a progressive increase in buoyancy, whereas plateau or hotspot subduction will induce a rapid change in crustal thickness and, hence, buoyancy. Subduction of plateaux and ridges has been proposed as a mechanism of orogenic growth in the Palaeoproterozoic Birimian terranes (Abouchami et al. 1990), and in the Archaean Zimbabwe craton (Kusky & Kidd 1992).

Ridge subduction is a diachronous process that typically involves a major change in plate convergence vectors between the upper plate and two subducting plates, with the change in plate convergence vectors across the subducting spreading ridge separated by a period of heating and igneous intrusion in the forearc and accretionary prism (above the slab window; Fig. 12). Such ridge–trench interaction played a major role in the development of the Tertiary North American Cordillera from Kodiak Island, Alaska, to Vancouver Island, British Columbia (Bradley et al. 2003; Sisson et al. 2003).

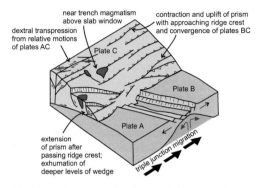

Fig. 12. Oblique view of hypothetical ridge–trench–trench triple junction, showing how structures in the upper plate will change with the passage of the triple junction, reflecting kinematic coupling between plates B and C before ridge subduction, and plates A and C after ridge subduction (adapted from Kusky et al. 1997a).

Deformation can be intense and is related not only to the plate convergence vectors (surface forces) but also to a change in dip of the subducting lithosphere as the ridge migrates along the trench (Kusky et al. 1997a; Haeussler et al. 2003; Pavlis et al. 2003; Roeske et al. 2003).

As a spreading ridge approaches a trench, progressively younger packages of offscraped sediments and volcanic rocks (Fig. 10) will be accreted to the overriding plate, and after a ridge is subducted, the accreted slabs will include progressively older packages of sediment and basalt along with the young sediments. Deformation becomes younger towards the trench and is superimposed on this complex pattern of ages of accreted sediments.

Accretion of buoyant lithosphere (terrane accretion)

If the lithosphere (oceanic or continental) is relatively thick (and buoyant) it may result in a temporary interruption to the subduction process through choking of the subduction zone, leading to the stepping out or flipping of the subduction zone (e.g. Ontong–Java Plateau; Petterson et al. 1999; Lister et al. 2001; Mann & Taira 2004). This flip will probably be associated with an interruption and/or migration of the magmatic arc (Fig. 11b). Terrane accretion was adopted by Coney et al. (1980) to explain the faulted juxtaposition of oceanic and convergent plate margin tectonostratigraphic units within the North American Cordillera. It is considered by many to constitute the main (sole) driving force for convergent margin orogenesis in that 'eventually a downgoing plate will carry continental or island arc crust into a subduction zone' (Moores & Twiss 1995, p. 212) to induce arc–arc or arc–continent collision, or terrane accretion (Dickinson 1977; Coney et al. 1980). Maxson & Tikoff (1996) argued that Cordilleran terrane accretion was the driving mechanism for the Laramide orogeny. Recent work in the Cordillera has emphasized that the number of terranes in the Cordillera is considerably less than originally envisaged by Coney and colleagues, and that many of the remaining terranes may not be suspect but are upper plate fragments that represent arcs and continental ribbons that lay outboard of, and along strike from, the Cordilleran margin (Monger & Knokleberg 1996; Johnston 2001; Colpron & Nelson 2006, 2009; Colpron et al. 2007). Seismic data across the northern Cordilleran orogen suggest that at least some of the accreted terranes are superficial with no deep crustal roots (Snyder et al. 2002, 2009) and may not be major impactors that drove orogenic events (Cawood & Buchan 2007); the terrane accretion model of orogenesis may therefore be 'suspect'.

Plate reorganization

Tectonic plate reorganization resulting from a change in the position and angular motion of Euler poles, perhaps related to termination of plate boundaries through collision or an increased spreading rate, will lead to a global readjustment in plate interactions and has been invoked as a potential cause of accretionary orogenesis (Colblentz & Richardson 1996). Vaughan (1995; see also Vaughan & Livermore 2005) proposed that pan-Pacific margin tectonic and metamorphic effects were a response to major plate reorganization associated with an increased spreading rate in the Pacific during the mid-Cretaceous (Sutherland & Hollis 2001). Cawood & Buchan (2007) highlighted evidence for deformation, mountain building and resultant crustal growth in accretionary orogens during phases of supercontinent assembly (Boger & Miller 2004; Foden et al. 2006). They undertook a detailed analysis of the timing of collisional orogenesis associated with supercontinent assembly compared with that for accretionary orogenesis along the margins of a supercontinent. They showed that age relations for assembly of Gondwana and Pangaea indicate that the timing of collisional orogenesis within the interior of the supercontinents was synchronous with subduction initiation and contractional orogenesis within the marginal Terra Australis orogen, which extended along the palaeo-Pacific margin of the these supercontinents.

Final assembly of Gondwana occurred at the end of the Neoproterozoic to early Palaeozoic, between about 590 and 510 Ma. This was coeval with a switch along the Pacific margin of the supercontinent from passive to convergent margin activity, followed by the Delamerian–Ross–Pampean orogenesis. Similarly, the final stages of assembly of the Pangaean supercontinent occurred during the end-Palaeozoic to early Mesozoic, between c. 320 and 250 Ma, and involved the accretion of Gondwana, Laurasia and Siberia. This phase of major plate boundary reorganization was accompanied by regional orogenesis along the Pacific margin of Gondwana–Pangaea (Gondwanide orogeny). The correspondence of this transitory coupling with, or immediately following, plate boundary reorganization, suggests that it may reflect plate kinematic readjustment involving increased relative convergence across the plate boundary. Cawood & Buchan (2007; see also Murphy & Nance 1991) suggested that this relationship probably reflects the global plate kinematic budget where termination of convergence during supercontinent assembly is compensated by subduction initiation and/or increased convergence along the exterior of the supercontinent. Transitory coupling across the plate boundary during subduction possibly accounts for the deformation and metamorphic pulses that develop in the accretionary orogens.

The analysis of Oncken et al. (2006b) suggests that Cenozoic orogenesis in the Andes is a response to global kinematic adjustment, in this case driven by opening of the Atlantic, resulting in an increase in westward drift of the South American plate relative to the Nazca plate (see Silver et al. 1998). Recently, Silver & Behn (2008) proposed that supercontinent collision may lead to a global loss of subduction. The data of Cawood & Buchan (2007), however, show that this concept is unlikely, at least in association with Gondwana and Pangaea assembly.

Accretionary orogens and plate tectonics; when did plate tectonics begin?

Because accretionary orogens require convergent plate margins, their appearance in the geological record heralds the initiation of horizontal plate interactions on the Earth. The question of when plate tectonics began, what criteria can be used to recognize it in the rock record and, once established, whether it was continuous, episodic and/or alternated with some alternative process are much debated (Sleep 2000; Hamilton 2003; Stern 2005; O'Neill et al. 2007; Condie & Kröner 2008). The consensus of opinion, however, is that convergent plate interaction and the recycling and subduction of material from the Earth's surface into the mantle has been active since at least 3.2–3.0 Ga (Cawood et al. 2006; Dewey 2007; Condie & Kröner 2008; Windley & Garde 2009) and possibly considerably earlier (Komiya et al. 1999; Harrison et al. 2005; Nemchin et al. 2008; Nutman et al. 2009; Polat et al. 2009). Well-constrained palaeomagnetic data demonstrate differential horizontal movements of continents in both Palaeoproterozoic and Archaean times, consistent with the lateral motion of lithospheric plates at divergent and convergent plate boundaries (Cawood et al. 2006). Well-preserved and unambiguous ophiolites associated with juvenile island-arc assemblages and modern-style accretion tectonics occur in the Palaeoproterozoic Trans-Hudson orogen of the Canadian shield (2.0 Ga Purtuniq ophiolite, Scott et al. 1992), the Svecofennian orogen of the Baltic shield (1.95 Ga Jormua ophiolite, Peltonen & Kontinen 2004), and in the Mazatzal–Yavapai orogens of southeastern Laurentia (1.73 Ga Payson ophiolite, Dann 1997). Furnes et al. (2007) proposed that ophiolite-related sheeted dykes and pillow basalts occur in the 3.8 Ga Isua greenstone belt in SW Greenland, but this has been disputed by Nutman & Friend (2007). Late Archaean mafic–ultramafic complexes in the North China craton have also been interpreted as ophiolites (Kusky et al. 2001; Kusky 2004), but have also been disputed (Zhao et al. 2007). Paired high-P–low-T and high-T–low-P tectonothermal environments (Miyashiro 1961, 1972, 1973b), requiring plate subduction, have been recognized as far back as the Neoarchaean (Brown 2006, 2009).

Inferred arc-related assemblages (magmatic arc, intra-arc basins and back-arc basins), indicative of subduction and convergent plate interaction occur within greenstone sequences in many Archaean cratons including Yilgarn, Pilbara, Superior, North China, Slave, and southern Africa. The lithological association in these greenstones, including calc-alkaline volcanic rocks, and locally boninite, shoshonite and high-Mg andesite, along with the associated geochemical signatures, are almost identical to those found in rocks of modern convergent plate margin arcs (Condie & Harrison 1976; Hallberg et al. 1976; de Wit & Ashwal 1997; Bai & Dai 1998; Polat & Kerrich 1999, 2004; Cousens 2000; Percival & Helmstaedt 2004; Smithies et al. 2004, 2005; Kerrich & Polat 2006; Polat et al. 2009).

Styles of Archaean and Proterozoic mineralization resemble Phanerozoic deposits related to subduction environments (Sawkins 1990; Kerrich et al. 2005), including a Palaeoarchaean porphyry Cu deposit (Barley 1992) and Archaean and Palaeoproterozoic volcanogenic massive sulphide Cu–Zn deposits (Barley 1992; Allen et al. 1996; Syme et al. 1999; Wyman et al. 1999a, b).

Condie & Kröner (2008) listed several distinctive petrotectonic assemblages such as accretionary prisms as well as arc–back-arc–forearc associations that argue for the existence of accretionary orogens since the early Archaean. For instance, the 3.2 Ga Fig Tree greywacke–shale sequence of the Barberton greenstone belt in South Africa has long been interpreted in terms of an accretionary prism (e.g. Lowe & Byerly 2007, and references therein), and there is geochronological, structural and geophysical evidence for terrane accretion in the late Archaean Abitibi greenstone belt, Superior Province, Canada through convergent plate interaction (Percival & Helmstaedt 2004, and references therein). The recognition of ocean plate stratigraphy in an orogen's rock record is a key indicator of both mid-ocean ridge spreading, required for its generation, and subduction accretion, necessary for its preservation. As such, it provides a key indicator for plate tectonics in the rock record. The proposal that an ocean plate stratigraphy is preserved in the Marble Bar greenstone belt in the Pilbara craton, supported by trace element geochemical data (Kato & Nakamura 2003), and in the Isua greenstone belt in Greenland (Komiya et al. 1999), suggests that ridge–trench movements and therefore plate tectonics were in operation in the Palaeoarchaean.

Deep seismic reflection profiling across a number of late Archaean and Palaeoproterozoic belts has identified dipping reflectors, in some cases extending into the mantle, which underlie arc assemblages in the preserved accretionary orogen and are interpreted as a frozen subduction surface (Calvert et al. 1995; Cook et al. 1999; Cook & Erdmer 2005; Korja & Heikkinen 2005; Percival et al. 2006; Lahtinen et al. 2009).

Nutman et al. (2009) and Polat et al. (2009) have presented data in support of convergent plate margin processes within the Eoarchaean accretionary orogens of Isua, Greenland and Anshan, China (c. 3.8–3.6 Ga). They showed that the lithotectonic assemblages in these regions and their geochemistry are similar to those in Phanerozoic convergent plate margins involving the subduction of young, hot lithosphere. Harrison et al. (2005) inferred that subduction may extend back to the earliest phases of Earth evolution. They suggested that the isotopic systematics of Jack Hills zircons, northern Yilgarn, indicate formation in a continental environment characterized by calc-alkaline magmatism and crustal anatexis, features seen in the modern Earth in convergent margin settings and implying that subduction was established by 4.4 Ga ago.

Accretionary orogens and continental growth

Continental growth involves the addition of mantle-derived (juvenile) material to the crust. Arc magmatism within accretionary orogens is invoked as the major source of this material but with additional input derived from mantle plumes. Geochemical and isotopic data have shown that the composition of continental crust resembles subduction-related igneous rocks and suggest that that there has been progressive growth of continental crust through time (Fig. 13; Taylor 1967; Taylor & McClennan 1985; McCulloch & Bennett 1994; Arculus 1999). Thus, accretionary orogens, with their subduction-related plate margins, are seen as the sites of net continental growth, rather than collisional orogens, which are envisaged as sites of crustal reworking (Dewey et al. 1986). Geochemical and isotopic studies from Neoproterozoic to Phanerozoic accretionary orogens in the Arabian–Nubian Shield, the Canadian Cordillera and the Central Asian Orogenic Belt indicate massive addition of juvenile crust during the period of 900–100 Ma (Samson et al. 1989; Kovalenko et al. 1996; Jahn et al. 2000a; Wu et al. 2000; Jahn 2004; Stern in press). However, recent whole-rock Nd and zircon Hf(–O) isotope data imply that continental crust formation was episodic, with significant pulses of juvenile magmatism and crustal growth in the Archaean and Palaeoproterozoic, and with no significant addition in the Phanerozoic (Fig. 13; Condie 1998, 2004; Hawkesworth & Kemp 2006; Kemp et al. 2006). Punctuated crustal growth may be related to mantle plume activity (Stein & Hofmann 1993; Condie 1998), and creates a paradox of global proportions, because plate tectonics can account for c. 90% of Earth's current heat loss with the remainder lost through plume activity (Davies 1999), and thus magmatic arcs should be the major site of continental growth. The isotopic age data outlined in Figure 13 are based on an analysis of the preserved rock record and the assumption that it is representative of the record of growth. In addition, improved microanalytical techniques are suggesting that the contribution of subduction processes to continental growth may have been masked. Hf–O isotopic analysis of zircons from the classic I-type granites of eastern Australia show that these rocks were formed by reworking of sedimentary materials through mantle-derived magmas rather than by melting of igneous rocks (Kemp et al. 2007) and thus are critical components of continental growth. Bulk rock isotopes will mask this component, and this suggests that the component of such material in continental growth may have been underestimated in the past.

The volume of continental crust that is added through time via juvenile magma addition appears to be effectively compensated by the return of continental and island arc crust to the mantle. Clift et al. (2009) and Scholl & von Huene (2009) have estimated that the long-term global average rate of arc magma additions is 2.8–3.0 $km^3 a^{-1}$. Crustal

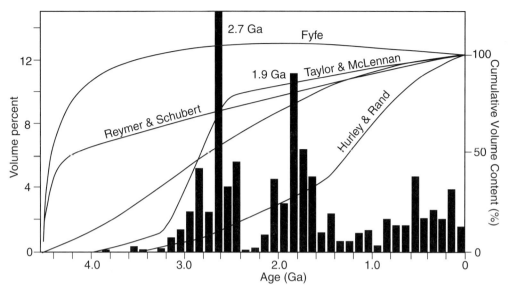

Fig. 13. Histogram of volume distribution of juvenile continental crust based on a compilation of U–Pb zircon ages integrated with Nd isotopic ages and lithological associations. Also plotted are models of continental growth, with 100% representing the present-day cumulative volume of crust (adapted from Condie 2005). Early models suggested slow initial growth followed by more rapid recent growth and were based on the geographical distribution of Rb–Sr and K–Ar isotope ages (Hurley & Rand 1969), which we now realize are probably reset by younger orogenic events. Some models suggested rapid early growth of continental crust, slowed by recycling of continental material and therefore resulting in a slower rate of growth or even a decrease in continental volume (Fyfe 1978; Reymer & Schubert 1984; Armstrong 1991). Other models have taken an intermediate approach to growth and require a more linear growth (unlabelled) or rapid growth during the late Archaean followed by steady-state growth driven by island arc magmatism (Taylor & McClennan 1985).

addition rates vary within arc systems over time, with the initial phase of arc construction (first 10 Ma) marked by rates that can be up to an order of magnitude larger than the long-term rate for an arc (Stern & Bloomer 1992; Stern 2004; Jicha et al. 2008).

Recycling of continental crust at convergent plate margins occurs by sediment subduction, subduction erosion and detachment of deeply underthrust crust (Scholl et al. 1980; Scholl & von Huene 2007, 2009; Clift et al. 2009). Sediment subduction entails the movement of lower plate sediment beneath the arc along the subduction channel (Fig. 9; Cloos & Shreve 1988a, b). Arc material may also be transported into the trench and then carried into the subduction channel. Material in the channel is carried beneath the frontal arc and if not underplated is carried into the mantle on the downgoing plate. Subduction erosion involves the transfer of material from the upper plate into the subduction channel and downward into the mantle. The loss of continental and arc crust through sediment subduction and subduction erosion has been estimated by Scholl & von Huene (2009) to be around 2.5 km^3 a^{-1} of which some 60% is due to erosion. Scholl & von Huene (2009) noted that continent and island arc crust is also carried into the mantle, where it can be detached, and is lost during final ocean closure and collision. They estimated that an additional 0.7 km^3 a^{-1} is recycled into the mantle by this process. Thus, the total volume of crustal material moved into the mantle at subduction zones is around 3.2 km^3 a^{-1}. This rate is sufficient that if plate tectonics has been operating since around 3.0 Ga (see Cawood et al. 2006) then a volume equal to the total current volume of continental crust would have been recycled into the mantle (Scholl & von Huene 2009).

Given uncertainties in these estimates for both the addition of crust and its removal from convergent plate margins, the net growth of continental crust is effectively zero, with crustal growth through magma addition effectively counterbalanced by removal of material. Thus, plate tectonics in general and convergent plate margins in particular, as represented by accretionary orogens, are not the sites of continental growth through time but rather sites of crustal reworking. Any single arc system can, however, show net addition or removal of material, hence allowing its preservation or removal from the rock record. For example, the South American margin has been undergoing long-term crustal loss such that the trench has migrated landward with respect to the upper plate with time, resulting in the magmatic arc younging away from the trench and Jurassic arc magmas forming the most seaward land outcrops in the current forearc (Stern 1991; Franz et al. 2006; Glodny et al. 2006; Kukowski & Oncken 2006). Areas of rapid accretion of material, either through arc magmatism during the early stages of arc development or through the accretion of already assembled continental (e.g. arc fragments of the North American Cordillera) and thickened oceanic (e.g. Ontong–Java) crustal fragments, are more likely to survive the effects of crustal reworking and are more likely to be preserved in the geological record. This may therefore lead to selective preservation of periods of continental growth in the rock record as exemplified in Figure 13. In addition, the thickened crust of oceanic islands and plateaux, once incorporated into convergent plate margins (such as Wrangellia, Sorachi, Sanbagawa and some Archaean greenstones), are selectively preserved even during periods of subduction erosion.

This is a contribution to the International Lithosphere Program, Task Force 1 (ERAS) and is Publication 499 of the Mainz Geocycles Cluster. We appreciate discussions with our ERAS colleagues during field workshops in Taiwan and Japan. R. M. Clowes, G. S. Fuis, B. M. Jahn, E. C. Leitch, G. S. Lister, S. Maruyama and Y. X. Wang have generously shared their many insights with us. B. Murphy and P. Leat are thanked for their reviews of the manuscript. This paper is dedicated to Akiho Miyashiro (1919–2008), pioneer of petrology and tectonics, father of paired metamorphic belts in accretionary orogens, and strong advocate of an island arc setting for ophiolites.

References

ABOUCHAMI, W., BOHER, M., MICHARD, A. & ALBARÈDE, F. 1990. A major 2.1 Ga old event of mafic magmatism in West Africa: An early stage of crustal accretion. *Journal of Geophysical Research*, **95**, 17605–17629.

ALLEN, R. L., WEIHED, P. & SVENSON, S.-A. 1996. Setting of Zn–Cu–Au–Ag massive sulfide deposits in the evolution and facies architecture of a 1.9 Ga marine volcanic arc, Skellefte District, Sweden. *Economic Geology*, **91**, 1022–1053.

ANSDELL, K. M. 2005. Tectonic evolution of the Manitoba–Saskatchewan segment of the Paleoproterozoic Trans-Hudson Orogen, Canada. *Canadian Journal of Earth Sciences*, **42**, 741–759.

ARCULUS, R. J. 1999. Origins of the continental crust. *Journal and Proceedings of the Royal Society of New South Wales*, **132**, 83–110.

ARMSTRONG, R. L. 1991. The persistent myth of crustal growth. *Australian Journal of Earth Sciences*, **38**, 613–630.

AUBOUIN, J. 1965. *Geosynclines*. Elsevier, Amsterdam.

BABEL WORKING GROUP. 1993a. Deep seismic reflection/refraction interpretation of crustal structure along BABEL profiles A and B in the southern Baltic Sea. *Geophysical Journal International*, **112**, 325–343.

BABEL WORKING GROUP. 1993b. Integrated seismic studies of the Baltic Shield using data in the Gulf of Bothnia region. *Geophysical Journal International*, **112**, 305–324.

BADARCH, G., CUNNINGHAM, W. D. & WINDLEY, B. F. 2002. A new terrane subdivision for Mongolia: Implications for the Phanerozoic crustal growth of Central Asia. *Journal of Asian Earth Sciences*, **21**, 87–110.

BAI, J. & DAI, F. Y. 1998. Archean Crust. *In*: MA, X. Y. & BAI, J. (eds) *Precambrian Crustal Evolution of China*. Springer, Berlin, 15–86.

BALLANCE, P. F., SCHOLL, D. W., VALLIER, T. L., STEVENSON, A. J., RYAN, H. & HERZER, R. H. 1989. Subduction of a Late Cretaceous seamount of the Louisville Ridge at the Tonga Trench: A model of normal and accelerated tectonic erosion. *Tectonics*, **8**, 953–962.

BALLY, A. W. 1981. Thoughts on the tectonics of folded belts. *In*: PRICE, N. J. & MCCLAY, K. (eds) *Thrust and Nappe Tectonics*. Geological Society, London, Special Publications, **9**, 13–32.

BANGS, N. L. & CANDE, S. C. 1997. The episodic development of a convergent margin inferred from structures and processes along the southern Chile margin. *Tectonics*, **16**, 489–505.

BARLEY, M. E. 1992. A review of Archean volcanic-hosted massive sulphide and sulfate mineralization in Western Australia. *Economic Geology*, **87**, 855–872.

BEAUMONT, C., ELLIS, S. & PFIFFNER, A. 1999. Dynamics of sediment subduction–accretion at convergent margins: Short-term modes, long-term deformation, and tectonic implications. *Journal of Geophysical Research*, **104**, 17573–17601.

BELTRANDO, M., HERMANN, J., LISTER, G. & COMPAGNONI, R. 2007. On the evolution of orogens: Pressure cycles and deformation mode switches. *Earth and Planetary Science Letters*, **256**, 372–388.

BIGNOLD, S. M. & TRELOAR, P. J. 2003. Northward subduction of the Indian plate beneath the Kohistan island arc, Pakistan Himalaya: New evidence from isotopic data. *Journal of the Geological Society, London*, **160**, 377–384.

BOGER, S. D. & MILLER, J. M. 2004. Terminal suturing of Gondwana and the onset of the Ross–Delamerian Orogeny: The cause and effect of an Early Cambrian reconstruction of plate motions. *Earth and Planetary Science Letters*, **219**, 35–48.

BRADLEY, D. C., KUSKY, T. M. ET AL. 2003. Geologic signature of early Tertiary ridge subduction in Alaska. *In*: SISSON, V. B., ROESKE, S. M. & PAVLIS, T. L. (eds) *Geology of a Transpressional Orogen Developed During Ridge–Trench Interaction along the North Pacific Margin*. Geological Society of America, Special Papers, **371**, 19–49.

BROWN, M. 1998a. Ridge–trench interactions and high-T–low-P metamorphism, with particular reference to the Cretaceous evolution of the Japanese Islands. *In*: TRELOAR, P. J. & O'BRIEN, P. J. (eds)

What Drives Metamorphism and Metamorphic Reactions? Geological Society, London, Special Publications, **138**, 137–169.

BROWN, M. 1998b. Unpairing metamorphic belts: P–T paths and a tectonic model for the Ryoke Belt, southwest Japan. *Journal of Metamorphic Geology*, **16**, 3–22.

BROWN, M. 2006. Duality of thermal regimes is the distinctive characteristic of plate tectonics since the Neoarchean. *Geology*, **34**, 961–964.

BROWN, M. 2009. Metamorphic patterns in orogenic systems and the geological record. *In*: CAWOOD, P. A. & KRÖNER, A. (eds) *Earth Accretionary Systems in Space and Time*. Geological Society, London, Special Publications, **318**, 37–74.

BURKE, K. C. A. 1977. Aulacogens and continental breakup. *Annual Review of Earth and Planetary Sciences*, **5**, 371–396.

CALVERT, A. J., SAWYER, E. W., DAVIS, W. J. & LUDDEN, J. N. 1995. Archaean subduction inferred from seismic images of a mantle suture in the Superior Province. *Nature*, **375**, 670–674.

CAWOOD, P. A. 1982. Structural relations in the subduction complex of the Paleozoic New England fold belt, eastern Australia. *Journal of Geology*, **90**, 381–392.

CAWOOD, P. A. 1983. Modal composition and detrital clinopyroxene geochemistry of lithic sandstones from the New England fold belt (East Australia): A Paleozoic forearc terrane. *Geological Society of America Bulletin*, **94**, 1199–1214.

CAWOOD, P. A. 1984. A geochemical study of metabasalts from a subduction complex in eastern Australia. *Chemical Geology*, **43**, 29–47.

CAWOOD, P. A. 1990. Provenance mixing in an intraoceanic subduction zone; Tonga Trench–Louisville Ridge collision zone, Southwest Pacific. *Sedimentary Geology*, **67**, 35–53.

CAWOOD, P. A. 1991a. Characterization of intra-oceanic magmatic arc source terranes by provenance studies of derived sediments. *New Zealand Journal of Geology and Geophysics*, **34**, 347–358.

CAWOOD, P. A. 1991b. Nature and record of igneous activity in the Tonga Arc, SW Pacific, deduced from the phase chemistry of derived detrital grains. *In*: MORTON, A. C., TODD, S. P. & HAUGHTON, P. D. W. (eds) *Developments in Sedimentary Provenance Studies*. Geological Society, London, Special Publications, **57**, 305–321.

CAWOOD, P. A. 2005. Terra Australis Orogen: Rodinia breakup and development of the Pacific and Iapetus margins of Gondwana during the Neoproterozoic and Paleozoic. *Earth-Science Reviews*, **69**, 249–279.

CAWOOD, P. A. & BUCHAN, C. 2007. Linking accretionary orogenesis with supercontinent assembly. *Earth-Science Reviews*, **82**, 217–256.

CAWOOD, P. A. & KORSCH, R. J. 2008. Assembling Australia: Proterozoic building of a continent. *Precambrian Research*, **166**, 1–35.

CAWOOD, P. A. & NEMCHIN, A. A. 2001. Source regions for Laurentian margin sediments: Constraints from U/Pb dating of detrital zircon in the Newfoundland Appalachians. *Geological Society of America Bulletin*, **113**, 1234–1246.

CAWOOD, P. A. & SUHR, G. 1992. Generation and obduction of ophiolites; constraints for the Bay of Islands Complex, western Newfoundland. *Tectonics*, **11**, 884–897.

CAWOOD, P. A. & TYLER, I. M. 2004. Assembling and reactivating the Proterozoic Capricorn Orogen: Lithotectonic elements, orogenies, and significance. *Precambrian Research*, **128**, 201–218.

CAWOOD, P. A., NEMCHIN, A. A., LEVERENZ, A., SAEED, A. & BALLANCE, P. F. 1999. U/Pb dating of detrital zircons: Implications for the provenance record of Gondwana margin terranes. *Geological Society of America Bulletin*, **111**, 1107–1119.

CAWOOD, P. A., NEMCHIN, A. A., STRACHAN, R. A., KINNY, P. D. & LOEWY, S. 2004. Laurentian provenance and an intracratonic tectonic setting for the Moine Supergroup, Scotland, constrained by detrital zircons from the Loch Eil and Glen Urquhart successions. *Journal of the Geological Society, London*, **161**, 861–874.

CAWOOD, P. A., KRÖNER, A. & PISAREVSKY, S. 2006. Precambrian plate tectonics: Criteria and evidence. *GSA Today*, **16**, 4–11.

CAWOOD, P. A., NEMCHIN, A. A., STRACHAN, R. A., PRAVE, A. R. & KRABBENDAM, M. 2007. Sedimentary basin and detrital zircon record along East Laurentia and Baltica during assembly and breakup of Rodinia. *Journal of the Geological Society, London*, **164**, 257–275.

CLAESON, D. T. & MEURER, W. P. 2004. Fractional crystallization of hydrous basaltic 'arc-type' magmas and the formation of amphibole-bearing gabbroic cumulates. *Contributions to Mineralogy and Petrology*, **147**, 288–304.

CLIFT, P., SCHOUTEN, H. & VANNUCCHI, P. 2009. Arc–continent collisions, sediment recycling and the maintenance of the continental crust. *In*: CAWOOD, P. A. & KRÖNER, A. (eds) *Earth Accretionary Systems in Space and Time*. Geological Society, London, Special Publications, **318**, 75–103.

CLOOS, M. & SHREVE, R. L. 1988a. Subduction-channel model of prism accretion, mélange formation, sediment subduction, and subduction erosion at convergent plate margins: 1. Background and description. *Pure and Applied Geophysics*, **128**, 456–500.

CLOOS, M. & SHREVE, R. L. 1988b. Subduction-channel model of prism accretion, mélange formation, sediment subduction, and subduction erosion at convergent plate margins: 2. Implications and discussion. *Pure and Applied Geophysics*, **128**, 501–545.

CLOWES, R. M., BRANDON, M. T., GREEN, A. G., YORATH, C. J., SUTHERLAND BROWN, A., KANASEWICH, E. R. & SPENCER, C. 1987. LITHOPROBE—southern Vancouver Island: Cenozoic subduction complex imaged by deep seismic reflections. *Canadian Journal of Earth Sciences*, **24**, 31–51.

CLOWES, R. M., ZELT, C. A., AMOR, J. R. & ELLIS, R. M. 1995. Lithospheric structure in the southern Canadian Cordillera from a network of seismic refraction lines. *Canadian Journal of Earth Sciences*, **32**, 1485–1513.

CLOWES, R. M., BAIRD, D. J. & DEHLER, S. A. 1997. Crustal structure of the Cascadian subduction zone, southwestern British Columbia, from potential field and seismic studies. *Canadian Journal of Earth Sciences*, **34**, 317–335.

COLBLENTZ, D. D. & RICHARDSON, R. M. 1996. Analysis of the South American intraplate stress field. *Journal of Geophysical Research*, **101**, 8643–8657.

COLLINS, W. J. 2002a. Hot orogens, tectonic switching, and creation of continental crust. *Geology*, **30**, 535–538.

COLLINS, W. J. 2002b. Nature of extensional accretionary orogens. *Tectonics*, **21**, 6-1–6-12, doi: 10.1029/2000TC001272.

COLLINS, W. J. & TEYSSIER, C. 1989. Crustal scale ductile fault systems in the Arunta Inlier, central Australia. *Tectonophysics*, **158**, 49–66.

COLPRON, M. & NELSON, J. L. (eds) 2006. *Paleozoic Evolution and Metallogeny of Pericratonic Terranes at the Ancient Pacific Margin of North America, Canadian and Alaskan Cordillera*. Geological Association of Canada, Special Paper, **45**.

COLPRON, M. & NELSON, J. L. 2009. A Palaeozoic Northwest Passage: incursion of Caledonian, Baltican and Siberian terranes into eastern Panthalassa, and the early evolution of the North American Cordillera. *In*: CAWOOD, P. A. & KRÖNER, A. (eds) *Earth Accretionary Systems in Space and Time*. Geological Society, London, Special Publications, **318**, 273–307.

COLPRON, M., NELSON, J. L. & MURPHY, D. C. 2007. Northern Cordilleran terranes and their interactions through time. *GSA Today*, **17**, 4–10.

CONDIE, K. C. 1998. Episodic continental growth and supercontinents: A mantle avalanche connection? *Earth and Planetary Science Letters*, **163**, 97–108.

CONDIE, K. C. 2004. Supercontinents and superplume events: Distinguishing signals in the geologic record. *Physics of the Earth and Planetary Interiors*, **146**, 319–332.

CONDIE, K. C. 2005. *Earth as an Evolving Planetary System*. Elsevier, Amsterdam.

CONDIE, K. C. & HARRISON, N. M. 1976. Geochemistry of the Archean Bulawayan Group, Midlands greenstone belt, Rhodesia. *Precambrian Research*, **3**, 253–271.

CONDIE, K. C. & KRÖNER, A. 2008. When did plate tectonics begin? Evidence from the geologic record. *In*: CONDIE, K. C. & PEASE, V. (eds) *When did Plate Tectonics Begin?* Geological Society, America, Memoirs, **440**, 281–295.

CONEY, P. J. 1973. Non-collision tectogenesis in western North America. *In*: TARLING, D. H. & RUNCORN, S. H. (eds) *Implications of Continental Drift to the Earth Sciences*. Academic Press, New York, 713–727.

CONEY, P. J. 1992. The Lachlan Fold Belt of eastern Australia and Circum-Pacific tectonic evolution. *Tectonophysics*, **214**, 1–25.

CONEY, P. J., JONES, D. L. & MONGER, J. W. H. 1980. Cordilleran suspect terranes. *Nature*, **288**, 329–333.

CONEY, P. J., EDWARDS, A., HINE, R., MORRISON, F. & WINDRIM, D. 1990. The regional tectonics of the Tasman orogenic system, eastern Australia. *Journal of Structural Geology*, **12**, 519–543.

COOK, F. A. & ERDMER, P. 2005. An 1800 km cross section of the lithosphere through the northwestern North American plate: Lessons from 4.0 billion years of Earth History. *Canadian Journal of Earth Sciences*, **42**, 1295–1311.

COOK, F. A., VAN DER VELDEN, A. J., HALL, K. W. & ROBERTS, B. J. 1999. Frozen subductions in Canada's Northwest Territories: Lithoprobe deep lithospheric reflection profiling of the western Canadian Shield. *Tectonics*, **18**, 1–24.

COOMBS, D. S., LANDIS, C. A., NORRIS, R. J., SINTON, J. M., BORNS, D. J. & CRAW, D. 1976. The Dun Mountain Ophiolite Belt, New Zealand, its tectonic setting, constitution and origin, with special reference to the southern portion. *American Journal of Science*, **276**, 561–603.

COUSENS, B. L. 2000. Geochemistry of the Archean Kam Group, Yellowknife greenstone belt, Slave province, Canada. *Journal of Geology*, **108**, 181–197.

COWAN, D. S. & PAGE, B. M. 1975. Recycled Franciscan material in Franciscan mélange, west of Paso Robles, California. *Geological Society of America Bulletin*, **85**, 1623–1634.

CROOK, K. A. W. 1974. Kratonisation of west Pacific-type geosynclines. *Journal of Geology*, **87**, 24–36.

CURRIE, C. A. & HYNDMAN, R. D. 2006. The thermal structure of subduction zone back arcs. *Journal of Geophysical Research*, **111**, B08404, doi:10.1029/2005JB004024.

CURRIE, C. A., HUISMANS, R. S. & BEAUMONT, C. 2008. Thinning of continental backarc lithosphere by flow-induced gravitational instability. *Earth and Planetary Science Letters*, **269**, 436–447.

DALZIEL, I. W. D. 1991. Pacific margins of Laurentia and east Antarctica–Australia as a conjugate rift pair: Evidence and implications for an Eocambrian supercontinent. *Geology*, **19**, 598–601.

DALZIEL, I. W. D., LAWVER, L. A. & MURPHY, J. B. 2000. Plumes, orogenesis, and supercontinental fragmentation. *Earth and Planetary Science Letters*, **178**, 1–11.

DANN, J. C. 1997. Pseudostratigraphy and origin of the Early Proterozoic Payson ophiolite, central Arizona. *Geological Society of America Bulletin*, **109**, 347–365.

DAVIES, G. F. 1999. *Dynamic Earth Plates, Plumes and Mantle Convection*. Cambridge University Press, Cambridge.

DE FRANCO, R., GOVERS, R. & WORTEL, R. 2008. Nature of the plate contact and subduction zones diversity. *Earth and Planetary Science Letters*, **271**, 245–253.

DEWEY, J. F. 1969. Evolution of the Appalachian–Caledonian orogen. *Nature*, **222**, 124–129.

DEWEY, J. F. 1980. Episodicity, sequence and style at convergent plate boundaries. *In*: STRANGEWAY, D. W. (ed.) *The Continental Crust and Its Mineral Deposits*. Geological Association of Canada, Special Papers, **20**, 553–576.

DEWEY, J. F. 2007. The secular evolution of plate tectonics and the continental crust: An outline. *In*: HATCHER, R. D. JR, CARLSON, M. P., MCBRIDE, J. H. & CATALÁN, J. M. (eds) *4-D Framework of Continental Crust*. Geological Society of America, Memoirs, **200**, 1–8.

DEWEY, J. F. & SPALL, H. 1975. Pre-Mesozoic plate tectonics. *Geology*, **3**, 422–424.

DEWEY, J. F., HEMPTON, M. R., KIDD, W. S. F., SAROGLU, F. & SENGOR, A. M. C. 1986. Shortening of continental lithosphere: The neotectonics of Eastern

Anatolia—a young collision zone. *In*: COWARD, M. P. & RIES, A. C. (eds) *Collision Tectonics*. Geological Society, London, Special Publications, **19**, 3–36.

DE WIT, M. J. & ASHWAL, L. D. 1997. *Greenstone Belts*. Oxford University Press, Oxford.

DICKINSON, W. R. 1977. Tectono-stratigraphic evolution of subduction-controlled sedimentary assemblages. *In*: TALWANI, M. & PITMAN, W. C., III (eds) *Island Arcs, Deep-Sea Trenches and Back-Arc Basins*. Maurice Ewing Series 1, American Geophysical Union, 33–40.

DICKINSON, W. R. 1985. Interpreting provenance relations from detrital modes of sandstones. *In*: ZUFFA, G. G. (ed.) *Provenance of Arenites*. NATO-ASI Series, **148**, 333–361.

DICKINSON, W. R. 1995. Forearc basins. *In*: BUSBY, C. J. & INGERSOLL, R. V. (eds) *Tectonics of Sedimentary Basins*. Blackwell Science, Cambridge, MA, 221–261.

DICKINSON, W. R. & SNYDER, W. S. 1978. Geometry of subducted slabs related to the San Andreas transform. *Journal of Geology*, **87**, 609–627.

DICKINSON, W. R. & SUCZEK, C. A. 1979. Plate tectonics and sandstone compositions. *AAPG Bulletin*, **63**, 2164–2182.

DICKINSON, W. R. & VALLONI, R. 1980. Plate settings and provenance of sands in modern ocean basins. *Geology*, **8**, 82–86.

DIRKS, P., HAND, M., COLLINS, W. J. & OFFLER, R. 1992. Structural metamorphic evolution of the Tia Complex, New England Fold Belt—thermal overprint of an accretion subduction complex in a compressional back-arc setting. *Journal of Structural Geology*, **14**, 669–688.

DIRKS, P. H. G. M., OFFLER, R. & COLLINS, W. J. 1993. Timing of emplacement and deformation of the Tia Granodiorite, southern New England Fold Belt, NSW: Implications for the metamorphic history. *Australian Journal of Earth Sciences*, **40**, 103–108.

ELSASSER, W. M. 1971. Sea-floor spreading as thermal convection. *Journal of Geophysical Research*, **76**, 1101–1112.

ERNST, W. G. 2005. Alpine and Pacific styles of Phanerozoic mountain building: Subduction-zone petrogenesis of continental crust. *Terra Nova*, **17**, 165–188.

FERGUSSON, C. L. 1984. The Gundahl Complex of the New England Fold Belt, eastern Australia: A tectonic mélange formed in the Paleozoic subduction complex. *Journal of Structural Geology*, **6**, 257–271.

FERGUSSON, C. L. 1985. Trench floor sedimentary sequences in a Palaeozoic subduction complex, eastern Australia. *Sedimentary Geology*, **42**, 181–200.

FERGUSSON, C. L. & CONEY, P. J. 1992. Convergence and intraplate deformation in the Lachlan Fold Belt of southeastern Australia. *Tectonophysics*, **214**, 417–439.

FODEN, J. D., ELBURG, M. A., DOUGHERTY-PAGE, J. & BURT, A. 2006. The timing and duration of the Delamerian Orogeny: Correlation with the Ross Orogen and implications for Gondwana assembly. *Journal of Geology*, **114**, 189–210.

FOSTER, D. A. & GRAY, D. R. 2000. The structure and evolution of the Lachlan Fold Belt (Orogen) of eastern Australia. *Annual Review of Earth and Planetary Sciences*, **28**, 47–80, doi:10.1146/annurev.earth.28.1.47.

FOSTER, D. A. & GRAY, D. R. 2007. Strain rate in Paleozoic thrust sheets, the western Lachlan Orogen, Australia: Strain analysis and fabric geochronology. *In*: SEARS, J. W., HARMS, T. A. & EVENCHICH, C. A. (eds) *Whence the Mountains? Inquiries into the Evolution of Orogenic Systems: A Volume in Honor of Raymond A. Price*. Geological Society of America, Boulder, CO, 349–368, doi:10.1130/2007.2433(17).

FOSTER, D. A., GRAY, D. R., SPAGGIARI, C., KAMENOV, G. & BIERLEIN, F. P. 2009. Palaeozoic Lachlan orogen, Australia; accretion and construction of continental crust in a marginal ocean setting: Isotopic evidence from Cambrian metavolcanic rocks. *In*: CAWOOD, P. A. & KRÖNER, A. (eds) *Earth Accretionary Systems in Space and Time*. Geological Society, London, Special Publications, **318**, 329–349.

FRANZ, G., KRAMER, F. W. ET AL. 2006. Crustal evolution at the central Andean continental margin: A geochemical record of crustal growth, recycling and destruction. *In*: ONCKEN, O., CHONG, G. ET AL. (eds) *The Andes: Active Subduction Orogeny*. Frontiers in Earth Sciences, **1**, 45–64.

FUIS, G. S. 1998. West margin of North America—a synthesis of recent seismic transects. *Tectonophysics*, **288**, 265–292.

FUIS, G. S. & CLOWES, R. M. 1993. Comparison of deep structure along three transects of the western North American continental margin. *Tectonics*, **12**, 1420–1435.

FUIS, G. S. & MOONEY, W. D. 1990. Lithosphere structure and tectonics from seismic refraction and other data. *In*: WALLACE, R. E. (ed.) *The San Andreas Fault System*. US Geological Survey, Professional Papers, **1515**, 206–236.

FUIS, G. S. & PLAFKER, G. 1991. Evolution of deep structure along the Trans-Alaska Crustal Transect, Chugach Mountains and Copper River basin, southern Alaska. *Journal of Geophysical Research*, **96**, 4229–4253.

FUIS, G. S., MURPHY, J. M., LUTTER, W. J., MOORE, T. E., BIRD, K. J. & CHRISTENSEN, N. I. 1997. Deep seismic structure and tectonics of northern Alaska: Crustal-scale duplexing with deformation extending into the upper mantle. *Journal of Geophysical Research*, **102**, 20873–20896, doi:10.1029/96JB03959.

FUIS, G. S., MOORE, T. E. ET AL. 2008. Trans-Alaska Crustal Transect and continental evolution involving subduction underplating and synchronous foreland thrusting. *Geology*, **36**, 267–270, doi:10.1130/G24257A.1.

FURNES, H., DE WIT, M., STAUDIGEL, H., ROSING, M. & MUEHLENBACHS, K. 2007. A vestige of Earth's oldest ophiolite. *Science*, **315**, 1704–1707.

FYFE, W. S. 1978. The evolution of the Earth's crust: Modern plate tectonics or ancient hot spot tectonics? *Chemical Geology*, **23**, 89–114.

GARDE, A. A., CHADWICK, B., GROCOTT, J., HAMILTON, M. A., MCCAFFREY, K. J. W. & SWAGER, C. P. 2002. Mid-crustal detachment during oblique convergence: Implications of partitioned transpression in the magmatic arc and proximal forearc of the Palaeoproterozoic Ketilidian orogen, southern Greenland. *Journal of the Geological Society, London*, **159**, 247–261.

GLODNY, J., ECHTLER, H. ET AL. 2006. Long-term geological evolution and mass-flow balance of the South–Central Andes. In: ONCKEN, O., CHONG, G. ET AL. (eds) The Andes: Active Subduction Orogeny. Frontiers in Earth Science, **1**, 401–428.

GOWER, C. F. 1996. The evolution of the Grenville Province in eastern Labrador, Canada. In: BREWER, T. S. (ed.) Precambrian Crustal Evolution in the North Atlantic Region. Geological Society, London, Special Publications, **112**, 197–218.

GOWER, C. F., RIVERS, T. & RYAN, A. B. 1990. Mid-Proterozoic Laurentia–Baltica: An overview of its geological evolution and a summary of the contributions made by this volume. In: GOWER, C. F., RIVERS, T. & RYAN, A. B. (eds) Mid-Proterozoic Laurentia–Baltica. Geological Association of Canada, Special Papers, **38**, 1–22.

GRAY, D. R. & FOSTER, D. A. 2006. Crust restoration for the western Lachlan Orogen using the strain-reversal, area-balancing technique: Implications for crustal components and original thicknesses. Australian Journal of Earth Sciences, **53**, 329–341.

GREENLY, E. 1919. The Geology of Anglesey. Memoir of the Geological Survey of Great Britain.

GROVES, D. I. & BIERLEIN, F. P. 2007. Geodynamic settings of mineral deposit systems. Journal of the Geological Society, London, **164**, 19–30.

GUTSCHER, M.-A. 2002. Andean subduction styles and their effect on thermal structure and interplate coupling. Journal of South American Earth Sciences, **15**, 3–10.

GUTSCHER, M.-A., SPAKMAN, W., BIJWAARD, H. & ENGDAHL, E. 2000. Geodynamics of flat subduction: Seismicity and tomographic constraints from the Andean margin. Tectonics, **19**, 814–833.

HAEUSSLER, P., BRADLEY, D. C. & GOLDFARB, R. 2003. Brittle deformation along the Gulf of Alaska margin in response to Palaeocence–Eocene triple junction migration. In: SISSON, V. B., ROESKE, S. M. & PAVLIS, T. L. (eds) Geology of a Transpressional Orogen Developed during Ridge–Trench Interaction along the North Pacific Margin. Geological Society of America, Special Papers, **371**, 119–140.

HALL, R. 2002. Cenozoic geological and plate tectonic evolution of SE Asia and the SW Pacific: Computer-based reconstructions, model and animations. Journal of Asian Earth Sciences, **20**, 353–431.

HALL, R. 2009. The Eurasian SE Asian margin as a modern example of an accretionary orogen. In: CAWOOD, P. A. & KRÖNER, A. (eds) Earth Accretionary Systems in Space and Time. Geological Society, London, Special Publications, **318**, 351–372.

HALLBERG, J. A., CARTER, D. N. & WEST, K. N. 1976. Archaean volcanism and sedimentation near Meekatharra, Western Australia. Precambrian Research, **3**, 577–595.

HAMILTON, W. 1979. Tectonics of the Indonesian Region. US Geological Survey, Professional Papers, **1078**.

HAMILTON, W. B. 2003. An alternative Earth. GSA Today, **13**, 4–12.

HAND, M. & SANDIFORD, M. 1999. Intraplate deformation in central Australia, the link between subsidence and fault reactivation. Tectonophysics, **305**, 121–140.

HARRISON, T. M., BLICHERT-TOFT, J., MÜLLER, W., ALBARÈDE, F., HOLDEN, P. & MOJZSIS, S. J. 2005. Heterogeneous Hadean hafnium: Evidence of continental crust at 4.4 to 4.5 Ga. Science, **310**, 1947–1950.

HAWKESWORTH, C. J. & KEMP, A. I. S. 2006. Evolution of continental crust. Nature, **443**, 811–817.

HAWKINS, J. W. 1994. Petrologic synthesis: Lau Basin transect (Leg 135). In: HAWKINS, J., PARSON, L. & ALLAN, J. (eds) Proceedings of the Ocean Drilling Program, Scientific Results, 135. Ocean Drilling Program, College Station, TX, 879–905.

HOFFMAN, P. F. 1973. Evolution of an early Proterozoic continental margin: The Coronation Geosyncline and associated aulacogens of the NW Canadian Shield. Philosophical Transactions of the Royal Society of London, Series A, **273**, 547–581.

HOFFMAN, P. F. 1991. Did the breakout of Laurentia turn Gondwanaland inside-out? Science, **252**, 1409–1412.

HOFFMAN, P. F., BURKE, K. C. A. & DEWEY, J. F. 1974. Aulacogens and their genetic relation to geosynclines, with a Proterozoic example from the Great Slave Lake, Canada. In: DOTT, R. H. & SHAVER, R. H. (eds) Modern and Ancient Geosyncline Sedimentation. Society of Economic Paleontologists and Mineralogists, Special Publications, **19**, 38–55.

HOLM, D. K., VAN SCHMUS, W. R., MACNEILL, L. C., BOERBOOM, T. J., SCHWEITZER, D. & SCHNEIDER, D. 2005. U–Pb zircon geochronology of Paleoproterozoic plutons from the northern midcontinent, USA: Evidence for subduction flip and continued convergence after geon 18 Penokean orogenesis. Geological Society of America Bulletin, **117**, 259–275.

HSÜ, K. J. 1971. Franciscan mélanges as a model for eugeosynclinal sedimentation and underthrusting tectonics. Journal of Geophysical Research, **76**, 1162–1170.

HURLEY, P. M. & RAND, J. R. 1969. Predrift continental nuclei. Science, **164**, 1229–1242.

HUSSON, L., CONRAD, C. P. & FACCENNA, C. 2008. Tethyan closure, Andean orogeny, and westward drift of the Pacific Basin. Earth and Planetary Science Letters, **271**, 303–310.

HYNDMAN, R. D., YORATH, C. J., CLOWES, R. M. & DAVIS, E. E. 1990. The northern Cascadia subduction zone at Vancouver Island: Seismic structure and tectonic history. Canadian Journal of Earth Sciences, **27**, 313–329.

HYNDMAN, R. D., WANG, K. & YAMANO, M. 1995. Thermal constraints on the seismogenic portion of the southwestern Japan subduction thrust. Journal of Geophysical Research, **100**, 15373–15392.

HYNDMAN, R. D., CURRIE, C. A. & MAZZOTTI, S. P. 2005. Subduction zone backarcs, mobile belts and orogenic heat. GSA Today, **15**, 4–10.

IAFFALDANO, G. & BUNGE, H.-P. 2008. Strong plate coupling along the Nazca–South American convergent margin. Geology, **36**, 443–446.

IAFFALDANO, G., BUNGE, H.-P. & DIXON, T. H. 2006. Feedback between mountain belt growth and plate convergence. Geology, **34**, 893–896.

INGERSOLL, R. V., GRAHAM, S. A. & DICKINSON, W. R. 1995. Remnant-ocean basins. In: BUSBY, C. J. &

INGERSOLL, R. V. (eds) *Tectonics of Sedimentary Basins*. Blackwell Science, Cambridge, MA, 363–391.

ISOZAKI, Y. 1996. Anatomy and genesis of a subduction-related orogen: A new view of geotectonic subdivision and evolution of the Japanese Islands. *Island Arc*, **5**, 289–320.

ISOZAKI, Y. 1997. Jurassic accretion tectonics of Japan. *Island Arc*, **6**, 25–51.

ISOZAKI, Y. & BLAKE, M. C. 1994. Biostratigraphic constraints on formation and timing of accretion in a subduction complex. An example from the Franciscan complex in northern California. *Journal of Geology*, **102**, 283–296.

ISOZAKI, Y., MARUYAMA, S. & FURUOKA, F. 1990. Accreted oceanic materials in Japan. *Tectonophysics*, **181**, 179–205.

JAHN, B. M. 2004. The Central Asian Orogenic Belt and growth of the continental crust in the Phanerozoic. *In*: MALPAS, J., FLETCHER, C. J. N., ALI, J. R. & AITCHESON, J. C. (eds) *Aspects of the Tectonic Evolution of China*. Geological Society, London, Special Publications, **226**, 73–100.

JAHN, B. M., WU, F. Y. & CHEN, B. 2000a. Granitoids of the Central Asian orogenic belt and continental growth in the Phanerozoic. *Transactions of the Royal Society of Edinburgh, Earth Sciences*, **91**, 181–193.

JAHN, B. M., WU, F. Y. & CHEN, B. 2000b. Massive granitoid generation in Central Asia: Nd isotope evidence and implication for continental growth in the Phanerozoic. *Episodes*, **23**, 82–92.

JAMIESON, R. A., BEAUMONT, C., FULLSACK, P. & LEE, B. 1998. Barrovian regional metamorphism: Where's the heat? *In*: TRELOAR, P. J. & O'BRIEN, P. J. (eds) *What Drives Metamorphism and Metamorphic Reactions?* Geological Society, London, Special Publications, **138**, 23–51.

JENNER, G. A., CAWOOD, P. A., RAUTENSCHLEIN, M. & WHITE, W. M. 1987. Composition of back-arc basin volcanics, Valu Fa Ridge, Lau Basin; evidence for a slab-derived component in their mantle source. *Journal of Volcanology and Geothermal Research*, **32**, 209–222.

JICHA, B. R., SCHOLL, D. W., SINGER, B. S., YOGODZINSKI, G. M. & KAY, S. M. 2008. Revised age of Aleutian Island Arc formation implies high rate of magma production. *Geology*, **34**, 661–664, doi: 10.1130/G22433.1.

JOHNSON, P. R. & WOLDEHAIMANOT, B. 2003. Development of the Arabian–Nubian Shield: Perspectives on accretion and deformation in the northern East African orogen and the assembly of Gondwana. *In*: YOSHIDA, M., WINDLEY, B. F. & DASGUPTA, S. (eds) *Proterozoic East Gondwana: Supercontinent Assembly and Breakup*. Geological Society, London, Special Publications, **206**, 289–325.

JOHNSTON, S. T. 2001. The Great Alaskan Terrane Wreck: Reconciliation of paleomagnetic and geological data in the northern Cordillera. *Earth and Planetary Science Letters*, **193**, 259–272.

JORDAN, T. E. 1995. Retroarc foreland and related basins. *In*: BUSBY, C. J. & INGERSOLL, R. V. (eds) *Tectonics of Sedimentary Basins*. Blackwell Science, Cambridge, MA, 331–362.

KARLSTROM, K. E., AHALL, K.-I., HARLAN, S. S., WILLIAMS, M. L., MCLELLAND, J. & GEISSMAN, J. W. 2001. Long-lived (1.8–1.0 Ga) convergent orogen in southern Laurentia, its extensions to Australia and Baltica, and implications for refining Rodinia. *Precambrian Research*, **111**, 5–30.

KATO, Y. & NAKAMURA, K. 2003. Origin and global tectonic significance of Early Archaean cherts from the Marble Bar greenstone belt, Pilbara craton, Western Australia. *Precambrian Research*, **125**, 191–293.

KATO, Y., OHTA, I., TSUNEMATSU, T., WATANABE, Y., ISOZAKI, Y., MARUYAMA, S. & IMAI, N. 1998. Rare earth element variations in mid-Archean banded iron formations: Implications for the chemistry of ocean and continent and plate tectonics. *Geochimica et Cosmochimica Acta*, **62**, 3475–3497.

KATO, Y., NAKAO, K. & ISOZAKI, Y. 2002. Geochemistry of late Permian to early Triassic pelagic cherts from southwest Japan: Implications for an oceanic redox change. *Chemical Geology*, **182**, 15–34.

KAWAI, T., WINDLEY, B. F., TERABAYASHI, M., YAMAMOTO, H., MARUYAMA, S. & ISOZAKI, Y. 2006. Mineral isograds and metamorphic zones of the Anglesey blueschist belt, UK: Implications for the metamorphic development of a Neoproterozoic subduction–accretion complex. *Journal of Metamorphic Geology*, **24**, 591–602.

KAWAI, T., WINDLEY, B. F. ET AL. 2007. Geotectonic framework of the Blueschist Unit on Anglesey–Lleyn, UK, and its role in the development of a Neoproterozoic accretionary orogen. *Precambrian Research*, **153**, 11–28.

KAY, M. 1951. *North American Geosynclines*. Geological Society of America, Memoirs, **48**.

KAY, S. M. & MPODOZIS, C. 2001. Central Andean ore deposits linked to evolving shallow subduction systems and thickening crust. *GSA Today*, **11**, 4–9.

KAY, S. M. & MPODOZIS, C. 2002. Magmatism as a probe to the Neogene shallowing of the Nazca plate beneath the modern Chilean flat-slab. *Journal of South American Earth Sciences*, **15**, 39–57.

KEMP, A. I. S., HAWKESWORTH, C. J., PATERSON, B. A. & KINNY, P. D. 2006. Episodic growth of the Gondwana supercontinent from hafnium and oxygen isotopes in zircon. *Nature*, **439**, 580–583.

KEMP, A. I. S., HAWKESWORTH, C. J. ET AL. 2007. Magmatic and crustal differentiation history of granitic rocks from Hf–O isotopes in zircon. *Science*, **315**, 980–983.

KERRICH, R. & POLAT, A. 2006. Archean greenstone–tonalite duality: Thermochemical mantle convection models or plate tectonics in the early Earth global dynamics? *Tectonophysics*, **415**, 141–165.

KERRICH, R., GOLDFARB, R. & RICHARDS, J. P. 2005. Metallogenic provinces in an evolving geodynamic framework. *Economic Geology*, **100**, 1097–1136.

KIMURA, G., SAKAKIBARA, M. & OKAMURA, M. 1994. Plumes in central Panthalassa? Deductions from accreted oceanic fragments in Japan. *Tectonics*, **13**, 905–916.

KIMURA, K. & HORI, R. 1993. Offscraping accretion of Jurassic chert–clastic complexes in the Mino-Tamba belt, central Japan. *Journal of Structural Geology*, **15**, 145–161.

KODAIRA, S., TAKAHASHI, N., PARK, J.-O., MOCHBAKI, K., SHINOHANA, M. & KIMMA, S. 2000. Western Nankai Trough seismogenic: Results from a wide-angle ocean bottom seismic survey. *Journal of Geophysical Research*, **105**, 5882–5906, doi:10.1029/1999JB900394.

KOMIYA, T., MARUYAMA, S., MASUDA, T., NOBDA, S., HAYASHI, M. & OKAMOTO, K. 1999. Plate tectonics at 3.8–3.7 Ga: Field evidence from the Isua Accretionary Complex, southern West Greenland. *Journal of Geology*, **107**, 515–554.

KORJA, A. & HEIKKINEN, P. 2005. The accretionary Svecofennian orogen—insight from the BABEL profiles. *Precambrian Research*, **136**, 241–268.

KORJA, A., KORJA, T., LUOSTO, U. & HEIKKINEN, P. 1993. Seismic and geoelectric evidence for collisional and extensional events in the Fennoscandian Shield—implications for Precambrian crustal evolution. *Tectonophysics*, **219**, 129–152.

KOVALENKO, V. I., YARMOLYUK, V. V., KOVACH, V. P., KOTOV, A. B., KOZAKOV, I. K. & SALNIKOVA, E. B. 1996. Sources of Phanerozoic granitoids in Central Asia: Sm–Nd isotope data. *Geochemical International*, **34**, 628–640.

KRÖNER, A., BRANDL, G., NEMCHIN, A. A. & PIDGEON, R. T. 1999. Single zircon ages for granitoid gneisses in the Central Zone of the Limpopo Belt, Southern Africa and geodynamic significance. *Precambrian Research*, **93**, 299–337.

KRÖNER, A. & WINDLEY, B. F. ET AL. 2007. Accretionary growth and crust-formation in the central Asian Orogenic Belt and comparison with the Arabian–Nubian shield. *In*: HATCHER, R. D., JR. CARLSON, M. P., MCBRIDE, J. H. & CATALAN, J. M. (eds) *The 4-D framework of the continental crust-Integrating crustal processes through time*. Geological Society of America, Memoirs, **200**, 181–210.

KRÖNER, A., HEGNER, E., LEHMANN, B., HEINHORST, J., WINGATE, M. T. D., LIU, D. Y. & ERMELOV, P. 2008. Palaeozoic arc magmatism in the Central Asian Orogenic Belt of Kazakhstan: SHRIMP zircon ages and whole-rock Nd isotopic systematics. *Journal of Asian Earth Sciences*, **32**, 118–130.

KUKOWSKI, N. & ONCKEN, O. 2006. Subduction erosion—the 'normal' mode of fore-arc, material transfer along the Chile margin? *In*: ONCKEN, O., CHONG, G. ET AL. (eds) *The Andes: Active Subduction Orogeny*. Frontiers in Earth Science, **1**, 217–236.

KUSKY, T. & YOUNG, C. 1999. Emplacement of the Resurrection Peninsula ophiolite in the southern Alaska forearc during a ridge–trench encounter. *Journal of Geophysical Research*, **104**, 29025–29054.

KUSKY, T. M. 1991. Structural development of an Archean orogen, western Point Lake, Northwest Territories. *Tectonics*, **10**, 820–841.

KUSKY, T. M. 2004. *Precambrian Ophiolites and Related Rocks*. Developments in Precambrian Geology, **13**.

KUSKY, T. M. & BRADLEY, D. C. 1999. Kinematics of mélange fabrics: Examples and applications from the McHugh Complex, Kenai Peninsula, Alaska. *Journal of Structural Geology*, **21**, 1773–1796.

KUSKY, T. M. & KIDD, W. S. F. 1992. Remnants of an Archean oceanic plateau, Belingwa greenstone belt, Zimbabwe. *Geology*, **20**, 43–46.

KUSKY, T. M. & POLAT, A. 1999. Growth of granite–greenstone terranes at convergent margins, and stabilization of Archean cratons. *Tectonophysics*, **305**, 43–73.

KUSKY, T. M., BRADLEY, D. C. & HAEUSSLER, P. 1997a. Progressive deformation of the Chugach accretionary complex, Alaska, during a Palaeogene ridge–trench encounter. *Journal of Structural Geology*, **19**, 139–157.

KUSKY, T. M., BRADLEY, D. C., HAEUSSLER, P. & KARL, S. 1997b. Controls on accretion of flysch and mélange belts at convergent margins: Evidence from Chugach Bay thrust and Iceworm Mélange, Chugach Terrane, Alaska. *Tectonics*, **16**, 855–878.

KUSKY, T. M., LI, J. G. & TUCKER, R. T. 2001. The Archean Dongwanzi ophiolite complex, North China Craton: 2.505 billion year old oceanic crust and mantle. *Science*, **292**, 1142–1145.

KUSKY, T. M., BRADLEY, D., DONLEY, D. T., ROWLEY, D. & HAEUSSLER, P. J. 2003. Controls on intrusion of near-trench magmas of the Sanak–Barabof belt, Alaska, during Palaeogene ridge subduction, and consequences for forearc evolution. *In*: SISSON, V. B., ROESKE, S. M. & PAVLIS, T. L. (eds) *Geology of a Transpressional Orogen Developed During a Ridge–Trench Interaction along the North Pacific Margin*. Geological Society of America, Special Papers, **371**, 269–292.

KUSKY, T. M., GLASS, A. & BRADLEY, D. C. 2007a. Structure, Cr-chemistry, and age of the Border Ranges ultramafic/mafic complex: A suprasubduction zone ophiolite complex. *In*: RIDGWAY, K. D., TROP, J. M., GLEN, J. M. G. & O'NEILL, J. M. (eds) *Tectonic Growth of a Collisional Continental Margin: Crustal Evolution of Southern Alaska*. Geological Society of America, Special Papers, **431**, 207–225.

KUSKY, T. M., WINDLEY, B. F. & ZHAI, M. G. 2007b. Tectonic evolution of the North China Block: From orogen to craton to orogen. *In*: ZHAI, M. G., WINDLEY, B. F., KUSKY, T. M. & MENG, Q. R. (eds) *Lithospheric Thinning under Eastern Asia*. Geological Society, London, Special Publications, **280**, 1–34.

LAHTINEN, R., KORJA, A. & NIRONEN, M. 2005. Palaeoproterozoic tectonic evolution of the Fennoscandian Shield. *In*: LEHTINEN, M., NURMI, P. & RÄMÖ, T. (eds) *The Precambrian Bedrock of Finland—Key to the evolution of the Fennoscandian Shield*. Elsevier, Amsterdam, 418–532.

LAHTINEN, R., KORJA, A., NIRONEN, M. & HEIKKINEN, P. 2009. Palaeoproterozoic accretionary processes in Fennoscandia. *In*: CAWOOD, P. A. & KRÖNER, A. (eds) *Earth Accretionary Systems in Space and Time*. Geological Society, London, Special Publications, **318**, 237–256.

LALLEMAND, S., HEURET, A. & BOUTELIER, D. 2005. On the relationships between slab dip, back-arc stress, upper plate absolute motion, and crustal nature in subduction zones. *Geochemistry, Geophysics, Geosystems*, **6**, article number Q09006.

LALLEMAND, S., HEURET, A., FACCENNA, C. & FUNICIELLO, F. 2008. Subduction dynamics as revealed by trench migration. *Tectonics*, **27**, doi:10.1029/2007/TC002212.

LEE, C.-T.-A., MORTON, D. M., KISTLER, R. W. & BAIRD, A. K. 2007. Petrology and tectonics of Phanerozoic continent formation: From island arcs to accretion and continental arc magmatism. *Earth and Planetary Science Letters*, **263**, 370–387.

LEITCH, E. C. 1984. Marginal basins of the SW Pacific and the preservation and recognition of their ancient analogues: A review. *In*: KOKELAAR, B. P. & HOWELLS, M. F. (eds) *Marginal Basin Geology*. Geological Society, London, Special Publications, **16**, 97–108.

LI, Z. X. & LI, X. H. 2007. Formation of a 1300-km-wide intracontinental orogen and post-orogenic magmatic province in Mesozoic South China: A flat-slab subduction model. *Geology*, **35**, 179–182.

LISTER, G. S. & FORSTER, M. A. 2006. The plate-tectonic significance of inversion cycles during orogenesis. *Geological Society of America, Abstracts with Programs*, Speciality Meeting No. 2, 119.

LISTER, G. S., FORSTER, M. A. & RAWLINGS, T. J. 2001. Episodicity during orogenesis. *In*: MILLER, J. A., HOLDSWORTH, R. E., BUICK, I. S. & HAND, M. (eds) *Continental Reactivaton and Reworking*. Geological Society, London, Special Publications, **184**, 89–113.

LOWE, D. R. & BYERLY, G. R. 2007. An overview of the geology of the Barberton greenstone belt and vicinity: Implications for early crustal development. *In*: VAN KRANENDONK, M. J., SMITHIES, R. H. & BENNETT, B. (eds) *Earth's Oldest Rocks*. Elsevier, Amsterdam, 481–526.

LUCAS, S. B., SYME, E. C. & ASHTON, K. E. 1999. New perspectives on the Flin Flon Belt, Trans-Hudson orogen, Manitoba and Saskatchewan: An introduction to the special issue on the NATMAP Shield Margin Project, Part 1. *Canadian Journal of Earth Sciences*, **36**, 135–140.

LUO, Z. 1991. The dynamical model for lithospheric evolution in the Longmenshan orogenic belt. *Journal of Chengdu College Geology*, **18**, 1–7.

MACKENZIE, J. M., CANIL, D., JOHNSTON, S. T., ENGLISH, J., MIHALYNUK, M. G. & GRANT, B. 2005. First evidence for ultrahigh-pressure garnet peridotite in the North American Cordillera. *Geology*, **33**, 105–108.

MAEDA, J. & KAGAMI, H. 1996. Interaction of a spreading ridge an accretionary prism: Implications from MORB magmatism in the Hidaka magmatic zone, Hokkaido, Japan. *Geology*, **24**, 31–34.

MANN, P. & TAIRA, A. 2004. Global tectonic significance of the Solomon Islands and Ontong Java Plateau convergent zone. *Tectonophysics*, **389**, 137–190.

MARSAGLIA, K. M. 1995. Interarc and backarc basins. *In*: BUSBY, C. J. & INGERSOLL, R. V. (eds) *Tectonics of Sedimentary Basins*. Blackwell Science, Cambridge, MA, 299–329.

MARUYAMA, S. 1997. Pacific-type orogeny revisited: Miyashiro-type orogeny proposed. *Island Arc*, **6**, 91–120.

MARUYAMA, S., KAWAI, T. & WINDLEY, B. F. in press. Ocean plate stratigraphy and its imbrication in an accretionary orogen: The Mona complex, Anglesey–Lleyn, Wales, UK. *In*: KUSKY, T. M., ZHAI, M. G. & XIAO, W. J. (eds) *The Evolved Continents: Understanding the Processes of Continental Growth*. Geological Society, London, Special Publications.

MATSUDA, T. & ISOZAKI, Y. 1991. Well-documented travel history of Mesozoic pelagic chert in Japan: From remote ocean to subduction zone. *Tectonics*, **10**, 475–499.

MATSUDA, T. & UYEDA, S. 1971. On the Pacific-type orogeny and its model: Extension of the paired belts concept and possible origin of marginal seas. *Tectonophysics*, **11**, 5–27.

MAXSON, J. & TIKOFF, B. 1996. Hit-and-run collisional model for the Laramide orogeny, western United States. *Geology*, **24**, 968–972.

MCCULLOCH, M. T. & BENNETT, V. C. 1994. Progressive growth of the Earth's continental crust and depleted mantle: Geochemical constraints. *Geochimica et Cosmochimica Acta*, **58**, 4717–4738.

METCALFE, I. 1996a. Gondwanaland dispersion, Asian accretion and evolution of eastern Tethys. *Australian Journal of Earth Sciences*, **43**, 605–623.

METCALFE, I. 1996b. Pre-Cretaceous evolution of SE Asian Terranes. *In*: HALL, R. & BLUNDELL, D. J. (eds) *Tectonic Evolution of SE Asia*. Geological Society, London, Special Publications, **106**, 97–122.

METCALFE, I. 2002. Permian tectonic framework and palaeogeography of SE Asia. *Journal of Asian Earth Sciences*, **20**, 551–566.

MIYASHIRO, A. 1961. Evolution of metamorphic belts. *Journal of Petrology*, **2**, 277–311.

MIYASHIRO, A. 1972. Metamorphism and related magmatism in plate tectonics. *American Journal of Science*, **272**, 629–656.

MIYASHIRO, A. 1973a. *Metamorphism and Metamorphic Belts*. Halstead Press, New York.

MIYASHIRO, A. 1973b. Paired and unpaired metamorphic belts. *Tectonophysics*, **17**, 241–254.

MIYASHIRO, A., AKI, K. & ŞENGÖR, A. M. C. 1982. *Orogeny*. Wiley, Chichester.

MONGER, J. & KNOKLEBERG, W. H. 1996. Evolution of the northern North America Cordillera: Generation, fragmentation, displacement and accretion of successive North American plate margin arcs. *In*: COYNER, A. R. & FAHAY, P. L. (eds) *Geology and Ore Deposits of the American Cordillera*. Geological Society of Nevada, 1133–1152.

MOORES, E. M. 1991. Southwest U.S.–East Antarctica (SWEAT) connection: A hypothesis. *Geology*, **19**, 425–428.

MOORES, E. M. & TWISS, R. J. 1995. *Tectonics*. W. H. Freeman, New York.

MORLEY, C. K. 2009. Evolution from an oblique subduction back-arc mobile belt to a highly oblique collisional margin: the Cenozoic tectonic development of Thailand and eastern Myanmar. *In*: CAWOOD, P. A. & KRÖNER, A. (eds) *Earth Accretionary Systems in Space and Time*. Geological Society, London, Special Publications, **318**, 373–403.

MORTIMER, N. 2004. New Zealand's geological foundations. *Gondwana Research*, **7**, 261–272.

MÜNTENER, O., KELEMEN, P. B. & GROVE, T. L. 2001. The role of H_2O during crystallization of primitive arc magmas under uppermost mantle conditions and genesis of igneous pyroxenites: An experimental study. *Contributions to Mineralogy and Petrology*, **141**, 643–658.

MURPHY, J. B. & NANCE, R. D. 1991. Supercontinent model for the contrasting character of Late Proterozoic orogenic belts. *Geology*, **19**, 469–472.

MURPHY, J. B., OPPLIGER, G. L. & BRIMHALL, G. H. 1998. Plume-modified orogeny: An example from the western United States. *Geology*, **26**, 731–734.

MURPHY, J. B., VAN STAAL, C. R. & KEPPIE, J. D. 1999. Middle to late Palaeozoic Acadian orogeny in the northern Appalachians: A Laramide-style plume-modified orogeny? *Geology*, **27**, 653–656.

MURPHY, J. B., HYNES, A. J., JOHNSTON, S. T. & KEPPIE, J. D. 2003. Reconstructing the ancestral Yellowstone plume from accreted seamounts and its relationship to flat-slab subduction. *Tectonophysics*, **365**, 185–194.

NAKANISHI, A., TAKAHASHI, N. *ET AL.* 2002. Crustal structure across the coseismic rupture zone of the 1944 Tonankai earthquake, the central Nankai Trough seismogenic zone. *Journal of Geophysical Research*, **107**, EMP 2-1-21, doi:10.1029/2001JB000424.

NEMCHIN, A. A., WHITEHOUSE, M. J., MENNEKEN, M., GEISLER, T., PIDGEON, R. T. & WILDE, S. A. 2008. A light carbon reservoir recorded in zircon-hosted diamond from the Jack Hills. *Nature*, **454**, 92–95.

NIE, S., YIN, A., ROWLEY, D. B. & JIN, Y. 1994. Exhumation of the Dabie Shan ultra-high pressure rocks and accumulation of the Songpan–Ganzi flysch sequence, central China. *Geology*, **22**, 999–1002.

NUTMAN, A. P. & FRIEND, C. R. L. 2007. Comment on 'A Vestige of Earth's Oldest Ophiolite'. *Science*, **318**, 746c, doi:10.1126/science.1144148.

NUTMAN, A. P., BENNET, V. C., FRIEND, C. R. L., JENNER, F., WAN, Y. & LIU, D.-Y. 2009. Eoarchaean crustal growth in West Greenland (Itsaq Gneiss Complex) and in northeastern China (Anshan area): review and synthesis. *In*: CAWOOD, P. A. & KRÖNER, A. (eds) *Earth Accretionary Systems in Space and Time*. Geological Society, London, Special Publications, **318**, 127–154.

OLDOW, J. S., BALLY, A. W. & AVÉ LALLEMANT, H. G. 1990. Transpression, orogenic float, and lithospheric balance. *Geology*, **18**, 991–994.

ONCKEN, O., CHONG, G. *ET AL.* 2006a. *The Andes: Active Subduction Orogeny*. Springer, Berlin.

ONCKEN, O., HINDLE, D., KLEY, J., ELGER, K. & SCHEMMANN, K. 2006b. Deformation of the Central Andean upper plate system—facts, fiction, and constraints for plateau models. *In*: ONCKEN, O., CHONG, G. *ET AL.* (eds) *The Andes: Active Subduction Orogeny*. Springer, Berlin, 3–27.

O'NEILL, C., LENARDIC, A., MORESI, L., TORSVIK, T. H. & LEE, C.-T. A. 2007. Episodic Precambrian subduction. *Earth and Planetary Science Letters*, **262**, 552–562.

OSOZAWA, S. 1988. Ridge subduction-induced orogeny, a case study of the Cretaceous to Palaeogene in southwest Japan. *In*: FLOWER, M. F. J., CHUNG, S.-L., LO, C.-H. & LEE, T.-Y. (eds) *Mantle Dynamics and Plate Interactions in East Asia*. American Geophysical Union, Geodynamic Series, **27**, 331–336.

OXBURGH, E. R. & TURCOTTE, D. L. 1970. Thermal structure of island arcs. *Geological Society of America Bulletin*, **81**, 1665–1688.

OXBURGH, E. R. & TURCOTTE, D. L. 1971. Origin of paired metamorphic belts and crustal dilation in island arc regions. *Journal of Geophysical Research*, **76**, 1315–1327.

PACKHAM, G. H. & LEITCH, E. C. 1974. The role of plate tectonic theory in the interpretation of the Tasman Orogenic Zone. *In*: DENMEAD, A. K., TWEEDALE, G. W. & WILSON, A. F. (eds) *The Tasman Geosyncline—A Symposium*. Geological Society of Australia, Brisbane, 129–154.

PAGE, R. A. & BROCHER, T. M. 1993. Thrusting of the central California margin over the edge of the Pacific plate during the transform regime. *Geology*, **21**, 635–638.

PARRA, M., FAUGERES, J.-C., GROUSSET, F. & PUJOL, C. 1997. Sr–Nd isotopes as tracers of fine-grained detrital sediments: The South-Barbados accretionary prism during the last 150 kyr. *Marine Geology*, **136**, 225–243.

PAVLIS, T. L., MARTY, K. & SISSON, V. B. 2003. Constrictional flow within the Eocene forearc of southern Alaska: An effect of dexteral shear during ridge subduction. *In*: SISSON, V. B., ROESKE, S. M. & PAVLIS, T. L. (eds) *Geology of a Transpressional Orogen Developed during Ridge–Trench Interaction along the North Pacific Margin*. Geological Society of America, Special Papers, **371**, 171–190.

PELTONEN, P. & KONTINEN, A. 2004. The Jormua ophiolite: A mafic–ultramafic complex from an ancient ocean–continent transition zone. *In*: KUSKY, T. M. (ed.) *Precambrian Ophiolites and Related Rocks*. Elsevier, Amsterdam, 35–71.

PERCIVAL, J. A. & HELMSTAEDT, H. 2004. Insights on Archean continent–ocean assembly, western Superior Province, from new structural, geochemical and geochronological observations: Introduction and summary. *Precambrian Research*, **132**, 209–212.

PERCIVAL, J. A., SANBORN-BARRIE, M., SKULSKI, T., STOTT, G. M., HELMSTAEDT, H. & WHITE, D. J. 2006. Tectonic evolution of the western Superior Province from NATMAP and Lithoprobe studies. *Canadian Journal of Earth Sciences*, **43**, 1085–1117.

PETFORD, N. & ATHERTON, M. P. 1995. Cretaceous–Tertiary volcanics and syn-subduction crustal extension in northern central Peru. *In*: SMELLIE, J. L. (ed.) *Volcanism Associated with Extension at Consuming Plate Margins*. Geological Society, London, Special Publications, **81**, 233–248.

PETTERSON, M. G., BABBS, T. *ET AL.* 1999. Geological-tectonic framework of Solomon Islands, SW Pacific: Crustal accretion and growth within an intra-oceanic setting. *Tectonophysics*, **301**, 35–60.

POLAT, A. & KERRICH, R. 1999. Formation of an Archean tectonic melange in the Schreiber–Hemlo greenstone belt, Superior Province, Canada: Implications for Archean subduction–accretion processes. *Tectonics*, **18**, 733–755.

POLAT, A. & KERRICH, R. 2004. Precambrian arc associations: Boninites, adakites, magnesian andesites, and Nb-enriched basalts. *In*: KUSKY, T. M. (ed.) *Precambrian Ophiolites and Related Rocks*. Elsevier, Amsterdam, 567–597.

POLAT, A., KERRICH, R. & WINDLEY, B. 2009. Archaean crustal growth processes in southern West Greenland

and the southern Superior Province: geodynamic and magmatic constraints. *In*: CAWOOD, P. A. & KRÖNER, A. (eds) *Earth Accretionary Systems in Space and Time*. Geological Society, London, Special Publications, **318**, 155–191.

RAMOS, V. A., CRISTALLINI, E. O. & PEREZ, D. J. 2002. The Pampean flat-slab of the Central Andes. *Journal of South American Earth Sciences*, **15**, 59–78.

RATSCHBACHER, L., FRISCH, W., CHEN, C. & PAN, G. 1996. Cenozoic deformation, rotation, and stress patterns in eastern Tibet and western Sichuan, China. *In*: YIN, A. & HARRISON, T. M (eds) *The Tectonic Evolution of Asia*. Cambridge University Press, Cambridge, 227–249.

REYMER, A. & SCHUBERT, G. 1984. Phanerozoic addition rates to the continental crust and crustal growth. *Tectonics*, **3**, 63–77.

ROESKE, S. M., SNEE, L. W. & PAVLIS, T. L. 2003. Dextral-slip reactivation of an arc–forearc boundary during Late Cretaceous–Early Eocene oblique convergence in the northern Cordillera. *In*: SISSON, V. B., ROESKE, S. M. & PAVLIS, T. L. (eds) *Geology of a Transpressional Orogen Developed during Ridge–Trench Interaction along the North Pacific Margin*. Geological Society of America, Special Papers, **371**, 141–169.

ROUSSE, S., GILDER, S., FARBER, D., MCNULTY, B., PATRIAT, P., TORRES, V. & SEMPERE, T. 2003. Palaeomagnetic tracking of mountain building in the Peruvian Andes since 10 Ma. *Tectonics*, **22**, 1048–1069.

ROYDEN, L. H. 1993*a*. Evolution of retreating subduction boundaries formed during continental collision. *Tectonics*, **12**, 629–638.

ROYDEN, L. H. 1993*b*. The tectonic expression of slab pull at continental convergent boundaries. *Tectonics*, **12**, 303–325.

RUSSO, R. M. & SILVER, P. G. 1996*a*. Cordillera formation, mantle dynamics, and the Wilson cycle. *Geology*, **24**, 511–514.

RUSSO, R. M. & SILVER, P. G. 1996*b*. Trench-parallel flow beneath the Nazca Plate from seismic anisotropy. *Science*, **263**, 1105–1111.

SAGIYA, T. & THATCHER, W. 1999. Coseismic slip resolution along a plate boundary megathrust: The Nankai Trough, Southwest Japan. *Journal of Geophysical Research*, **104**, 1111–1130.

SALEEBY, J. 2003. Segmentation of the Laramide Slab—evidence from the southern Sierra Nevada region. *Geological Society of America Bulletin*, **115**, 655–668.

SAMSON, S. D. & PATCHETT, P. J. 1991. The Canadian cordillera as a modern analogue of Proterozoic crustal growth. *Australian Journal of Earth Sciences*, **38**, 595–611.

SAMSON, S. D., MCCLELLAND, W. C., PATCHETT, P. J., GEHRELS, G. E. & ANDERSON, R. G. 1989. Evidence from neodymium isotopes for mantle contributions to Phanerozoic crustal genesis in the Canadian Cordillera. *Nature*, **337**, 705–709.

SANDIFORD, M. & HAND, M. 1998. Controls on the locus of intraplate deformation in central Australia. *Earth and Planetary Science Letters*, **162**, 97–110.

SANDIFORD, M., HAND, M. & MCLAREN, S. 2001. Tectonic feedback, intraplate orogeny and the geochemical structure of the crust: A central Australian perspective. *In*: MILLER, J. A., HOLDSWORTH, R. E., BUICK, I. S. & HAND, M. (eds) *Continental Reactivation and Reworking*. Geological Society, London, Special Publications, **184**, 195–218.

SAWAKI, Y., SHIBUYA, T. ET AL. in press. Imbricated ocean plate stratigraphy and U–Pb zircon ages from tuff beds in cherts in the Ballantrae Complex, SW Scotland. *Geological Society of America Bulletin*.

SAWKINS, F. J. 1990. *Metal Deposits in Relation to Plate Tectonics*. Springer, Berlin.

SCHELLART, W. P. 2008. Overriding plate shortening and extension above subduction zones: A parametric study to explain formation of the Andes Mountains. *Geological Society of America Bulletin*, **120**, 1441–1454.

SCHELLART, W. P. & LISTER, G. S. 2004. Tectonic models for the formation of arc-shaped convergent zones and backarc basins. *In*: SUSSMAN, A. J. & WEIL, A. B. (eds) *Orogenic Curvature: Integrating Palaeomagnetic and Structural Analyses*. Geological Society of America, Special Papers, **383**, 237–258.

SCHELLART, W. P., LISTER, G. S. & TOY, V. G. 2006. A Late Cretaceous and Cenozoic reconstruction of the Southwest Pacific region: Tectonics controlled by subduction and slab rollback processes. *Earth-Science Reviews*, **76**, 191–233.

SCHELLART, W. P., FREEMAN, J., STEGMAN, D. R., MORESI, L. & MAY, D. 2007. Evolution and diversity of subduction zones controlled by slab width. *Nature*, **446**, 308–311.

SCHOLL, D. W. & VON HUENE, R. 2007. Exploring the implications for continental basement tectonics if estimated rates of crustal removal (recycling) at Cenozoic subduction zones are applied to Phanerozoic and Precambrian convergent ocean margins. *In*: HATCHER, R. D. JR, CARLSON, M. P., MCBRIDE, J. H. & CATALÁN, J. M. (eds) *4-D Framework of Continental Crust*. Geological Society of America, Memoirs, 9–32.

SCHOLL, D. W. & VON HUENE, R. 2009. Implications of estimated magmatic additions and recycling losses at the subduction zones of accretionary (non-collisional) and collisional (suturing) orogens. *In*: CAWOOD, P. A. & KRÖNER, A. (eds) *Earth Accretionary Systems in Space and Time*. Geological Society, London, Special Publications, **318**, 105–125.

SCHOLL, D. W., VON HUENE, R., VALLIER, T. L. & HOWELL, D. G. 1980. Sedimentary masses and concepts about tectonic processes at underthrust ocean margins. *Geology*, **8**, 564–568.

SCOTT, D. J., HELMSTAADT, H. & BICKLE, M. J. 1992. Purtuniq ophiolite, Cape Smith belt, northern Quebec, Canada: A reconstructed section of Early Proterozoic oceanic crust. *Geology*, **20**, 173–176.

SEDLOCK, R. L. & ISOZAKI, Y. 1990. Lithology and biostratigraphy of Franciscan-like chert and associated rocks in west–central Baja California, Mexico. *Geological Society of America Bulletin*, **102**, 852–864.

SEELEY, D. R., VAIL, P. R. & WALTON, G. G. 1974. Trench slope model. *In*: BURK, C. A. & DRAKE, C. L. (eds) *The Geology of Continental Margins*. Springer, Berlin, 249–260.

ŞENGÖR, A. M. C. 1976. Collision of irregular continental margins: Implications for foreland deformation of Alpine-type orogens. *Geology*, **4**, 779–782.

ŞENGÖR, A. M. C. 1993. Turkic-type orogeny in the Altaids: Implications for the evolution of continental crust and methodology of regional tectonic analysis (34th Bennett Lecture). *Transactions of the Leicester Literature and Philosophical Society*, **87**, 37–54.

ŞENGÖR, A. M. C. 1995. Sedimentation and tectonics of fossil rifts. *In*: BUSBY, C. J. & INGERSOLL, R. V. (eds) *Tectonics of Sedimentary Basins*. Blackwell, Cambridge, MA, 53–117.

ŞENGÖR, A. M. C. & NATAL'IN, B. A. 1996. Paleotectonics of Asia: Fragments of a synthesis. *In*: YIN, A. & HARRISON, T. M. (eds) *The Tectonic Evolution of Asia*. Cambridge University Press, Cambridge, 486–640.

ŞENGÖR, A. M. C., BURKE, K. & DEWEY, J. F. 1978. Rifts and high angles to orogenic belts: Tests for their origin and the upper Rhine graben as an example. *American Journal of Science*, **278**, 24–40.

SHREVE, R. L. & CLOOS, M. 1986. Dynamics of sediment subduction, mélange formation, and prism accretion. *Journal of Geophysical Research*, **91**, 10229–10245.

SILVER, P. G. & BEHN, M. D. 2008. Intermittent Plate Tectonics? *Science*, **319**, 85–88.

SILVER, P. G., RUSSO, R. M. & LITHGOW–BERTELLONI, C. 1998. Coupling of South American and African plate motion and plate deformation. *Science*, **279**, 60–63.

SISSON, V. B., ROESKE, S. & PAVLIS, T. L. (eds) 2003. *Geology of a Transpressional Orogen Developed During Ridge–Trench Interaction along the North Pacific Margin*. Geological Society of America, Special Papers, **371**.

SLEEP, N. H. 2000. Evolution of the mode of convection within terrestrial planets. *Journal of Geophysical Research*, **105**, 17563–17578.

SMELLIE, J. L. (ed.) 1994. *Volcanism Associated with Extension at Consuming Plate Margins*. Geological Society, London, Special Publications, **84**.

SMITH, G. A. & LANDIS, C. A. 1995. Intra-arc basins. *In*: BUSBY, C. J. & INGERSOLL, R. V. (eds) *Tectonics of Sedimentary Basins*. Blackwell Science, Cambridge, MA, 263–298.

SMITHIES, R. H., CHAMPION, D. C. & SUN, S.-S. 2004. The case for Archean boninites. *Contributions to Mineralogy and Petrology*, **147**, 705–721.

SMITHIES, R. H., CHAMPION, D. C. & VAN KRANENDONK, M. J. 2005. Modern-style subduction processes in the Mesoarchean: Geochemical evidence from the 3.12 Ga Whundo intra-oceanic arc. *Earth and Planetary Science Letters*, **231**, 221–237.

SNYDER, D. B., LUCAS, S. B. & MCBRIDE, J. H. 1996a. Crustal and mantle reflectors from Palaeoproterozoic orogens and their relation to arc–continent collisions. *In*: BREWER, T. S. (ed.) *Precambrian Crustal Evolution in the North Atlantic Region*. Geological Society, London, Special Publications, **112**, 1–23.

SNYDER, D. B., PRASETYO, H., BLUNDELL, D. J., PIGRAM, C. J., BARBER, A. J., RICHARDSON, A. & TJOKOSAPROETRO, S. 1996b. A dual doubly vergent orogen in the Banda Arc continent–continent collision zone as observed on deep seismic reflection profiles. *Tectonics*, **15**, 34–53.

SNYDER, D. B., CLOWES, R. M., COOK, F. A., ERDMER, P., EVENCHICK, C. A., VAN DER VELDEN, A. J. & HALL, K. W. 2002. Proterozoic prism arrests suspect terranes: Insights into the ancient Cordilleran margin from seismic reflection data. *GSA Today*, **12**, 4–9.

SNYDER, D. B., PILKINGTON, M., CLOWES, R. M. & COOK, F. A. 2009. The underestimated Proterozoic component of the Canadian Cordillera accretionary margin. *In*: CAWOOD, P. A. & KRÖNER, A. (eds) *Earth Accretionary Systems in Space and Time*. Geological Society, London, Special Publications, **318**, 257–271.

SOH, W., NAKAYAMA, K. & KIMURA, T. 1998. Arc–arc collision in the Izu collision zone, central Japan, deduced from the Ashigara basin and adjacent Tanzawa mountains. *Island Arc*, **7**, 330–341.

STEIN, M. & HOFMANN, A. W. 1993. Mantle plumes and episodic crustal growth. *Nature*, **372**, 63–68.

STERN, C. R. 1991. Role of subduction erosion in the generation of Andean magmas. *Geology*, **19**, 78–81, doi:10.1130/0091-7613(1991)0192.3.CO:2.

STERN, R. J. 2004. Subduction initiation: Spontaneous and induced. *Earth and Planetary Science Letters*, **226**, 275–292.

STERN, R. J. 2005. Evidence from ophiolites, blueschists and ultrahigh-pressure metamorphic terranes that the modern episode of subduction tectonics began in the Neoproterozoic. *Geology*, **33**, 557–560.

STERN, R. J. in press. Neoproterozoic crustal growth: The solid Earth system during a critical episode of Earth history. *Gondwana Research*.

STERN, R. J. & BLOOMER, S. H. 1992. Subduction zone infancy: Examples from the Eocene Izu–Bonin–Mariana and Jurassic California arcs. *Geological Society of America Bulletin*, **104**, 1621–1636.

STERN, R. J., JOHNSON, P. R., KRÖNER, A. & YIBAS, B. 2004. Neoproterozoic ophiolites of the Arabian–Nubian Shield. *In*: KUSKY, T. M. (ed.) *Precambrian Ophiolites and Related Rocks*. Elsevier Developments in Precambrian Geology, **13**, 95–128.

ST-ONGE, M. R., VAN GOOL, J. A. M., GARDE, A. A. & SCOTT, D. J. 2009. Correlation of Archaean and Palaeoproterozoic units between northeastern Canada and western Greenland: constraining the pre-collisional upper plate accretionary history of the Trans-Hudson orogen. *In*: CAWOOD, P. A. & KRÖNER, A. (eds) *Earth Accretionary Systems in Space and Time*. Geological Society, London, Special Publications, **318**, 193–235.

SUTHERLAND, R. & HOLLIS, C. 2001. Cretaceous demise of the Moa plate and strike–slip motion at the Gondwana margin. *Geology*, **29**, 279–282.

SYME, E. C., LUCAS, S. B., BAILES, A. H. & STERN, R. A. 1999. Contrasting arc and MORB-like assemblages in the Palaeoproterozoic Flin Flon Belt, Manitoba, and the role of intra-arc extension in localizing volcanic-hosted massive sulphide deposits. *Canadian Journal of Earth Sciences*, **36**, 1767–1788.

TAPPONNIER, P., XU, Z., ROGER, F., MEYER, B., ARNAUD, N., WITTLINGER, G., JINGSUI, Y. & YANG, J. 2001. Oblique stepwise rise and growth of the Tibet Plateau. *Science*, **294**, 1671–1677.

TATSUMI, Y. 2005. The subduction factory: How it operates on Earth. *GSA Today*, **15**, 4–10.

TATSUMI, Y. & EGGINS, S. 1995. *Subduction Zone Magmatism*. Blackwell, Cambridge, MA.

TATSUMI, Y., MARUYAMA, S. & NOHDA, S. 1990. Mechanism of backarc opening in the Japan Sea: Role of asthenospheric injection. *Tectonophysics*, **181**, 299–306.

TAYLOR, B. & KARNER, G. D. 1983. On the evolution of marginal basins. *Reviews of Geophysics and Space Physics*, **21**, 1727–1741.

TAYLOR, S. R. 1967. The origin and growth of continents. *Tectonophysics*, **4**, 17–34.

TAYLOR, S. R. & MCCLENNAN, S. M. 1985. *The Continental Crust: Its Composition and Evolution*. Blackwell Scientific, Oxford.

THOMPSON, A. B., SCHULMANN, K., JEZEK, J. & TOLAR, V. 2001. Thermally softened continental extensional zones (arcs and rifts) as precursors to thickened orogenic belts. *Tectonophysics*, **332**, 115–141.

TROUW, R. A. J. & DE WIT, M. J. 1999. Relation between Gondwanide Orogen and contemporaneous intracratonic deformation. *Journal of African Earth Sciences*, **28**, 203–213.

UNDERWOOD, M. B. (ed.) 1993. *Thermal Evolution of the Tertiary Shimanto Belt, Southwest Japan: An Example of Ridge–Trench Interaction*. Geological Society of America, Special Papers, **273**.

UYEDA, S. & KANAMORI, H. 1979. Back-arc opening and the mode of subduction. *Journal of Geophysical Research*, **84**, 1049–1061.

VAN GOOL, J. A. M. & CAWOOD, P. A. 1994. Frontal vs. basal accretion and contrasting particle paths in metamorphic thrust belts. *Geology*, **22**, 51–54.

VAN HUNEN, J., VAN DEN BERG, A. P. & VLAAR, N. J. 2002. On the role of subducting oceanic plateaus in the development of shallow flat subduction. *Tectonophysics*, **352**, 317–333.

VAN KRANENDONK, M. J., HUGH SMITHIES, R., HICKMAN, A. H. & CHAMPION, D. C. 2007. Review: Secular tectonic evolution of Archean continental crust: Interplay between horizontal and vertical processes in the formation of the Pilbara Craton, Australia. *Terra Nova*, **19**, 1–38.

VAN STAAL, C. R., DEWEY, J. F., MAC NIOCALL, C. & MCKERROW, W. S. 1998. The Cambrian–Silurian tectonic evolution of the northern Appalachian and British Caledonides: history of a complex west and southwest Pacific-type segment of Iapetus. *In*: BUNDELL, D. J. & SCOTT, A. C. (eds) *Lyell: The Past is the Key to the Present*. Geological Society, London, Special Publications, **143**, 199–242.

VAUGHAN, A. P. M. 1995. Circum-Pacific mid-Cretaceous deformation and uplift: A superplume-related event? *Geology*, **23**, 491–494.

VAUGHAN, A. P. M. & LIVERMORE, R. A. 2005. Episodicity of Mesozoic terrane accretion along the Pacific margin of Gondwana: Implications for superplume–plate interactions. *In*: VAUGHAN, A. P. M., LEAT, P. T. & PANKHURST, R. J. (eds) *Terrane Processes at the Margins of Gondwana*. Geological Society, London, Special Publications, **246**, 143–178.

WALTER, M. R., VEEVERS, J. J., CALVER, C. R. & GREY, K. 1995. Neoproterozoic stratigraphy of the Centralian Superbasin, Australia. *Precambrian Research*, **73**, 173–195.

WANG, K., HYNDMAN, R. D. & YAMANO, M. 1995. Thermal regime of the Southwest Japan subduction zone: Effects of age history of the subducting plate. *Tectonophysics*, **248**, 53–69.

WANG, Y. X., MOONEY, W. D., YUAN, X. C. & OKAYA, N. in press. Crustal structure of the northeastern Tibetan Plateau from the southern Tarim basin to the Sichuan basin. *Journal of Geophysical Research*.

WANNAMAKER, P. E., BOOKER, J. R., JONES, A. G., CHAVE, A. D., FILLOUX, J. H., WAFF, H. S. & LAW, L. K. 1989. Resistivity cross-section through the Juan de Fuca subduction system and its tectonic implications. *Journal of Geophysical Research*, **94**, 14127–14144.

WELLS, R. E., BLAKELY, R. J., SUGIYAMA, Y., SCHOLL, D. W. & DINTERMAN, P. A. 2003. Basin-centered asperities in great subduction zone earthquakes: A link between slip, subsidence, and subduction erosion? *Journal of Geophysical Research*, **108**, 2507, doi:10.1029/2002JB002072.

WILLIAMS, I. S. & COLLINS, W. J. 1990. Granite–greenstone terranes in the Pilbara Block, Australia, as coeval volcano-plutonic complexes; Evidence from U–Pb zircon dating of the Mount Edgar Batholith. *Earth and Planetary Science Letters*, **97**, 41–53.

WILSON, J. T. 1966. Did the Atlantic close and then re-open? *Nature*, **211**, 676–681.

WINDLEY, B. F. 1992. Proterozoic collisional and accretionary orogens. *In*: CONDIE, K. C. (ed.) *Proterozoic Crustal Evolution*. Developments in Precambrian Geology, **10**, 419–446.

WINDLEY, B. F. & GARDE, A. A. 2009. Arc-generated blocks with crustal sections in the North Atlantic craton of West Greenland: New mechanism of crustal growth in the Archaean with modern analogues. *Earth-Science Reviews*, **93**, 1–30.

WINDLEY, B. F. & SMITH, J. V. 1976. Archaean high-grade complexes and modern continental margins. *Nature*, **260**, 671–675.

WINDLEY, B. F., WHITEHOUSE, M. J. & BA-BA-BTTAT, M. A. O. 1996. Early Precambrian gneiss terranes and Pan-African island arcs in Yemen: Crustal accretion of the eastern Arabian Shield. *Geology*, **24**, 131–134.

WINDLEY, B. F., ALEXEIEV, D., XIAO, W., KRÖNER, A. & BADARCH, G. 2007. Tectonic models for accretion of the Central Asian Orogenic Belt. *Journal of the Geological Society, London*, **164**, 31–47.

WU, F.-Y., JAHN, B.-M., WILDE, S. & SUN, D.-Y. 2000. Phanerozoic crustal growth: U–Pb and Sr–Nd isotopic evidence from the granites in northeastern China. *Tectonophysics*, **328**, 89–113.

WYMAN, D. A., BLEEKER, W. & KERRICH, R. 1999*a*. A 2.7 Ga plume, proto-arc, to arc transition and the geodynamic setting of the Kidd Creek deposit: Evidence from precise ICP MS trace element data. *Economic Geology Monograph*, **10**, 511–528.

WYMAN, D. A., KERRICH, R. & GROVES, D. I. 1999*b*. Lode gold deposits and Archean mantle–island arc interaction, Abitibi Subprovince, Canada. *Journal of Geology*, **107**, 715–725.

XIAO, W., WINDLEY, B. F., HAO, J. & ZHAI, M. 2003. Accretion leading to collision and the Permian Solonker suture, Inner Mongolia, China: Termination of the Central Asian Orogenic Belt. *Tectonics*, **22**, 1069, doi:10.1029/2002TC001484.

XIAO, W., WINDLEY, B. F., BADARCH, G., SUN, S., LI, J., QIN, K. & WANG, Z. 2004. Palaeozoic accretionary and convergent tectonics of the southern Altaids: Implications for the growth of Central Asia. *Journal of the Geological Society, London*, **161**, 339–342.

XIAO, X., LI, T., LI, G. & YUAN, X. 1998. *Tectonic evolution of the lithosphere of the Himalayas (General principle).* Structural Geology and Geomechanics, Series 5, 7, Geological Publishing House, Beijing [in Chinese].

ZHAO, G., WILDE, S. A., LI, S., SUN, M., GRANT, M. L. & LI, X. 2007. U–Pb zircon age constraints on the Dongwanzi ultramafic–mafic body, North China, confirm it is not an Archean ophiolite. *Earth and Planetary Science Letters*, **255**, 85–93.

ZHOU, D. & GRAHAM, S. A. 1996. The Songpan–Ganzi complex of the West Qinling Shan as a Triassic remnant ocean basin. *In*: YIN, A. & HARRISON, T. M. (eds) *The Tectonic Evolution of Asia.* Cambridge University Press, Cambridge, 281–299.

Metamorphic patterns in orogenic systems and the geological record

MICHAEL BROWN

Laboratory for Crustal Petrology, Department of Geology, University of Maryland, College Park, MD 20742-4211, USA (e-mail: mbrown@umd.edu)

Abstract: Regional metamorphism occurs in plate boundary zones. Accretionary orogenic systems form at subduction boundaries in the absence of continent collision, whereas collisional orogenic systems form where ocean basins close and subduction steps back and flips (arc collisions), simply steps back and continues with the same polarity (block and terrane collisions) or ultimately ceases (continental collisions). As a result, collisional orogenic systems may be superimposed on accretionary orogenic systems. Metamorphism associated with orogenesis provides a mineral record that may be inverted to yield apparent thermal gradients for different metamorphic belts, which in turn may be used to infer tectonic setting. Potentially, peak mineral assemblages are robust recorders of metamorphic P and T, particularly at high $P-T$ conditions, because prograde dehydration and melting with melt loss produce nominally anhydrous mineral assemblages that are difficult to retrogress or overprint without fluid influx. Currently on Earth, lower thermal gradients are associated with subduction (and early stages of collision) whereas higher thermal gradients are characteristic of back-arcs and orogenic hinterlands. This duality of thermal regimes is the hallmark of asymmetric or one-sided subduction and plate tectonics on modern Earth, and a duality of metamorphic belts will be the characteristic imprint of asymmetric or one-sided subduction in the geological record. Accretionary orogenic systems may exhibit retreating trench–advancing trench cycles, associated with high ($>750\,°\text{C GPa}^{-1}$) thermal gradient type of metamorphism, or advancing trench–retreating trench cycles, associated with low ($<350\,°\text{C GPa}^{-1}$) to intermediate ($350-750\,°\text{C GPa}^{-1}$) thermal gradient types of metamorphism. Whether the subducting boundary advances or retreats determines the mode of evolution. Accretionary orogenic systems may involve accretion of allochthonous and/or para-autochthonous elements to continental margins at subduction boundaries. Paired metamorphic belts, *sensu* Miyashiro, comprising a low thermal gradient metamorphic belt outboard and a high thermal gradient metamorphic belt inboard, are characteristic and may record orogen-parallel terrane migration and juxtaposition by accretion of contemporary belts of contrasting type. A wider definition of 'paired' metamorphism is proposed to incorporate all types of dual metamorphic belts. An additional feature is ridge subduction, which may be reflected in the pattern of high dT/dP metamorphism and associated magmatism. Apparent thermal gradients derived from inversion of age-constrained metamorphic $P-T$ data are used to identify tectonic settings of ancient metamorphism, to evaluate the age distribution of metamorphism in the rock record from the Neoarchaean Era to the Cenozoic Era, and to consider how this relates to the supercontinent cycle and the process of terrane export and accretion. In addition, I speculate about metamorphism and tectonics before the Mesoarchaean Era.

Forty years ago, the introduction of the plate tectonics paradigm provided a robust framework within which to understand the tectonics of the lithosphere, the strong outer layer of Earth above the softer asthenosphere (Isacks *et al.* 1968), during the Cenozoic and Mesozoic Eras (the maximum lifespan of the ocean floors before return to the mantle via subduction). Orogenesis, the process of forming mountains, was one of a number of fundamental geological processes that became understandable once placed within a plate tectonics context (Dewey & Bird 1970). Within a few years, Dewey *et al.* (1973) had demonstrated that the evolution of young orogenic systems could be unravelled by inverting geological data in combination with ocean-floor magnetic anomaly maps by following the kinematic principles of plate tectonics.

At the same time, the relationship between plate tectonics and metamorphism was addressed by Ernst (1971, 1973, 1975), Oxburgh & Turcotte (1971), Miyashiro (1972) and Brothers & Blake (1973).

Currently, the circum-Pacific and Alpine–Himalayan–Indonesian orogenic systems define two orthogonal great circle distributions of the continents (Fig. 1), each of which has a different type of orogenic system along the convergent plate boundary zone (Dickinson 2004). These orogenic systems record the two main zones of active subduction into the mantle, the circum-Pacific and the Alpine–Himalayan–Indonesian subduction systems (Collins 2003). Complementary to these are two major P- and S-wave low-velocity structures (superswells or superplumes) in the lower mantle, under southern Africa and the South Pacific

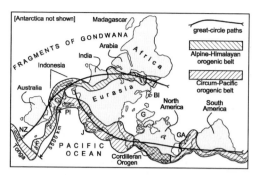

Fig. 1. Distribution of continents in relation to the Alpine–Himalayan and Circum-Pacific orogenic belts (Cordilleran orogen is cross-hatched) in 'circular' projection. BI, British Isles; F, Fiji; G, Greenland; GA, Greater Antilles; J, Japan; NZ, New Zealand; PI, Philippine Islands. Reprinted, with permission, from Dickinson, W. R. 2004. Evolution of the North American Cordillera. *Annual Review of Earth and Planetary Sciences*, **32**, 13–45. © 2004 by Annual Reviews (www.annualreviews.org).

(Montelli *et al.* 2006; Tan & Gurnis 2007). They are interpreted to record upwelling responsible for advective heat transport through the lower into the upper mantle (Nolet *et al.* 2006). This arrangement of subduction and upwelling reflects a simple pattern of long-wavelength mantle convection, a view that is supported by a variety of geophysical data (Richards & Engebretson 1992).

Accretionary orogenic systems form above a subduction boundary during continuing plate convergence in the absence of continental collision, as exemplified by the evolution of the Pacific Ocean rim during the Phanerozoic Eon (Coney 1992). These systems vary according to whether the subduction boundary or trench is retreating, neutral or advancing. They may exhibit cyclic behaviour in which a retreating trench changes to an advancing trench or in which an advancing trench changes to a retreating trench (Lister *et al.* 2001; Lister & Forster 2006). Accretion of arcs and/or allochthonous terranes, some of which may be far-travelled, is a common feature of accretionary orogenic systems, making the distinction from collisional orogenic systems somewhat arbitrary with the exception of terminal continent–continent collisions. Previously these orogenic systems have been called 'Pacific-type' (e.g. Matsuda & Uyeda 1971) or 'Cordilleran-type' (e.g. Coney *et al.* 1980), and more recently Maruyama (1997) has proposed that they should be called 'Miyashiro-type'.

Collisional orogenic systems are those in which an ocean is closed and arcs and/or allochthonous terranes and/or continents collide, as exemplified by the Tethysides (Dercourt *et al.* 1985, 1986; Savostin *et al.* 1986). The collision of an island arc with a rifted continental margin involves choking the subduction zone, subducting slab breakoff and a reversal or flip in the subduction polarity if subduction is initiated behind the arc. These 'soft' collisions generate only a short period of orogenesis because the forces opposing shortening are relieved by the renewed subduction (Dewey 2005). A similar process occurs where allochthonous blocks and terranes are sutured to an active continental margin, except that the subduction boundary steps back and continues to subduct towards the overriding plate and the associated orogenic event is minimal (e.g. van Staal *et al.* 2008). In contrast, orogens in which two continents become sutured involve significant thickening of the continental lithosphere over a wide zone, perhaps up to 1500 km across, and far-field shortening of the crust, which generates a mountain front and a wide orogenic plateau surrounded by internal basins, which themselves are bordered by thrust belts (Dewey 2005). Features such as deformation, metamorphism and magmatism may vary in intensity along and across the length and breadth of these systems because the continental margin being subducted need not be rectilinear.

Previously these orogenic systems have been called 'Himalayan-type' (Liou *et al.* 2004) or 'Turkic-type' (Şengör & Natal'in 1996). Turkic-type collisional orogenic systems are characterized by incorporation into the suture of large subduction–accretion complexes with associated magmatic arcs and terrane elements that evidence a complex history prior to terminal collision (Şengör *et al.* 1993; Moix *et al.* 2008). In some orogenic systems, such as the Alpine–Himalayan–Cimmerian orogenic system, the principal suture, in this case the Palaeo-Tethyan suture, divides the system into two parts on either side of which the subducting boundary zones, which were active accretionary orogens prior to collision, had very different character (Şengör 1992).

Accretionary orogenic systems may experience multiple 'soft' collisions. The modern example of such a system involves the Cenozoic evolution of southeastern Asia and the southwestern Pacific (Hall 2002). Here there were three system-wide events, at *c*. 45, 25 and 5 Ma, each involving changes in plate boundaries and motions as a result of collisions along some part of the subduction boundary system that led to subducting slab breakoff (Cloos *et al.* 2005).

In the older geological record, the early history of the Appalachian–Caledonian orogenic system involved peri-Laurentian arcs accreted during the Taconic (e.g. Newfoundland: Lissenberg *et al.* 2005; Zagorevski *et al.* 2006) or Grampian (e.g. western Ireland: Friedrich *et al.* 1999; Dewey 2005) orogenies. In these cases, the period from initial subduction of continental margin sediments to the end of shortening was no more than 20 Ma. The peak of metamorphism occurred within a few million years after initial collision, suggesting that emplacement of mafic magma below and into the evolving orogen advected the necessary heat. The generation of mafic magma might be achieved via subducting slab breakoff and decompression melting of asthenospheric mantle as it upwelled through a tear (Davies & von Blanckenburg 1995) or through gaps between drips (Lister *et al.* 2008), as shown, for example, for the evolution of New Guinea during the Neogene Era (Cloos *et al.* 2005).

Two of the largest accretionary orogenic systems on Earth are the Terra Australis orogenic system, which was active from the Edicaran Period into the Triassic Period, and the Central Asian Orogenic Belt (CAOB), which was active from the beginning of the Neoproterozoic Era to the end of the Mesozoic Era. The Terra Australis orogenic system, which extended some 18 000 km along the Gondwanan margin and up to 1600 km inboard, includes the Tasman, Ross and Tuhua orogens of Australia, Antarctica and New Zealand in East Gondwana, and the Cape Basin of Southern Africa and the Andean Cordillera of South America in West Gondwana (Cawood 2005; Cawood & Buchan 2007). Subduction along the Pacific margin of Gondwana was established *c.* 580–550 Ma (Cawood & Buchan 2007) contemporaneously with a major global plate reorganization associated with the last steps in the assembly of Gondwana by final closure of ocean basins and termination of intra-Gondwana subduction, and with opening of the Iapetus Ocean between Laurentia and Baltica. The Terra Australis orogenic system has a protracted history of continuing subduction and associated episodic plate boundary orogenesis, but terrane accretion and arc collisions are rare and continent–continent collisions are absent. In contrast, the CAOB formed by accretion of ophiolitic mélange zones, accretionary wedges, oceanic plateaux, island arcs and microcontinents in a manner comparable with circum-Pacific Mesozoic–Cenozoic accretionary orogens (Windley *et al.* 2007). The CAOB evolved over a period of more than 700 Ma culminating in terminal collision between the Siberian and North China cratons at around 250 Ma (Windley *et al.* 2007; Kröner *et al.* 2008). Closure of small ocean basins between accretion events means that ridge–trench interactions were likely (Windley *et al.* 2007). Furthermore, the CAOB represents a major site of juvenile magmatic additions to the crust during the Neoproterozoic and Palaeozoic Eras (Şengör *et al.* 1993; Jahn *et al.* 2000; Jahn 2004).

The Himalayan orogenic system is generally regarded as the 'type' example of a collisional orogenic system. Here the collisional phase has been evolving for *c.* 50 Ma (Rowley 1996) and was preceded by a longer period of subduction-related orogenesis involving accretion of arc terranes. This is by no means a long period of orogeny. In the older geological record, the orogenic system that sutured Laurentia with Amazonia (Grenville) evolved over *c.* 500 Ma (Rivers & Corrigan 2000). A continental-margin magmatic arc existed on the southeastern margin of Laurentia from *c.* 1500 to 1230 Ma, with part of the arc subsequently incorporated into the 1190–990 Ma collisional orogen. The arc oscillated between extension and shortening several times with back-arc deposits of several ages; closure of the back-arc basins occurred during two accretionary orogenies at *c.* 1495–1445 and 1250–1190 Ma, as well as during three crustal shortening events associated with the 1190–990 Ma collisional orogeny.

Young mountain chains on Earth are clearly associated with linear belts of distinctive sedimentation, deformation, magmatism and metamorphism, as well as the high relief. In contrast, ancient orogenic belts must be identified based on sedimentation and stratigraphy, particularly the occurrence of unconformities in the rock record, the type of associated volcanism and plutonism, commonly using chemical fingerprinting, and the style of tectonic deformation and regional metamorphism, simply because the mountains were eroded long ago.

A characteristic feature of subduction boundary zones is the development of dual thermal environments (Oxburgh & Turcotte 1970, 1971), representing the subduction zone or collisional suture (cooler) and the arc–back-arc system or orogenic hinterland (warmer). This feature is the hallmark of asymmetric or one-sided subduction on modern Earth (one-sided subduction is defined in Fig. 2). Brown (2006) showed that different types of metamorphism would be registered in each of these thermal environments, and proposed that the record of metamorphism in ancient orogens may be inverted to determine when this style of subduction boundary zone first was registered in the geological record. This simple approach to the geological record of metamorphism may appear incompatible with the complexity of evolution implied by the discussion of orogenic systems above. However, prograde metamorphism involves dehydration of the crust leading ultimately to nominally anhydrous peak mineral assemblages. In the absence of rehydration, which is generally linked with localized deformation (Boundy *et al.* 1992; Austrheim & Boundy 1994; Blattner 2005; Camacho *et al.* 2005; Clarke *et al.* 2005; Fitzherbert *et al.* 2005; Bjornerud & Austrheim 2006; Glodny *et al.* 2008), the initial record of metamorphism is likely to be preserved, at least partially, through

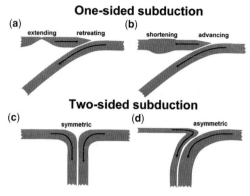

Fig. 2. Various one-sided (**a**, **b**) and two-sided (**c**, **d**) subduction geometries. In the case of the one-sided geometries, the overriding plate does not subduct. Depending on the relative plate motions, the overriding plate may be extending (a) or shortening (b). In the case of two-sided geometries, both plates subduct together. Depending on the relative thicknesses (ages) of the plates, two-sided subduction may be symmetric (c) or asymmetric (d). Reprinted from Gerya, T. V., Connolly, J. A. D. & Yuen, D. A. 2008. Why is terrestrial subduction one-sided? *Geology*, **36**, 43–46, with permission. © 2008 The Geological Society of America (GSA).

overprinting events such as may occur during subduction-to-collision orogenesis. Also, as shown by Hollis *et al.* (2006), the prograde history may be preserved through subsequent high-temperature events.

The $P-T-t$ evolution of orogens, the formation of paired metamorphic belts, and the temporal distribution of different types of metamorphism have been reviewed in a series of papers by Brown (1993, 1998a, b, 2001, 2002b, 2007a) that complement those by Ernst & Liou (1995), Maruyama *et al.* (1996), Maruyama & Liou (1998, 2005), Liou *et al.* (2004) and Ernst (2005). Here I consider metamorphism in orogenic systems in general, giving this paper a broader perspective than a simple discussion of metamorphism in accretionary orogenic systems. This is justified by the complexity induced in many ancient orogens by collision-related phenomena overprinting subduction-related phenomena, where the collisions vary from terrane accretion to arc accretion to continental margin subduction. The additional complexity that may result from this overprinting, which may vary with the style of collision involved, provides the foundation of fertile disagreements over the interpretation of ancient orogenic systems.

First, I review the tectonics of accretionary orogenic systems and consider when this style of orogenic system first appeared in the geological record. I also consider how to classify metamorphic belts for evaluating tectonic setting back through time. Next, I review types of metamorphism in relation to accretionary and subduction-to-collision orogenic systems as they occur on Earth since the Cretaceous Period, before considering the imprint of different types of metamorphism in the rock record back to the Neoarchaean Era using a dataset for high-temperature and high-pressure metamorphism compiled in 2005 (Brown 2007a). Using this dataset, I define three types of metamorphism characterized by different ranges of apparent thermal gradient, which I relate to the changing tectonics of orogenic systems through Earth history. Finally, I evaluate the age distribution of metamorphism from the Neoarchaean Era to the Cenozoic Era and consider how this relates to the supercontinent cycle and the process of terrane export and accretion (see Brown 2007b, 2008). Excluded from consideration in this review are intraplate orogenies such as the Ediacaran–Cambrian Petermann and the Devonian–Carboniferous Alice Springs orogenies in Central Australia (e.g. Sandiford & Hand 1998; Hand & Sandiford 1999; Sandiford 2002).

Tectonics of orogenic systems on Earth

At a global scale on modern Earth, subduction at convergent plate boundaries is continuous. Orogenic systems occur along convergent plate boundaries, where they act as buffer zones (Lister *et al.* 2001) to accommodate deformation (Kreemer *et al.* 2003). Although deformation is continuing at convergent plate boundaries, generally involving strain accumulation at low strain rates, orogenic systems may also exhibit episodic behaviour of various types at all scales (Dewey 1981); some of these episodic events have been inferred to be globally significant (Lister *et al.* 2001). Global events may be associated with switches in tectonic mode, from contraction to extension and vice versa (Lister *et al.* 2001). These switches will effect change in the thermal structure of orogenic systems, and therefore they should be registered in the metamorphic record.

Orogenesis in accretionary orogenic systems may be driven by reorganization of plate motions (Dewey 1975), ridge–trench interactions (Brown 1998b; Wakabayashi 2004), subduction of ocean-floor debris (Cloos 1993; Collins 2002), or terrane accretion (Jones *et al.* 1983; Howell *et al.* 1985; Coney 1992). Processes that change the state of stress along one plate boundary will affect the balance of torques on all other plate boundaries (Coblentz & Sandiford 1994; Coblentz *et al.* 1994, 1995, 1998; Richardson & Coblentz 1994; Coblentz & Richardson 1996). To maintain the torque balance, kinematic adjustments must be made elsewhere along the plate boundary system in response to the change along the first plate boundary. For this reason processes such as ridge–trench interactions, subduction of ocean-floor debris and terrane accretion that have a significant impact along one plate boundary will affect the plate boundary torque balance and promote changes in plate kinematics that may be imprinted on the rock record as a global event. Of course, this is also true for terminal continent collision in subduction-to-collision orogenesis.

For accretionary orogenic systems, the behaviour of the subduction boundary or trench in relation to the overriding plate is important (Uyeda & Kanamori 1979; Uyeda 1982; Royden 1993a, b; Lister et al. 2001; Lister & Forster 2006; Schellart 2007b, 2008a, b; Schellart et al. 2007; Lallemand et al. 2008; but see Doglioni (2008) for an alternative perspective). The mode of evolution of the system is primarily determined by whether the trench advances (is pushed back by the advance of the overriding plate) or retreats (is pulled back faster than the overriding plate is able to adjust). Trench migration is also influenced by the velocity of the subducting plate (Fig. 3), which largely depends on its age at the trench (Lallemand et al. 2008), and proximity to lateral slab edges (Schellart et al. 2007; Schellart 2008b). Shortening of the overriding plate and formation of an orogenic plateau in the hinterland behind a mountain chain

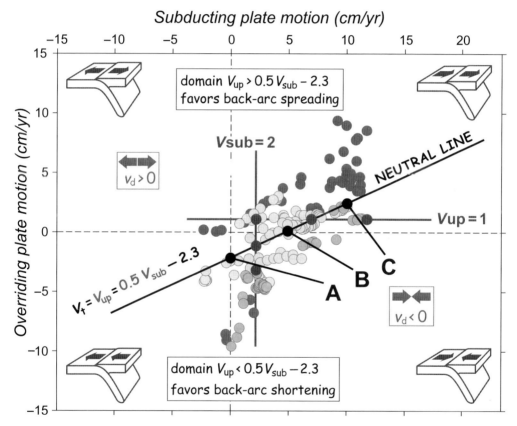

Fig. 3. Overriding plate absolute motion v. subducting plate absolute motion for 166 transects across trenches considered by Lallemand et al. (2008). V_{up}, V_t and V_{sub} are the absolute upper plate, trench and subducting plate motions counted positive landward (velocities in HS3 reference frame; references frames were discussed by Lallemand et al. 2008). The regression line $V_{up} = 0.5 V_{sub} - 2.3$ in cm a^{-1} (or $V_{sub} - 2 V_{up} = 4.6$) is valid only for the neutral subduction zone transects (light grey dots). Quality factor $R^2 = 0.37$. Along the neutral line, the trench velocity V_t equals the upper plate velocity V_{up} because there is no deformation across the arc system. All transects characterized by active shortening (medium grey dots) are located below this line, and all transects characterized by active spreading (= extension; dark grey dots) are located above the line (with the single exception of four transects across the New Hebrides trench, discussed by Lallemand et al. 2008, p. 7). Republished from Lallemand, S., Heuret, A., Faccenna, C. & Funiciello, F. 2008. Subduction dynamics as revealed by trench migration. *Tectonics*, **27**, TC3014, doi:10.1029/2007TC002212, with permission. © 2008 American Geophysical Union.

in the absence of terminal continent collision, such as occurs in the modern Andes, requires resistance to trench retreat, a condition that is met far from lateral slab edges (Schellart et al. 2007; Schellart 2008b), and overriding plate motion towards the trench (van Hunen et al. 2002), which is most probably driven by ridge-push forces. In addition, in these advancing systems, extensional collapse of the orogenic suprastructure may be a factor caused by the gravitational potential energy of the orogen, as described for collisional orogenic systems by Dewey (1988).

Thompson et al. (1997) treated orogenic systems along subduction boundaries as complex transpressive systems in which the degree of obliquity or ratio of pure shear to simple shear components controls the style of metamorphism, whereas Lister & Forster (2006) argued that switches in tectonic mode are primarily responsible for differences in style of metamorphism. Thompson et al. (1997) proposed that a low ratio of pure shear to simple shear, typical for wrench-dominated plate boundaries, implies a larger component of horizontal transport from a position deep in a transpressive orogenic system, which allows for a longer time during which heating may occur. In contrast, a high ratio of pure shear to simple shear, typical for convergence-dominated plate boundaries, implies a smaller component of horizontal transport from a position deep in a transpressive orogenic system, which drives more rapid exhumation and allows less time during which heating may occur. Thompson et al. (1997) proposed that high dT/dP, intermediate dT/dP and low dT/dP metamorphism are associated with an increasing angle of subduction obliquity, respectively. However, the complexity of natural obliquely subducting plate boundary zones suggests that this analysis may be far too simple (e.g. Baldwin et al. 2004).

In contrast, Lister & Forster (2006) identified two end-member types of orogenic system based on cyclicity of tectonic mode switches. Some systems may exhibit a cycle in which a retreating trench changes to an advancing trench ('pull–push inversion cycles' of Lister & Forster 2006), and these systems are associated with high thermal gradient type of metamorphism ($>750\ °C\ GPa^{-1}$). Other systems may exhibit a cycle in which an advancing trench changes to a retreating trench ('push–pull inversion cycles' of Lister & Forster 2006), and these systems are associated with low ($<350\ °C\ GPa^{-1}$) to intermediate ($350-750\ °C\ GPa^{-1}$) thermal gradient types of metamorphism.

Metamorphic imprints in orogens are likely to be the result of discrete events caused locally by ridge–trench interactions, subduction of ocean-floor debris, terrane accretion and terminal continent collision or globally by changes in plate kinematics to regain the plate boundary torque balance. Low thermal gradient type metamorphic rocks in accretionary orogenic systems dominated by cycles in which an advancing trench changes to a retreating trench come back to the surface relatively soon after their burial at rates comparable with plate boundary velocities. Rapid return to shallow crustal levels is also the case for low thermal gradient type metamorphic rocks exhumed by extrusion, such as in the Thompson et al. (1997) model.

In the Tethysides, the occurrence of low thermal gradient type metamorphic rocks in several belts related to a single subduction system suggests that these cycles represent recurrent transient events (Lister et al. 2001). Exhumation of these metamorphic rocks appears to have been associated in time and space with subduction–accretion of continental ribbon terranes that triggered slab rollback or slab step-back outboard of the accreted terrane and created the necessary space for exhumation (Brun & Faccenna 2008). This is not the only mechanism by which low thermal gradient type metamorphic rocks are exhumed, but a detailed discussion is outside the scope of this review and the interested reader is referred to papers by Platt (1993), Gerya et al. (2002) and Warren et al. (2008a, b) and references therein.

How far back in time are we able to recognize these orogenic systems?

Since the early days of the new global tectonics, there has been much debate about when Earth might have adopted a mobile lid mode of convection, whether once one-sided subduction was established it was maintained to the present day, and whether the formation and stabilization of the continental lithosphere and the change to one-sided subduction were related to each other. Alternative geodynamic scenarios for the early history of the Earth include an alternation between plate tectonics and some other mode (Sleep 2000) and episodic (O'Neill et al. 2007) or intermittent (Silver & Behn 2008) plate tectonics.

Consensus is emerging that Earth had adopted, either partially or completely, one-sided subduction and a form of plate tectonics akin to that on modern Earth by sometime in the Archaean Eon (Brown 2006, 2007a, b, 2008). However, there are those who argue for plate tectonics as early as the Hadean Eon (Davies 2006; Harrison et al. 2006; Shirey et al. 2008), consistent with the null hypothesis that plate tectonics was the mode of convection throughout Earth history. There are also those who argue against worldwide modern-style subduction before the Neoproterozoic Era (Stern 2005, 2007, 2008), requiring an alternative paradigm to plate tectonics for the period from the Hadean Eon to the Mesoproterozoic Era.

The style of subduction is the key. It is the change from symmetric to asymmetric or one-sided subduction sometime in the Archaean Eon (Fig. 2; Gerya et al. 2008) that marks the beginning of what I have called a 'Proterozoic plate-tectonics regime'; that is to say, a regime with dual thermal environments, one on the oceanward side that is cooler and another on the hinterland side that is warmer (Brown 2006, 2007a, b, 2008). The Neoarchaean Era is when many of the characteristic features of plate tectonics become widespread in the geological record (Condie & Pease 2008), the continental crust became stabilized, and the continental lithosphere was created as crustal fragments aggregated into several supercratons (Bleeker 2003).

Since the formation of supercratons in the Neoarchaean Era (Bleeker 2003), the continents have grown laterally through time. This growth has occurred via the processes of magmatism in arc–back-arc systems, accretion at continental margin trenches of ocean plate materials, including offscraping and underplating of ocean-floor topographic features, and collision with incoming terranes and arc systems that also may have been displaced laterally by orogen-parallel motions at active continental margins. These processes all contribute to the complexity recorded by accretionary orogenic systems at convergent plate margins (Dewey et al. 1990; Şengör et al. 1993; Collins 2002; Dickinson 2004; Cawood 2005), prior to eventual incorporation of these accretionary orogenic systems into collisional orogenic systems during supercontinent formation. Whether accretionary orogenic systems are confined to the post-Mesoarchaean period or whether they predate the stabilization of the continental lithosphere during the Mesoarchaean to Neoarchaean Eras periods remains an open question (Condie & Kröner 2008).

The classification of metamorphic belts

The principal tectonic settings for regional metamorphism are zones of continent extension, continent margins associated with subduction and collision, island arcs and ocean plateaux (excluding the ocean ridge systems for the purpose of this review). In a classic paper, Miyashiro (1961) classified metamorphic facies series characteristic of regional metamorphic belts into the following five types with increasing pressure: andalusite–sillimanite type; low-pressure intermediate group; kyanite–sillimanite type; high-pressure intermediate group; jadeite–glaucophane type. Miyashiro (1961) observed that in some regions a metamorphic belt of the andalusite–sillimanite type and/or low-pressure intermediate group and a metamorphic belt of the jadeite–glaucophane type and/or high-pressure intermediate group occurred together as a pair. The latter belt was inferred to develop at a site of low heat flow such as a subduction zone whereas the former belt was inferred to develop at a site of high heat flow such as that beneath an associated volcanic arc or back-arc basin. Miyashiro (1961) suggested that these belts formed in different phases of the same cycle of orogenesis, whereas subsequently Brown (1998a, b, 2002b) proposed they might have formed along different sectors of the same subducting margin, and became juxtaposed only by later orogen-parallel translation. Miyashiro (1961) perspicaciously observed (in relation to secular change) 'regional metamorphism under higher rock pressures appears to have taken place in later geological times'.

Since 1961 not only has the plate tectonics paradigm revolutionized our view of orogenesis and its attendant processes but also we have seen the limits of regional metamorphism pushed to pressures of up to 10 GPa (ultrahigh-pressure metamorphism, UHPM) and temperatures of at least 1050 °C (ultrahigh-temperature metamorphism, UHTM); the field between these types of metamorphism is one in which eclogite and high-pressure granulite mineralogies overlap according to the dictates of bulk composition in relation to $P-T$ conditions (Brown 2007a, 2008). In 2007 I highlighted three types of metamorphism (namely, granulite–ultrahigh-temperature metamorphism (G–UHTM), eclogite–high-pressure granulite metamorphism (E–HPGM) and high-pressure–ultrahigh-pressure metamorphism (HPM–UHPM)) because I was interested in the different thermal regimes they record (Brown 2007a). Close-to-peak $P-T$ conditions for single samples from granulites and ultrahigh-temperature metamorphic belts and high-pressure to ultrahigh-pressure metamorphic belts define two well-delineated sectors of $P-T$ space (Fig. 4), with a gap that is represented in the rock record by eclogite–high-pressure granulite metamorphic belts (e.g. O'Brien & Rotzler 2003).

Miyashiro used the metamorphic field gradients determined for a limited number of metamorphic belts to place each of them into one of his five types of metamorphism based on increasing pressure (Miyashiro 1961, fig. 4). The concept of grouping metamorphic belts by pressure is a useful one. However, the lower grade and/or prograde history of many higher temperature belts commonly is not recorded. For this reason, in this review I use apparent thermal gradient defined by close-to-peak $P-T$ conditions for classification rather than metamorphic field gradients. As with Miyashiro's approach, the resulting classification into high dT/dP (synonymous with low-P–high-T or LP–HT), intermediate dT/dP (similar to 'Barrovian')

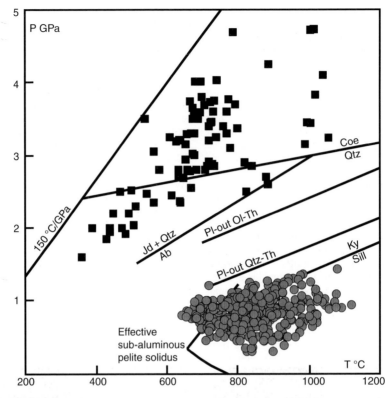

Fig. 4. $P-T$ diagram to show the results of internally consistent thermobarometry of single rock samples for high-pressure and ultrahigh-pressure metamorphism (black squares; $P-T$ calculated using Grt–Cpx–Ky–Pag–Qtz/Coe thermobarometry, mostly from Ravna & Terry (2004) with additional data from Hacker (2006) and Tsujimori et al. (2006)) and granulite and ultrahigh-temperature metamorphism (grey circles; $P-T$ calculated by Pattison et al. (2003) using Grt–Opx thermobarometry with Al-solubility in Opx corrected for retrograde Fe–Mg exchange). Uncertainties on the $P-T$ data are likely to be of the order of 0.2–0.3 GPa and 30–60 °C. Thermal gradients of <150 °C GPa^{-1} are not recorded in crustal rocks on Earth.

and low dT/dP (synonymous with high-P–low-T or HP–LT) metamorphism coincides with major changes in facies series. These three types of metamorphism culminate in ultrahigh-temperature, eclogite–high-pressure granulite or ultrahigh-pressure metamorphism, respectively. They register the imprint of three different thermal environments in the geological record, each with a characteristic range of dP/dT or apparent thermal gradients.

Of course, metamorphism is a dynamic and not a static process and metamorphic rocks record the evidence of this dynamic process in changes in mineralogy and/or mineral chemistry that track the pressure (P) and temperature (T) with time (t). Once this encoded information is decoded and inverted the result is commonly represented as a $P-T-t$ path. This is the record of burial and exhumation in a particular thermal environment or combination of thermal environments in which the minerals crystallized and/or re-equilibrated (e.g. Brown 1993, 2001). In effect, the $P-T-t$ path records crustal evolution across the changing spectrum of transient metamorphic geotherms characteristic of one or more thermal environments.

This observation about the dynamic evolution of metamorphic belts is important because classification is based on an apparently static view of metamorphism based on the metamorphic facies concept. The close-to-peak $P-T$ conditions recorded by the equilibrium mineralogy and the implied apparent thermal gradients derived by linear extrapolation back to the surface do not preclude a $P-T$ evolution that transgresses from one thermal environment to another. Thus, a metamorphic belt may evolve from a low dT/dP to intermediate dT/dP type or from an intermediate dT/dP to a high dT/dP type. This may or may not be registered in the rock by features such as disequilibrium microstructures or mineral chemistries that identify the change from one environment to the other; for example, from the HPM–UHPM facies series to the E–HPGM facies or from the E–HPGM facies to the

G–UHTM facies series, respectively. Thus, the metamorphic facies concept, the apparent thermal gradient and the metamorphic $P-T-t$ path record different characteristics of a metamorphic belt.

The metamorphic realm

The $P-T$ field of conditions recorded by crustal rocks traditionally has been divided into facies, each represented by a group of mineral assemblages associated in space and time that are inferred to register equilibration within a limited range of $P-T$ conditions (Fig. 5a). The limits of P and T recorded by crustal rocks are higher than in 1961. Some rocks characteristic of the granulite facies record temperatures >900 °C at pressures of 0.6–1.3 GPa (ultrahigh-temperature metamorphism (UHTM); e.g. Harley 1998, 2008; Brown 2007a; Kelsey 2008) and some eclogite facies rocks record pressures up to 10 GPa at temperatures of 600–1000 °C (ultrahigh-pressure metamorphism (UHPM); e.g. Chopin 2003; Liu et al. 2007; Brown 2007a). In addition, we now recognize a transition between these two facies, referred to as medium-temperature eclogite–high-pressure granulite metamorphism (E–HPGM; e.g. O'Brien & Rötzler 2003; Brown 2007a). On modern Earth, the different types of metamorphic facies series leading to granulite–ultrahigh-temperature metamorphism (G–UHTM), medium-temperature eclogite–high pressure granulite metamorphism (E–HPGM) and high-pressure metamorphism–ultrahigh-pressure metamorphism (HPM–UHPM) are generated in different tectonic settings with contrasting thermal regimes at convergent plate boundary zones (Brown 2006, 2007a). Based on apparent thermal gradients (discussed further below), these metamorphic facies series correspond to high dT/dP, intermediate dT/dP and low dT/dP metamorphism, respectively.

Recently, Stüwe (2007) has introduced an alternative division of $P-T$ space, based on whether thermal conditions implied by close-to-peak metamorphic mineral assemblages in orogenic crust were warmer or cooler than a normal (conductive) continental geotherm (Fig. 5b). On this $P-T$ diagram, we may distinguish $P-T$ fields that are reached as a function of different tectonic processes.

For thermal conditions warmer than a normal continental geotherm, a thermal gradient of

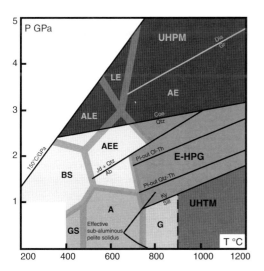 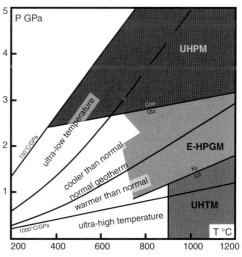

Fig. 5. (Left) $P-T$ diagram to show the principal metamorphic facies in $P-T$ space and the $P-T$ ranges of different types of extreme metamorphism. HP–UHP metamorphism includes the following: BS, blueschist; AEE, amphibole–epidote eclogite facies; ALE, amphibole lawsonite eclogite facies; LE, lawsonite eclogite facies; AE, amphibole eclogite facies; UHPM, ultrahigh-pressure metamorphism; GS, greenschist facies; A, amphibolite facies; E-HPG, medium-temperature eclogite–high-pressure granulite metamorphism; G, granulite facies, whereas UHTM is the ultrahigh-temperature metamorphic part of the granulite facies. (Right) An alternative division of $P-T$ space based on whether thermal conditions implied by peak metamorphic mineral assemblages were warmer or cooler than a normal (conductive) continental geotherm, which is constructed to pass through 500 °C at 1 GPa and 1200 °C at 3 GPa (see Stüwe 2007). This diagram distinguishes $P-T$ fields that are reached as a function of different tectonic processes. The boundary between the warmer than normal and ultrahigh temperature fields reflects the conductive limit; it has dT/dP of 1000 °C GPa^{-1}. The boundary between the cooler than normal and ultra-low-temperature fields represents half the normal geothermal gradient. It should be noted that the field boundaries in this figure are simpler than the facies boundaries based on equilibrium mineral assemblages.

c. 1000 °C GPa^{-1} is a practical limit for a conductive response, based on the unlikely scenario of asthenospheric mantle being located at the base of normal thickness crust (Stüwe 2007). Metamorphic belts that record apparent thermal conditions hotter than this limit require a component of advection associated with mantle-derived magma in addition to thinning the subcrustal lithospheric mantle (e.g. Sandiford & Powell 1991), consistent with an arc–back-arc setting (Hyndman et al. 2005a, b; Currie & Hyndman 2006, 2007; Schellart 2007a; Currie et al. 2008), or internal heat generation by radioactive decay in over-thickened crust (Le Pichon et al. 1997; McKenzie & Priestley 2008), consistent with an orogenic hinterland setting.

For thermal conditions cooler than a normal continental geotherm, Stüwe (2007) suggested a two-fold division into a 'cooler than normal' and an 'ultra-low-temperature' P–T field. The cooler than normal P–T field is a thermal regime that may be reached by thickening both the crust and mantle components of the lithosphere for an acceptable range of thermal parameters (Sandiford & Dymoke 1991; Stüwe 2007). This observation suggests that at least some low dT/dP metamorphic belts within orogenic systems may reflect the thermal response to lithospheric thickening and may not necessarily reflect thermal regimes attendant on subduction or in response to unusually rapid exhumation. In contrast, processes other than lithospheric thickening are required to reach the ultra-low-temperature P–T field; subduction is one possible process by which rocks may enter this thermal regime (Stüwe 2007).

In this review, the range of P–T conditions recorded by HPM–UHPM, E–HPGM and G–UHTM is the same as those specified by Brown (2007a). High-pressure metamorphism (HPM) and granulite-facies (G) metamorphism are well-understood terms in the literature and are not redefined here.

For UHPM, pressure must have exceeded the stability field of quartz, recognized by the presence of coesite or diamond in crustal rocks or by equivalent P–T conditions determined using robust thermobarometry. Examples of robust thermobarometry currently in use, as recently evaluated by Hacker (2006), include the method of Ravna & Terry (2004) and the average P–T mode in THERMOCALC (Powell et al. 1998) with the internally consistent thermodynamic dataset of Holland & Powell (1998, most recent update).

For UHTM, I follow Harley (1989, 1998) in using 900 °C as an arbitrary lower temperature limit and Brown (2007a) in using the stability field of sillimanite in rocks of appropriate composition to define the pressure. UHTM in crustal rocks may be recognized either by robust thermobarometry or by the presence of a diagnostic mineral assemblage in an appropriate bulk composition and oxidation state, such as assemblages with sapphirine + quartz, spinel + quartz or corundum + quartz. The current state-of-the-art in robust thermobarometry for granulite metamorphism in general for a variety of bulk compositions is either the method proposed by Pattison et al. (2003) or the average P–T mode in THERMOCALC (Powell et al. 1998) with the internally consistent thermodynamic dataset of Holland & Powell (1998, most recent update).

The transition from granulite- to high-pressure granulite-facies conditions is not well defined, but I recommend the sillimanite to kyanite transformation in metapelitic rocks as a simple rule of thumb. Furthermore, the field of high-pressure granulite-facies metamorphism overlaps the transformation of amphibolite to eclogite in basaltic rocks, which occurs over more than 1 GPa according to the chemical composition of the protolith (O'Brien & Rötzler 2003). These two features define the intermediate E–HPGM field.

Peak mineral assemblages are potentially robust recorders of metamorphic P and T, particular at high P–T conditions (HPM–UHPM, E–HPGM and G–UHTM), because prograde dehydration and melting with melt loss produce nominally anhydrous mineral assemblages that are difficult to retrogress or overprint without fluid influx. None the less, the effects of overprinting of younger orogenic events on older orogens leading to polymetamorphism must be avoided wherever possible, or where present or suspected must be distinguished (e.g. Hensen & Zhou 1995; Hensen et al. 1995). Without attention to the possible degradation of the dataset by incorporation of potentially overprinted P–T information, use of the geological record of metamorphism in relation to secular change will be potentially flawed. Furthermore, in evaluating secular change in patterns of metamorphism it is essential to use precise P–T–age relations. The P–T conditions should be assessed based on robust thermobarometry or the presence of a diagnostic mineral assemblage in an appropriate bulk composition and oxidation state, and the age should be determined using a robust chronometer and should be related to a specific P–T point along the P–T–t evolution; if possible, close to peak P–T.

Today we are confident in our ability as forensic scientists correctly to interpret the evidence recovered from eroded orogens (e.g. O'Brien 1997, 1999; Brown 2001, 2002a, 2007a; White et al. 2002; O'Brien & Rötzler 2003; Johnson & Brown 2004; Baldwin et al. 2005; Powell et al. 2005); this confidence was lacking in the early days of studies of metamorphism under extreme P–T conditions (Green 2005). Our confidence is in part due to improvements in our ability to interrogate rocks and to recognize the effects of overprinting; for example, by using chemical mapping of mineral grains and high spatial resolution *in situ* geochronology.

The wider availability of high-precision ages on accessory phases linked to specific microstructural sites and robustly determined $P-T$ conditions, commonly through the identification of mineral assemblages preserved as micro-inclusions in the core, mantle or rim of zircon grains, has increased our confidence that we know the true 'age' of close-to-peak metamorphism or at least a particular $P-T$ point or range in many metamorphic belts. However, for UHTM rocks accessory phase geochronology most commonly records crystallization along a segment of the high-temperature cooling path (Harley et al. 2007; Baldwin & Brown 2008; Harley 2008; Kelsey 2008). With these caveats in mind we may now assess reliably the imprint of metamorphism through Earth history.

Metamorphism since the Cretaceous Period

In this section, I discuss the relationship between the three types of metamorphism based on ranges of apparent thermal gradient identified above and the variety of tectonic settings that are associated with convergent plate boundary zones during the interval of Earth history for which information about plate kinematics may be retrieved from the magnetic anomaly patterns in the ocean basins. In addition, I review likely sources of heat to drive metamorphism in general, and to drive high dT/dP metamorphism in particular.

High-pressure metamorphism–ultrahigh-pressure metamorphism (low dT/dP metamorphism)

High-pressure metamorphism–ultrahigh-pressure metamorphism (blueschist–eclogite–ultrahigh-pressure facies series; Fig. 5) most commonly occurs associated with subduction, particularly in the subduction zone prior to and during collision. Evidence includes samples of glaucophane schist entrained in serpentine mud volcanoes in the Mariana forearc (e.g. Shipboard Scientific Party 2002), where incipient blueschists with the assemblage lawsonite–pumpellyite–hematite yield T of 150–250 °C at P of 0.5–0.6 GPa (Maekawa et al. 1993) and dT/dP of 300–400 °C GPa^{-1} (approximately equivalent to 9–11 °C km^{-1}). This apparent thermal gradient is consistent with thermal models for subduction zones (e.g. Hacker et al. 2003) and apparent thermal gradients derived from Eocene low-grade blueschist-facies series rocks (e.g. Potel et al. 2006).

Ultrahigh-pressure metamorphic rocks generally register deep subduction and exhumation of continental crust (lithosphere) during the early stage of collision, as evidenced by coesite-bearing eclogites of Eocene age at Tso Morari in the frontal Himalayas (Sachan et al. 2004; Leech et al. 2005). However, strong coupling also may drag down hanging-wall lithosphere during the subduction process, as might be suggested for the coesite-bearing eclogites of Pliocene age in Eastern Papua New Guinea (Baldwin et al. 2004). Additionally, Searle et al. (2001) have argued that deep earthquakes of the Hindu Kush seismic zone represent a tracer of contemporary coesite and diamond formation in crustal rocks, and Rondenay et al. (2008) have used seismic imaging to reveal the release of metamorphic fluids during subduction zone metamorphism.

Lawsonite blueschists and lawsonite eclogites generally record ultra-low-temperature subduction in comparison with typical HPM–UHPM. We know this from thermodynamic modelling of Eocene lawsonite blueschists and lawsonite eclogites exhumed to the Earth's surface in the Pam Peninsular, New Caledonia (Clarke et al. 1997, 2006), and xenoliths of lawsonite eclogite of Eocene age inferred to record ultra-low-temperature subduction to >3.5 GPa brought up in Tertiary kimberlite pipes at Garnet ridge in the Colorado Plateau, USA (Usui et al. 2003, 2006).

Medium-temperature eclogite–high-pressure granulite metamorphism (intermediate dT/dP)

Modern examples of medium-temperature eclogite–high-pressure granulite metamorphism are not common at outcrop in orogens of Cenozoic age. One example within the appropriate range of apparent thermal gradients occurs in the upper units of the Nevado–Filabride Complex in the Betic Cordillera of southern Spain (Platt et al. 2006), where eclogite lenses sheathed by blueschist within pelitic schist yield $P-T$ of 1.1–1.8 GPa and 550–700 °C yielding dP/dT around 390 °C GPa^{-1}. These units record clockwise $P-T$ paths, and the evolution from HPM conditions to E–HPGM conditions records the transition from subduction to collision. Another example from a collision zone is provided by Miocene age crustal xenoliths from the southern Pamir (Hacker et al. 2005), where basaltic eclogite, sanidine eclogite and high-P felsic granulite record $P-T$ of 2.5–2.8 GPa and 1000–1100 °C yielding dP/dT around 400 °C GPa^{-1}.

High-P granulites also are reported from two exhumed Cretaceous arcs. The first example is the

Jijal complex in northern Pakistan, part of the Kohistan arc, which cooled from magmatic conditions to E–HPGM conditions of 2.0–1.5 GPa and 1100–800 °C, based on relict magmatic assemblages within overprinting metamorphic assemblages (Ringuette et al. 1999; Yamamoto et al. 2005). The second example is the Arthur River complex in Fiordland, New Zealand, which yields E–HPGM conditions of 1.4 GPa and >750 °C (Clarke et al. 2000; Daczko et al. 2001; Hollis et al. 2003, 2004). Both arcs yield apparent thermal gradients of 530–550 °C GPa^{-1}.

I interpret E–HPGM to register collisional orogenesis, most probably identifying the suture where elements of the subduction complex and the overriding plate become tectonically juxtaposed and commonly become structurally attenuated. Thus, E–HPGM registers the subduction-to-collision process. In Phanerozoic orogens E–HPGM may be closely associated with HPM–UHPM.

Granulite–ultrahigh-temperature metamorphism (high dT/dP)

Granulite–ultrahigh-temperature metamorphism (granulite–ultrahigh-temperature facies series) occurs in the deeper parts of Mesozoic oceanic plateaux (e.g. Gregoire et al. 1994; Shafer et al. 2005) and in exposed middle to lower crust of young continental arcs (e.g. Lucassen & Franz 1996). In addition, metapelitic xenoliths retrieved from Neogene volcanoes in central Mexico (Hayob et al. 1989, 1990) and at El Joyazo in SE Spain (Cesare & Gomez-Pugnaire 2001) record evidence of Cenozoic G–UHTM during crustal extension, and evidence of melt-related processes in lower crustal garnet granulite xenoliths from Kilbourne Hole, Rio Grande Rift, suggests contemporary G–UHTM in the lower crust of rifts (Scherer et al. 1997). It is likely that G–UHTM conditions are being generated today in orogenic hinterlands, such as under Tibet and the Altiplano, based on the interpretation of multiple geophysical datasets (e.g. Nelson et al. 1996; Schilling & Partzsch 2001; Unsworth et al. 2005) that suggest the presence of melt, inferred to have been derived from mica breakdown melting, and new inferences from seismic data and expectations from numerical modelling (Le Pichon et al. 1997; McKenzie & Priestley 2008). The rarity of young G–UHTM rocks at the Earth's surface probably reflects the general expectation that granulites are exhumed in a subsequent orogenic cycle (Harley 1989). Metasedimentary protoliths may attain G–UHTM $P-T$ conditions by internal heating if sufficient time is available (Le Pichon et al. 1997; McKenzie & Priestly 2008) or by inversion of thinned lithosphere with high heat flow such as occurs in back-arcs (Hyndman et al. 2005a, b), as suggested for the Variscan evolution of the Moldanubian Zone in the Vosges Mountains by Schulmann et al. (2002), or by extension of the lithosphere and advection of heat with magma (Dewey et al. 2006). This is discussed further below.

What is the source of the heat? The mechanism that provides the heat necessary to drive high dT/dP metamorphism remains elusive in most circumstances. Radiogenic heating is important and sometimes is argued to be sufficient, particularly if time for incubation is available between thickening and reaching peak metamorphic conditions (Le Pichon et al. 1997; Gerdes et al. 2000; Jamieson et al. 2002; Andreoli et al. 2006; McLaren et al. 2006; McKenzie & Priestley 2008). The observation that belts of regional metamorphism typically contain abundant crust-derived intrusive rocks leads to the postulate that intracrustal magmatism increases the regional thermal gradient at shallow crustal levels, consistent with observations from several metamorphic belts that medium-pressure regional metamorphism grades with decreasing crustal depth into regional-scale contact metamorphism (e.g. Brown & Solar 1999). In addition, in accretionary orogenic systems ridge subduction may introduce hot asthenospheric mantle to the base of the overriding plate to generate anomalous high dT/dP conditions in the forearc leading to anatexis at shallow crustal levels (Brown 1998a, b; Groome & Thorkelson 2008).

Apparent thermal gradients that significantly exceed 750 °C GPa^{-1} are retrieved from G–UHTM terranes (see below), but these gradients cannot be sustained to mantle depths in overthickened crust without exceeding the peridotite solidus. Two alternatives are implied by this observation. The first is that the asthenosphere was close to the Moho, which could lead to transient thermal gradients up to the conductive limit (1000 °C GPa^{-1}, potentially higher over a plume head), which might occur, for example, if a lithospheric root was removed as a result of convective instability (Platt & England 1994; Platt et al. 1998, 2003) or if the subducting slab breaks off (Davies & von Blanckenburg 1995). The second is that there was sufficient advection of heat into the crust with mantle-derived magma (Sandiford & Powell 1991; Stüwe 2007), for which evidence is generally scarce (Harley 2004). An example occurs in the Bohemian Massif, where there appears to be a relationship between the late Variscan ultrapotassic magmatism and G–UHTM in the Moldanubian Zone (Janousek & Holub 2007; Leichmann et al. 2007). Underplating by basaltic magma commonly is implicated to provide heat (e.g. Dewey et al. 2006), but the temporal relations between extension, crustal melting and emplacement of basic magma sometimes are ambiguous (e.g. Barboza et al. 1999; Peressini et al. 2007). Modelling studies suggest that heating by multiple intraplating by dykes and sills may be effective at melting the lower crust in continental arcs (e.g. Jackson et al. 2003; Dufek & Bergantz 2005; Annen et al. 2006).

Heating by viscous dissipation may be important in some subduction-to-collision orogenic systems. Although viscous dissipation may generate heat (e.g. Kincaid & Silver 1996; Stüwe 1998; Leloup et al. 1999; Burg & Gerya 2005), it requires initially strong lithosphere (e.g. a differential stress of 100–300 MPa); England & Houseman (1988) suggested that a differential stress of 100–200 MPa is necessary to generate a plateau, consistent with this requirement. The strong positive correlation between overall intensity of viscous heating in crustal rocks and the instantaneous convergence rate suggests that a significant contribution (>0.1 $\mu W\,m^{-3}$) of viscous heating into the crustal heat balance may be expected when the convergence rate exceeds 1 cm a^{-1}, particularly if the lower crust is strong (Burg & Gerya 2005). Therefore, heating by viscous dissipation may become a dominant heat source in the early stage of subduction-to-collisional orogenesis if convergence rates are rapid (e.g. the Himalayas), and it may be significant in the heat budget of orogens (e.g. Stüwe 2007).

Currently, the major mountain belts of the circum-Pacific orogenic systems are located in former subduction zone back-arcs, which are characterized by high heat flow of 69 ± 16 to 85 ± 16 mW m^{-2} for continental crust with average radiogenic heat production (Hyndman et al. 2005a; Currie & Hyndman 2006; Currie et al. 2008), similar to the average value of 65 ± 10 mW m^{-2} for Variscan crust of the Iberian mainland (Fernandez et al. 1998) but lower than the range of 90–150 mW m^{-2} for a limited number of measurements from Tibet (Francheteau et al. 1984). Currie & Hyndman (2006) reported observations that indicate Moho temperatures of 800–900 °C, uniformly high temperatures of 1200 °C in the shallow mantle and a thin lithosphere c. 60 km thick over back-arc widths of 250 to >900 km, compared with Moho temperatures of 400–500 °C and lithosphere 200–300 km thick for cratons; the difference results in back-arc lithosphere being at least an order of magnitude weaker than cratons. Similar high temperatures are inferred for extensional back-arcs of the western Pacific and southern Europe, but the thermal structures are complicated by extension and spreading (Currie & Hyndman 2006).

Following termination of subduction by collision, the high temperatures decay over a time scale of about 300 Ma (Currie & Hyndman 2006). The reason subduction zone back-arcs are hot may be principally related to thin lithosphere and shallow convection in the mantle wedge asthenosphere, where convection is inferred to result from a reduction in viscosity induced by water from the underlying subducting plate, with additional effects from local extension (Currie & Hyndman 2006, 2007; Schellart 2007a; Currie et al. 2008). Most circum-Pacific mountain belts are broad zones of long-lived tectonic activity because they remain sufficiently weak to deform by the forces developed at plate boundaries.

Summary

I conclude that metamorphism since the Cretaceous Period occurs in several different thermal environments. One environment is characterized by low dT/dP, corresponding to the subduction zone or to the transition from subduction to collision. The subduction-to-collision setting also appears to be appropriate for intermediate dT/dP metamorphism, where some components are related to subduction (medium-temperature eclogite) but other components record thickening of the crust in arcs (high-pressure granulite). These elements have become juxtaposed during the collision process. Another environment is characterized by high dT/dP, corresponding to crustal extensional settings or the back-arc or the orogenic hinterland. The back-arc and orogenic hinterland may be inverted or thickened during collision and may be modified subsequently by orogenic collapse or extension.

Metamorphism earlier in the Phanerozoic Eon

In this section, I discuss the relationship between the different types of metamorphism and orogenic systems during the interval of Earth history for which there is no disagreement about the operation of modern plate tectonics (Stern 2005; Brown 2006).

Accretionary orogenic systems

High dT/dP metamorphism in accretionary orogenic systems associated with retreating trench–advancing trench tectonics, such as occurred in the Tasmanides of eastern Australia (Collins 2002) or the Acadian of northeastern North America (Solar & Brown 2001a, b), commonly follows counterclockwise $P-T-t$ paths in which deformation (thickening) and metamorphism (heating) proceed contemporaneously and peak metamorphism is late syntectonic (e.g. Solar & Brown 1999). Evidence of extreme metamorphism generally is absent from the level of erosion. None the less, blueschists may occur sporadically early in the orogenic cycle (e.g. the southern Lachlan orogen in the Tasmanides; Spaggiari et al. 2002a, b) or may be exposed during core complex formation late in the orogenic cycle (e.g. the New England orogen in the Tasmanides; Little et al. 1992, 1993, 1995; Holcombe & Little 1994). G–UHTM may be implied at depth (e.g. the Lachlan orogen; Collins 2003). Short-lived phases of orogenesis may relate to interruptions in the continuity of subduction by topographic features on the ocean plate, particularly oceanic plateaux (Cloos 1993; Koizumi & Ishiwatari 2006), as has been suggested for the Lachlan orogen (Collins 2002). Conversely, orogen-wide change from a retreating trench to an advancing trench may relate to changes in plate kinematics and advance of the subduction hinge (e.g. the Acadian orogen in the Appalachians; Solar et al. 1998; Solar & Brown 2002b).

In contrast, we may consider the Andes, where the Late Palaeozoic accretionary prism of the Coastal Cordillera in Chile (Willner 2005) and the Cretaceous Diego de Almagro Metamorphic Complex in Chilean Patagonia (Willner et al. 2004) are characterized by low dT/dP metamorphism, including blueschists, registering a thermal environment with an apparent thermal gradient around $350\,°C\,GPa^{-1}$. Such conditions may be more typical of an advancing subduction hinge in the early stage of an advancing trench–retreating trench cycle. It should be noted, however, that the modern Andes probably was developed as a consequence of two factors (Schellart 2008b), the large slab width (i.e. trench-parallel extent) and the trenchward motion of the overriding plate (i.e. trench advance).

In accretionary orogenic systems that involve terrane accretion, allochthonous and/or para-autochthonous elements become accreted to continental margins in convergent systems that involved oblique relative plate motion vectors (e.g. Dewey et al. 1990). Accretion at trenches has played a major role in the growth and evolution of many segments of the circum-Pacific margin, as revealed by investigations of the geology around the Pacific rim since the 1970s. These studies identified the importance of tectonostratigraphic terranes in understanding Pacific rim geology (Jones et al. 1983; Howell et al. 1985; Coney 1992) and the metamorphic patterns in accretionary orogens (Ernst et al. 1990). Much of the Cordillera in western North America comprises a collage in which terranes have been either accreted to the continent or displaced along the trench by orogen-parallel motions during the Mesozoic and Cenozoic (Coney et al. 1980; Jones et al. 1983; Ernst et al. 1990; Johnston 2001, 2008; Dickinson 2004; Johnston & Borel 2007; Redfield et al. 2007;

Ridgway & Flesch 2007). The Mesozoic–Cenozoic Cordilleran orogen of the western USA (Fig. 1) includes multiple tectonostratigraphic terranes principally composed of subduction complexes and interoceanic island arcs that were accreted sequentially to the continental margin since the Triassic Period (Dickinson 2008). Because of this terrane accretion, the continental margin of the western USA my have grown by as much as 800 km during Mesozoic–Cenozoic Cordilleran orogenesis.

Paired metamorphic belts *sensu* Miyashiro (1961) are a characteristic that results from orogen-parallel terrane migration and juxtaposition by accretion of contemporary belts of contrasting type (Brown 1998a, b, 2002b). These paired belts comprise a low dT/dP metamorphic belt outboard and a high dT/dP metamorphic belt inboard, as exemplified by the Mesozoic metamorphic belts of Japan (Miyashiro 1961) and the Palaeozoic–Mesozoic metamorphic belts of the western USA (Ernst et al. 1990; Patrick & Day 1995). In some systems, an important additional feature was ridge subduction, which may be reflected in both the pattern of high dT/dP metamorphism and the associated magmatism (Hole & Larter 1993; Thorkelson 1996; Brown 1998b; Sisson et al. 2003; Farris & Paterson 2009), although the thermal effect of ridge subduction may vary with orientation of the ridge segment in relation to the trench and whether separation continues as the ridge is subducted (e.g. Daniel et al. 2001; Farris & Paterson 2009; Groome & Thorkelson 2008).

Granulites may occur at the highest grade of metamorphism in the high dT/dP belt (Osanai et al. 1998; Miyazaki 2004). Although high-pressure rocks such as lawsonite eclogites are common (Tsujimori et al. 2006), UHPM generally appears to be absent in the outboard low dT/dP belt. Examples with UHPM registered include the Sambagawa Belt in Japan (Ota et al. 2004) and the western USA at Garnet ridge, on the Colorado Plateau (Watson & Morton 1969; Helmstaedt & Schulze 1988; Usui et al. 2003, 2006); additional studies elsewhere may reveal UHPM as the norm in terrane accretion orogenic systems formed during the Phanerozoic Eon. Rapid loading and unloading in some sectors of contractional (transpressive) continental arcs is an additional feature in some terrane accretion orogenic systems, identified by close-to-isothermal increase followed by close-to-isothermal decrease in P, which yields a distinctive P–T–t path (Hiroi et al. 1998; Whitney et al. 1999).

Subduction-to-collision orogenic systems

In subduction-to-collision orogenic systems, metamorphism associated with continental subduction and terminal continental collision follows standard subduction-related metamorphism of accreted materials; initial subduction of the continental margin may generate HPM–UHPM of continental crust as it is being subducted until the subduction zone is choked. Conductive relaxation associated with thickening across the suture may lead to E–HPGM of suture-related rocks, and deformation of the hinterland may generate clockwise metamorphic $P-T-t$ paths in granulite-facies series rocks, possibly leading to UHTM conditions, according to the mechanisms and relative rates of thickening and thinning versus heat transfer by conduction and advection. In many subduction-to-collision orogenic systems penetrative deformation largely precedes the peak of metamorphism because rates of continental deformation typically are about an order of magnitude faster than rates of thermal equilibration over the length scale of the crust (Stüwe 2007).

The European Variscides represent a classic example of a subduction-to-collision orogenic system. Here early (Devonian) subduction-to-collision-related HPM–UHPM is registered in the Galicia–Trás-os-Montes allochthon of NW Iberia (Martínez Catalán et al. 1997), the Champtoceaux and Essarts complexes of southern Brittany (Bosse et al. 2000; Godard 2001), the Monts du Lyonnais unit in the eastern French Massif Central (Lardeaux et al. 2001) and the Saxothuringian belt across the German–Czech border (Massonne 2001; Konopásek & Schulmann 2005). This is followed by later (Carboniferous) high dT/dP metamorphism and anatexis in the foreland in the Central Iberian Zone (Pereira & Bea 1994), the southern Brittany migmatite belt (Johnson & Brown 2004) and the central French Massif Central (Macaudière et al. 1992; Montel et al. 1992), and E–HPGM to G–UHTM in the Saxothuringian belt and the Bohemian Massif (e.g. Willner et al. 1997; O'Brien & Rötzler 2003; Štípská & Powell 2005). At a larger scale, the variation in metamorphic style of subduction-to-collision orogenic systems is well illustrated by the variation along the length of both the Appalachian/Caledonian–Variscide–Altaid and Alpine–Himalayan–Cimmerian chains.

Summary

Each of these orogenic systems that formed before the Cretaceous Period in the Phanerozoic Eon preserves metamorphic belts with contrasting types of metamorphism that may be inferred to record contrasting thermal regimes. These examples are consistent with the premise that plate tectonics may be extrapolated back through at least three ocean lithosphere turnover cycles (to c. 600 Ma); they also highlight the variable imprint in the rock record generated by the complex interactions and diversity of processes along plate boundary zones. My aim now is to extend this analysis back through the Precambrian using the record of metamorphism exposed at the surface in continents.

Metamorphism since the Neoarchaean Era

As discussed above, metamorphic belts are classified into three types according to the characteristic metamorphic facies series and peak $P-T$ conditions registered by the belt. In this section, I review the geological record of metamorphism since the Neoarchaean Era in relation to this typology using the dataset from Brown (2007a).

HPM–UHPM is characterized by lawsonite blueschist- to lawsonite eclogite-facies series rocks and blueschist- to eclogite- to ultrahigh-pressure-facies series rocks, where T plotted in Figure 6a is that registered at maximum P. E–HPGM is characterized by facies series that reach peak $P-T$ in the eclogite–high-pressure granulite facies (O'Brien & Rötzler 2003), where maximum P and T generally are achieved sequentially, but close enough to be considered as contemporaneously for this analysis (Fig. 6a). G–UHTM is characterized by granulite-facies series rocks that may reach ultrahigh-temperature metamorphic conditions, where P plotted in Figure 6a is that registered at maximum T. Each metamorphic belt is represented by a single datum that is my best estimate of representative peak $P-T$ conditions, which defines an apparent thermal gradient, and close-to-peak age as defined above. These arbitrary simplifications ignore the dynamic nature of metamorphism, recorded by $P-T-t$ paths, and the evolution of geotherms in orogens with time. A future analysis could be extended to take account of the spatial and temporal variation within each metamorphic belt. Bearing in mind these caveats, let us examine the rock record of metamorphism.

The $P-T$ value for each terrane shown in Figure 6b records a point on a metamorphic (transient) geotherm, and different apparent thermal gradients are implied by each type of metamorphism. These apparent thermal gradients are inferred to reflect different tectonic settings. HPM–UHPM is characterized by apparent thermal gradients of 150–350 °C GPa^{-1} (approximately equivalent to 4–10 °C km^{-1}), and plots across the boundary between the 'cooler than normal' and 'ultra-low-temperature' fields. About half of these terranes require a process other than simple thickening to achieve such cold gradients. We know from the global context that all of these terranes were associated with subduction, so it is likely that subduction was the process that created the

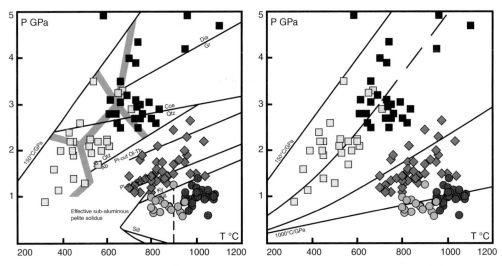

Fig. 6. Left. Metamorphic patterns based on representative 'peak' metamorphic $P-T$ conditions of metamorphic belts in relation to the metamorphic facies. Common granulite and granulite-facies ultrahigh-temperature metamorphism (G–UHTM belts; P at maximum T; light grey circles are common granulite belts and dark grey circles are granulite-facies ultrahigh-temperature metamorphic belts; data from tables 1 and 2 of Brown 2007a); medium-temperature eclogites–high-pressure granulites (E–HPGM belts; peak $P-T$; diamonds; data from table 3 of Brown 2007a); lawsonite blueschists–lawsonite eclogites and ultrahigh-pressure metamorphic rocks (HPM–UHPM belts; T at maximum P; light grey squares are lawsonite-bearing rocks and black squares are ultrahigh-pressure belts; data from tables 4 and 5 of Brown 2007a). Right. Peak $P-T$ values for each terrane (from Fig. 3a) in relation to a normal (conductive) continental geotherm. Each of these data records a point on a metamorphic (transient) geotherm, and the different apparent thermal gradients implied by each type of metamorphism are inferred to reflect different tectonic settings and thermal regimes.

ultra-low-temperature environment. E–HPGM is characterized by apparent thermal gradients of 350–750 °C GPa^{-1} (10–20 °C km^{-1}), and plots across the normal continental geotherm but mostly in the field where heating may be a conductive response to thickening. G–UHTM is characterized by apparent thermal gradients ≫750 °C GPa^{-1} (≫20 °C km^{-1}), and mostly plots across the boundary into the field that requires an advective component of heating.

Figure 7 illustrates apparent thermal gradient, which is inferred to relate to tectonic setting, plotted against age of peak metamorphism. Each type of metamorphism has a distinct range of apparent thermal gradient, as anticipated from Figure 6, and HPM–UHPM is restricted to the later part of the Neoproterozoic Era and the Phanerozoic Eon. However, what is now clear is the dual nature of the thermal regimes represented in the metamorphic record since the Neoarchaean Era. The period from the Neoarchaean Era to the Neoproterozoic Era is characterized by G–UHTM and E–HPGM, whereas the period from late in the Neoproterozoic Era and through the Phanerozoic Eon is characterized by HPM–UHPM and E–HPGM together with G–UHTM, although the latter crops out only sporadically after the Cambrian Period. We might expect that the high dT/dP metamorphism would be of similar age to or younger than the low to intermediate dT/dP metamorphism, as the latter records the transition from subduction to collision whereas the former records inversion of a back-arc basin (Hyndman et al. 2005a, b) or heating by radioactive decay in a thickened orogenic hinterland (Le Pichon et al. 1997; McKenzie & Priestley 2008). At present we do not have data to test this prediction.

Analysis of the data in Figure 7 provides a set of compelling first-order observations from which to argue that the modern era of ultra-low-temperature subduction began in the Neoproterozoic Era, as registered by the occurrence of HPM–UHPM, but that ultra-low-temperature subduction alone is not the hallmark of plate tectonics. In contrast, G–UHTM and E–HPGM are present in the exposed rock record back to at least the Neoarchaean Era, registering a duality of thermal regimes, which has been argued to represent the hallmark of plate tectonics (Brown 2006). Based on this observation, plate tectonics processes probably were operating in the Neoarchaean Era as recorded by the imprints of dual types of metamorphism in the rock record, and this may manifest the first record of a global plate tectonics mode on Earth.

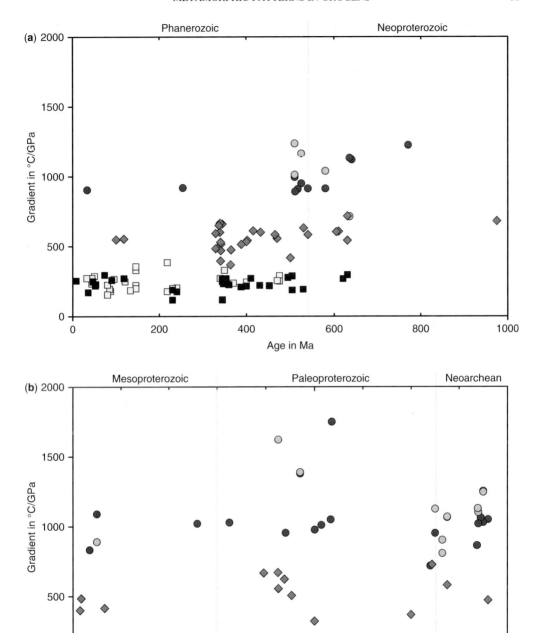

Fig. 7. Apparent thermal gradient (in °C GPa^{-1}) plotted against age of peak metamorphism (in Ma) (data from tables 1–5 of Brown 2007a) for the three main types of extreme metamorphic belt (G–UHTM, circles; E–HPGM, diamonds; HPM–UHPM, squares; for further details see Fig. 6) for two time intervals: (**a**) Phanerozoic and Neoproterozoic; (**b**) Mesoproterozoic to Neoarchaean.

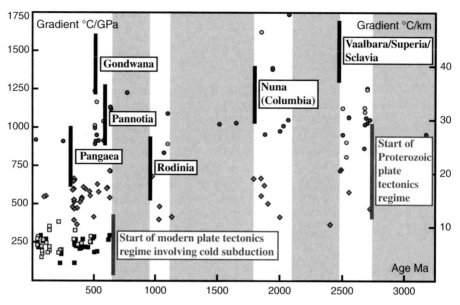

Fig. 8. Plot of apparent thermal gradient (in °C GPa^{-1}) v. age of peak metamorphism (in Ma) for the three main types of extreme metamorphic belt (G–UHTM, circles; E–HPGM, diamonds; HPM–UHPM, squares; for further details see Fig. 6); an approximate conversion to °C km^{-1} is also shown (data from tables 1–5 of Brown 2007a). Also shown are the timing of supercraton amalgamation (Vaalbara, Superia and Sclavia) and supercontinent amalgamation (Nuna (Columbia), Rodinia and Pannotia and Gondwana as steps to Pangaea), and the start of the Proterozoic plate tectonics and modern plate-tectonics regimes in relation to thermal gradients for the three main types of extreme metamorphic belt.

The distribution of these types of metamorphism throughout Earth history based on the age of close to peak metamorphism as defined above is displayed in Figure 8 together with periods of supercontinent amalgamation. Changes in the metamorphic record broadly coincide with the transitions from the Archaean to Proterozoic Eons and the Proterozoic to Phanerozoic Eons, and imply a different style of tectonics in the Archaean Eon in comparison with the Proterozoic Eon and in the Proterozoic Eon in comparison with the Phanerozoic Eon. Overall, the restricted timespan of different types of metamorphism through Earth history and the periods of metamorphic quiescence during the Proterozoic Eon suggest a link with the supercontinent cycle and major events in the mantle. This issue is discussed in more detail later in this review.

Caveats There are several caveats about possible bias in this record. It is commonly argued that going back through time increases loss of information by erosion of the older record. However, the data in Figures 7 and 8 plot in particular periods and there is a clear distinction between the period before the Neoproterozoic Era, where UHPM does not occur, and during the Neoproterozoic Era and the Phanerozoic Eon, where UHPM is common. These observations are inconsistent with a progressively degraded record with increasing age. Extrapolation back in time also raises questions about partial to complete overprinting by younger orogenic events on older orogens, which is a concern in any metamorphic study. However, our ability to recognize the effects of overprinting has improved significantly (discussed above), and overprinting has been avoided in compiling the dataset used for this analysis (Brown 2007a). Finally, as discussed above it is likely that some Phanerozoic G–UHTM rocks have not yet been exposed at Earth's surface, leading to possible bias in the younger part of the record.

The transition from the Archaean Eon to the Proterozoic Eon

Granulite-facies ultrahigh-temperature metamorphism (G–UHTM) is documented in the rock record predominantly from the Neoarchaean Era to the Cambrian Period, although it may be inferred at depth in some younger orogenic systems (Le Pichon *et al.* 1997; Brown 2007a; McKenzie & Priestley 2008). The first occurrence of G–UHTM in the rock record signifies a change in geodynamics that generated transient sites of very high heat flow, perhaps analogous to modern subduction zone back-arcs (and possible deep in arcs) or orogenic hinterlands. In the Neoarchaean Era, G–UHTM occurs within the Kaapvaal Craton and the Southern Marginal Zone of the Limpopo Belt, southern Africa, and in the Lewisian Complex of the Assynt terrane, Scotland (the Badcallian event), at $c.$ 2.72–2.69 Ga, and within the Napier Complex, East Antarctica, and the Andriamena Unit, Madagascar, at $c.$ 2.56–2.46 Ga (references in Brown 2007a, 2008).

Although well-characterized examples are rare in the transition from the Neoarchaean Era to the Palaeoproterozoic Era, medium-temperature eclogite–high-pressure granulite metamorphism (E–HPGM) also is first recognized in the Neoarchaean Era and occurs at intervals throughout the rock record during the Proterozoic and Palaeozoic Eons (Brown 2007a). E–HPGM belts are inferred to record subduction-to-collision orogenesis and are complementary to G–UHTM belts. However, there are not yet sufficient localities for these types of metamorphism in the Neoarchaean Era to determine whether they are contemporary or not in any particular orogen. The oldest occurrence of E–HPGM is represented by eclogite blocks within mélange in the Gridino Zone of the Eastern Domain of the Belomorian Province, White Sea, Russia. The eclogite-facies metamorphism appears to have been reliably dated at $c.$ 2.72 Ga and the P–T data of 1.40–1.75 GPa and 740–865 °C are well characterized. This occurrence is one of the earliest records of E–HPGM within a suture zone and is critical in evaluating the start of plate tectonics, but the fact that the ages come from blocks in a mélange cannot be avoided and does not necessarily undermine the antiquity of these eclogites.

The occurrence of both G–UHTM and E–HPGM belts since the Neoarchaean Era manifests the onset of a 'Proterozoic plate-tectonics regime', which may have begun locally during the Mesoarchaean and Neoarchaean Eras and may have become global only during the transition to the Palaeoproterozoic Eon (Brown 2007b, 2008). This premise is consistent with aggregation of continental crust into progressively larger units to form supercratons, perhaps indicating a change in the pattern of mantle convection during the transition to the Proterozoic Eon.

The emergence of plate tectonics requires forces sufficient to initiate and drive subduction, and lithosphere with sufficient strength to subduct coherently. These requirements probably were met as basalt became able to transform to eclogite. Secular change in thermal regimes to allow this transformation appears to have been gradual, occurring regionally first during the Mesoarchaean to Neoarchaean Eras, leading to the successive formation of the supercratons Vaalbara, Superia and Sclavia, and worldwide during the Palaeoproterozoic Era as evidenced in orogenic belts that suture Nuna (Columbia), unless this distribution is an artefact of (lack of) preservation or thorough overprinting of eclogite in the exposed Archaean cratons.

The western Superior craton in Laurentia represents a prime example of craton formation by terrane accretion. Although elements of the geology extend back perhaps to the Eoarchaean Era (e.g. Böhm et al. 2000), the western Superior craton was created by a sequence of five orogenic events that assembled continental and oceanic terranes into a coherent unit, part of the supercraton Sclavia (Bleeker 2003), during the Mesoarchaean and Neoarchaean Eras (Percival et al. 2006). The constituent terranes exhibit similar sequences of events comprising termination of arc magmatism, early deformation, synorogenic sedimentation, sanukitoid magmatism, main-phase shortening deformation and regional metamorphism to granulite facies (e.g. Pikwitonei granulite domain; Mezger et al. 1990; Böhm et al. 2004), late transpression, and emplacement of crust-derived granites. Seismic reflection and refraction images indicate north-dipping structures, interpreted as a stack of discrete 10–15 km thick terranes. A model of progressive accretion by early plate tectonics is suggested by the presence of a slab of high-velocity material, inferred to represent subcreted oceanic lithosphere, and Moho offsets, consistent with shallow subduction (Percival et al. 2006).

The transition to a Proterozoic plate-tectonics regime resulted in stabilized lithosphere in which cratons form the cores of continents that subsequently grew dominantly by marginal accretion. Furthermore, the transition coincides with the first occurrence of ophiolite-like complexes in suture zones during the Proterozoic Eon (Moores 2002) and with the increase in $\delta^{18}O$ of magmas through the Palaeoproterozoic Era, which may reflect maturation of the crust, the beginning of recycling of supracrustal rocks and their increasing involvement in magma genesis via subduction (Valley et al. 2005). Although the style of subduction during the Proterozoic Eon remains cryptic, the change in tectonic regime whereby interactions between discrete lithospheric plates generated tectonic settings with contrasting thermal regimes was a landmark event in Earth history (Brown 2006).

As an example of accretion during the Palaeoproterozoic Era, we may consider the metasedimentary belts within the Lewisian complex of northwestern Scotland. Although the Lewisian complex has been correlated with the intercontinental collisional belts of Palaeoproterozoic age that suture Laurentia, identifying an appropriate tectonic setting for amalgamation of the various components of the Lewisian complex has been elusive because of uncertainty about the age and origin of the supracrustal rocks (the Loch Maree Group on the mainland and the Leverburgh Belt in South Harris, Outer Hebrides) and the paucity of subduction-related intrusive rocks. The Loch Maree Group is made up of two components (Park et al. 2001), one oceanic (plateau basalts or primitive arcs, plus associated abyssal sediments, ferruginous hydrothermal deposits, and platform carbonates) and the other continental (deltaic flysch, greywacke shale), and probably represents an accretionary complex similar to the Shimanto belt in Japan. Subsequently, the various elements of the Loch Maree Group became tectonically intermixed and subject to extreme deformation under amphibolite-facies conditions during accretion to an overriding Lewisian terrane in the Orosirian Period. Metamorphism in the Leverburgh Belt was more extreme, reaching high-pressure granulite-facies conditions (transitional UHTM to E–HPGM), also in the Orosirian Period (Baba 1998, 1999; Hollis et al. 2006). The preservation of evidence of prograde increase in P, high P–T mineral assemblages, and chemical disequilibrium features in these rocks is attributed to a short-lived tectonometamorphic cycle that terminated in rapid exhumation after a collisional event. Overall, the Lewisian complex is best interpreted as a collage of terranes resulting from subduction-to-collision orogenesis, with the accretionary complexes (all probably part of the same unit) sandwiched between the terrane elements. However, the complex as a whole contains comparatively little juvenile material of Palaeoproterozoic age, limited to possible arc remnants in the Outer Hebrides (Roineabhal and Nis terranes) and the volcano-sedimentary assemblage at Loch Maree (Park et al. 2005).

The absence of HPM–UHPM terranes before the Ediacaran may relate to weakness of the subducting lithosphere, which might have been strong enough to allow shallow subduction but the rheology may have been too weak to allow deep subduction of coherent slabs and/or provide a mechanism for eduction of continental crust if, indeed, it was ever subductable before the Ediacaran (Burov & Watts 2006; Brun & Faccenna 2008; van Hunen & van den Berg 2008). This does not conflict with subduction of ocean lithosphere late during the Archaean Eon and during the Proterozoic Eon, as early slab breakoff and slab breakup still might have transported materials with a surface chemical signature deep into the mantle as recorded in some diamonds. Thus, subduction zone metamorphism and the transition from subduction to collision are recorded by eclogites and high-pressure granulites (intermediate dT/dP metamorphism) during the Neoarchaean Era and the Proterozoic Eon.

Paired metamorphic belts revisited

Orogens may be composed of belts with contrasting types of metamorphism that record different apparent thermal gradients. Classic paired metamorphic belts in which an inboard high dT/dP metamorphic belt is juxtaposed against an outboard low dT/dP metamorphic belt along a tectonic contact are found in accretionary orogens of the circum-Pacific (Miyashiro 1961). Generally, they appear to result from tectonic juxtaposition of terranes with different metamorphic facies series that may or may not be exactly contemporaneous and that may or may not be far-travelled (Brown 1998a, b, 2002b; Tagami & Hasebe 1999). This is an inevitable consequence of the difference between contemporaneous subduction, generating low dT/dP blueschists and eclogites in the subduction zone and extension in an arc–back-arc system with formation of depositional basins, compared with the timing of events such as ridge subduction (e.g. Brown 1998b, 2002b) or deformation and thickening of the arc–back-arc system (e.g. Collins 2002), generating high dT/dP metamorphism. Thus, metamorphic belts of contrasting type may be formed during a single orogenic cycle along different sectors of a common convergent margin by multiple processes, but juxtaposition may have been due to the obliquity of convergence and lateral translation along the margin late in the orogenic event (e.g. Brown 1998a).

In Miyashiro's original classification of types of metamorphism (Miyashiro 1961), an intermediate type of metamorphism was included for unpaired belts such as those in the Scottish Highlands and the Northern Appalachians, although in both cases the intermediate dT/dP metamorphic belt (Barrovian type) is juxtaposed against a high dT/dP metamorphic belt (Buchan type). Miyashiro (1973) subsequently suggested that

'paired and unpaired (single) metamorphic belts form by the same mechanism, and an unpaired belt represents paired belts in which the contrast between the two belts is obscure, or in which one of the two belts is undeveloped or lost'.

One issue to consider is whether to extend the concept of 'paired metamorphic belts' more widely than accretionary orogens, outside the original usage by Miyashiro (1961), to subduction-to-collision orogenic systems, as perhaps was implied by Miyashiro in his later publication (Miyashiro 1973). This issue is raised in part by the use of the term 'paired metamorphic mountain belt' by Goscombe & Hand (2000) to refer to the eastern Himalaya in Nepal, where two sectors of the mountain belt are characterized by contrasting types of P–T path, clockwise versus counter-clockwise in P–T space, but the metamorphism is of 'Barrovian' (intermediate dT/dP) and 'Buchan' (high dT/dP) type, respectively, without a low dT/dP metamorphic belt at all. This is similar to the Scottish Highlands, from where these two types of metamorphism were identified, and the two types of metamorphic belt in the Northern Appalachians (Spear *et al.* 2002).

The modern plate-tectonics regime is characterized by a duality of thermal environments, the subduction zone and the suture zone of subduction-to-collision orogens, and the back-arc or orogenic hinterland, in which contrasting types of regional-scale metamorphic belts are being formed contemporaneously. I consider this duality to be the hallmark of one-sided subduction. Thus, the characteristic imprint of one-sided subduction and perhaps of plate tectonics in the geological record will be the broadly contemporaneous occurrence of two contrasting types of metamorphism reflecting this duality of thermal environments (Brown 2006). On this basis, I propose the following:

paired metamorphic belts are penecontemporaneous belts of contrasting type of metamorphism that record different apparent thermal gradient, one warmer and the other colder, juxtaposed by plate tectonics processes.

By this definition, the combination of G–UHTM with E–HPGM (from the Neoarchaean Era to the Neoproterozoic Era) or HPM–UHPM and/or E–HPGM with G–UHTM (during the Phanerozoic Eon) may be described as a paired metamorphic belt. This extends the original concept of Miyashiro (1961) beyond the simple pairing of high dT/dP and low dT/dP metamorphic belts in circum-Pacific accretionary orogens, and makes it more useful in the context of our better understanding of the relationship between thermal regimes and tectonic setting. This is particularly useful in subduction-to-collision orogenic systems, where an accretionary phase is overprinted by a collision phase that will be registered by the imprint of penecontemporaneous low to intermediate dT/dP and high dT/dP metamorphism in the rock record (Brown 2006).

The supercontinent cycle

The continental lithosphere has been amalgamated into a supercontinent at several intervals during Earth history with dramatic effects on both surface and deep Earth processes. Aggregated continental lithosphere influences mantle heat loss by acting as an insulator to the convecting mantle and by imposing its own wavelength on mantle convection by impeding downwelling (Le Pichon & Huchon 1984; Gurnis 1988; Guillou & Jaupart 1995; Cooper *et al.* 2006; Coltice *et al.* 2007). Therefore, during periods of aggregated continental lithosphere, we may anticipate longer wavelength convection and global variations in mantle heat loss and mantle potential temperature (Grigné *et al.* 2005; Lenardic *et al.* 2005, 2006; Coltice *et al.* 2007).

Correlations

The temporal relationship between metamorphism at extreme $P-T$ conditions, plate tectonics and the supercontinent cycle is shown in Figure 8. There are five principal periods of crustal amalgamation since the Neoarchaean Era (Bleeker 2003), as follows: formation of the supercratons Vaalbara, Superia and Sclavia from older superterranes and juvenile crustal additions during the Mesoarchaean and Neoarchaean Eras (Vaalbara slightly older and Sclavia slightly younger); and, formation of the supercontinents Nuna (Columbia; 2.0–1.8 Ga), Rodinia (1.2–1.0 Ga) and Pangaea (320–250 Ma; via the intermediate steps of Pannotia (650–570 Ma) and Gondwana (570–510 Ma)). There is a correspondence between the timing of extreme metamorphism in the rock record and suturing of continental lithosphere by subduction-to-collision orogenesis into supercratons and supercontinents.

The Phanerozoic Eon

The Phanerozoic Eon has been distinguished by rearrangement of the continental lithosphere within a continent-dominated hemisphere in a series of steps leading to the assembly of Pangaea during the Palaeozoic Era, followed by the early stage of supercontinent breakup and dispersal during the Mesozoic–Cenozoic (Brown 2008). The formation of Pangaea involved successive subduction-to-collision orogenic systems (both accretionary orogens, involving ribbon continent terranes and/or arcs, and collisional orogens, involving multiple continents) within a single (Pangaean) convection cell, resulting in the Appalachian/Caledonian–Variscide–Altaid and the Alpine–Himalayan–Cimmerian orogenic systems. In contrast, a complementary (Panthalassan) convection cell, centred on the South Pacific low-velocity structure, is composed of ocean lithosphere and defined at its outer edge by the circum-Pacific accretionary orogenic systems (Coney 1992; Collins 2003).

The Panthalassan convection cell became established towards the end of the Neoproterozoic Era, centred on the former Rodinia and most probably reflecting a slab 'graveyard' as a result of Rodinia assembly, and reached a maximum horizontal extent around the Devonian–Carboniferous boundary. It has been in decline since, as the Pangaean cell, centred on the former Gondwana and most probably reflecting a slab 'graveyard' as a result of

Gondwana assembly, began to expand in the Jurassic–Cretaceous Eras as excessive heating of the upper mantle beneath Pangaea led to breakup (Le Pichon & Huchon 1984; Gurnis 1988).

Slab 'graveyards' might form after continent aggregation by collision if stagnant slabs break off and avalanche into the lower mantle. They are thought to be associated with P- and S-wave low-velocity structures (LVS; Christensen & Hofmann 1994; McNamara & Zhong 2004; Tan & Gurnis 2007). LVS coincide with positive anomalies in the geoid, are inferred to be warmer than surrounding mantle and are interpreted to be high-bulk-modulus domes composed of pyroxenite (perovskite-rich material in the lower mantle), supporting the link with subducted mid-ocean ridge basalt (MORB) (Tan & Gurnis 2007). Confirmation of crustal recycling in the Pacific mantle comes from Sr, Nd, Hf and Pb isotope data from garnet-clinopyroxenite xenoliths in lavas from Malaita, Solomon Islands (Ishikawa *et al.* 2007).

Wilson-type cycles and Hoffman-type breakups

There are two contrasting models for the behaviour of supercontinents (Brown 2008), based on different concepts introduced by Wilson (1966, closing and reopening an ocean) and Hoffman (1991, turning a supercontinent inside out), respectively. In one model, continental lithosphere simply fragments and reassembles along the same (internal) contacts or introverts (the 'Wilson cycle' *sensu stricto*; Wilson 1966), whereas in the alternative model a supercontinent fragments, disperses and reassembles by closure of the complementary superocean (Hoffman 1991). Wilson-type cycles operated during the Phanerozoic Era, when the continental lithosphere was restricted to one hemisphere. In contrast, a Hoffman-type breakup was the process by which the Gondwanan elements of Rodinia were dispersed and reassembled, possibly first as part of the ephemeral Pannotia (Dalziel 1997) before final amalgamation as Gondwana (Collins & Pisarevsky 2005; Tohver *et al.* 2006). In addition, a Hoffman-type breakup was most probably the process by which Nuna (Columbia) was reconfigured into Rodinia, although in this case the continental fragments were fewer and larger (Condie 2002). Whether any Wilson-type cycles occurred within Rodinia or Nuna remains to be confirmed, but there is no *a priori* reason to suppose that Wilson-type cycles did not occur during the Proterozoic Eon and the Trans-Hudson orogen in North America has been interpreted in terms of a Wilson-type cycle (Murphy & Nance 2005).

Examples of Wilson-type cycles and intrasupercontinent orogenesis during the Phanerozoic Eon

During Wilson-type cycles, 'expansion' of a continent-dominated hemisphere may be limited by the persistence of an opposed ocean-dominated hemisphere, as has been the case for the Pacific Ocean through the Phanerozoic Eon. In this case, modification of the way the continents are aggregated requires splitting the continental lithosphere to generate an internal ocean, such as the Iapetus Ocean (Murphy & Nance 2005), which separated Laurentia from Baltica and Gondwana, or the Rheic Ocean, which separated the Avalonian and Hunic terranes from Gondwana (von Raumer & Stampfli 2008), or the closure of the Palaeotethys and the formation of the Neotethys as 'Cimmeria' was sutured to Eurasia (Metcalfe 1996).

In the case of splitting off a terrane, the terrane migrates across a closing ocean before suturing at a new location on the opposite margin (commonly referred to as 'terrane export'). Calving of another terrane from the parent continent by splitting inboard from the margin and initiation of subduction behind the newly accreted terrane on the opposite side of the new ocean in turn allows a second terrane to migrate across a closing ocean to be accreted on the opposite margin. The process, which may continue to repeat, is best exemplified by the dispersion of terranes from Gondwana across the eastern Tethys to form Asia by accretion (e.g. Metcalfe 1996). It is the successive closure of these newly generated but relatively short-lived oceans by subduction and terrane export from one continental aggregate to another that leads to a new arrangement of the continental lithosphere within the continental hemisphere (e.g. Stampfli *et al.* 2002; Collins 2003; von Raumer *et al.* 2003; von Raumer & Stampfli 2008).

HPM–UHPM, which may be linked with or transitional to E–HPGM, is the metamorphic style generally associated with Wilson-type cycles during the Phanerozoic Eon (Carswell & Compagnoni 2003; Liou *et al.* 2004), reflecting the short-lived subduction and terrane collision style of tectonics characteristic of the Caledonian–Variscide–Altaid and the Alpine–Himalayan–Cimmerian orogenic systems (Brun & Faccenna 2008). These also correspond to the 'push–pull' cycles of Lister & Forster (2006).

Numerical modelling of the minimum subduction necessary to generate thermal conditions suitable for blueschist metamorphism is consistent with limited subduction of short-lived ocean basins (Maresch & Gerya 2005). Because of the geometric interplay between the nose of the subducting ocean lithosphere and the evolving array of isotherms, subduction of younger and hotter ocean lithosphere may lead to earlier formation of blueschists than subduction of older and cooler ocean lithosphere. Maresch & Gerya (2005) identified an optimum age of 40–60 Ma for subducted ocean lithosphere to generate blueschist conditions. The thermal structure also explains the rarity of arc volcanic rocks in these orogens because the necessary flux of water into the mantle wedge to drive melting will be weak or even absent.

The Appalachian orogen is rather different, particularly in the northern sector, where older Barrovian and younger high dT/dP metamorphism may be a consequence of subduction and arc–back-arc formation on one or sometimes both margins of sutures between peri-Laurentian and peri-Gondwanan terranes and intervening Iapetan elements. UHPM rocks have not yet been identified in the southern Appalachians, but HPM rocks are present (Dilts et al. 2006) and are a characteristic feature of the final (Alleghanian) phase of orogenesis.

The Iapetus Ocean was closed along the Appalachian–Caledonian orogenic system; the intervening internally generated lithosphere was consumed during assembly of Laurentia, Avalonia and Baltica in the Early Devonian forming Laurussia (Murphy & Nance 2005). UHPM rocks are a feature of the Norwegian Caledonides (e.g. Carswell & Compagnoni 2003; Liou et al. 2004). The Rheic Ocean was closed along the Variscide–Altaid orogenic system. In the Variscide sector, the intervening internally generated lithosphere was consumed by subduction during clockwise rotation of Gondwana and collision with Laurussia. UHPM rocks are a common feature of the European Variscides (e.g. Carswell & Compagnoni 2003; Liou et al. 2004). The Altaid Central Asian Orogenic Belt is something of an enigma; although it includes some of the oldest formed and most extreme UHPM rocks known (at Kotchatev; Kaneko et al. 2000) it also includes high dT/dP metamorphic belts, inferred to record ridge–trench interactions, and arc collisions (Windley et al. 2007; see also Filippova et al. 2001; Collins 2003). Within the Mesozoic–Cenozoic Tethysides, the Central Orogenic Belt of China (the Tienshan–Qinling–Dabie–Sulu orogenic system; e.g. Yang et al. 2005) and the Alpides (the Alpine–Zagros–Himalayan orogenic system; e.g. Şengör 1987; Hafkenscheid et al. 2006) are products of multiple accretion events or collisions. In each case, the orogenic event involved the destruction of a relatively short-lived ocean, and the sutures commonly are decorated with occurrences of HPM and/or UHPM rocks.

Supercontinent fragmentation, dispersal and reassembly by Hoffman-type breakups

Continental lithosphere tends to amalgamate over cold downwellings to form a supercontinent, which inhibits subduction and mantle cooling (Gurnis 1988). The mantle beneath the supercontinent may eventually overheat and become the site of a new upwelling that fragments the overlying continental lithosphere under tension to dissipate the thermal anomaly. Thus, the breakup of Rodinia was centred over the Pacific LVS and the breakup of the Gondwanan sector of Pangaea was centred over the Africa LVS. The next supercontinent amalgamates above areas of mantle downwelling by subduction-to-collision orogenesis, as did the Gondwanan elements of Rodinia by the end of the Cambrian Period. This is the basic principle that leads to breakup of one supercontinent and formation of another. As a result of Hoffman-type breakup, the internal rifted margins of the old supercontinent become the external margins of the new supercontinent, and the external accretionary margins of the old supercontinent become deformed and smeared out along the internal sutures of the new supercontinent.

Let us consider the transformation of Rodinia to Gondwana, which is the original example of a Hoffman-type breakup (Hoffman 1991). This transformation involved the fragmentation, dispersal and reassembly of the continental lithosphere by subduction-to-collision orogenesis to form the network of Brasiliano and Pan-African belts that suture Gondwana (e.g. Cordani et al. 2003; Collins & Pisarevsky 2005), possibly via the ephemeral supercontinent Pannotia (Dalziel 1997). The orphaned Laurasian continental fragments combined with each other at a later time and then with Gondwana to form Pangaea by the Permian Period. The internal geometry of Rodinia changed considerably during its few hundred million years of existence. Geological and palaeomagnetic data suggest that the supercontinent consolidated at 1100–1000 Ma and most probably disintegration began between 850 and 800 Ma (e.g. Cordani et al. 2003; Torsvik 2003). Reassembly to form Gondwana occurred by destruction of parts of the complementary superocean as Rodinia progressively disintegrated, although the exact configuration and global location of the different continental lithosphere fragments in Rodinia at the time of breakup is uncertain (e.g. Murphy et al. 2004). Although G–UHTM and E–HPGM are associated with the Brasiliano and Pan-African orogenic belts, the Trans-Saharan segment of the Pan-African also records the first coesite-bearing eclogites, and sutures within the Anti-Atlas and the South China block record the first blueschists. These features point to a transition caused by secular cooling and the changeover to the 'modern plate-tectonics regime' characterized by colder subduction.

Ancient Earth

Metamorphism before the Neoarchaean

The rock record from the Eoarchaean Era through the Mesoarchaean Era generally records $P-T$ conditions characteristic of low-to-moderate-P–moderate-to-high-T metamorphic facies, yielding dT/dP in the range for G–UHTM but without achieving the extreme temperatures that appear in the record during the Neoarchaean Era. In greenstone belts metamorphic grade varies from prehnite–pumpellyite facies through greenschist facies to amphibolite facies and, rarely, into the granulite facies; ocean-floor metamorphism of the protoliths is common. In high-grade terranes, granulite-facies metamorphism at temperatures below 900 °C and multiple episodes of anatexis are the norm.

In southern West Greenland, in the Isua Supracrustal Belt the metamorphism is polyphase, occurring during the Eoarchaean and Neoarchaean Eras, and an age of c. 3.7 Ga has been argued for the early metamorphism. $P-T$ conditions of 0.5–0.7 GPa and 500–550 °C (or up to 600 °C) have been retrieved from the Isua Supracrustal Belt (Hayashi et al. 2000; Rollinson 2002). These data yield warm apparent thermal gradients of 800–1000 °C GPa^{-1} (c. 23–28 °C km^{-1}), which is just within the range for a purely conductive response to thickening, but may reflect thinner lithosphere with higher heat flow than later in Earth history. Also in southern West Greenland, in mafic rocks on the eastern side of Innersuartuut Island in the Itsaq Gneiss Complex of the Færingehavn terrane, where the gneisses record several events in the interval c. 3.67–3.50 Ga, early orthopyroxene + plagioclase assemblages record poorly defined but rather ordinary granulite-facies conditions whereas overprinting garnet + clinopyroxene + quartz assemblages probably record widespread high-pressure granulite-facies metamorphism associated with final terrane assembly in the interval 2.715–2.650 Ga (Friend & Nutman 2005b, 2007). This is generally consistent with an evolution during the Mesoarchaean and Neoarchaean Eras involving: a postulated island arc complex in the eastern Akia terrane of southern West Greenland, built on a volcanic substrate in the interval c. 3.07–3.00 Ga and subsequently thickened and metamorphosed in the interval c. 3.00–2.97 Ga (Garde 2007); progressive terrane assembly during the interval c. 2.95–2.65 Ga (Friend & Nutman

2005a); and plate tectonics processes operating on Earth (Brown 2006, 2007b).

Although part of the Isua Supracrustal Belt appears to represent early ocean-floor and interoceanic island arc material (Furnes et al. 2007), it is by no means clear that claims that the Isua Supracrustal Belt is dominantly a subduction–accretion complex are valid (Komiya et al. 2002). Early plate-like behaviour may have allowed crust to float to form thick stacks above zones of boundary-layer downwelling that 'drip' into the mantle, or, for a slightly cooler Earth, thick stacks above zones of 'sub-lithospheric' subduction, in which the mantle part of the 'lithosphere' is rigid enough to subduct (Fig. 9; Davies 1992; van Hunen et al. 2008). These thick stacks might be called 'pseudo-accretionary complexes', but neither boundary-layer downwelling nor 'sub-lithospheric' subduction is likely to have created the dual thermal environments at the site of tectonic stacking argued to be characteristic of plate tectonics (Brown 2006). It is possible, however, that thickening (with or without delamination of (?) eclogite sinkers)

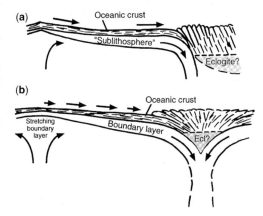

Fig. 9. (a) Sketch of a conjectured regime that might precede plate tectonics proper (from Davies 1992). The mantle part of the ocean lithosphere is thick enough to behave rigidly. Although the ocean lithosphere as a whole is buoyant, the lower mantle part of the lithosphere is negatively buoyant and detaches from the crust to subduct into the deeper mantle (sublithospheric subduction). The crust may thicken and differentiate internally, and the deeper part may transform to eclogite that may drip into the mantle. (b) Sketch of a conjectured pre-plate regime (from Davies 1992). The buoyant crust decouples from a denser mantle boundary layer, and accumulates above the downwelling mantle boundary layer as it drips into the deeper mantle. The crust may thicken and differentiate internally, and the deeper part may transform to eclogite that may drip into the mantle. Reprinted from Davies, G. F. 1992. On the emergence of plate tectonics. Geology, **20**, 963–966, with permission. © 1992 The Geological Society of America (GSA).

might have induced melting deep in the pile to generate tonalite–trondhjemite–granite-type magmas in a manner that might resemble future arcs. In principle, the metamorphic imprint imposed by either of these behaviours during the early part of Earth history should be distinguishable from that imposed by a plate tectonics mode of convection.

In South Africa, recent work on the c. 3.23 Ga metamorphism of the Barberton and related greenstone belts has yielded the following P–T data: in the Onverwacht Group greenstone remnants, P–T conditions of 0.8–1.1 GPa and 650–700 °C (Dziggel et al. 2002); in the southern Barberton greenstone belt, P–T conditions of 0.9–1.2 GPa and 650–700 °C (Diener et al. 2005, 2006); and, in amphibolite-dominated blocks of supracrustal rocks in tectonic mélange from the Inyoni Shear Zone, P–T conditions of 1.2–1.5 GPa and 600–650 °C (Moyen et al. 2006). Apparent thermal gradients for these belts are in the range 450–700 °C GPa^{-1} (c. 13–20 °C km^{-1}), which are not outside the limit for a conductive response to thickening. Consequently, an earlier conclusion that this intermediate dT/dP metamorphism represents evidence for subduction-driven tectonic processes during the evolution of the early Earth may be premature.

Tectonics before the Neoarchaean Era

Blueschists are not documented in the Archaean Eon and there is no metamorphic imprint of subduction of continental crust to mantle conditions and return to crustal depths. However, the chemistry of eclogite xenoliths of Mesoarchaean to Neoarchaean age and the chemistry and paragenesis of diamonds of Neoarchaean age in kimberlites within cratons suggest that some process associated with supercraton formation at convergent margins was operating to take basalt and other supracrustal material into the mantle by the Neoarchaean Era (Brown 2008). This may have been some form of subduction. However, subduction as we know it on modern Earth requires strong lithosphere and appropriate driving forces. The hotter lithosphere in the Archaean Eon is likely to have been weaker because of lower viscosity, perhaps weakening coupling across the subduction boundary and perhaps leading to more frequent slab breakoff and/or slab breakup (e.g. Burov & Watts 2006). Thus, it may be that weaker lithosphere prevented exhumation of subducted crust by mechanisms related to subduction and plate kinematics (e.g. Brun & Faccenna 2008; van Hunen & van den Berg 2008). Regardless of the tectonic style, an essential requirement for the return of ocean crust into the mantle is the efficient transformation of basalt to eclogite (van Hunen & van den Berg 2008). The interested reader is referred to the paper by van Hunen et al. (2008), where a variety of possible models for tectonics during the Archaean Eon are summarized and discussed.

Tonalite–trondhjemite–granite suites (TTGs) dominate the rock record of the Archaean Eon, but their origin has been controversial. One recent view suggests that many older TTGs formed by melting of garnet amphibolite of broadly basaltic composition hydrated by interaction with seawater, whereas melting of eclogite increased in importance through the Mesoarchaean and Neoarchaean Eras, as shown by an increase in Nb/Ta in TTGs (Foley et al. 2002, 2003). Melting of garnet amphibolite may be achieved in the lower part of thickened basaltic crust or by subduction on a warmer Earth. Conceptually, this is consistent with early plate-like behaviour that allowed crust to float to form thick stacks above zones of boundary-layer downwelling, or, for a slightly cooler Earth, thick stacks above zones of 'sub-lithospheric' subduction, and this indeed may have been the convective mode during the Eoarchaean and Palaeoarchaean Eras (Fig. 9; van Hunen et al. 2008). By the Mesoarchaean and Neoarchaean Eras, and possibly as early as the Palaeoarchaean Era, subduction was operating in some regions of the Earth; this subduction was probably characterized by warmer geotherms that allowed melting of subducting garnet amphibolite and, with secular cooling, a change to melting of subducting eclogite.

Worldwide, the oldest surviving crustal remnants from the Palaeoarchaean Era generally are composed of juvenile TTGs formed prior to a period of polyphase granulite-facies metamorphism in the interval 3.65–3.60 Ga; extreme thermal conditions typical of G–UHTM are not generally registered before the Neoarchaean Era. This appears to be counterintuitive, as we might expect that the higher abundance of heat-producing elements might have led to higher crustal heat production and hotter orogens. However, it is possible that contemporary heat loss through oceans and continents was higher and lithosphere rheology was generally weaker on early Earth. This is permissive for a 'crème brûlée' lithosphere rheology structure. Also, it is consistent with: (1) modelling suggesting dominance of unstable subduction for plate collision regimes with very hot geotherms in which convergence is accommodated by pure shear thickening and development of gravitational (Rayleigh–Taylor) instabilities (Burov & Watts 2006); (2) modelling that rules out the commonly proposed flat subduction model for early Earth tectonics (van Hunen et al. 2004); (3) the proposition that early plate-like behaviour may have allowed crust to float to form thick stacks above zones of boundary-layer downwelling, or, for a slightly cooler Earth, thick stacks above zones of 'sub-lithospheric' subduction (Davies

1992; van Hunen et al. 2008); (4) modelling of thinner ocean lithosphere suggesting that plate motions in the Archaean Eon might have been faster than on modern Earth, which may explain the scarcity of accretionary prisms in the rock record during the Archaean Eon, as high convergence rates will favour subduction erosion over subduction accretion (Hynes 2008); and (5) the absence of E–HPGM and HPM–UHPM in the early Earth rock record, inferred herein to be more typical of plate collisions in the Proterozoic and Phanerozoic Eons (Brown 2006).

Concluding discussion

Metamorphism associated with orogenesis provides a mineral record that may be inverted to yield ambient apparent thermal gradients. Apparent thermal gradients derived from inversion of age-constrained metamorphic P–T data are used to identify tectonic settings of ancient metamorphism and evaluate geodynamic regimes. On modern Earth, tectonic settings with lower thermal gradients are characteristic of subduction zones whereas those with higher thermal gradients are characteristic of back-arcs and orogenic hinterlands. If a duality of thermal regimes is the hallmark of plate tectonics, then a duality of metamorphic belts is the characteristic imprint of plate tectonics in the record. Ideally, these belts should be 'paired' broadly in time and space, although the spatial arrangement may be due to late orogenic juxtaposition. Given this characteristic imprint, I propose broadening the definition of paired metamorphic belts to the following: paired metamorphic belts are penecontemporaneous belts of contrasting type of metamorphism that record different apparent thermal gradient, one warmer and the other colder, juxtaposed by plate tectonics processes.

The first global occurrences of granulite-facies ultrahigh-temperature metamorphism (G–UHTM) are recorded during the Neoarchaean Era and contemporaneous but more sparse occurrences of medium-temperature eclogite–high-pressure granulite metamorphism (E–HPGM), signifying changes in global geodynamics that generated transient sites of higher and lower heat flow than required by apparent thermal gradients recovered from the earlier record. G–UHTM is dominantly a phenomenon limited to the period from the Neoarchaean Era to the Cambrian Period and is documented during four distinct periods, the Neoarchaean Era, the Orosirian Period, the Ectasian and Stenian Periods, and the Ediacaran and Cambrian Periods, synchronous with formation of supercratons, supercontinents and the Gondwana segment of Pangaea. Many G–UHTM belts may have developed in settings analogous to modern back-arcs and orogenic hinterlands. The oldest record of E–HPGM also coincides with the formation of supercratons; in addition to occurrences formed during the Proterozoic Era, E–HPGM is common in the Caledonides and Variscides of the Palaeozoic Era. E–HPGM belts complement G–UHTM belts, and are generally inferred to record subduction-to-collision orogenesis.

The occurrence of G–UHTM and E–HPGM belts since the Neoarchaean Era signifies change to globally 'subductable' oceanic lithosphere, possibly beginning as early as the Palaeoarchaean Era, and transfer of water into upper mantle. The first appearance of dual metamorphic belts also registers operation of a global 'Proterozoic plate-tectonics regime', which evolved during a transition in the Neoproterozoic Era to the 'modern plate-tectonics regime' characterized by subduction of continental crust deep into the mantle and its (partial) return from depths of up to 300 km. This transition in the Neoproterozoic Era has implications for transfer of water deeper into mantle. Age distribution of metamorphic belts that record extreme conditions is not uniform; they occur in periods that correspond to amalgamation of continental lithosphere into supercratons (e.g. Superia/Sclavia) or supercontinents (e.g. Nuna (Columbia), Rodinia, and the Gondwana segment of Pangaea). Accretionary orogenic systems form at sites of subduction of oceanic lithosphere; these systems dominate during the lifetime of a supercontinent and during break-up and dispersal. Collisional orogenic systems form where ocean basins close and subduction ultimately ceases; these systems dominate during crustal aggregation and assembly of supercontinents. It follows that collisional orogenic systems may be superimposed on accretionary orogenic systems.

For accretionary orogenic systems where a retreating plate boundary is dominant, high dT/dP metamorphism is typical, commonly with counter-clockwise P–T–t paths and peak metamorphic mineral growth syn- to late in relation to tectonic fabrics. G–UHTM and E–HPGM rocks generally are not exposed, and although rare, blueschists may occur early to record HPM, but UHPM is not found. Short-lived contractional phases of orogenesis probably relate to interruptions in the continuity of subduction caused by topographic features on the ocean plate, particularly ocean plateaux. Extensive granite magmatism accompanies metamorphism. Examples include the Lachlan orogen, Australia, the Acadian orogen, NE USA and Maritime Canada, and the orogens of the Proterozoic Eon in the SW USA. In contrast, for accretionary orogenic systems where an advancing plate boundary is dominant, such as the accretionary prism of the Coastal Cordillera in Chile, which formed late during the Late Palaeozoic Era, and the Diego de Almagro

Metamorphic Complex in Chilean Patagonia, which formed during the Cretaceous Period, low dT/dP metamorphism is characteristic, as recorded by blueschists in these forearc complexes.

At many plate boundaries, oblique convergence leads to orogen-parallel transport in terrane accretion orogenic systems. Commonly a low dT/dP metamorphic belt occurs outboard and is separated from a high dT/dP metamorphic belt by a terrane boundary. In some terrane accretion orogenic systems an additional feature is ridge subduction, which may be reflected in the pattern of high dT/dP metamorphism and the magmatism. Granulites may occur at the highest grade of metamorphism in the high dT/dP belt, where granite magmatism is common, but UHPM occurs only rarely in the outboard low dT/dP belt. Examples include the Mesozoic metamorphic belts of Japan and the North American Cordillera, where successively accreted younger metamorphic belts occur on the outboard side. Terrane accretion orogenic systems may be transitional to collisional orogenic systems. If this transitional type includes large subduction–accretion complexes, the orogen is referred to as Turkic-type, with examples in both the Appalachian/Caledonian–Variscide–Altaid and the Alpine–Himalayan–Cimmerian orogenic systems.

I thank I. Fitzsimons and G. Zhao for their stimulating review comments, some of which were challenging to meet! Of course, any remaining infelicities are mine. I thank P. Cawood for his editorial patience.

References

ANDREOLI, M. A., HART, R. J., ASHWAL, L. D. & COETZEE, H. 2006. Correlations between U, Th content and metamorphic grade in the western Namaqualand Belt, South Africa, with implications for radioactive heating of the crust. *Journal of Petrology*, **47**, 1095–1118.

ANNEN, C., BLUNDY, J. D. & SPARKS, R. S. J. 2006. The genesis of intermediate and silicic magmas in deep crustal hot zones. *Journal of Petrology*, **47**, 71–95.

AUSTRHEIM, H. & BOUNDY, T. M. 1994. Pseudotachylytes generated during seismic faulting and eclogitization of the deep crust. *Science*, **265**, 82–83.

BABA, S. 1998. Proterozoic anticlockwise $P-T$ path of the Lewisian complex of South Harris, Outer Hebrides, NW Scotland. *Journal of Metamorphic Geology*, **16**, 819–841.

BABA, S. 1999. Sapphirine-bearing orthopyroxene–kyanite/sillimanite granulites from South Harris, NW Scotland: Evidence for Proterozoic UHT metamorphism in the Lewisian. *Contributions to Mineralogy and Petrology*, **136**, 33–47.

BALDWIN, J. A. & BROWN, M. 2008. Age and duration of ultrahigh-temperature metamorphism in the Anápolis–Itauçu Complex, southern Brasília Belt, Central Brazil—constraints from U–Pb geochronology, mineral rare earth element chemistry and trace-element thermometry. *Journal of Metamorphic Geology*, **26**, 213–233.

BALDWIN, J. A., POWELL, R., BROWN, M., MORAES, R. & FUCK, R. A. 2005. Mineral equilibria modelling of ultrahigh-temperature metamorphism: An example from the Anápolis–Itauçu Complex, central Brazil. *Journal of Metamorphic Geology*, **23**, 511–531.

BALDWIN, S. L., MONTELEONE, B. D., WEBB, L. E., FITZGERALD, P. G., GROVE, M. & HILL, E. J. 2004. Pliocene eclogite exhumation at plate tectonic rates in eastern New Guinea. *Nature*, **431**, 263–267.

BARBOZA, S. A., BERGANTZ, G. W. & BROWN, M. 1999. Regional granulite facies metamorphism in the Ivrea zone: Is the Mafic Complex the smoking gun or a red herring? *Geology*, **27**, 447–450.

BJORNERUD, M. G. & AUSTRHEIM, H. 2006. Hot fluid or rock in eclogite metamorphism? *Nature*, **440**, E4–E4.

BLATTNER, P. 2005. Transport of low-aH_2O dehydration products to melt sites via reaction-zone networks, Milford Sound, New Zealand. *Journal of Metamorphic Geology*, **23**, 569–578.

BLEEKER, W. 2003. The late Archaean record: A puzzle in ca. 35 pieces. *Lithos*, **71**, 99–134.

BÖHM, C. O., HEAMAN, L. M., CREASER, R. A. & CORKERY, M. T. 2000. Discovery of pre-3.5 Ga exotic crust at the northwestern Superior Province margin, Manitoba. *Geology*, **28**, 75–78.

BÖHM, C. O., HEAMAN, L. M. & MACHADO, N. 2004. *Neoarchaean high-grade metamorphism of the Pikwitonei Granulite Domain, NW Superior Craton*. GAC–MAC Annual Meeting, St. Catharines, Program with Abstracts, **29**, 330.

BOSSE, V., FERAUD, G., RUFFET, G., BALLÈVRE, M., PEUCAT, J.-J. & DE JONG, K. 2000. Late Devonian subduction and early-orogenic exhumation of eclogite-facies rocks from the Champtoceaux Complex (Variscan belt, France). *Geological Journal*, **35**, 297–325.

BOUNDY, T. M., FOUNTAIN, D. M. & AUSTRHEIM, H. 1992. Structural development and petrofabrics of eclogite facies shear zones, Bergen Arcs, Western Norway—Implications for deep crustal deformational processes. *Journal of Metamorphic Geology*, **10**, 127–146.

BROTHERS, R. N. & BLAKE, M. C. 1973. Tertiary plate tectonics and high-pressure metamorphism in New Caledonia. *Tectonophysics*, **17**, 337–358.

BROWN, M. 1993. $P-T-t$ evolution of orogenic belts and the causes of regional metamorphism. *Journal of the Geological Society, London*, **150**, 227–241.

BROWN, M. 1998a. Unpairing metamorphic belts: $P-T$ paths and a tectonic model for the Ryoke Belt, southwest Japan. *Journal of Metamorphic Geology*, **16**, 3–22.

BROWN, M. 1998b. Ridge–trench interactions and high-T–low-P metamorphism, with particular reference to the Cretaceous evolution of the Japanese Islands. *In*: TRELOAR, P. J. & O'BRIEN, P. J. (eds) *What Drives Metamorphism and Metamorphic Reactions?* Geological Society, London Special Publications, **138**, 131–163.

BROWN, M. 2001. From microscope to mountain belt: 150 years of petrology and its contribution to understanding geodynamics, particularly the tectonics of orogens. *Journal of Geodynamics*, **32**, 115–164.

BROWN, M. 2002a. Prograde and retrograde processes in migmatites revisited. *Journal of Metamorphic Geology*, **20**, 25–40.

BROWN, M. 2002b. Plate margin processes and 'paired' metamorphic belts in Japan: Comment. *Earth and Planetary Science Letters*, **199**, 483–492.

BROWN, M. 2006. A duality of thermal regimes is the distinctive characteristic of plate tectonics since the Neoarchean. *Geology*, **34**, 961–964.

BROWN, M. 2007a. Metamorphic conditions in orogenic belts: A record of secular change. *International Geology Review*, **49**, 193–234.

BROWN, M. 2007b. Metamorphism, plate tectonics, and the supercontinent cycle. *Earth Science Frontiers*, **14**, 1–18.

BROWN, M. 2008. Characteristic thermal regimes of plate tectonics and their metamorphic imprint throughout Earth history. *In*: CONDIE, K. & PEASE, V. (eds) *When Did Plate Tectonics Begin?* Geological Society of America, Special Papers, **440**, 97–128.

BROWN, M. & SOLAR, G. S. 1999. The mechanism of ascent and emplacement of granite magma during transpression: A syntectonic granite paradigm. *Tectonophysics*, **312**, 1–33.

BRUN, J.-P. & FACCENNA, C. 2008. Exhumation of high pressure rocks driven by slab rollback. *Earth and Planetary Science Letters*, doi:10.1016/j.epsl.2008.02.038.

BURG, J.-P. & GERYA, T. V. 2005. The role of viscous heating in Barrovian metamorphism of collisional orogens: Thermomechanical models and application to the Lepontine Dome in the Central Alps. *Journal of Metamorphic Geology*, **23**, 75–95.

BUROV, E. B. & WATTS, A. B. 2006. The long-term strength of continental lithosphere: 'Jelly sandwich' or 'crème brûlée'? *GSA Today*, **16**, 4–10.

CAMACHO, A., LEE, J. K. W., HENSEN, B. J. & BRAUN, J. 2005. Short-lived orogenic cycles and the eclogitization of cold crust by spasmodic hot fluids. *Nature*, **435**, 1191–1196.

CARSWELL, D. A. & COMPAGNONI, R. 2003. *Ultrahigh Pressure Metamorphism*. Eötvös University Press, Budapest.

CAWOOD, P. A. 2005. Terra Australis Orogen: Rodinia breakup and development of the Pacific and Iapetus margins of Gondwana during the Neoproterozoic and Paleozoic. *Earth-Science Reviews*, **69**, 249–279.

CAWOOD, P. A. & BUCHAN, C. 2007. Linking accretionary orogenesis with supercontinent assembly. *Earth-Science Reviews*, **82**, 217–256.

CESARE, B. & GOMEZ-PUGNAIRE, M. T. 2001. Crustal melting in the Alboran domain: Constraints from xenoliths of the Neogene Volcanic Province. *Physics and Chemistry of the Earth, Part A*, **26**, 255–260, doi:10.1016/S1464-1895(01)00053-9.

CHOPIN, C. 2003. Ultrahigh-pressure metamorphism: Tracing continental crust into the mantle. *Earth and Planetary Science Letters*, **212**, 1–14, doi:10.1016/S0012-821X(03)00261-9.

CHRISTENSEN, U. R. & HOFMANN, A. W. 1994. Segregation of subducted oceanic crust in the convecting mantle. *Journal of Geophysical Research*, **99**, 19867–19884.

CLARKE, G. L., AITCHISON, J. C. & CLUZEL, D. 1997. Eclogites and blueschists of the Pam Peninsula, NE New Caledonia: A reappraisal. *Journal of Petrology*, **38**, 843–876.

CLARKE, G. L., KLEPEIS, K. A. & DACZKO, N. R. 2000. Cretaceous high-P granulites at Milford Sound, New Zealand: Metamorphic history and emplacement in a convergent margin setting. *Journal of Metamorphic Geology*, **18**, 359–374.

CLARKE, G. L., DACZKO, N. R., KLEPEIS, K. A. & RUSHMER, T. 2005. Roles for fluid and/or melt advection in forming high-P mafic migmatites, Fiordland, New Zealand. *Journal of Metamorphic Geology*, **23**, 557–567.

CLARKE, G. L., POWELL, R. & FITZHERBERT, J. A. 2006. The lawsonite paradox: A comparison of field evidence and mineral equilibria modelling. *Journal of Metamorphic Geology*, **24**, 715–725.

CLOOS, M. 1993. Lithospheric buoyancy and collisional orogenesis: Subduction of oceanic plateaus, continental margins, island arcs, spreading ridges, and seamounts. *Geological Society of America Bulletin*, **105**, 715–737.

CLOOS, M., SAPIIE, B., VAN UFFORD, A. Q., WEILAND, R. J., WARREN, P. Q. & MCMAHON, T. P. 2005. Collisional delamination in New Guinea: The geotectonics of subducting slab breakoff. *Geological Society of America, Special Papers*, **400**, 1–51.

COBLENTZ, D. D. & RICHARDSON, R. M. 1996. Analysis of the South American intraplate stress field. *Journal of Geophysical Research—Solid Earth*, **101**, 8643–8657.

COBLENTZ, D. D. & SANDIFORD, M. 1994. Tectonic stresses in the African Plate—constraints on the ambient lithospheric stress state. *Geology*, **22**, 831–834.

COBLENTZ, D. D., RICHARDSON, R. M. & SANDIFORD, M. 1994. On the gravitational potential of the Earth's lithosphere. *Tectonics*, **13**, 929–945.

COBLENTZ, D. D., SANDIFORD, M., RICHARDSON, R. M., ZHOU, S. H. & HILLIS, R. 1995. The origins of the intraplate stress-field in continental Australia. *Earth and Planetary Science Letters*, **133**, 299–309.

COBLENTZ, D. D., ZHOU, S. H., HILLIS, R. R., RICHARDSON, R. M. & SANDIFORD, M. 1998. Topography, boundary forces, and the Indo-Australian intraplate stress field. *Journal of Geophysical Research—Solid Earth*, **103**, 919–931.

COLLINS, A. S. & PISAREVSKY, S. A. 2005. Amalgamating eastern Gondwana: The evolution of the Circum-Indian Orogens. *Earth-Science Reviews*, **71**, 229–270.

COLLINS, W. J. 2002. Nature of extensional accretionary orogens. *Tectonics*, **21**, doi:10.1029/2000TC001272.

COLLINS, W. J. 2003. Slab pull, mantle convection, and Pangaean assembly and dispersal. *Earth and Planetary Science Letters*, **205**, 225–237.

COLTICE, N., PHILLIPS, B. R., BERTRAND, H., RICARD, Y. & REY, P. 2007. Global warming of the mantle at the origin of flood basalts over supercontinents. *Geology*, **35**, 391–394.

CONDIE, K. C. 2002. Breakup of a Paleoproterozoic supercontinent. *Gondwana Research*, **5**, 41–43.

CONDIE, K. & KRÖNER, A. 2008. When did plate tectonics begin? *In*: CONDIE, K. & PEASE, V. (eds) *When Did Plate Tectonics Begin on Planet Earth?* Geological Society of America, Special Papers, **440**, 281–294.

CONDIE, K. & PEASE, V. 2008. When did plate tectonics begin on planet earth? *Geological Society of America, Special Papers*, **440**, 1–294.

CONEY, P. J. 1992. The Lachlan Belt of eastern Australia and Circum-Pacific tectonic evolution. *Tectonophysics*, **214**, 1–25.

CONEY, P. J., JONES, D. L. & MONGER, J. W. H. 1980. Cordilleran suspect terranes. *Nature*, **28**, 329–333.

COOPER, C. M., LENARDIC, A. & MORESI, L. 2006. Effects of continental insulation and the partitioning of heat producing elements on the Earth's heat loss. *Geophysical Research Letters*, **33**, L13313, doi:10.1029/2006GL026291.

CORDANI, U. G., D'AGRELLA-FILHO, M. S., BRITO-NEVES, B. B. & TRINDADE, R. I .F. 2003. Tearing up Rodinia: The Neoproterozoic palaeogeography of South American cratonic fragments. *Terra Nova*, **15**, 350–359, doi:10.1046/j.1365-3121.2003.00506.x.

CURRIE, C. A. & HYNDMAN, R. D. 2006. The thermal structure of subduction zone back arcs. *Journal of Geophysical Research*, **111**, B08404, doi:10.1029/2005JB004024.

CURRIE, C. A. & HYNDMAN, R. D. 2007. Reply to comment by W. P. Schellart on 'The thermal structure of subduction zone back arcs'. *Journal of Geophysical Research—Solid Earth*, **112**, B11408, doi:10.1029/2007JB005415.

CURRIE, C. A., HUISMANS, R. S. & BEAUMONT, C. 2008. Thinning of continental back-arc lithosphere by flow-induced gravitational instability. *Earth and Planetary Science Letters*, **269**, 436–447.

DACZKO, N. R., KLEPEIS, K. A. & CLARKE, G. L. 2001. Evidence of Early Cretaceous collisional-style orogenesis in northern Fiordland, New Zealand and its effects on the evolution of the lower crust. *Journal of Structural Geology*, **23**, 693–713.

DALZIEL, I. W. D. 1997. Neoproterozoic–Paleozoic geography and tectonics: Review, hypothesis, environmental speculation. *Geological Society of America Bulletin*, **109**, 16–42.

DANIEL, A. J., KUSZNIR, N. J. & STYLES, P. 2001. Thermal and dynamic modeling of deep subduction of a spreading center: Implications for the fate of the subducted Chile Rise, southern Chile. *Journal of Geophysical Research*, **106**, 4293–4304.

DAVIES, G. F. 1992. On the emergence of plate tectonics. *Geology*, **20**, 963–966.

DAVIES, G. F. 2006. Gravitational depletion of the early Earth's upper mantle and the viability of early plate tectonics. *Earth and Planetary Science Letters*, **243**, 376–382, doi:10.1016/j.epsl.2006.01.053.

DAVIES, J. H. & VON BLANCKENBURG, F. 1995. Slab breakoff: A model of lithosphere detachment and its test in the magmatism and deformation of collisional orogens. *Earth and Planetary Science Letters*, **129**, 85–102.

DERCOURT, J., ZONENSHAIN, L. P. *ET AL*. 1985. Presentation of 9 paleographic maps at 20 millions scale from Atlantic to Pamir between Lias and present. *Bulletin de la Société Géologique de France*, **5**, 637–652.

DERCOURT, J., ZONENSHAIN, L. P. *ET AL*. 1986. Geological evolution of the Tethys belt from the Atlantic to the Pamirs since the Lias. *Tectonophysics*, **123**, 241–315.

DEWEY, J. F. 1975. Finite plate implications—some implications for evolution of rock masses at plate margins. *American Journal of Science*, **A275**, 260–284.

DEWEY, J. F. 1981. Episodicity, sequence and style at convergent plate boundaries. *In*: STRANGWAY, D. W. (ed.) *Episodicity, Sequence and Style at Convergent Plate Boundaries*. Geological Society of Canada, Special Publications, **19**, 3–36.

DEWEY, J. F. 1988. Extensional collapse of orogens. *Tectonics*, **7**, 1123–1139.

DEWEY, J. F. 2005. Orogeny can be very short. *Proceedings of the National Academy of Sciences*, **102**, 15286–15293.

DEWEY, J. F. & BIRD, J. M. 1970. Mountain belts and new global tectonics. *Journal of Geophysical Research*, **75**, 2625–2647.

DEWEY, J. F., PITMAN, W. C., RYAN, W. B. F. & BONNIN, J. 1973. Plate tectonics and evolution of alpine system. *Geological Society of America Bulletin*, **84**, 3137–3180.

DEWEY, J. F., GASS, I. G., CURRY, G. B., HARRIS, N. B. W. & ŞENGÖR, A. M. C. 1990. Allochthonous terranes. *Philosophical Transactions of the Royal Society of London*, **331**, 455–647.

DEWEY, J. F., ROBB, L. & VAN SCHALKWYK, L. 2006. Did Bushmanland extensionally unroof Namaqualand? *Precambrian Research*, **150**, 173–182.

DICKINSON, W. R. 2004. Evolution of the North American Cordillera. *Annual Review of Earth and Planetary Sciences*, **32**, 13–45.

DICKINSON, W. R. 2008. Accretionary Mesozoic–Cenozoic expansion of the Cordilleran continental margin in California and adjacent Oregon. *Geosphere*, **4**, 329–353.

DIENER, J. F. A., STEVENS, G., KISTERS, A. F. M. & POUJOL, M. 2005. Metamorphism and exhumation of the basal parts of the Barberton greenstone belt, South Africa: Constraining the rates of Mesoarchaean tectonism. *Precambrian Research*, **143**, 87–112.

DIENER, J., STEVENS, G. & KISTERS, A. 2006. High-pressure intermediate-temperature metamorphism in the southern Barberton granitoid–greenstone terrain. *In*: BENN, K., MARESCHAL, J.-C. & CONDIE, K. C. (eds) *Archean Geodynamics and Environments*. American Geophysical Union, Geophysical Monographs, **164**, 239–254.

DILTS, S. L., STEWART, K. G. & LOEWY, S. L. 2006. Analysis of mineral inclusions in zircon from the eclogite-bearing Ashe Metamorphic Suite, North Carolina: Implications for exhumation history. *EOS Transactions, American Geophysical Union*, **87**, Joint Assembly Supplement, Abstract T41A-04.

DOGLIONI, C. 2008. Comment on 'The potential influence of subduction zone polarity on overriding plate deformation, trench migration and slab dip angle' by W. P. Schellart. *Tectonophysics*, doi:1016/j.tecto.2008.02.012.

DUFEK, J. & BERGANTZ, G. W. 2005. Lower crustal magma genesis and preservation: A stochastic framework for the evaluation of basalt–crust interaction. *Journal of Petrology*, **46**, 2167–2195.

DZIGGEL, A., STEVENS, G., POUJOL, M., ANHAEUSSER, C. R. & ARMSTRONG, R. A. 2002. Metamorphism of the granite–greenstone terrane south of the Barberton greenstone belt, South Africa: An insight into the tectono-thermal evolution of the 'lower' portions of the Onverwacht Group. *Precambrian Research*, **114**, 221–247.

ENGLAND, P. C. & HOUSEMAN, G. A. 1988. The mechanics of the Tibetan Plateau. *Philosophical Transactions of the Royal Society of London, Series A*, **326**, 301–320.

ERNST, W. G. 1971. Metamorphic zonations on presumably subducted lithospheric plates from Japan, California and Alps. *Contributions to Mineralogy and Petrology*, **34**, 43–59.

ERNST, W. G. 1973. Blueschist metamorphism and $P–T$ regimes in active subduction zones. *Tectonophysics*, **17**, 255–272.

ERNST, W. G. 1975. Systematics of large-scale tectonics and age progressions in Alpine and Circum-Pacific blueschist belts. *Tectonophysics*, **26**, 229–246.

ERNST, W. G. 2005. Alpine and Pacific styles of Phanerozoic mountain building: Subduction-zone petrogenesis of continental crust. *Terra Nova*, **17**, 165–188.

ERNST, W. G. & LIOU, J. G. 1995. Contrasting plate-tectonic styles of the Qinling–Dabie–Sulu and Franciscan Metamorphic Belts. *Geology*, **23**, 353–356.

ERNST, W. G., GASS, I. G. & ROBERTSON, H. F. 1990. Metamorphism in allochthonous and autochthonous terranes of the western United States [and Discussion]. *Philosophical Transactions of the Royal Society of London, Series A*, **331**, 540–570.

FARRIS, D. W. & PATERSON, S. R. 2009. Subduction of a segmented ridge along a curved continental margin: Variations between the western and eastern Sanak–Baranof belt, southern Alaska. *Tectonophysics*, **464**, 100–117, doi:10.1016/jtecto.2007.10.008.

FERNANDEZ, M., MARZAN, I., CORREIA, A. & RAMALHO, E. 1998. Heat flow, heat production, and lithospheric thermal regime in the Iberian Peninsula. *Tectonophysics*, **292**, 29–53.

FILIPPOVA, I. B., BUSH, V. A. & DIDENKO, A. N. 2001. Middle Paleozoic subduction belts: The leading factor in the formation of the central Asian fold-and-thrust belt. *Russian Journal of Earth Sciences*, **3**, 405–426.

FITZHERBERT, J. A., CLARKE, G. L. & POWELL, R. 2005. Preferential retrogression of high-P metasediments and the preservation of blueschist to eclogite facies metabasite during exhumation, Diahot terrane, NE New Caledonia. *Lithos*, **83**, 67–96.

FOLEY, S., TIEPOLO, M. & VANNUCCI, R. 2002. Growth of early continental crust controlled by melting of amphibolite in subduction zones. *Nature*, **417**, 837–840.

FOLEY, S. F., BUHRE, S. & JACOB, D. E. 2003. Evolution of the Archaean crust by delamination and shallow subduction. *Nature*, **421**, 249–252.

FRANCHETEAU, J., JAUPART, C. *ET AL.* 1984. High heat-flow in southern Tibet. *Nature*, **307**, 32–36.

FRIEDRICH, A. M., BOWRING, S. A., MARTIN, M. W. & HODGES, K. V. 1999. Short-lived continental magmatic arc at Connemara, western Irish Caledonides: Implications for the age of the Grampian orogeny. *Geology*, **27**, 27–30.

FRIEND, C. R. L. & NUTMAN, A. P. 2005a. New pieces to the Archaean terrane jigsaw puzzle in the Nuuk region, southern West Greenland: Steps in transforming a simple insight into a complex regional tectonothermal model. *Journal of the Geological Society, London*, **162**, 147–162.

FRIEND, C. R. L. & NUTMAN, A. P. 2005b. Complex 3670–3500 Ma orogenic episodes superimposed on juvenile crust accreted between 3850 and 3690 Ma, Itsaq Gneiss Complex, southern West Greenland. *Journal of Geology*, **113**, 375–397.

FRIEND, C. R. L., KINNY, P. D. & LOVE, G. J. 2007. Timing of magmatism and metamorphism in the Gruinard Bay area of the Lewisian Gneiss Complex: Comparison with the Assynt Terrane and implications for terrane accretion: Reply. *Contributions to Mineralogy and Petrology*, **153**, 489–492.

FURNES, H., DE WIT, M., STAUDIGEL, H., ROSING, M. & MUEHLENBACHS, K. 2007. A vestige of Earth's oldest ophiolite. *Science*, **315**, 1704–1707.

GARDE, A. A. 2007. A mid-Archaean island arc complex in the eastern Akia terrane, Godthåbsfjord, southern West Greenland. *Journal of the Geological Society, London*, **164**, 565–579.

GERDES, A., WORMER, G. & HENK, A. 2000. Post-collisional granite generation and HT–LP metamorphism by radiogenic heating: The Variscan South Bohemian Batholith. *Journal of the Geological Society, London*, **157**, 577–587.

GERYA, T. V., STOCKHERT, B. & PERCHUK, A. L. 2002. Exhumation of high pressure metamorphic rocks in a subduction channel: A numerical simulation. *Tectonics*, **21**, TC1056, doi:10.1029/2002TC001406.

GERYA, T. V., CONNOLLY, J. A. D. & YUEN, D. A. 2008. Why is terrestrial subduction one-sided? *Geology*, **36**, 43–46.

GLODNY, J., KUHN, A. & AUSTRHEIM, H. 2008. Geochronology of fluid-induced eclogite and amphibolite facies metamorphic reactions in a subduction–collision system, Bergen Arcs, Norway. *Contributions to Mineralogy and Petrology*, **156**, 27–48.

GODARD, G. 2001. The Les Essarts eclogite-bearing metamorphic Complex (Vendée, Southern Armorican Massif, France). *Géologie de la France*, **1-2**, 19–51.

GOSCOMBE, B. & HAND, M. 2000. Contrasting $P–T$ paths in the Eastern Himalaya, Nepal: Inverted isograds in a paired metamorphic mountain belt. *Journal of Petrology*, **41**, 1673–1719.

GREEN, H. 2005. Psychology of a changing paradigm: 40 + years of high-pressure metamorphism. *International Geology Review*, **47**, 439–456.

GREGOIRE, M., MATTIELLI, N. *ET AL.* 1994. Oceanic mafic granulite xenoliths from the Kerguelen archipelago. *Nature*, **367**, 360–363, doi:10.1038/367360a0.

GRIGNÉ, C., LABROSSE, S. & TACKLEY, P. J. 2005. Convective heat transfer as a function of wavelength: Implications for the cooling of the Earth. *Journal of Geophysical Research—Solid Earth*, **110**, B03409.

GROOME, W. G. & THORKELSON, D. J. 2008. The three-dimensional thermo-mechanical signature of ridge subduction and slab window migration. *Tectonophysics*, doi:10.1016/j.tecto.2008.07.003.

GUILLOU, L. & JAUPART, C. 1995. On the effect of continents on mantle convection. *Journal of Geophysical Research—Solid Earth*, **100**, 24217–24238.

GURNIS, M. 1988. Large-scale mantle convection and the aggregation and dispersal of supercontinents. *Nature*, **332**, 695–699.

HACKER, B. R. 2006. Pressures and temperatures of ultrahigh pressure metamorphism: Implications for UHP tectonics and H$_2$O in subducting slabs. *International Geology Review*, **48**, 1053–1066.

HACKER, B. R., ABERS, G. A. & PEACOCK, S. M. 2003. Subduction factory—1. Theoretical mineralogy, densities, seismic wave speeds, and H$_2$O contents. *Journal of Geophysical Research—Solid Earth*, **108**, article no. 2029, doi:10.1029/2001JB001127.

HACKER, B., LUFFI, P. ET AL. 2005. Near-ultrahigh pressure processing of continental crust: Miocene crustal xenoliths from the Pamir. *Journal of Petrology*, **44**, 1833–1865.

HAFKENSCHEID, E., WORTEL, M. J. R. & SPAKMAN, W. 2006. Subduction history of the Tethyan region derived from seismic tomography and tectonic reconstructions. *Journal of Geophysical Research*, **111**, B08401, doi:10.1029/2005JB003791.

HALL, R. 2002. Cenozoic geological and plate tectonic evolution of SE Asia and the SW Pacific: Computer-based reconstructions, model and animations. *Asian Earth Sciences*, **20**, 353–431.

HAND, M. & SANDIFORD, M. 1999. Intraplate deformation in central Australia, the link between subsidence and fault reactivation. *Tectonophysics*, **305**, 121–140.

HARLEY, S. L. 1989. The origins of granulites—A metamorphic perspective. *Geological Magazine*, **126**, 215–247.

HARLEY, S. L. 1998. On the occurrence and characterisation of ultrahigh-temperature (UHT) crustal metamorphism. *In*: TRELOAR, P. J. & O'BRIEN, P. J. (eds) *What Drives Metamorphism and Metamorphic Reactions?* Geological Society, London, Special Publications, **138**, 75–101.

HARLEY, S. L. 2004. Extending our understanding of ultrahigh temperature crustal metamorphism. *Journal of Mineralogical and Petrological Sciences*, **99**, 140–158.

HARLEY, S. L. 2008. Refining the P–T records of UHT crustal metamorphism. *Journal of Metamorphic Geology*, **26**, 125–154.

HARLEY, S. L., KELLY, N. M. & MOLLER, A. 2007. Zircon behaviour and the thermal histories of mountain chains. *Elements*, **3**, 25–30.

HARRISON, T. M., BLICHERT-TOFT, J., MÜLLER, W., ALBARÈDE, F., HOLDEN, P. & MOJZSIS, S. J. 2006. Heterogeneous Hadean hafnium: Evidence of continental crust at 4.4 to 4.5 Ga: Reply. *Science*, **312**, 1139.

HAYASHI, M., KOMIYA, T., NAKAMURA, Y. & MARUYAMA, S. 2000. Archean regional metamorphism of the Isua supracrustal belt, southern West Greenland: Implications for a driving force for Archean plate tectonics. *International Geology Review*, **42**, 1055–1115.

HAYOB, J. L., ESSENE, E. J., RUIZ, J., ORTEGA-GUTIÉRREZ, F. & ARANDA-GÓMEZ, J. J. 1989. Young high-temperature granulites from the base of the crust in central Mexico. *Nature*, **342**, 265–268, doi:10.1038/342265a0.

HAYOB, J. L., ESSENE, E. J. & RUIZ, J. 1990. Reply. *Nature*, **347**, 133–134.

HELMSTAEDT, H. & SCHULZE, D. J. 1988. Eclogite-facies ultramafic xenoliths from Colorado Plateau diatreme breccias: Comparison with eclogites in crustal environments, evaluation of the subduction hypothesis, and implications for eclogite xenoliths from diamondiferous kimberlites. *In*: SMITH, D. C. (ed.) *Eclogites and Eclogite-Facies Rocks*. Elsevier, New York, 387–450.

HENSEN, B. J. & ZHOU, B. 1995. Retention of isotopic memory in garnets partially broken down during an overprinting granulite-facies metamorphism—implications for the Sm–Nd closure temperature. *Geology*, **23**, 225–228.

HENSEN, B. J., ZHOU, B. & THOST, D. E. 1995. Are reaction textures reliable guides to metamorphic histories? Timing constraints from garnet Sm–Nd chronology for 'decompression' textures in granulites from Sostrene Island, Prydz Bay, Antarctica. *Geological Journal*, **30**, 261–271.

HIROI, Y., KISHI, S., NOHARA, T., SATO, K. & GOTO, J. 1998. Cretaceous high-temperature rapid loading and unloading in the Abukuma metamorphic terrane, Japan. *Journal of Metamorphic Geology*, **16**, 67–81.

HOFFMAN, P. E. 1991. Did the breakout of Laurentia turn Gondwanaland inside-out? *Science*, **252**, 1409–1412.

HOLCOMBE, R. J. & LITTLE, T. A. 1994. Blueschists of the New-England orogen—structural development of the Rocksberg Greenstone and associated units near Mt-Mee, southeastern Queensland. *Australian Journal of Earth Sciences*, **41**, 115–130.

HOLE, M. J. & LARTER, R. D. 1993. Trench-proximal volcanism following ridge crest–trench collision along the Antarctic Peninsula. *Tectonics*, **12**, 897–910.

HOLLAND, T. J. B. & POWELL, R. 1998. An internally consistent thermodynamic data set for phases of petrological interest. *Journal of Metamorphic Geology*, **16**, 309–343.

HOLLIS, J. A., CLARKE, G. L., KLEPEIS, K. A., DACZKO, N. R. & IRELAND, T. R. 2003. Geochronology and geochemistry of high-pressure granulites of the Arthur River Complex, Fiordland, New Zealand: Cretaceous magmatism and metamorphism on the palaeo-Pacific margin. *Journal of Metamorphic Geology*, **21**, 299–313.

HOLLIS, J. A., CLARKE, G. L., KLEPEIS, K. A., DACZKO, N. R. & IRELAND, T. R. 2004. The regional significance of Cretaceous magmatism and metamorphism in Fiordland, New Zealand, from U–Pb zircon geochronology. *Journal of Metamorphic Geology*, **22**, 607–627.

HOLLIS, J. A., HARLEY, S. L., WHITE, R. W. & CLARKE, G. L. 2006. Preservation of evidence for prograde metamorphism in ultrahigh-temperature, high-pressure kyanite-bearing granulites, South Harris, Scotland. *Journal of Metamorphic Geology*, **24**, 263–279.

HOWELL, D. G., JONES, D. L. & SCHERMER, E. R. 1985. Tectonostratigraphic terranes of the Circum-Pacific region: Principles of terrane analysis. *In*: HOWELL, D. G. (ed.) *Tectonostratigraphic Terranes of the Circum-Pacific Region*. Circum-Pacific Council for Energy and Mineral Resources, Earth Science Series, **1**, 3–31.

HYNDMAN, R. D., CURRIE, C. A. & MAZZOTTI, S. P. 2005a. Subduction zone back-arcs, mobile belts, and orogenic heat. *GSA Today*, **15**, 4–10.

HYNDMAN, R. D., CURRIE, C. & MAZZOTTI, S. 2005b. The origin of global mountain belts: Hot subduction zone backarcs. *Eos Transactions, American Geophysical Union*, **86**, T11B-0367.

HYNES, A. 2008. Some effects of a warmer Archean mantle on the geological record of plate tectonics. *In*: CONDIE, K. & PEASE, V. (eds) *When Did Plate Tectonics Begin?* Geological Society of America, Special Papers, **440**, 149–156.

ISACKS, B., SYKES, L. R. & OLIVER, J. 1968. Seismology and the new global tectonics. *Journal of Geophysical Research*, **73**, 5855–5899.

ISHIKAWA, A., KURITANI, T., MAKISHIMA, A. & NAKAMURA, E. 2007. Ancient recycled crust beneath the Ontong Java plateau: Isotopic evidence from the garnet clinopyroxenite xenoliths, Malaita, Solomon Islands. *Earth and Planetary Science Letters*, **259**, 134–148.

JACKSON, M. D., CHEADLE, M. J. & ATHERTON, M. P. 2003. Quantitative modeling of granitic melt generation and segregation in the continental crust. *Journal of Geophysical Research—Solid Earth*, **108**, article no. 2332.

JAHN, B. M. 2004. The Central Asian orogenic belt and growth of the continental crust in the Phanerozoic. *In*: MALPAS, J., FLETCHER, C. J. N., ALI, J. R. & AITCHINSON, J. C. (eds) *Aspects of the Tectonic Evolution of China*. Geological Society, London, Special Publications, **226**, 73–100.

JAHN, B. M., WU, F. Y. & CHEN, B. 2000. Massive granitoid generation in Central Asia: Nd isotope evidence and implication for continental growth in the Phanerozoic. *Episodes*, **23**, 82–92.

JAMIESON, R. A., BEAUMONT, C., NGUYEN, M. H. & LEE, B. 2002. Interaction of metamorphism, deformation and exhumation in large convergent orogens. *Journal of Metamorphic Geology*, **20**, 9–24.

JANOUSEK, V. & HOLUB, F. V. 2007. The casual link between HP–HT metamorphism and ultrapotassic magmatism in collisional orogens: Case study from the Moldanubian Zone of the Bohemian Massif. *Proceedings of the Geologists' Association*, **118**, 75–86.

JOHNSTON, S. T. 2001. The great Alaskan terrane wreck: Reconciliation of paleomagnetic and geological data in the northern Cordillera. *Earth and Planetary Science Letters*, **193**, 259–272.

JOHNSON, S. T. 2008. The Cordilleran Ribbon Continent of North America. *Annual Review of Earth and Planetary Sciences*, **36**, 495–530.

JOHNSTON, S. T. & BOREL, G. D. 2007. The odyssey of the Cache Creek terrane, Canadian Cordillera: Implications for accretionary orogens, tectonic setting of Panthalassa, the Pacific superswell, and break-up of Pangea. *Earth and Planetary Science Letters*, **253**, 415–428.

JOHNSON, T. & BROWN, M. 2004. Quantitative constraints on metamorphism in the Variscides of southern Brittany—a complementary pseudosection approach. *Journal of Petrology*, **38**, 1237–1259.

JONES, D. L., HOWELL, P. G., CONEY, P. J. & MONGER, J. W. 1983. Recognition, character and analysis of tectonostratigraphic terranes in western North America. *In*: HASHIMOTO, M. & UYEDA, S. (eds) *Accretion Tectonics in the Circum-Pacific region*. Terra Scientific, Tokyo, 21–35.

KANEKO, Y., MARUYAMA, S. *ET AL.* 2000. Geology of the Kokchetav UHP–HP metamorphic belt, Northern Kazakhstan. *Island Arc*, **9**, 264–283.

KELSEY, D. E. 2008. On ultrahigh-temperature crustal metamorphism. *Gondwana Research*, **13**, 1–29.

KINCAID, C. & SILVER, P. 1996. The role of viscous dissipation in the orogenic process. *Earth and Planetary Science Letters*, **142**, 271–288.

KOIZUMI, K. & ISHIWATARI, A. 2006. Oceanic plateau accretion inferred from Late Paleozoic greenstones in the Jurassic Tamba accretionary complex, southwest Japan. *Island Arc*, **15**, 58–83.

KOMIYA, T., HAYASHI, M., MARUYAMA, S. & YURIMOTO, H. 2002. Intermediate-P/T type Archaean metamorphism of the Isua supracrustal belt: Implications for secular change of geothermal gradients at subduction zones and for Archaean plate tectonics. *American Journal of Science*, **302**, 806–826.

KONOPÁSEK, J. & SCHULMANN, K. 2005. Contrasting Early Carboniferous field geotherms: Evidence for accretion of a thickened orogenic root and subducted Saxothuringian crust (Central European Variscides). *Journal of the Geological Society, London*, **162**, 463–470.

KREEMER, C., HOLT, W. E. & HAINES, A. J. 2003. An integrated global model of present-day plate motions and plate boundary deformation. *Geophysical Journal International*, **154**, 8–34.

KRÖNER, A., WINDLEY, B. F. *ET AL.* 2008. Accretionary growth and crust-formation in the Central Asian Orogenic Belt and comparison with the Arabian–Nubian shield. *In*: HATCHER, R. (ed.) *4-D Framework of the Continental Crust—Integrating Crustal Processes through Time*. Geological Society of America, Memoirs, **200**, 181–210.

LALLEMAND, S., HEURET, A., FACCENNA, C. & FUNICIELLO, F. 2008. Subduction dynamics as revealed by trench migration. *Tectonics*, **27**, TC3014, doi: 10.1029/2007TC002212.

LARDEAUX, J. M., LEDRU, P., DANIELA, I. & DUCHENE, S. 2001. The Variscan French Massif Central—a new addition to the ultra-high pressure metamorphic 'club': Exhumation processes and geodynamic consequences. *Tectonophysics*, **332**, 143–167.

LEECH, M. L., SINGH, S., JAIN, A. K., KLEMPERER, S. L. & MANICKAVASAGAM, R. M. 2005. The onset of India–Asia continental collision: Early, steep subduction required by the timing of UHP metamorphism in the western Himalaya. *Earth and Planetary Science Letters*, **234**, 83–97, doi:10.1016/j.epsl. 2005.02.038.

LEICHMANN, J., NOVAK, M., BURIANEK, D. & BURGER, D. 2007. High-temperature to ultrahigh-temperature metamorphism related to multiple ultrapotassic intrusions: Evidence from garnet–sillimanite–cordierite kinzigite and garnet–orthopyroxene migmatites in the eastern part of the Moldanubian Zone (Bohemian Massif). *Geologica Carpathica*, **58**, 415–425.

LELOUP, PH. D., RICARD, Y., BATTAGLIA, J. & LACASSIN, R. 1999. Shear heating in continental strike-slip shear zones: Model and field examples. *Geophysical Journal International*, **136**, 19–40.

LENARDIC, A., MORESI, L.-N., JELLINEK, M. & MANGA, M. 2005. Continental insulation, mantle cooling, and the surface area of oceans and continents. *Earth and Planetary Science Letters*, **234**, 317–333.

LENARDIC, A., RICHARDS, M. A. & BUSSE, F. H. 2006. Depth-dependent rheology and the horizontal length scale of mantle convection. *Journal of Geophysical Research—Solid Earth*, **111**, B07404.

LE PICHON, X. & HUCHON, P. 1984. Geoid, Pangaea and convection. *Earth and Planetary Science Letters*, **67**, 123–135.

LE PICHON, X., HENRY, P. & GOFFE, B. 1997. Uplift of Tibet: From eclogites to granulites—Implications for the Andean Plateau and the Variscan belt. *Tectonophysics*, **273**, 57–76.

LIOU, J. G., TSUJIMORI, T., ZHANG, R. Y., KATAYAMA, I. & MARUYAMA, S. 2004. Global UHP metamorphism and continental subduction/collision: The Himalayan model. *International Geology Review*, **46**, 1–27.

LISSENBERG, C. J., ZAGOREVSKI, A., MCNICOLL, V. J., VAN STAAL, C. R. & WHALEN, J. B. 2005. Assembly of the Annieopsquotch Accretionary Tract, Newfoundland Appalachians: Age and geodynamic constraints from syn-kinematic intrusions. *Journal of Geology*, **113**, 553–570.

LISTER, G. & FORSTER, M. 2006. The plate-tectonic significance of inversion cycles during orogenesis. *Geological Society of America, Abstracts with Programs*, Speciality Meeting No. 2, 119.

LISTER, G. S., FORSTER, M. A. & RAWLING, T. J. 2001. Episodicity during orogenesis. Geological Society, London, Special Publications, **184**, 89–113.

LISTER, G., KENNETT, B., RICHARDS, S. & FORSTER, M. 2008. Boudinage of a stretching slablet implicated in earthquakes beneath the Hindu Kush. *Nature Geoscience*, **1**, 196–201.

LITTLE, T. A., HOLCOMBE, R. J., GIBSON, G. M., OFFLER, R., GANS, P. B. & MCWILLIAMS, M. O. 1992. Exhumation of Late Paleozoic blueschists in Queensland, Australia, by extensional faulting. *Geology*, **20**, 231–234.

LITTLE, T. A., HOLCOMBE, R. J. & SLIWA, R. 1993. Structural evidence for extensional exhumation of blueschist-bearing serpentinite matrix melange, New-England orogen, southeast Queensland, Australia. *Tectonics*, **12**, 536–549.

LITTLE, T. A., MCWILLIAMS, M. O. & HOLCOMBE, R. J. 1995. Ar-40/Ar-39 thermochronology of epidote blueschists from the North Daguilar Block, Queensland, Australia—timing and kinematics of subduction complex unroofing. *Geological Society of America Bulletin*, **107**, 520–535.

LIU, L., ZHANG, J., GREEN, H. W., JIN, Z. & BOZHILOV, K. N. 2007. Evidence of former stishovite in metamorphosed sediments, implying subduction to >350 km. *Eos Transactions, American Geophysical Union*, **86**, V51E-08.

LUCASSEN, F. & FRANZ, G. 1996. Magmatic arc metamorphism: Petrology and temperature history of metabasic rocks in the Coastal Cordillera of northern Chile. *Journal of Metamorphic Geology*, **14**, 249–265, doi:10.1046/j.1525-1314.1996.59011.x.

MACAUDIÈRE, J., BARBEY, P., JABBORI, J. & MARIGNAC, C. 1992. Le stade initial de fusion des dômes anatectiques: Le dome du Velay (Massif Central Francais). *Comptes Rendus de l'Académie des Sciences, Série II*, **315**, 1761–1767.

MAEKAWA, H., SHOZUI, M., ISHII, T., FRYER, P. & PEARCE, J. A. 1993. Blueschist metamorphism in an active subduction zone. *Nature*, **364**, 520–523, doi: 10.1038/364520a0.

MARESCH, W. V. & GERYA, T. V. 2005. Blueschists and blue amphiboles: How much subduction do they need? *International Geology Review*, **47**, 688–702.

MARTÍNEZ CATALÁN, J. R., ARENAS, R., DÍAZ GARCÍA, F. & ABATI, J. 1997. Variscan accretionary complex of northwest Iberia: Terrane correlation and succession of tectonothermal events. *Geology*, **25**, 1103–1106.

MARUYAMA, S. 1997. Pacific-type orogeny revisited: Miyashiro-type orogeny proposed. *Island Arc*, **6**, 91–120.

MARUYAMA, S. & LIOU, J. G. 1998. Initiation of ultrahigh-pressure metamorphism and its significance on the Proterozoic–Phanerozoic boundary. *Island Arc*, **7**, 6–35.

MARUYAMA, S. & LIOU, J. G. 2005. From snowball to Phanerozoic Earth. *International Geology Review*, **47**, 775–791.

MARUYAMA, S., LIOU, J. G. & TERABAYASHI, M. 1996. Blueschists and eclogites of the world and their exhumation. *International Geology Review*, **38**, 490–596.

MASSONNE, H.-J. 2001. First find of coesite in the ultrahigh-pressure metamorphic area of the central Erzgebirge, Germany. *European Journal of Mineralogy*, **13**, 565–570.

MATSUDA, T. & UYEDA, S. 1971. On the Pacific-type orogeny and its model—extension of the paired belts concept and possible origin of marginal seas. *Tectonophysics*, **11**, 5–27.

MCKENZIE, D. M. & PRIESTLEY, K. 2008. The influence of lithospheric thickness variations on continental evolution. *Lithos*, **102**, 1–11.

MCLAREN, S., SANDIFORD, M., POWELL, R., NEUMANN, N. & WOODHEAD, J. 2006. Palaeozoic intraplate crustal anatexis in the Mount Painter province, South Australia: Timing, thermal budgets and the role of crustal heat production. *Journal of Petrology*, **47**, 2281–2302.

MCNAMARA, A. K. & ZHONG, S. 2004. Thermochemical structures within a spherical mantle: Superplumes or piles? *Journal of Geophysical Research*, **109**, B07402, doi:10.1029/2003JB002847.

METCALFE, I. 1996. Gondwanaland dispersion, Asian accretion and evolution of eastern Tethys. *Australian Journal of Earth Sciences*, **43**, 605–623.

MEZGER, K., BOHLEN, S. R. & HANSON, G. N. 1990. Metamorphic history of the Archean Pikwitonei granulite domain and the Cross Lake subprovince, Superior

Province, Manitoba, Canada. *Journal of Petrology*, **31**, 483–517.

MIYASHIRO, A. 1961. Evolution of metamorphic belts. *Journal of Petrology*, **2**, 277–311.

MIYASHIRO, A. 1972. Metamorphism and related magmatism in plate tectonics. *American Journal of Science*, **272**, 629–656.

MIYASHIRO, A. 1973. Paired and unpaired metamorphic belts. *Tectonophysics*, **17**, 241–254.

MIYAZAKI, K. 2004. Low-*P*–high-*T* metamorphism and the role of heat transport by melt migration in the Higo Metamorphic Complex, Kyushu, Japan. *Journal of Metamorphic Geology*, **22**, 793–809.

MOIX, P., BECCALETTO, L., KOZUR, H. W., HOCHARD, C., ROSSELET, F. & STAMPFLI, G. M. 2008. A new classification of the Turkish terranes and sutures and its implication for the paleotectonic history of the region. *Tectonophysics*, **451**, 7–39.

MONTEL, J. M., MARIGNAC, C., BARBEY, P. & PICHAVANT, M. 1992. Thermobarometry and granite genesis: The Hercynian low-*P* high-*T* Velay anatectic dome (French Massif Central). *Journal of Metamorphic Geology*, **10**, 1–15.

MONTELLI, R., NOLET, G., DAHLEN, F. A. & MASTERS, G. 2006. A catalogue of deep mantle plumes: New results from finite-frequency tomography. *Geochemistry, Geophysics, Geosystems*, **7**, Q11007, doi:10.1029/2006GC001248.

MOORES, E. M. 2002. Pre 1 Ga (pre-Rodinian) ophiolites: Their tectonic and environmental implications. *Geological Society of America Bulletin*, **114**, 80–95.

MOYEN, J.-F., STEVENS, G. & KISTERS, A. 2006. Record of mid-Archaean subduction from metamorphism in the Barberton terrain of South Africa. *Nature*, **442**, 559–562.

MURPHY, J. B. & NANCE, R. D. 2005. Do supercontinents turn inside-in or inside-out? *International Geology Review*, **47**, 591–619.

MURPHY, J. B., PISAREVSKY, S. A., NANCE, R. D. & KEPPIE, J. D. 2004. Neoproterozoic–early Paleozoic evolution of peri-Gondwanan terranes: Implications for Laurentia–Gondwana connections. *International Journal of Earth Sciences*, **93**, 659–682.

NELSON, K. D., ZHAO, W. J. ET AL. 1996. Partially molten middle crust beneath southern Tibet: Synthesis of project INDEPTH results. *Science*, **274**, 1684–1688.

NOLET, G., KARATO, S. I. & MONTELLI, R. 2006. Plume fluxes from seismic tomography. *Earth and Planetary Science Letters*, **248**, 685–699

O'BRIEN, P. J. 1997. Garnet zoning and reaction textures in overprinted eclogites, Bohemian Massif, European Variscides: A record of their thermal history during exhumation. *Lithos*, **41**, 119–133.

O'BRIEN, P. J. 1999. Asymmetric zoning profiles in garnet from HP–HT granulite and implications for volume and grain boundary diffusion. *Mineralogical Magazine*, **63**, 227–238.

O'BRIEN, P. J. & RÖTZLER, J. 2003. High-pressure granulites: Formation, recovery of peak conditions and implications for tectonics. *Journal of Metamorphic Geology*, **21**, 3–20.

O'NEILL, C., LENARDIC, A., MORESI, L., TORSVIK, T. H. & LEE, C.-T. A. 2007. Episodic Precambrian subduction. *Earth and Planetary Science Letters*, doi:10.1016/j.epsl.2007.04.056.

OSANAI, Y., HAMAMOTO, T., MAISHIMA, O. & KAGAMA, H. 1998. Sapphirine-bearing granulites and related high-temperature metamorphic rocks from the Higo metamorphic terrane, west–central Kyushu, Japan. *Journal of Metamorphic Geology*, **16**, 53–66.

OTA, T., TERABAYASHI, M. & KATAYAMA, I. 2004. Thermobaric structure and metamorphic evolution of the Iratsu eclogite body in the Sanbagawa belt, central Shikoku, Japan. *Lithos*, **73**, 95–126.

OXBURGH, E. R. & TURCOTTE, D. L. 1970. Thermal structure of island arcs. *Geological Society of America Bulletin*, **81**, 1665–1688.

OXBURGH, E. R. & TURCOTTE, D. L. 1971. Origin of paired metamorphic belts and crustal dilation in island arc regions. *Journal of Geophysical Research*, **76**, 1315–1327.

PARK, R. G., TARNEY, J. & CONNELLY, J. N. 2001. The Loch Maree Group: Palaeoproterozoic subduction–accretion complex in the Lewisian of NW Scotland. *Precambrian Research*, **105**, 205–226.

PARK, R. G., KINNY, P. D., FRIEND, C. R. L. & LOVE, G. J. 2005. Discussion on a terrane-based nomenclature for the Lewisian Gneiss Complex of NW Scotland. *Journal of the Geological Society, London*, **162**, 893–895.

PATRICK, B. E. & DAY, H. W. 1995. Cordilleran high-pressure metamorphic terranes: Progress and problems. *Journal of Metamorphic Geology*, **13**, 1–8.

PATTISON, D. R. M., CHACKO, T., FARQUHAR, J. & MCFARLANE, C. R. M. 2003. Temperatures of granulite-facies metamorphism: Constraints from experimental phase equilibria and thermobarometry corrected for retrograde exchange. *Journal of Petrology*, **44**, 867–900.

PERCIVAL, J. A., SANBORN-BARRIE, M., SKULSKI, T., STOTT, G. M., HELMSTAEDT, H. & WHITE, D. J. 2006. Tectonic evolution of the western superior province from NATMAP and Lithoprobe studies. *Canadian Journal of Earth Sciences*, **43**, 1085–1117.

PEREIRA, M. D. & BEA, F. 1994. Cordierite-producing reactions at the Peña Negra Complex, Avila Batholith, central Spain: The key role of cordierite in low-pressure anatexis. *Canadian Mineralogist*, **32**, 763–780.

PERESSINI, G., QUICK, J. E., SINIGOI, S., HOFMANN, A. W. & FANNING, M. 2007. Duration of a large mafic intrusion and heat transfer in the lower crust: A SHRIMP U–Pb zircon study in the Ivrea–Verbano Zone (western Alps, Italy). *Journal of Petrology*, **48**, 1185–1218.

PLATT, J. P. 1993. Exhumation of high-pressure rocks—A review of concepts and processes. *Terra Nova*, **5**, 119–133.

PLATT, J. P. & ENGLAND, P. C. 1994. Convective removal of lithosphere beneath mountain belts—thermal and mechanical consequences. *American Journal of Science*, **294**, 307–336.

PLATT, J. P., SOTO, J. I., WHITEHOUSE, M. J., HURFORD, A. J. & KELLEY, S. P. 1998. Thermal evolution, rate of exhumation, and tectonic significance of metamorphic rocks from the floor of the Alboran extensional basin, western Mediterranean. *Tectonics*, **17**, 671–689.

PLATT, J. P., WHITEHOUSE, M. J., KELLEY, S. P., CARTER, A. & HOLLICK, L. 2003. Simultaneous extensional exhumation across the Alboran Basin: Implications for the causes of late orogenic extension. *Geology*, **31**, 251–254.

PLATT, J. P., ANCZKIEWICZ, R., SOTO, J. I., KELLEY, S. P. & THIRLWALL, M. 2006. Early Miocene continental subduction and rapid exhumation in the western Mediterranean. *Geology*, **34**, 981–984.

POTEL, S., FERREIRO MÄHLMANN, R., STERN, W. B., MULLIS, J. & FREY, M. 2006. Very low-grade metamorphic evolution of pelitic rocks under high-pressure/low-temperature conditions, NW New Caledonia (SW Pacific). *Journal of Petrology*, **47**, 991–1015, doi:10.1093/petrology/egl001.

POWELL, R., HOLLAND, T. & WORLEY, B. 1998. Calculating phase diagrams involving solid solutions via non-linear equations, with examples using THERMOCALC. *Journal of Metamorphic Geology*, **16**, 577–588.

POWELL, R., GUIRAUD, M. & WHITE, R. W. 2005. Truth and beauty in metamorphic phase equilibria: Conjugate variables and phase diagrams. *Canadian Mineralogist*, **43**, 21–33.

RAVNA, E. J. KROGH & TERRY, M. P. 2004. Geothermobarometry of UHP and HP eclogites and schists—an evaluation of equilibria among garnet–clinopyroxene–kyanite–phengite–coesite/quartz. *Journal of Metamorphic Geology*, **22**, 570–592.

REDFIELD, T. F., SCHOLL, D. W., FITZGERALD, P. G. & BECK, M. E. JR. 2007. Escape tectonics and the extrusion of Alaska: Past, present, and future. *Geology*, **35**, 1039–1042.

RICHARDS, M. A. & ENGEBRETSON, D. E. 1992. Large-scale mantle convection and the history of subduction. *Nature*, **355**, 437–440.

RICHARDSON, R. M. & COBLENTZ, D. D. 1994. Stress modelling in the Andes—constraints on the South American intraplate stress magnitudes. *Journal of Geophysical Research—Solid Earth*, **99**, 22015–22025.

RIDGWAY, K. D. & FLESCH, L. M. 2007. Cenozoic tectonic processes along the southern Alaska convergent margin. *Geology*, **35**, 1055–1056.

RINGUETTE, L., MARTIGNOLE, J. & WINDLEY, B. F. 1999. Magmatic crystallization, isobaric cooling, and decompression of the garnet-bearing assemblages of the Jijal sequence (Kohistan terrane, western Himalayas). *Geology*, **27**, 139–142.

RIVERS, T. & CORRIGAN, D. 2000. Convergent margin on southeastern Laurentia during the Mesoproterozoic: Tectonic implications. *Canadian Journal of Earth Sciences*, **37**, 359–383.

ROLLINSON, H. R. 2002. The metamorphic history of the Isua Greenstone Belt, West Greenland. *In*: FOWLER, C. M. R., EBINGER, C. J. & HAWKESWORTH, C. J. (eds) *The Early Earth*. Geological Society, London, Special Publications, **199**, 329–350.

RONDENAY, S., ABERS, G. A. & VAN KEKEN, P. E. 2008. Seismic imaging of subduction zone metamorphism. *Geology*, **36**, 275–278.

ROWLEY, D. B. 1996. Age of initiation of collision between India and Asia: A review of stratigraphic data. *Earth and Planetary Science Letters*, **145**, 1–13.

ROYDEN, L. H. 1993a. The tectonic expression of slab pull at continental convergent boundaries. *Tectonics*, **12**, 303–325.

ROYDEN, L. H. 1993b. Evolution of retreating subduction boundaries formed during continental collision. *Tectonics*, **12**, 629–638.

SACHAN, H. K., MUKHERJEE, B. K., OGASAWARA, Y., MARUYAMA, S., ISHIDA, H., MUKO, A. & YOSHIOKA, N. 2004. Discovery of coesite from Indus Suture Zone (ISZ), Ladakh, India: Evidence for deep subduction. *European Journal of Mineralogy*, **16**, 235–240.

SANDIFORD, M. 2002. Low thermal Peclet number intraplate orogeny in central Australia. *Earth and Planetary Science Letters*, **201**, 309–320.

SANDIFORD, M. & DYMOKE, P. 1991. Some remarks on the stability of blueschists and related high P–low T assemblages in continental orogens. *Earth and Planetary Science Letters*, **102**, 14–23.

SANDIFORD, M. & HAND, M. 1998. Controls on the locus of intraplate deformation in central Australia. *Earth and Planetary Science Letters*, **162**, 97–110.

SANDIFORD, M. & POWELL, R. 1991. Some remarks on high-temperature–low-pressure metamorphism in convergent orogens. *Journal of Metamorphic Geology*, **9**, 333–340.

SAVOSTIN, L. A., SIBUET, J. C., ZONENSHAIN, L. P., LEPICHON, X. & ROULET, M. J. 1986. Kinematic evolution of the Tethys Belt from the Atlantic Ocean to the Pamirs since the Triassic. *Tectonophysics*, **123**, 1–35.

SCHELLART, W. P. 2007a. Comment on 'The thermal structure of subduction zone back arcs'. *Journal of Geophysical Research—Solid Earth*, **112**, B11407, doi:10.1029/2007/JB005287.

SCHELLART, W. P. 2007b. The potential influence of subduction zone polarity on overriding plate deformation, trench migration and slab dip angle. *Tectonophysics*, **445**, 363–372.

SCHELLART, W. P. 2008a. Reply to comment on 'The potential influence of subduction zone polarity on overriding plate deformation, trench migration and slab dip angle'. *Tectonophysics*, doi:10.1016/j.tecto.2008.03.016.

SCHELLART, W. P. 2008b. Overriding plate shortening and extension above subduction zones: A parametric study to explain formation of the Andes Mountains. *Geological Society of America, Bulletin*, **120**, 1441–1454, doi:10.1130/B26360.1.

SCHELLART, W. P., FREEMAN, J., STEGMAN, D. R., MORESI, L. & MAY, D. 2007. Evolution and diversity of subduction zones controlled by slab width. *Nature*, **446**, 308–311.

SCHERER, E. E., CAMERON, K. L., JOHNSON, C. M., BEARD, B. L., BAROVICH, K. M. & COLLERSON, K. D. 1997. Lu–Hf geochronology applied to dating Cenozoic events affecting lower crustal xenoliths from Kilbourne Hole, New Mexico. *Chemical Geology*, **142**, 63–78, doi:10.1016/S0009-2541(97)00076-4.

SCHILLING, F. R. & PARTZSCH, G. M. 2001. Quantifying partial melt fraction in the crust beneath the central Andes and the Tibetan plateau. *Physics and Chemistry of the Earth: Part A: Solid Earth and Geodesy*, **26**, 239–246, doi:10.1016/S1464-1895(01)00051-5.

SCHULMANN, K., SCHALTEGGER, U., JEZEK, J., THOMPSON, A. B. & EDEL, J. B. 2002. Rapid burial and exhumation during orogeny: Thickening and synconvergent exhumation of thermally weakened and thinned crust (Variscan orogen in Western Europe). *American Journal of Science*, **302**, 856–879.

SEARLE, M., HACKER, B. R. & BILHAM, R. 2001. The Hindu Kush seismic zone as a paradigm for the creation of ultrahigh-pressure diamond- and coesite-bearing continental rocks. *Journal of Geology*, **109**, 143–153.

ŞENGÖR, A. M. C. 1987. Tectonics of the Tethysides—orogenic collage development in a collisional setting. *Annual Review of Earth and Planetary Sciences*, **15**, 213–244.

ŞENGÖR, A. M. C. 1992. The Paleo-Tethyan suture: A line of demarcation between two fundamentally different architectural styles in the structure of Asia. *Island Arc*, **1**, 78–91.

ŞENGÖR, A. M. C. & NATAL'IN, B. A. 1996. Turkic-type orogeny and its role in the making of the continental crust. *Annual Review of Earth and Planetary Sciences*, **24**, 263–337.

ŞENGÖR, A. M. C., NATALIN, B. A. & BURTMAN, V. S. 1993. Evolution of the Altaid Tectonic Collage and Paleozoic crustal growth in Eurasia. *Nature*, **364**, 299–307.

SHAFER, J. T., NEAL, C. R. & MAHONEY, J. J. 2005. Crustal xenoliths from Malaita, Solomon Islands: A window to the lower crust of the Ontong Java Plateau. *EOS Transactions, American Geophysical Union*, **86**, T11C-0397.

SHIPBOARD SCIENTIFIC PARTY 2002. Leg 195 summary. In: SALISBURY, M. H., SHINOHARA, M., & RICHTER, C. ET AL. (eds) *Proceedings of the Ocean Drilling Program, Initial Reports*, **195**. World Wide Web Address: http://www-odp.tamu.edu/publications/195_IR/VOLUME/CHAPTERS/IR195_01.PDF

SHIREY, S. B., KAMBER, B. S., WHITEHOUSE, M. J., MUELLER, P. A. & BASU, A. R. 2008. Isotopic constraints on mantle evolution and plate tectonics. In: CONDIE, K. & PEASE, V. (eds) *When Did Plate Tectonics Begin?* Geological Society of America, Special Papers, **440**, 1–29.

SILVER, P. G. & BEHN, M. D. 2008. Intermittent plate tectonics? *Science*, **319**, 85–88.

SISSON, V. B., PAVLIS, T. L., ROESKE, S. M. & THORKELSON, D. J. 2003. Introduction: An overview of ridge–trench interactions in modern and ancient settings. In: SISSON, V. B., ROESKE, S. M. & PAVLIS, T. L. (eds) *Geology of a Transpressional Orogen Developed During Ridge–Trench Interaction along the North Pacific Margin*. Geological Society of America, Special Papers, **371**, 1–18.

SLEEP, N. H. 2000. Evolution of the mode of convection within terrestrial planets. *Journal of Geophysical Research*, **105**, 17563–17578.

SOLAR, G. S. & BROWN, M. 1999. The classic high-T–low-P metamorphism of west–central Maine, USA: Is it post-tectonic or syn-tectonic? Evidence from porphyroblast–matrix relations. *Canadian Mineralogist*, **37**, 289–311.

SOLAR, G. S. & BROWN, M. 2001a. Petrogenesis of migmatites in Maine, USA: Possible source of peraluminous leucogranite in plutons. *Journal of Petrology*, **42**, 789–823.

SOLAR, G. S. & BROWN, M. 2001b. Deformation partitioning during transpression in response to Early Devonian oblique convergence, Northern Appalachian orogen, USA. *Journal of Structural Geology*, **23**, 1043–1065.

SOLAR, G. S., PRESSLEY, R. A., BROWN, M. & TUCKER, R. D. 1998. Granite ascent in convergent orogenic belts: Testing a model. *Geology*, **26**, 711–714.

SPAGGIARI, C. V., GRAY, D. R., FOSTER, D. A. & FANNING, C. M. 2002a. Occurrence and significance of blueschist in the southern Lachlan orogen. *Australian Journal of Earth Sciences*, **49**, 255–269.

SPAGGIARI, C. V., GRAY, D. R. & FOSTER, D. A. 2002b. Blueschist metamorphism during accretion in the Lachlan orogen, south-eastern Australia. *Journal of Metamorphic Geology*, **20**, 711–726.

SPEAR, F. S., KOHN, M. J., CHENEY, J. T. & FLORENCE, F. 2002. Metamorphic, thermal, and tectonic evolution of central New England. *Journal of Petrology*, **43**, 2097–2120.

STAMPFLI, G. M., VON RAUMER, J. & BOREL, G. 2002. The Palaeozoic evolution of pre-Variscan terranes: From Gondwana to the Variscan collision. In: MARTINEZ CATALAN, J. R., HATCHER, R. D., ARENAS, R. & DIAZ GARCIA, F. (eds) *Variscan–Appalachian Dynamics: The Building of the Late-Palaeozoic Basement*. Geological Society of America, Special Papers, **364**, 263–280.

STERN, R. J. 2005. Evidence from ophiolites, blueschists, and ultrahigh-pressure metamorphic terranes that the modern episode of subduction tectonics began in Neoproterozoic time. *Geology*, **33**, 557–560.

STERN, R. J. 2007. When and how did plate tectonics begin? Theoretical and empirical considerations. *Chinese Science Bulletin*, **52**, 578–591.

STERN, R. J. 2008. Modern-style plate tectonics began in Neoproterozoic time: An alternate interpretation of Earth tectonic history. In: CONDIE, K. & PEASE, V. (eds) *When did Plate Tectonics Begin?* Geological Society of America, Special Papers, **440**, 265–280.

ŠTÍPSKÁ, P. & POWELL, R. 2005. Does ternary feldspar constrain the metamorphic conditions of high-grade meta-igneous rocks? Evidence from orthopyroxene granulites, Bohemian Massif. *Journal of Metamorphic Geology*, **23**, 627–647.

STÜWE, K. 1998. Heat sources of Cretaceous metamorphism in the Eastern Alps—a discussion. *Tectonophysics*, **287**, 251–269.

STÜWE, K. 2007. *Geodynamics of the Lithosphere*. Springer, Berlin.

TAGAMI, T. & HASEBE, N. 1999. Cordilleran-type orogeny and episodic growth of continents: Insights from the circum-Pacific continental margins. *Island Arc*, **8**, 206–217.

TAN, E. & GURNIS, M. 2007. Compressible thermochemical convection and application to lower mantle structures. *Journal of Geophysical Research—Solid Earth*, **112**, B06304, doi:10.1029/2006JB004505.

THOMPSON, A. B., SCHULMANN, K. & JEZEK, J. 1997. Thermal evolution and exhumation in obliquely convergent (transpressive) orogens. *Tectonophysics*, **280**, 171–184.

THORKELSON, D. J. 1996. Subduction of diverging plates and the principles of slab window formation. *Tectonophysics*, **255**, 47–63.

TOHVER, E., D'AGRELLA-FILHO, M. S. & TRINDADE, R. I. F. 2006. Paleomagnetic record of Africa and South America for the 1200–500 Ma interval, evaluation of Rodinia and Gondwana assemblies. *Precambrian Research*, **147**, 193–222.

TORSVIK, T. H. 2003. The Rodinia jigsaw puzzle. *Science*, **300**, 1379–1381.

TSUJIMORI, T., SISSON, V. B., LIOU, J. G., HARLOW, G. E. & SORENSEN, S. S. 2006. Very-low-temperature record of the subduction process: A review of worldwide lawsonite eclogites. *Lithos*, **92**, 609–624.

UNSWORTH, M. J., JONES, A. G., WEI, W., MARQUIS, G., GOKARN, S. G. & SPRATT, J. E. 2005. Crustal rheology of the Himalaya and Southern Tibet inferred from magnetotelluric data. *Nature*, **438**, 78–81.

USUI, T., NAKAMURI, E., KOBAYASHI, K., MARUYAMA, S. & HELMSTAEDT, H. 2003. Fate of the subducted Farallon plate inferred from eclogite xenoliths in the Colorado Plateau. *Geology*, **31**, 589–592.

USUI, T., NAKAMURA, E. & HELMSTAEDT, H. 2006. Petrology and geochemistry of eclogite xenoliths from the Colorado Plateau: Implications for the evolution of subducted oceanic crust. *Journal of Petrology*, **47**, 929–964.

UYEDA, S. 1982. Subduction zones—An introduction to comparative subductology. *Tectonophysics*, **81**, 133–159.

UYEDA, S. & KANAMORI, H. 1979. Back-arc opening and the mode of subduction. *Journal of Geophysical Research*, **84**, 1049–1061.

VALLEY, J. W., LACKEY, J. S. ET AL. 2005. 4.4 billion years of crustal maturation: Oxygen isotope ratios of magmatic zircon. *Contributions to Mineralogy and Petrology*, **150**, 561–580, doi:10.1007/s00410-005-0025-8.

VAN HUNEN, J. & VAN DEN BERG, A. P. 2008. Plate tectonics on the early Earth: Limitations imposed by strength and buoyancy of subduction lithosphere. *Lithos*, **103**, 217–235.

VAN HUNEN, J., VAN DEN BERG, A. P. & VLAAR, N. J. 2002. The impact of the South American plate motion and the Nazca Ridge subduction on the flat subduction below south Peru. *Geophysical Research Letters*, **29**, Article Number. 1690.

VAN HUNEN, J., VAN DEN BERG, A. P. & VLAAR, N. J. 2004. Various mechanisms to induce present-day shallow flat subduction and implications for the younger Earth: A numerical parameter study. *Physics of the Earth and Planetary Interiors*, **146**, 179–194.

VAN HUNEN, J., DAVIES, G., HYNES, A. & VAN KEKEN, P. E. 2008. Mantle geodynamics and when plate tectonics began. *In*: CONDIE, K. & PEASE, V. (eds) *When Did Plate Tectonics Begin?* Geological Society of America, Special Papers, **440**, 157–171.

VAN STAAL, C. R., CURRIE, K. L., ROWBOTHAN, G., GOODFELLOW, W. & ROGERS, N. 2008. Pressure–temperature paths and exhumation of Late Ordovician–Early Silurian blueschists and associated metamorphic nappes of the Salinic Brunswick subduction complex, northern Appalachians. *Geological Society of America Bulletin*, doi:10.1130/B26324.1.

VON RAUMER, J. & STAMPFLI, G. M. 2008. The birth of the Rheic Ocean—Early Paleozoic subsidence patterns and subsequent tectonic plate scenarios. *Tectonophysics*, doi:10.1016/j.tecto.2008.04.012.

VON RAUMER, J., STAMPFLI, G. M. & BUSSEY, F. 2003. Gondwana-derived microcontinents—the constituents of the Variscan and Alpine collisional orogens. *Tectonophysics*, **365**, 7–22.

WAKABAYASHI, J. 2004. Tectonic mechanisms associated with $P-T$ paths of regional metamorphism: alternatives to single-cycle thrusting and heating. *Tectonophysics*, **392**, 193–218.

WARREN, C. J., BEAUMONT, C. & JAMIESON, R. A. 2008a. Formation and exhumation of ultra-high-pressure rocks during continental collision: Role of detachment in the subduction channel. *Geochemistry, Geophysics, Geosystems*, **9**, Article Number Q04019.

WARREN, C. J., BEAUMONT, C. & JAMIESON, R. A. 2008b. Modelling tectonic styles and ultra-high pressure (UHP) rock exhumation during the transition from oceanic subduction to continental collision. *Earth and Planetary Science Letters*, **267**, 129–145.

WATSON, K. D. & MORTON, D. M. 1969. Eclogite inclusions in kimberlite pipes at Garnet Ridge, northeastern Arizona. *American Mineralogy*, **54**, 267–285.

WHITE, R. W., POWELL, R. & CLARKE, G. L. 2002. The interpretation of reaction textures in Fe-rich metapelitic granulites of the Musgrave Block, central Australia: Constraints from mineral equilibria calculations in the system $K_2O-FeO-MgO-Al_2O_3-SiO_2-H_2O-TiO_2-Fe_2O_3$. *Journal of Metamorphic Geology*, **20**, 41–55.

WHITNEY, D. L., MILLER, R. B. & PATERSON, S. R. 1999. $P-T-t$ evidence for mechanisms of vertical tectonic motion in a contractional orogen: North-western US and Canadian Cordillera. *Journal of Metamorphic Geology*, **17**, 75–90.

WILLNER, A. P. 2005. Pressure–temperature evolution of a Late Palaeozoic paired metamorphic belt in north–central Chile (34°00′–35°30′S). *Journal of Petrology*, **46**, 1805–1833.

WILLNER, A. P., RÖTZLER, J. & MARESCH, W. V. 1997. Pressure–temperature and fluid evolution of quartzofeldspathic metamorphic rocks with a relic high-pressure, granulite-facies history from the central Erzgebirge (Saxony, Germany). *Journal of Petrology*, **38**, 307–336.

WILLNER, A. P., HERVÉ, F., THOMSON, S. N. & MASSONNE, H. J. 2004. Converging $P-T$ paths of Mesozoic HP–LT metamorphic units (Diego de Almagro Island, Southern Chile): Evidence for juxtaposition during late shortening of an active continental margin. *Mineralogy and Petrology*, **81**, 43–84.

WILSON, J. T. 1966. Did the Atlantic close and then re-open? *Nature*, **211**, 676–681.

WINDLEY, B. F., ALEXEIEV, D., XIAO, W., KRÖNER, A. & BADARCH, G. 2007. Tectonic models for accretion of the Central Asian Orogenic Belt. *Journal of the Geological Society, London*, **164**, 31–47.

YAMAMOTO, H., KOBAYASHI, K., NAKAMURA, E., KANEKO, Y. & KAUSAR, A. B. 2005. U–Pb zircon dating of regional deformation in the lower crust of the Kohistan arc. *International Geology Review*, **47**, 1035–1047.

YANG, J., LIU, F. ET AL. 2005. Two ultrahigh-pressure metamorphic events recognized in the central orogenic belt of China: Evidence from the U–Pb dating of coesite-bearing zircons. *International Geology Review*, **47**, 327–343.

ZAGOREVSKI, A., ROGERS, N., VAN STAAL, C. R., MCNICOLL, B., LISSENBERG, C. J. & VALVERDE-VAQUERO, P. 2006. Lower to Middle Ordovician evolution of peri-Laurentian arc and back-arc complexes in Iapetus: Constraints from the Annieopsquotch accretionary tract, central Newfoundland. *Geological Society of America Bulletin*, **118**, 324–342.

Arc–continent collisions, sediment recycling and the maintenance of the continental crust

PETER D. CLIFT[1,2]*, HANS SCHOUTEN[3] & PAOLA VANNUCCHI[4]

[1]*School of Geosciences, University of Aberdeen, Meston Building, Kings College, Aberdeen AB24 3UE, UK*

[2]*DFG-Research Centre Ocean Margins (RCOM), Universität Bremen, Klagenfurter Strasse, 28359 Bremen, Germany*

[3]*Department of Geology and Geophysics, Woods Hole Oceanographic Institution, Woods Hole, MA 02543, USA*

[4]*Dipartimento di Scienze della Terra, Università deli Studi di Firenze, Via La Pira, 4, 50121 Firenze, Italy*

*Corresponding author (e-mail: pclift@abdn.ac.uk)

Abstract: Subduction zones are both the source of most new continental crust and the locations where crustal material is returned to the upper mantle. Globally the total amount of continental crust and sediment subducted below forearcs currently lies close to 3.0 Armstrong Units (1 AU = 1 km^3 a^{-1}), of which 1.65 AU comprises subducted sediments and 1.33 AU tectonically eroded forearc crust, compared with an average of *c.* 0.4 AU lost during subduction of passive margins during Cenozoic continental collision. Margins may retreat in a wholesale, steady-state mode, or in a slower way involving the trenchward erosion of the forearc coupled with landward underplating, such as seen in the central and northern Andean margins. Tephra records of magmatism evolution from Central America indicate pulses of recycling through the roots of the arc. While this arc is in a state of long-term mass loss this is achieved in a discontinuous fashion via periods of slow tectonic erosion and even sediment accretion interrupted by catastrophic erosion events, probably caused by seamount subduction. Crustal losses into subduction zones must be balanced by arc magmatism and we estimate global average melt production rates to be 96 and 64 km^3 Ma^{-1} km^{-1} in oceanic and continental arc, respectively. Critical to maintaining the volume of the continental crust is the accretion of oceanic arcs to continental passive margins. Mass balancing across the Taiwan collision zones suggests that almost 90% of the colliding Luzon Arc crust is accreted to the margin of Asia in that region. Rates of exhumation and sediment recycling indicate that the complete accretion process spans only 6–8 Ma. Subduction of sediment in both erosive and inefficient accretionary margins provides a mechanism for returning continental crust to the upper mantle. Sea level governs rates of continental erosion and thus sediment delivery to trenches, which in turn controls crustal thicknesses over 10^7–10^9 years. Tectonically thickened crust is reduced to normal values (35–38 km) over time scales of 100–200 Ma.

The origins of the continental crust and the timing of its generation have been contentious issues in the Earth sciences for many years, largely revolving around arguments about whether the vast majority of the crust was generated during the Archaean (Armstrong & Harmon 1981; Bowring & Housh 1995; Elliott *et al.* 1999; Goldstein *et al.* 1997), or whether growth has been more gradual since that time (Moorbath 1978; O'Nions *et al.* 1979; Albarède & Brouxel 1987; Ellam & Hawkesworth 1988). More recently it has been suggested that continental crust has largely been generated in a series of rapid bursts of production between 1 and 3 Ga that were linked to rapid convection events in the mantle (Hawkesworth & Kemp 2006). In this model new upper crust is formed by differentiation from melt, leaving a large volume of dense residue, which may then be recycled back into the upper mantle via gravitational delamination. Generation of significant volumes of new crust does not appear to have been associated with areas of greatly thickened continental crust.

Although such an explanation implies that mantle plumes may have been fundamental in the generation of some early crust (e.g. West African and Arabian Shields; Boher *et al.* 1992; Stein & Hofmann 1994), most geochemical evidence instead indicates that subduction magmatism was the dominant source of new crustal material, especially during the Phanerozoic. Although continental crust is generally more andesitic than most modern arc magmatism (Rudnick 1995), a combination of geochemical and tectonic evidence indicates that active margins are probably the principal source of the continental crust (Dewey & Windley 1981; Rudnick & Fountain 1995; Taylor & McLennan 1995; Barth et al. 2000). Exactly how new crust is generated and how rapidly this is recycled through the subduction zones is, however, still enigmatic. In this paper we examine the rates of mass recycling through subduction zones and assess the role of oceanic arc accretion in building and maintaining the continental crust. We specifically focus on the Phanerozoic, when tectonic processes were the same as in modern plate tectonics. In this respect we do not try to understand how crust is recycled on the gigayear time scale, but rather how the volume of continental crust is maintained in the present tectonic regime. We do this by generating mass budgets for sediment and crustal flux to the global trench systems and comparing this with other possible mechanisms for crustal loss.

Generation of the continental crust in the first place is only part of the process of how the modern continents were formed and is not the focus of this paper, which is the role that subduction processes play in the maintenance of the crust during Phanerozoic times. Some constraints on long-term crustal volumes have been gained through consideration of variations in global sea level. Because the volume of the ocean basins is controlled by the relative proportions of continental versus oceanic crust a progressive loss or gain of continental crust would necessarily result in long-term variations in the global sea level, assuming that the volume of water on the Earth's surface has remained roughly constant. Although sea level has varied in the geological past, the magnitude of the change over a variety of time scales is modest compared with the total depth of the ocean basins (c. 200 m compared with mostly 3–6 km; Haq et al. 1987) and there has been little net change since at least the start of the Phanerozoic. Schubert & Reymer (1985) argued that because of the generally constant degree of continental freeboard (i.e. the average height of the continents above sea level) since the Precambrian, the continental crust must have remained close to constant volume since that time. In fact, those workers argued that gradual cooling of the Earth has resulted in slight deepening of the ocean basins, implying slow crustal growth since that time. This slow growth model is supported by Nd isotopic evidence for continental extraction from the upper mantle (Jacobsen 1988), as well as new Hf–O isotope data from zircons in cratonic rocks (Hawkesworth & Kemp 2006). Thus there appears to be a long-term balance between growth of new continental crust through arc magmatism and recycling of this crust back into the mantle via subduction zones.

Mass budgets for the continental crust

Subduction of continental crust

For many years it was believed that large-scale subduction of continental material was impossible because of the density differences between continental crust, oceanic crust and the upper mantle. None the less, if continental crust is generated at convergent margins then this requires some type of return flow to the upper mantle for the volume balance inferred from the freeboard argument to be maintained. Whereas some have argued for delamination and loss of lower crustal blocks (Gao et al. 2004), others have suggested that large volumes of crust can be subducted during major continental collision events (Hildebrand & Bowring 1999; Johnson 2002).

An estimate of the degree of crustal recycling possible during continental collisions can be derived by considering the Earth's mountain belts formed by continent–continent and arc–continent (passive margin) collisions during the Cenozoic. In practice, this means the Neotethyan belts of the Mediterranean, the Middle East and Himalayas, as well as Taiwan and the Ryukyu Arc, northern Borneo and the Australia–Papua New Guinea region. In total, these amount to almost 16 000 km of margin (Table 1). We assume a gradual thinning of the crust across these margins as a result of the progressive extension prior to break-up, and thus an average crustal thickness of 18 km (half normal continental crustal thickness; Christensen & Mooney 1995). We further assume that, as in the modern oceans, around 50% of the subducted margins were volcanic (50 km wide continent–ocean transition (COT)) and the remainder non-volcanic (150 km wide COT). The sedimentary cover to colliding margins is usually imbricated into the orogen but apart from small occurrences of high-grade rock subducted and then re-exhumed to the surface (e.g. Leech & Stockli 2000; Ratschbacher et al. 2000) the crystalline crust is potentially lost to the mantle. We calculate that around 28×10^6 km^3 have been subducted since the Mesozoic, resulting in a long-term average of 0.43 Armstrong Units (1 AU = 1 km^3 a^{-1}). Table 1 shows the basis of this estimate, involving the subduction of around 2.8×10^9 km^3 of passive margin crust since 65 Ma. If we assume that all the subducted margins were non-volcanic then the total subducted would increase to 4.2×10^9 km^3 and the long-term recycling rate to 0.65 AU.

Mechanisms of subduction erosion

This study focuses on quantifying the processes by which crust is returned to the mantle via subduction

Table 1. *Summary of the lengths of Cenozoic mountain belts along which passive margins have been subducted since 65 Ma, with an estimate of the subducted crustal volumes assuming 50% of the margins are volcanic and 50% non-volcanic*

Cenozoic orogen	Length of subducted margin (km)	Margin type	% of total	Length of margin (km)	Width of COT (km)	Average crustal thickness (km)	Volume (km³)
Alboran Sea	1300	Volcanic	50	7875	50	18	7 087 500
Pyrenees	450	Non-volcanic	50	7875	150	18	21 262 500
Alps–Apennines–Hellenides	3000	**Total**					**28 350 000**
Turkey–Zagros	3100						
Himalaya	2700						
Taiwan–Ryukyu	1100						
Borneo	1300						
Australia–Papua New Guinea	2800						
Total	**15 750**						

zones. Von Huene & Scholl (1991) synthesized a series of earlier studies and proposed that large volumes of sediment and forearc crust are being subducted in modern trenches. The concept that convergent margins not only accrete trench sediments in forearc prisms (accretionary wedges) but also allow deep subduction of continental material within a 'subduction channel' was pioneering. Tectonic erosion of some plate margins had been recognized before (Miller 1970; Murauchi 1971; Scholl *et al.* 1977; Hilde 1983), but the proposal by von Huene & Scholl (1991) that this was a process common in many of the Earth's subduction zones helped geologists reconcile the apparent mismatch between a stable crustal volume and the continuing arc magmatism. Their study helped properly establish the fact that there are two end-member types of active margin, accretionary and erosive,

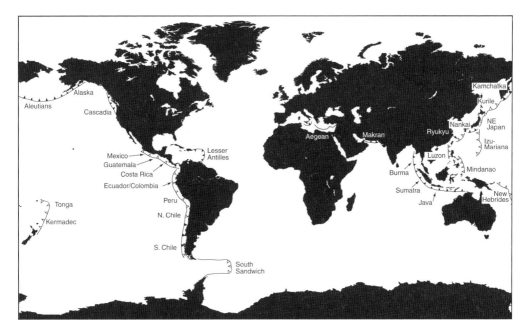

Fig. 1. Schematic map of the global subduction system showing the distribution of accreting v. eroding subduction zones. Accretionary margins are shown with filled barbs on the plate boundary, whereas empty barbs mark erosive margins. Modified after Clift & Vannucchi (2004), reproduced with permission of the American Geophysical Union.

and that the latter are common, especially in the circum-Pacific region (Figs 1 and 2).

It is still not yet agreed how tectonic erosion is accomplished, with some models suggesting mechanical abrasion by fault blocks on the underside of the subducting plate acting like a 'buzz saw' (Hilde 1983), whereas others emphasize the importance of subducting aseismic ridges in abrading material from the forearc (Clift & MacLeod 1999; Ranero & von Huene 2000; Clift et al. 2003a; Vannucchi et al. 2003; von Huene & Ranero 2003). Alternatively, the importance of fluids in this process was first suggested by von Huene & Lee (1982). In this type of model water expelled from the subducting plate causes hydrofracturing and disintegration of the base of the overlying forearc (Platt 1986; von Huene et al. 2004). The disrupted fragments would then be released into the subduction channel and transported to depth, either to the roots of the volcanic arc or even to the

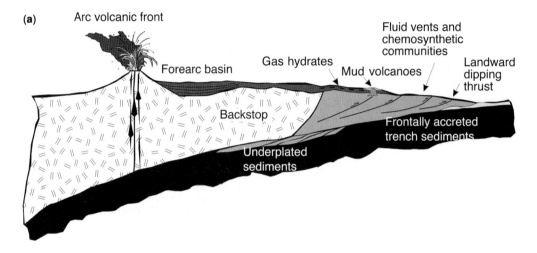

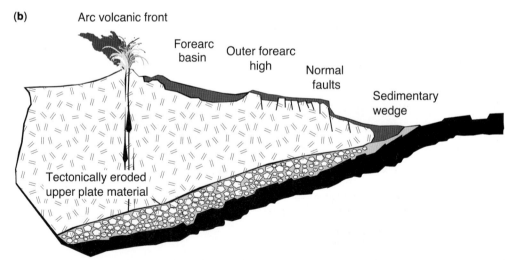

Fig. 2. Schematic illustrations of the features common to the two basic types of active margin from Clift & Vannucchi (2004). (**a**) Accretionary margins, such as Cascadia, are characterized by forearc regions composed of thrust and penetratively deformed trench and oceanic sediments that often develop mud diapirism and volcanism as a result of sediment overpressuring. Gas hydrate zones are also commonly associated with structures in the wedge. (**b**) In contrast, erosive plate margins, such as Tonga, are marked by steep trench slopes, composed of volcanic, plutonic and mantle rocks. Sedimentary rocks are typically limited to the forearc basin, where they may be faulted but are not strongly sheared in the fashion of an accretionary wedge. In the Marianas serpentinite mud volcanism is recorded. Reprinted with permission from the American Geophysical Union.

upper mantle. Evidence from the first recognized palaeo-subduction channel in the Italian Apennines shows that material was being returned to the deep Earth via a 500 m wide zone and that tectonic erosion of the forearc wedge was occurring both on its underside and in the frontal 'toe' region (Vannucchi et al. 2008).

Clift & Vannucchi (2004) used large-scale, long duration (>10 Ma) estimates of forearc mass change to improve the estimated rates of accretion or mass loss in all convergent margins. They predicted that around 3.6 AU were being subducted globally, rather higher than the 1.6 AU estimated by von Huene & Scholl (1991). They recorded 57% of modern trenches as being in a long-term tectonically erosive mode (Fig. 1) and noted that there was a correlation between margins where there was less than 1 km of sediment in the trench and tectonic erosion. In contrast, margins with >1 km of trench sediment were sites of net accretion (Fig. 3). Because the last 2–4 Ma are generally recognized as being a period of faster than normal continental surface erosion (Zhang et al. 2001), and presumably faster clastic trench sedimentation, it seems likely that 1 km is an overestimate of this threshold over long periods of geological time. This is because the Plio-Pleistocene pulse of faster sediment supply would not have had sufficient time to affect the long-term growth rates of convergent margins.

However, an absolute lower bound to the accretionary threshold is 400 m, because this is the thickness of purely pelagic sediment found in many trenches in both the eastern and western Pacific Ocean, where there are no clastic trench sediments and where long-term tectonic erosion is recorded. Because of the increased superficial erosion linked to glacial cycles the Earth's subduction systems are likely to now be in a particularly accretionary mode, and prior to northern hemispheric glaciation tectonic erosion would have been more dominant.

An important feature of accretionary margins is that although they are regions of net continental growth this does not mean that no continental crust is being recycled to the mantle in these margins (von Huene & Scholl 1991). Indeed, over long periods of geological time the proportion of sediment reaching the trench and preserved in accretionary margins averages only c. 19% globally (Clift & Vannucchi 2004). It is not clear whether this is because subduction accretion is a very inefficient process, or because even accretionary margins are affected by phases of tectonic erosion, perhaps triggered by subduction of seamounts or aseismic ridges. Drilling and modelling of the accretionary wedge in the Nankai Trough of SW Japan indicates that décollement in the trench sediment pile preferentially occurs near the base of the coarser grained trench sands, whereas the finer grained hemipelagic

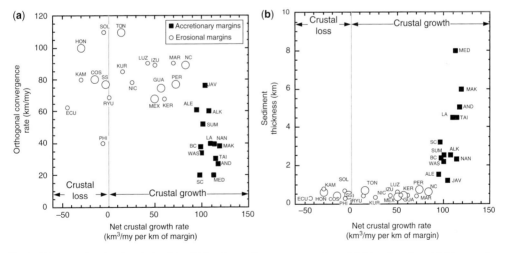

Fig. 3. Diagrams showing the relationship of (**a**) orthogonal convergence rates and (**b**) trench sediment thicknesses to the net crustal growth or loss along the global active margins, from Clift & Vannucchi (2004). Modified with permission from American Geophysical Union. Large circles show erosive plate margins for which a trench retreat is well defined; small circles represent margins for which tectonic erosion rates are inferred. ALE, Aleutians; ALK, Alaska; AND, Andaman; BC, British Columbia; COS, Costa Rica; ECU, Ecuador; GUA, Guatemala; HON, Honshu; IZU, Izu; JAV, Java; KAM, Kamchatka; KER, Kermadec; KUR, Kurile; LA, Lesser Antilles; LUZ, Luzon–Philippine; MAK, Makran; MAR, Marianas; MED, Mediterranean; MEX, Mexico; NAN, Nankai; NC, Northern Chile; NIC, Nicaragua; PER, Peru; PHI, Philippines; RYU, Ryukyu; SC, Southern Chile; SOL, Solomons; SS, South Sandwich; SUM, Sumatra; TAI, Taiwan; TON, Tonga; WAS, Washington–Oregon.

sediments below seem to be underthrust, at least below the toe region of the wedge (Le Pichon et al. 1993). This means that a significant proportion of the trench sediment column is underthrust at that margin, and presumably at most other margins, even when no ridge or seamount collision is occurring. Mass loss rates increase during subduction of aseismic ridges. Three-dimensional seismic reflection data have shown that such ridges have resulted in temporary periods of tectonic erosion, followed by a reversion to more efficient sediment accretion (Bangs et al. 2006).

Non-steady-state tectonic erosion

Observations of shallow-water or subaerial sedimentary rocks now found in trenches indicate long-term crustal loss over periods of $>10^7$ years, but do not preclude shorter intervals of net accretion. Conversely, margins in a state of long-term net accretion may also experience periods of tectonic erosion, usually linked to the passage of seamounts or other topographic features on the subducting plate. Some estimates of the time scale of these alternations can be derived from reconstructions of vertical tectonic motions. Melnick & Echtler (2006) used seismic data from the southern Andean margin to show that the marine forearc in this area has experienced basin inversion and uplift since the Pliocene, reversing a long-term trend to subsidence and crustal loss. In that study inversion was attributed to the increased sediment flux to the trench as a result of faster continental erosion driven by the glacial–interglacial cycles. Increased sediment flux was inferred to lubricate the base of the forearc wedge, reducing friction along that contact and driving a readjustment of the critical angle of the wedge. In this respect the model followed the suggestion by Lamb & Davis (2003) that reduced sediment flux to the central Andean trench increased coupling between overriding and subducting plates and triggered uplift of the main Andean ranges. If this were true then the total crustal flux to the upper mantle would increase as continental erosion increased.

Further north variability in the flux of material through the subduction zone is better constrained thanks to linked onshore–offshore datasets. The Andean margin of Peru and northern Chile provides a good example of an active margin in a state of long-term mass loss (von Huene & Ranero 2003). About 250 km of crust is estimated to have been lost from the western edge of South America in northern Chile and southern Peru since 150 Ma (Rutland 1971; Scheuber & Reutter 1992; Stern 2004), with as much as 30 km of this loss occurring since 10 Ma (Laursen et al. 2002). A landward step of 30–40 km in the trench axis north of the Juan Fernandez Ridge since 10 Ma resulted in trench retreat rate estimates of about 3.0 km Ma^{-1} during the Miocene–Recent, with similar values estimated from offshore Mejillones (von Huene & Ranero 2003). In such an environment forearc subsidence is predicted and has been observed in seismic and drilling data (Laursen et al. 2002; Clift et al. 2003a; von Huene & Ranero 2003; Fig. 4), with temporary periods of uplift at the point where aseismic ridges are in collision with the trench.

Comparison of onshore and offshore data in the central Andes, however, indicates that the margin is not in a continuous state of wholesale landward retreat. Sedimentary evidence from basins exposed along the coast demonstrates that the shoreline has been relatively stationary since 16 Ma and has not migrated inshore, as might be expected. Moreover, the exposure of these basins and the presence of marine terraces over long stretches of the margin (Fig. 5) point to recent uplift since at least the start of the Pleistocene (c. 1.8 Ma). Subsidence analysis backstripping methods were applied by Clift & Hartley (2007) to measured sections from these onshore coastal basins to isolate the tectonically driven subsidence of the basement, after correcting for sediment loading and sea-level variability. Although rates of tectonic erosion and subsidence are much higher close to the trench the continental basins are useful because the shallow shelf depths involved provide relatively high-precision constraints on vertical motions compared with deep-water trench or slope sediments.

Figure 6 shows the results of these analyses for the Pisco, Salinas, Mejillones, Caldera and Carrizalillo Basins. What is remarkable is that all these basins show a coherent, albeit slow basement subsidence, as might be predicted for a tectonically erosive margin, since 16 Ma, at the same time as the trench is known to retreat landwards and the marine forearc to subside (Fig. 4). However, like the southern Andean basins (Melnick & Echtler 2006) they also all show a late Pliocene unconformity, followed by terrestrial and coastal sedimentation preserved in a series of Pleistocene terraces. Although uplift can be generated by flexure in the footwall of normal faults, this is necessarily localized. However, the regional style of uplift observed along long stretches of the Andean coast and in the form of extensive terracing across a zone 5–10 km wide requires regional crustal thickening. This is most probably caused by subduction accretion and basal underplating of the forearc crust. The observation of active extensional faulting in much of the forearc precludes thickening by horizontal compression and thrust faulting. The central Andean margin experiences tectonic erosion but instead of the wholesale retreat of the plate margin the

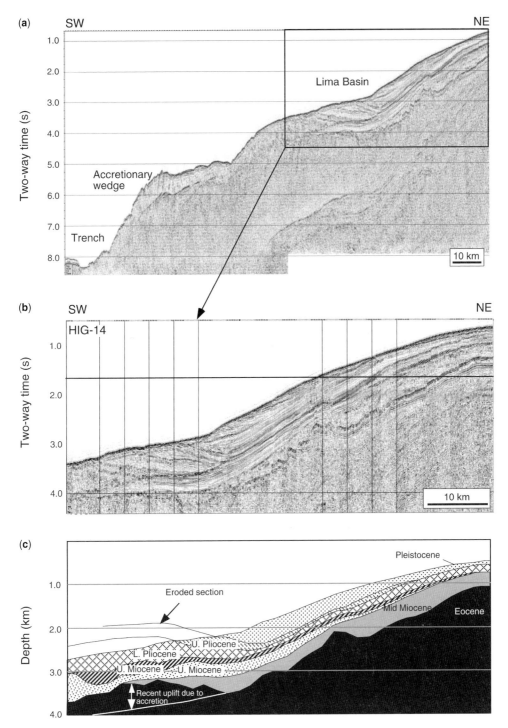

Fig. 4. (a) Multichannel seismic reflection profile HIG-14, with (b) close-up of Lima Basin and (c) after depth conversion and interpretation of age structure, largely correlated from ODP Site 679 located at the landward end of the profile.

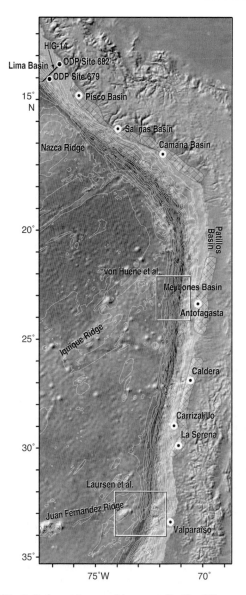

Fig. 5. Bathymetric map of the eastern Pacific offshore Peru and Chile showing the locations of the basins discussed in the text, as well as the major bathymetric ridges now in collision with the Andean margin. Depth data are from GEBCO. Depth contours in 500 m intervals. Boxes show regions of detailed offshore geophysical (seismic and bathymetric mapping) surveys. Reprinted with permission from Geological Society of America (Clift & Hartley 2007).

forearc was constrained to steepen since 16 Ma (Fig. 7a) and then was subject to underplating since 2 Ma. As a result, mass flow rates are much lower here than would be predicted for a steady-state erosive margin.

It is tempting to relate the Pleistocene shift to limited accretion under much of the terrestrial forearc as a result of faster trench sedimentation forced by climatic changes during the onset of northern hemispheric glaciation, much as suggested for the southern Andes (Melnick & Echtler 2006). However, this model does not hold up in the central Andes, where there are effectively no trench clastic sediments even today, reflecting the very arid climate of the continental interior. We conclude that the recent change in trench tectonics in the Andes is probably driven by a tectonic change in the way that the Pacific and South American plates interact.

Analysis of seismic and drilling data from the Andean forearc offshore the Lima Basin (Figs 4 and 5) also suggests that tectonic erosion is a temporally and spatially discontinuous process. Backstripping analysis of Ocean Drilling Program (ODP) Site 682 on the trench slope has been used to demonstrate long-term mass loss (Fig. 6; Clift et al. 2003a), accelerating after the collision of that part of the trench with the aseismic Nazca Ridge. Like the onshore basins to the south, ODP Site 679, also located in the Lima Basin but in shelf water depths, shows slow basement subsidence, with temporary uplift during passage of the Nazca Ridge along this part of the trench at 11–4 Ma (Fig. 5; Hampel 2002). However, interpretation of seismic lines across the forearc (Clift et al. 2003a; Fig. 4) shows that the trenchward part of the Lima Basin has been uplifted >500 m relative to the landward portions of the basin since the end of the Pliocene. This suggests Pleistocene underplating of the outer part of the forearc, where typically subduction erosion is most rapid. Uplift is consistent with the observation of a small accretionary complex at the foot of the trench slope in this region (Fig. 4a). The Aleutian forearc provides an additional example of this kind of process because although a significant accretionary wedge is developed at the trench the mid-forearc is dominated by a deep basin whose subsidence can be explained only by major basal erosion of the wedge (Ryan & Scholl 1993; Wells et al. 2003). Clearly, accretion and erosion are spatially and temporally variable, even on a single length of margin.

Geochemical evidence for temporal variations

Changes in the rates of sediment subduction and forearc tectonic erosion affect forearc vertical motions but must also affect the chemistry of arc volcanism. Geochemical evidence exists in Central America to suggest variations on the 1–4 Ma time scale in tectonic erosion, as well as related but

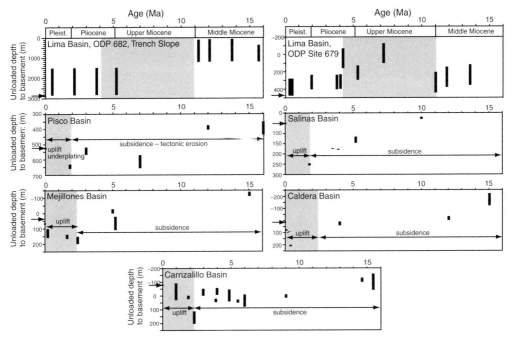

Fig. 6. Reconstructed depth to basement for a series of Andean forearc sedimentary basins during the Cenozoic, showing the general slow rate of subsidence consistent with long-term net mass loss during subduction erosion. Lima Basin (ODP Site 679 on the inner shelf, and ODP Site 682 on the trench slope) reconstruction is from Clift et al. (2003a). Pisco Basin data are taken from Dunbar et al. (1990) and Tsuchi (1992). Mejillones Peninsula data are from Krebs et al. (1992), Hartley & Jolley (1995), Ortlieb et al. (1996) and Ibaraki (2001). Caldera and Carrizalillo reconstructed are from Marquardt et al. (2004) and Le Roux et al. (2005), respectively. Height of black vertical bars shows the magnitude of the uncertainties in palaeo-water depth. Altered from Clift & Hartley (2007) and with permission from Geological Society of America.

independent changes in sediment subduction. In Costa Rica two different views of mass flux have been proposed. Tectonicists have shown that the margin and trench slope are in a state of long-term subsidence and presumed mass loss as a result of subduction erosion (Meschede et al. 1999; Vannucchi et al. 2001, 2003). Geochemical data from the active arc (e.g. ^{10}Be isotopes) indicate that the sedimentary cover in the modern trench cannot at present be contributing significantly to petrogenesis. This has been interpreted to imply either that the sediment is being offscraped and accreted to the margin (Valentine et al. 1997; Morris et al. 2002) or that tectonic subduction erosion is adding large volumes of additional material to the subduction channel so that the sedimentary signal to the arc magma is strongly diluted (Vannucchi et al. 2003). Early geophysical surveys had proposed that the Costa Rican forearc was largely composed of an accretionary wedge (Shipley et al. 1992), yet drilling of the region by ODP Leg 170 demonstrated that in fact the slope was formed by an extension of the onshore igneous Nicoya Complex, mantled by a relatively thin sedimentary apron of mass-wasted continental detritus and hemipelagic sediments (Kimura et al. 1997). Although the ODP drilling ruled out the possibility of much accretion of oceanic sediments at the toe of the forearc wedge during the recent geological past, it was not able to show whether sediments have been transferred to the overriding plate by underplating at greater depths. Clearly, competing models advocating accretion and erosion cannot both be correct, at least over the same time scales.

Study of the tephras deposited offshore the Nicoya Peninsula, Costa Rica, offer a possible explanation to these contradictory lines of evidence. Tephras can be used to reconstruct the magmatic evolution of the adjacent arc (the prevailing winds blow directly offshore) because the isotopic composition of the arc magmas (from which the tephras are derived) reflects the degree of sediment and forearc recycling in the subduction zone. As a result, the isotope geochemistry of the tephras can be used to quantify the degree of recycling at the time of their deposition. The cause of the tectonic erosion cannot be constrained from the tephras alone, but is inferred in this case from modern geophysical

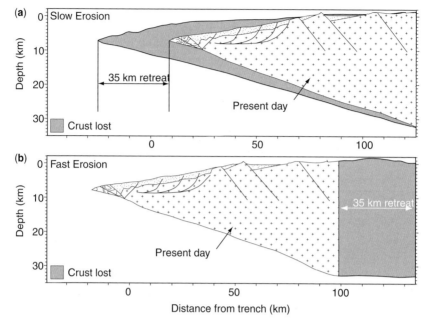

Fig. 7. Schematic illustration of the alternative modes of subduction erosion. (**a**) Shows the slower and non-steady state style of erosion as typified by Neogene northern and central Andes in which erosion is concentrated close to the trench, whereas the coast remains approximately stationary. Trench retreat is accommodated by steepening of the forearc taper. (**b**) Shows the fast, steady-state mode of erosion, apparently operating in the western Pacific, where taper angle remains constant and trench retreat rate corresponds to loss of complete crustal thickness.

data that indicate this process to be the primary cause of erosion in Costa Rica (Ranero & von Huene 2000).

Tephra glasses younger than 2.5 Ma were analysed by microprobe and found to be essentially unaltered, allowing them to be used to trace the temporal evolution of the onshore volcanic arc (Clift et al. 2005a). Trace element characteristics were compared with the active arc and a generally good fit was recognized between the tephra and the Costa Rican part of the Central American volcanic zone, a match that is possible because of the major along-strike variability noted between Costa Rica, Guatemala and Nicaragua (Leeman et al. 1994; Reagan et al. 1994; Patino et al. 2000; Carr et al. 2003). Clift et al. (2005a) employed Li and Nd isotopes to quantify the influence of sediment subduction and tectonic erosion on arc petrogenesis, an approach that was possible because of the different isotopic characteristics of trench sediments, the Nicoya Complex forearc basement and the altered oceanic crust entering the trench (Fig. 8a). These end-members allow the relative contributions of these sources to be resolved and unmixed. Figure 8b shows that at 1.45 Ma an unusual tephra was deposited showing an anomalous composition in Nd isotopes (i.e. relatively negative ε_{Nd}). This composition departs sharply from that of the nearest Costa Rican arc volcano, Arenal, as well as the other tephra in the stratigraphy, thus indicating a short-lived pulse of enhanced continental sediment subduction. The Li isotope composition also shows strong temporal variability, with high δ^7Li values being achieved at and for around 0.5 Ma after the ε_{Nd} spike. Subsequently, δ^7Li reduced to low values at the present day. These trends were interpreted to show that both sediment subduction and forearc erosion are currently at low levels, as inferred from study of the volcanoes themselves, but that this pattern is atypical of the recent geological past. The patterns of isotope evolution may reflect collision of a seamount with the trench before 1.45 Ma, driving subduction of a sediment wedge formed there, and subsequently increasing erosion and subduction of forearc materials. The whole cycle from accretion to erosion and back to accretion appears to span <2 Ma.

Rates of mass recycling

Rates of sediment and crustal subduction

Table 2 presents revised estimates of the rates of sediment accretion, tectonic erosion and arc

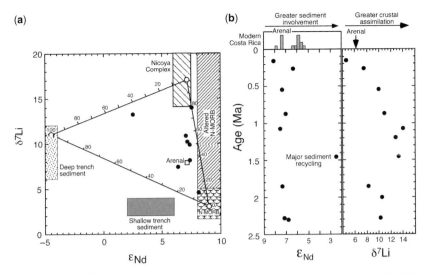

Fig. 8. (a) Plot of ε_{Nd} and δ^7Li shows that although the majority of tephra glasses at ODP Sites 1039, 1041 and 1043 could be explained by a petrogenesis mixing recycled MORB crust and subducted sediments, an additional, probably forearc component is required. Sediment data from Chan & Kastner (2000) and Kelly (2003). (b) Diagrams showing the evolving Li and Nd isotopic composition of Costa Rican tephras since 2.5 Ma. Histogram of ε_{Nd} values for modern Costa Rica is from GEOROC. Modern δ^7Li value for Arenal volcano is from Chan et al. (1999); Nd data are from Feigenson et al. (2004). Sediment proportion of petrogenesis is calculated from end-member mixing model based on Nd isotopes. Diagram shows the strong temporal variability in the degree of sediment and forearc crustal recycling, probably linked to seamount collision.

magmatism for global active margins based on the study of Clift & Vannucchi (2004), but updated with new constraints for some crucial margins. These values indicate less crustal subduction than proposed in the earlier study, mostly as a result of reduced estimates for tectonic erosion in the Andean margin of Peru and northern Chile (Clift & Hartley 2007). Even so, we estimate that worldwide total continental mass subduction is close to 3.0 AU, compared with around 3.6 AU in the synthesis of Clift & Vannucchi (2004), but still more than the 1.6 AU of von Huene & Scholl (1991). This value is almost an order of magnitude more than the 0.4 AU estimate for recycling of subducted passive margins in orogenic belts (Table 3).

The total amount of sediment being subducted at the world's trenches is relatively well constrained, because trench sediment thicknesses and convergence rates are relatively well known. In contrast, rates of subduction erosion of forearc crust are less well documented and appear to be variable in time. Of the total continental crust subducted globally at least as deeply as the magmatic roots of the arc systems we estimate that 1.65 AU comprise subducted sediments, with 1.33 AU of eroded forearc. Thus the latitude for error in the total estimate principally lies in the 1.33 AU of tectonic erosion. Regions where this value is not well known are themselves becoming scarcer. Although the degree of tectonic erosion may change as a result of further research, we believe that the amount of change in the total sum cannot be too large and that the 3.0 AU recycling rate is unlikely to be more than c. 20% in error.

Rates of trench retreat are the most common source of tectonic erosion estimates and many are based on the identification of terrestrial or shallow-water sediments overlying basement on the trench slope of modern systems (von Huene et al. 1982; Clift & MacLeod 1999; Vannucchi et al. 2001, 2003, 2004; Laursen et al. 2002). Rates of loss are then calculated based on the distance between the modern location of the shallow-water sedimentary rock and the location where similar facies sediments are now being deposited in the modern forearc. Rates of trench retreat are then derived by dividing this across-strike distance by the age of the sedimentary rock. Trench retreat rates in turn can be converted to rates of crustal volume loss if the thickness of the forearc crust is known and if the mode of subduction erosion is assumed. In the simplest and most conservative model trench retreat occurs along a margin in which the geometry remains effectively constant through time. Figure 7b shows that in this case a retreat of any given distance must involve loss of a complete crustal section of this width, as the forearc wedge itself remains constant. However, over short timespans tectonic

Table 2. *Summary of some of the major tectonic characteristics of the world's subduction zones, shown in Figure 1*

Arc	Length (km)	Orthogonal rate of convergence (km Ma^{-1})	Trench retreat rate (km Ma^{-1})	Trench sediment thickness (km)	Rate of sediment delivery (km^3 Ma^{-1} km^{-1})	Rate of forearc erosion (km^3 Ma^{-1} km^{-1})	Rate of sediment accretion (km^3 Ma^{-1} km^{-1})	Rate of crustal subduction (km^3 Ma^{-1} km^{-1})	Efficiency of sediment accretion (%)	Magmatic productivity rate (km^3 Ma^{-1} km^{-1})	Net crustal growth rate (km^3 Ma^{-1} km^{-1})
Erosive margins											
North Chile	2000	89	3.0	0.3	13	15		28		99	84
Peru	2200	77	3.1	0.7	29	15		44		86	71
Ecuador–Colombia	1100	63	3.0	0.6	20	15		35		71	56
Costa Rica	450	80	3.0	0.4	17	105		122		89	−16
Nicaragua	275	78	2.0	0.3	12	64		76		86	22
Guatemala	500	74	0.9	0.3	12	34		46		82	48
Mexico	1700	68	1.0	0.6	22	40		62		75	35
Kurile	1100	85	3.0	0.5	20	75		95		94	19
Kamchatka	1100	80	3.0	0.8	36	120		156		89	−31
NE Japan	1000	100	3.0	0.8	45	120		165		111	−9
Mariana	1600	90	1.0	0.4	19	20		39		100	80
Izu–Bonin	1300	89	2.0	0.4	19	40		59		99	59
Ryukyu	1000	69	3.0	0.4	14	90		104		77	−13
South Luzon	400	90	1.5	0.4	19	48		67		100	52
Philippine	1000	40	3.0	0.6	13	96		109		45	−51
Tonga	1500	110	3.8	0.4	23	76		99		122	46
Kermadec	1250	68	1.5	0.4	14	30		44		75	45
Solomons	2750	110	3.8	0.3	17	95		112		122	27
South Sandwich	700	77	4.7	0.4	16	94		110		86	−8
Accretionary margins											
South Chile	2000	20		3.2	43		7	36	16	22	29
Lesser Antilles	850	40		4.5	137		19	118	14	44	64
Oregon–Washington	850	34		2.2	48		10	38	21	38	48
British Columbia	550	38		2.5	61		10	51	16	42	52
Aleutians	1500	61		1.5	54		4	50	7	68	72
Alaska	2050	60		2.5	96		17	79	18	67	84
Taiwan–North Luzon	700	30		4.5	95		25	70	26	33	58
SW Japan–Nankai	900	39		2.3	65		24	41	37	44	68
Sumatra	1800	52		2.5	83		11	72	13	58	69
Java	2100	76		1.2	54		14	40	26	84	98
Burma–Andaman	1800	27		5.0	99		27	71	28	31	58
Makran	1000	38		6.0	179		29	150	16	42	71
Aegean	1200	20		8.0	131		22	109	17	22	45

Values are modified after those presented by Clift & Vannucchi (2004), especially with regard to the North Chile and Peruvian margins. Magmatic production rates are set to balance subducted losses and are scaled to reflect convergence velocity.

Table 3. *Rates of mass recycling for major crustal mass repositories*

	Rate (AU)
Arc magma production	2.98
Sediment subduction	1.65
Tectonic erosion	1.33
Subduction accretion	0.28
Total river discharge	6.70
Mainland Asian erosion	0.75
MORB production	20.00
OIB production	0.5
Subduction of passive margins in orogens	0.4

AU = Armstrong Unit (1 km^3 a^{-1}). Arc magmatic production rate is calculated to match the degree of crustal subduction (sediment and forearc crust). Subduction accretion rate is from Clift & Vannucchi (2004). Mid-ocean ridge basalt (MORB) production rate is from Dick *et al.* (2003). Ocean island basalt (OIB) production is from S. Hart (pers. comm.). Total river discharge is from Milliman (1997). Asian Cenozoic erosion rate is from Clift *et al.* (2004).

erosion can result in steepening of the forearc slope and a retreat of the trench with much lower total volumes of crustal subduction (Clift & Hartley 2007).

Rates of arc magmatism

Although short-term variations in mass flux through any given subduction zone can be driven by ridge and seamount collisions, or even climatically induced changes in trench sedimentation, the recognition that some margins have been in a long-term state of erosion or accretion allows mass budgets for plate margins to be made over periods of geological time 10^7 years or longer. As argued above, this loss has been balanced by arc magmatism, assuming a relatively constant volume of water at the Earth's surface and in the presence of a stable sea level. Although Clift & Vannucchi (2004) calculated that this rate of production averaged *c.* 90 km^3 Ma^{-1} km^{-1} of trench this value can now be reduced in light of the much lower crustal recycling rates for the central and northern Andes (Clift & Hartley 2007). Because of the length of the Andean margin and the thickness of the continental crust here, this adjustment makes a significant difference to the rate of loss and brings the global average rate of arc production down to 74 km^3 Ma^{-1} km^{-1}. The rate of recycling is likely to change as more geophysical information because available from the long stretches of active margin that have not yet been investigated.

Determining if this value is realistic or not can be difficult because rates of melt production under continental arcs are hard to determine. This is because only a small fraction of the melt is actually erupted, whereas most is intruded into or underplated onto the base of the crust. This is typically hard to resolve seismically and even harder to date. The situation is easier in oceanic arcs, where the whole crustal section is generated by subduction magmatism. If the age of subduction initiation is known then an average rate of production can be calculated. Holbrook *et al.* (1999) estimated long-term magmatic growth rates of 55–82 km^3 Ma^{-1} km^{-1} of margin for the Aleutians, whereas Suyehiro *et al.* (1996) indicated long-term average accretion rates of 66 km^3 Ma^{-1} km^{-1} of margin in the Izu Arc. Both these estimates do not account for the gradual loss of crust by subduction erosion, meaning that the true estimates of magmatic output for these arcs would be higher. However, it should also be recognized that most of the forearc crust in the Tonga, Mariana, Izu, and Bonin arcs is boninitic and was produced rapidly after the initiation of subduction, *c.* 45 Ma (Bloomer & Hawkins 1987; Stern & Bloomer 1992). This means that the steady-state average rate of crustal production is probably somewhat below the long-term average derived from seismic measurements. We conclude that the inferred rate of long-term crustal productivity is within error of the best constrained arc magmatic production rates.

Melting in continental active margins necessarily adds new material directly to the continental crustal mass, but the same is not necessarily true for oceanic arcs. Because these are built on oceanic lithosphere this material may be subducted when the arc collides with another trench system along with the lithosphere on which it is built. If oceanic arc crust is not accreted to continental margins during arc–continent collision then the continental mass balance would be disturbed and this subducted arc crust would need to be compensated for by greater production in continental arcs. This seems unlikely to be occurring because continental arc production is typically estimated to be lower than that in oceanic arcs (Plank & Langmuir 1988). For example, Atherton & Petford (1996) estimated production in the Andes be only 8 km^3 Ma^{-1} km^{-1}. Although oceanic arcs account for only 31% of the total active margins in the world our mass budget suggests that 40% (1.2 AU) of global arc melt production occurs along these margins. We estimate global average melt production rates to be 96 and 64 km^3 Ma^{-1} km^{-1} in oceanic and continental arc, respectively. If all the production needed to balance global subduction losses (3.0 AU) occurred only in the continental arcs then the average rate of melt production in these settings would increase to an unrealistic 106 km^3 Ma^{-1} km^{-1}. Thus efficient accretion of oceanic arc crust is important to the maintenance of the continental crust, but whether this really occurs or not is open to question. It is noteworthy that there are very few accreted oceanic island arc sections on Earth. Kohistan, Himalayas (Treloar et al. 1996; Khan et al. 1997), Alisitos, Baja California (Busby et al. 2006) and Talkeetna, Alaska (DeBari & Coleman 1989) are noteworthy for being the most complete, whereas many other examples are small and fragmentary, typically comprising only lavas and volcaniclastic sedimentary rocks. The fate of the oceanic Luzon Arc in the classic arc–continent collision zone of Taiwan is unclear, as only a fragment can be seen in the Coastal Ranges. Here we explore its accretion.

Arc accretion as a steady-state process

If arc accretion is to be understood as an integral part of the plate-tectonic cycle, and as a key process in maintaining the continental crustal volume, then the classic example of Taiwan may be used to quantify the processes that occur when an oceanic arc (Luzon) collides with a passive continental margin (Suppe 1981; Chemenda et al. 1997). This type of collision must be relatively common within the lifespan of the Earth, and it has been suggested that it is this process that transforms mafic, depleted arc crust into more siliceous, enriched continental crust (Draut et al. 2002). In this scenario high-silica, light rare earth element (LREE)-enriched magmas are injected into the mafic oceanic arc crust as it collides with a passive continental margin. At the same time the mafic–ultramafic lower crust may be subducted along with the mantle lithosphere on which the arc is constructed, made possible by detachment in the weak middle crust.

Arcs may also be accreted to active continental margins and there are examples (e.g. Dras–Kohistan and Talkeetna in southern Alaska) where a complete oceanic crustal section has been transferred to a continental margin when subduction eliminates the oceanic crust between two arcs of the same subduction polarity (DeBari & Coleman 1989; Treloar et al. 1996; Aitchison et al. 2000; Clift et al. 2005b). In both these examples the Moho itself can be observed in the field and there is little doubt that crustal accretion has been efficient. However, the fate of the arc crust during arc–passive margin collision is less clear because such collisions result in the formation of active continental margins following subduction polarity reversal, so that the oceanic plate on which the oceanic arc was constructed is then subducted (Casey & Dewey 1984).

In Taiwan the topographic massif of the North Luzon Arc disappears beneath the SE flank of Taiwan and it is unclear how much arc crust is accreted because the Taiwan ranges themselves expose thick sequences of deformed and metamorphosed Chinese passive margin sedimentary rocks, with only small scattered exposures of volcanic and volcaniclastic rocks found in the eastern Coastal Ranges (Lundberg et al. 1997; Song & Lo 2002). In this study we attempt to mass balance the collision to assess the efficiency of the accretion process. In doing so we model the collision of Luzon and China as being a steady-state process that is migrating progressively to the SW with time and that started to the NE of the present collision point at some point in the geological past. We note that some models propose initial Luzon collision to have only initiated around 6–9 Ma, effectively just to the east of the present collision zone, with a transform boundary between the Manila and Ryukyu Trenches prior to collision (Suppe 1984; Sibuet & Hsu 1997; Huang et al. 2000, 2006). Here we favour a more steady-state collision that may have started much earlier and potentially a long way to the east on modern Taiwan (Suppe 1984; Teng 1990; Clift et al. 2003b).

Migration of the arc collision along the margin is then controlled by the speed and obliquity of convergence (Fig. 9). In this scenario along-strike variability in the orogen represents different phases in the arc accretion process, with a pre-collisional Luzon Arc south of Taiwan, peak collision in the centre of the island and post-collisional orogenic collapse and subduction polarity reversal to the NE of Taiwan. In effect, the Luzon Arc can be considered to act in a rigid fashion like a snow plough deforming the passive margin of China into an accretionary stack, which then collapsed into the newly formed Ryukyu Trench as the collision point migrated to the SW along the margin (Fig. 10). In this scenario the collapsed orogen comprises slices of Chinese passive margin and the accreted Luzon Arc crust, which are then overlain in the Okinawa Trough by sediments eroded from the migrating orogen.

By taking cross-sections through the arc in the pre-, syn- and post-collisional stages we can assess to what extent the transfer of arc crust is an efficient process or not. Fortunately, gravity and seismology data allow the large-scale crustal structure of the orogen to be constrained, especially the depth and geometry of the main detachment on which the thrust sheets of passive margin sedimentary rock are carried (Fig. 11; Carena et al. 2002; Lee et al. 2002; Chen et al. 2004). Similarly, the depth and width of the foreland basin is well characterized by seismic and gravity data (Lin et al. 2003). Combined gravity, seismic reflection and refraction surveys constrain the crustal structure in the western Okinawa Trough and Ryukyu Arc, where Moho depths appear to be 25 and 30 km, respectively

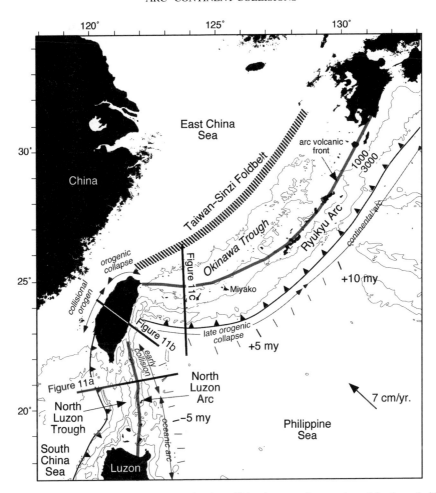

Fig. 9. Bathymetric map of the Taiwan region showing the collisional orogen, the opposing subduction polarities and the Okinawa Trough opening in the wake of orogenic collapse. The numbered lines adjacent to the plate boundary show the inferred time of peak arc collision between the Luzon Arc and the passive margin of China in the past and future. Map is labelled to show the stage of arc–continent collision along strike. Wide dashed line shows location of the Taiwan–Sinzi Foldbelt, interpreted as remnants of the former collisional orogen; the grey line shows the location of the modern arc volcanic front, focused by extension in the Okinawa Trough close to Taiwan (reprinted from Clift et al. 2003b).

(Kodaira et al. 1996; Wang et al. 2004). The volume of crust in the colliding arc is less well constrained, but the regional structure across the arc and forearc is known from seismic surveys (Hayes & Lewis 1984), and the crustal thickness of the igneous arc can be estimated at reaching a maximum of c. 25 km from gravity data and comparison with seismic data from other oceanic island arcs (Suyehiro et al. 1996; Holbrook et al. 1999). Sediment volumes on the northern margin of the South China Sea are best characterized from the Pearl River Mouth Basin to the SW of Taiwan, where abundant seismic reflection and drilling data provide a good estimate of the total amount of sediment available to be imbricated into the thrust sheets (Clift et al. 2004). Our estimates of the crustal volumes are derived from the sections shown in Figure 11 and are listed in Table 4.

By summing the volume of the crustal blocks at each stage of the collision we are able to assess how much material might be subducted during the collision process. This budget is shown graphically in Figure 12. What is most striking is that the great crustal thickness under Taiwan can be accounted for only if the metasedimentary thrust sheets that characterize the exposed ranges are underlain by a great thickness of arc crust. The total volume of sediment on the Chinese margin, together with recycled sediments in the accretionary complex west of the Luzon Arc, do not come close to

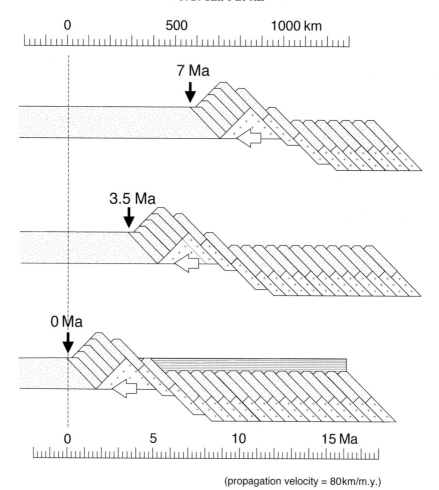

Fig. 10. Schematic illustration of the proposed tectonic model for Luzon Arc collision in Taiwan. Arc crust ('+' pattern) acts much like a snow plough, deforming and uplifting sediments of the Chinese passive margin (stippled). As the collision point migrates along the margin the compressive stress is released and the orogen collapses to form a basin, subsequently infilled by sediment, largely eroded from the orogen (horizontal shading).

explaining the volume of material lying above the basal detachment in the orogen. Although the cross-section could be drawn in alternative ways that would change the precise balance between crustal blocks, the need to accrete most of the Luzon Arc crust to the Chinese margin to make the sections balance is a common feature of all the reasonable reconstructions. Reducing the volume of accreted arc crust requires that this volume be made up by metasedimentary thrust sheets, yet the total volume of sediment on the Chinese margin or in the Manila Accretionary Complex is insufficient to account for this. It is this relative lack of material from which to form the Central Range thrust sheets, together with a relatively voluminous colliding arc, that requires the boundary between arc and orogen to dip westwards under the Central Ranges from the Longitudinal Valley. If this boundary is close to vertical then this would require the volume of the thrust sheets to be much greater than all the sediments on the Chinese passive margin and Manila Accretionary Complex from which the range is apparently built. A west-dipping thrust is none the less consistent with the structure of the eastern Central Ranges.

In our estimate c. 87% of the incoming arc crust is accreted. In addition, some sedimentary material appears to be lost, probably through erosion and recycling back into the pre-collisional accretionary complex, in which only part of the incoming section is imbricated, whereas some is lost to depth below the arc. The collapse thrust sheets and accreted Luzon Arc crust is then required to form the basement to the new Ryukyu Arc. The deformed

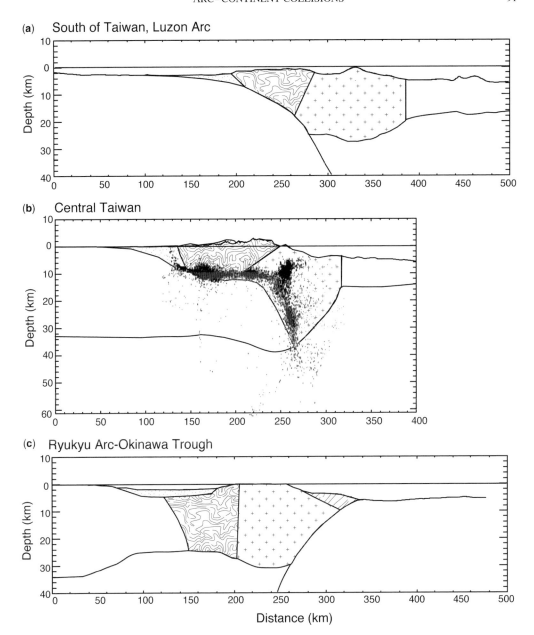

Fig. 11. Cross-sections through the Taiwan collision zone, south of the collision zone, at the collisional maximum in central Taiwan and across the Okinawa Trough and Ryukyu forearc, interpreted to be in a post-collisional state. Crustal structure under Taiwan is inferred from the seismic evidence of Carena *et al.* (2002), shown as black and grey dots projected onto the section from a number of major faults in the central Taiwan region. (See Fig. 9 for locations of the sections.)

edge of the Chinese margin is marked by the Taiwan–Sinzi Folded Belt (Fig. 9), located landward of the Okinawa Trough. This means that the Okinawa Trough itself and the arc ridge must be composed of accreted sediments and arc crust.

Sediment recycling during arc accretion

The degree of sediment recycling during arc accretion can be estimated through a similar mass-balancing exercise and through understanding of

Table 4. *Estimates of the volumes of crustal material found in the colliding Luzon Arc, Taiwan orogen and post-collisional Ryukyu–Okinawa Arc system, and erosion estimates for the Taiwan orogen with different ages for the start of collision*

Tectonic unit	Volume ($km^3\ km^{-1}$ of trench)
Taiwan thrust sheets	1076
Sediment in Taiwan foreland	250
Trench sediment in Luzon Trench	125
Accretionary prism offshore Luzon	825
Arc under Taiwan	2155
Luzon Arc igneous crust	2475
Ryukyu accretionary prism	210
Sediment in Okinawa Trough	300
Sediment on South China margin	750
Sediment eroded from Taiwan (5 Ma collision)	880
Sediment eroded from Taiwan (4 Ma collision)	720
Sediment eroded from Taiwan (3 Ma collision)	550

the rates of rock uplift and exhumation along the strike of Taiwan. Regional trends in rock uplift rates can be determined from the current elevations and the age of the collision, together with estimates for the amount of exhumation derived from the metamorphic grade and fission-track data (Tsao *et al.* 1992; Liu *et al.* 2000; Dadson *et al.* 2003; Willett *et al.* 2003; Fuller *et al.* 2006). Although in some locations modern rates of uplift have been determined by dating terraces (Peng *et al.* 1977; Vita-Finzi & Lin 1998) these are necessarily limited to the coastal regions, mostly in the Coastal Ranges of eastern Taiwan, and are not useful for estimating rates of the main accretionary stack. We follow Fuller *et al.* (2006) in placing peak erosion rates at $8\ mm\ a^{-1}$, falling to $6\ mm\ a^{-1}$ over much of the Central Ranges. These values are consistent with the suspended sediment load calculations of Fuller *et al.* (2003) that predicted rates of $2.2-8.3\ mm\ a^{-1}$. Similarly, Dadson *et al.* (2003) used data from river gauging stations to yield an average Central Range erosion rate of $c.\ 6\ mm\ a^{-1}$.

Exhumation rates driven by erosion reach a peak in the south of the island because rates of rock uplift are highest during the most intense period of collisional compression between Luzon and China; these are partly balanced by erosion driven largely by precipitation, but also by tectonic extension (Crespi *et al.* 1996; Teng *et al.* 2000). Exhumation

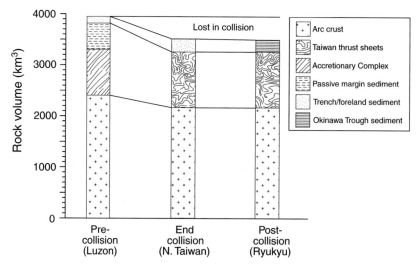

Fig. 12. Diagram showing the relative volumes of accreted and subducted arc crust v. eroded sediment in the Taiwan collision zone. The figure shows the relatively high efficiency of the accretion process in transferring oceanic arc crust to the Asian margin (88%). Crustal volumes are derived from the balanced cross-sections shown in Figure 11.

and vertical uplift rates decrease rapidly toward the northern end of the Central Ranges, and especially around the Ilan Basin (Fig. 13), although active motion along a detachment reversing the Lishan Fault causes increased exhumation in the Hsüehshan Range. Because the rates of exhumation are known from the fission-track analysis the total amount of exhumation can be calculated by knowing the duration of the collision (i.e. the speed of arc migration). We show three possible models for sediment accretion and erosion in Figure 14, based on collision starting at 5 Ma,

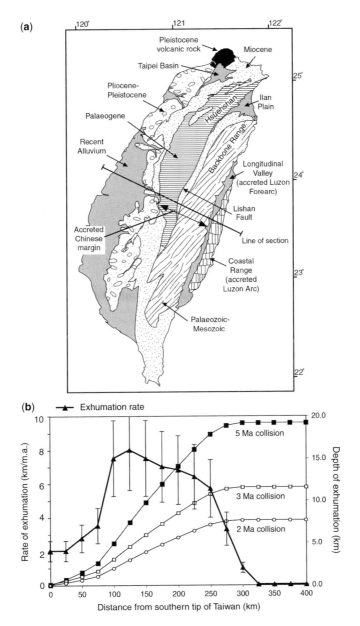

Fig. 13. (a) Simplified geological map of Taiwan showing the map tectonic units that comprise the island and the location of the Ilan Plain along the strike of the Lishan Fault, which separates the Hsüehshan and Backbone Ranges. (b) Diagram showing the rates of erosion along the Central Ranges of Taiwan, running south to north. Rates are based on fission-track analysis of Willett *et al.* (2003). Modelled depths of erosion imply a start of collision at 5, 3 and 2 Ma at the northern end of Taiwan and a propagating collisional orogen youngindg to the south.

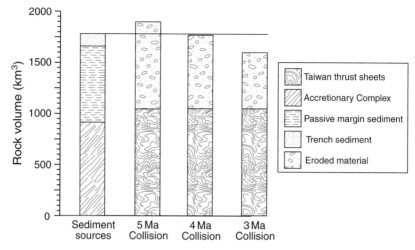

Fig. 14. Diagram showing the balance between the sediments and sedimentary rocks available to form the Taiwan orogen within the Luzon forearc and Chinese passive margin and the measured volume of the Taiwan thrust sheets and eroded volumes predicted for collision starting at different times for that point on the Luzon Arc now directly under the northern tip of Taiwan. The generally accepted 5 Ma collision age appears to require more material than is actually available in the system.

4 Ma and 3 Ma. The 5 Ma collision age would suggest c. 20 km of exhumation, somewhat more than indicated by the upper greenschist-facies rocks that characterize the Central Ranges (c. 12–15 km burial), yet this apparent mismatch does not preclude an earlier collision because of the continuous nature in which passive margin sediments may be added to the thrust stack, and recycled at shallower levels, so that although 20 km of erosion occurs this does not result in the progressive unroofing of deeper buried rocks (Willett et al. 1993; Fig. 15).

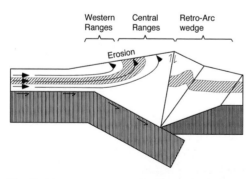

Fig. 15. Diagram redrawn from Willett et al. (1993) showing that crust from the subducting plate (China) is fluxed through the orogen, so that erosion at the mountain front can continue indefinitely without exposing any deeper level material at any given point. This type of horizontal transport must be occurring if the 20 km of erosion has occurred in the Central Ranges without higher grade metamorphic rocks being exposed.

The 5 Ma collision age is favoured by most recent syntheses (Huang et al. 2006), typically based on the ages of cooling, but also on the age of syncollisional sedimentary rocks exposed in the Coastal Ranges. The Shuilien Conglomerate is dated to start deposition around 3 Ma (Chi et al. 1981), yet even this can be considered a minimum age because of the potential for along-strike transport from the location of erosion to the depositional basin.

The most striking result of the sedimentary mass balance is that the volume of metasedimentary rocks together with the eroded volumes when compared with the total volumes of possible sediment source seem to require that the rocks now exposed in northern Taiwan were uplifted and became emergent in a palaeo-southern Taiwan no later than 4 Ma. There is simply insufficient sediment on the northern margin of the South China Sea to account for the orogen and its eroded sediment if uplift and subaerial erosion in the current area of Taiwan started at 5 Ma or before. This means that the speed of collision requires only 4 Ma for a rock to travel from the southern coast of Taiwan to the collapsing mountains around the Ilan Plain. Although some of the eroded sediment is transported into the Okinawa Trough the bulk is recycled back through the foreland basin and south into the Luzon Trench. Because the distal passive margin begins to collide with the Manila Trench south of southern Taiwan we estimate 6–8 Ma for the duration of the entire accretionary process.

Erosional control on crustal recycling and thickness

The image we have of the global subduction system is that it both generates new crust principally via arc magmatism and to a lesser extent via subduction accretion, and also returns crustal material to the mantle in tectonically erosive and accretionary active margins. Trench sediment thickness is the primary control on whether a margin is in a state of long-term accretion or tectonic erosion, and this in turn must be controlled by rates of continental erosion. What controls continental erosion is still debated, yet it is clear that tectonically driven rock uplift and climate are key processes (Zhang et al. 2001; Burbank et al. 2003; Dadson et al. 2003). Thus orogeny, uplifting blocks of crust above sea level, increases erosion and feeds sediment to the continental margins where some of it can be recycled into the mantle.

The topography of Earth is dominated by deep oceanic basins that contrast with continents that mostly lie close to sea level, except in regions of recent tectonic deformation (Fig. 16). An analysis of global topography shows that most ocean basins lie 3–7 km below sea level, whereas the vast majority of continental crust lies <500 m above sea level (Fig. 17). Sedimentary records indicate that the volume of water on the planet's surface has approximately filled the oceanic basins since Precambrian times, neither underfilling nor overspilling by more than c. 200 m (Haq et al. 1987). This state has lasted probably since the Archaean (McLennan & Taylor 1982; Campbell & Taylor 1983; Nisbet 1987; Harrison 1994). The volume of those basins is determined by the volume, density and thickness of equilibrium continental crust relative to the oceanic crust, but what controls crustal thickness?

Clift & Vannucchi (2004) calculated that at modern rates of plate motion arc magmatism is incapable of producing crust >35 km in tectonically erosive settings (oceanic or continental) because melt production would be incapable of keeping up

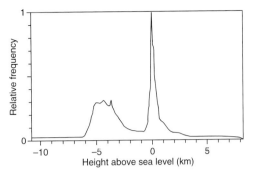

Fig. 16. Hypsometry of the Earth, showing the bipolar distribution between ocean basins and continental crust, which shows a particular sharp maximum close to sea level, representing the equilibrium state of the continental crust and equivalent to a thickness of c. 36 km with normal density structure (Rudnick & Fountain 1995).

with mass removal rates. In addition, delamination of dense lower crustal gabbronorite and pyroxenite in active arcs limits the thickness that can be formed by magmatism. Behn & Kelemen (2006) demonstrated that such lithologies are convectively unstable relative to the underlying mantle. Seismic velocity data show that most lower crust in modern arcs has seismic velocities (V_p) of more than 7.4 km s^{-1}, compared with the $V_p > 7.4$ km s^{-1} of the dense gabbronorite. This observation suggests that gravitationally unstable material must founder rapidly on geological time scales. As a result, greater crustal thicknesses would thus require additional tectonically driven horizontal compression.

Although trench processes act as an initial limit on crustal thicknesses, the narrow range of 36–41 km found in most continental crust (Christensen & Mooney 1995) must in part reflect the erosion of subaerially exposed crust. Except in regions of long-term extreme aridity (e.g. Atacama Desert; Hartley et al. 2005) orographic rain at mountain fronts will necessarily drive erosion and move crustal material from areas of thickened crust onto continental

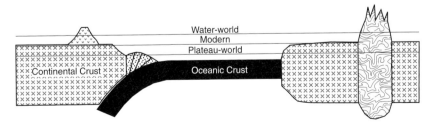

Fig. 17. Schematic illustration of the two states of continental crust at subduction margins. Thin crust lying below sea level is not eroded and is built up by voluminous melt production, whereas subaerial continental arcs have low melt production and are continuously eroded to sea level, even when active.

margins in the form of sediment. Even where climate inhibits erosion the process of continental drift will tend to move crustal blocks out of arid regions and into wetter ones over time periods of 10^7-10^8 years, so that eventually tectonically or magmatically thickened crust must be eroded and freeboard reduced to sea level. The 'excess' continental crust (i.e. eroded sediment) is returned to the mantle by subduction, either directly to a trench or to a passive margin that is eventually involved in collision with a trench.

No long-term trend in global sea level has been recorded over 10^7-10^9 years (Haq et al. 1987; Hallam 1992) and there appears to be good evidence for ocean depths in the Archaean being close to modern values (Wise 1974; Nisbet 1987; Harrison 1999). However, there seems no a priori reason why the volume of water should not cause sea level to overspill the oceanic basins and flood the continents to a depth of 1 km or more. Conversely, if the volume of water were much less, or the oceanic basins much larger, then sea level would lie well below the level of the continents (Fig. 17). None the less, evidence from the stratigraphic record shows that this has not happened. Harrison (1999) argued that continental freeboard above sea level is controlled by the volume of the water at Earth's surface, which varies over long periods of time (10^7-10^9 years). Erosion of elevated topography as a result of precipitation and glaciation will tend to reduce elevated terrain to sea level (Harrison 1994), thinning the crust as it does so. Conversely, thin, submarine arc crust will tend to be built up to sea level by the compressive tectonic forces favoured by sediment-starved trenches and where coupling between subducting and overriding plates is stronger (Wells et al. 2003), as well as by magmatic accretion. The process is self-limiting because a thickening, uplifting margin will tend to result in faster erosion, more sediment in the trench and a reduction in tectonic coupling. Thus continental crustal thickness reflects the combined influence of seawater volume and the ability to subduct significant volumes of crust at both accretionary and erosive margins.

Erosion of orogenic topography

The rate at which continental crust is recycled through subduction zones is dependent on how fast orogenic topography can be reduced to sea level. Mountain belts are prone to reactivation after their initial peneplanation, yet they rarely regain major altitude after this initial phase of crustal thinning. The Urals reach heights >1800 m more than 250 Ma after their formation (Brown et al. 2006), yet generally Mesozoic and especially Palaeozoic belts rarely exceed 1 km elevations. The highest mountains on Earth represent Cenozoic and modern plate collisions. Arc–continent collisional orogens reduce to sea level as a result of gravitational collapse into the new trench immediately after the collision point has passed (Teng 1996; Clift et al. 2003b), but continent–continent collisions are often different and require extended periods of erosion before equilibrium is attained (i.e. before they are reduced in altitude close to sea level). For example, the mountains formed by the Acadian orogeny after closure of the Iapetus Ocean reached peak metamorphism around 395 Ma (Armstrong et al. 1992) but are truncated by a peneplain erosion unconformity of Late Carboniferous age (c. 320 Ma) in western Ireland (Graham 1981), suggesting c. 75 Ma to remove the excess crust by erosion.

Estimates for the erosion of the Himalaya and Tibet provide an end-member example of a very large, long-lived orogen. The volume of eroded rock in the basins around Tibet can be used to determine long-term erosion rates. Eroded rock volumes, with a correction for sediment porosity, are shown in Table 5. These values are determined from regional seismic profiles and are largely from the studies of Métivier et al. (1999) and Clift et al. (2004). As well as the sediments now seen in the major depocentres we estimate those sediments already lost by subduction, especially in the Andaman Arc, where sediment has been subducted at a rate of 1.3×10^5 km^3 Ma^{-1} in the recent geological past, although this would have been less earlier as the fan has grown significantly in the Neogene (Métivier et al. 1999). We estimate that c. 3.2×10^6 km^3 has been subducted in this way since 50 Ma, assuming no Bengal Fan at that time and progressive subsequent growth. Subducted losses in the Indus Fan via subduction in the Makran Accretionary Complex would be somewhat less, because that fan has been cut off from the Gulf of Oman since 20 Ma by uplift of Murray Ridge (Mountain & Prell 1990), so that only the Palaeogene erosional record is being lost. The basins of continental Central Asia, including the Tarim, Junngar and Hexi Corridor Basins, account for 1.9×10^6 km^3 (Métivier et al. 1999). The excess crustal mass in Tibet is estimated using a value of 3.5×10^6 km^2 as the area of the plateau and 70 km for the crustal thickness, around double the standard equilibrium continental crust.

The average long-term rate of erosion is 0.56 AU since 50 Ma; requiring 220 Ma to reduce Tibet crust from 70 to 35 km thickness, if the plateau ceased growing today and that long-term rate of erosion was maintained (Fig. 18; Table 6). Whether this is a reasonable value is debatable because mass accumulation rates in East Asian marginal seas appear to have increased rapidly after 33 Ma

Table 5. *Volumes of basins in Asia that have largely been filled by erosion of the Himalaya and Tibetan Plateau*

Basin	Eroded rock volume (km³)	Percentage of total
Bengal Fan	10271159	36.8
Indus Fan	4108463	14.7
Sediment subducted in Andaman Trench	3200000	11.5
Himalayan Foreland Basin	2042711	7.3
Central Asian basins (Tarim, Junggar, Qaidam)	1900000	6.8
Irrawaddy Fan	1472473	5.3
Sediment subducted in Makran	1300000	4.7
Song Hong–Yinggehai Basin	451931	1.6
Katawaz Basin	389972	1.4
Sulaiman–Kirthar Ranges	389972	1.4
East China Sea	345111	1.2
Pearl River Mouth Basin	328677	1.2
Malay Basin	315530	1.1
Nam Con Son Basin	287592	1.0
SE Hainan Basin (Qiongdongnan Basin)	246508	0.9
Burma Basin	237697	0.9
Mekong–Cuu Long Basin	177486	0.6
Indo-Burma Ranges	129991	0.5
Bohai Gulf	123254	0.4
Pengyu Basin	65735	0.2
Pattani Trough	65078	0.2
West Natuna Basin	36976	0.1
Hanoi Basin	8124	0.0
Total eroded	**27894442**	

Data compiled from Clift *et al.* (2004) for the marine and foreland basin and from Métivier *et al.* (1999) for the basins of Central Asia. This volume of sediment can be used to calculate long-term average rates of sediment production. Given the modern size of Tibet and the known 'excess' crustal thickness above average continental values, the time needed to remove Tibet can be estimated. Erosion has been much faster since c. 33 Ma and thus a range of possible values can be calculated if we assume erosion starting after collision c. 50 Ma or only since 33 Ma.

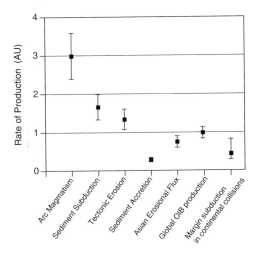

Fig. 18. Diagram showing the comparative rates of sediment recycling in subduction zones relative to tectonic erosion, arc magmatism, ocean island basalt (OIB) magmatism, subduction accretion and the erosional flux from Asia.

(Clift *et al.* 2004), and recent palaeo-weathering studies from South and East Asia suggest an initial intensification of monsoon summer rains after 22 Ma (Clift & Plumb 2008). As rains are a major trigger for continental erosion, much of the sediment now seen around Asia has been eroded since that time. Using 33 and 22 Ma as the start dates for the bulk of the erosion yields average erosion rates of 0.9 and 1.3 AU, respectively, and consequently durations of 145 and 97 Ma to reduce Tibet back to sea level. Thus it appears that little orogenic topography can survive at the Earth's surface beyond c. 200 Ma, without the excess crust being recycled to the mantle via subduction zones.

Summary

Geochemical arguments strongly suggest that much of the modern continental crust has been formed in or at least recycled via subduction zones. Subduction zones are also the locations where crustal material is returned to the upper mantle both via tectonic erosion of forearc crust and subduction of

Table 6. *Estimates of average rates of erosion for the Himalaya and Tibetan Plateau assuming different ages for the start of uplift and erosion*

	Rate of erosion (km^3 Ma^{-1})	Armstrong units
Average erosion rate since 50 Ma	557889	0.56
Average erosion rate since 33 Ma	845286	0.85
Average erosion rate since 22 Ma	1267929	1.27
Time required to erode modern Tibet at 50 Ma rate	220	
Time required to erode modern Tibet at 33 Ma rate	145	
Time required to erode modern Tibet at 22 Ma rate	97	

Data from Clift *et al.* (2004).

sediment in erosive margins, and via the inefficient process of subduction accretion that allows c. 83% of the incoming sedimentary column to be subducted over long periods of geological time (Clift & Vannucchi 2004). Total continental mass subduction rates are c. 3.0 Armstrong Units (1 AU = 1 km^3 a^{-1}), of which 1.65 AU comprises subducted sediments and 1.33 AU tectonically eroded forearc crust (Fig. 18). Recycling of crust by subduction of passive margins in continent–continent and arc–passive margin collisions is substantially less, averaging 0.4 AU during the Cenozoic.

The process of subduction erosion is still not well understood and may involve both mechanical abrasion of the forearc wedge and/or fluid-induced fracturing of the forearc (von Huene *et al.* 2004; Vannucchi *et al.* 2008). What is clear is that erosion can operate in several fashions. Margins may retreat landwards in a wholesale, steady-state mode, such as apparently is the case in the South Sandwich, Tonga and Marianas Arcs. Alternatively, trench retreat may occur by erosion of the trenchward forearc coupled with landward underplating, such as seen in the central and northern Andean margin, both associated with trench sediment thicknesses of <1 km. Tectonic erosion is probably a discontinuous process, as suggested by tephra records from offshore Costa Rica (Clift *et al.* 2005a), a region generally associated with long-term tectonic erosion (Vannucchi *et al.* 2001). Geochemical data indicate that subduction erosion in the Central American arc is achieved via periods of slow tectonic erosion interrupted by shorter periods of accelerated erosion, probably driven by seamount subduction, and by periods of sediment accretion.

Subducted crustal losses must be balanced by new production via arc magmatism. Average global melt production rates are 96 and 64 km^3 Ma^{-1} km^{-1} in oceanic and continental arc, respectively, close to those estimated by seismic methods, after correction for subduction erosion losses (Suyehiro *et al.* 1996; Holbrook *et al.* 1999). The accretion of oceanic arc crust to continental passive margins during arc–continent collision is crucial to maintaining the volume of the continental crust without excessive melt production in continental arcs. Mass balancing across the Taiwan collision zone suggests that almost 90% of the colliding Luzon Arc crust is accreted to the margin of Asia in that region. The accretion is seen as a steady-state migrating process, with the collapsed arc and deformed passive margin of China underlying the new Ryukyu forearc ridge and Okinawa Trough. Rates of exhumation and sediment recycling indicate that the subaerial phase of orogenesis spans only c. 4 Ma, although initial collision with the deep-water passive margin must have predated that time by 2–3 Ma.

Subduction of sediment in both erosive and inefficient accretionary margins provides a mechanism for returning continental crust to the upper mantle and appears to be fundamental in governing the thickness of the continental crust. Sea level controls rates of continental erosion, reducing topography caused by tectonically overthickened crust to sea level over time scales of 100–200 Ma. Much of this eroded sediment is delivered directly or indirectly to trenches, allowing its return to the upper mantle. Thus sea level and the volume of water on the Earth's surface control crustal thicknesses over periods of 10^7–10^9 years.

P.C. wishes to thank the Alexander von Humboldt Foundation for support during the writing of this paper at the University of Bremen, as well as the College of Physical Sciences, University of Aberdeen for its generous support. PRIN 2005 Subduction complex dynamics: mass transfer in fossil systems and comparisons with modern examples, funded P.V. We thank D. Brown and D. Tappin for their constructive comments in review for improving the manuscript.

References

AITCHISON, J. C., BADENGZHU, D. *ET AL*. 2000. Remnants of a Cretaceous intra-oceanic subduction system within the Yarlung–Zangbo suture (southern Tibet). *Earth and Planetary Science Letters*, **183**, 231–244.

ALBARÈDE, F. & BROUXEL, M. 1987. The Sm/Nd secular variation of the continental crust and the depleted mantle. *Earth and Planetary Science Letters*, **82**, 25–35.

ARMSTRONG, R. L. & HARMON, R. S. 1981. Radiogenic isotopes: The case for crustal recycling on a near-steady-state no-continental-growth Earth. *Philosophical Transactions of the Royal Society of London, Series A*, **301**, 443–472.

ARMSTRONG, T. R., TRACY, R. J. & HAMES, W. E. 1992. Contrasting styles of Taconian, Eastern Acadian and Western Acadian metamorphism, central and western New England. *Journal of Metamorphic Geology*, **10**, 415–426.

ATHERTON, M. P. & PETFORD, N. 1996. Plutonism and the growth of Andean crust at 9 degrees S from 100 to 3 Ma. *Journal of South American Earth Sciences*, **9**, 1–9.

BANGS, N. L. B., GULICK, S. P. S. & SHIPLEY, T. H. 2006. Seamount subduction erosion in the Nankai Trough and its potential impact on the seismogenic zone. *Geology*, **34**, 701–704, doi: 10.1130/G22451.1.

BARTH, M. G., MCDONOUGH, W. F. & RUDNICK, R. L. 2000. Tracking the budget of Nb and Ta in the continental crust. *Chemical Geology*, **165**, 197–213.

BEHN, M. & KELEMEN, P. 2006. Stability of arc lower crust; insights from the Talkeetna Arc section, south central Alaska, and the seismic structure of modern arcs. *Journal of Geophysical Research*, **111**, doi:10.1029/2006JB004327.

BLOOMER, S. H. & HAWKINS, J. W. 1987. Petrology and geochemistry of boninite series volcanic rocks from the Mariana trench. *Contributions to Mineralogy and Petrology*, **97**, 361–377.

BOHER, M., ABOUCHAMI, W., MICHARD, A., ALBAREDE, F. & ARNDT, N. T. 1992. Crustal growth in West Africa at 2.1 Ga. *Journal of Geophysical Research*, **97**, 345–369.

BOWRING, S. A. & HOUSH, T. 1995. The Earth's early evolution. *Science*, **269**, 1535–1540.

BROWN, D., PUCHKOV, V., ALVAREZ-MARRON, J., BEA, F. & PEREZ-ESTAUN, A. 2006. Tectonic processes in the Southern and Middle Urals; an overview. *In*: GEE, D. G. & STEPHENSON, R. A. (eds) *European Lithosphere Dynamics*. Geological Society, London, Memoirs, **32**, 407–419.

BURBANK, D. W., BLYTHE, A. E. *ET AL.* 2003. Decoupling of erosion and precipitation in the Himalayas. *Nature*, **426**, 652–655.

BUSBY, C., FACKLER ADAMS, B., MATTINSON, J. & DEOREO, S. 2006. View of an intact oceanic arc, from surficial to mesozonal levels: Cretaceous Alisitos arc, Baja California. *Journal of Volcanology and Geothermal Research*, **149**, 1–46.

CAMPBELL, I. H. & TAYLOR, S. R. 1983. No water, no granites; no oceans, no continents. *Geophysical Research Letters*, **10**, 1061–1064.

CARENA, S., SUPPE, J. & KAO, H. 2002. Active detachment of Taiwan illuminated by small earthquakes and its control of first-order topography. *Geology*, **30**, 935–938.

CARR, M. J., FEIGENSON, M. D., PATINO, L. C. & WALKER, J. A. 2003. Volcanism and geochemistry in Central America: Progress and problems. *In*: EILER, J. (ed.) *Inside the Subduction Factory*. American Geophysical Union, Geophysical Monograph, **138**, 153–174.

CASEY, J. F. & DEWEY, J. F. 1984. Initiation of subduction zones along transform and accreting plate boundaries, triple-junction evolution, and forearc spreading centres—implications for ophiolitic geology and obduction. *In*: GASS, I. G., LIPPARD, S. J. & SHELTON, A. W. (eds) *Ophiolites and Oceanic Lithosphere*. Geological Society, London, Special Publications, **13**, 269–290.

CHAN, L.-H. & KASTNER, M. 2000. Lithium isotopic compositions of pore fluids and sediments in the Costa Rica subduction zone; implications for fluid processes and sediment contribution to the arc volcanoes. *Earth and Planetary Science Letters*, **183**, 275–290.

CHAN, L.-H., LEEMAN, W. P. & YOU, C. F. 1999. Lithium isotopic composition of Central American Volcanic Arc lavas: Implications for modification of subarc mantle by slab-derived fluids. *Chemical Geology*, **160**, 255–280.

CHEMENDA, A. I., YANG, R. K., HSIEH, C. H. & GROHOLSKY, A. L. 1997. Evolutionary model for the Taiwan collision based on physical modeling. *Tectonophysics*, **274**, 253–274.

CHEN, P.-F., HUANG, B.-S. & LIANG, W.-T. 2004. Evidence of a slab of subducted lithosphere beneath central Taiwan from seismic waveforms and travel times. *Earth and Planetary Science Letters*, **229**, 61–71.

CHI, W. R., NAMSON, J. & SUPPE, J. 1981. Stratigraphic record of plate interactions in the Coastal Range of eastern Taiwan. *Geological Society of China Memoir*, **4**, 155–194.

CHRISTENSEN, N. I. & MOONEY, W. D. 1995. Seismic velocity structure and composition of the continental crust; a global view. *Journal of Geophysical Research*, **100**, 9761–9788.

CLIFT, P. D. & HARTLEY, A. 2007. Slow rates of subduction erosion along the Andean margin and reduced global crustal recycling. *Geology*, **35**, 503–506.

CLIFT, P. D. & MACLEOD, C. J. 1999. Slow rates of subduction erosion estimated from subsidence and tilting of the Tonga forearc. *Geology*, **27**, 411–414.

CLIFT, P. & VANNUCCHI, P. 2004. Controls on tectonic accretion versus erosion in subduction zones; implications for the origin and recycling of the continental crust. *Reviews of Geophysics*, **42**, (RG2001), doi: 10.1029/2003RG000127.

CLIFT, P. D. & PLUMB, R. A. 2008. *The Asian Monsoon: Causes, History and Effects*. Cambridge, Cambridge University Press.

CLIFT, P. D., PECHER, I., KUKOWSKI, N. & HAMPEL, A. 2003a. Tectonic erosion of the Peruvian forearc, Lima Basin, by subduction and Nazca Ridge collision. *Tectonics*, **22**, 1023, doi:10.1029/2002TC001386.

CLIFT, P. D., SCHOUTEN, H. & DRAUT, A. E. 2003b. A general model of arc–continent collision and subduction polarity reversal from Taiwan and the Irish Caledonides. *In*: LARTER, R. D. & LEAT, P. T. (eds) *Intra-Oceanic Subduction Systems: Tectonic and Magmatic Processes*. Geological Society, London, Special Publications, **219**, 81–98.

CLIFT, P. D., LAYNE, G. D. & BLUSZTAJN, J. 2004. Marine sedimentary evidence for monsoon strengthening, Tibetan uplift and drainage evolution in east Asia. In: CLIFT, P., KUHNT, W., WANG, P. & HAYES, D. (eds) *Continent–Ocean Interactions in the East Asian Marginal Seas.* American Geophysical Union, Geophysical Monograph, **149**, 255–282.

CLIFT, P. D., CHAN, L.-H., BLUSZTAJN, J., LAYNE, G. D., KASTNER, M. & KELLY, R. K. 2005a. Pulsed subduction accretion and tectonic erosion reconstructed since 2.5 Ma from the tephra record offshore Costa Rica. *Geochemistry, Geophysics, Geosystems*, **6**, doi:10.1029/2005GC000963.

CLIFT, P. D., PAVLIS, T., DEBARI, S. M., DRAUT, A. E., RIOUX, M. & KELEMEN, P. B. 2005b. Subduction erosion of the Jurassic Talkeetna–Bonanza Arc and the Mesozoic accretionary tectonics of western North America. *Geology*, **33**, 881–884.

CRESPI, J., CHAN, Y. C. & SWAIM, M. 1996. Synorogenic extension and exhumation of the Taiwan hinterland. *Geology*, **24**, 247–250.

DADSON, S., HOVIUS, N. ET AL. 2003. Links between erosion, runoff variability and seismicity in the Taiwan orogen. *Nature*, **426**, 648–651.

DEBARI, S. M. & COLEMAN, R. G. 1989. Examination of the deep levels of an island arc: Evidence from the Tonsina ultramafic–mafic assemblage, Tonsina, Alaska. *Journal of Geophysical Research*, **94**, 4373–4391.

DEWEY, J. F. & WINDLEY, B. F. 1981. Growth and differentiation of the continental crust. *Philosophical Transactions of the Royal Society, London, Series A*, **301**, 189–206.

DICK, H. J. B., LIN, J. & SCHOUTEN, H. 2003. An ultraslow-spreading class of ocean ridge. *Nature*, **426**, 405–412.

DRAUT, A. E., CLIFT, P. D., HANNIGAN, R. E., LAYNE, G. & SHIMIZU, N. 2002. A model for continental crust genesis by arc accretion; rare earth element evidence from the Irish Caledonides. *Earth and Planetary Science Letters*, **203**, 861–877.

DUNBAR, R. B., MARTY, R. C. & BAKER, P. A. 1990. Cenozoic marine sedimentation in the Sechura and Pisco basins, Peru. *Palaeogeography, Palaeoclimatology, Palaeoecology*, **77**, 235–261.

ELLAM, R. M. & HAWKESWORTH, C. J. 1988. Is average continental crust generated at subduction zones? *Geology*, **16**, 314–317.

ELLIOTT, T., ZINDLER, A. & BOURDON, B. 1999. Exploring the kappa conundrum; the role of recycling in the lead isotope evolution of the mantle. *Earth and Planetary Science Letters*, **169**, 129–145.

FEIGENSON, M. D., CARR, M. J., MAHARAJ, S. V., JULIANO, S. & BOLGE, L. L. 2004. Lead isotope composition of Central American volcanoes: Influence of the Galapagos plume. *Geochemistry, Geophysics, Geosystems*, **5**, doi:10.1029/2003GC000621.

FULLER, C. W., WILLETT, S. D., HOVIUS, N. & SLINGERLAND, R. 2003. Erosion rates for Taiwan mountain basins: New determinations from suspended sediment records and a stochastic model of their variation. *Journal of Geology*, **111**, 71–87.

FULLER, C. W., WILLETT, S. D., FISHER, D. & LU, C. Y. 2006. A thermomechanical wedge model of Taiwan constrained by fission-track thermochronometry. *Tectonophysics*, **425**, 1–24.

GAO, S., RUDNICK, R. L. & YUAN, H.-L. 2004. Recycling lower continental crust in the North China craton. *Nature*, **432**, 892–897.

GOLDSTEIN, S. L., ARNDT, N. T. & STALLARD, R. F. 1997. The history of a continent from U–Pb ages of zircons from Orinoco River sand and Sm–Nd isotopes in Orinoco Basin river sediments. *Chemical Geology*, **139**, 271–286.

GRAHAM, J. R. 1981. Fluvial sedimentation in the Lower Carboniferous of Clew Bay, County Mayo, Ireland. *Sedimentary Geology*, **30**, 195–211.

HALLAM, A. 1992. *Phanerozoic Sea-level Changes.* Columbia University Press, New York.

HAMPEL, A. 2002. The migration history of the Nazca Ridge along the Peruvian active margin: A re-evaluation. *Earth and Planetary Science Letters*, **203**, 665–679.

HAQ, B. U., HARDENBOL, J. & VAIL, P. R. 1987. Chronology of fluctuating sea levels since the Triassic. *Science*, **235**, 1156–1167.

HARRISON, C. G. A. 1994. Rates of continental erosion and mountain building. *Geologische Rundschau*, **83**, 431–437.

HARRISON, C. G. A. 1999. Constraints on ocean volume change since the Archean. *Geophysical Research Letters*, **26**, 1913–1916.

HARTLEY, A. J. & JOLLEY, E. J. 1995. Tectonic implications of Late Cenozoic sedimentation from the Coastal Cordillera of northern Chile (22–24°S). *Journal of the Geological Society, London*, **152**, 51–63.

HARTLEY, A. J., CHONG, G., HOUSTON, J. & MATHER, A. E. 2005. 150 million years of climatic stability; evidence from the Atacama Desert, northern Chile. *Journal of the Geological Society, London*, **162**, 421–424.

HAWKESWORTH, C. J. & KEMP, A. I. S. 2006. Evolution of the continental crust. *Nature*, **443**, 811–817, doi:10.1038/nature05191.

HAYES, D. E. & LEWIS, S. D. 1984. A geophysical study of the Manila Trench, Luzon, Philippines; 1, Crustal structure, gravity, and regional tectonic evolution. *Journal of Geophysical Research*, **89**, 9171–9195.

HILDE, T. W. C. 1983. Sediment subduction versus accretion around the Pacific. *Tectonophysics*, **99**, 381–397.

HILDEBRAND, R. S. & BOWRING, S. A. 1999. Crustal recycling by slab failure. *Geology*, **27**, 11–14.

HOLBROOK, W. S., LIZARRALDE, D., MCGEARY, S., BANGS, N. & DIEBOLD, J. 1999. Structure and composition of the Aleutian island arc and implications for continental crustal growth. *Geology*, **27**, 31–34.

HUANG, C. Y., YUAN, P. B., LIN, C. W. & WANG, T. K. 2000. Geodynamic processes of Taiwan arc–continent collision and comparison with analogs in Timor, Papua New Guinea, Urals and Corsica. *Tectonophysics*, **325**, 1–21, doi:10.1016/S0040-1951(00)00128-1.

HUANG, C. Y., YUAN, P. B. & TSAO, S. H. 2006. Temporal and spatial records of active arc–continent collision in Taiwan: A synthesis. *Geological Society of America Bulletin*, **118**, 274–288.

IBARAKI, M. 2001. Neogene planktonic foraminifera of the Caleta Herradura de Mejillones section in northern Chile: Biostratigraphy and paleoceanographic implications. *Micropalaeontology*, **47**, 257–267.

JACOBSEN, S. B. 1988. Isotopic constraints on crustal growth and recycling. *Earth and Planetary Science Letters*, **90**, 315–329.

JOHNSON, M. R. W. 2002. Shortening budgets and the role of continental subduction during the India–Asia collision. *Earth-Science Reviews*, **59**, 101–123.

KELLY, R. K. 2003. *Subduction dynamics at the Middle America Trench: New constraints from swath bathymetry, multichannel seismic data, and ^{10}Be*. PhD thesis, Massachusetts Institute of Technology–Woods Hole Oceanographic Institution, Cambridge, MA–Woods Hole, MA.

KHAN, M. A., STERN, R. J., GRIBBLE, R. F. & WINDLEY, B. F. 1997. Geochemical and isotopic constraints on subduction polarity, magma sources, and palaeogeography of the Kohistan intra-oceanic arc, northern Pakistan Himalaya. *Journal of the Geological Society London*, **154**, 935–946.

KIMURA, G., SILVER, E. A. & BLUM, P. ODP LEG 170 SCIENTIFIC PARTY (eds) 1997. *Proceedings of the Ocean Drilling Program, Initial Reports, 170*. Ocean Drilling Program, College Station, TX.

KODAIRA, S., IWASAKI, T., URABE, T., KANAZAWA, T., EGLOFF, F., MAKRIS, J. & SHIMAMURA, H. 1996. Crustal structure across the middle Ryukyu Trench obtained from ocean bottom seismographic data. *Tectonophysics*, **263**, 39–60.

KREBS, W. N., ALEMAN, A. M., PADILLA, H., ROSENFELD, J. H. & NIEMEYER, H. 1992. Age and paleoceanographic significance of the Caleta Herradura diatomite, Peninsula de Mejillones, Antofagasta, Chile. *Revista Geologica de Chile*, **19**, 75–81.

LAMB, S. & DAVIS, P. 2003. Cenozoic climate change as a possible cause for the rise of the Andes. *Nature*, **425**, 792–797.

LAURSEN, J., SCHOLL, D. W. & VON HUENE, R. 2002. Neotectonic deformation of the central Chile margin: Deepwater forearc basin formation in response to hot spot ridge and seamount subduction. *Tectonics*, **21**, 1038(5), doi:10.1029/2001TC901023.

LEE, J. C., CHU, H.-T., ANGELIER, J., CHAN, Y.-C., HU, J.-C., LU, C.-Y. & RAU, R.-J. 2002. Geometry and structure of northern surface ruptures of the 1999 Mw = 7.6 Chi-Chi, Taiwan earthquake: Influence from inherited fold belt structures. *Journal of Structural Geology*, **24**, 173–192.

LEECH, M. & STOCKLI, D. F. 2000. The late exhumation history of the ultrahigh-pressure Maksyutov Complex, south Ural Mountains, from new apatite fission track data. *Tectonics*, **19**, 153–167.

LEEMAN, W. P., CARR, M. J. & MORRIS, J. D. 1994. Boron geochemistry of the Central American volcanic arc; constraints on the genesis of subduction-related magmas. *Geochimica et Cosmochimica Acta*, **58**, 149–168.

LE PICHON, X., HENRY, P. & LALLEMANT, S. 1993. Accretion and erosion in subduction zones: The role of fluids. *Annual Review of Earth and Planetary Sciences*, **21**, 307–331.

LE ROUX, J. P., GÓMEZA, C. V. ET AL. 2005. Neogene–Quaternary coastal and offshore sedimentation in north central Chile: Record of sea-level changes and implications for Andean tectonism. *Journal of South American Earth Sciences*, **19**, 83–98.

LIN, A. T., WATTS, A. B. & HESSELBO, S. P. 2003. Cenozoic stratigraphy and subsidence history of the South China Sea margin in the Taiwan region. *Basin Research*, **15**, 453–478.

LIU, T. K., CHEN, Y. G., CHEN, W. S. & JIANG, S. H. 2000. Rates of cooling and denudation of the Early Penglai Orogeny, Taiwan, as assessed by fission-track constraints. *Tectonophysics*, **320**, 69–82.

LUNDBERG, N., REED, D. L., LIU, C.-S. & LIESKE, J. 1997. Forearc–basin closure and arc accretion in the submarine suture zone south of Taiwan. *Tectonophysics*, **274**, 5–23.

MARQUARDT, C., LAVENU, A., ORTLIEB, L., GODOYA, E. & COMTE, D. 2004. Coastal neotectonics in southern Central Andes; uplift and deformation of marine terraces in northern Chile (27°S). *Tectonophysics*, **394**, 193–219.

MCLENNAN, S. M. & TAYLOR, S. R. 1982. Geochemical constraints on the growth of the continental crust. *Journal of Geology*, **90**, 347–361.

MELNICK, D. & ECHTLER, H. P. 2006. Inversion of forearc basins in south–central Chile caused by rapid glacial age trench fill. *Geology*, **34**, 709–712.

MESCHEDE, M., ZWEIGEL, P. & KIEFER, E. 1999. Subsidence and extension at a convergent plate margin: Evidence for subduction erosion off Costa Rica. *Terra Nova*, **11**, 112–117.

MÉTIVIER, F., GAUDEMER, Y., TAPPONNIER, P. & KLEIN, M. 1999. Mass accumulation rates in Asia during the Cenozoic. *Geophysical Journal International*, **137**, 280–318.

MILLER, H. 1970. *Vergleichende Studien an praemesozoischen Gesteinen Chiles unter besonderer Beruecksichtigung ihrer Kleintektonik*. Schweizerbart Stuttgart.

MILLIMAN, J. D. 1997. Fluvial sediment discharge to the sea and the importance of regional tectonics. *In*: RUDDIMAN, W. F. (ed.) *Tectonic Uplift and Climate Change*. Plenum, New York, 239–257.

MOORBATH, S. 1978. Age and isotope evidence for the evolution of the continental crust. *Philosophical Transactions of the Royal Society of London, Series A*, **288**, 401–413.

MORRIS, J., VALENTINE, R. & HARRISON, T. 2002. ^{10}Be imaging of sediment accretion and subduction along the northeast Japan and Costa Rica convergent margins. *Geology*, **30**, 59–62.

MOUNTAIN, G. S. & PRELL, W. L. 1990. A multiphase plate tectonic history of the southeast continental margin of Oman. *In*: ROBERTSON, A. H. F., SEARLE, M. P. & RIES, A. C. (eds) *The Geology and Tectonics of Oman*. Geological Society, London, Special Publications, **49**, 725–743.

MURAUCHI, S. 1971. The renewal of island arcs and the tectonics of marginal seas. *In*: ASANO, S. & UDINTSEV, G. B. (eds) *Island Arc and Marginal Seas*. Tokai University Press, Tokai, 39–56.

NISBET, E. G. 1987. *The Young Earth; an Introduction to Archaean Geology*. Allen & Unwin, London.

O'NIONS, R. K., EVENSEN, N. M. & HAMILTON, P. J. 1979. Geochemical modelling of mantle differentiation and crustal growth. *Journal of Geophysical Research*, **84**, 6091–6101.

ORTLIEB, L., ZAZO, C., GOY, J. L., DABRIO, C. & MACHARÉ, J. 1996. Pampa del Palo; ananomalous composite marine terrace on the uprising coast of southern Peru. *Journal of South American Earth Sciences*, **9**, 367–379.

PATINO, L. C., CARR, M. J. & FEIGENSON, M. D. 2000. Local and regional variations in Central American arc lavas controlled by variations in subducted sediment input. *Contributions to Mineralogy and Petrology*, **138**, 265–283.

PENG, T.-H., LI, Y.-H. & WU, F. T. 1977. Tectonic uplift rates of the Taiwan Island since the early Holocene. *Geological Society of China Memoir*, **2**, 57–69.

PLANK, T. & LANGMUIR, C. H. 1988. An evaluation of the global variations in the major element chemistry of arc basalts. *Earth and Planetary Science Letters*, **90**, 349–370.

PLATT, J. P. 1986. The mechanics of frontal imbrication: A first-order analysis. *Geologische Rundschau*, **77**, 577–589.

RANERO, C. R. & VON HUENE, R. 2000. Subduction erosion along the Middle America convergent margin. *Nature*, **404**, 748–752.

RATSCHBACHER, L., HACKER, B. R. ET AL. 2000. Exhumation of the ultrahigh-pressure continental crust in east central China; Cretaceous and Cenozoic unroofing and the Tan-Lu Fault. *Journal of Geophysical Research*, **105**, 13303–13338.

REAGAN, M., MORRIS, J., HERRSTROM, E. & MURRELL, M. 1994. Uranium series and beryllium isotopic evidence for an extended history of subduction modification of the mantle below Nicaragua. *Geochimica et Cosmochimica Acta*, **58**, 4199–4212.

RUDNICK, R. L. 1995. Making continental crust. *Nature*, **378**, 573–578.

RUDNICK, R. L. & FOUNTAIN, D. M. 1995. Nature and composition of the continental crust; a lower crustal perspective. *Reviews of Geophysics*, **33**, 267–309.

RUTLAND, R. W. R. 1971. Andean orogeny and ocean floor spreading. *Nature*, **233**, 252–255.

RYAN, H. F. & SCHOLL, D. W. 1993. Geologic implications of great interplate earthquakes along the Aleutian arc. *Journal of Geophysical Research*, **98**, 22135–22146.

SCHEUBER, E. & REUTTER, K. J. 1992. Magmatic arc tectonics in the Central Andes between 21° and 25°S. *Tectonophysics*, **205**, 127–140, doi:10.1016/0040-1951(92)90422-3.

SCHOLL, D. W., MARLOW, M. S. & COOPER, A. K. 1977. Sediment subduction and offscraping at Pacific margins. *In*: TALWANI, M. & PITMAN, W. C. (eds) *Island Arcs, Deep Sea Trenches and Back-arc Basins*. American Geophysical Union, Maurice Ewing Series, **1**, 199–210.

SCHUBERT, G. & REYMER, A. P. S. 1985. Continental volume and freeboard through geological time. *Nature*, **316**, 336–339.

SHIPLEY, T. H., MCINTOSH, K. D., SILVER, E. A. & STOFFA, P. L. 1992. Three-dimensional imaging of the Costa Rica accretionary prism: Structural diversity in a small volume of the lower slope. *Journal of Geophysical Research*, **97**, 4439–4459.

SIBUET, J.-C. & HSU, S.-K. 1997. Geodynamics of the Taiwan arc–arc collision. *Tectonophysics*, **274**, 221–251.

SONG, S.-R. & LO, H.-J. 2002. Lithofacies of volcanic rocks in the central Coastal Range, eastern Taiwan; implications for island arc evolution. *Journal of Asian Earth Sciences*, **21**, 23–38.

STEIN, M. & HOFMANN, A. W. 1994. Mantle plumes and episodic crustal growth. *Nature*, **372**, 63–68.

STERN, C. R. 2004. Active Andean volcanism: Its geologic and tectonic setting. *Revista Geologica de Chile*, **31**, 161–206.

STERN, R. J. & BLOOMER, S. H. 1992. Subduction zone infancy: Examples from the Eocene Izu–Bonin–Mariana and Jurassic California arcs. *Geological Society of America Bulletin*, **104**, 1621–1636.

SUPPE, J. 1981. Mechanics of mountain building and metamorphism in Taiwan. *Geological Society of China, Memoir*, **4**, 67–89.

SUPPE, J. 1984. Kinematics of arc–continent collision, flipping of subduction, and backarc spreading near Taiwan. *In*: TSAN, S. F. (ed.) *A special volume dedicated to Chun-Sun Ho on the occasion of his retirement*. Geological Society of China, Memoir, **6**, 21–33.

SUYEHIRO, K., TAKAHASHI, N. ET AL. 1996. Continental crust, crustal underplating, and low-Q upper mantle beneath an oceanic island arc. *Science*, **272**, 390–392.

TAYLOR, S. R. & MCLENNAN, S. M. 1995. The geochemical evolution of the continental crust. *Reviews of Geophysics*, **33**, 241–265.

TENG, L. S. 1990. Geotectonic evolution of late Cenozoic arc–continent collision in Taiwan. *Tectonophysics*, **183**, 57–76.

TENG, L. S. 1996. Extensional collapse of the northern Taiwan mountain belt. *Geology*, **24**, 949–952.

TENG, L. S., LEE, C.-T., TSAI, Y.-B. & HSIAO, L.-Y. 2000. Slab breakoff as a mechanism for flipping of subduction polarity in Taiwan. *Geology*, **28**, 155–158.

TRELOAR, P. J., PETTERSON, M. G., JAN, M. Q. & SULLIVAN, M. A. 1996. A re-evaluation of the stratigraphy and evolution of the Kohistan arc sequence, Pakistan Himalaya: Implications for magmatic and tectonic arc-building processes. *Journal of the Geological Society, London*, **153**, 681–693.

TSAO, S. H., LI, T.-C., TIEN, J.-L., CHEN, C.-H., LIU, T.-K. & CHEN, C.-H. 1992. Illite crystallinity and fission track ages along the east central cross-island highway of Taiwan. *Acta Geologica Taiwanica*, **30**, 45–64.

TSUCHI, R. 1992. Neogene events in Japan and on the Pacific coast of South America. *Revista Geologica de Chile*, **19**, 67–73.

VALENTINE, R. B., MORRIS, J. D., DUNCAN, D., JR. & ODP SCIENTIFIC PARTY LEG. 170 1997. Sediment subduction, accretion, underplating and arc volcanism along the margin of Costa Rica: Constraints from Ba, Zn, Ni, and ^{10}Be concentrations. *EOS Transactions, American Geophysical Union*, **78**, 673.

VANNUCCHI, P., SCHOLL, D. W., MESCHEDE, M. & MCDOUGALL-REID, K. 2001. Tectonic erosion and consequent collapse of the Pacific margin of Costa Rica: Combined implications from ODP Leg 170, seismic offshore data, and regional geology of the Nicoya Peninsula. *Tectonics*, **20**, 649–668.

VANNUCCHI, P., RANERO, C. R., GALEOTTI, S., STRAUB, S. M., SCHOLL, D. W. & MCDOUGALL-RIED, K. 2003. Fast rates of subduction erosion along the Costa Rica Pacific margin: Implications for nonsteady rates of crustal recycling at subduction zones. *Journal of Geophysical Research—Solid Earth*, **108**, 2511, doi:10.1029/2002JB002207.

VANNUCCHI, P., GALEOTTI, S., CLIFT, P. D., RANERO, C. R. & VON HUENE, R. 2004. Long-term subduction–erosion along the Guatemalan margin of the Middle America Trench. *Geology*, **32**, 617–620.

VANNUCCHI, P., REMITTI, F. & BETTELLI, G. 2008. Geological record of fluid flow and seismogenesis along an erosive subducting plate boundary. *Nature*, **451**, 699–704, doi:10.1038/nature06486.

VITA-FINZI, C. & LIN, J. C. 1998. Serial reverse and strike slip on imbricate faults: The Coastal Range of east Taiwan. *Geology*, **26**, 279–282.

VON HUENE, R. & LEE, H. 1982. The possible significance of pore fluid pressures in subduction zones. *In*: WATKINS, J. S. & DRAKE, C. L. (eds) *Studies in Continental Margin Geology*. American Association of Petroleum Geologists, Memoirs, **34**, 781–791.

VON HUENE, R. & RANERO, C. R. 2003. Subduction erosion and basal friction along the sediment-starved convergent margin off Antofagasta, Chile. *Journal of Geophysical Research*, **108**, 2079, doi:10.1029/2001JB001569.

VON HUENE, R. & SCHOLL, D. W. 1991. Observations at convergent margins concerning sediment subduction, subduction erosion, and the growth of continental crust. *Reviews of Geophysics*, **29**, 279–316.

VON HUENE, R., LANGSETH, M., NASU, N. & OKADA, H. 1982. A summary of Cenozoic tectonic history along the IPOD Japan trench transect. *Geological Society of America Bulletin*, **93**, 829–846.

VON HUENE, R., RANERO, C. & VANNUCCHI, P. 2004. Generic model of subduction erosion. *Geology*, **32**, 913–916.

WANG, T.-K., LIN, S.-F., LIU, C.-S. & WANG, C.-S. 2004. Crustal structure of southernmost Ryukyu subduction zone; OBS, MCS and gravity modelling. *Geophysical Journal International*, **157**, 147–163.

WELLS, R. E., BLAKELY, R. J., SUGIYAMA, Y., SCHOLL, D. W. & DINTERMAN, P. A. 2003. Basin-centered asperities in great subduction zone earthquakes: A link between slip, subsidence, and subduction erosion? *Journal of Geophysical Research*, **108**, doi:10.1029/2002JB002072.

WILLETT, S., BEAUMONT, C. & FULLSACK, P. 1993. Mechanical model for the tectonics of doubly vergent compressional orogens. *Geology*, **21**, 371–374.

WILLETT, S. D., FISHER, D., FULLER, C., CHAO, Y.-E. & YU, L.-C. 2003. Erosion rates and orogenic-wedge kinematics in Taiwan inferred from fission-track thermochronometry. *Geology*, **31**, 945–948.

WISE, D. U. 1974. Continental margins, freeboard and the volumes of continents and oceans through time. *In*: BURK, C. A. & DRAKE, C. L. (eds) *The Geology of Continental Margins*. Springer, New York, 45–58.

ZHANG, P., MOLNAR, P. & DOWNS, W. R. 2001. Increased sedimentation rates and grain sizes 2–4 Myr ago due to the influence of climate change on erosion rates. *Nature*, **410**, 891–897.

Implications of estimated magmatic additions and recycling losses at the subduction zones of accretionary (non-collisional) and collisional (suturing) orogens

DAVID W. SCHOLL[1,2]* & ROLAND VON HUENE[3]

[1]*University of Alaska Fairbanks, Fairbanks, AK 99775, USA*
[2]*U.S. Geological Survey, 345 Middlefield Road, Menlo Park, CA 94025, USA*
[3]*University of California Davis, Davis, CA 95616, USA*
*Corresponding author (e-mail: dscholl@usgs.gov)

Abstract: Arc magmatism at subduction zones (SZs) most voluminously supplies juvenile igneous material to build rafts of continental and intra-oceanic or island arc (CIA) crust. Return or recycling of accumulated CIA material to the mantle is also most vigorous at SZs. Recycling is effected by the processes of sediment subduction, subduction erosion, and detachment and sinking of deeply underthrust sectors of CIA crust. Long-term (>10–20 Ma) rates of additions and losses can be estimated from observational data gathered where oceanic crust underruns modern, long-running (Cenozoic to mid-Mesozoic) ocean-margin subduction zones (OMSZs, e.g. Aleutian and South America SZs). Long-term rates can also be observationally assessed at Mesozoic and older crust-suturing subduction zone (CSSZs) where thick bodies of CIA crust collided in tectonic contact (e.g. Wopmay and Appalachian orogens, India and SE Asia). At modern OMSZs arc magmatic additions at intra-oceanic arcs and at continental margins are globally estimated at $c.$ 1.5 AU and $c.$ 1.0 AU, respectively (1 AU, or Armstrong Unit, $= 1 \text{ km}^3 \text{ a}^{-1}$ of solid material). During collisional suturing at fossil CSSZs, global arc magmatic addition is estimated at 0.2 AU. This assessment presumes that in the past the global length of crustal collision zones averaged $c.$ 6000 km, which is one-half that under way since the early Tertiary. The average long-term rate of arc magmatic additions extracted from modern OMSZs and older CSSZs is thus evaluated at 2.7 AU. Crustal recycling at Mesozoic and younger OMSZs is assessed at $c.$ 60 km^3 Ma^{-1} km^{-1} ($c.$ 60% by subduction erosion). The corresponding global recycling rate is $c.$ 2.5 AU. At CSSZs of Mesozoic, Palaeozoic and Proterozoic age, the combined upper and lower plate losses of CIA crust via subduction erosion, sediment subduction, and lower plate crustal detachment and sinking are assessed far less securely at $c.$ 115 km^3 Ma^{-1} km^{-1}. At a global length of 6000 km, recycling at CSSZs is accordingly $c.$ 0.7 AU. The collective loss of CIA crust estimated for modern OMSZs and for older CSSZs is thus estimated at $c.$ 3.2 AU. SZ additions (+2.7 AU) and subtractions (−3.2 AU) are similar. Because many uncertainties and assumptions are involved in assessing and applying them to the deep past, the net growth of CIA crust during at least Phanerozoic time is viewed as effectively nil. With increasing uncertainty, the long-term balance can be applied to the Proterozoic, but not before the initiation of the present style of subduction at $c.$ 3 Ga. Allowing that since this time a rounded-down rate of recycling of 3 AU is applicable, a startlingly high volume of CIA crust equal to that existing now has been recycled to the mantle. Although the recycled volume ($c.$ 9 × 10^9 km^3) is small ($c.$ 1%) compared with that of the mantle, it is large enough to impart to the mantle the signature of recycled CIA crust. Because subduction zones are not spatially fixed, and their average global lengths have episodically been less or greater than at present, recycling must have contributed significantly to creating recognized heterogeneities in mantle geochemistry.

Juvenile igneous material supplied largely by arc, hotspot, spreading ridge, and rift-related magmatism accumulates magmatically and tectonically to form rafts of continental and island or intra-oceanic arc (CIA) crust (Table 1). Although the tectonic accretion of oceanic crust to the edges of CIA masses adds juvenile material, in particular the large igneous provinces (LIPs) of oceanic plateaux (Table 1), the great bulk of growth occurs through arc magmatism at subduction zones (SZs) (Reymer & Schubert 1984; Rudnick & Gao 2003; Condie 2005, 2007; O'Neill *et al.* 2007; Whitmeyer & Karlstrom 2007; Stern 2008). Arc magmatism has been under way since at least the initiation of modern subduction systems at $c.$ 3 Ga (van Kranendonk 2004; van Kranendonk & Cassidy 2004; Condie 2005, 2007; van der Velden & Cook 2005; Cawood *et al.* 2006). Forms of subduction (i.e. crustal foundering) also prevailed prior to this time (Condie 2005).

Table 1. *Terms of reference*

Accretionary (non-collisional) orogen: An orogen formed in the upper plate at OMSZs that is not caused by regional-scale collision of large masses of CIA crust. A modern example of an accretionary orogen is the Andean system; a fossil example is the Tasman orogen of SE Australia (Fig. 4; Collins & Richards 2008). Localized collisional tectonism is typical of accretionary orogens because they commonly involved along-margin transpressive processes and subduction zone collision between upper and lower plate crustal blocks (e.g. the Yakutat block with SE Alaska, Fig. 4).

AU: Armstrong Unit. A solid volume of $1 \text{ km}^3 \text{ a}^{-1}$ of juvenile igneous rock added to the continental mass or of older crustal material removed and recycled to the mantle. $1 \text{ AU} = 1 \text{ km}^3 \text{ a}^{-1}$.

CIA crust: Thick (>20–30 km) sequences of continental crust and the magmatic constructs of offshore, intra-oceanic arcs.

Collisional orogen: An orogen formed at an SZ as a consequence of the collision of thick sections of upper and lower plate CIA crust or of CIA and LIP crustal bodies. Modern examples at CSSZs include the continental-scale India–Asia collisional orogen of the Himalaya. In comparison, modern examples at OMSZs include the laterally less extensive Yakutat orogen of SE Alaska and the offshore Taiwan orogen of south–central Asia (Fig. 4).

Continental margin (Andean) arc–subduction zone: An arc massif–subduction zone formed at a continental margin. Modern examples include the SZs and growing arc massifs of Kamchatka, Alaska, Cascadia and Mexico–America (Fig. 4).

Crustal growth at subduction zones (SZs): The arithmetic, long-term sum or net of arc magmatic additions less recycling losses at ocean-margin and crust-suturing subduction zones (OMSZs and CSSZs). Not included in these estimates are tectonic additions by the accretion of LIPs and fragments of oceanic crust or the delamination of thickened upper plate crust.

CSSZ: Crust-suturing subduction zone. CSSZs are where thick sections of upper and lower plate crust meet in collisional contact. Examples of modern CSSZs include the colliding and suturing crustal masses of Australia with eastern Indonesia and India with SE Asia (Fig. 4). Less laterally extensive CSSZs are, potentially, where thick bodies of lower plate crust enter modern OMSZs; for example, where the far western block of the Aleutian Arc collides with Kamchatka (Fig. 4).

Frontal prism: Landward thickening, prism-shaped mass of actively deforming (chiefly shortening) material forming the lower or seaward toe of the forearc. Frontal prisms are typically 10–40 km wide and constructed dominantly of either accreted lower plate ocean-floor sediment or downslope displaced debris shed from the forearc.

IBM: The Izu–Bonin–Mariana SZ and arc massif system of the Pacific's western rim (Fig. 4).

Inner prism: The margin's older, landward body of little deforming basement rock. Inner prisms are always present at convergent margin and commonly exposed in the coastal area. Younger frontal or middle prisms are added to the seaward terminus of the inner prism (Figs 2 and 3).

Intra-oceanic arc–subduction zone: An arc massif–subduction zone that forms in an offshore position within older, pre-existing oceanic crust (e.g. the Mariana and Aleutian systems). Most modern intra-oceanic arc–trench systems formed in the early to middle Tertiary.

LIP: Large igneous province, in particular of the crustally thick (20–30 km) magmatic piles of oceanic plateaux and their trailing aseismic ridges. Most LIPs and linked aseismic ridges are produced at plume-nourished hotspots (HS in Fig. 4). Modern examples of LIPs entering OMSZs include the Ontong–Java Plateau of the SW Pacific and the Hikurangi Plateau off NE New Zealand (Fig. 4). Modern examples of their subducting aseismic ridges include the Emperor Ridge at the Kamchatka SZ, the Nazca Ridge at the Peru SZ, and the Juan Fernandez Ridge at the Chile SZ (Fig. 4).

Middle prism: Tectonic accumulations of older, little deforming accreted lower plate rock and sediment added to the seaward terminus of the margin's inner prism. The middle prism, where formed, separates the inner and frontal prisms (Fig. 3). Middle prism evidently forms only at accreting margins (at present *c*. 25% of all OMSZs).

OMSZ: Ocean-margin subduction zone. OMSZs are convergent continental or intra-oceanic margins underthrust by a lower plate of subducting oceanic lithosphere. Modern examples include the intra-oceanic Aleutian and the continental margin Andean SZs. Accretionary orogens form at OMSZs.

(Continued)

Table 1. *Continued*

Sediment subduction: Describes the landward, subsurface transport of ocean-floor sediment below the margin's middle and inner prisms. Subducted sediment is transported within the subduction channel and is either underplated landward of the frontal prism or carried into the mantle with the subducting lower plate the (Figs 1–3).

Subduction channel: Describes a relatively thick, typically 0.5–1.0 km or greater, section of sediment and tectonically eroded forearc debris that separates the stronger material of the upper and lower plates. Material in the subduction channel is driven downward and toward the mantle by the subducting lower plate (Figs 1–3; Cloos & Shreve 1988a, b). Examples of seismic images of the subduction channel have been shown for the Costa Rica margin by Ranero & von Huene (2000), for the Ecuador and Colombia subduction zones by Sage *et al.* (2006) and Collot *et al.* (2008), for the Nankai margin by Park *et al.* (2002) and Moore *et al.* (2007), and for the southern Sumatra margin by Kopp *et al.* (2001).

Subduction erosion: Describes the frontal (seaward edge of margin) or basal (beneath the margin) removal of upper plate material by the underthrusting action of the lower plate. Landward and subsurface transport of the dislodged debris is effected by the subduction channel. Detached forearc material can either underplate the forearc in the coastal area or be recycled to the mantle (von Huene & Lallemand 1990).

SZ: Subduction zone.

Because of concern about their meaningfulness, rates of additions and losses extracted from modern-style active and fossil SZs are not applied to the bulk of Archaean time.

The dominant process of generating new CIA crust occurs at ocean-margin subduction zones (OMSZs). Convergent margins of OMSZs are underrun by subducting oceanic crust (Fig. 1).

JUVENILE MAGMATIC PRODUCTIVITY AT OCEAN-MARGIN SUBDUCTION ZONES (OMSZs)

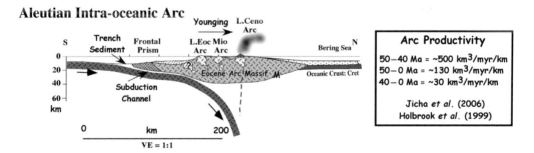

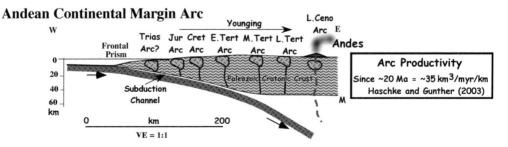

Fig. 1. Diagrams showing arc magmatic additions at ocean-margin subduction zones (OMSZs). Arc building is highest at intra-oceanic arcs, and in particular for the first c. 10 Ma after the establishment of a new subduction zone (Stern & Bloomer 1992; Holbrook *et al.* 1999; Jicha *et al.* 2006). Episodes of widespread initiation of new subduction zones will correspond to periods of rapid growth in the Earth's inventory of continental and island arc crust (Condie 2005; O'Neill *et al.* 2007), which is at presently $c.\ 7 \times 10^9$ km^3.

OMSZ, occur at intra-oceanic arcs (e.g. the modern Mariana and Aleutian SZs) and at continental margins (e.g. the modern Peru–Chile and Japan SZs) (see Fig. 4). Accretionary or non-collisional orogens commonly form at OMSZs; for example, at the modern and growing Andean orogen, and at fossil OMSZs that constructed the Mesozoic and early Tertiary North America Cordillera, and, in the Palaeozoic, the Tasman orogen of SE Australia (Fig. 4; Collins & Richards 2008) and the highly obliquely deformed Devonian to late Jurassic Pacific margin of western North America (Ernst et al. 2008). Arc magmatism also occurs, although less voluminously, at crust-suturing subduction zones (CSSZs) where large bodies of CIA crust are in tectonic contact and construct collisional orogens (e.g. Africa and Eurasia, India and southern Asia) (Figs 3 and 4; Table 1).

Subduction zones are the principal setting not only for additions of CIA crust but also for its removal and recycling to the mantle (Figs 2 and 3). Recycling at OMSZs is caused by the linked, co-tectonic processes of sediment subduction and subduction erosion (von Huene & Scholl 1991; Clift & Vannucchi 2004; Scholl & von Huene 2007, in press; Clift et al. 2009). Sediment subduction describes the transport of ocean-floor sediment entering the subduction zone landward beneath the margin's older rock framework and toward the mantle (Table 1, Figs 2 and 3). Subduction erosion describes the removal of upper plate material and its subsurface transport landward and toward the mantle. Transport is carried out by the

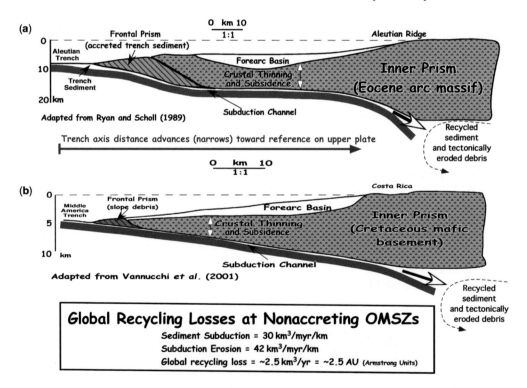

Fig. 2. Diagrams of long-term, nonaccreting margins of ocean-margin subduction zones (OMSZs) listing recycling losses attributed to sediment subduction and subduction erosion (see Table 2). Nonaccreting margins are structurally characterized by the close approach (within 5–30 km) of the rock framework of the inner prism to the base of the forearc slope (trench floor). Over periods >5–10 Ma the base of the slope migrates (narrows) toward a fixed, onshore reference (Clift & Vannucchi 2004). Nonaccreting margins include (**a**) those that have a non-widening frontal prism of accreted trench deposits of glacial age (adapted from Ryan & Scholl 1989), and (**b**) margins with a narrow frontal prism constructed of slope material and debris (adapted from Vannucchi et al. 2001). Nonaccreting margins make up at least c. 75% of all OMSZs (Scholl & von Huene 2007).

LONG-TERM RECYCLING AT ACCRETING MARGINS

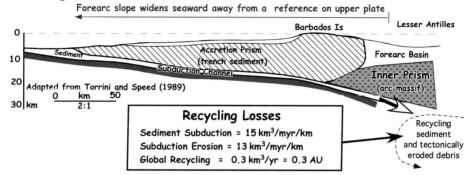

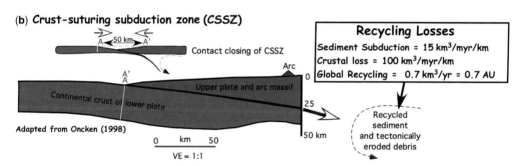

Fig. 3. Diagrams of an accreting (**a**) ocean-margin subduction zone (OMSZ) and (**b**) accreting crust-suturing subduction zone (CSSZ), listing recycling losses attributed to sediment subduction, subduction erosion, and slab breakoff (**b**). Accreting margins of OMSZs (a) are structurally characterized by a wide (>50 km) separation between the seaward edge of the inner prism of basement rock and a well-sedimented trench floor. The separating rock body, the middle prism, is evidently constructed of accumulated masses of older frontal prisms and underplated sediment. Over periods >5–10 Ma the submerged forearc migrates (widens) away from a fixed onshore reference (Clift & Vannucchi 2004). Accreting margins construct c. 25% of all OMSZs (Scholl & von Huene 2007). All are characterized by relatively slow convergence (c. 50 km Ma^{-1}) and the long-sustained (30–50 Ma) tectonic injection of a thick (2–3 km) sequence of lower plate sediment into the SZ. Accreting CSSZs are those at which contact closure of the thick crustal bodies of upper and lower plates results in the construction of a collisional orogen. Closure, which eventually occludes the subduction zone, typically takes several tens of millions of year. (b) is based on a mid-European collision zone of late Carboniferous age that exhibits evidence of forearc thinning and subsidence, both of which were attributed by Oncken (1998) to subduction erosion. The likelihood that a lengthy sector of underthrust lower plate crust was dragged into the mantle with the subducted ocean lithosphere is based on studies of the Palaeoproterozoic Wopmay orogen of NW Canada by Hildebrand & Bowring (1999) (Fig. 4).

subduction channel that separates the rock framework of the upper and lower plates (Table 1; Figs 1 and 2; Cloos & Shreve 1988a, b).

At CSSZs, during the initial phase of colliding CIA bodies, material losses by subduction erosion and probably also sediment subduction occur (Wagreich 1993; Oncken 1998). However, at these subduction zones larger tracts of CIA crust are destroyed by deep underthrusting, detachment, and foundering into the deeper mantle (Hildebrand & Bowring 1999).

Because the greater additions and losses of CIA crust occur at SZs, the question arises of the long-term (over tens to hundreds of millions of years) material balance between arc massif building and destructive crustal losses. It is debated whether, since the late Archaean, the Earth's volumetric store of CIA crust remained much the same, as posited by Armstrong (1981, 1991), or has been progressively expanding or even diminishing, as hypothesized by many others (see discussions by Sylvester (2000) and Condie (2005)).

Based on offshore and onshore observations, this paper attempts to derive a working estimate of the long-term addition and loss of CIA material at OMSZs and at CSSZs. These assessment are used to explore the implications for the net growth of CIA crust since c. 3 Ga. Estimates are reported in Armstrong Units (AU), as defined by Kay & Kay (2009), where 1 AU equals a solid-volume rate of 1 km^3 a^{-1} (Table 1).

The estimated long-term or average volumetric magmatic gains (+2.7 AU) and recycling losses (−3.2 AU) registered at Cenozoic and late Mesozoic subduction zones are similar. In view of the many uncertainties and linked assumptions involved in estimating these global rates, the difference of 0.5 AU is not considered significant. The perspective is thus gained that a long-term balance has been maintained at least during the Phanerozoic and most probably also the Proterozoic, if not to the inception of the modern subduction system at c. 3 Ga.

Extending the assessment of long-term balance back in time prior to the mid-Mesozoic calls for two major assumptions: (1) that the Cenozoic distribution and length of subduction zone types (OMSZs and CSSZs) is representative of the average Phanerozoic and Proterozoic conditions; (2) that arc production and recycling rates assessed at currently active Cenozoic OMSZs (Scholl & von Huene 2007) and three fossil CSSZs (Wagreich 1993; Oncken 1998; Hildebrand & Bowring 1999) are meaningfully typical for the period since modern subduction began, at c. 3 Ga (or whenever this happened; see Cawood et al. 2006).

However, it is increasingly clear that global arc additions did not proceed at a steady pace, but proceeded episodically at extremely high or much reduced rates (Condie 1998, 2004, 2005; O'Neill et al. 2007; Stern 2008; Silver & Behn 2008). Recycling losses at SZs also do not proceed at a steady pace but episodically at elevated and reduced rates (von Huene & Scholl 1991; Clift & Vannucchi 2004; Kay et al. 2005).

Why a global balance might exist can speculatively be tied to the circumstance that rates of plate movements, creation of oceanic crust, and convergence at subduction zones are proportional to rates of arc magmatism and recycling losses there. Continental additions by hotspot and rift volcanism and accretion of oceanic, LIP or back-arc crust and losses by lower-crust delamination are not included in the assessment that a long-term balance has been struck at SZs.

Volume of arc magmatic material added at subduction zones

At modern OMSZs

In recent years, estimates of arc magmatic productivity that summed magmatic growth at newly formed (Eocene and younger) intra-oceanic SZs and longer-established continental margin or Andean SZs, centred about c. 30 km^3 Ma^{-1} km^{-1} of arc (Reymer & Schubert 1984). The corresponding global rate for OMSZs c. 42 000 km in length is c. 1.3 AU. However, initial arc magmatism at newly formed intra-ocean SZs is initially at least an order of magnitude higher at 300 km^3 Ma^{-1} km^{-1}, a rate equivalent to that of a slow-spreading ridge (Stern & Bloomer 1992; Stern 2004; Jicha et al. 2006). A global estimate of arc productivity over long time periods thus requires separate assessments of productivity of long-established or mature SZs of continental margins or edges and that of newly formed SZs, in particular of intra-oceanic arcs that exhibit prodigious rates of start-up magmatism (Figs 1 and 4). These include, for example, the Aleutian, Izu–Bonin–Mariana (IBM), and Tonga–Kermadec systems. Longer established arc systems are chiefly those of Pacific-rim continental margins; for example, the Alaska, Mexico, western South America and Japan margins (Fig. 4).

Drawing upon recently acquired geophysical data and advanced dating techniques, during the past 50 Ma the rate of magmatic growth at modern intra-oceanic arc massifs has averaged 100–160 km^3 Ma^{-1} km^{-1} (Holbrook et al. 1999; Jicha et al. 2006; Takahasi et al. 2007). These high construction speeds, 3–5 times greater than the commonly accepted value of c. 30 km^3 Ma^{-1} km^{-1}, were first recognized and documented by Stern & Bloomer (1992). The high, long-term average of intra-oceanic arcs is strongly biased by the initial phase of vigorous arc magmatism (Hawkins et al. 1984; Stern & Bloomer 1992). The Aleutian SZ, for example, produced in c. 50 Ma some 4900 km^3 of igneous crust for each kilometre of arc. Including material removed from the arc massif by subduction erosion increases the volume produced to c. 6400 km^3 Ma^{-1} km^{-1}. The long-term or lifetime production rate is thus estimated at c. 130 km^3 Ma^{-1} km^{-1} of arc (Fig. 1; Jicha et al. 2006). Because Eocene igneous units are exposed and geophysically mapped across the width of the Aleutian Ridge, the great bulk of arc building occurred between about 40 and 50 Ma (or less) at a rate estimated at c. 500 km^3 Ma^{-1} km^{-1} (Fig. 1). This magmatic outpouring is equivalent to that produced across a spreading centre opening at c. 80 km Ma^{-1} (sum of both sides).

Similar initial and longer-term rates of arc growth are characteristic of the IBM system (Stern & Bloomer 1992). Takahasi et al. (2007) estimated that juvenile magmatism during the past c. 50 Ma produced the combined arc massifs of the West Mariana and Mariana Ridges of c. 4800 km^3 km^{-1} (Fig. 3). Adding a volume of 1000 km^3 for the remnant arc of the Palau–Kyushu Ridge and subduction erosion losses of c. 1000 km^3 brings the total

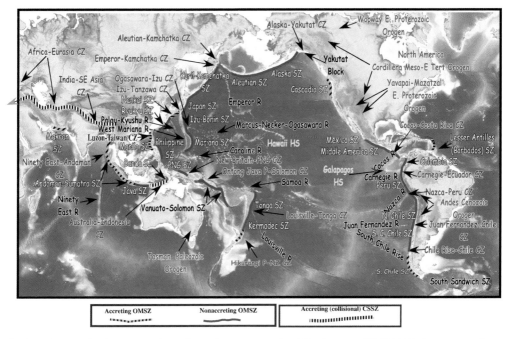

Fig. 4. Index map of major nonaccreting and accreting ocean-margin subduction zones (OMSZs). Modern nonaccreting margins have narrowed during the past 10 Ma or longer in response to subduction erosion, whereas accreting margins have widened seaward in response to subduction accretion of sea-floor sediment. Also shown are major oceanic ridges and rises (R) and plateaux or LIPs (P) entering subduction zones (SZs). Subduction of these bathymetric elements enhances rates of subduction erosion by several fold to an order of magnitude above long-term background rates (Clift & Vannucchi 2004; Clift et al. 2009). Collision sectors of OMSZs at which crustal suturing is likely are indicated (CZ) (e.g. Yakutat CZ of SE Alaska, Ontong–Java Plateau and Solomon CZ) and also the far more laterally extensive crust-suturing subduction zones (CSSZs) of colliding continental bodies (e.g. Australia–Indonesia and India–SE Asia CZs). HS, hotspot; PNG, Papua New Guinea.

volume to 6800 km^3 km^{-1}, which is virtually that of the Aleutian arc massif, and probably also the Tonga–Kermadec system (Scholl et al. 1985; Scholl & Herzer 1992, 1994). The long-term or average rate of IBM arc magmatism is thus c. 135 km^3 Ma^{-1} km^{-1}. Stern & Sherman (1992) assessed the initial subduction rate at 300 km^3 Ma^{-1} km^{-1}. However, adding stranded masses of the Eocene arc massif across the width of the Philippine Sea and subduction erosion losses raises the rate for the IBM systems to c. 500 km^3 Ma^{-1} km^{-1}, which is that of the Aleutian arc massif.

Once the great bulk of the arc massif accumulates, arc magmatism wanes and becomes organized along the magmatic front tracked by aligned trends of eruptive centres (Fig. 1; Stern & Bloomer 1992). The generally accepted value of productivity is around 30 km^3 Ma^{-1} km^{-1} of arc, a rate based on the partitioning factor that the volume of arc extrusive material is c. 25% and intrusive additions 75% of the volume of juvenile magma supplied by mantle melting (Kay & Kay 1985; Stern & Bloomer 1992). This rate is 15 times less than the initial rate based on the geophysically measured volume of the early formed arc massif (Fig. 1).

Magmatism at intra-oceanic arcs can be extinguished when they collide at CSSZs to build rafts of CIA crust; for example, the Jurassic Talkeetna–Bonanza arc system of southern Alaska (Clift et al. 2005). Arc extinction also occurs when their subduction zone is jammed by a colliding, underthrusting mass of CIA or thick oceanic crust (e.g. an LIP). The subduction zone can then shift elsewhere, as happened after the collision of the Ontong–Java Plateau with the Pacific-facing subduction zone of the Solomon arc massif (Mann & Taira 2004; Stern 2004). Similarly, the Pacific-rim collision of the offshore Olyutorsky arc complex with Kamchatka evidently initiated the formation of the offshore Aleutian subduction zone (Garver et al. 2004; Scholl 2007). Other potential examples include the continuing collision of the New Britain arc with Papua New Guinea (Cloos et al. 2005), the Luzon arc with SE Asia at Taiwan (Byrne & Liu 2002), and the Australia continental mass with Indonesia (Stern 2004) (Fig. 4). The average global rate of formation of CIA crust at intra-oceanic arcs can thus remain high if tectonic collision forces a new subduction zone, at either an intra-oceanic or a continental margin setting, to form somewhere else.

Modern intra-ocean arcs have been building their arc massifs for the past $c.$ 50 Ma. Older ones, for example, the accreted early Jurassic Talkeetna and Bonanza arcs of Alaska, were magmatically active for at least this long (Clift et al. 2005). Condie (2007) assessed the typical lifespan of oceanic arc systems before their collisional extinction at <100 Ma. Adopting an average lifespan of 70 Ma, the long-term rate of magmatic productivity for an intra-ocean arc before accretion would be $c.$ 100 km^3 Ma^{-1} km^{-1} (500 km^3 Ma^{-1} for the first 10 Ma plus 30 km^3 Ma^{-1} for 60 Ma thereafter). At this long-term rate, the global output for the present 14 500 km of intra-oceanic arcs (principally the Aleutian, South Sandwich, Kermadec, Tonga, Vanuatu, Solomon, New Britain, IBM, east Java or Banda and Antilles systems) would alone average $c.$ 1.5 AU (Fig. 4).

For the older-established arc systems of continental margins, initial high rates of arc magmatic productivity were progressively muted by their long-continued activity. For example, since the mid-Cretaceous, productivity estimates for the Andean system range from 4 to 35 km^3 Ma^{-1} km^{-1} (Clift et al. 2003; Kukowski & Oncken 2006). The estimate of Haschke & Gunter (2003) of 35 km^3 Ma^{-1} km^{-1} for Andean magmatism since the middle Miocene, which is effectively the globally general rate identified by Reymer & Schubert (1984), is adopted as a useful long-term rate applicable to continental margin productivity. This rate is applied to continental margins characterized by non-collisional or accretionary orogenesis, such as the North America Cordillera. The combined length of continental margin arc systems at modern OMSZs is $c.$ 27 500 km (Scholl & von Huene 2007). The corresponding global addition rate is thus $c.$ 1.0 AU. Combined with the estimates for offshore arcs (1.5 AU), the long-term addition of arc magma at OMSZs of dominantly Cenozoic age is thus assessed at $c.$ 2.5 AU.

At CSSZs

Arc magmatic additions also accompany contact suturing at CSSZs, where regional-scale contact or collisional orogenesis occurs. During the Cenozoic, these subduction zones principally included the $c.$ 10 000 km long Mediterranean–Indian or neo-Tethyan suturing of Cenozoic age and the current $c.$ 2000 km long Australia–Indonesian collision belt. Along OMSZs far less lengthy CSSZs sectors include where the Yakutat block plunges northward into the eastern end of the Alaska SZ (Eberhart-Phillips et al. 2006; Gulick et al. 2007; Fuis et al. 2008; Wallace 2008), and the Ontong–Java Plateau, the remnants of Pacific-facing or northern Solomon SZ (Mann & Taira 2004). Other potential CSSZ sectors of OMSZs are, as mentioned above, the New Britain arc massif with Papua New Guinea, the Luzon arc with Taiwan, the Izu arc body with the Tanzawa massif of central Japan, and the Aleutian arc with Kamchatka (Fig. 4).

In general, arc volcanism is not voluminous where suturing collision is under way, a process that can magmatically continue for 50 Ma (Hoffman & Bowring 1984; Cook et al. 1998; Oncken 1998). Applying the Andean arc production rate of $c.$ 35 km^3 Ma^{-1} km^{-1} for each kilometre of lengthy CSSZs, and allowing that the time-average length of CSSZs is 6000 km (at present $c.$ 12 000 km), then a volume of arc productivity of $c.$ 0.2 AU should be added to the $c.$ 1.0 AU estimated for Andean magmatism at OMSZs. The global output at continental margin arcs is thus assessed at $c.$ 1.2 AU.

Compiled additions

In total, at OMSZs at which non-collisional, accretionary orogenesis occurs, arc productivity is estimated to have globally averaged $c.$ 1.5 AU (at 100 km^3 Ma^{-1} km^{-1}), and 1.0 AU (at 35 km^3 Ma^{-1} km^{-1}) for intra-oceanic and continental margin arcs, respectively. These estimates are extracted from, and most applicable to, Cretaceous and Cenozoic margins. During crustal bonding at mostly older CSSZs, a global arc addition of $c.$ 0.2 AU (at 35 km^3 Ma^{-1} km^{-1}) is assumed. The combined global addition rate is thus 2.7 AU.

If this rate is broadly applicable since at least the late Archaean ($c.$ 3 Ga), it is in keeping with the implications of geochemical modelling by Hawkesworth & Kemp (2006) that older estimates of the magmatic growth of CIA crust ($c.$ 1 AU) are too low by a factor of two or three.

Volume of crustal material removed (recycled) at subduction zones

Losses at OMSZs

Recycling losses occur at OMSZs dominated by both nonaccreting and accreting styles of evolution (Fig. 4; Table 1). Accreted material is the sedimentary and igneous bodies scraped off the underthrusting lower plate and tectonically added, accreted, frontally or by underplating to the upper plate.

All convergent margins have a frontal prism that forms at the base of the landward trench slope just inboard of the trench floor (Figs 1–3). The frontal prism, typically <30–40 km in width and commonly <10 km, is an actively deforming (principally by thrusting), structurally weak body of either young accreted trench-floor sediment or tectonized slope sediment and mass-wasting debris shed from the margin. All nonaccreting and accreting margins also have an inner or landward prism of older basement or framework rock that is commonly coastally exposed (Figs 1–3). Whereas the age of the frontal prism is late Cenozoic, the age of the rock framework of the inner prism is typically early Tertiary or much older (e.g. the Palaeozoic and Proterozoic cratonic crust of western South America).

Accreting margin differs in that older frontal prisms and underplated sediment are stored in a middle prism that separates the actively deforming frontal prism and the margin's inner rock framework. Little is known about the material make-up of the middle prism, perhaps best exposed at the island of Barbados (Fig. 3; Torrini et al. 1985; Torrini & Speed 1989).

Nonaccreting margins narrow over long periods of time (tens of millions of years) with respect to a fixed spot on the upper plate (Clift & Vannucchi 2004; Table 1). At nonaccreting margins 80–100% of the sub-trench sedimentary section entering the SZ is subducted beneath the frontal prism and the inner prism (Scholl & von Huene 2007; Fig. 2a). Nonaccreting margins are structurally signified by the seaward extension of the framework rock of the inner prism to within a few tens of kilometres of the trench and thus to the back of the frontal prism (Table 1; Figs 1 and 2).

For the purpose of this paper's long-term view, nonaccreting margins also include those that have accumulated a young (c. <5 Ma old) frontal prism of accreted sediment supplied by enhanced rates of glacial-age turbiditic sedimentation in high-latitude trenches; for example, those of the Aleutian and south–central Chile SZs (Fig. 2; von Huene & Scholl 1991; Bangs & Cande 1997; Scholl & von Huene 2007). These frontal prisms are considered to be either geological ephemeral or existing in a dynamic balance of accretionary gains and subduction losses and thus not widening with time beyond c. 30–35 km (von Huene et al. 2009). They are attached to the seaward edge of the inner prism. Nonaccreting subduction zones at present constitute at least 75% of the global length of OMSZs (Fig. 4; Scholl & von Huene 2007).

Accreting margins of OMSZs are those that by progressive tectonic accumulation of trench-floor sediment and fragments of igneous oceanic crust have, with respect to a fixed point on the upper plate, quasi-continuously widened seaward since the early or middle Cenozoic (Table 1). At accreting margins the seaward terminus of the forearc's older or inner rock framework lies >50 km landward of the trench and the actively deforming accretionary frontal prism (Fig. 3). Between the frontal prism and the inner prism are stored older, only slightly deforming frontal prisms and deeply underplated material (Schulter et al. 2002; Kopp & Kukowski 2003).

Modern accreting margins, such as those of the Lesser Antilles (Barbados), Sumatra, Makran and Luzon(?) SZs, at present account for c. 25% of all OMSZs (Fig. 4). Accreting margins are most characteristic of subduction zones underthrust at low orthogonal speeds (c. 50 km Ma^{-1}) and in particular those dominated by the long-sustained input of thick sections of terrigenous sediment, either as trench-axis turbidite deposits or those of adjacent deep-sea fans. Sustained high rates of ocean-floor sedimentation must have been true of past accretionary margins, and possibly some non-accreting margins, to nourish the growth of the large accretionary complexes of late Jurassic to earliest Tertiary age that are widely exposed around the northern rim of the Pacific (see, e.g. Ernst 1970; Ridgway et al. 2007; Trop & Ridgway 2007).

Sediment subduction. During the past four decades offshore seismic and drilling investigations established that a large quantity of ocean-floor sediment conveyed tectonically into the subduction zone does not accumulate as a frontal prism of accreted material but continues landward in the subsurface beneath the frontal prism and the forearc's middle and inner prisms (Figs 1–3). The tectonic process involved is that of sediment subduction (Table 1; Scholl et al. 1980; von Huene & Scholl 1991; Clift & Vannucchi 2004; Clift et al. 2009).

Sediment subduction inserts clastic material into the subduction channel that separates the upper plate of the forearc from the underthrusting lower plate (Figs 1 and 2; Cloos & Shreve 1988a, b; Sage et al. 2006). During the late Cenozoic (i.e. the past 5–10 Ma), the missing proportion of frontally 'accretable' ocean-floor sediment is estimated to be 30% at c. 25% of the global length of OMSZs (c. 42 000 km), c. 80% at c. 25% of OMSZs, and 100% at the remaining c. 50% of modern OMSZs (von Huene & Scholl 1991). Sediment recycling is highest, c. 30 km^3 Ma^{-1} km^{-1}, at nonaccreting margins where c. 85% of all subducted sediment is recycled or geologically temporarily underplated; Table 2; Figs 2a and 3). However, good information about the volume of sediment subducted at accreting margin is lacking and it could be much higher than the estimate, 15 km^3 Ma^{-1} km^{-1}, listed in Table 2 and shown in Figure 3.

Table 2. *Solid volumes of subducted sediment and tectonically eroded forearc debris transported at OMSZs beneath the submerged forearc and toward the mantle*

	Average recycled for each km of subduction zone (km^3 Ma^{-1})	Globally recycled	
		in AU (km^3 a^{-1})	%
Subducted sediment			
Subducted and recycled to mantle depths at nonaccreting margins*	30	0.9	85
Subducted and recycled to mantle depths at accreting margins†	15	0.2	15
Global volume recycled to mantle depths at all margins	26	1.1	100
Subduction erosion			
Subduction erosion @ 40 km^3 Ma^{-1} km^{-1} at nonaccreting margins‡	42	1.3	90
Subduction erosion @ 12 km^3 Ma^{-1} at accreting margins‡	13	0.1	10
Global recycled to mantle	34	1.4	100
Total subducted material			
At nonaccreting margins (c. 31 000 km)			
Subducted sediment	30	0.9	42
Subduction erosion	42	1.3	58
Global sum subducted toward mantle	72	2.2	100
At accreting margins (c. 11 000 km)			
Subducted sediment	15	0.2	54
Subduction erosion	13	0.1	46
Global sum subducted toward mantle	28	0.3	100
At all margins (c. 42 000 km)			
Subducted sediment	26	1.1	43
Subduction erosion	34	1.4	57
Global sum subducted toward mantle	60	2.5	100

*Nonaccreting or erosive ocean-margin subduction zones are those that with respect to a fixed reference on the upper plate have narrowed over the past 10–20 Ma or longer (see Fig. 3; Clift & Vannucchi 2004). Modern nonaccreting margins, which include those with a recently added glacial-age accretionary frontal prism, total c. 31 000 km in global length (e.g. the Aleutian, Peru–Chile, IBM and Japan systems; Fig. 2; Scholl & von Huene 2007).
†Accreting ocean-margin subduction zones are those that with respect to a fixed reference on the upper plate have widered over the past c. 10–20 Ma or longer (see Fig. 3; Clift & Vannucchi 2004). Modern accreting margins, which do not include those with a recently added glacial-age accretionary frontal prism, total c. 11 000 km in global length (e.g. the Barbados or Antilles, Sumatra and Makran systems; Fig. 2; Scholl & von Huene 2007).
‡Scholl & von Huene (2007).
Transport occurs within the subduction channel (Figs 1–3). Adapted from Scholl & von Huene (2007, table 4). The bulk of the removed material is, eventually, thought to be recycled to the mantle.

In the late Cenozoic, the combined global loss of subducted sediment at accreting and nonaccreting SZs is estimated at c. 1.1 AU, which is equivalent to an average recycling of subducted sediment of c. 26 km^3 Ma^{-1} for each kilometre of OMSZ (Table 2; see also Clift et al. 2009).

Subduction erosion. Offshore and coastal studies of OMSZs document that sediment subduction is accompanied by the removal of large masses of upper plate CIA crust (Table 1). Removal is effected by processes of subduction erosion that truncate the seaward edge of the submerged forearc and thin and subside the crust of the submerged forearc (von Huene et al. 2004). Nearly 90% of tectonically eroded material at OMSZs occurs at nonaccreting margins, which are commonly termed erosive margins (Fig. 2; Kukowski & Oncken 2006; Ranero et al. 2006). Removal of upper plate CIA crust thins the submerged forearc and progressively truncates, and thus narrows, its seaward extent relative to a fixed onshore reference (Clift & Vannucchi 2004). The average linear rate of truncation at nonaccreting subduction zones is c. 2.5 km Ma^{-1}, a slow process that is only about 3% of the rate of orthogonal convergence (Scholl & von Huene 2007).

Franz et al. (2006), Glodny et al. (2006) and Kukowski & Oncken (2006) observed that recycling of upper plate crust along the western South America margin by subduction erosion has since the Mesozoic averaged c. 35–45 km^3 Ma^{-1} km^{-1} of margin. The volume of recycled upper plate material increases to c. 90 km^3 Ma^{-1} km^{-1} if subducted sediment and eroded forearc crust are combined (see tables 2 and 3 of Scholl & von Huene 2007). For the Andean system, this combined volume exceeds that estimated for the long-term magmatic productivity by a factor of at least two.

The long-term average rate of forearc destruction that combines that assessed for nonaccreting (42 km^3 Ma^{-1} km^{-1}) and accreting (13 km^3 Ma^{-1} km^{-1}) margins is estimated at c. 34 km^3 Ma^{-1} km^{-1} of forearc, a rate equivalent to a global recycling of c. 1.4 AU (Figs 2 and 3; Table 2; von Huene & Scholl 1991; Clift & Vannucchi 2004; Scholl & von Huene 2007; Clift et al. 2009). As an approximation, at OMSZs recycling effected by subduction erosion is about 60% of the total solid-volume of CIA material lost to the mantle (Table 2).

Recycling losses at CSSZs

Physical evidence that deep subduction of sedimentary material occurs at OMSZs is provided by the widespread exposure of packets of high P–T metamorphic rock (Ernst 1970; Jacobson et al. 1996, 2007; Grove et al. 2003, 2008; Ernst 2005). Similarly, the common occurrence of ultrahigh-pressure terranes at collisional orogens is prima facie evidence that during their formation at CSSZs material of CIA origin is carried well into the upper mantle (Hacker et al. 2006; Griffin 2007; Hacker 2007). The perspective gained from these relations is that crustal recycling to deep mantle circulation occurs at all subduction zones.

With respect to lower plate crust, for the early Proterozoic Wopmay collisional orogen of NW Canada, Hildebrand & Bowring (1999) estimated that crustal losses caused by detachment and sinking of deeply underthrust lower plate crust were c. 4000 km^3 km^{-1} of SZ (Fig. 4). Allowing that initial contact closing at this CSSZ required c. 50 Ma, then c. 80 km^3 Ma^{-1} km^{-1} of lower plate material was recycled to the mantle as suturing commenced. For the upper plate, Oncken (1998) estimated that subduction erosion thinned a central European Variscan CSSZ at a rate of c. 30 km^3 Ma^{-1} km^{-1}. Upper plate tectonic erosion has been recognized by Wagreich (1993) during the collisional building of the northern Calcareous Alps.

A conservative estimate is made that during underthrusting closure of the Wopmay CSSZ the combined crustal destruction by detachment and subduction erosion was about 100 km^3 Ma^{-1} km^{-1} (i.e. 50 km^3 Ma^{-1} km^{-1} for each plate). Adopting a c. 6000 km length of closing CSSZs as typical of the Proterozoic and Phanerozoic, a crustal volume of 0.6 km^3 a^{-1} or 0.6 AU of CIA crust was destroyed at CSSZs.

Sediment subduction should also attend the initial contact closing of a CSSZ system. Information about this likelihood is sparse, but large bodies of accreted sediment are not commonly described at Palaeozoic and Proterozoic collisional orogens; this implies that some or much of the sediment reaching the sea floor between closing blocks is either directly subducted or uplifted and later removed by subaerial erosion. If the eroded debris reaches active trench systems, then sediment recycling by follow-on sediment subduction will ensue. Modern examples are the sediment-charged Manila and Alaska trenches fed from the collisional uplifted coastal mountains of Taiwan and SE Alaska, respectively.

It is assumed, as a default value, that sediment subduction at CSSZs occurs at a rate currently estimated for accreting OMSZs (c. 15 km^3 Ma^{-1} km^{-1} Fig. 3; Table 2. At a long-term average length of 6000 km for past CSSZs, the global recycling rate of sediment at them averaged c. 0.1 AU. The total crustal loss at CSSZs of subducted sediment and eroded and detached forearc crust is thus estimated at c. 115 km^3 Ma^{-1} km^{-1}. The global rate of recycling at CSSZs is accordingly c. 0.7 AU (Fig. 3).

Balance of additions and losses

The long-term rate of destruction of upper plate CIA crust at each kilometre of OMSZs is estimated to be c. 60 km^3 Ma^{-1} km^{-1} (Table 2); that is, c. 25 km^3 of subducted sediment and 35 km^3 of tectonically eroded material from the submerged forearc. The corresponding global loss for the present length of OMSZs (c. 42 000 km) is thus c. 2.5 AU (i.e. c. 2.5 km^3 a^{-1}).

The corresponding rate of crustal recycling at CSSZs is far less securely assessed at 115 km^3 Ma^{-1} km^{-1} (c. 15 km^3 for sediment subduction and c. 100 km^3 for tectonically eroded upper plate and loss of detached, deeply subducted lower plate crust). At an average global length of 6000 km for CSSZs, an additional c. 0.7 AU is recycled. The assessed volume rate of crustal destruction at all subduction zones is thus c. 3.2 AU.

In comparison with the estimated long-term rate of loss of 3.2 AU, the global addition rate of juvenile magmatic crust is assessed at c. 2.7 AU (2.5 AU for OMSZs and 0.2 AU for CSSZs). Because arc additions and crustal losses are similar, and many uncertainties are involved in arriving at these estimations, the net growth of CIA crust during at least the Phanerozoic is assessed as effectively nil. With less certainty, this working conclusion of balance can be applied to the Proterozoic, or since the inception of the modern subduction zone. This inference of balance may not change if estimates of subduction losses and additions are determined to be higher or lower than stated above.

Uncertainties and testing of assumptions

Considerable uncertainties arise in applying to the whole of the Phanerozoic and Proterozoic estimated rates of SZ additions and subtractions extracted from the interpreted rock record of OMSZs of mostly Cenozoic age and only a few older fossil CSSZs.

For the recycling process of subduction erosion, long-term estimates based on the 10 best-constrained SZs range over nearly an order of magnitude, from 11 to 90 $km^3 Ma^{-1} km^{-1}$, and the lower values correspond to margins with the longest observational time base (50–160 Ma) (Scholl & von Huene 2007, table 4; Scholl & von Huene in press, fig. 4). Applications of the average of these estimates to the Phanerozoic and the Proterozoic must thus be viewed as working or exploratory values until more robustly measured ones become available.

Sediment subduction is proportional to the thickness of trench-floor sediment entering the subduction zone (von Huene & Scholl 1991). Drawing on marine geophysical data, the late Cenozoic insertion of ocean-floor sediment into the global network of subduction zones can be observationally assessed. To calculate longer-term rates of sediment subduction, the current thickness (2–2.5 km) of turbiditic sediment in high-latitude trenches requires a correction for elevated, late Cenozoic sedimentation provided by glaciated drainages (von Huene & Scholl 1991). However, regardless of latitude, trench axial sections are also thick to exceptionally thick (5–7 km) where collisional or accretionary orogenesis enhances the delivery of terrigenous sediment to SZs. Examples of such trenches are those of the Barbados (Lesser Antilles), Makran, Sumatra, West Java and Manila subduction zones (Fig. 4). For the Alaska–Aleutian Trench system, both coastal orogenesis and glaciated alpine drainages have combined to sustain a sediment-flood trench axis c. 3000 km in length since at least the late Miocene (Ryan & Scholl 1989; Rea & Snoeckx 1995; Kelemen et al. 2003; Gulick et al. 2007).

Assessing the delivery of continent-derived debris (terrigenous and biogenic) to Mesozoic, Palaeozoic and Proterozoic trench systems cannot be based on direct sea-floor observations. However, the abundance of accretionary underplates exposed at continental sectors of the Pacific rim (see, e.g. Kimura et al. 1992; Plafker & Berg 1994; Jacobson et al. 1996; Grove et al. 2003; Ernst 2005; Glodny et al. 2006) implies that continental debris was delivered at high rates to late Palaeozoic, Mesozoic and early Tertiary trenches. Presumably, the estimated rates of sediment subduction listed in Table 2, which is most reflective of the Cenozoic, a time of both collisional and non-collisional or accretionary orogenesis, are not uselessly off scale with respect to assessing long-term rates for the longer geological past.

However, applying them requires that the present length of OMSZs is representative of the Palaeozoic and Proterozoic as well as the global average rate of orthogonal subduction at them. Evidence is compelling that past rates of arc magmatic productivity were episodically high and linked in a complicated way to the assembly and breakup of supercontinents (Condie 1998, 2004, 2005; O'Neill et al. 2007; Stern 2008). It is also likely that during the closing phase of Wilson-cycle assembly of supercontinents the global length of subduction zones was greatly diminished (Silver & Behn 2008). At these times both recycling and arc magmatism may have been greatly reduced.

At collisional CSSZs, the rates of gains and losses are based on the interpreted rock record of only a few Mesozoic and older examples known to the authors. The assessment of crustal losses at CSSZs are thus poorly constrained. Also, the average global length of pre-Cenozoic CSSZs has been estimated rather than determined. Hildebrand & Bowring (1999) estimated that a typical global length of past collisional CSSZs ranged from 1000 to 20 000 km. A typical global length of 6000 km, half the present value, was assumed in the present paper. Field research at the suture zones of ancient collisional orogens is required to provide much-needed improvements in the gains and losses of CIA that occurred at them beyond the exploratory estimates offered in this paper.

With respect to cycles of supercontinent formation and linked issues of global subduction vigour, after the breakup of Pangaea c. 200 Ma, reassembly of dispersing continental-scale fragments along c. 12 000 km of CSSZs has been occurring since the early Tertiary. At the same time, older continental margin subduction zones of the northern and western Pacific were abandoned and replaced offshore with newly established, rapid arc-building intra-oceanic SZs, (e.g. the Aleutian, IBM and Tonga systems; Fig. 4).

Except for past periods of assembled Wilson-cycle supercontinents and potential stagnant-lid convection and low subduction zone vitality, the present global distribution of OMSZs and CSSZs and the orogens they built might be reasonably characteristic of the Phanerozoic and Proterozoic. Also, because many past events of superplume activity are recognized (Condie 2004; Neal et al. 2008), the population of our present ocean basins with plume-constructed eruptive relief of Cretaceous and Cenozoic age could well be typical of many past periods. The passage of high relief down the subduction zone greatly increases the recycling rate of CIA crust, thus countering pulses of rapid arc growth at newly established SZs.

The above considerations warrant at least exploring the implications of our estimates of SZ additions and recycling losses to the Phanerozoic and Proterozoic, if not since the organization of modern subduction zones at $c.$ 3 Ga (Cawood et al. 2006).

Global addition estimates. The average productivity at modern intra-oceanic arcs ($c.$ 130–135 km^3 Ma^{-1} km^{-1} for the Aleutian and IBM systems) appears reasonably well constrained for the past 50 Ma, and a long-term rate of $c.$ 100 km^3 km^{-1} Ma^{-1} seems both conservative and reasonable to apply before arc magmatic activity is extinguished by accretionary collision, which typically occurs within 100 Ma (Condie 2007). Estimates of magmatic productivity along continental margin arcs are made with more difficulty. The generally agreed upon rate, and the one used in this paper (35 km^3 Ma^{-1} km^{-1}) is based on the findings of Stern & Bloomer (1992) and Haschke & Gunter (2003). This rate is effectively that proposed earlier by Reymer & Schubert (1984).

The accuracy of the estimated rates of oceanic arc productivity and their applicability to the geological past can be examined by considering the growth rate of large tracts of Precambrian crust; for example, in the southwestern USA, the early Proterozoic Yavapai–Mazatzal orogenic crust of southern Laurentia accumulated in $c.$ 200 Ma between $c.$ 1800 and 1600 Ma (Condie 2007; Whitmeyer & Karlstrom 2007; Amato et al. 2008; Fig. 3). Field and age relations and positive Nb isotopic ratios imply that a substantial proportion of the Palaeoproterozoic body is juvenile arc crust generated at intra-oceanic and continental margin SZs. However, widespread bimodal magmatism, abundant pre-orogenic zircon cores, and Sm–Nd and zircon Hf isotopic data document an abundance of remelted and mobilized older Palaeoproterozoic and Archaean crust (Bickford & Hill 2007; Bickford et al. 2008).

The Yavapai–Mazatzal orogen is $c.$ 800 km wide, and, although deeply eroded now, its original crustal thickness was probably close to 50 km (Condie 2007). A 1 km wide corridor of the orogen thus originally involved a crustal volume of $c.$ 40 000 km^3. This volume was produced in $c.$ 200 Ma at an average corresponding rate of $c.$ 200 km^3 Ma^{-1} km^{-1} of orogen.

Not included in the crustal volume is that lost by subduction erosion. Presuming that at least one SZ was constantly operating during crustal assembly, subduction erosion would have recycled older CIA crust at a rate at least typical of that of modern OMSZs (i.e. $c.$ 34 km^3 Ma^{-1} km^{-1}; Table 2). Correspondingly, over 200 Ma, a crustal volume of $c.$ 6800 km^3 km^{-1} of SZ would have been demolished. Added to the thickness-based volume, the restored volume of igneous material of the Yavapai and Mazatzal crustal body was $c.$ 50 000 km^3. The estimated long-term production rate is thus $c.$ 250 km^3 Ma^{-1} km^{-1} of orogen. This rough calculation includes volumetric contributions of accreted bodies of juvenile back-arc crust and oceanic plateaux, and large tracts of pre-existing CIA and younger non-juvenile igneous crust (Bickford & Hill 2007; Bickford et al. 2008). This volume-rate estimate is similar to the total accretion rate of 181 km^3 Ma^{-1} km^{-1} assessed by Condie (2007); this lower rate does not include recycling losses.

Allowing that at least one-half of the restored Yavapai–Mazatzal igneous mass was contributed by arc magmatism (see Bickford & Hill 2007, fig. 3), then arc addition was at 125 km^3 Ma^{-1} km^{-1}. Condie (2007) estimated more carefully that juvenile crust was added to the Yavapai–Mazatzal cratonic complex at 165 km^3 Ma^{-1} km^{-1}. Magma production at Cenozoic, intra-oceanic SZs has during the past $c.$ 50 Ma operated at similar high rates of 130–135 km^3 Ma^{-1} km^{-1} (Fig. 1; Holbrook et al. 1999; Jicha et al. 2006; Takahasi et al. 2007). The comparable rates of magmatic productivity of Cenozoic intra-ocean arcs and during the fabrication of Palaeoproterozoic crust imply that juvenile additions at high rates have remained little changed since the early Proterozoic. High construction rates of Proterozoic crust also signal that massif building was at newly established intra-oceanic and continental margin SZs (see also Condie & Chomiak 1996). It seems likely that elevated rates of juvenile arc additions contributed substantially to the production of Neoproterozoic crust, which forms 17–20% of the Earth's preserved continental mass (Stern 2008).

Losses. With respect to losses, evidence of uplift and patterns of deformation and sedimentation imply that part of the material subducted beneath the submerged forearc is reattached to the upper plate beneath the coastal region (Hartley et al. 2000; Clift & Hartley 2007). The volume of CIA material sequestered as underplates is difficult to assess. Based on the increasing recognition of exhumed bodies of subduction channel accumulations

(Vannucchi et al. 2008), and in particular of the widespread exposure of Mesozoic and early Tertiary underplates at North Pacific margins and detected there seismically, lower crustal rocks formed by underplating may be voluminous (Sample & Moore 1987; Moore et al. 1991; Kimura et al. 1992; Gutscher et al. 1998; Calvert 2004; Fuis et al. 2008; Grove et al. 2008). The volume of high-pressure rock in older Phanerozoic and Precambrian rock sequences is none the less globally small in comparison with the amount of CIA material estimated to have been subducted at OMSZs (at least 2.5×10^9 km^3 since 1 Ga; Table 2). It is thus supposed that in the long term (many tens to hundreds of millions of years) the bulk of subducted sediment and tectonically eroded material stored initially as underplates is eventually flushed to the mantle (von Huene & Scholl 1991; Scholl & von Huene 2007; Clift et al. 2009). As evidenced by episodes of inboard migration of the arc magmatic front, rapid subduction-erosion flushing of stored or older material may occur periodically rather than at a slower but more continuous background rate (Kay et al. 2005; also see Fig. 4).

Although few observation-based estimates of upper plate losses have been made at the CSSZs of collisional margins, the rates obtained are comparable with those much better estimated at OMSZs. For example, the global average rate of subduction erosion at modern OMSZs (c. 34 km^3 Ma^{-1} km^{-1}, Table 2) and those for the upper plate estimated by Wagreich (1993) for the northern Calcareous Alps and by Oncken (1998) for a Variscan orogen (c. 30 km^3 Ma^{-1} km^{-1}) are similar.

For the lower plate, regional-scale examples for comparison of losses at modern and ancient CSSZs are also sparse. However, Johnson (2002) estimated that large volumes of lower plate Indian crust have been subducted at the India–Asia CSSZ. Based on the observations of Hildebrand & Bowring (1999), lower plate losses of c. 80 km^3 Ma^{-1} km^{-1} for the Wopmay accretionary orogen of NW Canada are qualitatively in keeping with subduction losses of >50% of lower plate crust at the collisional closures of the Yakutat terrane with the eastern Alaska SZ (Eberhart-Phillips et al. 2006; Gulick et al. 2007; Fuis et al. 2008; Wallace 2008) and the Ontong–Java Plateau with the Solomon arc massif (Mann & Taira 2004; Fig. 4).

Pulses of addition and losses

Silver & Behn (2008) posited that a lengthy lull in the global rate of subduction could be a likely consequence of the closing of Pacific-type ocean basins to form an areally expansive supercontinental mass. Closing would occlude many thousands of kilometres of circum-Pacific SZs, ending both arc volcanism and crustal recycling at them. If new subduction zones did not rapidly form peripheral to the supercontinental body, or at distance elsewhere, then for periods 100 Ma or longer a balance between SZ gains and losses, if maintained, would be at much reduced levels. Similarly, palaeomagnetic data support the notion that during the Precambrian subduction vigour was episodic. Exceptional periods of rapid subduction matched episodes of rapid crustal growth by arc magmatism (O'Neill et al. 2007).

Rapid growth of CIA crust, at a long-term rate at least as high as 100 km^3 Ma^{-1} km^{-1} and for some periods of time between 500 and 300 km^3 Ma^{-1} km^{-1}, occurs after the initiation of new intra-oceanic subduction zones (Fig. 1). It is not known what elevated rates attend the initiation of a new continental margin or Andean-type subduction zones. If breakup of a supercontinent and dispersal of its fragments initiates new offshore subduction zones, an episode of greatly heightened global productivity would ensue. For example, the rapid growth (c. 165 km^3 Ma^{-1} km^{-1}; Condie 2007) of juvenile crust to form the early Proterozoic Yavapai and Mazatzal arc complexes occurred prior to their joining with Rodinia, which was finally assembled by 1.3–1.0 Ga (Condie 2005, 2007). This implies that rapid growth of arc-magmatic crust might be a consequence of rapidly moving (mobile) crustal blocks (plates) during assembly of a supercontinental mass and dispersal of its parts after breakup. Arc productivity was probably lower, possibly significantly so, during the lifespan of the supercontinent, when plate immobility and stagnant lid conditions prevailed (O'Neill et al. 2007; Silver & Behn 2008).

Global episodes of rapid recycling are stimulated by increased discharge of turbiditic sediment to the trench axis and adjacent sea floor and the subduction of high sea-floor relief. For example, sediment subduction beneath the sediment-nourished south–central Chile margin (33–45°S) is c. 100 km^3 Ma^{-1} km^{-1}, that beneath the trench-charged Aleutian margin is c. 75 km^3 Ma^{-1} km^{-1}, whereas along the sediment-starved IBM system the rate drops to c. 25 km^3 Ma^{-1} km^{-1} (von Huene & Scholl 1991; Scholl & von Huene 2007; Table 2).

Underthrusting of high relief enhances rates of subduction erosion several fold to an order of magnitude or higher (von Huene & Scholl 1991; Vannucchi et al. 2003; Clift & Vannucchi 2004; Ranero et al. 2006). Elevated rates are sustained where tracks of seamounts or lengthy ridges enter the subduction zone, and in particular where its regional trend is oblique to tracks of subducting relief (Ballance *et al.* 1989; Behrmann *et al.* 1994; Bourgois *et al.* 1996; von Huene *et al.* 2000; Laursen *et al.* 2002; Vannucchi *et al.* 2003; von Huene & Ranero in press). The western margin of South America has in particular been affected by subducted relief (Nur & Ben-Avraham 1981; Larson 1991; Gutscher *et al.* 1999, 2000; Hampel *et al.* 2003, 2004; Kukowski *et al.* 2008). In part because of a long history of high-relief subduction, since the mid-Mesozoic this margin registers a higher rate of crustal loss than arc magmatic additions (Stern 1991; Franz *et al.* 2006; Glodny *et al.* 2006; Kukowski & Oncken 2006).

Elevated sea-floor relief is constantly in construction at spreading ridges and by hotspot volcanism (e.g. Hawaii and the Galapagos centres, Fig. 4). During episodes of abundant growth of oceanic plateaus large areas of the ocean floor are populated with LIPs and their trailing aseismic ridges (Neal *et al.* 2008). Examples of these edifices entering eastern Pacific SZs at present include, from north to south, the Cocos, Carnegie, Nazca and Juan Fernandez Ridges, and also the spreading ridge of the South Chile Rise (Fig. 3). Prominent examples of LIPs and aseismic ridges colliding with western Pacific SZs include, from south to north, the Hikurangi Plateau, the Louisville, Samoa and Carolina Ridges, the Ontong–Java Plateau, and the Marcus–Necker–Ogasawara and Emperor Ridges (Fig. 3). Many if not most of these igneous constructs were initially generated during a mid-Cretaceous (120–100 Ma) episode of superplume activity (Larson 1991, 1997). Earlier episodes of plume-supported plateau formation were posited by Condie (2005, 2007) to have occurred in the early Proterozoic and late Archaean at 1.9 Ga and 2.7 Ga, respectively.

The perspective is gained that past episodes of widespread formation of new intra-oceanic or continental margin subduction zone would enhance arc magmatic productivity by a factor at least 3–4 time greater than widely thought typical of currently active continental or intra-oceanic arcs (*c.* 30–35 $km^3 Ma^{-1} km^{-1}$). Increased rates of recycling losses would coincide with the prolonged subduction of tracks of high sea-floor relief produced by episodes of intraplate magmatism. Relief arming of the Pacific basin occurred widely in the early and mid-Cretaceous, when opening of the Atlantic was under way (Sager *et al.* 1988; Atwater 1989; Larson 1991, 1997; Tarduno *et al.* 2003; Sager 2005; Norton 2007).

The festooning of the Pacific's margins with its present set of rapidly grown intra-oceanic arcs (e.g. Aleutian, Kuril, IBM, Vanuatu–Solomon, New Britain, Tonga–Kermadec and South Sandwich) did not occur until after the early Eocene. However, older accreted offshore, intra-oceanic massifs of Jurassic and Cretaceous arcs are widely reported from the Pacific margins; for example, for the sectors of California (Ernst *et al.* 2008), Alaska (Plafker & Berg 1994; Plafker *et al.* 1994; Condie & Chomiak 1996; Clift *et al.* 2005) and Kamchatka (Garver *et al.* 2004; Sukhov *et al.* 2004; Chekhovich & Sukhov 2006; Chekhovich *et al.* 2006). It seems likely that episodes of rapid formation of new subduction zones and the growth of CIA crust and rapid subduction zone recycling could occur very differently in time and space. If a global balance is struck between growth and loss, it would only be realized over many hundreds of millions of years.

It is not so obvious why a long-term balance might exist between arc magmatic production and crustal recycling. However, it could be linked to the fact that rates of arc magmatism and crystal recycling at SZs are both proportional to average global production of oceanic crust, plate movements and subduction vigour. If a balance has been struck, it has to be established across times of supercontinent assembly and perhaps depressed global subduction rates (Silver & Behn 2008), episodes of rapid plate movements and creation of new SZs at which prodigious rates of arc magmatism would be expected (Stern & Bloomer 1992; O'Neill *et al.* 2007; Fig. 4), and times of superplume production of high sea-floor relief (Larson 1991, 1997; Condie 1998, 2004; Neal *et al.* 2007), which, when subducted, greatly enhances recycling rates (von Huene & Scholl 1991; Clift & Vannuchi 2004). In the background are cycles of increased continental erosion and delivery of sediment to subduction zones.

Whole-Earth perspectives and implications

Hildebrand & Bowring (1999) emphasized that $(3-6) \times 10^3 km^3$ of lower plate passive margin crust were recycled at each kilometer of the Wopmay collisional orogen, an early Proterozoic CSSZ. For a global length of CSSZs equivalent to the present value (*c.* 12 000 km), the subducted volume of lower plate crust alone would be *c.* $50 \times 10^6 km^3$, a mass corresponding to nearly 1% of the Earth's present volume of CIA crust (*c.* $7 \times 10^9 km^3$). As Hildebrand & Bowring noted, summed over long periods of time, the volume of subducted lower plate crust at collisional origins is very large. For example, a protracted 50 Ma closing of a global length of CSSZ of 6000 km would at a recycling rate of $(3-6) \times 10^3 km^3 km^{-1}$ of CSSZ return *c.* $0.5 \times 10^6 km^3 Ma^{-1}$ of lower plate crust to the mantle. If this value is characteristic of the past, since 3 Ga a volume of CIA crust equivalent to 1.5×10^9 km, or c. 20% of the existing volume, has at CSSZs alone been subducted and probably largely recycled to the mantle.

A similar conclusion is reached by considering the implications of the long-term rate of crustal loss at the combined length of OMSZs and CSSZs. The global rate of destruction of CIA crust from both the upper and lower plates is estimated in this paper at 3.2 AU (3.2 km^3 a^{-1}), and earlier was estimated by Scholl & von Huene (2007) at 2.5 AU for OMSZs alone (see Clift et al. 2009, for a similar evaluation). At either of these rates, during the past 3 Ga, the volume of deeply subducted continental and island arc crust is equal to or somewhat larger than the existing volume of the Earth's crust. This amount seems astonishingly high.

Except for relatively small proportions of subducted material permanently returned to the upper plate by crustal underplating, or returned via arc magmatism, the bulk of subducted material is presumed to be entrained into mantle circulation. Since 3 Ga, when continental and island arc subduction systems are thought to have been initiated, the mass of CIA crust recycled to the mantle is estimated at c. 3 km^3 a^{-1} is c. 10^{10} km^3, a mass equivalent to c. 1% of the mantle's volume (c. 10^{12} km^3). This fraction, although small, is capable of imparting a CIA crustal signature to the mantle (Scholl & von Huene 2007). Because subduction zones at any time are not equally distributed across the Earth's surface (Condie 2005), and at times had greater or much reduced global length (O'Neill et al. 2007: Silver & Behn 2008), recycling at subduction zones must have been a substantial contributor to establishing mantle geochemical heterogeneities; for example, the large continental recycling anomaly recorded in Samoan lavas (Workman et al. 2008).

Summary

The volumetric growth of juvenile continental and island arc (CIA) crust is principally produced by arc volcanism at ocean-margin subduction zones (OMSZs) underthrust by oceanic crust, and less abundantly at collisional or crust-suturing subduction zones (CSSZs). Global crustal recycling is also concentrated at OMSZs via the processes of sediment subduction and subduction erosion, and by these means and the breakoff and foundering of deeply underthrust sectors of CIA crust at collisional CSSZs.

For the Cenozoic and late Mesozoic, onshore and offshore field observations lead to estimates of the global long-term rate of juvenile magmatic productivity at 2.7 AU (2.5 AU at OMSZs and 0.2 AU at CSSZs), where 1 AU (Armstrong Unit) is 1 km^3 a^{-1}. This rate can with more difficulty be applied to the older Phanerozoic and Proterozoic, and perhaps since modern subduction zones were established at c. 3 Ga. Arc magmatic additions to form CIA crust are highest at newly formed, intra-oceanic subduction zones. Presumably, high rates also attend the initiation of subduction at continental margins.

The estimate of additions (+2.7 AU) and losses (−3.2 AU) at subduction zones are similar. The implication is that since c. 3 Ga juvenile crustal gains and crustal losses at subduction zones have been approximately in balance. However, periods of enhanced rate of growth occur in particular when new subduction zones form widely, presumably during rapid microplate assembly of supercontinents and their subsequent breakup and dispersal of crustal fragments.

Rates of recycling losses are boosted at OMSZs by the subduction of lower plate high relief; for example, the entrance into the subduction zone of oceanic plateaux, aseismic ridges, and groups of seamounts and, at CSSZs, the leading edge of underthrusting CIA crust. The development of sea-floor high relief occurs in particular during periods of widespread hotspot volcanism or superplume activity. These episodes are apparently linked to rapid rates of sea-floor generation (Larson 1991, 1997) that, presumably, would stimulate both increased rates of arc productivity and recycling. Because ocean basins are large, long periods of time may separate the building of elements of high sea-floor relief and their arrival at a a subduction zone. It is likely, therefore, that episodes of rapid growth of juvenile CIA crust are separated in time and space from episodes of enhanced rates of crustal recycling.

An astonishing outcome of the estimated rate of crust recycling is that since 3 Ga a volume equal to that of the standing continental mass (c. 7×10^9 km^3) has been recycled to the mantle. Although small in comparison with the volume of the mantle, the amount of CIA material injected into the mantle is adequate to impart a continental crustal signature and to contribute importantly to forming mantle geochemical heterogeneities.

Our exploration of issues and observations concerning crustal additions and recycling losses at subduction zones benefited hugely from discussions with colleagues familiar with modern subduction zones and the rock record of ancient ones. For the generous sharing of their time and knowledge, we wish to thank in particular Nathan Bangs, Kevin Burke, Peter Cawood, Peter Clift, Mark Cloos, Jean-Yves Collot, Kent Condie, Gary Ernst, Marty Grove, Brad Hacker, Bob Hildebrand, Sue and Bob Kay, Steve Kirby, Simon Klemperer, Heidrun Kopp, Nina Kukowski, Paul Mann, Bob McLaughlin, Casey Moore, Onno Oncken, Terry Pavlis, Victor Ramos, Cesar Ranero, Holly Ryan, Bob Stern, Tracy Vallier, Paola Vannucchi and Ray Wells.

References

AMATO, J. M., BOULLION, A. O., FARMER, G. L., SANDERS, A. E., GEHRELS, G. E. & WOODEN, J. L. 2008. Evolution of the Mazatzal province and the timing of the Mazatzal orogeny: Insights from U–Pb geochronology and geochemistry of igneous and metasedimentary rocks in southern New Mexico. *Geological Society of America Bulletin*, **120**, 328–346, doi: 10.1130/B26200.1.

ARMSTRONG, R. L. 1981. Radiogenic isotopes: The case for crustal recycling on a near-steady-state no-continental growth Earth. *Philosophical Transactions of the Royal Society of London*, **301**, 443–472.

ARMSTRONG, R. L. 1991. The persistent myth of crustal growth. *Australian Journal of Earth Sciences*, **38**, 613–630.

ATWATER, T. 1989. Plate tectonic history of the northeast Pacific and western North America. *In*: WINTERER, E. L., HUSSONG, D. M. & DECKER, R. W. (eds) *Eastern Pacific Ocean and Hawaii*. The Geology of North America, **N**, 21–72.

BALLANCE, P. F., SCHOLL, D. W., VALLIER, T. L., STEVENSON, A. J., RYAN, H. F. & HERZER, R. H. 1989. Subduction of a Late Cretaceous seamount of the Louisville chain at the Tonga Trench: A model of normal and accelerated tectonic erosion. *Tectonics*, **8**, 953–962.

BANGS, N. L. & CANDE, S. C. 1997. The episodic development of a convergent margin inferred from structures and processes along the southern Chile margin. *Tectonics*, **16**, 489–505.

BEHRMANN, J. H., LEWIS, S. D., CANDE, S. C. & ODP LEG 141 SCIENTIFIC PARTY. 1994. Tectonics and geology of spreading ridge subduction at the Chile Triple Junction: A synthesis of results from Leg 141 of the Ocean Drilling Program. *Geologische Rundschau*, **83**, 832–852.

BICKFORD, M. E. & HILL, B. M. 2007. Does the arc accretion model adequately explain the Paleoproterozoic evolution of southern Laurentia? An expanded interpretation. *Geology*, **35**, 167–170, doi:10.1130/G23174A.1.

BICKFORD, M. E., MUELLER, P. A., KAMENOV, G. D. & HILL, B. M. 2008. Crustal evolution of southern Laurentia during the Paleoproterozoic: Insights from zircon Hf isotopic studies of *ca.* 1.75 Ga rocks in central Colorado. *Geology*, **36**, 555–558, doi: 10.1130/G24700A.1.

BOURGOIS, J., MARTIN, H., LAGABRIELLE, Y., LE MOIGNE, J. & FRUTOS, J. 1996. Subduction erosion related to spreading-ridge subduction: Taitao peninsula (Chile margin triple junction area). *Geology*, **24**, 723–726.

BYRNE, T. B. & LIU, C.-S. (eds) 2002. *Geology and Geophysics of an Arc–Continent Collision, Taiwan*. Geological Society of America, Special Papers, **358**.

CALVERT, A. 2004. Seismic reflection imaging of two megathrust shear zones in the northern Cascadia subduction zone. *Nature*, **428**, 163–167.

CAWOOD, P. A., KRONER, A. F. & PSAREVSKY, S. 2006. Precambrian plate tectonics: Criteria and evidence. *GSA Today*, **16**, doi:10.1130/GSAT01607.1.

CHEKHOVICH, V. D. & SUKHOV, A. N. 2006. Breakup of the Late Cretaceous Achaivayam–Valagin volcanic arc in the Paleocene (terranes of southern Koryakia and eastern Kamchatka). *Doklady Earth Sciences*, **497A**, 893–896.

CHEKHOVICH, V. D., SUKHOV, A. N., FILATOVA, N. I., VISHNEVSKAYA, V. S. & BASOV, I. A. 2006. New data on Cretaceous volcanic arcs of the northeastern Asian margin. *Doklady Earth Sciences*, **407A**, 381–384.

CLIFT, P. D. & HARTLEY, A. J. 2007. Slow rates of subduction erosion and coastal underplating along the Andean margin of Chile and Peru. *Geology*, **35**, 503–506, doi:10.1130/G23584A.1.

CLIFT, P. D. & VANNUCCHI, P. 2004. Controls on tectonic accretion versus erosion in subduction zones: Implications for the origin and recycling of the continental crust. *Reviews of Geophysics*, **42**, RG2001, doi: 10.1029/2003RG000127.

CLIFT, P. D., PECHER, I., KUKOWSKI, N. & HAMPEL, A. 2003. Tectonic erosion of the Peruvian forearc, Lima Basin, by subduction and Nazca Ridge collision. *Tectonics*, **22**, 1023, doi:10.1029/2002TC001386.

CLIFT, P. D., PAVLIS, T., DEBARI, S. M., DRAUT, A. E., RIOUX, M. & KELEMEN, P. B. 2005. Subduction erosion of the Jurassic Talkeetna–Bonanza arc and the Mesozoic accretionary tectonics of western North America. *Geology*, **33**, 881–884, doi:10.1130/G21822.1.

CLIFT, P., SCHOUTEN, H. & VANNUCCHI, P. 2009. Arc–continent collisions, sediment recycling and the maintenance of the continental crust. *In*: CAWOOD, P. A. & KRÖNER, A. (eds) *Earth Accretionary Systems in Space and Time*. Geological Society, London, Special Publications, **318**, 75–103.

CLOOS, M. & SHREVE, R. L. 1988a. Subduction-channel model of prism accretion, melange formation, sediment subduction, and subduction erosion at convergent plate margins: 1. Background and description. *Pure and Applied Geophysics*, **128**, 456–500.

CLOOS, M. & SHREVE, R. L. 1988b. Subduction-channel model of prism accretion, melange formation, sediment subduction, and subduction erosion at convergent plate margins: 2. Implications and discussion. *Pure and Applied Geophysics*, **128**, 501–545.

CLOOS, M., SAPIIE, B., VAN UFFORD, A. Q., WEILAND, R. J., WARREN, P. Q. & MCMAHON, T. P. 2005. *Collisional Delamination in New Guinea: The Geotectonics of Subducting Slab Breakoff*. Geological Society of America, Special Papers, **400**.

COLLINS, W. J. & RICHARDS, S. W. 2008. Geodynamic significance of S-type granites in Circum-Pacific orogens. *Geology*, **36**, 559–562, doi:10.1130/G24658A.1.

COLLOT, J.-Y., AGUDELO, W. & RIBODETTI, A. 2008. Origin of crustal splay fault and its relation to the seismogenic zone and underplating at the erosional north Ecuador and southwest Colombia margin. *Journal of Geophysical Research*, **113**, B12102, doi:10.1029/2008JB005691.

CONDIE, K. C. 1998. Episodic continental growth and supercontinents: A mantle avalanche connection? *Earth and Planetary Science Letters*, **163**, 97–108.

CONDIE, K. C. 2004. Supercontinents and superplume events: Distinguishing signals in the geologic record.

Physics of the Earth and Planetary Interiors, **146**, 319–332.

CONDIE, K. C. 2005. *Earth as an Evolving Planetary System*. Elsevier Academic Press, Amsterdam, Boston.

CONDIE, K. C. 2007. Accretionary orogens in space and time. *In*: HATCHER, R. D. JR, CARLSON, M. P., MCBRIDE, J. H. & MARTINEZ CATALÁN, J. R. (eds) *4D Framework of Continental Crust*. Geological Society of America, Memoirs, **200**, 145–158.

CONDIE, K. C. & CHOMIAK, B. 1996. Continental accretion: Contrasting Mesozoic and Early Proterozoic tectonic regimes in North America. *Tectonophysics*, **265**, 101–126.

COOK, F. A., VAN DER VELDEN, A. J., HALL, K. W. & ROBERTS, B. J. 1998. Tectonic delamination and subcrustal imbrication of the Precambrian lithosphere in northwestern Canada mapped by LITHOPROBE. *Geology*, **26**, 839–842.

EBERHART-PHILLIPS, D., CHRISTENSEN, D. H., BROCHER, T. M., HANSEN, R., RUPPERT, N. A., HAEUSSLER, P. J. & ABERS, G. A. 2006. Imaging the transition from Aleutian subduction to Yakutat collision in central Alaska with local earthquakes and active source data. *Journal of Geophysical Research*, **111**, B11303, doi:10.1029/2005JB004240.

ERNST, W. G. 1970. Tectonic contact between the Franciscan melange and the Great Valley sequence, crustal expression of a Late Mesozoic Benioff zone. *Journal of Geophysical Research*, **75**, 886–901.

ERNST, W. G. 2005. Alpine and Pacific styles of Phanerozoic mountain building: Subduction-zone petrogenesis of continental crust. *Terra Nova*, **17**, 165–188.

ERNST, W. G., SNOW, C. A. & SCHERER, H. H. 2008. Contrasting early and late Mesozoic petrotectonic evolution of northern California. *Geological Society of America Bulletin*, **120**, 179–194, doi:10.1130/B26173.1.

FRANZ, G. F. & KRAMER, W. ET AL. 2006. Crustal evolution at the central Andean continental margin: A geochemical record of crustal growth, recycling and destruction. *In*: ONCKEN, O., CHONG, G. ET AL. (eds) *The Andes, Active Subduction Orogeny*. Frontiers in Earth Science, **1**, 45–64.

FUIS, G. S., MOORE, T. E. ET AL. 2008. The Trans-Alaska Crustal Transect and continental evolution involving subduction underplating and synchronous foreland thrusting. *Geology*, **36**, 267–270, doi:10.1130/G24257A.1.

GARVER, J. I., HOURIGAN, J., SOLOVIEV, A. & BRANDON, M. T. 2004. Collision of the Olyutorsky terrane, Kamchatka. *Geological Society of America, Abstracts with Programs*, **36**, 121.

GLODNY, J., ECHTLER, H. ET AL. 2006. Long-term geological evolution and mass-flow balance of the south-central Andes. *In*: ONCKEN, O., CHONG, G. ET AL. (eds) *The Andes, Active Subduction Orogeny*. Frontiers in Earth Science, **1**, 401–428.

GRIFFIN, W. L. 2007. Major transformations reveal Earth's deep secrets. *Geology*, **36**, 95–96, doi: 10.1130/focus012008.1.

GROVE, M., JACOBSON, C. E., BARTH, A. P. & VUCIC, A. 2003. Temporal and spatial trends of Late Cretaceous–early Tertiary underplating of Pelona and related schist beneath southern California and southwestern Arizona. *In*: JOHNSON, S. E., PATERSON, S. R., FLETCHER, J. M., GIRTY, G. & KIMBROUGH, D. L. (eds) *Tectonic Evolution of Northwestern Mexico and the Southwestern USA*. Geological Society of America, Special Papers, **374**, 381–406.

GROVE, M., BEBOUT, G. E., JACOBSON, C. E., KIMBROUGH, D. L., KING, R. L. & LOVERA, O. M. 2008. Medial Cretaceous subduction erosion of southwestern North America: New hypothesis for the formation of the Catalinas schist. *In*: DRAUT, A. E., CLIFT, P. D. & SCHOLL, D. W. (eds) *Formation and Applications of the Sedimentary Record in Arc Collision Zones*. Geological Society of America, Special Papers, **436**, 335–362.

GULICK, S. P. S., LOWE, L. A., PAVLIS, T. L., GARDNER, J. V. & MAYER, L. A. 2007. Geophysical insights into the transition fault debate (2007): Propagating strike slip in response to stalling Yakutat block subduction in the Gulf of Alaska. *Geology*, **35**, 763–766, doi: 10.1130/G23585A.1.

GUTSCHER, M.-A., KUKOWSKI, N., MALAVIEILLE, J. & LALLEMAND, S. 1998. Episodic imbricate thrusting and underthrusting: Analog experiments and mechanical analysis applied to the Alaskan accretionary wedge. *Journal of Geophysical Research*, **103**, 10161–10176.

GUTSCHER, M. A., MALAVIELLE, J., LALLEMAND, S. & COLLOT, J.-Y. 1999. Tectonic segmentation of the North Andean margin: Impact of the Carnegie Ridge collision. *Earth and Planetary Science Letters*, **168**, 255–270, doi:10.1016/S0012-821X(99)00060-6.

GUTSCHER, M. A., SPACKMAN, W., BIJWAARD, H. & ENGDAHL, E. R. 2000. Geodynamics of flat subduction: Seismicity and tomographic constraints from the Andean margin. *Tectonics*, **19**, 814–833.

HACKER, B. R. 2007. Ascent of ultrahigh-pressure western gneiss region, Norway. *In*: CLOOS, M., CARLSON, W. D., GILBERT, M. C., LIOU, J. G. & SORENSEN, S. S. (eds) *Convergent Margin Terranes and Associated Regions: A Tribute to W. G. Ernst*. Geological Society of America, Special Papers; **419**, 171–184.

HACKER, B. R., MCCLELLAND, W. C. & LIOU, J. G. (eds) 2006. *Ultrahigh-pressure Metamorphism: Deep Continental Subduction*. Geological Society of America, Special Papers, **403**.

HAMPEL, A., ADAM, J. & KUKOWSKI, N. 2003. Response of the tectonically erosive south Peruvian forearc to subduction of the Nazca Ridge: Analysis of three-dimensional analogue experiments. *Tectonics*, **23**, TC5003, doi:101029/2003TC001585.

HAMPEL, A., KUKOWSKI, N., BIALAS, J., HUEBSCHER, C. & HEINBOCKEL, R. 2004. Ridge subduction at an erosive margin: The collision zone of the Nazca Ridge in southern Peru. *Journal of Geophysical Research*, **109**, B02101, doi:10.1029/2003JB002593.

HARTLEY, A. J., CHONG, G., TURNER, P., KAPE, S. J., MAY, G. & JOLLEY, E. J. 2000. Development of a continental forearc: A Cenozoic example from the Central Andes, northern Chile. *Geology*, **28**, 331–334.

HASCHKE, M. & GUNTHER, A. 2003. Balancing crustal thickening in arcs by tectonic vs magmatic means. *Geology*, **31**, 933–936.

HAWKESWORTH, C. J. & KEMP, A. I. S. 2006. Evolution of the continental crust. *Nature*, **443**, 811–817.

HAWKINS, J. W., BLOOMER, S. H., EVANS, C. A. & MECHIOR, J. T. 1984. Evolution of intra-oceanic arc–trench systems. *Tectonophysics*, **102**, 175–205.

HILDEBRAND, R. S. & BOWRING, S. A. 1999. Crustal recycling by slab failure. *Geology*, **27**, 11–14.

HOFFMAN, P. F. & BOWRING, S. A. 1984. Short-lived continental margin and destruction, Wopmay orogen, northwest Canada. *Geology*, **12**, 68–72.

HOLBROOK, W. S., LIZARRALDE, D., MCGEARY, S., BANGS, N. & DIEBOLD, J. 1999. Structure and composition of the Aleutian island arc and implications for continental crustal growth. *Geology*, **27**, 31–34.

JACOBSON, C. E., OYARZABAL, F. R. & HAXEL, G. B. 1996. Subduction and exhumation of the Pelona–Orocopia–Rand schists, southern California. *Geology*, **24**, 547–550.

JACOBSON, C. E., GROVE, M., VUCIC, A., PEDRICK, J. N. & EBERT, K. A. 2007. Exhumation of the Orocopia schist and associated rocks of southeastern California: Relative roles of erosion, synsubduction tectonic denudation, and middle Cenozoic extension. *In*: CLOOS, M., CARLSON, W. D., GILBERT, M. C., LIOU, J. G. & SORENSEN, S. S. (eds) *Convergent Margin Terranes and Associated Regions: A Tribute to W. G. Ernst*. Geological Society of America, Special Papers, **419**, 1–37.

JICHA, B. R., SCHOLL, D. W., SINGER, B. S., YOGODZINSKI, G. M. & KAY, S. M. 2006. Revised age of Aleutian Island Arc formation implies high rate of magma production. *Geology*, **34**, 661–664; doi: 10.1130/G22433.1.

JOHNSON, M. R. W. 2002. Shortening budget and the role of continental subduction during the India–Asia collision. *Earth-Science Reviews*, **59**, 101–123.

KAY, S. M. & KAY, R. W. 1985. Role of crustal accumulates and the oceanic crust in the formation of the lower crust of the Aleutian arc. *Geology*, **13**, 4612–464.

KAY, R. W. & KAY, S. M. 2009. The Armstrong Unit (AU = km^3/yr) and processes of crust–mantle mass flux. *Abstracts for Goldschmidt 2008—From Sea to Sky Conference, 13–18 July, Vancouver*, abstract 1879.

KAY, S. M., GODOY, E. & ANDREW, K. 2005. Episodic arc migration, crustal thickening, subduction erosion, and magmatism in the south–central Andes. *Geological Society of America Bulletin*, **117**, 67–88, doi:10.1130/B25431.1.

KELEMEN, P. B., YOGODZINSKI, G. M. & SCHOLL, D. W. 2003. Along-strike variation in lavas of the Aleutian Island Arc: Genesis of high Mg# andesite and implications for continental crust. *In*: EILER, J. (ed.) *Inside the Subduction Factory*. American Geophysical Union, Geophysical Monograph, **138**, 223–276.

KIMURA, G. & SAKAKIBARA, M. *ET AL.* 1992. A deep section of accretionary complex: Susunai Complex in Sakhalin Island, Northwest Pacific Margin. *Island Arc*, **1**, 166–175.

KOPP, H. & KUKOWSKI, N. 2003. Backstop geometry and accretionary mechanics of the Sunda margin. *Tectonics*, **22**, 1072, doi:10.1029/2002TC001420.2003.

KOPP, H., FLUEH, E. R., KLAESCHEN, D., BIALAS, J. & REICHERT, C. 2001. Crustal structure of the central Sunda Margin at the onset of oblique subduction. *Geophysical Journal International*, **147**, 449–474.

KUKOWSKI, N. & ONCKEN, O. 2006. Subduction erosion—the 'normal' mode of fore-arc, material transfer along the Chile margin? *In*: ONCKEN, O., CHONG, G. *ET AL.* (eds) *The Andes, Active Subduction Orogeny*. Frontiers in Earth Science, **1**, 217–236.

KUKOWSKI, N., HAMPEL, A., HOTH, S. & BIALAS, J. 2008. Morphotectonic and morphometric analysis of the Nazca plate and the adjacent offshore Peruvian continental slope: Implications for submarine landscape evolution. *Marine Geology*, **254**, 107–120.

LARSON, R. L. 1991. Latest pulse of Earth: Evidence for a mid-Cretaceous superplume. *Geology*, **19**, 547–550.

LARSON, R. L. 1997. Superplumes and ridge interactions between Ontong Java and Manihiki Plateaus and the Nova–Canton Trough. *Geology*, **25**, 779–782.

LAURSEN, J., SCHOLL, D. W. & VON HUENE, R. 2002. Neotectonic deformation of the central Chile margin: Deepwater forearc basin formation in response to hot spot ridge and seamount subduction. *Tectonics*, **21**, 1038, doi:10.1029/2001TC901023.

MANN, P. & TAIRA, A. 2004. Global tectonic significance of the Solomon Islands and Ontong Java Plateau convergent zone. *Tectonophysics*, **389**, 137–190.

MOORE, J. C., DIEBOLD, J. *ET AL.* 1991. EDGE deep seismic reflection transect of the eastern Aleutian arc–trench layered lower crust reveals underplating and continental growth. *Geology*, **19**, 420–424.

MOORE, G. F., BANGS, N. L., TAIRA, A., KURAMOTO, S., PANGBORN, E. & TOBIN, H. J. 2007. Three-dimensional splay fault geometry and implications for Tsunami generation. *Science*, **318**, 1128–1131.

NEAL, C. R., COFFIN, M. F. *ET AL.* 2008. Investigatiing large igneous provinces formation and associated paleoenvironmental events: A white paper for scientific drilling. *Scientific Drilling (IODP)*, **6**, 4–18, doi:10.2204/iodp.sd.6.01.

NORTON, I. O. 2007. Speculations on Cretaceous tectonic history of the northwest Pacific and a tectonic origin for the Hawaii hotspot. *In*: FOULGER, G. R. & JURDY, D. M. (eds) *Plates, Plumes, and Planetary Processes*. Geological Society of America, Special Papers, **530**, 451–470.

NUR, A. & BEN-AVRAHAM, Z. 1981. Volcanic gaps and the consumption of aseismic ridges in South America. *In*: KULM, L., DYMOND, J., DASCH, E. J. & HUSSONG, D. M. (eds) *Nazca Plate: Crustal Formation and Andean Convergence*. Geological Society of America, Memoirs, **154**, 729–740.

ONCKEN, O. 1998. Evidence for precollisional subduction erosion in ancient collisional belts: The case of the Mid-European Variscides. *Geology*, 1075–1078.

O'NEILL, C., LENARDIC, A., MORES, L., TORSVIK, T. H. & LEE, C.-T. A. 2007. Episodic Precambrian subduction. *Earth and Planetary Science Letters*, **262**, 552–562.

PARK, J.-O., TSURU, T., KODAIRA, S., CUMMINS, P. R. & KANEDA, Y. 2002. Splay fault branching along the Nankai subduction zone. *Science*, **297**, 1157–1160.

PLAFKER, G. & BERG, H. C. 1994. The geology of Alaska. *In*: PLAFKER, G. & BERG, H. C. (eds) *The Geology of Alaska*. The Geology of North America, **G-1**, 1–1055.

PLAFKER, G., MOORE, J. C. & WINKLER, G. 1994. Geology of the southern Alaska margin. *In*: PLAFKER, G. & BERG, H. C. (eds) *The Geology of Alaska*. The Geology of North America, **G-1**, 389–449.

RANERO, C. R. & VON HUENE, R. 2000. Subduction erosion along the Middle America convergent margin. *Nature*, **404**, 748–752.

RANERO, C. R., VON HUENE, R., WEINREBE, W. & REICHERT, C. 2006. Tectonic processes along the chile convergent margin. *In*: ONCKEN, O., CHONG, G. ET AL. (eds) *The Andes, Active Subduction Orogeny*. Frontiers in Earth Science, **1**, 91–119.

REA, D. K. & SNOECKX, H. 1995. Sediment fluxes in the Gulf of Alaska; paleoceanographic record from Site 887 on the Patton–Murray Seamount Platform. *In*: REA, D. K., BASOV, I. A., SCHOLL, D. W. & ALLAN, J. F. (eds) *Proceedings of the Ocean Drilling Program, Scientific results, 145*. Ocean Drilling Program, College Station, TX, 247–256.

REYMER, A. & SCHUBERT, G. 1984. Phanerozoic addition rates to the continental crust and crustal growth. *Tectonics*, **3**, 63–77.

RIDGWAY, K. D., TROP, J. M., GLEN, J. M. G. & O'NEILL, J. M. (eds) 2007. *Tectonic Growth of a Collisional Continental Margin: Crustal Evolution of Southern Alaska*. Geological Society of America, Special Papers, **431**.

RUDNICK, R. L. & GAO, S. 2003. Composition of Continental Crust. *In*: HOLLAND, H. D. & TUREKIAN, K. K. (eds) *Treatise on Geochemistry*, **3**, 1–64. Elsevier-Pergamon, Oxford.

RYAN, H. F. & SCHOLL, D. W. 1989. The evolution of forearc structures along an oblique convergent margin, central Aleutian Arc. *Tectonics*, **8**, 497–516.

SAGE, F., COLLOT, J.-Y. & RANERO, C. 2006. Interplate patchiness and subduction-erosion mechanisms: Evidence from depth-migrated seismic images at the central Ecuador convergent margin. *Geology*, **34**, 997–1000, doi:10.1130/G22790A.1.

SAGER, W. W. 2005. What built Shatsky Rise, a mantle plume or ridge tectonics? *In*: FOULGER, G. R., NATLAND, J. H., PRESNAL, D. C. & ANDERSON, D. L. (eds) *Plates, Plumes and Paradigms*. Geological Society of America, Special Papers, **388**, 721–733.

SAGER, W. W., HANDSCHUMACHER, D. W., HILDE, T. W. C. & BRACEY, D. R. 1988. Tectonic evolution of the northern Pacific plate and Pacific–Farallon–Izanagi triple junction in the Late Jurassic and Early Cretaceous (M21–M10). *Tectonophysics*, **155**, 345–364.

SAMPLE, J. C. & MOORE, J. C. 1987. Structural style and kinematics of an underplated slate belt, Kodiak and adjacent islands, Alaska. *Geological Society of America Bulletin*, **99**, 7–20.

SCHOLL, D. W. 2007. Viewing the tectonic evolution of the Kamchatka–Aleutian (KAT) connection with an Alaska crustal extrusion perspective. *In*: EICHELBERGER, J. C., GORDEEV, E., IZBEKOV, P., KASAHARA, M. & LEES, J. (eds) *Volcanism and Subduction in the Kamchatka Region*. American Geophysical Union, Geophysical Monograph, **172**, 3–35.

SCHOLL, D. W. & HERZER, R. H. 1992. Geology and resource potential of the southern Tonga platform, 1992. *In*: WATKINS, J. S., ZHIQIANG, FENG & MCMILLEN, K. J. (eds) *Geology and Geophysics of Continental Margins*. American Association of Petroleum Geologists, Memoirs, **53**, 139–156.

SCHOLL, D. W. & HERZER, R. H. 1994. Geology and resource potential of the Tonga–Lau region. *In*: STEVENSON, A. J., HERZER, H. R. & BALLANCE, P. F. (eds) *Geology and Submarine Resources of the Tonga–Lau Region*. SOPAC Technical Bulletin, **8**, 329–350.

SCHOLL, D. W. & VON HUENE, R. 2007. Crustal recycling at modern subduction zones applied to the past—issues of growth and preservation of continental basement, mantle geochemistry, and supercontinent reconstruction. *In*: HATCHER, R. D. JR, CARLSON, M. P., MCBRIDE, J. H. & MARTÍNEZ CATALÁN, J. R. (eds) *4D Framework of Continental Crust*. Geological Society of America, Memoirs, **200**, 9–32.

SCHOLL, D. W. & VON HUENE, R. In press. Subduction zone recycling and truncated or missing tectonic and depositional domains expected at continental suture zones. *In*: CLOWES, R. & SKULSKI, T. (eds) *LITHOPROBE—Parameters, Processes and the Evolution of a Continent. LITHOPROBE Synthesis Volume II*. National Research Council Canada, Research Press.

SCHOLL, D. W., VON HUENE, R., VALLIER, T. L. & HOWELL, D. G. 1980. Sedimentary masses and concepts about tectonic processes at underthrust ocean margins. *Geology*, **8**, 564–568.

SCHOLL, D. W., VALLIER, T. L. & PACKHAM, G. 1985. Framework geology and resource potential of the southern Tonga platform and adjacent terranes: A synthesis. *In*: SCHOLL, D. W. & VALLIER, T. L. (eds) *Resources of Island Arcs—Tonga Region*. Circum-Pacific Council for Energy and Mineral Resources, Earth Science Series, **2**, 457–475.

SCHULTER, H. U., GAEDICKE, C. ET AL. 2002. Tectonic features of the southern Sumatra–western Java forearc of Indonesia. *Tectonics*, **21**, 1047, doi:10.1029/2001TC901048.

SILVER, P. G. & BEHN, M. D. 2008. Intermittent plate tectonics. *Science*, **319**, 85–88.

STERN, C. R. 1991. Role of subduction erosion in the generation of Andean magmas. *Geology*, **19**, 78–81, doi:10.1130/0091-7613(1991)0192.3.CO:2.

STERN, R. J. 2004. Subduction initiation: Spontaneous and induced. *Earth and Planetary Science Letters*, **226**, 275–292.

STERN, R. J. 2008. Neoproterozoic crustal growth: The solid Earth system during a critical episode of Earth history. *Gondwana Research*, **14**, 33–50.

STERN, R. J. & BLOOMER, S. H. 1992. Subduction zone infancy: Example from the Eocene Izu–Bonin–Mariana and Jurassic California arc. *Geological Society of America Bulletin*, **104**, 1621–1636.

SUKHOV, A. N., BOGDANOV, N. A. & CHEKHOVICH, V. D. 2004. Geodynamics and paleogeography of the northwestern Pacific continental margin in the Late Cretaceous. *Geotectonics*, **38**, 61–71.

SYLVESTER, P. J. 2000. *Continental Formation, Growth and Recycling*. Elsevier, Amsterdam.

TAKAHASI, N., KODAIRA, S., KLEMPERER, S. L., TATSUMI, Y., KANEDA, Y. & SUYEHIRO, K. 2007. Crustal structure and evolution of the Mariana intra-oceanic island arc. *Geology*, **35**, 203–206, doi:10.1130/G23212A.1.

TARDUNO, J. A., DUNCAN, R. A. ET AL. 2003. The Emperor Seamounts: Southward motion of the Hawaiian hotspot plume in Earth's mantle. *Science*, **301**, 1064–1069.

TORRINI, R. JR & SPEED, R. C. 1989. Tectonic wedging in the forearc basin—Accretionary prism transition, Lesser Antilles Forearc. *Journal of Geophysical Research*, **94**, 10549–10584.

TORRINI, R. JR, SPEED, R. C. & MATTIOLI, G. S. 1985. Tectonic relationships between forearc-basin strata and the accretionary complex at Bath, Barbados. *Geological Society of America Bulletin*, **96**, 861–874.

TROP, J. M. & RIDGWAY, K. D. 2007. Mesozoic and Cenozoic tectonic growth of southern Alaska: A sedimentary basin perspective. *In*: RIDGWAY, K. D., TROP, J. M., GLEN, J. M. G. & O'NEILL, J. M. (eds) *Tectonic Growth of a Collisional Margin: Crustal Evolution of Southern Alaska*. Geological Society of America, Special Papers, **431**, 55–93.

VAN DER VELDEN, A. J. & COOK, F. A. 2005. Relict subduction zones in Canada. *Journal of Geophysical Research*, **110**, B08403, doi:10.1029/2004/JB003333.

VAN KRANENDONK, M. J. 2004. Archean tectonics 2004: A review. *Precambrian Research*, **131**, 143–151.

VAN KRANENDONK, M. J. & CASSIDY, K. 2004. Comment: An alternative Earth, Warren B. Hamilton, GSA Today, **13**, 4–12. *GSA Today*, **14**, 14–14.

VANNUCCHI, P., SCHOLL, D. W. & MESCHEDE, M. 2001. Tectonic erosion and consequent collapse of the Pacific margin of Costa Rica: Combined implications from ODP Leg 170, seismic offshore data, and regional geology of the Nicoya Peninsula. *Tectonics*, **20**, 649–668.

VANNUCCHI, P., RANERO, C. R., GALEOTTI, S., STRAUB, S. M., SCHOLL, D. W. & MCDOUGALL-RIED, K. 2003. Fast rates of subduction erosion along the Costa Rica Pacific margin: Implications for nonsteady rates of crustal recycling at subduction zones. *Journal of Geophysical Research*, **108**, 2511, doi:10.1029/2002JB002207.

VANNUCCHI, P., REMITTI, F. & BETTELLI, G. 2008. Geological record of fluid flow and seismogenesis along an erosive subducting plate boundary. *Nature*, **451**, 699–703, doi:10.1038/nature06486.

VON HUENE, R. & LALLEMAND, S. 1990. Tectonic erosion along the Japan and Peru convergent margins. *Geological Society of America Bulletin*, **102**, 704–720.

VON HUENE, R. & RANERO, C. R. In press. Neogene collision and deformation of convergent margins along the Americas backbone. *In*: KAY, S., RAMOS, V. & DICKINSON, W. (eds) *Backbone of the Americas: Shallow Subduction, Plateau Uplift, and Ridge Collision*. Geological Society of America, Memoirs

VON HUENE, R. & SCHOLL, D. W. 1991. Observations at convergent margins concerning sediment subduction, subduction erosion, and the growth of continental crust. *Reviews of Geophysics*, **29**, 279–316.

VON HUENE, R., RANERO, C. R., WEINREBE, W. & HINZ, K. 2000. Quaternary convergent margin tectonics of Costa Rica, segmentation of the Cocos plate and Central America volcanism. *Tectonics*, **19**, 314–334.

VON HUENE, R., RANERO, C. R. & VANNUCCHI, P. 2004. Generic model of subduction erosion. *Geology*, **32**, 913–916, doi:10.1130/G20563.1.

VON HUENE, R., RANERO, C. R. & SCHOLL, D. W. 2009. Convergent margin structure in high quality images and current models: A review. *In*: LALLEMAND, S. (ed.) *Subduction Zones Geodynamics*. Springer, Berlin.

WAGREICH, M. 1993. Subcrustal tectonic erosion in orogenic belts: A model for the Late Cretaceous subsidence of the northern Calcareous Alps (Austria). *Geology*, **21**, 941–944.

WALLACE, W. K. 2008. Yakataga fold-and-thrust belt: Structural geometry and tectonic implications of a small continental collision zone. *In*: FREYMUELLER, J., HAEUSSLER, P., WESSON, R. & EKSTROM, G. (eds) *Active Tectonics and Seismic Potential of Alaska*. American Geophysical Union, Geophysical Monograph, **179**, 1–431.

WHITMEYER, S. J. & KARLSTROM, K. E. 2007. Tectonic model for the Proterozoic growth of North America. *Geosphere*, **3**, 220–259, doi:10.1130/GES00055.1.

WORKMAN, R. K., EILER, J. M., HART, S. R. & JACKSON, M. G. 2008. Oxygen isotopes in Samoan lavas: Confirmation of continent recycling. *Geology*, **36**, 551–554.

Eoarchaean crustal growth in West Greenland (Itsaq Gneiss Complex) and in northeastern China (Anshan area): review and synthesis

ALLEN P. NUTMAN[1,2]*, VICKIE C. BENNETT[2], CLARK R. L. FRIEND[1,3], FRANCES JENNER[2], YUSHENG WAN[1] & DUNYI LIU[1]

[1]*Institute of Geology and Chinese International Centre for Precambrian Research, Chinese Academy of Geological Sciences, 26 Baiwanzhuang Road, Beijing, 100037, China*

[2]*Research School of Earth Sciences, Australian National University, Canberra, ACT 0200, Australia*

[3]*45, Stanway Road, Headington, Oxford OX3 8HU, UK*

*Corresponding author (e-mail: nutman@bjshrimp.cn)

Abstract: Eoarchaean crust in West Greenland (the Itsaq Gneiss Complex, 3870–3600 Ma) is >80% by volume orthogneisses derived from plutonic tonalite–trondhjemite–granodiorite (TTG) suites, <10% amphibolites derived from basalts and gabbros, <10% crustally derived granite, <1% metasedimentary rocks and ≪1% tectonic slices of upper mantle peridotite. Amphibolites at >3850, c. 3810 and c. 3710 Ma have some compositional similarities to modern island arc basalts (IAB), suggesting their origin by hydrous fluxing of a suprasubduction-zone upper mantle wedge. Most of the Eoarchaean tonalites match in composition high-silica, low-magnesian adakites, whose petrogenesis is dominated by partial melting of garnetiferous mafic rocks at high pressure. However, associated with the tonalites are volumetrically minor more magnesian quartz diorites, whose genesis probably involved melting of depleted mantle to which some slab-derived component had been added. This assemblage is evocative of suites of magmas produced at Phanerozoic convergent plate boundaries in the case where subducted crust is young and hot. Thus, Eoarchaean 'subduction' first gave rise to short-lived episodes of mantle wedge melting by hydrous fluxing, yielding IAB-like basalts ± boninites. In the hotter Eoarchaean Earth, flux-dominated destructive plate boundary magma generation quickly switched to slab melting of ('subducted') oceanic crust. This latter process produced the voluminous tonalites that were intruded into the slightly older sequences consisting of tectonically imbricated assemblages of IAB-like pillow lavas + sedimentary rocks, gabbros and upper mantle peridotite slivers. Zircon dating shows that Eoarchaean TTG production in the Itsaq Gneiss Complex was episodic (3870, 3850–3840, 3820–3810, 3795, 3760–3740, 3710–3695 and 3660 Ma). In each case, emplacement of small volumes of magma was probably followed by 10–40 Ma quiescence, which allowed the associated thermal pulse to dissipate. This explains why Greenland Eoarchaean crustal growth did not have granulite-facies metamorphism directly associated with it. Instead, 3660–3600 Ma granulite-facies metamorphism(s) in the Itsaq Gneiss Complex were consequential to collisional orogeny and underplating, upon termination of crustal growth. Similar Eoarchaean crustal history is recorded in the Anshan area of China, where a few well-preserved rocks as old as 3800 Ma have been found including high-MgO quartz diorites. For 3800 Ma rocks, this is a rare, if not unique, situation outside of the Itsaq Gneiss Complex. The presence of volumetrically minor 3800 Ma mantle-derived high-MgO quartz diorites in both the Itsaq Gneiss Complex and the Anshan area indicates either that Eoarchaean 'subduction' zones were overlain by a narrow mantle wedge or that the shallow subduction trapped slivers of upper mantle between the conserved and consumed plates.

The first Rb–Sr isotopic studies (Moorbath *et al.* 1972; Moorbath 1975) of the then oldest-known Eoarchaean orthogneisses from West Greenland recognized that their primitive (i.e. low initial $^{87}Sr/^{86}Sr$) signatures meant the gneisses represented predominantly 'juvenile' magmas derived from material separated from the mantle shortly before their genesis. These juvenile magmas were accreted into existing crust, and the gneisses were not 'reworked' from appreciably older Hadean (>4000 Ma) rocks. (Accreted is used here in an Archaean sense, to mean growth of crust by emplacement of magmas, probably at convergent plate boundaries, as exemplified by Moorbath *et al.* (1972) and Wells (1980).) This interpretation was coupled with the recognition that these ancient rocks were derived largely from tonalitic plutonic protoliths (McGregor 1973, 1979) with a likely

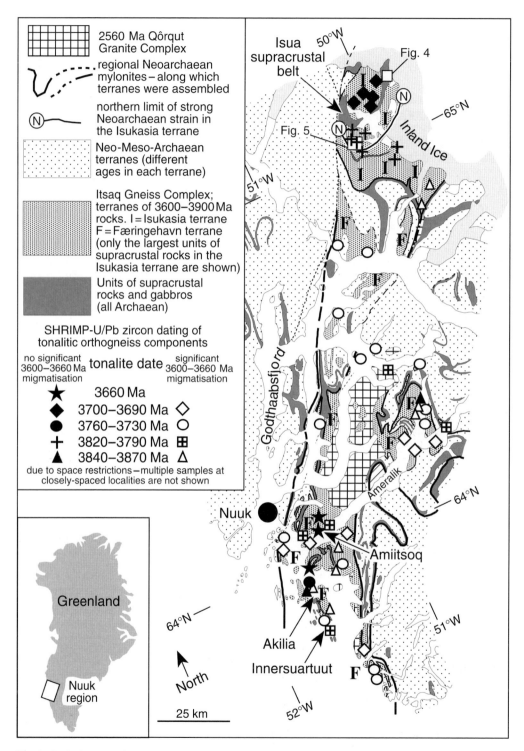

Fig. 1. Geological map of the Nuuk region, displaying the Eoarchaean Itsaq Gneiss Complex, and summary of SHRIMP U/Pb zircon results on TTG rocks. Localities mentioned in the text are indicated.

origin by melting of mafic rocks (O'Nions & Pankhurst 1974; McGregor 1979). Rocks of such compositions are overwhelmingly produced at convergent plate boundaries (see Martin 1999, for summary). Structural and mapping evidence shows that Eoarchaean tonalite–trondhjemite–granodiorite (TTG) suite rocks were intruded into already tectonically imbricated packages of metavolcanic rocks, gabbros, mantle-derived harzburgites and dunites and banded iron formation–chert (e.g. Nutman & Collerson 1991; Nutman et al. 1996, 2007a; Friend et al. 2002). Thus in this paper we coin the phrase 'crustal accretion–growth', to describe the process responsible for imbrication of unrelated rocks followed by intrusion of isotopically juvenile TTG.

Thus since the 1970s, it has been proposed that crustal accretion–growth processes with some resemblance to modern plate boundary processes, operated at least as far back in time as the Eoarchaean (3500–4000 Ma). Although there are many differences in the evidence cited and methods used, subsequent studies of Greenland Eoarchaean crustal evolution concur with Moorbath (1975) that Eoarchaean rocks are dominantly the juvenile additions to the crust, and it has been proposed that they formed at convergent plate boundaries (e.g. Nutman & Collerson 1991; Bennett et al. 1993; Nutman et al. 1993, 1996, 2000, 2002a, b; Komiya et al. 1999; Polat et al. 2002; Hanmer & Greene 2002; Polat & Hofmann 2003; Steenfelt et al. 2005; Furnes et al. 2007).

This paper examines the nature of very ancient crustal accretion–growth by focusing on the rich Eoarchaean geological record of West Greenland, preserved within the Itsaq Gneiss Complex (Fig. 1). This record is then compared with the less extensive Eoarchaean geological record of China (Liu et al. 1992; Song et al. 1996; Wan et al. 2005). Despite being more limited, this record is important, as this is perhaps the only other place in the world where single-component meta-igneous rocks as old as 3800 Ma have been found (Wan et al. 2005). These two ancient rock suites that are interpreted to form at >3600 Ma convergent plate boundaries are then briefly contrasted with the products of Phanerozoic crustal accretionary systems.

Eoarchaean geological record of West Greenland (Itsaq Gneiss Complex) and understanding ancient crustal accretion–growth

Study of modern versus ancient systems

Subduction zones and other related components of modern convergent plate boundaries where crustal accretion is currently taking place can be studied in great detail, using numerous methods, ranging from deep seismic imaging (e.g. tomography), charting centimetre-scale relative crustal motions using global positioning system (GPS), to structural geology and numerous geochemical and isotopic analytical techniques. However, the Eoarchaean crustal rocks that are interpreted as the products of crustal accretion–growth at convergent plate boundaries are preserved only as small fragments in younger Meso-Neoarchaean orogens, where they underwent repeated later deformation and metamorphism (see Nutman 2006, for summary). Thus compared with modern settings, a much narrower range of information can be obtained from these Eoarchaean rocks. Information from these rocks is heavily biased towards geochemical and isotopic data, with use of structural information limited to very rare areas where total strain was low. Thus in studying the Eoarchaean geological record, there is no hope of identifying either intact fossil subduction zones (as revealed by seismic studies of the Neoarchaean Superior Province greenstone–granite terrane of Canada; Calvert et al. 1995) or extensive original plate boundaries and palaeogeographies.

Geological setting of the Itsaq Gneiss Complex

In West Greenland, Eoarchaean rocks occur in the Nuuk region (Fig. 1), where they crop out over about 3000 km^2, and are known collectively as the Itsaq Gneiss Complex (Nutman et al. 1996). Although this complex constitutes at least a quarter of the preserved global Eoarchaean geological record, it is smaller than many single islands in a modern island arc system, giving an idea of the scale of the surviving Eoarchaean geological record. Additionally, the Itsaq Gneiss Complex Eoarchaean crustal accretion products were affected by latest Eoarchaean (<3660 Ma) high-grade metamorphism and deformation (Griffin et al. 1980; Nutman et al. 1996; Friend & Nutman 2005a). We interpret these latest Eoarchaean events to be consequential to a collisional orogeny that terminated crustal accretion–growth (Friend & Nutman 2005a). Thus the products of likely Eoarchaean crustal accretion–growth are preserved mostly as strongly deformed amphibolite–granulite-facies gneisses (Fig. 2a), from which it is hard to extract any detailed information concerning their origins. However, locally there are low-strain domains where some more detailed information on Eoarchaean crustal accretion can be gathered (Fig. 2b). Most of these domains are in and around the Isua supracrustal belt (Fig. 1), an area that since the

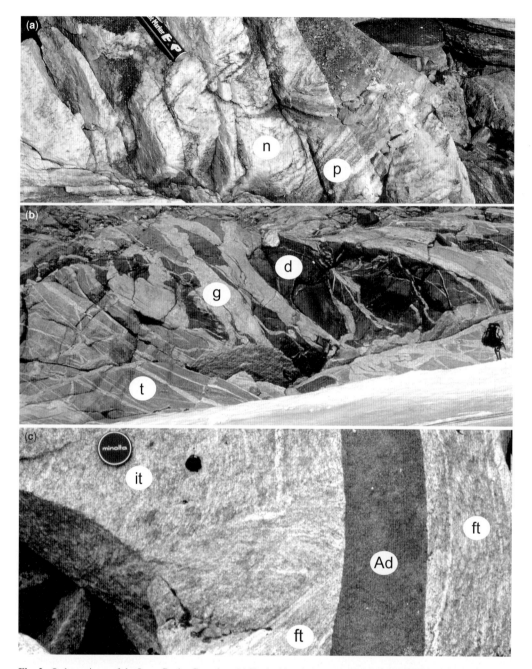

Fig. 2. Orthogneisses of the Itsaq Gneiss Complex. (a) Typical banded grey gneiss from the Færingehavn terrane. Reconnaissance SHRIMP U/Pb zircon dating (H. Horie, A. P. Nutman & H. Hidaka, unpubl. data) indicates that this rock contains predominantly c. 3750 Ma palaeosome (p) and c. 3650 Ma neosome (n). Because of the intimate association of components of different age caused by some *in situ* melting and superimposed strong polyphase Archaean deformation, such rocks are unsuitable for geochemical studies. (b) Intrusive relationships preserved in plutonic rocks north of the Isua supracrustal belt. Weakly deformed c. 3700 Ma tonalites (t) are first cut by a dioritic c. 3660 Ma Inaluk dyke (d) and then by marginally younger (3660–3650 Ma) leucogranite sheets (g). Photo courtesy of J. L. Crowley. (c) Complexities of c. 3800 Ma tonalites in rare low-strain zone south of the Isua supracrustal belt. Isotropic biotite tonalite (it) at top left forms matrix to jumbled blocks of weakly foliated or banded tonalite (ft). Tonalites are cut by an amphibolitized, but undeformed Ameralik dyke (Ad) with likely age of 3500 Ma.

early 1970s (Moorbath et al. 1973; Bridgwater & McGregor 1974; Allaart 1976) has been an enduring focus of research into Eoarchaean geology.

The Itsaq Gneiss Complex occurs as mylonite-bounded tectonic slivers within a collage of younger terranes (each terrane has its own rock ages and internal evolution). The Itsaq Gneiss Complex in the Nuuk region is represented by the Isukasia terrane to the north (with the Isua supracrustal belt) and the Færingehavn terrane to the south (Fig. 1). These terranes were assembled with younger rocks during several Neoarchaean tectonothermal events (see Friend et al. 1988; Crowley 2002; Friend & Nutman 2005b). Thus the Eoarchaean rocks are allochthons found within a Neoarchaean orogen (McGregor et al. 1991; Friend & Nutman 2005b).

The Itsaq Gneiss Complex contains >95% quartzo-feldspathic rocks of intrusive origin, now mostly strongly deformed into orthogneisses. The rare areas of relatively little deformation show that these gneisses usually formed from older tonalite and younger granite components (Fig. 2b). All researchers (e.g. Steenfelt et al. 2005) agree that the tonalites are compositionally like Archaean TTG suite rocks. True (potassic) granites also occur (Nutman et al. 1984a, 1996; Nutman & Bridgwater 1986; Crowley et al. 2002). These were formed largely by melting of crust dominated by tonalite (Baadsgaard et al. 1986). Volcanic and sedimentary (supracrustal) rocks form <5% of the complex and are found scattered through the banded gneisses as enclaves and tectonic slivers, and range in size from the 35 km long Isua supracrustal belt (e.g. Bridgwater & McGregor 1974; Allaart 1976; Nutman et al. 1984b, 1997, 2002b; Komiya et al. 1999) down to metre-sized pods (e.g. McGregor & Mason 1977; Nutman et al. 2002a). Volcanic and sedimentary rocks are dominated by associations of banded and commonly skarn-bearing amphibolites, with lesser amounts of quartz–magnetite banded iron formation (BIF), marbles, siliceous rocks and calc-silicate rocks. In the amphibolites, rare occurrences of relict pillow lava structure have been found (Fig. 3a; Komiya et al. 1999; Solvang 1999; Furnes et al. 2007). Several workers have interpreted the BIF, siliceous and calc-silicate lithologies together as a variegated suite of chemical sediments (Fig. 3b; e.g. Nutman et al. 1984b; Dymek & Klein 1988; Dauphas et al. 2004; Bohlar et al. 2004), albeit these have often been heavily metasomatized. Less common are felsic schists and pelites, with all of those that are agreed to have a volcano-sedimentary rather than metasomatic origin occurring in the Isua supracrustal belt (e.g. Nutman et al. 1984b, 1997, 2002b; Kamber et al. 2005). Metagabbros, locally grading into anorthosites (Fig. 3c) also occur, and are usually spatially associated with layered meta-peridotites (Chadwick & Crewe 1986; Nutman et al. 1996; Friend et al. 2002). These are interpreted as fragments of layered (basic) intrusions. Other ultramafic rocks are highly magnesian, with very low alumina and lime. Most of these are metasomatized amphibole ± phlogopite-bearing schists. However, very rarely, they are much better preserved, and occur as fine-grained olivine + orthopyroxene ± aluminous spinel ± amphibole dunites and harzburgites (Nutman et al. 1996; Friend et al. 2002).

The Isukasia terrane occurs in the northern part of the Nuuk region and contains the Isua supracrustal belt (Fig. 1). In and around the Isua supracrustal belt there are most of the world's occurrences of Eoarchaean rocks preserved in a low-strain state. Additionally, the Isua area as a whole underwent polyphase metamorphism at only amphibolite-facies conditions, rather than granulite-facies conditions as in other parts of the complex (Griffin et al. 1980; Nutman et al. 1996). In the northern parts of the Isukasia terrane (north of line 'N' in Fig. 1), c. 3500 Ma Ameralik dykes are weakly to non-deformed, showing that the amount of Neoarchaean deformation is generally low (Bridgwater & McGregor 1974; Allaart 1976). The low-strain domains show that in the Isukasia terrane, in situ melting was very limited (Nutman et al. 1996), which greatly aids the interpretation of these rocks. The southern edge of the Isukasia terrane is more deformed, and consists of banded gneisses resembling in appearance those of the Færingehavn terrane (Bridgwater & McGregor 1974; Nutman et al. 1996).

The Færingehavn terrane is where terrestrial >3600 Ma rocks were first identified (Black et al. 1971; McGregor 1973). The type locality of these rocks was Amîtsoq (using the old Greenlandic orthography) on the north shore of Ameralik (fjord), and hence the Eoarchaean banded orthogneisses became known as the Amîtsoq gneisses (McGregor 1973). McGregor and his collaborators abandoned this term in 1996, when instead these rocks became regarded as the dominant part of the Itsaq Gneiss Complex (Nutman et al. 1996). The Færingehavn terrane occurs in the central and southern part of the Nuuk region (Fig. 1) and overall its rocks are much more modified by deformation and high-grade metamorphism than the northern parts of the Isukasia terrane, making it harder to obtain unequivocal interpretations of the rocks. In rare domains of lower strain in the Færingehavn terrane, tonalitic protoliths are locally well preserved (Nutman et al. 2000, 2007a, b), and it is seen that granitic components are derived both from in situ melting of the tonalites and from intrusions. These anatectic and emplacement events

Fig. 3. Supracrustal rocks and gabbros of the Itsaq Gneiss Complex. (**a**) Relict pillow structure in amphibolite with IAB chemistry in the western part of the Isua supracrustal belt. Pillows face to the left of the picture but have been substantially flattened orthogonally to their original layering. Tops (T) and bases (B) of some pillows are indicated. Triangular domains of silica-rich interpillow hyaloclastite are locally preserved (iph). (**b**) BIF from the eastern end of the Isua supracrustal belt. In the right-hand side of the photograph the original sedimentary layering is preserved, but is

have been equated with petrographic evidence for Eoarchaean granulite-facies metamorphism in the Færingehavn terrane (McGregor & Mason 1977; Griffin et al. 1980; Friend & Nutman 2005a). Units of c. 3640 Ma (Baadsgaard 1973) mixed Fe-rich augen granites, monzonites and ferro-gabbros are a distinct component of the southern part of the Færingehavn terrane (McGregor 1973). These mixed rocks are the product of hybridization of magmas derived from the mantle and deep crust, and resemble A-type or within-plate granites with high Nb, Zr, TiO_2 and P_2O_5 (Nutman et al. 1984a, 1996).

The orthogneisses also contain lenses of amphibolites, ultramafic and siliceous rocks, named the Akilia association by McGregor & Mason (1977). Although it is agreed that these largely represent enclaves of mafic volcanic rocks, gabbros and chemical sediments, there is lack of agreement concerning their age. This is because in most places the original relationship with the surrounding orthogneisses has been obliterated by migmatization and high strain (compare Myers & Crowley (2000) with Nutman et al. (2000, 2002a)). Nutman et al. (1993, 1996, 2000) and Friend & Nutman (2005a) presented evidence that the Akilia association represents the remains of several supracrustal sequences, ranging from $\geq$3850 Ma to 3650–3600 Ma. However, Whitehouse et al. (1999) and Whitehouse & Kamber (2005) considered that there is not yet strong enough evidence for any of the Akilia association being $\geq$3850 Ma.

Isua supracrustal belt and surrounding gneisses

Isua supracrustal belt

Despite the fact that the Isua supracrustal belt escaped much deformation in the Neoarchaean, most of it is strongly deformed as a result of Eoarchaean deformation (Nutman et al. 1984b, 2002a; Myers 2001). Thus, in most places, primary volcanic and sedimentary structures were obliterated, and outcrop-scale compositional layering is dominantly of transposed tectonic origin. It is only in rare augen of total low strain that volcanic and sedimentary structures are preserved. Unequivocal pillow structures in Isua supracrustal belt amphibolites (Komiya et al. 1999; Solvang 1999; Furnes et al. 2007) are important, because, by following them into high-strain equivalents, they demonstrate that many of the Isua amphibolites were derived from subaqueous volcanic rocks (Fig. 4a). This contrasts with previous ideas held in the 1970s and 1980s that amphibolites derived from gabbros were an equally important component in the belt as metavolcanic ones (e.g. Nutman et al. 1984b).

Rocks of chemical sedimentary origin such as banded iron formation (Moorbath et al. 1973; Dymek & Klein 1988) have quartz and either calc-silicate or magnetite layering. This is mostly a transposed layering, and only locally does it represent original (albeit still deformed) sedimentary structure (Fig. 4b). There are two prominent felsic schist units. The first one is restricted to the north-eastern end of the belt and has yielded several zircon dates of c. 3710 Ma, and there is agreement that it is a felsic volcano-sedimentary unit (Nutman et al. 1997, 2002b; Kamber et al. 2005). The second unit crops out throughout the length of the belt, and has in several places yielded zircon dates of c. 3800 Ma (e.g. Baadsgaard et al. 1984; Compston et al. 1986; Nutman et al. 2002b). For this unit there is lack of agreement on whether it is a (metasomatized) felsic volcanic rock (Allaart 1976; Nutman et al. 1984b, 2002b), an altered tonalitic intrusion (Rosing et al. 1996) or metasomatized basalt (Myers 2001).

Nutman et al. (1984b) initially regarded the Isua supracrustal belt as fragments of one or possibly two volcano-sedimentary sequences ('A' and 'B'), disrupted by early tectonic breaks. At that time there was only zircon geochronology on one outcrop of felsic rocks in the belt, with ages of c. 3800 Ma (Michard-Vitrac et al. 1977; Baadsgaard et al. 1984). The only other age constraints came from a whole-rock Pb–Pb errorchron of 3710 ± 70 Ma on the BIF (Moorbath et al. 1973), an Sm–Nd errorchron of 3770 ± 42 Ma for a mixed suite of felsic and mafic rocks (Hamilton et al. 1978), and Rb–Sr whole-rock Eoarchaean errorchrons ($\pm >50$ Ma) for orthogneiss components invading and surrounding the belt (e.g. Moorbath et al. 1972, 1977). Within this early geochronological framework, it was considered that the Isua supracrustal rocks were all related, a concept then

Fig. 3. (Continued) still deformed. In the left-hand side of the photograph deformation is much stronger, with transposed quartz + magnetite layering and development of a new magnetite foliation. This is the more typical state of preservation of BIF in the Itsaq Gneiss Complex. (c) Relict graded layering in a layered gabbro–anorthosite enclave, in the Færingehavn terrane. Anorthosite (an) at the bottom of the exposure has a sharp contact with overlying melagabbro (mg), which then grades up into gabbro (g) and leucogabbro (lg). Mineralogy is now entirely metamorphic plagioclase + hornblende. There is no igneous plagioclase + pyroxenes $\pm$ olivine preserved. A segregation vein formed under granulite-facies conditions but now retrogressed under amphibolite-facies conditions (granu vein) cuts the igneous layering.

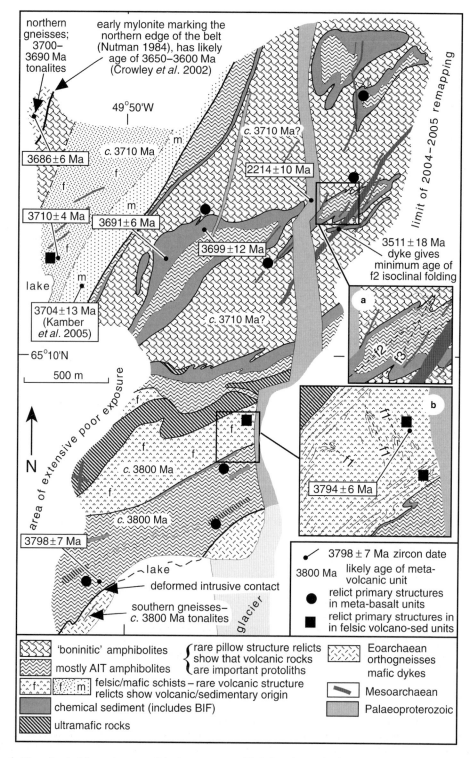

Fig. 4. Map of part of the eastern part of the Isua supracrustal belt, based on revised mapping by Nutman & Friend in 2004 and 2005. (a) Detail of structural relationships between AIT, boninites and chemical sediment. (b) Detail of early isoclinal folds in c. 3800 Ma felsic schist unit. All zircon age determinations without name indicated are by Nutman and coworkers (unpublished). AIT, (Archaean) Island arc tholeiities.

widely accepted for all other ancient volcano-sedimentary sequences, such as the Barberton Greenstone Belt. With the start of a broader sensitive high-resolution ion microprobe (SHRIMP) U–Pb zircon-dating programme with rock ages with uncertainties of $\leq \pm 5$ Ma, it was realized that the belt contained supracrustal rocks varying in age by 100 Ma, and hence the earlier 'stratigraphy' was abandoned (Nutman et al. 1997). Thus the belt's southern part is dominated by c. 3800 Ma rocks, whereas along its northern and central reaches c. 3700 Ma rocks are present (e.g. Fig. 4; Nutman et al. 1996, 1997, 2002b; Crowley 2003; Kamber et al. 2005). Nutman et al. (1997) first proposed that these unrelated sequences were separated by Eoarchaean mylonites and have expanded on this in more recent work (Nutman et al. 2002b, 2007a). Other workers have concurred with this interpretation, and Appel et al. (1998), Komiya et al. (1999) and Myers (2001) all indicated mylonites in their mapping, to divide the belt into Eoarchaean panels of unrelated rocks.

Orthogneisses adjacent to the Isua supracrustal belt

The Isua supracrustal belt is bounded to the north by orthogneisses, whose main components are 3700–3690 Ma tonalites, and several suites of 3660–3630 Ma granites and pegmatites (Nutman & Bridgwater 1986; Nutman et al. 1996, 2000, 2002b; Crowley et al. 2002). The contact between these tonalites and the eastern part of Isua supracrustal belt is an Eoarchaean shear zone (Nutman 1984; Nutman et al. 1997, 2002b), which probably formed at c. 3650 Ma (Crowley et al. 2002).

Although superficially similar in the field, the tonalites and their gneissic equivalents on the south side of the Isua supracrustal belt are older than those to the north, with ages of 3810–3795 Ma (Nutman et al. 1996, 1999, 2000; Crowley 2003). Furthermore, along the southern side of the belt metabasaltic amphibolites have been intruded by c. 3800 Ma tonalites, thus giving their minimum age (Nutman et al. 1996, 1997; Crowley 2003). In the southern 3800 Ma tonalite area there are 3660–3630 Ma granitic sheets, but they are less voluminous than granite sheets cutting the tonalites to the north. The southern 3800 Ma tonalites, however, contain more inclusions of mafic and ultramafic rocks than the northern 3700 Ma tonalites.

Both the northern and southern tonalites contain rare areas of very low total strain, where igneous textures (weakly plagioclase phyric) and structures (brecciation of weakly foliated tonalite by homogeneous 3800 Ma tonalite; Fig. 2c) are preserved. However, despite the low-strain, even these rocks have been thoroughly recrystallized, and no igneous phases (apart from zircons and in the northern tonalites rare centres of plagioclase) are preserved. Associated with both suites of tonalites are lesser volumes of quartz diorites, with the same age as the tonalites (Nutman et al. 1999).

Tectonic features related to Eoarchaean accretion–growth

In the Isua supracrustal belt, rocks of different age and origin are tectonically juxtaposed along mylonites that are then folded. As shown in the eastern part of the Isua supracrustal belt (Fig. 4) these mylonites must be Eoarchaean, because an amphibolitized mafic dyke cutting a mylonite has yielded a zircon age of c. 3500 Ma (Nutman et al. 2007a). Furthermore, the mylonite at the northern edge of the belt probably formed at c. 3650 Ma (Crowley et al. 2002). Thus the Isua supracrustal belt is clearly the site of Eo- to Palaeoarchaean tectonic intercalation of different packages of rocks. However, in the Isua supracrustal belt the absolute age constraints on Eoarchaean mylonites are sparse, and a relationship between these Isua Eoarchaean mylonites and the juvenile tonalite suites (e.g. clearly cross-cutting TTG sheets) is not found. Thus it is as yet unknown to what extent tectonic intercalation in the Isua supracrustal belt occurred during crustal accretion–growth, or shortly afterwards during collisional orogeny.

The 3800 Ma tonalitic gneisses south of the Isua supracrustal belt contain a mixed inclusion suite of skarn-bearing amphibolite with some BIF from the upper crust, plutonic layered gabbros with layered peridotites from deeper levels, and slivers and pods of dunite and harzburgite derived from the upper mantle (Fig. 5a; Nutman et al. 1996, 2002b; Friend et al. 2002). This inclusion assemblage is dissected by Eoarchaean mylonites. Granitic lithons in mylonite and deformed granite sheets cutting some mylonites have yielded zircon dates of 3630–3600 Ma (Friend et al. 2002; Nutman et al. 2002b). Hence, such Eoarchaean mylonites are >150 Ma younger than the tonalites that mark crustal accretion–growth in this area. Such mylonites must have formed during younger superimposed orogenic events. However, in some cases, earlier mylonites have been detected between mantle duniteor or harzburgite slivers and crustal rocks, such as layered gabbros and cumulate layered ultramafic rocks (Fig. 5b). These mylonites predate c. 3800 Ma tonalite–trondhjemites that engulf the ultramafic and mafic rocks (Nutman et al. 1996; Friend et al. 2002). These indicate that mantle rocks and rocks from different crustal levels were being tectonically intercalated prior to

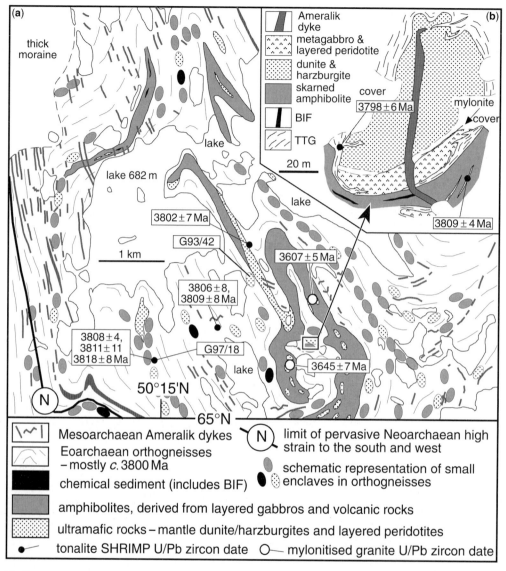

Fig. 5. (a) Map of an area dominated by c. 3800 Ma orthogneisses, south of the Isua supracrustal belt. Based on mapping by Nutman in 1981, with revisions in 1997. (b) Map showing mantle harzburgite in tectonic contact with layered peridotites and gabbros, prior to intrusion of c. 3800 Ma tonalites.

being engulfed and dissipated by large volumes of tonalite. This is strong field evidence (see McGregor 1979) for a synkinematic regime for tonalite emplacement during (c. 3800 Ma) Eoarchaean crustal accretion. This would mean that during crustal accretion–growth, many tonalite bodies could be already foliated and even gneissic, prior to deformation and high-grade metamorphism superimposed during later unrelated orogenic events. Rare examples of jumbled blocks of foliated tonalite dated at c. 3800 Ma (Nutman et al. 1999) set in a matrix of homogeneous tonalite (Fig. 2c), are also evidence for this.

Geochronological framework

Age of supracrustal and mantle rocks

Archaean volcano-sedimentary sequences dominated by mafic volcanic rocks and chemical sediments are notoriously difficult to date accurately and precisely, because they do not crystallize

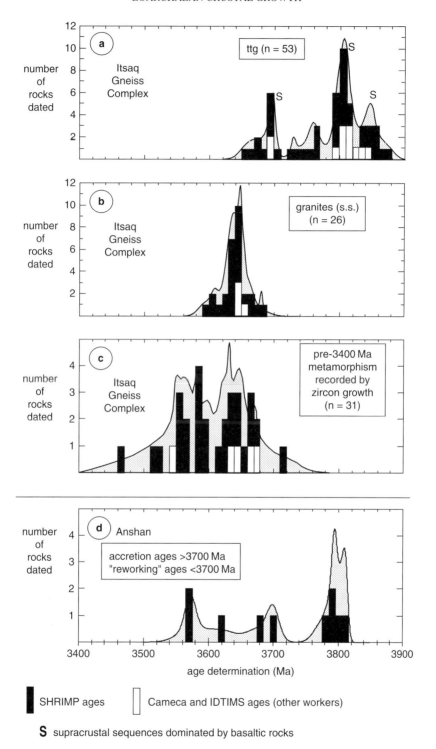

Fig. 6. Histograms summarizing the SHRIMP zircon geochronological framework for the Itsaq Gneiss Complex of Greenland and the Eoarchaean rocks of the Anshan area, NE China. IDTIMS, isotope dilution thermal ionization mass spectrometry.

igneous zircon when they form. Instead, some age constraints can be obtained by dating discordant tonalite sheets within them, as has been done for amphibolites from the southern edge of the Isua supracrustal belt and in amphibolites and ultramafic rocks further to the south (Nutman et al. 1996, 1997, 1999; Friend et al. 2002; Crowley 2003).

In other instances, rare, small oscillatory-zoned euhedral zircons have been extracted from chemical sediments in the northern side of the belt and have given ages of c. 3700 Ma (Nutman et al. 2002b, 2007a). These are interpreted as a small volcanogenic component that fell into the accumulating chemical sedimentary sequences, thereby given a proxy age of c. 3700 Ma. Some graded felsic volcano-sedimentary rocks in the NE of the belt contain abundant oscillatory-zoned euhedral igneous zircon, and have yielded direct dates of c. 3700 Ma (Fig. 6; Nutman et al. 1996, 1997, 2002b; Kamber et al. 2005).

Age of banded gneisses, tonalites and quartz diorites

Whitehouse et al. (1999) identified multiple generations of oscillatory-zoned igneous zircons in samples of banded gneisses from the Færingehavn terrane. They interpreted the youngest oscillatory-zoned zircons as giving the igneous age of single protoliths, with all older zircons being xenocrysts. Nutman et al. (2000, 2004) pointed out that as such rocks are actually complex banded gneisses, an equally permissible, if not more likely, interpretation is that several ages of igneous material are present. Such diversity of opinion shows that working on single-component rocks from low-strain zones rather than banded gneisses is important to avoid ambiguity in the interpretation of (expensive) geochronological data. By focusing geochronology on low-strain domains, Nutman et al. (1993, 1996, 2000, 2002a, 2007b) proposed that tonalites within the Itsaq Gneiss Complex are c. 3850, 3800, 3760–3730, 3700 and 3660 Ma old (Fig. 6). Detailed arguments for this have been given by Nutman et al. (2000, 2004, 2007a, b) and Mojzsis & Harrison (2002).

Eoarchaean island arc basalts, 'boninites', upper mantle peridotites, tonalites and quartz diorites

Gill et al. (1981) first noted that some Isua supracrustal belt banded amphibolites had compositional affinities with some modern arc basalts. More recent studies have focused on rare low-strain zones in Isua amphibolites, particularly where pillow centres can be sampled separately from pillow rims and interstitial material (Polat et al. 2002; Polat & Hofmann 2003; Jenner et al. 2006). Careful sampling allows chemical signatures that are original to the volcanic protoliths to be discriminated from those caused by sea-floor metasomatism and later alteration during high-grade metamorphism and deformation (Polat & Hofmann 2003). The trace element chemistry of recently collected suites of Itsaq Gneiss Complex amphibolites indicates that they were derived from rocks whose closest modern analogues are island arc basalts (IAB), rather than being plume related or mid-ocean ridge basalt (MORB; Table 1; Polat & Hofmann 2003). Diagnostic signatures are depletions in Ti and Nb and (variable) enrichments in Pb, Sr, Ba, Rb (Fig. 7). Within the Isua supracrustal belt, rocks at the southern edge of IAB affinity (Fig. 7a; Jenner et al. 2006) certainly have an age of c. 3800 Ma. However, other suites with IAB affinity in the northern side of the belt (Fig. 7b; Polat & Hoffman, 2003) are interlayered with BIF–metachert with c. 3710 Ma volcano-sedimentary zircons (Nutman et al. 2002b) and thus are probably c. 3710 Ma old. Amphibolites of IAB affinity also occur on the islands of Akilia and Innersuartuut, in the Færingehavn terrane (Fig. 1; Cates & Mojzsis 2006; Jenner et al. 2006; Jenner 2007). Although in the Færingehavn terrane volcanic structures are not preserved, careful sampling to avoid superimposed metasomatism has revealed geochemical traits similar to those seen in better-preserved rocks in Isua (Fig. 7c). From zircon dating of tonalitic components of the gneiss complex that invade and surround these mafic rocks, Nutman et al. (1996, 1997, 2002b, 2004, 2007a, b), Mojzsis & Harrison (2002) and Manning et al. (2006) interpreted such amphibolites on Akilia and Innersuartuut as >3850 Ma old. As such, these rocks' IAB geochemical signatures are with remnants of the world's oldest-preserved mafic crust. However, it should be noted that others (e.g. Whitehouse & Kamber 2005) dispute our evidence that these rocks are as old as 3850 Ma.

Within the Isua supracrustal belt, the prominent unit of garbenschiefer-textured amphibolite (Fig. 4) is distinguished from neighbouring amphibolites by its high MgO and lower high field strength element contents (HFSE) (TiO_2 <0.4 wt%; Zr <30 ppm; Nb <0.13 ppm; Y <14 ppm; Polat et al. 2002; Polat & Hofmann 2003). Polat & Hofmann (2003) have likened the chemistry of the garbenschiefer unit to that of boninites found in modern arc settings. These rocks also show pronounced enrichment in Pb and Sr; representative analyses are given in Table 1 and their trace element chemistry is shown in Figure 7d. Although the exact analogy with modern boninites might be debated, the high MgO contents of these rocks

Table 1. Representative whole-rock chemical analyses

	Mantle slivers		Layered intrusions					Mafic volcanic rocks			
Sample lithology age	G93/48 dunite >3800 Ma	G93/42 harzb >3800 Ma	G93/49 layered-p >3800 Ma	G93/71 layered-p >3800 Ma	G03/64 gabbro >3850 Ma	G03/62 gabbro >3850 Ma	G91/30 anorth >3850 Ma	G99/33 IAB-Akilia >3850 Ma	JG03/03 IAB-Akilia >3850 Ma	G91/47 IAB-Akilia >3850 Ma	JG03/45 IAB-Isua >3800 Ma
Greenland mafic and ultramafic rocks											
SiO_2	40.57	44.46	42.33	48.71	47.4	47.5	48.73	49.2	45.6	52.1	52.7
TiO_2	0.01	0.1	0.06	0.23	0.23	0.25	0.17	0.69	0.75	0.48	0.97
Al_2O_3	0.17	3.2	4.12	6.1	22.1	18.2	27.91	15.1	14.68	15.91	14.5
Fe_2O_3	10.01	9.36	12.81	11.7	6.58	8.48	3.47	12.46	14.51	9.78	11.5
FeO											
MnO	0.15	0.11	0.16		0.12	0.16	0.07	0.29	0.3	0.13	0.21
MgO	48.14	41.47	36.42	27.35	7.11	10.6	4.43	6.14	6.77	8.68	5.3
CaO	0.02	0.15	3.43	5.51	13.6	11.3	13.97	13.79	14.9	7.97	8
Na_2O			0.3	0.5	2.03	2.17	1.7	1.14	1.45	3.44	4.67
K_2O			0.06	0.11	0.67	1.43	0.1	0.35	0.58	0.57	0.13
P_2O_5	0.01		0.01		0.01	0.02	0.02	0.08	0.03	0.04	0.1
LOI					0.04	0	1.2	0.55	0.19	0.72	
Cr	1286	2233	2194	3185	225	455	139	571	662	184	83
Co	150	146	137	97							37.5
Ni	2697	2560	1595	865	174	225	103	183	219	81.2	70.7
Rb					11	86.9	8.7	5	10.4	16.8	0.53
Sr					108	74.8	367	91.8	73.4	104	265
Ba					39.7	44.5	111	6.6	15	11.6	11.5
V								233	252	233	
Ta					0.051	0.036		0.695	0.062	0.093	0.24
Nb					0.739	0.533		2.8	1.17	0.935	3.54
Zr					12.5	14	17.8	35.1	31.3	32.4	87.3
Th					0.065	0.032		0.256	0.015	0.351	1.22
U					0.042	0.064	174	0.615	0.019	0.089	0.26
Y					5.61	6.22	27	17	16.1	13.5	21.2
Pb											
La	0.025	0.027	0.26	0.4	1.49	1.76	14.4	3.64	1.4	2.02	8.29
Ce	0.064	0.087	0.786	1.271	3.51	3.55	35.7	8.39	4.26	4.75	19.9
Pr	0.011	0.017	0.122	0.235	0.494	0.492	5.1	1.23	0.753	0.692	2.79
Nd	0.06	0.1	0.542	1.347	2.38	2.28	23.8	6.06	4.31	3.43	12.9
Sm	0.024	0.046	0.148	0.495	0.705	0.648	5.95	2	1.66	1.12	3.39
Eu	0.003	0.009	0.09	0.144	0.311	0.304	2	0.682	0.593	0.399	1.09
Gd	0.036	0.08	0.209	0.783	0.856	0.843	6.23	2.75	2.46	1.64	3.97
Tb	0.007	0.017	0.04	0.151	0.154	0.164	0.984	0.487	0.449	0.313	0.68
Dy	0.045	0.116	0.251	0.961	1.03	1.07	5.32	2.95	2.83	1.59	3.96
Ho	0.011	0.029	0.056	0.209	0.215	0.238	1.02	0.641	0.605	0.5	0.81
Er	0.031	0.1	0.172	0.632	0.618	0.684	2.56	1.95	1.79	1.68	2.36
Tm											
Yb	0.035	0.132	0.173	0.601	0.653	0.764	2.16	1.72	1.7	1.68	2.26
Lu	0.007	0.026	0.028	0.092	0.096	0.112		0.264	0.27	0.256	0.32

(Continued)

Table 1. *Representative whole-rock chemical analyses*

	Mafic volcanic rocks									
Sample	JG03/46	JG03/48	JG03/49	2000-6	2000-10	2000-14	2000-19	462904	462945	462946	462948
Lithology	IAB-Isua	IAB-Isua	IAB-Isua	IAB-Isua	IAB-Isua	IAB-Isua	IAB-Isua	bon-Isua	bon-Isua	bon-Isua	bon-Isua
Age	>3800 Ma	>3800 Ma	>3800 Ma	c. 3700 Ma	c. 3700 Ma	c. 3700 Ma	c. 3700 Ma	c. 3700 Ma	c. 3700 Ma	c. 3700 Ma	c. 3700 Ma

Greenland mafic and ultramafic rocks

	JG03/46	JG03/48	JG03/49	2000-6	2000-10	2000-14	2000-19	462904	462945	462946	462948
SiO_2	50.8	51.7	51.2	51.7	51.2	48.76	50.95	48.62	50.13	51.57	54
TiO_2	0.78	0.76	0.75	0.78	0.62	0.67	0.59	0.2	0.32	0.36	0.4
Al_2O_3	9.2	9.3	9.3	10.28	8.38	8.89	7.99	14.09	17.96	18.34	17.74
Fe_2O_3	13.6	13	12.9	13.63	12.84	14.61	13.11	10.68	11.12	10.62	9.3
FeO											
MnO	0.22	0.22	0.22	0.25	0.19	0.23	0.17	0.24	0.19	0.17	0.13
MgO	12	11.4	11.4	9.36	14.42	13.98	15.28	12.46	8.97	8.46	7.05
CaO	10.6	10.6	10.7	11.4	10.03	11.39	10.05	12.82	8.54	7.11	8.39
Na_2O	1.61	1.77	1.81	2.37	2.19	1.27	1.72	0.76	2.6	3.24	2.85
K_2O	0.16	0.16	0.16	0.15	0.07	0.12	0.05	0.11	0.14	0.1	0.12
P_2O_5	0.05	0.05	0.05	0.07	0.06	0.07	0.07	0.01	0.02	0.02	0.02
LOI				2.11	0.98	1.36	1.03	1.7	2.63	1.72	1.76
Cr	974	960	987	594	1621	1559	1648	1922	196	204	141
Co	64.8	63.2	64.6	57	80	82	81	68	44	49	35
Ni	257	250	258	127	516	554	597	560	103	96	69
Rb	1.43	1.52	1.49	0.6	1.9	3.5	0.8	1.9	5.1	1.2	4.3
Sr	101	122	127	56	131	38	102	40	76	118	116
Ba	41	45.3	41.8	9.1	8.2	10.5	3.5	6	12	17	14
V				210	149	174	147	161	212	189	195
Ta	0.17	0.16	0.17	0.13	0.12	0.12	0.12	0.243	0.533	0.584	0.552
Nb	2.35	2.33	2.27	1.54	1.4	1.33	1.37				
Zr	65.2	63.7	62.4	49	55.6	50.9	53.7	16.5	23.6	23.4	27.5
Th	0.95	0.92	0.89	0.4	0.63	0.63	0.69	0.135	0.168	0.197	0.229
U	0.19	0.19	0.19		0.1	0.11	0.11				
Y	16.3	15.9	15.8	16.6	12.1	15.9	12.3	9	10	10.8	13
Pb				1.26	1.3	1.02		2.3		4.2	3.91
La	4.74	4.29	4.64	2.96	3.84	4.18	4.7	1.032	0.676	1.039	1.262
Ce	12.5	11.6	12.1	7.73	9.24	10.49	10.69	2.476	1.823	2.54	3.084
Pr	1.85	1.75	1.8	1.17	1.35	1.48	1.38	0.349	0.273	0.367	0.459
Nd	9.03	8.52	8.82	5.85	6.53	7.45	7.29	1.664	1.401	1.857	2.278
Sm	2.7	2.6	2.6	1.84	1.79	2.51	2.1	0.615	0.552	0.687	0.912
Eu	0.89	0.88	0.88	0.64	0.55	0.76	0.64	0.196	0.229	0.22	0.283
Gd	3.25	3.2	3.17	2.81	2.09	3.12	2.4	0.879	0.856	1.057	1.318
Tb	0.57	0.55	0.55	0.46	0.31	0.44	0.37	0.177	0.174	0.209	0.262
Dy	3.23	3.19	3.14	2.91	2.23	3	2.33	1.358	1.431	1.721	2.039
Ho	0.64	0.64	0.63	0.63	0.47	0.59	0.48	0.349	0.357	0.462	0.537
Er	1.83	1.77	1.7	1.58	1.2	1.54	1.22	1.303	1.278	1.679	1.914
Tm				0.25	0.17	0.22	0.18	0.206	0.192	0.249	0.281
Yb	1.58	1.55	1.58	1.67	1.24	1.48	1.12	1.447	1.344	1.653	1.832
Lu	0.23	0.23	0.22	0.24	0.15	0.24	0.18	0.217	0.22	0.245	0.278

(*Continued*)

Table 1. *Representative whole-rock chemical analyses*

								Tonalitic and quartz-diorite components						
Sample	G01/113	G93/05	G91/49	G93/07	G99/22	G97/18	G97/31	G97/39	G97/38	G93/44	292128	292127	G97/30	292483	292489
Lithology	tonalite	tonalite	tonalite	tonalite	tonalite	tonalite	tonalite	tonalite	tonalite	tonalite	qtz-diorite	qtz-diorite	qtz-diorite	qtz-diorite	qtz-diorite
Age	3849 Ma	3842 Ma	3850 Ma	3852 Ma	3862 Ma	3808 Ma	3809 Ma	3818 Ma	3811 Ma	3806 Ma	3806 Ma	c. 3800 Ma	c. 3800 Ma	c. 3800 Ma	c. 3800 Ma
Greenland grey gneisses															
SiO$_2$	69.59	67.65	69.04	67.29	66.92	69.79	66.79	69.81	70.13	70.02	54.73	56.58	60.21	61.94	62.5
TiO$_2$	0.27	0.41	0.31	0.49	0.51	0.25	0.31	0.26	0.25	0.27	0.91	0.81	1.03	0.57	0.47
Al$_2$O$_3$	15.54	16.03	15.52	15.55	16.09	16.7	17.93	16.67	16.74	16.45	19.1	18.2	5.36	15.21	14.56
Fe$_2$O$_3$	2.24	3.76	2.84	4.1	3.5	2.12	2.36	2.15	2.05	1.95	7.16	1.78	7.64	1.28	1.85
FeO												4.67		4.2	4.68
MnO	0.03	0.06	0.04	0.07	0.06	0.03	0.03	0.03	0.03	0.02	0.1	0.07	0.12	0.07	0.06
MgO	2.5	2.39	2.59	1.53	1.32	0.94	1.2	0.94	0.89	0.96	3.37	3.35	3.8	1.55	1.27
CaO	4.38	4.53	5.82	4.45	4.61	3.94	5.02	3.95	3.93	3.57	7.71	6.4	4.57	4.32	4.66
Na$_2$O	3.12	3.7	1.5	4.78	4.95	5.22	5.4	5.17	5.33	5.14	4.87	4.78	3.28	4.05	3.55
K$_2$O	1.57	1.45	1.79	1.17	1.03	0.93	0.79	0.87	0.98	1.08	1.28	1.62	2.85	2.15	1.97
P$_2$O$_5$	0.07	0.12	0.06	0.14	0.14	0.01	0.07	0.08	0.07	0.08	0.29	0.27	0.44	0.16	0.18
LOI												1.64		1.39	1.19
Cr	4.3	11	31	30.9	6.7	32	37	30	36	54	36	77		22	26
Co	30.8	43	34	26	31			17	15	12	44	8		40	41
Ni	66.7	94	58	25.1	16.5	31	26	27	37	53	35	50		17	10
Rb	96	128	123	337.7	350.4	429	469	424	431	394	642	53		103	93
Sr	120.7	90	97	115.2	93.5	100	87	93	122	136	163	669		417	392
Ba		40	19			21	26	19	19	19	115	224		257	197
V												78		56	49
Ta	0.27	0.3	0.52												
Nb	3.2	4.4	3.1	3.9	4.5	1.3	1.2	1.2	1.3	1.6	5.6	6.9		10	9
Zr	141.3	118	122	142.6	157.8	121	115	122	125	109	54	53		194	197
Th	2.1	2.02	0.42	0.18	0.2	0.77	0.38	0.63	0.79	0.43	1.31	3			
U	0.3	0.41	0.25	0.12	0.1	0.32	0.24	0.27	0.33	0.23	1.05				
Y	3.3	8.1	2.1	8.1	9.5	2.9	3.5	2.9	3	1.4	18.2	17		33	21
Pb	11.1	10.6	4.4		8.4	6.5	10.3	6.8	9	12	15.2	15		20	22
La		13.4	10.1	14.5	28.8	5.81	4.83	5.17	5.17	4.23	20.2				
Ce	24.2	29.8	19.5	32.2	3.8	11.4	9.8	9.7	10.3	7.7	53.6				
Pr	2.7	3.94	2.32	4.1	16.1	1.48	1.38	1.54	1.31	0.9	8.29				
Nd	10	15.3	8.5	16.5	3.12	6.05	6.18	5.95	5.89	3.37	37.4				
Sm	1.69	2.58	1.4	3.09	1	1.3	1.23	1.2	1.38	0.64	7.41				
Eu	0.54	0.65	0.58	0.89	2.74	0.48	0.54	0.42	0.45	0.37	1.94				
Gd	1.25	2.09	0.84	2.42	0.34	0.91	1.14	0.84	0.98	0.47	5.19				
Tb	0.13	0.29	0.1	0.32	1.98	0.13	0.15	0.14	0.19	0.05	0.66				
Dy	0.71	1.6	0.54	1.64	0.35	0.63	0.75	0.65	0.56	0.29	3.63				
Ho	0.12	0.33	0.11	0.29	0.96	0.12	0.11	0.11	0.1	0.06	0.68				
Er	0.31	0.83	0.27	0.75		0.27	0.3	0.29	0.26	0.15	1.74				
Tm															
Yb	0.29	0.84	0.27	0.63	0.82	0.28	0.29	0.33	0.3	0.17	1.52				
Lu	0.044	0.11	0.036	0.093	0.12	0.039	0.045	0.048	0.035	0.027	0.2				

(*Continued*)

Table 1. *Representative whole-rock chemical analyses*

Tonalitic and quartz-diorite components

Sample	G97/97	G97/98	VM90/02	229404	229403	237000	225892	225893	236999	236991	225989	225990	VM90/08	248054	248046A
Lithology	tonalite	tonalite	tonalite	qtz-diorite	qtz-diorite	qtz-diorite	tonalite	tonalite	tonalite	tonalite	qtz-diorite	qtz-diorite	qtz-diorite	qtz-diorite	tonalite
Age	c. 3795 Ma	3795 Ma	3760 Ma	c. 3700	c. 3700	c. 3700	c. 3700	c. 3700	c. 3700	c. 3700	c. 3700	c. 3700	c. 3650 Ma	c. 3650 Ma	c. 3650 Ma
SiO_2	71.75	69.23	70.38	59.86	61.7	64.12	69.27	69.29	65.97	68.02	60.89	61.62	50.42	53.44	72.14
TiO_2	0.3	0.33	0.31	0.84	0.73	0.47	0.42	0.43	0.45	0.41	0.67	0.69	0.57	0.55	0.26
Al_2O_3	15.1	16.19	15.89	16.33	15.79	14.76	15.91	15.74	14.41	15.22	14.88	15.38	16.89	17.43	15
Fe_2O_3	2.47	2.61	2.35	1.55	1.61	2.22	0.58	0.57	1.34	1.29	1.26	1.67	8.43	0.89	0.33
FeO				4.72	4.23	5.12	2.43	2.38	4.5	2.66	4.74	3.82		6.67	1.65
MnO	0.04	0.05	0.03	0.1	0.09	0.08	0.04	0.04	0.07	0.04	0.09	0.08	0.13	0.16	0.06
MgO	0.71	0.96	0.7	3.47	3.21	1.25	0.92	0.99	1.06	1.17	2.59	2.96	5.92	5.42	0.66
CaO	2.82	3.44	2.73	5.73	5.23	4.51	2.92	3.2	4.06	3.66	4.68	4.47	8.88	7.69	2.18
Na_2O	4.73	4.89	4.99	4.44	4.01	3.77	5.06	4.88	3.74	4.62	5.19	4.69	4.18	4.67	4.67
K_2O	1.86	1.66	1.41	1.45	1.68	1.92	1.66	1.51	1.94	1.41	0.63	1.13	0.89	1.34	2.5
P_2O_5	0.01	0.01	0.09	0.23	0.22	0.21	0.14	0.14	0.19	0.14	0.19	0.21	0.09	0.09	0.06
LOI			0.96	1.03	0.96	0.92	0.52	0.6	0.87	0.71	2.73	2.55	0.71	1.28	0.47
Cr	4	5		109	100	15	15	16	22	14	89	86	44.4	82	10
Co	12	17		60	64	49	57	58	46	52	43	32		31	9
Ni	78	72		56	53	6	4	4	7	2	37	41	97	98	68
Rb	236	294		79	106	64	73	54	75	87	26	49	25.9	71	80
Sr	212	204		433	380	267	416	490	244	263	258	319	170.9	222	199
Ba	17	22		325	344	334	313	373	325	199	106	248	39.9	112	232
V				95	86	46	28	23	41	35	67	69		120	36
Ta	4.3	3.8													
Nb	124	86		9	8	10	6	4	8	9	8	9	3.2	3	7
Zr	4.14	3.73		161	135	158	168	179	174	169	136	146	92.4	67	124
Th	0.82	1.2		4	6	1	8	7	5	4	3	11			
U	8.6	16.3													
Y	13.9	15.1		24	23	26	5	5	19	20	20	23	15	14	7
Pb	21.6	15		16	16	12	20	10	13	17	8	12		19	9
La	41.1	25					24.1		22.8	16.9			6.81		
Ce	4.68	3.34					44.1		44.7	36.6			12.1		
Pr	16.6	13.5											2.54		
Nd	2.81	2.92					3.19		3.94	4.24			11.4		
Sm	0.74	0.65					0.726		1.34	0.926			2.78		
Eu	1.93	2.79											0.672		
Gd	0.28	0.39							0.51	0.436			2.78		
Tb	1.58	2.86											0.454		
Dy	0.29	0.54											2.53		
Ho	0.92	1.58											0.507		
Er													1.39		
Tm															
Yb	0.88	1.62					0.202		1.32	1.33			1.35		
Lu	0.13	0.25					0.031		0.179	0.187					

(*Continued*)

Table 1. Representative whole-rock chemical analyses

				Tonalitic and quartz-dioritic components					
Sample	A9011	A0518	A9604	A0512	A0423	A0404	A0405-1	A0405-2	A9011
Lithology	trondhj	trondhj	qtz-diorite	trondhj	trondhj	trondhj	trondhj	trondhj	trondhj
Age	3804	3800	3794	3777	3680	3620	3573	3573	3804

Anshan area Eoarchaean orthogneisses

	A9011	A0518	A9604	A0512	A0423	A0404	A0405-1	A0405-2	A9011
SiO_2	74.29	74.39	57.88	70.53	71.10	76.10	73.03	76.18	74.29
TiO_2	0.10	0.08	0.61	0.28	0.37	0.09	0.18	0.03	0.10
Al_2O_3	15.62	14.73	18.99	15.61	15.34	13.39	14.83	13.40	15.62
Fe_2O_3	0.21	0.52	5.00	0.86	2.19	0.86	1.07	0.17	0.21
FeO	0.92	0.40		1.87	0.66	0.20	0.40	0.22	0.92
MnO	0.03	0.01	0.15	0.03	0.05	0.02	0.02	0.02	0.03
MgO	0.22	0.16	2.65	1.33	0.83	0.43	0.54	0.23	0.22
CaO	1.48	1.46	4.52	0.49	1.43	1.86	1.41	0.72	1.48
Na_2O	4.72	5.71	5.56	5.18	4.49	5.86	5.00	5.02	4.72
K_2O	1.88	1.04	2.53	2.00	2.39	0.48	2.72	3.11	1.88
P_2O_5	0.00	0.01	0.20	0.08	0.07	0.02	0.05	0.01	0.00
LOI		0.70		1.28	1.30	0.66	0.94	0.22	
		4.73		9.78	19.0	11.1	19.9	11.7	
Cr	18								18
Co									
Ni									
Rb	70	48.2	216	125	186	20.7	83.7	63.3	70
Sr	376	497	371	95.3	256	250	176	108	376
Ba	729	457	313	410	517	81.0	403	262	729
V									
Ta		0.25		1.71	1.24	0.26	0.89	0.41	
Nb	2	2.82	5	11.8	9.87	2.35	6.15	2.05	2
Zr	103	71.7	157	138	169	81.9	116	35.2	103
Th	5	4.82	5	4.22	6.02	10.6	12.4	9.13	5
U		1.63		1.13	0.46	0.97	1.93	2.33	
Y	3	2.84	39	10.8	34.4	4.60	12.2	11.2	3
Pb									
La	6.55	5.99	40.43	16.5	36.8	7.91	19.3	2.81	6.55
Ce	9.59	7.16	79.09	27.2	50.1	14.6	25.3	4.16	9.59
Pr	2.45	1.10	7.5	2.81	6.57	1.49	3.41	0.55	2.45
Nd	5.4	3.88	28.55	9.48	22.9	5.03	11.5	2.00	5.4
Sm	1.78	0.79	8	1.80	5.15	1.07	2.36	0.55	1.78
Eu	0.46	0.33	2.35	0.33	0.86	0.39	0.54	0.13	0.46
Gd	1.56	0.68	7.26	1.80	4.91	0.96	2.10	0.85	1.56

(*Continued*)

Table 1. Continued

		Tonalitic and quartz-dioritic components						
Sample	A9011	A0518	A9604	A0512	A0423	A0404	A0405-1	A0405-2	A9011
Lithology	trondhj	trondhj	qtz-diorite	trondhj	trondhj	trondhj	trondhj	trondhj	trondhj
Age	3804	3800	3794	3777	3680	3620	3573	3573	3804
Tb	0.3	0.10	1.3	0.30	0.85	0.13	0.33	0.21	0.3
Dy	0.76	0.60	7.44	1.80	4.84	0.88	1.84	1.68	0.76
Ho	0.2	0.11	1.45	0.37	0.92	0.16	0.42	0.39	0.2
Er	0.49	0.34	3.93	1.14	2.66	0.46	1.20	1.21	0.49
Tm	0.1	0.05	0.59	0.18	0.40	0.06	0.16	0.19	0.1
Yb	0.41	0.34	3.81	1.19	2.97	0.45	1.05	1.31	0.41
Lu	0.1	0.05	0.56	0.18	0.44	0.08	0.14	0.19	0.1

Data are from Nutman & Bridgwater (1986), Nutman et al. (1996, 1999, 2007b), Polat & Hofmann (2003), Jenner (2007) and the Geological Survey of Greenland (previously unpublished data of A. Nutman, and compilations of the late V. R. McGregor and the late D. Bridgwater). Lithology abbreviations in the fourth row are: layered-p, layered peridotite (from layered basic intrusions); anorth, anorthosite (from layered basic intrusions); IAB, metabasaltic amphibolite with chemical affinities with island arc basalt; bon, metabasaltic amphibolite with chemically affinities with boninites; qtz-diorite, quartz diorite; trondhj, trondhjemite. Age constraints in the fifth row are from SHRIMP U/Pb zircon geochronology: >XXXX Ma is where the minimum age for mafic or ultramafic rocks has been obtained from intrusive tonalite sheets; c. XXXX Ma is the likely age based on geological relationships and associated dated rocks. Other ages are direct ages on the samples for which whole-rock analyses are presented. (Analytical methods are described in the source references.)

clearly requires more extensive melting of a depleted mantle source compared with the melting responsible for the IAB (Polat et al. 2002; Jenner et al. 2006). This would indicate elevated mantle temperatures and/or increased fluxing of mantle by fluids, as corroborated by the enrichment in Pb and Sr (Polat et al. 2002). In a modern setting, this occurs in the most juvenile stages of island arc systems (see Shervais 2001, for review). There is no direct precise date for the 'boninitic' amphibolite unit(s). Attempts to date it by the Sm–Nd whole-rock method gave ages between 3700 and 3800 Ma with uncertainties $> \pm 40$ Ma (e.g. Hamilton et al. 1978; Rosing 1999). Felsic intermediate volcano-sedimentary rocks of likely arc derivation that are intimately interlayered or tectonically interdigitated with the 'boninitic' unit contain c. 3710 Ma detrital zircons (Nutman et al. 1997, 2002b). Thus, as a working hypothesis, we consider that c. 3700 Ma is the likely age for the 'boninitic' unit. Although amphibolites of 'boninitic' chemistry have been most widely documented from the Isua supracrustal belt, there are some amphibolites (>3850 Ma?) in the Færingehavn terrane on Akilia island with similar, but less extreme compositional traits (Jenner 2007).

Gabbros together with the related layered ultramafic rocks and anorthosites that they grade into occur in small amounts (Table 1). These rocks are now thoroughly recrystallized such that no igneous phases are preserved. However, compositional variation of geochemically well-preserved suites indicates control by plagioclase, pyroxene and olivine fractionation, not amphibole (Jenner 2007). Thus prior to metamorphism with amphibolitization, it is considered that these were anhydrous intrusions. As the gabbros are derived from layered complexes (Fig. 3c), their protoliths were mixtures of residual and trapped liquid + calcic plagioclase + pyroxene ± olivine ± chromian spinel. Despite being crystal–liquid mixtures, trace element variations such as negative Ti and Nb anomalies in trace-element spidergrams show that their parental liquids also had affinities with IAB, rather than komatiites or MORB (Jenner 2007). Thus these rocks have a close compositional affinity with the associated supracrustal amphibolites of the Itsaq Gneiss Complex.

Highly magnesian (up to 48 wt%) dunites and harzburgites with low alumina (<5.5 wt%, but generally much lower) and low lime (<2 wt%) are locally preserved as small relict cores within larger amounts of amphibole-rich ultramafic schists (Table 1; Friend et al. 2002). The dunites and harzburgites are generally massive and devoid of compositional layering, in contrast to the layered peridotites that grade into gabbros. Most commonly the ultramafic rocks occur as isolated pods

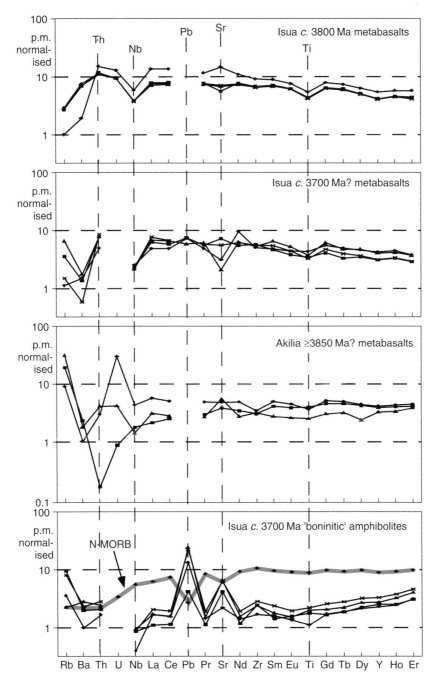

Fig. 7. Depleted mantle normalized trace element spidergrams for selected amphibolites from the Itsaq Gneiss Complex. Data are from Polat *et al.* (2002), Polat & Hofmann (2003) and Jenner (2007).

completely surrounded by orthogneisses or, more rarely, are in tectonic contact with crustal rocks, and then surrounded by tonalitic gneisses (Fig. 5b; Nutman *et al.* 1996; Friend *et al.* 2002). Olivine in these rare well-preserved rocks has compositions of Fo89–90 and can contain inclusions of aluminous spinel (Friend *et al.* 2002). Spinels from one of these dunites (G93/42 in Fig. 5a) have given

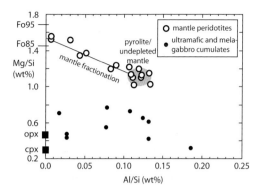

Fig. 8. Al v. Mg (wt. proportion) for some ultramafic rocks from the Itsaq Gneiss Complex. Homogeneous dunites and harzburgites follow the mantle trend, and have distinct compositions from layered peridotites representing cumulate components in layered gabbro complexes.

the most primitive measured terrestrial $^{187}Os/^{188}Os$ isotopic ratio (Bennett et al. 2003). In terms of bulk compositions (e.g. Al/Si v. Mg/Si; Fig. 8), the dunites and harzburgites follow the mantle fractionation trend, with many samples being strongly depleted. Hence, they bear no resemblance to either komatiites or cumulate peridotites from layered complexes (Fig. 8). Instead, they are interpreted as slivers of Eoarchaean upper mantle. South of the Isua supracrustal belt, these slivers had already been tectonically intercalated with crustal rocks when c. 3800 Ma tonalites were intruded (Nutman et al. 1996; Friend et al. 2002). Thus their emplacement into the crust was part of the crust formation process.

Tonalites are siliceous (typically 65–70 wt%) potassium-poor intrusive rocks that are the most voluminous of all Archaean rock types. Geochemical and experimental petrological research shows that Archaean tonalites were generated by melting of hydrated mafic crust, after transformation into garnet amphibolite, high-pressure granulite or eclogite (as summarized by Martin et al. 2005). Therefore, it is generally considered that Archaean tonalites formed in a broadly similar fashion at ancient convergent plate boundaries (e.g. Plank 2005). Sr, Nd and Hf isotopic studies show that the tonalites from the Itsaq Gneiss Complex were originally juvenile magmas whose source materials had been extracted from the mantle only shortly before melted occurred (e.g. Moorbath et al. 1972; Moorbath 1975; Bennett et al. 1993; Vervoort et al. 1996). In terms of major element chemistry (Table 1), the Itsaq Gneiss Complex tonalites mostly plot in the high-silica, low-magnesian field for older TTG (Fig. 9a). In

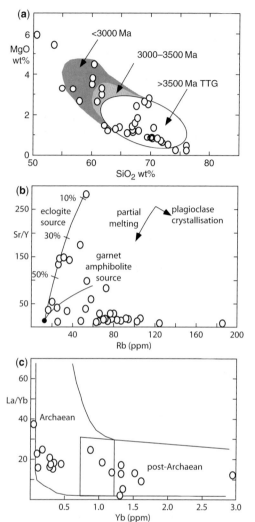

Fig. 9. Geochemistry of Itsaq Gneiss Complex and Anshan area tonalites–trondhjemites and quartz diorites. (a) SiO_2 v. MgO (wt%) plot, demonstrating that most samples plot in the >3500 Ma field, for compositions dominated by melting of a basaltic slab (Martin et al. 2005). The coeval, volumetrically minor quartz diorites are too magnesian to have formed solely by melting of a basaltic slab, and melting of (metasomatized) mantle must have been involved as well. These samples plot within the field of younger Archaean, and even post-Archaean compositions, despite their >3600 Ma ages. (b) Rb (ppm) v. Sr/Y plot. Shown on it are melting models, using eclogite and garnet amphibolite sources. (c) Yb (ppm) v. La/Yb plot.

terms of their trace element variation, they show enrichment of the light REE (LREE) over heavy REE (HREE), indicative of the role of residual garnet in their source region preferentially retaining the HREE (Fig. 9b; Arth & Hanson 1975; Martin 1986; Martin & Moyen 2002; Champion & Smithies 2007, and references therein). Trace element modelling (Nutman et al. 1999) of some c. 3800 Ma tonalites demonstrated that they could be produced by 40% partial melting of an eclogite source, whose protolith in turn was formed by 30% melting of depleted upper mantle (Rb v. Sr/Y, Fig. 9c). However, average Archaean tonalite and many Itsaq Gneiss Complex compositions lie to the right of this model eclogite melting curve (Fig. 9c). Such compositions can arise from different permutations of plagioclase fractionation, melting of garnet amphibolite rather than eclogite sources and contributions from melting in the upper mantle.

Associated with the c. 3800 Ma, 3700 Ma and 3660 Ma tonalite suites (Table 1) are much smaller volumes of quartz diorite, which have lower silica (55–60 wt%) and higher MgO (3 to 4 wt%; Fig. 9a), with Mg-number values approaching 50. As pointed out by Smithies & Champion (2000), the highly magnesian character of such rocks means that they cannot have been derived solely by melting of basic rocks, but must have involved some melting of high Mg-number mantle ultramafic rocks fluxed by fluids as well. Thus it appears that although crust formation represented by the Itsaq Gneiss Complex largely involved emplacement of tonalites whose petrogenesis was dominated by the melting of (hydrated) basaltic rocks at high pressure (Nutman et al. 1999; Steenfelt et al. 2005), there was also some contribution to siliceous magmas from the melting of (metasomatized) mantle rocks.

Thermal characteristics of the crustal accretion–growth that formed the Itsaq Gneiss Complex

The Itsaq Gneiss Complex comprises >80% by volume of tonalitic orthogneisses. Numerous experimental studies (e.g. Rapp 1997; Wyllie et al. 1997) indicate that parental tonalite magmas are hot (>1000 °C), water-undersaturated and have positive dP/dT for their solidi. This positive solidus slope permits magmas to rise from their source regions without freezing (Brown & Fyffe 1970). Additionally, tonalite magmas have a specific gravity of c. 2.4–2.5 (Grove & Baker 1983), which is considerably less than the 2.7–3.0 specific gravity for the assemblage of predominantly mafic host rocks. Therefore, tonalite magmas should be emplaced into high levels in the crust, at which point they attain neutral buoyancy. If each successive batch of magma is rapidly emplaced above the other, a mode of tonalite intrusion termed overaccretion by Wells (1980), there are predictable thermal consequences.

If the intrusion of TTG magmas occurs as several pulses in a short period it will result in more elevated crustal temperatures, with an increased likelihood of achieving temperatures of >800–850 °C with dehydration-melting reactions and hence granulite-facies metamorphism. Wells (1980) modelled this quantitatively using 1D conductive relaxation models, for continental crust undergoing rapid magmatic thickening in a predominantly recumbent tectonic regime. Intuitively, if the emplacement of a given volume of >900 °C tonalite magma as several pulses is spread over a longer time, then it is less likely to cause regional high-grade metamorphism, as the heat from one magma pulse can be dissipated by conduction before intrusion of the next pulse. Some overaccretion scenarios presented by Wells (1980) are reproduced here in Figure 10. They clearly demonstrate that rapid accretion–growth (<10 Ma) results in sustained elevated 800–850 °C temperatures at depth in the crust, which are appropriate for granulite-facies metamorphism. Slow episodic accretion–growth (using Wells' scenario of three intrusive pulses in 50 Ma) does not lead to >800 °C temperatures, but instead there is a 600–800 °C temperature range, more appropriate to amphibolite-facies metamorphism. At these lower temperatures, dehydration-melting reactions causing granulite-facies metamorphism are unlikely. This slow (c. 50 Ma) scenario is compatible for crustal accretion–growth in the Itsaq Gneiss Complex, because zircon geochronology shows that its tonalite components were emplaced episodically from c. 3870 to 3660 Ma, with 10–40 Ma interludes between each batch (Fig. 6).

Discussion

Interpretation of Eoarchaean juvenile magmatic suites and coeval tectonic regimes: crustal accretion–growth at Eoarchaean convergent plate boundaries?

For heat-loss efficiency reasons, the hotter Eoarchaean Earth must have had a convecting mantle, and thereby some form of plate-tectonic regime. The presence of BIF chemical sediments at the start of the geological record (described initially by Moorbath et al. 1973) shows that this convective system was already water-cooled at its top, providing hydrated crust for melting at convergent plate boundaries (Campbell & Taylor 1983). Thus by the start of the Eoarchaean, crust production and lithospheric processes had essential resemblance to those in the Phanerozoic. Beyond this, there is no consensus on whether the Eoarchaean regime

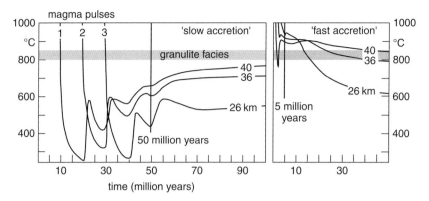

Fig. 10. One-dimensional thermal relaxation models of overaccretion models (Wells 1980). Overaccretion is used here as defined by Wells, as the successive emplacement of batches of magma above those previously emplaced. Figures from that paper have been reorganized to facilitate discussion in this paper.

involved thicker or thinner oceanic crust, more numerous, smaller plates, and similar or greater ridge length (for discussion and arguments, see McKenzie & Bickle 1988; Abbott et al. 1994; De Wit 1998; Davies 2006; Kerrich & Polat 2006). However, despite this debate, we consider Eoarchaean plate boundary magmatism as a certainty, not a speculation.

The Itsaq Gneiss Complex contains early sequences of metabasalts and gabbros with IAB and boninitic-like chemistry that are intruded by large volumes of TTG with minor quartz diorite. This represents an assemblage of diverse magmatic rocks that together have closest affinity with those found at Phanerozoic plate boundaries, rather than any other type of magmatic province. Furthermore, such sequences form concomitantly with tectonic imbrication of crustal mafic rocks, mantle ultramafic rocks and chemical sediments (Nutman & Collerson 1991; Nutman et al. 1996, 2007a; Friend et al. 2002). Together, these Eoarchaean assemblages resemble the Phanerozoic accretionary–crustal growth assemblages represented in suprasubduction-zone ophiolites (for review, see Shervais 2001).

In Phanerozoic systems, the destruction of the crust on one side of a newly formed convergent plate boundary when it founders into the mantle causes rapid extension of the conserved upper plate into the space created and is coupled with high heat flow as asthenospheric mantle migrates into the extensional domain. As explained by numerous researchers (e.g. Shervais 2001), fluid release from the destroyed crust into the hot overlying mantle wedge is the origin of island arc basalts (IAB) that characterize the juvenile stages of island arc systems–accretionary orogens. Boninites are magnesian, large ion lithophile element (LILE)-enriched mafic volcanic rocks that occur in the early stages of some arc systems. They are commonly attributed to higher degrees of fluxing of the asthenospheric mantle wedge above a subduction zone, permitting higher degrees of melting (hence higher MgO content) and higher LILE content than IAB.

If, later in the life of a modern arc system, the crust being destroyed by subduction is younger, it will be hotter as a result of more residual mid-ocean ridge heat. In this case, if that crust is <20 Ma old, it will start to melt (in the stability field of eclogite or high-pressure granulite) before it can release an LILE-bearing hydrous fluid into the overlying asthenospheric mantle wedge (Martin 1986). This scenario of melting of the destroyed or subducted slab leads to adakite magmas along convergent plate boundaries, with their characteristic geochemical signatures of melting of hydrous mafic rocks under high-pressure conditions, leaving garnet as a residual phase (e.g. Smithies & Champion 2000; Martin et al. 2005). Many researchers have in general terms likened the composition and genesis of adakites and Archaean tonalites, thereby proposing that tonalites formed at ancient convergent plate boundaries in accretionary orogens (as summarized by Martin et al. 2005).

In the Itsaq Gneiss Complex, crustal accretion–growth probably occurred in several cycles (Fig. 6), as is seen in complex Phanerozoic orogens. Earliest rocks are always sequences dominated by amphibolites with IAB-like ± boninite-like chemistries (e.g. Polat & Hofmann 2003). Although in detail the chemistry of these rocks differs from that of their modern analogues (e.g. Jenner 2007), overall the similarities are much more striking than the differences, and it is a plausible interpretation that they formed early in the development of ancient convergent plate boundaries (e.g. Polat & Hofmann 2003). The early IAB-dominated sequences are then always intruded by much larger volumes of tonalite, formed by melting of mafic crust at high pressure. This is because in the hotter early Earth, very soon after a destructive plate boundary initiated, hotter 'oceanic' crust was being destroyed at the boundary. This meant that in the Eoarchaean, IAB ± boninite magmatism invariably switched over to voluminous tonalite magmatism. Although the importance of the tonalite component in Eoarchaean crustal accretion–growth has long been recognized (Moorbath 1975; McGregor 1979; Martin 1986; Nutman et al. 1999; Steenfelt et al. 2005), it is only in recent years that robust evidence for Eoarchaean IAB ± boninite associations has been presented (e.g. Polat et al. 2002; Polat & Hoffman 2003; Cates & Mojzsis 2006; Jenner et al. 2006; Jenner 2007).

The episodic nature of Eoarchaean crustal accretion–growth; Greenland

The Itsaq Gneiss Complex contains associations of mafic rocks of broadly IAB and boninitic affinity that have been intruded by tonalite ± quartz diorite whose Sr, Nd and Hf isotopic signatures indicate they represent new additions to the crust. Thus the Itsaq Gneiss Complex is dominated by both mafic and felsic rocks that are widely interpreted to form by hydrous melting at destructive plate boundaries. We would also propose that the Itsaq Gneiss Complex shows repeated cycles of crustal accretion–growth of this mode, at c. 3850 Ma, 3800 Ma, maybe 3760 Ma, 3700 Ma and c. 3660 Ma, although we acknowledge that there is some dispute of our geochronological interpretations upon which this conclusion is based (e.g. Whitehouse et al. 1999; Whitehouse & Kamber 2005).

Given that the Itsaq Gneiss Complex consists of a small remaining fragment of a once much larger crustal accretion–growth complex, it is impossible to reconstruct details concerning crustal architecture and palaeogeography during accretion (in this respect we differ from the synthesis for the Itsaq Gneiss Complex of Komiya et al. (1999)). However, two points are noteworthy.

First, the presence of upper mantle peridotite slivers (Nutman et al. 1996; Friend et al. 2002) and high-MgO quartz diorites (data of McGregor 1979; Nutman et al. 1999; this paper) in the Itsaq Gneiss Complex places some constraints on the construction and architecture of the crust. Thus, upper mantle peridotite slivers tectonically intercalated with mafic crustal rocks prior to the time of tonalite emplacement rule out a model for ancient convergent plate boundaries as a crustal sagging or ductile flow domain driven by an eclogitized root, which then melted to produce tonalite. This rules out the least uniformitarian geometry for ancient convergent plate boundaries. Likewise, the presence of the high-MgO quartz diorites coeval with several generations of tonalite (Fig. 9) indicates that subordinate melting of metasomatized mantle occurred simultaneously with the tonalites whose origin involved slab melting. Melting could have occurred in the high-pressure root of tectonically imbricated mafic crustal rocks and upper mantle slivers (obduction model of De Wit 1998) or in the context of more uniformitarian shallow subduction, with a small amount of upper mantle trapped between the conserved and destroyed crustal plates (models discussed by Smithies & Champion 2000).

Second, certain ages of Itsaq Gneiss Complex tonalite are generally found in close association. Thus, parts of the Færingehavn terrane show an association of c. 3850 and 3750 Ma TTG, whereas in other parts around Amiitsoq only 3660 Ma TTG occur. In contrast, in the northern Isukasia terrane, extensive domains of c. 3800 and 3700 Ma TTG seem to have been tectonically juxtaposed in the Eoarchaean (Figs 1 and 4). Thus it is possible that the Itsaq Gneiss Complex consists of several juvenile crustal accretion–growth complexes formed at different convergent plate boundaries, which were later docked together in a collisional orogeny starting at c. 3660 Ma.

Collisional orogeny and the end of Itsaq Gneiss Complex Eoarchaean crustal accretion–growth (from 3660 Ma)

In the Itsaq Gneiss Complex, a c. 200 Ma history (>3850 to 3660 Ma) of crustal evolution is marked by different suites of IAB-like mafic rocks of different age invaded by voluminous juvenile tonalites with lesser amounts of quartz diorites. This evolution comprised the assembly of a number of different blocks that were then juxtaposed by Eoarchaean tectonic events that finished at c. 3660 Ma (Friend & Nutman 2005a). All researchers of the Itsaq Gneiss Complex agree that 3660–3600 Ma was a time of major tectonothermal activity in the complex, but there are differences in interpretation as to the significance and results (e.g. compare Kamber & Moorbath 1998 and Whitehouse et al. 1999 v. Nutman et al. 2000 and Friend & Nutman 2005a).

Friend & Nutman (2005a) proposed that following 3850–3690 Ma juvenile crustal accretion–growth events, the evolution of the Itsaq Gneiss Complex continued with numerous crustal-reworking events between 3660 and 3550 Ma. Examples of these events are: (1) intrusion of several generations of geochemically diverse granites (*sensu stricto*), monzonites, and subordinate ferro-gabbros and ferro-diorites; (2) superimposed, multiple episodes of amphibolite to granulite-facies metamorphism; (3) the occurrence in part of the complex around Amîitsoq and the mountains to the north of the youngest tonalitic gneisses at c. 3660 Ma, which are indistinguishable in age from the oldest recognized granites (*sensu stricto*) elsewhere; (4) the occurrence also in the mountains north of Amîitsoq of deformed remnants of volcano-sedimentary basins that were still being filled between 3650 and 3600 Ma (from dating youngest detrital zircons) and that were inverted and metamorphosed by c. 3570 Ma (dating of oldest *in situ* metamorphic overgrowths). These 3670–3550 Ma events are interpreted as reflecting a complex orogen (initially collisional) superimposed on the products of the earlier (3850–3690 Ma) juvenile crustal accretion–growth regimes. However, Kamber & Moorbath (1997), using the whole-rock Pb isotopic compositions of regional collections of orthogneisses ranging from granitic to dioritic in composition, have regarded this period to embrace important crustal accretion–growth events as well, essentially coeval with crustal reworking.

The large amount of zircon geochronology integrated with field studies from the Itsaq Gneiss Complex indicates that Eoarchaean granulite-facies metamorphism (McGregor & Mason 1977; Griffin *et al.* 1980) did not occur until 3660–3600 Ma (e.g. Friend & Nutman 2005a). This is in accord with the reappraisal of the thermal modelling integrated with zircon geochronology of the Itsaq Gneiss Complex TTG, which suggests that the 'slow' tonalite overaccretion could not have been responsible for the granulite-facies metamorphism. Instead, as pointed out by Friend & Nutman (2005a), the granulite-facies metamorphism in the Itsaq Gneiss Complex is most probably related to post-collisional orogenic events that marked the termination of crustal accretion–growth, when the underplating of the crust by mafic magma caused deep crustal melting, with production of various suites of granites, monzonites and Fe-rich mafic rocks.

Source compositions of the tonalites

A convergent plate margin–island arc setting for the Itsaq Gneiss Complex mafic rocks and the tonalites that engulf them means that it is unlikely that the preserved mafic rocks are the same as those that melted at depth to generate the tonalites. This parallels the situation in Phanerozoic arcs, where early mafic rocks are mostly of IAB affinity, whereas the mafic rocks destroyed by subduction are mostly MORB oceanic crust. Instead, we propose that allochthonous amphibolite units at depth melted to give rise to the Itsaq Gneiss Complex tonalites, as in modern arcs, where adakite magmas are generated from subducted MORB crust and not from mafic arc rocks that occur as enclaves within them. Thus, in the Itsaq Gneiss Complex, 'disappeared' Eoarchaean MORB might be the appropriate source material for the tonalites. Scenarios of tonalite generation from an eclogite representing c. 40% melting of a depleted mantle composition were found to be appropriate for some Itsaq Gneiss Complex TTG (Nutman *et al.* 1999).

Comparison of Eoarchaean crustal evolution in Greenland and China

The Anshan area of northern China is a significant part of the oldest rock record because although it is very small in extent, integrated field and laboratory work has identified relict single igneous phase components within the gneisses (Wan *et al.* 2005). This allows more precise statements about crust-forming processes than in cases where only ancient polyphase migmatitic gneisses have been recognized. Particularly noteworthy is the documentation of not only trondhjemitic–tonalitic phases, but also c. 3800 Ma high-MgO quartz diorite (Wan *et al.* 2005), which is unique outside the West Greenland Eoarchaean record. The geochronology and the lithological assemblage in the Itsaq Gneiss Complex are compared in Figure 6 with the Eoarchaean geological record of the Anshan area (Liu *et al.* 1992, 2007; Song *et al.* 1996; Wan *et al.* 2005). Around the city of Anshan, the amount of known Eoarchaean crust is at present only about 1 km^2 at three localities. Anshan Eoarchaean crustal evolution is marked first by emplacement of c. 3800 and 3700 Ma trondhjemites and lesser amounts of 3800 Ma quartz diorite, with positive initial ε_{Nd} values (Wan *et al.* 2005). These are then intruded by c. 3600 Ma granitoid sheets, and variably affected by c. 3600 Ma high-grade metamorphism with migmatization. Thus again, as in Greenland, tonalite–trondhjemite emplacement appears not to be directly responsible for granulite-facies metamorphism and granite formation. Given the small amount of surviving Archaean crust in the Anshan area, it is not surprising that even less is known about the supracrustal sequences into which the tonalites–trondhjemites must have been emplaced. However, at the Baijiafen quarry locality, Liu *et al.* (2007) described mafic, biotite-rich schists with silica-rich layers and a zircon age of 3727 ± 17 Ma, which might represent a metasomatized relict of volcano-sedimentary rocks. Overall, despite the much more scanty Eoarchaean geological record in NE China compared with that in Greenland, the two regions appear to show a general similarity in evolution (Fig. 6). This suggests that the well-documented crustal accretion–growth processes in the Itsaq Gneiss

Complex of Greenland are not unique, and might be more broadly applicable to the early Earth. Of particular importance then from Anshan is the occurrence of 3800 Ma high-MgO quartz diorite as also found in West Greenland, indicating the contribution of metasomatized depleted mantle in the production of juvenile continental crust at that time. This is contrary to recent opinions that production of Eoarchaean juvenile crust was solely by partial melting of eclogites (e.g. Martin & Moyen 2002).

Conclusions

(1) Eoarchaean accretion–growth, exemplified by the geological record in the Itsaq Gneiss Complex, started with formation of volcano-sedimentary sequences dominated by amphibolites with IAB-like $\pm$ boninite-like geochemistry, with associated chemical sediments, particularly chert and BIF. They are interpreted as juvenile arc assemblages, whose mafic volcanic components were formed by fluxing of a mantle wedge by fluids derived from destroyed ('subducted') mafic crust at convergent plate boundaries.

(2) Tonalites that are the dominant rocks in Eoarchaean terranes intrude the earlier IAB-dominated sequences. This occurred when hot (young) oceanic crust was being subducted, which caused melting within it at high pressures (residual garnet). Because of the generally hotter state of the Eoarchaean Earth, convergent plate boundary magmatism was typically dominated by tonalitic compositions.

(3) Both the dominant mafic and felsic components of Eoarchaean Itsaq Gneiss Complex have chemical signatures resembling igneous suites currently forming at some modern convergent plate boundaries. This supports 30-year-old notions that ancient crustal accretion–growth took place in settings with some similarities to modern convergent plate boundaries.

(4) Although a convergent plate boundary setting is preferred for the origin of the juvenile crustal accretion–growth components in the Itsaq Gneiss Complex, little of the complex is preserved, and most was strongly modified in later tectonothermal events. However, models for the geometry and dynamics of these boundaries must be able to explain the presence of slivers of upper mantle peridotite tectonically intercalated with crustal rocks during accretion, and the presence of small volumes of high-MgO quartz diorites that are coeval with the voluminous tonalites.

This research on the Itsaq Gneiss Complex was supported by NERC grant NER/A/S/1999/00024 and ARC grant DP0342798. Research on Eoarchaean crustal evolution in the Anshan area was supported by the Key Programmes of the Ministry of Land and Resources of China (1212010711815, 1212010561608). The paper is published with the permission of the Director of the Geological Survey of Denmark and Greenland.

References

ABBOTT, D., DRURY, R. & SMITH, W. H. F. 1994. Flat to steep transition in subduction style. *Geology*, **22**, 937–940.

ALLAART, J. H. 1976. The pre-3760 m.y. old supracrustal rocks of the Isua area, central West Greenland, and the associated occurrence of quartz-banded ironstone. *In*: WINDLEY, B. F. (ed.) *The Early History of the Earth*. Wiley, Chichester, 177–189.

APPEL, P. W. U., FEDO, C. M., MOORBATH, S. & MYERS, J. S. 1998. Recognisable primary volcanic and sedimentary features in a low-strain domain of the highly deformed, oldest-known (*ca*. 3.7–3.8 Gyr) greenstone belt, Isua, Greenland. *Terra Nova*, **10**, 57–62.

ARTH, J. G. & HANSON, G. N. 1975. Geochemistry and origin of the Early Precambrian crust of north-eastern Minnesota. *Geochimica et Cosmochimica Acta*, **39**, 325–362.

BAADSGAARD, H. 1973. U–Th–Pb dates on zircons from the early Precambrian Amîtsoq gneisses, Godthaab district, West Greenland. *Earth and Planetary Science Letters*, **19**, 22–28.

BAADSGAARD, H., NUTMAN, A. P., BRIDGWATER, D., MCGREGOR, V. R., ROSING, M. & ALLAART, J. H. 1984. The zircon geochronology of the Akilia association and the Isua supracrustal belt, West Greenland. *Earth and Planetary Science Letters*, **68**, 221–228.

BAADSGAARD, H., NUTMAN, A. P. & BRIDGWATER, D. 1986. Geochronology and isotope geochemistry of the early Archaean Amîtsoq gneisses of the Isukasia area, southern West Greenland. *Geochimica et Cosmochimica Acta*, **50**, 2173–2183.

BENNETT, V. C., NUTMAN, A. P. & MCCULLOCH, M. T. 1993. Nd isotopic evidence for transient, highly depleted mantle reservoirs in the early history of the Earth. *Earth and Planetary Science Letters*, **119**, 299–317.

BENNETT, V. C., NUTMAN, A. P. & ESAT, T. M. 2003. Constraints on mantle evolution and differentiation from $^{187}Os/^{188}Os$ isotopic compositions of Archaean ultramafic rocks from southern West Greenland (3.8 Ga) and Western Australia (3.45 Ga). *Geochimica et Cosmochimica Acta*, **66**, 2615–2630.

BLACK, L. P., GALE, N. H., MOORBATH, S., PANKHURST, R. J. & MCGREGOR, V. R. 1971. Isotopic dating of very early Precambrian amphibolite facies gneisses from the Godthåb district, West Greenland. *Earth and Planetary Science Letters*, **12**, 245–259.

BOHLAR, R., KAMBER, B. S., MOORBATH, S., FEDO, C. M. & WHITEHOUSE, M. J. 2004. Characterisation of early Archaean chemical sediments by trace element signatures. *Earth and Planetary Science Letters*, **222**, 43–60.

BRIDGWATER, D. & MCGREGOR, V. R. 1974. Field work on the very early Precambrian rocks of the Isua area, southern West Greenland. *Rapport Grønlands Geologiske Undersøgelse*, **65**, 49–54.

BROWN, G. C. & FYFFE, W. S. 1970. Production of granitic melts during ultrametamorphism. *Contributions to Mineralogy and Petrology*, **28**, 310–318.

CALVERT, A. J., SAWYER, E. W., DAVIS, W. J. & LUDDEN, J. N. 1995. Archaean subduction inferred from seismic images of a mantle suture in the Superior Province. *Nature*, **375**, 670–673.

CAMPELL, I. H. & TAYLOR, S. R. 1983. No water, no granites – no oceans, no continents. *Journal of Geophysical Research*, **10**, 1061–1064.

CATES, N. L. & MOJZSIS, S. J. 2006. Chemical and isotopic evidence for widespread Eoarchean metasedimentary enclaves in southern West Greenland. *Geochimica et Cosmochimica Acta*, **70**, 4229–4257.

CHADWICK, B. & CREWE, M. A. 1986. Chromite in the early Archaean Akilia association (*c.* 3,800 m.y.), Ivîsartoq region, inner Godthåbsfjord, southern West Greenland. *Economic Geology*, **81**, 184–191.

CHAMPION, D. C. & SMITHIES, R. H. 2007. Geochemistry of Palaeoarchaean granites of the east Pilbara terrane, Pilbara Craton, Western Australia: Implications for early Archaean crustal growth. *In*: VAN KRANENDONK, M. J., SMITHIES, R. H. & BENNETT, V. C. (eds) *Earth's Oldest Rocks*. Elsevier, Amsterdam, 369–410.

COMPSTON, W., KINNY, P. D., WILLIAMS, I. S. & FOSTER, J. J. 1986. The age and lead loss behaviour of zircons from the Isua supracrustal belt as determined by ion microprobe. *Earth and Planetary Science Letters*, **80**, 71–81.

CROWLEY, J. L. 2002. Testing the model of late Archean terrane accretion in southern West Greenland: A comparison of timing of geological events across the Qarliit Nunaat fault, Buksefjorden region. *Precambrian Research*, **116**, 57–79.

CROWLEY, J. L. 2003. U–Pb geochronology of 3810–3630 Ma granitoid rocks south of the Isua greenstone belt, southern West Greenland. *Precambrian Research*, **126**, 235–257.

CROWLEY, J. L., MYERS, J. S. & DUNNING, G. R. 2002. The timing and nature of multiple 3700–3600 Ma tectonic events in granitoid rocks north of the Isua greenstone belt, southern West Greenland. *Geological Society of America Bulletin*, **114**, 1311–1325.

DAUPHAS, N., VAN ZUILEN, M., WADHWA, M., DAVIS, A. M., MARTEY, B. & JANNEY, P. E. 2004. Clues from Fe isotope variations on the origin of early Archean BIFs from Greenland. *Science*, **302**, 2077–2080.

DAVIES, G. F. 2006. Gravitational depletion of early Earth's upper mantle and the viability of early plate tectonics. *Earth and Planetary Science Letters*, **243**, 376–382.

DE WIT, M. J. 1998. On Archean granites, greenstones, cratons and tectonics: Does the evidence demand a verdict? *Precambrian Research*, **91**, 181–226.

DYMEK, R. F. & KLEIN, C. 1988. Chemistry, petrology and origin of banded iron-formation lithologies from the 3800 Ma Isua supracrustal belt, West Greenland. *Precambrian Research*, **37**, 247–302.

FRIEND, C. R. L. & NUTMAN, A. P. 2005a. Complex 3670–3500 Ma orogenic episodes superimposed on juvenile crust accreted between 3850–3690 Ma, Itsaq Gneiss Complex, southern West Greenland. *Journal of Geology*, **113**, 375–398.

FRIEND, C. R. L. & NUTMAN, A. P. 2005b. New pieces to the Archaean terrane jigsaw puzzle in the Nuuk region, southern West Greenland: Steps in transforming a simple insight into a complex regional tectonothermal model. *Journal of the Geological Society, London*, **162**, 147–163.

FRIEND, C. R. L., NUTMAN, A. P. & MCGREGOR, V. R. 1988. Late Archaean terrane accretion in the Godthåb region, southern West Greenland. *Nature*, **335**, 535–538.

FRIEND, C. R. L., BENNETT, V. C. & NUTMAN, A. P. 2002. Abyssal peridotites >3,800 Ma from southern West Greenland: Field relationships, petrography, geochronology, whole-rock and mineral chemistry of dunite and harzburgite inclusions in the Itsaq Gneiss Complex. *Contributions to Mineralogy and Petrology*, **143**, 71–92.

FURNES, H., DE WIT, M., STAUDIGEL, H., ROSING, M. & MUEHLENBACHS, K. 2007. Vestige of Earth's oldest ophiolite. *Science*, **215**, 1704–1707.

GILL, R. C. O., BRIDGWATER, D. & ALLAART, J. H. 1981. The geochemistry of the earliest known basic metavolcanic rocks, West Greenland: a preliminary investigation. *Geological Society of Australia Special Publication*, **7**, 313–325.

GRIFFIN, W. L., MCGREGOR, V. R., NUTMAN, A. P., TAYLOR, P. N. & BRIDGWATER, D. 1980. Early Archaean granulite-facies metamorphism south of Ameralik. *Earth and Planetary Science Letters*, **50**, 59–74.

GROVE, T. L. & BAKER, M. B. 1983. Effects of melt density on magma mixing in calc-alkaline series magmas. *Nature*, **305**, 5933.

HAMILTON, P. J., O'NIONS, R. K., EVENSEN, N. H., BRIDGWATER, D. & ALLAART, J. H. 1978. Sm–Nd isotopic investigations of Isua supracrustals and implications for mantle evolution. *Nature*, **272**, 41–43.

HANMER, S. & GREENE, D. C. 2002. A modern structural regime in the Paleoarchean (~3.64 Ga); Isua Greenstone Belt, southern West Greenland. *Tectonophysics*, **346**, 201–222.

JENNER, F. 2007. Geochemistry and petrogenesis of Archaean mafic and ultramafic rocks, southern West Greenland. PhD thesis, Australian National University, Canberra.

JENNER, F. E., BENNETT, V. C., NUTMAN, A. P., FRIEND, C. R. L., NORMAN, M. D. & YAXLEY, G. 2008. Evidence for plate tectonics at 3.8 Ga: Geochemistry of arc-like metabasalts from the southern edge of the Isua supracrustal belt. *Chemical Geology*, doi: 10.1016/j.chem.geo.2008.09.016.

KAMBER, B. S. & MOORBATH, M. 1998. Initial Pb of the Amîtsoq gneiss revisited: Implication for the timing of early Archaean crustal evolution in West Greenland. *Chemical Geology*, **150**, 19–41.

KAMBER, B. S., WHITEHOUSE, M. J., BOLHAR, R. & MOORBATH, S. 2005. Volcanic resurfacing and the early terrestrial crust: Zircon U–Pb and REE constraints from the Isua Greenstone Belt, southern West Greenland. *Earth and Planetary Science Letters*, **240**, 276–290.

KERRICH, R. & POLAT, A. 2006. Archean greenstone-tonalite duality: thermochemical mantle convection models or plate tectonics in the early Earth global dynamics? *Tectonophysics*, **415**, 141–165.

KOMIYA, T., MARUYAMA, S., MASUDA, T., APPEL, P. W. U. & NOHDA, S. 1999. The 3.8–3.7 Ga plate tectonics on the Earth; Field evidence from the Isua accretionary complex, West Greenland. *Journal of Geology*, **107**, 515–554.

LIU, D., NUTMAN, A. P., COMPSTON, W., WU, J. S. & SHEN, Q. H. 1992. Remnants of ≥3800 Ma crust in the Chinese part of the Sino-Korean craton. *Geology*, **20**, 339–342.

LIU, D., WAN, Y., WU, J. S., WILDE, S. A., ZHOU, H. Y., DONG, C. Y. & YIN, X. Y. 2007. Eoarchean rocks and zircons in the North China Craton. *In*: VAN KRANENDONK, M. J., SMITHIES, R. H. & BENNETT, V. C. (eds) *Earth's Oldest Rocks*. Elsevier, Amsterdam, 251–274.

MANNING, C. E., MOJZSIS, S. J. & HARRISON, T. M. 2006. Geology, age and origin of supracrustal rocks, Akilia, Greenland. *American Journal of Science*, **306**, 303–366.

MARTIN, H. 1986. Effect of steeper Archaean geothermal gradient on geochemistry of subduction-zone magmas. *Geology*, **14**, 753–756.

MARTIN, H. 1999. Adakitic magmas: Modern analogues of Archaean granitoids. *Lithos*, **46**, 411–429.

MARTIN, H. & MOYEN, J.-F. 2002. Secular changes in tonalite–trondhjemite–granodiorite composition as markers of the progressive cooling of Earth. *Geology*, **30**, 319–322.

MARTIN, H., SMITHIES, R. H., RAPP, R., MOYEN, J.-F. & CHAMPION, D. 2005. An overview of adakite, tonalite–trondhjemite–granodiorite (TTG), and sanukitoid: Relationships and some implications for crustal evolution. *Lithos*, **79**, 1–24.

MCGREGOR, V. R. 1973. The early Precambrian gneisses of the Godthåb district, West Greenland. *Philosophical Transactions of the Royal Society of London, Series A*, **273**, 343–358.

MCGREGOR, V. R. 1979. Archean gray gneisses and the origin of continental crust: Evidence form the Godthåb region, West Greenland. *In*: BARKER, F. (ed.) *Trondhjemites, Dacites and Related Rocks*. Developments in Petrology, **6**, 169–204.

MCGREGOR, V. R. & MASON, B. 1977. Petrogenesis and geochemistry of metabasaltic and metasedimentary enclaves in the Amîtsoq gneisses, West Greenland. *American Mineralogist*, **62**, 887–904.

MCGREGOR, V. R., FRIEND, C. R. L. & NUTMAN, A. P. 1991. The late Archaean mobile belt through Godthåbsfjord, southern West Greenland: A continent–continent collision zone? *Bulletin of the Geological Society of Denmark*, **39**, 179–197.

MCKENZIE, D. P. & BICKLE, M. J. 1988. The volume and composition of melt generated by extension of the lithosphere. *Journal of Petrology*, **29**, 625–679.

MICHARD-VITRAC, A., LANCELOT, J., ALLEGRE, C. J. & MOORBATH, S. 1977. U–Pb ages on single zircons from early Precambrian rocks of West Greenland and the Minnesota River Valley. *Earth and Planetary Science Letters*, **35**, 449–453.

MOJZSIS, S. J. & HARRISON, T. M. 2002. Establishment of a 3.83-Ga magmatic age for the Akilia tonalite (southern West Greenland). *Earth and Planetary Science Letters*, **202**, 563–576.

MOORBATH, S. 1975. Evolution of Precambrian crust from strontium isotopic evidence. *Nature*, **254**, 395–398.

MOORBATH, S., O'NIONS, R. K., PANKHURST, R. J., GALE, N. H. & MCGREGOR, V. R. 1972. Further rubidium–strontium age determinations on the very early Precambrian rocks of the Godthåb district, West Greenland. *Nature*, **240**, 78–82.

MOORBATH, S., O'NIONS, R. K. & PANKHURST, R. J. 1973. Early Archaean age for the Isua iron formation, West Greenland. *Nature*, **245**, 138–139.

MOORBATH, S., ALLAART, J. H., BRIDGWATER, D. & MCGREGOR, V. R. 1977. Rb–Sr ages of early Archaean supracrustal rocks and Amîtsoq gneisses at Isua. *Nature*, **270**, 43–45.

MYERS, J. S. 2001. Protoliths of the 3.8–3.7 Ga Isua greenstone belt, West Greenland. *Precambrian Research*, **105**, 129–141.

MYERS, J. S. & CROWLEY, J. L. 2000. Vestiges of life in the oldest Greenland rocks? A review of early Archean geology in the Godthåbsfjord region, and reappraisal of field evidence for >3800 Ma life on Akilia. *Precambrian Research*, **103**, 101–124.

NUTMAN, A. P. 1984. Early Archaean crustal evolution of the Isukasia area, southern West Greenland. *In*: KRÖNER, A. & GREILING, R. (eds) *Precambrian Tectonics Illustrated*. Schweizerbart, Stuttgart, 79–93.

NUTMAN, A. P. 2006. Antiquity of the oceans and continents. *Elements*, **2**, 223–227.

NUTMAN, A. P. & BRIDGWATER, D. 1986. Early Archaean Amîtsoq tonalites and granites from the Isukasia area, southern West Greenland: Development of the oldest-known sial. *Contributions to Mineralogy and Petrology*, **94**, 137–148.

NUTMAN, A. P. & COLLERSON, K. D. 1991. Very early Archean crustal-accretion complexes preserved in the North Atlantic Craton. *Geology*, **19**, 791–794.

NUTMAN, A. P., BRIDGWATER, D. & FRYER, B. 1984a. The iron rich suite from the Amîtsoq gneisses of southern West Greenland: Early Archaean plutonic rocks of mixed crustal and mantle origin. *Contributions to Mineralogy and Petrology*, **87**, 24–34.

NUTMAN, A. P., ALLAART, J. H., BRIDGWATER, D., DIMROTH, E. & ROSING, M. T. 1984b. Stratigraphic and geochemical evidence for the depositional environment of the early Archaean Isua supracrustal belt, southern West Greenland. *Precambrian Research*, **25**, 365–396.

NUTMAN, A. P., FRIEND, C. R. L., KINNY, P. D. & MCGREGOR, V. R. 1993. Anatomy of an Early Archaean gneiss complex: 3900 to 3600 Ma crustal evolution in southern West Greenland. *Geology*, **21**, 415–418.

NUTMAN, A. P., MCGREGOR, V. R., FRIEND, C. R. L., BENNETT, V. C. & KINNY, P. D. 1996. The Itsaq Gneiss Complex of southern West Greenland; the world's most extensive record of early crustal evolution (3900–3600 Ma). *Precambrian Research*, **78**, 1–39.

NUTMAN, A. P., BENNETT, V. C., FRIEND, C. R. L. & ROSING, M. T. 1997. ~ 3710 and ≥3790 Ma volcanic sequences in the Isua (Greenland) supracrustal belt; structural and Nd isotope implications. *Chemical Geology*, **141**, 271–287.

NUTMAN, A. P., BENNETT, V. C., FRIEND, C. R. L. & NORMAN, M. 1999. Meta-igneous (non-gneissic) tonalites and quartz-diorites from an extensive ca. 3800 Ma terrain south of the Isua supracrustal belt, southern West Greenland: Constraints on early crust formation. *Contributions to Mineralogy and Petrology*, **137**, 364–388.

NUTMAN, A. P., FRIEND, C. R. L., BENNETT, V. C. & MCGREGOR, V. R. 2000. The early Archaean Itsaq Gneiss Complex of southern West Greenland: The importance of field observations in interpreting dates and isotopic data constraining early terrestrial evolution. *Geochimica et Cosmochimica Acta*, **64**, 3035–3060.

NUTMAN, A. P., MCGREGOR, V. R., SHIRAISHI, K., FRIEND, C. R. L., BENNETT, V. C. & KINNY, P. D. 2002a. ≥3850 Ma BIF and mafic inclusions in the early Archaean Itsaq Gneiss Complex around Akilia, southern West Greenland? The difficulties of precise dating of zircon-free protoliths in migmatites. *Precambrian Research*, **117**, 185–224.

NUTMAN, A. P., FRIEND, C. R. L. & BENNETT, V. C. 2002b. Evidence for 3650–3600 Ma assembly of the northern end of the Itsaq Gneiss Complex, Greenland: Implication for early Archean tectonics. *Tectonics*, **21**, article 5, doi: 10.1029/2000TC001203.

NUTMAN, A. P., FRIEND, C. R. L., BARKER, S. S. & MCGREGOR, V. R. 2004. Inventory and assessment of Palaeoarchaean gneiss terrains and detrital zircons in southern West Greenland. *Precambrian Research*, **135**, 281–314.

NUTMAN, A. P., FRIEND, C. R. L., HORIE, H. & HIDAKA, H. 2007a. Construction of pre-3600 Ma crust at convergent plate boundaries, exemplified by the Itsaq Gneiss Complex of southern West Greenland. *In*: VAN KRANENDONK, M. J., SMITHIES, R. H. & BENNETT, V. C. (eds) *Earth's Oldest Rocks*. Elsevier, Amsterdam, 187–218.

NUTMAN, A. P., BENNETT, V. C., FRIEND, C. R. L., HORIE, K. & HIDAKA, H. 2007b. ~3850 Ma tonalites in the Nuuk region, Greenland: Geochemistry and their reworking within an Eoarchaean gneiss complex. *Contributions to Mineralogy and Petrology*, **154**, 385–408.

O'NIONS, R. K. & PANKHURST, R. J. 1974. Rare-earth element distribution in Archaean gneisses and anorthosites, Godthåb area, West Greenland. *Earth and Planetary Science Letters*, **22**, 328–338.

PLANK, T. 2005. Constraints from thorium/lanthanum on sediment recycling at subduction zones and the evolution of continents. *Journal of Petrology*, **46**, 921–944.

POLAT, A. & HOFMANN, A. W. 2003. Alteration and geochemical patterns in the 3.7–3.8 Ga Isua greenstone belt, West Greenland. *Precambrian Research*, **126**, 197–218.

POLAT, A., HOFMANN, A. W. & ROSING, M. T. 2002. Boninite-like volcanic rocks in the 3.7–3.8 Ga Isua greenstone belt, West Greenland: Geochemical evidence for intra-oceanic subduction processes in the early Earth. *Chemical Geology*, **184**, 231–254.

RAPP, R. P. 1997. Heterogeneous source regions for Archean granitoids. *In*: DE WIT, M. J. & ASHWAL, L. D. (eds) *Greenstone Belts*. Oxford University Press, Oxford, 267–279.

ROSING, M. 1999. ^{13}C-depleted carbon microparticles in >3700 Ma seafloor sedimentary rocks from Greenland. *Science*, **283**, 674–676.

ROSING, M. T, ROSE, N. M., BRIDGWATER, D. & THOMSEN, H. S. 1996. Earliest part of the Earth's stratigraphic record: A reappraisal of the >3.7 Ga Isua (Greenland) supracrustal sequence. *Geology*, **24**, 43–46.

SHERVAIS, J. W. 2001. Birth, death, and resurrection: The life cycle of suprasubduction zone ophiolites. *Geochemistry, Geophysics, Geosystems*, paper number 2000GC000080.

SMITHIES, R. H. & CHAMPION, D. C. 2000. The Archaean high-Mg diorite suite: Links to tonalite–trondhjemite–granodiorite magmatism and implications for early Archaean crustal growth. *Journal of Petrology*, **41**, 1653–1671.

SOLVANG, M. 1999. *An investigation of metavolcanic rocks from the eastern part of the Isua greenstone belt, Western Greenland*. Geological Survey of Denmark and Greenland (GEUS) Internal Report.

SONG, B., NUTMAN, A. P., LIU, D. Y. & WU, J. S. 1996. 3800 to 2500 Ma crustal evolution in the Anshan area of Liaoning province, northeastern China. *Precambrian Research*, **78**, 79–94.

STEENFELT, A., GARDE, A. A. & MOYEN, J.-F. 2005. Mantle wedge involvement in the petrogenesis of Archaean grey gneisses in West Greenland. *Lithos*, **79**, 207–228.

VERVOORT, J. D., PATCHETT, P. J., GEHRELS, G. E. & NUTMAN, A. P. 1996. Constraints on early Earth differentiation from hafnium and neodymium isotopes. *Nature*, **379**, 624–627.

WAN, Y., LIU, D. Y., SONG, B., WU, J., YANG, C., ZHANG, Z. & GENG, Y. 2005. Geochemical and Nd isotopic compositions of 3.8 Ga meta-quartz dioritic and trondhjemitic rocks from the Anshan area and their geological significance. *Journal of Asian Earth Sciences*, **24**, 563–575.

WELLS, P. R. A. 1980. Thermal models for the magmatic accretion and subsequent metamorphism of continental crust. *Earth and Planetary Science Letters*, **46**, 253–265.

WHITEHOUSE, M. J. & KAMBER, B. S. 2005. Assigning dates to thin gneissic veins in high-grade metamorphic terranes: A cautionary tale from Akilia, southwest Greenland. *Journal of Petrology*, **46**, 291–318.

WHITEHOUSE, M. J., KAMBER, B. S. & MOORBATH, S. 1999. Age significance of U–Th–Pb zircon data from early Archaean rocks of west Greenland—a reassessment based on combined ion-microprobe and imaging studies. *Chemical Geology*, **160**, 201–224.

WYLLIE, P. J., WOLF, M. B. & VAN DER LAAN, S. R. 1997. Conditions for the formation of tonalites and trondhjemites. *In*: DE WIT, M. J. & ASHWAL, L. D. (eds) *Greenstone Belts*. Oxford University Press, Oxford, 250–266.

Archaean crustal growth processes in southern West Greenland and the southern Superior Province: geodynamic and magmatic constraints

ALI POLAT[1]*, ROB KERRICH[2] & BRIAN WINDLEY[3]

[1]*Department of Earth Sciences, University of Windsor, Windsor, ON, N9B 3P4, Canada*

[2]*Department of Geological Sciences, University of Saskatchewan, SK, S7N 5E2, Canada*

[3]*Department of Geology, University of Leicester, Leicester LE1 7RH, UK*

*Corresponding author (e-mail: polat@uwindsor.ca)

Abstract: Eo- to Mesoarchaean greenstone belts (e.g. 3800–3700 Ma Isua, c. 3075 Ma Ivisaartoq, 3071 Ma Qussuk) occur within orthogneisses of the southern West Greenland Craton. Greenstone belts are composed mainly of metavolcanic rocks with minor ultramafic and sedimentary schists. Compositionally, volcanic rocks are dominantly tholeiitic basalts, boninites, and picrites, with minor intermediate to felsic volcanic rocks. These greenstone belts appear to have formed in convergent margin geodynamic settings. Detailed field observations, contrasting ages, and metamorphic and structural histories suggest that this craton was assembled in several accretionary tectonothermal events, involving accretion of arcs, back-arcs, forearcs, and continental fragments by horizontal tectonics. The Superior Province of Canada was also built by the amalgamation of oceanic and continental fragments ranging in age from 3700 to 2650 Ma, during five discrete tectonothermal events over 40 Ma between 2720 and 2680 Ma. The Neoarchaean (2750–2670 Ma) Wawa greenstone belts are composed of tectonically juxtaposed fragments of oceanic plateaux, oceanic island arcs, back-arcs, and siliciclastic trench turbidites. Following juxtaposition, these diverse lithologies were collectively intruded by syn- to post-kinematic granitoids with subduction zone geochemical signatures. Oceanic island arc lavas are easily distinguished from oceanic plateau counterparts because they possess positively fractionated rare earth element ($La/Sm_{cn} > 1$ and $Gd/Yb_{cn} > 1$) and high field strength element depleted ($Nb/Th_{pm} < 1$; $Nb/La_{pm} < 1$) patterns. In addition, the island arc association includes pyroclastic rocks that are rare to absent in the oceanic plateau volcanic association. Structural studies indicate that the Wawa greenstone belts underwent a complex history of deformation including thrusting, strike-slip faulting, and asymmetric folding. These belts constitute part of a c. 1000 km scale subduction–accretion complex that formed along an intra-oceanic convergent plate margin during trenchward migration of the magmatic arc axis. Several first-order geological observations on Archaean greenstone belts of SW Greenland and the Superior Province suggest that Phanerozoic-style plate-tectonic models can provide an elegant explanation for their structural, lithological, metamorphic and geochemical characteristics.

The origin of Archaean greenstone belts is a matter of longstanding debate (Windley 1976, 1993; Condie 1981, 2005a; Goodwin 1996; de Wit 1998; Kusky 2004; Cawood et al. 2006; Kerrich & Polat 2006). Numerous models, including uniformitarian and non-uniformitarian tectonic models, have been proposed to explain their origin (Burke et al. 1976; Windley 1976, 1993; Desrochers et al. 1993; de Wit 1998; Hamilton 1998; Kusky & Polat 1999; Polat & Hofmann 2003; Stern 2005). Uniformitarian models have invoked Phanerozoic-like geodynamic settings like those of today, including intracontinental rifts, mid-ocean ridges, back-arc basins, island arcs, ocean plateaux, plumes in a continental setting, and subduction–accretion complexes, or some combination, to account for the variety of structural, lithological and geochemical characteristics of Archaean greenstone belts. Specifically, the Superior Province, the largest Archaean craton in the world, has been interpreted as a series of allochthonous, amalgamated, oceanic and continental fragments or terranes, ranging in age from 3700 to 2680 Ma, that accreted from north to south over 40 Ma between 2720 and 2680 Ma (Card 1990; Williams et al. 1991a; Stott 1997; Daigneault et al. 2004; Percival et al. 2006a).

Similarly, on the basis of detailed field observations, contrasting ages, and metamorphic histories, several workers have shown that the SW Greenland Archaean craton is a collage of Eoarchaean to Neoarchaean oceanic island arcs and continental fragments, assembled in several accretionary tectonothermal events by horizontal tectonics (Bridgwater et al. 1974; Friend et al. 1988, 1996;

McGregor et al. 1991; Nutman et al., 2002, 2004; Friend & Nutman 2005a, b; Garde 2007, and references therein; Windley & Garde 2009).

Geochemical studies in Archaean greenstone belts have documented two major types of volcanic rock associations: (1) an oceanic plateau association composed dominantly of compositionally uniform Mg- to Fe-rich tholeiitic basalts and komatiites erupted from mantle plumes; and (2) a compositionally diverse intra-oceanic island arc association (Polat et al. 1998; Condie 2005a; Kerrich & Polat 2006; Polat & Kerrich 2006; Ujike et al. 2007). The second is composed predominantly of tholeiitic to calc-alkaline basalts, andesites, dacites, and rhyolites (BADR).

Cenozoic island arc associations display great compositional diversity (Hawkins 2003; Pearce 2003; Kelemen et al. 2004). In addition to 'normal' tholeiitic to calc-alkaline BADR, and alkaline lavas, there are rare boninites, low-Ti tholeiites (LOTI), picrites, adakites, magnesian andesites (MA), and Nb-enriched basalts (NEB) (Table 1). The latter six types have been described from certain arcs in terms of hot subduction: high thermal gradients that may stem from subduction of young, hot oceanic lithosphere, from ridge subduction, or slab windows (Hawkins 2003; Pearce 2003; Schuth et al. 2004; Thorkelson & Breitsprecher 2005; Table 1). Notably, similar rocks occur in Palaeozoic to Mesozoic supra-subduction ophiolites and accreted juvenile island arcs (Table 1).

Until the 1990s several of these lithologies, such as boninites and LOTI, were thought to be confined to Phanerozoic arcs. On the basis of their compositional similarities, and spatial–temporal associations, to Cenozoic volcanic suites, Archaean boninites, LOTI, picrites, adakites, MA, and NEB have been referred to as 'hot subduction' volcanic rocks by Polat & Kerrich (2006). Given the convergent margin geodynamic regime in which Cenozoic

Table 1. *Archaean hot subduction volcanic rocks and their Phanerozoic analogues*

Rock type	Archaean example	Phanerozoic analogues
Boninite	Abitibi (Kerrich et al. 1998); Isua (Polat et al. 2002); Karelia (Shchipansky et al. 2004); Whundo (Smithies et al. 2005); Dharwar Craton (Manikyamba et al. 2005)	Troodos ophiolite (Beccaluva & Serri 1988); Papua New Guinea ophiolite (Crawford et al. 1989); Izu–Bonin–Mariana forearc (Pearce et al. 1992); Betts Cove ophiolite (Bédard 1999); Oman ophiolite (Ishikawa et al. 2002); Caribbean island arc (Escuder Viruete et al. 2006); Lau Basin (Deschamps & Lallemand 2003)
Low-Ti tholeiite (LOTI)	Abitibi (Kerrich et al. 1998; Wyman 1999); Karelia (Shchipansky et al. 2004)	Mariana trench, Lau Basin, New Guinea (Beccaluva & Serri 1988); Tasmania (Brown & Jenner 1989)
Picrite	Isua (Polat & Hofmann 2003); Wawa (Polat & Kerrich 1999); Zunhua (Polat et al. 2006); Ivisaartoq (Polat et al. 2007)	Solomon Islands (Schuth et al. 2004); Central Aleutians (Nye & Reid 1986); Vanuatu arc (Eggins 1993); Lesser Antilles (Thirlwall et al. 1996)
Adakite	Wabigoon (Tomlinson et al. 1999); Birch–Uchi (Hollings & Kerrich 2000); Wawa (Polat & Kerrich 2001a); Abitibi (Wyman et al. 2002); Wabigoon (Ujike et al. 2007); Vedlozero–Segozero (Svetov et al. 2004)	Central America (Defant et al. 1992; Escuder Viruete et al. 2007); Philippines (Sajona et al. 1996); Japan (Morris 1995); Southern and Northern Volcanic Zones, Andes (Stern & Kilian 1996; Bourdon et al. 2002); Baja California (Benoit et al. 2002)
Magnesian andesite (MA)	Birch–Uchi (Hollings & Kerrich 2000); Wawa (Polat & Kerrich 2001a); Abitibi (Wyman et al. 2002)	Chile (Rogers & Saunders 1989); Philippines (Sajona et al. 1996); Western Aleutians (Yogodzinski et al. 1995); Baja California (Calmus et al. 2003); Caribbean (Escuder Viruete et al. 2007)
Nb-enriched basalt (NEB)	Birch–Uchi (Hollings & Kerrich 2000); Wabigoon (Ujike et al. 2007); Wawa (Polat & Kerrich 2001a); Karelia (Shchipansky et al. 2004)	Central America, Panama (Defant et al. 1992); Northern Kamchatka (Kepezhinskas et al. 1996); Philippines (Sajona et al. 1996); Baja California (Benoit et al. 2002) Caribbean island arc (Escuder Viruete et al. 2007)

Modified from Polat & Kerrich (2006).

'hot subduction' lavas develop, the geochemical characteristics of Archaean counterparts are of particular interest for understanding Archaean subduction zone geodynamic and petrogenetic processes and the early evolution of the crust–mantle system (Wyman et al. 2002; Wyman 2003; Kerrich & Polat 2006). In addition, many Archaean greenstone terranes are structurally similar to Phanerozoic subduction–accretion complexes, being accretionary assemblages of fault-bounded lithotectonic terranes that record multiple phases of compressional, strike-slip, and extensional deformation (Table 2; Taira et al. 1992; Şengör & Natal'in 1996, 2004; Kusky & Polat 1999; Kerrich & Polat 2006). Collisional orogenies have been classified by Şengör (1993) and Şengör & Natal'in (1996) into three superfamilies: (1) Altaid-type (also termed Turkic, Accretionary, or Cordilleran); (2) Alpine-type; (3) Himalayan type, citing the Neoarchaean Yilgarn Craton and Neoproterozoic Altaids as examples of the Altaid type (Tables 3 and 4).

In this review, we summarize recent studies of the 3800–2800 Ma accretionary greenstone terranes in SW Greenland and of the 2750–2670 Ma greenstone belts of the Wawa subprovince, Superior Province, Canada (Figs 1 and 2). Specifically, we address the geodynamic significance of recently recognized 'hot subduction' volcanic rocks and the style of deformation, to better constrain the origin of Archaean greenstone–granitoid terranes. Given

Table 2. *Interpreted geodynamic processes in Archaean greenstone–granitoid terranes and their Phanerozoic analogues*

Geodynamic process	Proposed archaean examples	Phanerozoic analogues
Plume–subduction interaction	Superior Province (Dostal & Mueller 1997; Wyman 1999)	Tonga forearc (Danyushevsky et al. 1995)
Plume–lithosphere interaction	Superior Province (Tomlinson et al. 1999); Pilbara Craton (Blake et al. 2004)	Gondwanaland; Siberian traps, Iceland–North Atlantic, Basin and Range (Storey 1995, and references therein)
Subduction accretion (Arc–arc, arc–continent, continent–continent; arc–oceanic plateau accretion)	Superior Province (Mueller et al. 1996; Polat et al. 1998; Percival et al. 2006a); Baltic Shield (Shchipansky et al. 2004); Greenland (Garde 2007; Nutman & Friend 2007a); Kaapvaal Craton (de Ronde & de Wit 1994); Zimbabwe Craton (Kusky & Kidd 1992)	Himalayas, Alps, Altaids, Alaska, Appalachians, North American Cordillera (Burchfiel et al. 1992; Kearey & Vine 1996; Şengör & Natal'in 2004); Japan (Taira et al. 1992); SW Pacific (Mann & Taira 2004)
Thrusting and imbrication	Abitibi (Lacroix & Sawyer 1995); Wawa (Gill 1992; Corfu & Stott 1998); Rainy Lake (Poulsen et al. 1980); North Karelia (Shchipansky et al. 2004); Barberton (de Ronde & de Wit 1994); Yilgarn Craton (Davis & Maidens 2003)	Canadian Cordillera (Umhoefer & Schiarizza 1996; US Cordillera (Wakabayashi 1992); Taurides (Polat et al. 1996); Altaids (Şengör & Natal'in 1996, 2004); Ontong–Java (Mann & Taira 2004)
Strike-slip faulting	Superior Province (Mueller et al. 1996; Percival et al. 2006b); Dharwar Craton (Chadwick et al. 2000); Pilbara Craton (Zegers et al. 1998); Yilgarn Craton (Chen et al. 2003)	Alps and Himalayas (Şengör 1990); Altaids (Şengör & Natal'in 1996); Canadian Cordillera (Umhoefer & Schiarizza 1996); Andes (Dewey & Lamb 1992)
Continental rifting	Superior Province (Tomlinson et al. 1999); Yilgarn Craton (Swager 1997); Pilbara Craton (Blake et al. 2004); Slave Province (Bleeker 2002)	East Africa, South and North Atlantic, Basin and Range, Rio Grande (Kearey & Vine 1996, and references therein)
Orogenic collapse	Superior Province (Calvert et al. 2004); Greenland (Hanmer & Greene 2002); Barberton (Kisters et al. 2003); Yilgarn Craton (Davis & Maidens 2003)	Alps, Himalayas, Appalachians (Dewey 1988); North American Cordillera (Vanderhaeghe & Teyssier 2001); Basin and Range (Dilek & Moores 1999)

Modified from Polat & Kerrich (2006).

Table 3. *Comparison of the Altaid–Cordilleran and Alpine–Himalayan type orogens*

Altaid–Cordilleran orogeny	Alpine–Himalayan orogeny
Accretion of allochthonous juvenile oceanic island arcs, forearcs, and continental blocks	Continent–continent collision, presence of giant ophiolite nappes between two continents
Closure of external ocean	Closure of internal ocean, such as Tethys
Multiple sutures, ophiolites (e.g. ensimatic arc basement, forearc), numerous ophiolitic fragments (ophirags) in the accretionary prism	Long, narrow single suture zone with more or less complete ophiolite nappe emplaced onto passive continental margins (e.g. Oman, Kizildag)
Subduction–accretion complex	Deformed passive margin sedimentary rocks
Multiple deformation of subduction–accretion complex, large degree of structural shuffling by thrusting and strike-slip faulting	Reworking of pre-existing crust
Magmatic arc migrates through subduction–accretion complex	Subdued magmatism
Heat advected by magmas	Internal radiogenic heat production
Subduction erosion of lithospheric mantle	Delamination of thickened lithospheric mantle

Data are from Şengör & Natal'in (1996, 2004).

Table 4. *Examples of the Altaid–Cordilleran and Alpine Himalayan type orogens in Earth's history*

Altaid–Cordilleran orogeny	Alpine–Himalayan orogeny
Isua greenstone belt, Eoarchaean	Terminal collision in Trans-Hudson (c. 1.7 Ga)
Ivisaartoq and Qussuk greenstone belts, Mesoarchaean	Grenville (<1.2 Ga)
Superior Province and, Yilgarn Craton, Neoarchaean	Caledonides, Early Palaeozoic
Altaids, Late Proterozoic–Palaeozoic	Alpine–Himalayan, Mesozoic–Cenozoic
Nippinoids, Mesozoic–Cenozoic	
Tasmanids, Early Palaeozoic	
North American Cordillera, Mesozoic–Cenozoic	

Data are from Şengör & Natal'in (1996, 2004).

the problems with identifying Archaean geodynamic settings from geochemistry alone, we emphasize associations of volcanic rocks that are repeated in space and time. The prefix 'meta' is implicit for all lithologies.

Archaean hot subduction volcanic rocks

Boninites

The term 'boninite' refers to a variety of primary, or near-primary magmas defined by Crawford *et al.* (1989) as having $SiO_2 > 53$ wt% and Mg-number >60. These lavas have a wide range of CaO/Al_2O_3 ratios of <0.55 to >0.75 in conjunction with U-shaped rare earth element (REE) patterns. The classification of boninitic rocks was initially based mainly on major elements such as CaO, K_2O, Na_2O, Al_2O_3 and SiO_2, and mineralogical and textural features (Crawford *et al.* 1989). Given the fact that CaO, K_2O, Na_2O and SiO_2 are likely to be mobile during hydrothermal alteration and metamorphism, the terms 'boninitic series' or 'boninite-like' have been used for metamorphosed Archaean lavas that share low TiO_2, Zr, and Nb contents, high Al_2O_3/TiO_2 ratios, and U-shaped REE patterns of fresh boninites (Kerrich *et al.* 1998; Polat *et al.* 2002; Smithies *et al.* 2005; Fig. 3; Table 1), yet have lower SiO_2 contents than in the boninite definition of Crawford *et al.* (1989). LOTI are Light REE (LREE)-depleted, high-Mg tholeiitic rocks associated with boninites; they share the negatively fractionated REE patterns and high Al_2O_3/TiO_2 ratios of boninites but do not have the U-shaped patterns or negative Nb anomalies (Brown & Jenner 1989; Wyman *et al.* 1999).

The distinctive U-shaped REE patterns of most boninites, with negative anomalies of Nb, Ta, P, and Ti, are considered to result from a two-stage mantle melting process: high degree of partial melting, followed by second-stage melting of the refractory, depleted residue under hydrous conditions at convergent margins (Sun & Nesbitt 1978; Stern *et al.* 1991). Most boninites have been reported from western Pacific arcs but are also

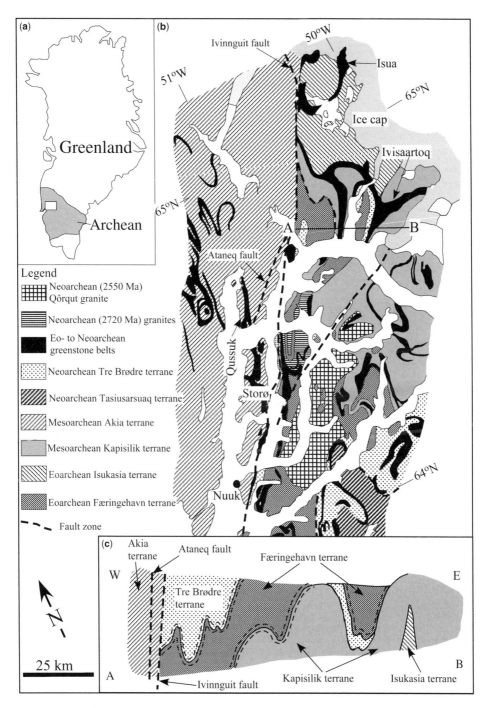

Fig. 1. (a) Location of the study area in SW Greenland. (b) A simplified geological map of the Nuuk region, showing the Eo- to Neoarchaean tectonic terranes and the location of the Ivisaartoq greenstone belt. (c) A simplified geological cross-section along line A–B; vertical and horizontal scales are exaggerated for illustration purposes. The map is modified from Friend & Nutman (2005a, b) and Nutman & Friend (2007a). The cross-section is modified from Friend & Nutman (2005a). Because of recent revisions in both terrane boundaries and terminology, there are some discrepancies between new and old terrane boundaries and names.

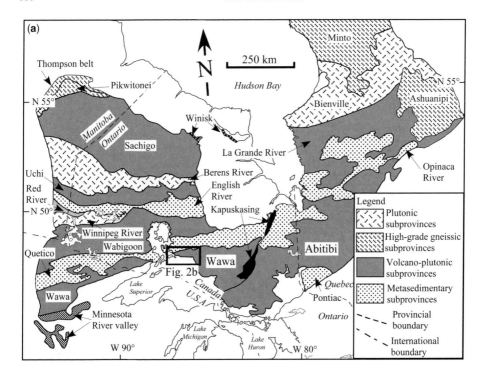

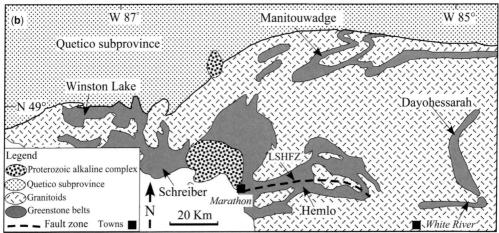

Fig. 2. (a) Simplified geological of the Superior Province (modified from Card & Ciesielski 1986). (b) Simplified geological map of the Schreiber, Hemlo, Manitouwadge, and White River areas, Wawa Subprovince, showing the location of the Winston Lake and Manitouwadge greenstone belts (modified after Williams et al. 1991b).

known from Phanerozoic ophiolites. LOTI may also be second-stage melts, but under anhydrous conditions (Table 1; Brown & Jenner 1989).

In the 3800–3700 Ma Isua greenstone belt, boninitic rocks are tectonically juxtaposed against picrites that have arc-like trace element patterns (Polat & Hofmann 2003). Boninitic rocks are associated with LOTI and sanukutoid intrusions in the 3010–2935 Ma Mallina Basin, Pilbara Craton, and with LOTI and tholeiitic to calc-alkaline basalts in the North Karelian and Abitibi greenstone belts (Shchipansky et al. 2004; Smithies et al. 2004; Table 1). The 3120 Ma Whundo boninitic lavas are associated with a calc-alkaline BADR suite (Smithies et al. 2005), whereas counterparts from the Frotet–Evans belt occur with adakites

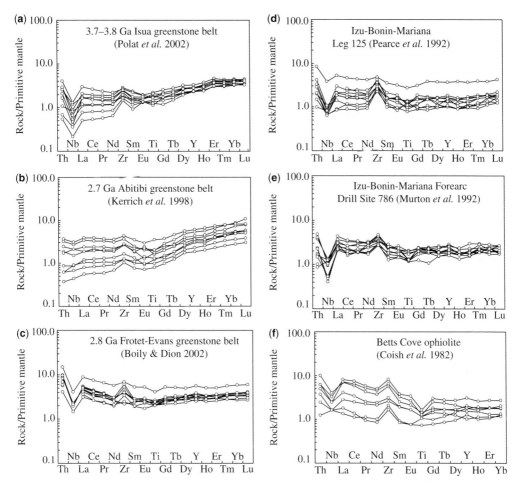

Fig. 3. Primitive mantle-normalized trace element diagrams for Archaean boninitic volcanic rocks and Phanerozoic counterparts, adapted from Polat & Kerrich (2006). Normalization values are from Hofmann (1988).

(Boily & Dion 2002). These are structurally overlain by tholeiitic to calc-alkaline BADR, with Th- and LREE-enriched, but Nb- and Ti-depleted, primitive mantle-normalized trace element patterns. In summary, all documented Archaean boninites occur with a variety of volcanic rock types that in the Phanerozoic are restricted to convergent margins.

Picrites

In contrast to boninitic rocks, Phanerozoic island arc picrites display positively fractionated heavy REE (HREE) patterns $(Gd/Yb)_{cn} > 1.5$, and the normalized negative anomalies of Nb, Ta, P, Zr, Hf and Ti are characteristic of most convergent margin magmas (Fig. 4). Picrites discussed in this paper differ from the Neoarchaean plume-derived Abitibi and Enemy Lake picrites (see Stone et al. 1995; Francis et al. 1999) in that they are spatially and temporally associated with the BADR magma series and have less fractionated HREE patterns, negative Nb and Ti anomalies, and lower Fe_2O_3 contents. Similarly, they are distinct from plume-derived Phanerozoic picrites in terms of high field strength element (HFSE)/REE ratios (see Kerr 2004).

Polat & Hofmann (2003) reported the presence of island arc picrites and tholeiitic basalts in the 3800–3700 Ma Isua greenstone belt (Table 1). Similar types of rocks have recently been recognized in the 3075 Ma Ivisaartoq, Greenland, and 2.5 Ga Zunhua, China, greenstone terranes (Polat & Hofmann 2003; Polat et al. 2006), and in the 2700 Ma Wawa greenstone terranes (Table 1; Polat & Kerrich 1999). A sheeted dyke complex, with supra-subduction zone geochemical characteristics, has been documented with the picrite–basalt

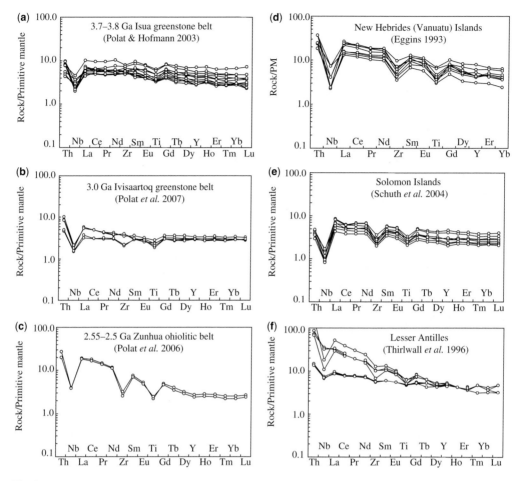

Fig. 4. Primitive mantle-normalized trace element diagrams for Archaean picritic volcanic rocks and Phanerozoic counterparts, adapted from Polat & Kerrich (2006). Normalization values are from Hofmann (1988).

association of the Isua greenstone belt, consistent with the formation of the belt by sea-floor spreading processes (Furnes *et al.* 2007), possibly in a forearc setting. However the presence of a sheeted dyke complex in the Isua greenstone belt has been questioned by Nutman & Friend (2007). All these rocks are characterized by primitive mantle-normalized trace element patterns similar to Phanerozoic arc picrites, but distinct from all known Archaean komatiites or boninites.

Adakites

Adakites are Al- and Na-enriched, intermediate to felsic calc-alkaline volcanic rocks having high Sr/Y and La/Yb$_{cn}$ ratios (Drummond *et al.* 1996; Martin 1999; Prouteau *et al.* 1999; Martin *et al.* 2005; Fig. 5). Rapp *et al.* (1999) and Martin *et al.* (2005) distinguished two major types of adakites on the basis of SiO$_2$ content, namely high-SiO$_2$ adakites (HSA) and low-SiO$_2$ adakites (LSA), respectively, synonymous with adakites and magnesian andesites (MA). In the Cenozoic, arc adakites occur as small volume flows and intrusions, often associated with MA and NEB. Adakites are distinctive in the conjunction of extremely fractionated REE but low Yb contents, coupled with high Mg-number, Cr and Ni compared with 'normal' dacites. They are generally interpreted as slab melts that hybridized with peridotite when traversing the mantle wedge (Martin 1999; Rapp *et al.* 1999; Smithies 2000; Condie 2005*b*; Martin *et al.* 2005). Adakites have been described from the Western Aleutians, Japan, the Philippines, Central America, Baja California, Ecuador, and Chile (Table 1).

Recent studies have documented the presence of adakitic intrusive and volcanic rocks in several

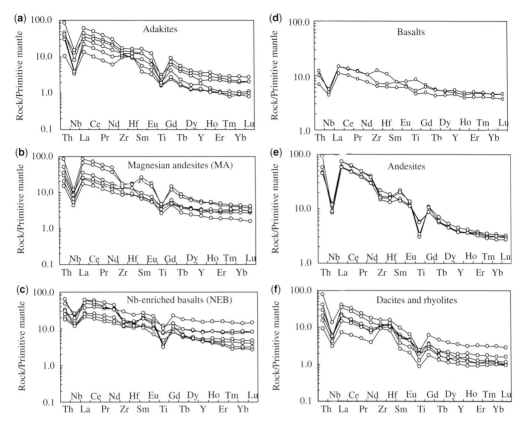

Fig. 5. Primitive mantle-normalized trace element diagrams for Wawa island arc volcanic rocks, adapted from Polat & Münker (2004). Normalization values are from Hofmann (1988).

Phanerozoic terranes (Xu et al. 2002; Gao et al. 2004; Wang et al. 2006; Jiang et al. 2007). Formation of these rocks appears to have not been related to any subduction zone geodynamic processes; they were emplaced long after the cessation of subduction. Their origin has been attributed to melting of lower continental crust during post-collisional tectonothermal events. Delamination and melting of lower crust in the upper mantle may have played an important role in petrogenesis of some of these adakitic rocks (Gao et al. 2004). In a recent review, Richards & Kerrich (2007) suggested several approaches for distinguishing slab melts from crustal melts having some adakite-like characteristics.

The presence of adakites, sometimes with MA and NEB, within 'normal' calc-alkaline BADR arc sequences has recently been reported from several Neoarchaean greenstone belts, notably the Superior Province (Hollings & Kerrich 2000; Polat & Kerrich 2001a; Boily & Dion 2002; Percival et al. 2003; Table 1), and Dharwar craton (Manikyamba & Khanna 2007; Manikyamba et al. 2007). Similarly, adakites, without MA or NEB, have been documented from bimodal arc sequences of the Yellowknife greenstone belt of the Slave Province, Wabigoon subprovince of the Superior Province, and the Vedlozero–Segozero greenstone belt of the Baltic Shield (Cousens et al. 2002; Svetov et al. 2004; Ujike et al. 2007).

MA and NEB association

High-magnesian andesites are near-primary, intermediate calc-alkaline volcanic rocks, having high Mg-number (>50), and Cr and Ni contents (Kelemen 1995; Sajona et al. 1996; Kelemen et al. 2004). Nb-enriched basalts are mafic volcanic arc rocks possessing high Nb concentrations (>7 ppm compared with <2 ppm for 'normal' arc basalts) and higher Nb/Th_{pm} and Nb/La_{pm} ratios than 'normal' arc basalts (Defant et al. 1992; Kepezhinskas et al. 1996; Sajona et al. 1996; Fig. 5).

The association of MA and NEB with adakites in Cenozoic arcs is generally explained as follows: (1) adakites are slab melts that are mildly hybridized with peridotitic mantle wedge; (2) MA are generated by more extensive hybridization of an adakite melt, or melting of a sub-arc mantle wedge that was previously metasomatized by adakitic melts; and (3) NEB are melts of the residue of MA hybridization as it is dragged to greater depth by induced convection in the mantle wedge (Sajona et al. 1996; Tatsumi 2001; Martin et al. 2005, and references therein). MA have been documented from the Aleutians, Philippines, Central America, Baja California, and Caribbean islands (Table 1). The presence of the MA and NEB association, with or without adakites, within 'normal' tholeiitic to calc-alkaline BADR arc sequences has recently been reported from several Neoarchaean greenstone belts of the Superior Province (Table 1).

Alternative models to slab melting, some of which are also alternatives to Archaean plate tectonics, have been proposed for high-Al tonalite–trondhjemite–granodiorite (TTG) batholiths of greenstone belts that compositionally resemble adakites. Such models include melting of thick crustal basalt sequences (Whalen et al. 2002), melting in thick plume-derived crust (Condie 2005b), resurfacing by thermochemical mantle convection (van Thienen et al. 2004), and catalytic delamination of eclogite (Bédard 2006). These batholiths are syn- to late-tectonic, rather than syn-arc, are volumetrically large, and are not spatially or temporally associated with BARD series volcanic rocks, boninites, MA or NEB. Accordingly, these alternative models are not further considered here.

Eo- to Neoarchaean greenstone belts in SW Greenland

The Archaean craton of West Greenland largely consists of Eo- to Neoarchean (c. 3800–2700 Ma) orthogneisses with TTG compositions (Steenfelt et al. 2005). These gneisses contain many conformable layers of amphibolite and anorthosite, both of which are up to c. 2 km thick, together with important but rare metasediments; taken together these rocks occur in alternating granulite- and amphibolite-facies belts (Bridgwater et al. 1976; Fig. 1). Since the 1980s terrane accretion models have been invoked to explain the geodynamic evolution of the Archaean crust in SW Greenland (Friend et al. 1988, 1996; Crowley 2002), and the accretion models have been refined as new geochronological, field, and geochemical data have been amassed over the last few years (Friend & Nutman 2005a; Garde 2007; Nutman & Friend 2007a). From their extensive sensitive high-resolution ion microprobe (SHRIMP) U/Pb zircon dating, and on the basis of their differing metamorphic, structural, and isotopic histories, Nutman and co-workers defined four Eoarchaean terranes, namely Færingehavn, Isukasia, Qarliit Taserssuat, and Aasivik, in the Itsaq Gneiss Complex; two Mesoarchaean terranes (Kapisilik and Akia) and two Neoarchaean terranes (Tre Brødre and Tasiusarsuaq) were also recognized (Nutman et al. 1996, 2004; Friend & Nutman 2005a, b; Nutman & Friend 2007a).

In this section we summarize the main geological and geochemical characteristics of the Eoarchaean (3800–3700 Ma) Isua and Mesoarchaean (c. 3075 Ma) Ivisaartoq and Qussuk greenstone belts, SW Greenland (Fig. 1). The Isua and Ivisaartoq greenstone belts contain both metasedimentary and metavolcanic supracrustal rocks and are characterized by polyphase deformation and metamorphism (Figs 6 and 7). The Isua and Ivisaartoq belts are located within the recently defined Eoarchaean Isukasia and Mesoarchaean Kapisilik tectonic terranes, respectively (Friend & Nutman 2005a; Nutman & Friend 2007a). Field relationships indicate that the Isukasia terrane is structurally overlain by the Kapisilik terrane to the south; the Kapisilik terrane, in turn, is structurally overlain by the Færingehavn and Tre Brødre terranes to the SSW (Fig. 1c). The latter two terranes were juxtaposed by 2950 Ma. The contact between the Isukasia and Kapisilik terranes is characterized by mylonites, asymmetric folds and migmatites (Fig. 7). Geochronological data and field observations constrain the collision between the Tre Brødre and the Kapisilik terranes to shortly after 2800 Ma. Geodynamic settings of these belts are interpreted on the basis of field relationships and recently obtained extensive trace element geochemical data (Polat et al. 2002, 2007, 2008; Polat & Hofmann 2003).

Eoarchaean (3800–3700 Ma) Isua greenstone belt

The Eoarchaean Isua greenstone belt is about 30 km long and up to 4 km wide (Bridgwater et al. 1976; Appel et al. 1998; Nutman et al. 2002, and references therein; Fig. 1). The belt occurs within Eoarchaean (c. 3800–3600 Ma) quartzo-feldspathic orthogneisses (Friend & Nutman 2005a, and references therein). On the basis of recent mapping, the eastern part of the belt has been divided into three lithotectonic domains: northwestern, central, and southeastern (Appel et al. 1998; Myers 2001). Each domain is characterized by different lithological associations and intensities of deformation. Similar lithotectonic domains also exist in the western part of the belt, which has been informally divided into three arc-shaped lithotectonic domains: the outer, central, and inner arcs. According to Nutman et al. (2002), the outer arc domain is composed of c. 3800 Ma volcano-sedimentary

Fig. 6. Field photographs of the Isua greenstone belt. (**a**) Pillow basalts. (**b**) Deformed cherts and amphibolites. (**c**) Deformed banded iron formations and cherts. (**d**) Deformed felsic ocelli in picritic flows and pillow basalts.

rocks, whereas the central arc domain is characterized by c. 3700 Ma volcano-sedimentary rocks. Recent geochronological studies suggest that the inner arc lithotectonic domain consists of 3720–3700 Ma rocks (A. P. Nutman, pers. comm.).

The Isua greenstone belt contains the oldest known rocks deposited on the surface of Earth, comprising volcanic, volcaniclastic, clastic and chemical sedimentary rocks (Komiya et al. 1999; Myers 2001; Polat & Hofmann 2003; Fig. 6). These supracrustal rocks have been repeatedly metamorphosed up to amphibolite facies and folded during several phases of deformation. Despite the polyphase deformation and metamorphism some low-strain zones display a wealth of well-preserved primary volcanic structures such as pillow lavas, minor debris flows, and pillow breccias (Fig. 6a). Polymictic conglomerates and pelites, mainly staurolite–mica schists, can be traced for more than 1 km along strike. Layers of chert-banded iron formation (BIF) are interbedded with volcanic rocks (Fig. 6b and c). In the low-strain domains, the variably deformed pillow basalts are intercalated with ultramafic units. The original stratigraphic relationships between mafic (pillow lavas) and ultramafic units (serpentinites) have been disrupted and complicated throughout the belt.

On the basis of extensive geochemical data obtained from the least altered volcanic rocks in the western part of the belt, Polat & Hofmann (2003) recognized the presence of two geochemically distinct volcanic associations in the Isua greenstone belt: a low-HFSE and high-HFSE association. Similar volcanic associations also exist in the eastern part of the belt. Volcanic rocks from the central arc have consistently lower HFSE (e.g. TiO_2 0.20–0.40 wt%; Zr 12–30 ppm; Nb 0.13–0.80 ppm) contents than those from the inner or outer arcs (TiO_2 0.50–1.14 wt%; Zr 34–77 ppm; Nb 1.2–2.7 ppm) at a given Mg-number value, and MgO and Ni contents (Figs 3a and 4a; Table 5). Given their respective geochemical coherence of many alteration-insensitive elements, differences between the two suites cannot be attributed to post-depositional alteration, fractional crystallization, the degree of partial melting, or crustal contamination processes, but could be explained by two geochemically distinct sources in the early Archaean mantle (Polat & Hofmann 2003). The high MgO, Ni and Cr contents in the least altered

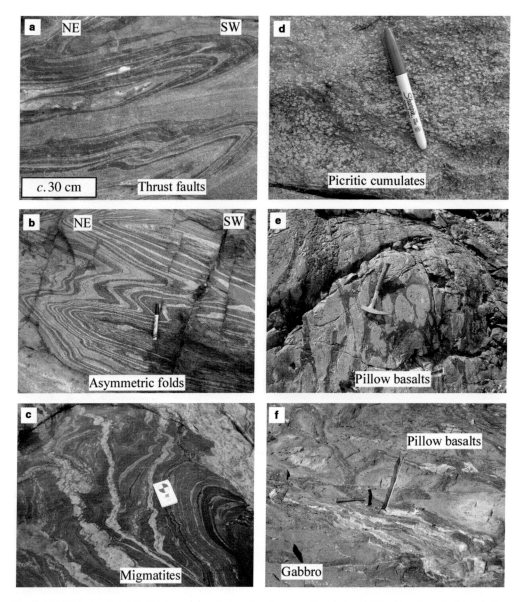

Fig. 7. Field photographs of the Ivisaartoq greenstone belt and surrounding region. (**a**) Mesoscopic-scale thrust fault at the contact between the Kapisilik and Tre Brødre terranes (to the north of the Ivisaartoq greenstone belt). (**b**) Mesoscopic-scale asymmetric folds at the contact between the Kapisilik and Tre Brødre terranes (to the north of the Ivisaartoq greenstone belt). (**c**) Migmatites at the contact between the Isukasia and Kapislik terranes (to the north of the Ivisaartoq greenstone belt). (**d**) Cumulate layer (picrite) in the Ivisaartoq greenstone belt. (**e**) Pillow basalts in the Ivisaartoq greenstone belt. (**f**) Altered pillow basalts and gabbros in the Ivisaartoq greenstone belt.

volcanic rocks of both associations are consistent with high degrees of partial melting. The low-HFSE association of the central lithotectonic sequence is geochemically comparable with Phanerozoic boninitic rocks (Polat *et al.* 2002); whereas the high-HFSE association of the inner and outer arc domains is similar to those of Phanerozoic island arc basaltic to picritic volcanic rocks (Figs 3 and 4). Both associations plot in the field of modern subduction zone rocks on a Ta/Yb v. Th/Yb diagram (Fig. 8; boninites are not shown).

Cenozoic boninites and picrites tend to form in restricted intra-oceanic supra-subduction zone settings associated, in space and time, with the

Table 5. *Summary of the ranges of significant compositional values and element ratios for picritic (outer and inner arc domains) and boninitic (central arc domain) volcanic rocks in the 3.8–3.7 Ga Isua greenstone belt*

	Outer arc tectonic domain basalts to picrites	Central arc tectonic domain boninites	Inner arc tectonic domain basalts to picrites
(wt%)			
SiO_2	48–53	47–54	48–53
MgO	8.9–19.6	6.8–16.1	4.5–21.0
TiO_2	0.50–1.05	0.17–0.40	0.52–1.14
Al_2O_3	6.5–11.9	13.9–20.2	7.9–14.1
Fe_2O_3	12.0–17.7	8.2–11.9	12.2–15.0
Mg-no.	58–77	61–77	40–77
(ppm)			
Cr	194–2916	60–1920	33–2268
Co	60–100	35–81	40–95
Ni	88–952	60–645	31–826
Sc	16–46	26–49	20–50
V	150–305	160–310	140–470
Nb	1.46–2.46	0.13–0.80	1.21–2.75
Zr	34–71	12.1–29.5	46–77
Th	0.2–2.12	0.04–0.29	0.40–0.79
La	1.80–3.95	0.31–1.83	2.4–6.3
Yb	1.04–2.01	0.94–1.84	1.1–2.7
Y	12.3–21.7	6.0–13.7	10.9–27.6
La/Sm_{cn}	0.7–1.1	0.56–1.39	1.0–1.7
La/Yb_{cn}	1.2–2.1	0.16–0.79	1.3–3.0
Gd/Yb_{cn}	1.4–1.7	0.26–0.61	1.4–1.9
Zr/Y	2.6–3.3	1.3–2.5	2.8–5.1
Ti/Zr	80–97	60–116	62–95
Al_2O_3/TiO_2	11.3–13.5	45–94	12.5–15.0
Nb/La_{pm}	0.37–0.82	0.23–0.77	0.29–0.60
Th/La_{pm}	0.7–5.81	0.57–2.07	0.8–1.9
Nb/Th_{pm}	0.14–0.80	0.31–0.76	0.22–0.57

Modified from Polat & Hofmann (2003).

subduction of young, hot, oceanic crust; for example, the Izu–Bonin–Mariana and Solomon arcs. In the Solomon arc, picrites occur above the subducting Woodlark spreading centre (Schuth *et al.* 2004). If the geochemical characteristics of the Isua picritic volcanic rocks had similar geodynamic significance to Phanerozoic counterparts, then they were possibly the products of early Archaean subduction zone geodynamic processes. The picritic and boninitic volcanic associations of the Isua greenstone belt were likely to have been juxtaposed as a consequence of Phanerozoic-style plate-tectonic processes operating in the early Earth.

Our recent geochemical and mineralogical analyses of felsic volcanic ocelli (Fig. 6d; Appel *et al.* 2009) in the high-HFSE volcanic association of basalts and picrites are comparable with oceanic plagiogranites (tonalites, trondhjemites) occurring in Phanerozoic ophiolites. Plagiogranites are particularly common in supra-subduction zone ophiolites (Hawkins 2003). In conclusion, picritic, boninitic, and plagiogranitic geochemical trends are consistent with the Isua greenstone belt having formed by similar geodynamic processes to Phanerozoic supra-subduction ophiolites.

Mesoarchaean (c. 3075 Ma) Ivisaartoq greenstone belt

The Ivisaartoq greenstone belt contains the largest Mesoarchaean supracrustal assemblage in SW Greenland (Chadwick 1990; Friend & Nutman 2005a, b; Polat *et al.* 2007, 2008, and references therein; Fig. 1). Its maximum age is constrained by a mean U–Pb zircon age of 3075 ± 15 Ma for felsic volcaniclastic rocks (Friend & Nutman 2005a; Polat *et al.* 2007), whereas a minimum age is given by weakly deformed 2961 ± 12 Ma granites to the north (Chadwick 1990; Friend & Nutman 2005a).

Notwithstanding isoclinal folding and amphibolite-facies metamorphism, this greenstone belt contains well-preserved primary magmatic structures in low-strain domains, such as pillow lavas,

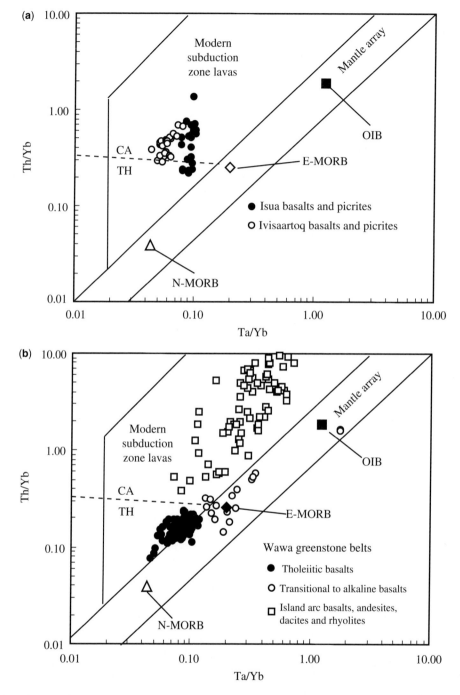

Fig. 8. Ta/Yb v. Th/Yb variation diagram showing the mantle array and field for modern subduction zone lavas. Both Isua and Ivisaartoq picritic to basaltic lavas plot in the field of modern subduction zone lavas. Wawa calc-alkaline BADR, adakites, MA, and NEB also plot in the field of modern subduction volcanic rocks. Wawa transitional to alkaline basalts plot in the mantle array between the oceanic plateau tholeiitic basalts and modern OIB (modified after Pearce 1982, 2003). Data for Isua and Ivisaartoq lavas are from Polat & Hofmann (2003) and Polat et al. (2007, 2008). Data for Wawa rocks are from Polat et al. (1999), Polat & Kerrich (2000) and Polat (2007). Normalization values and data for modern N-MORB, E-MORB and OIB are from Sun & McDonough (1989). CA, calc-alkaline; TH, tholeiitic.

volcanic breccias, and clinopyroxene cumulate in picrite layers, and sedimentary units (Chadwick 1990; Polat et al. 2007; Fig. 7d and e). There are also variably deformed gabbroic to dioritic dykes (Fig. 7f), actinolite schists, and serpentinites.

Chadwick (1990) subdivided the belt into lower and upper amphibolite units. These units are separated by a thin layer (up to 50 m thick) of magnetite-rich ultramafic schists, called the 'magnetic marker'. Hydrothermal alteration of the 'magnetic marker' and volcanic rocks in its vicinity resulted in the formation of calc-silicate rocks hosting strata-bound scheelite mineralization (Appel 1997). The intensity of deformation appears to increase towards the boundary between the two amphibolite units, suggesting that they are tectonically juxtaposed.

Cumulate picrites have uniform major and trace element compositions. They have sub-chondritic Nb/Ta (11–17) and Zr/Y (1.8–2.1) ratios and slightly super-chondritic Al_2O_3/TiO_2 (23–25) ratios (Table 6). On chondrite- and primitive mantle-normalized diagrams, they have the following trace element characteristics: (1) moderately enriched LREE patterns; and (2) flat HREE patterns; (3) negative Eu, Nb, Zr, and Ti anomalies (Fig. 9). Pillow basalts have the following trace element characteristics: (1) flat to depleted LREE patterns; (2) flat to slightly enriched HREE patterns; and (3) negative Nb, Zr, and Ti anomalies (Fig. 9; Table 6). Trace element patterns of gabbros and diorites are similar to those of pillow basalts (Fig. 9). The LREE-enriched and Nb-depleted trace element patterns of the Ivisaartoq volcanic association and intrusive rocks are consistent with a subduction zone geochemical signature (see Saunders et al. 1991; Hawkesworth et al. 1993; Murphy 2007). In addition, on the Ta/Yb v. Th/Yb variation diagram, the Ivisaartoq volcanic rocks plot in the field of modern subduction-derived magmas (see Pearce 1982, 2003, 2008; Fig. 8a).

The trace element systematics of the Ivisaartoq rocks are consistent with subduction zone geochemical signatures, but crustal contamination cannot be completely ruled out to account for the origin of the negative Nb anomalies in these rocks. Accordingly, the following geodynamic settings are possible: (1) a Red Sea-type proto-ocean setting (Fig. 10a); (2) a Rocas Verdes-type (southern Andes) continental arc rift or back-arc basin (see Stern & de

Table 6. *Summary of major (wt%) and trace (ppm) element concentrations and significant element ratios for the Ivisaartoq rocks*

	Cumulates	Pillow lavas	Gabbros	Diorites	Anorthositic inclusions
SiO_2	48.7–50.3	47.7–55.7	47.6–51.0	55.2–57.1	47.4–49.0
MgO	22.3–23.5	4.5–18.8	7.8–14.1	3.8–7.6	1.1–1.6
TiO_2	0.27–0.28	0.37–0.75	0.50–1.00	0.64–1.14	0.04–0.08
Al_2O_3	6.3–7.1	8.7–15.2	13.6–16.0	15.0–16.9	29.1–30.3
Fe_2O_3	9.2–10.4	7.7–13.9	9.0–13.2	6.2–8.0	2.4–3.2
Mg-no.	81.7–83.2	53.6–76.9	53.9–75.6	50.6–70.1	43.6–49.5
Cr	1575–1670	62–5700	230–1060	180–1030	6–11
Co	80–84	48–96	45–61	36–48	6–9
Ni	730–830	125–705	87–230	60–120	<2
Sc	24–26	26.5–43.4	33–41	35–48	1–3
V	100–170	136–485	190–475	230–300	14–21
Nb	0.69–0.77	0.09–1.61	0.10–1.77	1.33–3.06	0.10–0.21
Zr	12.5–15.6	22.3–42.4	28.2–48.5	36.3–68.4	3.6–21.1
Th	0.22–0.36	0.37–0.79	0.22–0.74	0.15–0.36	0.04–0.12
La	1.59–1.83	2.25–3.56	2.21–7.98	2.44–6.63	0.68–0.98
Yb	0.80–1.00	1.08–1.69	1.49–2.20	1.08–2.58	0.16–0.21
Y	7.0–8.4	9.4–15.2	13.6–19.8	9.5–22.2	1.3–2.2
La/Sm_{cn}	1.50–1.90	0.97–2.31	0.90–2.30	0.89–1.89	2.60–3.35
La/Yb_{cn}	1.27–1.60	1.09–2.19	0.80–3.30	1.05–2.05	2.80–3.32
Gd/Yb_{cn}	0.97–1.02	1.00–1.20	1.00–1.40	1.10–1.30	0.94–1.28
Al_2O_3/TiO_2	23.5–25.6	20.3–26.3	15.8–27.0	14.8–23.5	390–673
Zr/Y	1.8–2.1	2.0–2.9	1.2–2.7	1.2–3.8	1.8–9.8
Ti/Zr	103–128	76–116	94–123	95–106	13–90
Nb/La_{pm}	0.36–0.45	0.30–0.50	0.21–0.76	0.36–0.53	0.13–0.22
Th/La_{pm}	0.94–1.52	1.2–2.11	0.50–2.48	1.20–1.90	0.29–1.34
Nb/Th_{pm}	0.25–0.39	0.21–0.42	0.20–1.08	0.22–0.35	0.13–0.71

Data are from Polat et al. (2007, 2008).

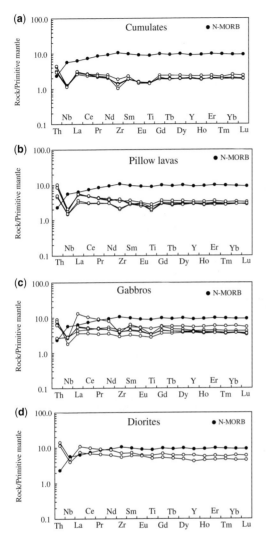

Fig. 9. Primitive mantle-normalized trace element diagrams for the Ivisaartoq volcanic rocks (picritic cumulate flows, pillow basalts) and intrusive rocks (gabbros and diorites). Normalization and N-MORB values are from Hofmann (1988).

Wit 1997; Fig. 10b); or (3) a Western Pacific-type oceanic arc–back-arc–forearc system (e.g. Izu–Bonin–Mariana, Solomon–New Hebrides, Tonga–Kermadec–Lau) (see Pearce & Peate 1995; Hawkins 2003; Schuth et al. 2004; Fig. 10c).

Given the presence of pillow structures and evidence for sea-floor hydrothermal alteration, the belt appears to have originated in an oceanic rather than a continental environment. There is no field evidence such as basal conglomerates for deposition on older continental basement as an autochthonous continental flood basalt province (e.g. Deccan, Karoo, Parana basalts) or on an Andean-type continental margin. The presence of gabbroic and dioritic dykes and siliciclastic sedimentary rocks is inconsistent with a typical oceanic plateau (e.g. Ontong–Java, Caribbean) setting. No komatiitic flows, the origin of which has typically been attributed to mantle plumes (Storey et al. 1991; Herzberg 1992; Xie et al. 1993; Condie 2004; Ernst et al. 2004), have been documented in the belt to suggest a mantle plume origin. Accordingly, we suggest that the Ivisaartoq greenstone belt formed in a Mesoarchaean convergent margin geodynamic setting.

On the basis of geological similarities between the Ivisaartoq greenstone belt, Phanerozoic forearc ophiolites, and intra-oceanic island arcs, Polat et al. (2007, 2008) suggested that this greenstone belt represents a relic of dismembered Mesoarchaean supra-subduction zone oceanic crust.

Mesoarchaean (c. 3071 Ma) Qussuk greenstone belt

The 3071 Ma Qussuk greenstone belt (informal name) is located in the upper part of the Akia tectonic block (Fig. 1). Like the other tectonic terranes in the Nuuk region, the Akia terrane underwent polyphase deformation and records multiple metamorphic and magmatic events (Garde 2007, and references therein). The belt was subjected to amphibolite-facies metamorphism at about 2980 Ma. The geochemical and geochronological characteristics of supracrustal rocks exposed in the Qussuk Peninsula have recently been investigated by Garde (2007). Amphibolites and associated gabbros have LREE-depleted trace element patterns, with minor negative Nb anomalies, whereas mafic to intermediate pyroclastic rocks display LREE-enriched patterns with large negative anomalies. According to Garde (2007), the Qussuk volcanic sequence originated at a convergent plate margin at c. 3071 Ma and was intruded by slab-derived granitoids around 3060 Ma, forming an intra-oceanic island arc system. Volcanic and intrusive events were followed by thrusting and folding at about 3030 Ma. Both the c. 3071 Ma Qussuk and the c. 3075 Ma Ivisaartoq volcanic associations formed at similar times and in comparable geodynamic settings.

Neoarchaean Abitibi–Wawa superterrane

The Wawa subprovince is located in the southern Superior Province. Linear map patterns of the approximately east–west-striking plutonic, volcanic–plutonic (greenstone–granitoid), high-grade gneissic, and sedimentary subprovinces reflect the lateral accretion of oceanic and continental

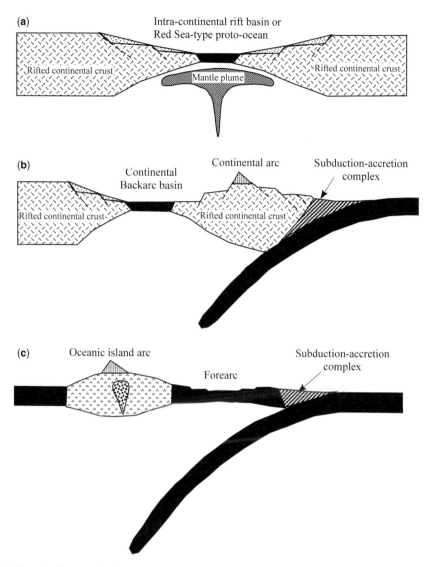

Fig. 10. Schematic diagrams showing possible geodynamic settings for the Ivisaartoq greenstone belt. (**a**) A Red Sea-type proto-ocean resulting from continental rifting where plume–lithosphere interaction could have resulted in negative Nb anomalies. (**b**) A Rocas Verdes-type continental back-arc basin where a combination of crustal assimilation and subduction zone process would have been the cause of the negative Nb anomalies. (**c**) An oceanic arc–forearc setting where the negative Nb anomalies resulted from subduction zone petrogenetic processes.

fragments along Neoarchaean convergent plate boundaries over a length of more than 1000 km (Williams *et al.* 1991*a*; Stott 1997; Polat & Kerrich 2000, 2001*b*; Percival *et al.* 2006*a*, *b*).

The Superior Province was built by the accretionary amalgamation of oceanic and continental fragments, ranging in age from 3700 to 2650 Ma, during five discrete tectonothermal events over 40 Ma between 2720 and 2680 Ma (Percival *et al.* 2006*a*, and references therein). These tectonothermal events began with the collision of the Northern superterrane and the Northern Caribou superterrane in the north at about 2720 Ma, continued progressively southward, and ended with accretion of the Minnesota River Valley and Pontiac terranes to the Abitibi–Wawa superterrane in the south at about 2680 Ma.

The Abitibi–Wawa superterrane evolved in a mainly oceanic setting until its collision with the composite Superior superterrane to the north at about 2695 Ma (Corfu & Stott 1998; Polat et al. 1998; Calvert & Ludden 1999; Percival et al. 2006a), and the Pontiac superterrane to the south at c. 2690 Ma (Feng & Kerrich 1992). Polat & Kerrich (2000, 2001b) interpreted the Neoarchaean Wawa greenstone belts as part of a c. 1000 km scale subduction–accretion complex that formed along an intra-oceanic convergent plate margin during trenchward migration of the magmatic arc axis. The geological characteristics of the Wawa and Abitibi subprovinces are summarized in Table 7.

The Wawa subprovince extends from the Vermilion district of Minnesota in the west to the Kapuskasing structural zone in the east (Williams et al. 1991b; Fig. 2a). The Kapuskasing structural zone separates the formerly contiguous Abitibi subprovince to the east. The subprovince is composed of two linear trends of greenstone belts: the northern trend at the border with the Quetico subprovince comprises the Vermilion district, Shebandowan, Winston Lake, Schreiber–Hemlo, Manitouwadge, Hornepayne, White River–Dayohessarah, and

Table 7. *Summary of the major lithological and structural characteristics of the Neoarchaean Wawa and Abitibi subprovinces*

		Wawa	Abitibi	Geodynamic and kinematic interpretation	Possible Phanerozoic analogues
Rock types					
Volcanic rocks		Tholeiitic basalt and komatiite association, with minor transitional to alkaline basalts	Tholeiitic basalt and komatiite association	Plume-derived oceanic plateaux	Ontong–Java; Caribbean; Wrangellia (western Canadian Cordillera)
		Tholeiitic to calc-alkaline bimodal volcanic association, including basalts, andesites, dacites, and rhyolites, with minor adakites, magnesian andesites, Nb-enriched basalts, and picrites; pyroclastic deposits	Tholeiitic to calc-alkaline bimodal volcanic association, including basalts, andesites, dacites, and rhyolites, with minor adakites and magnesian andesites; Low-Ti tholeiite and boninite association	Oceanic island arcs; back-arcs; plume–subduction zone interaction	Southwestern Pacific island arcs (e.g. Izu–Bonin–Mariana system; Japan; Solomon Islands; Lau Basin); Altaids (Central Asia); Papua New Guinea; Central America
Sedimentary rocks		Siliciclastic turbidites, banded iron formations, cherts	Siliciclastic turbidites; banded iron formations; alluvial fan deposits (conglomerates)	Intra-arc basins; forearc basins; back-arc basins; trenchs; pull-apart basins	Shimanto belt (Japan); Altaids (Central Asia); Lake Hazar basin; Erzincan basin (Turkey)
Intrusive rocks		Tonalites–trondhjemites–granodiorites (TTG), granites, diorites, sanukitoids, and syenites, and minor anothorsites and ultramafic sills	Tonalites–trondhjemites–granodiorites (TTG), granites, diorites, syenites, ultramafic sills	Magmatic arc	Western North America (Alaska; Canadian Cordillera; Sierra Nevada); Altaids (Central Asia); Japan
Ore deposits		Volcanogenic massive sulphide deposits, lode gold deposits, banded iron formations	Volcanogenic massive sulphide deposits, lode gold deposits, banded iron formations	Magmatic arc; back-arc; forearc; strike-slip fault zone	Lau Basin; Japan; Izu–Bonin–Mariana system; Altaids (Central Asia)

(*Continued*)

Table 7. *Continued*

	Wawa	Abitibi	Geodynamic and kinematic interpretation	Possible Phanerozoic analogues
Deformation				
D_1	Asymmetric isoclinal to closed folds, sheath folds; mesoscopic-scale reverse faults; regional-scale thrust faults; stretching lineations; slaty cleavage; metamorphic foliation	E–W-trending asymmetric isoclinal to closed folds; sheath folds; regional scale thrust faults; slaty cleavage; metamorphic layering; stretching lineations; S–C planar fabrics	NNW–SSE compression; collision–accretion structures; tectonic juxtaposition of trench turbidites with oceanic plateaus and island arcs	Altaids (Central Asia); Shimanto belt (Japan); Chugach accretionary complex (Alaska); western Canadian Cordillera; Andes
D_2	Right-lateral strike-slip faults; isoclinal to closed upright folds; mesoscopic-scale reverse faults; rhombohedral and scaly cleavages; stretching lineations; asymmetric sigmoidal boudins; kink bands; Riedel shears; S–C planar fabrics	Regional-scale right-lateral strike-slip faults; thrust faults; upright Z-shaped folds; steeply plunging isoclinal folds; steeply dipping foliations; crenulation cleavages; steeply plunging stretching lineations	NNW–SSE transpression; orogen-parallel strike-slip faulting; Timiskaming-type volcanism and sedimentation	West Antarctica; Altaids (Central Asia); Philippines; Japan; Western Canadian Cordillera
D_3	E–W-trending regional-scale strike-slip faults; pull-apart basins?	E–W-trending regional-scale strike-slip faults; pull-apart basins	Orogen-parallel strike-slip faulting developed during island arc accretion	Altyn Tag fault (Central Asia); North Anatolian fault (Turkey); Border Range fault (Alaska)

Modified from Polat & Kerrich (2001*b*). New data added from Şengör & Natal'in (1996, 2004) and Daigneault *et al.* (2004).

Kabinakagami greenstone belts; the southern trend includes the Mishibishu, Michipicoten, and Gamitagama greenstone belts (Fig. 2b). In this study, we summarize the geochemical and structural characteristics of the Schreiber–Hemlo, Manitouwadge, and Winston Lake greenstone belts.

Structural studies indicate that greenstone belts in the northern Wawa subprovince underwent a complex history of faulting, folding, and fabric development, resulting in the destruction of the original stratigraphic relationships between the various lithological units (Zaleski & Peterson 1995; Corfu & Stott 1998; Polat *et al.* 1998; Polat & Kerrich 1999; Zaleski *et al.* 1999; Davis & Lin 2003; Muir 2003). Consequently, the present exposure of each unit or volcanic association may not reflect the original relative abundances.

In the Schreiber–Hemlo greenstone belt, which is representative of typical greenstone belts in the Wawa subprovince, volcanism extended from c. 2750 to 2688 Ma; a granitoid intrusive phase is recorded from 2720 to 2677 Ma, and siliciclastic turbidites were deposited from 2705 to 2685 Ma (Corfu & Muir 1989*a*, *b*; Williams *et al.* 1991*b*; Beakhouse & Davis 2005). The greenstone belt was divided into three lithotectonic assemblages by Williams *et al.* (1991*b*): Schreiber, Hemlo–Black River, and Heron Bay. The Schreiber and Hemlo–Black River assemblages are separated by the Proterozoic Coldwell alkaline complex (Fig. 2b). The Hemlo–Black River and Heron Bay assemblages are located to the north and south of the right-lateral Lake Superior–Hemlo fault zone, respectively (Fig. 2b). These three assemblages are composed of similar volcanic and siliciclastic sedimentary rocks.

The Schreiber–Hemlo greenstone belt contains well-preserved primary volcanic and sedimentary features, such as pillow lavas, spinifex-textured komatiites, hyaloclastite, pyroclastic deposits, and thinly layered cherts in low-strain domains (Fig. 11). Early sea-floor hydrothermal alteration and spheroidal variolitic structures have been well preserved mainly in oceanic plateau tholeiitic basalts (Fig. 11). Cherts fill the spaces between pillows (Fig. 11b). They also occur as 20–30 cm thick, 1–3 m long discontinuous layers between flows. Many pillow drainage cavities are filled with quartz. Carbonates (calcite) occur primarily as cement in pillow breccia.

The Schreiber–Hemlo greenstone belt underwent at least three phases of ductile deformation,

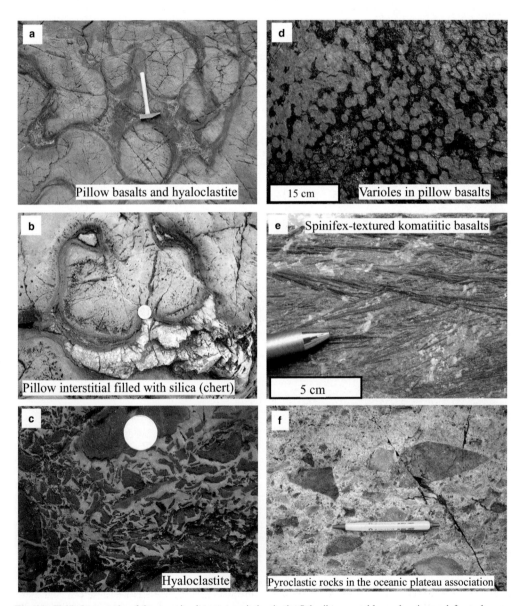

Fig. 11. Field photographs of the oceanic plateau association in the Schreiber assemblage, showing undeformed primary volcanic structures. (**a**) Pillow basalts characterized by chlorite alteration in the rims. Hyaloclastite also occurs between pillows. (**b**) Chert formation between pillows. (**c**) Hyaloclastite displaying chlorite and carbonate alteration. (**d**) Variolites in pillow basalts spatially associated with komatiites and komatiitic basalt. (**e**) Spinifex texture in komatiitic basalts (16 wt% MgO). (**f**) Pyroclastic deposits occurring between pillow basaltic flows.

resulting in folding, faulting, transposition and regional foliation (Polat *et al.* 1998; Davis & Lin 2003; Beakhouse & Davis 2005; Fig. 12; Table 7). The first phase of deformation (D_1) occurred between 2695 and 2688 Ma and the second phase (D_2) between 2690 and 2680 Ma (Beakhouse & Davis 2005). D_1, is defined by reverse (possibly rotated thrust faults) faults, SSE-verging, tight asymmetric folds, and steeply dipping foliation and associated mineral elongation lineations

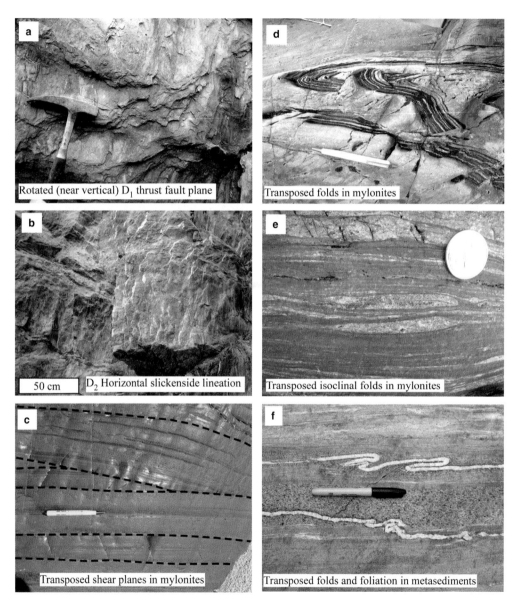

Fig. 12. Field photographs showing the nature of deformation in the Schreiber–Hemlo greenstone belt. (**a**) A near-vertical (rotated) D_1 thrust fault. (**b**) A D_2 strike-slip fault, displaying horizontal lineation on the slickenside. (**c**) Transposed mylonite layers. The outcrop includes numerous shear zones superficially resembling sedimentary layers. (**d**) Transposed fold including siliceous thin mylonite layers. (**e**) Transposed mylonitic isoclinal rootless folds. The outcrop contains numerous asymmetric sigma (σ) structures and fold hooks. (**f**) Strongly deformed sedimentary rocks with transposed folds and foliation. (a)–(e) are from the Schreiber assemblage; (f) is from the Hemlo–Black River assemblage. All these structures are comparable with those found in Phanerozoic orogenic belts.

(Fig. 12a; Table 7). D_2 is dominated by ENE-striking, right-lateral, orogen parallel strike-slip faults, and steeply dipping reverse faults. The presence of coeval right-lateral strike-slip and compressional D_2 structures is consistent with a right-lateral transpressional deformation phase (Polat & Kerrich 1999; Fig. 12b). Transposed D_1 and D_2 folds, foliations, and shear planes occur throughout the belt, but are exposed mainly in the mylonites of the McKellar Harbour area (Fig. 12c–e). Similar structures are also seen in the Hemlo area (Fig. 12f). The age of the third phase of deformation is not well constrained. Regional metamorphism occurred between 2690 and 2670 Ma (Davis & Lin 2003; Beakhouse & Davis 2005, and references therein). Metamorphic grade generally increases from greenschist to amphibolite facies towards some of the intrusion–supracrustal contacts.

The Manitouwadge greenstone belt is located on the northern margin of the Wawa subprovince and is tectonically juxtaposed against the metasedimentary Quetico subprovince (Fig. 2b). This belt is characterized by east–west-trending, variably dipping, intensely deformed and metamorphosed volcanic and sedimentary rocks aged 2720–2680 Ma. The oldest magmatic rocks in the Manitouwadge belt are 2720 Ma felsic volcanic flows and associated TTG plutons (see Zaleski et al. 1999); 2720 Ma barren and mineralized felsic rocks are spatially associated with tholeiitic basalts. Metamorphic grade of most of the supracrustal rocks is greenschist to amphibolite facies.

The Winston Lake greenstone belt is the western continuation of the Manitouwadge belt, near the Quetico subprovince (Williams et al. 1991b; Fig. 2b). The western part of the belt is separated from the Schreiber–Hemlo belt to the south by the Crossman Lake batholith. The nature of the contact between the two belts to the east is not well constrained. Like its Manitouwadge counterpart, the Winston Lake belt is characterized by mafic to felsic volcanic and siliciclastic sedimentary rocks, which are collectively intruded by TTG and gabbroic plutons (Williams et al. 1991b). There is polyphase deformation, and greenschist- to amphibolite-facies metamorphism. The greenstone belt is composed of three major units: (1) a lower volcano-sedimentary unit consisting mainly of calc-alkaline felsic flows, and minor pyroclastic and siliciclastic (greywacke and shale) sedimentary rocks; (2) a central volcanic unit, composed dominantly of Mg- to Fe-tholeiitic basalts; and (3) an upper volcano-sedimentary unit of intercalated mafic to felsic volcanic and turbiditic siliciclastic sedimentary rocks. Contacts between these spatially associated units are often marked by deformation, intense metasomatic alteration, and/or intrusions, obscuring the original stratigraphic relationships.

Our studies in the Wawa greenstone belts have documented two major types of volcanic rock association: (1) an oceanic plateau association composed dominantly of compositionally uniform Mg- to Fe-rich tholeiitic basalts (oceanic plateau basalts; OPB), and komatiites erupted from mantle plumes in an intra-oceanic setting; and (2) a compositionally diverse intra-oceanic island arc association consisting mainly of tholeiitic to calc-alkaline basalts, andesites, dacites, and rhyolites (BADR). In the oceanic plateau association, there are volumetrically minor transitional to alkaline basalts that plot with Recent ocean island basalts on the trend of low-pressure high degrees of melt for tholeiites to high-pressure low degrees of melt for alkali basalts (Polat et al. 1999; Greenough et al. 2005; Polat 2009). The latter association contains volumetrically minor picrites, adakites, MA, and NEB in an island arc association. The intra-oceanic island arc association is paired with trench turbidites. The geochemical characteristics of oceanic plateaux and island arc associations are summarized below.

Summary of Archaean terranes

Southern West Greenland: Isua and Ivisaartoq greenstone belts

It has been demonstrated that the 3800–2800 Ma granitoid–greenstone terranes of southern West Greenland record distinct deformation–metamorphic–magmatic histories that can be best explained by modern-style convergent margin plate-tectonic processes (Bridgwater et al. 1974; Hanmer et al. 2002; Friend & Nutman 2005a, b; Nutman 2006; Garde 2007; Nutman & Friend 2007a). Low-strain domains preserve primary magmatic textures and near-primary geochemical signatures in the Eoarchaean (3800–3700 Ma) Isua and Mesoarchaean (c. 3075 Ma) Ivisaartoq greenstone belts, providing important information on the nature of Archaean oceanic crust and geodynamic processes (Rosing et al. 1996; Nutman et al. 2002; Polat & Hofmann 2003; Furnes et al. 2007).

Isua volcanic rocks are characterized by compositionally distinct, tectonically juxtaposed, boninitic and island arc picritic sequences. If the geochemical characteristics of these sequences have the same geodynamic significance as Phanerozoic counterparts, then they were the products of higher geothermal gradients, stemming from subduction of young, hot oceanic lithosphere, including ridge subduction and development of slab windows. Trench rollback following the initiation of intra-oceanic subduction may have caused the extension of the overriding oceanic crust, creating a sheeted dyke complex in the picritic to basaltic association (Furnes *et al.* 2007).

Picrites of the Mesoarchaean Ivisaartoq belt require a depleted, shallow (<80 km) upper mantle source in a subduction zone setting. Following the initiation of the subduction zone, the forearc region of the overriding plate may have undergone a significant amount of extension in response to a slab rollback, resulting in a large degree of partial melting of the mantle wedge. Such extensive partial melting could have generated thick (>20 km) basaltic to picritic flows, with minor gabbroic to dioritic dykes in the forearc region. As the extension of the overriding plate continued, the adiabatically upwelling ultramafic melts intruded the volcanic rocks as dunitic to wehrlitic sills, generating the Mesoarchaean forearc oceanic crust (Fig. 10).

Wawa subprovince

The original stratigraphic relationships of the oceanic plateau and island arc volcanic associations of the Wawa subprovince have been disrupted and complicated by multiple phases of ductile deformation. However, in a few areas (e.g. Hemlo), it is clear that the (crustally uncontaminated) oceanic plateau association is overlain by the calc-alkaline basalt, andesite, dacite, and rhyolite suite. In addition, in the Hemlo and Heron Bay areas the oceanic plateau association is intruded by pre-tectonic subduction-derived 2720–2697 Ma TTG plutons (Polat *et al.* 1998; Davis & Lin 2003; Beakhouse & Davis 2005, and references therein).

Primitive mantle-normalized trace element patterns of the Wawa oceanic plateau association vary from depleted normal mid-ocean ridge basalt (N-MORB)-like, through flat (oceanic plateau-like), to enriched (E-MORB, ocean island basalt (OIB)-like) (Figs 13 and 14). Similar trace element patterns have been reported from Phanerozoic plume-derived oceanic plateaux and ocean islands (see Hofmann 2004; Kerr 2004; Greenough *et al.* 2005; Kerr & Mahoney 2007). Accordingly, it is proposed that geochemical diversity in the Wawa ocean plateau association probably resulted from mixing of enriched and depleted melts within an upwelling heterogeneous mantle plume. Enriched and depleted basaltic melts may reflect low and high, respectively, percentage of partial melting of a depleted mantle source at variable depths in the mantle plume (see Sproule *et al.* 2002). It is clear that the Recent MORB–OIB array had become established in the Archaean mantle by 2.7 Ga.

Oceanic island arc volcanic rocks are predominantly calc-alkaline BADR with volumetrically minor adakite, MA and NEB (Table 8). They are

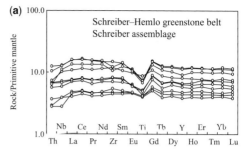
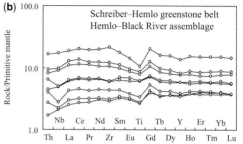
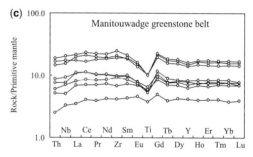

Fig. 13. Primitive mantle-normalized trace element diagrams for the Wawa oceanic plateau tholeiitic basalts. Normalization values are from Sun & McDonough (1989).

distinct from the oceanic plateau Mg- to Fe-tholeiitic sequences in terms of large variations in both major and trace element contents (Figs 8b and 15; Table 8). They are characterized by: (1) positively fractionated REE patterns as reflected in high $La/Sm_{cn}=1.2–6.0$ and $Gd/Yb_{cn}=1.2–6.5$ ratios; and (2) enrichment of Th and La over Nb, generating negative Nb anomalies (Fig. 5). All these geochemical features are typical of intra-oceanic subduction zone magmatism (see Saunders *et al.* 1991; Hawkesworth *et al.* 1993; Murphy 2007).

The presence of both the oceanic plateau and island arc associations in the Schreiber–Hemlo greenstone belt can be explained by the initiation of a subduction zone at the edge of an oceanic plateau. Compositionally diverse calc-alkaline lavas including basalts, andesites, dacites, rhyolites, adakites, MA, and NEB erupted from an island arc

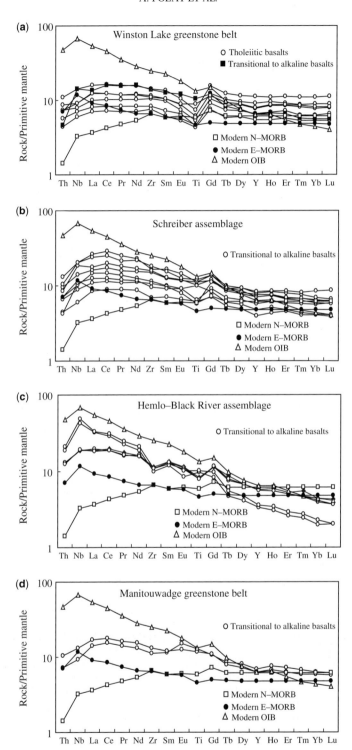

Fig. 14. Primitive mantle-normalized diagrams for Wawa oceanic plateau tholeiitic and transitional to alkaline basalts. Data are from Polat *et al.* (1999) and Polat (2007). Normalization values and data for modern N-MORB, E-MORB and OIB are from Sun & McDonough (1989).

Table 8. Summary of the significant compositional values and element ratios for the 2.7 Ga Wawa oceanic plateau basalts and oceanic island arc adakites, magnesian andesites (MA), Nb-enriched basalts (NEB), basalts, andesites, and dacites–rhyolites

	Oceanic plateau association		Oceanic island arc association				
	Tholeiitic basalts	Adakites	MA	NEB	Basalts	Andesites	Dacites–rhyolites
(wt%)							
SiO_2	48–54	64–73	55–64	50–57	46–54	57–64	65–74
MgO	3.5–10.6	1.0–3.8	3.4–7.5	3.8–5.7	3.6–6.4	1.5–2.9	0.5–3.8
TiO_2	0.60–2.40	0.28–0.80	0.47–1.83	0.63–2.24	0.6–2.1	0.52–1.72	0.16–0.80
Al_2O_3	12.5–17.6	15–18	14–17	12.3–17.1	13.0–17.6	14–18	14–18
Fe_2O_3	7.2–17.0	2.5–6.7	5.7–10.5	8.2–16.6	8.6–18.3	5.6–14.9	1.5–7.0
Mg-no.	34–73	39–58	50–70	41–61	35–50	29–52	36–60
(ppm)							
Cr	10–680	11–166	106–531	14–221	14–362	9–165	6–165
Co	36–108	8–27	21–39	36–52	33–52	15–45	4–27
Ni	20–175	3.0–88	21–229	9–90	27–72	4.0–81	2–88
Sc	27–61	4–19	12–27	24–45	17–50	11–49	2–17
V	186–435	36–143	78–185	150–386	132–368	78–308	21–143
Nb	2.0–14.4	2.1–10.9	2.6–12.9	7.3–16.2	2.7–5.2	3.9–6.9	1.6–9.9
Zr	50–280	91–204	81–221	110–278	62–138	99–165	75–204
Th	0.12–1.40	0.72–9.9	1.08–6.90	1.42–3.83	0.56–3.51	1.0–4.6	0.7–9.9
La	1.7–13.7	8.2–51.5	9.6–61.6	11.9–37.2	6.7–24.6	8.3–44.6	5–55
Yb	1.5–8.0	0.34–1.4	0.72–1.81	1.2–5.8	1.0–3.8	0.81–3.40	0.2–1.4
Y	15–75	5.1–23.3	8.8–21.6	18.4–59.4	13.3–35.0	9.5–31.4	3.0–23.3
La/Sm_{cn}	0.80–1.17	2.6–4.2	2.34–3.83	1.42–2.79	1.37–2.93	1.54–3.92	2.2–5.2
La/Yb_{cn}	0.77–1.40	10.7–50.0	5.26–27.9	2.65–18.2	1.9–17.3	2.7–22.9	8.7–50.0
Gd/Yb_{cn}	0.94–1.30	1.8–5.3	1.50–4.87	1.25–4.05	1.28–3.07	1.14–3.32	1.5–5.3
Zr/Y	1.9–3.9	9.3–26.9	5.86–11.68	3.76–8.9	3.2–7.9	3.4–12.2	9.3–28.0
Ti/Zr	46–127	12.2–36.9	26–135	22–100	37–92	20–90	11–40
Al_2O_3/TiO_2	5.7–25.3	21–70	9.2–30.1	6.6–24.1	6.1–25.1	8.5–30.1	21–104
Nb/La_{pm}	0.60–1.10	0.09–0.33	0.06–0.53	0.23–0.67	0.14–0.51	0.14–0.61	0.09–0.53
Th/La_{pm}	0.57–1.10	0.81–2.96	0.51–1.77	0.54–1.43	0.59–1.10	0.63–1.0	0.76–1.89
Nb/Th_{pm}	0.77–1.50	0.08–0.40	0.09–0.69	0.39–1.03	0.13–0.86	0.15–0.67	0.08–0.51

Data are from Polat & Kerrich (2000, 2001a) and Polat & Münker (2004).

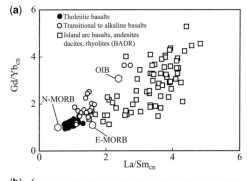
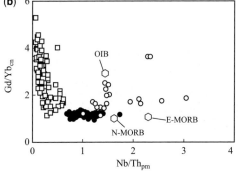
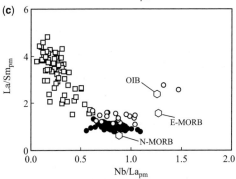

Fig. 15. (**a**) La/Sm_{cn} v. Gd/Yb_{cn}, (**b**) Nb/Th_{pm} v. Gd/Yb_{cn} and (**c**) Nb/La_{pm} v. La/Sm_{cn} variation diagrams for Wawa oceanic plateau tholeiitic basalts and transitional to alkaline basalts, and calc-alkaline island arc basalts, andesites, dacites and rhyolites. Data are from Polat et al. (1999), Polat & Kerrich (2000) and Polat (2007). Data for calc-alkaline island arc basalts, andesites, dacites and rhyolites are from Polat & Kerrich (2000). Normalization values are from Sun & McDonough (1989).

developed at the edge of the oceanic plateau. Extension of the overriding plate may have led to rifting of the arc, resulting in the formation of a back-arc basin (Fig. 16). Hydrothermal alteration of back-arc oceanic crust produced the volcanogenic massive sulphide deposits in the Winston Lake and Manitouwadge greenstone belts (Fig. 16).

Independent lines of evidence are all consistent with the Neoarchaean Wawa–Abitibi greenstone belts as a subduction–accretion complex that formed along an Archaean convergent plate margin: (1) geochronologically distinct, tectonically juxtaposed volcanic and sedimentary rocks; (2) tectonic mélange; and (3) regional-scale transpressional structure (Fig. 16). These greenstone belts originated as a subduction–accretion complex that formed along a NNW-dipping subduction zone. In this geodynamic framework, komatiite and associated Mg- to Fe-tholeiitic basalt sequences are interpreted as dismembered fragments of an Archaean oceanic plateau(x), whereas LREE-enriched and HFSE-depleted TTG plutons and associated bimodal volcanic sequences are interpreted as fragments of oceanic island arcs. Similarly, siliciclastic turbidites, having positively fractionated REE patterns and HFSE depletion, are defined as trench turbidites. Volumetrically minor adakites, MA and NEB occur within the island arc association. The adakites represent melts from subducted oceanic crust and all other suites were derived from the mantle wedge above the subducting oceanic lithosphere.

Implications and conclusions

Significance of hot subduction rocks

In the Western Pacific Ocean, Cenozoic boninites, picrites, adakites, MA, and NEB were erupted in supra-subduction zone settings, including forearcs, arcs, and back-arcs (Table 1). Formation of these rocks appears to have resulted from the following five major geodynamic processes: (1) interaction between a subducting young plate, or spreading centre, and an arc, resulting in the formation of adakites, MA, NEB (e.g. Philippines) and picrites (e.g. Solomon arc); (2) interaction of a back-arc spreading centre with an arc, generating boninites (e.g. Bonin islands, North Tonga, Valu Fa Ridge) and picrites (e.g. Vanuatu arc); (3) intersection of a back-arc spreading centre with a transform fault, producing boninites (e.g. Mariana islands; South Vanuatu); (4) subduction initiation either along a spreading centre or transform fault (Izu–Bonin islands); and (5) mantle plume–arc interaction (Izu–Bonin–Mariana forearc; North Tonga) (Casey & Dewey 1984; Pearce et al. 1992; Stern & Bloomer 1992; Danyushevsky et al. 1995; Sajona et al. 1996; Macpherson & Hall 2001; Deschamps & Lallemand 2003; Hawkins 2003; Pearce 2003; Schuth et al. 2004). The above volcanic rock types formerly thought to be restricted to Phanerozoic or contemporary convergent margin settings have been identified in several Archaean greenstone belts (Polat & Kerrich 2006).

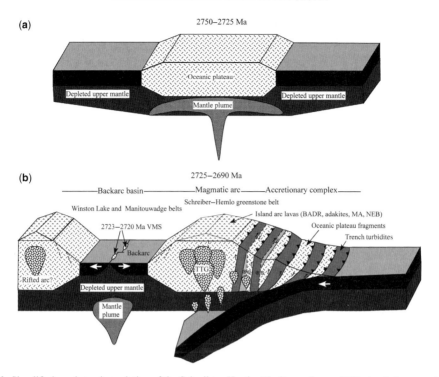

Fig. 16. Simplified geodynamic evolution of the Schreiber–Hemlo, Manitouwadge, and Winston Lake greenstone belts. (**a**) At 2750–2725 Ma komatiites, tholeiitic pillow basalts, and transitional to alkaline basalts erupted from a mantle plume, forming a oceanic plateau. (**b**) The initiation of a subduction zone at the edge of the oceanic plateau resulting in the formation of a magmatic arc, and production of calc-alkaline basalts, andesites, dacites, rhyolites, adakites, Nb-enriched basalts, and magnesian andesites. Rifting of the magmatic arc gave rise to the opening of a back-arc basin, resulting in the formation of volcanogenic massive sulphide deposits in the Manitouwadge and Winston Lake greenstone belts (modified from Polat & Kerrich 1999).

Archaean equivalents of boninites, picrites, adakites, NEB and MA probably formed in comparable geodynamic settings given similar associations (Fig. 17). These rocks are rare in both the Archaean (<5%) and Phanerozoic geological records; however, their geochemical signatures provide significant insights into petrological, thermal, and geodynamic processes (Drummond *et al.* 1996; Pearce 2003; Schuth *et al.* 2004). We suggest that vigorous mantle convection and subduction of younger, hotter oceanic slabs, including ridge subduction, and slab windows, may have provided optimized conditions for the production of boninites, picrites, adakites, MA, and NEB in many Archaean subduction zones (see Martin 1999; Kerrich & Polat 2006; Fig. 17).

Phanerozoic analogues for Archaean greenstone belts and style of crustal growth

It has been shown that map patterns, lithological association, structural characteristics, and style of geodynamic evolution of Archaean orogens (greenstone–granitoid terranes) are comparable with those in the Altaids, circum-Pacific (e.g. Japan, Alaska, Solomon arc–Ontong–Java), and Caribbean (e.g. Greater Antilles) accretionary orogenic systems (Tables 1 and 2; Williams *et al.* 1991*a*; Kimura *et al.* 1993; Şengör & Natal'in 1996, 2004; Polat *et al.* 1998; Kusky & Polat 1999; Mann & Taira 2004). These accretionary orogens are predominantly composed of juvenile crust in subduction–accretion complexes, with extensive vertical accretion above a subducting slab (Şengör & Natal'in 1996, 2004; Windley *et al.* 2007). In these orogens accretion of various lithotectonic units is dominantly oblique, with the partitioned compressional and strike-slip components responsible for much of the deformation. Trenchward growth of these orogens involves multiple allochthonous lithotectonic units, bounded by diachronous sutures, over long durations of 100–400 Ma. Rather than a single curvilinear magmatic arc, as in continent–continent collision orogens (see Şengör, 1990), broad (1000 km scale) arcs develop. These orogenic systems show a

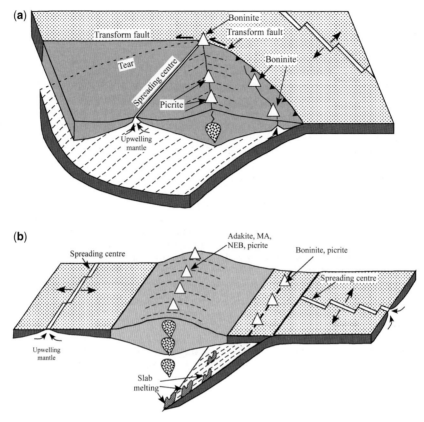

Fig. 17. Simplified geodynamic models illustrating how boninites, picrites, adakites, magnesian andesites (MA), and Nb-enriched basalts (NEB) form in a supra-subduction zone setting during interactions between transform fault, and convergent and divergent plate boundaries. Boninites originate either at a back-arc spreading centre–transform fault intersection at the termination of a subduction zone or in an extending forearc during slab rollback (e.g. Izu–Bonin–Mariana, Tethyan boninites). Picrites are formed during subduction of young, hot oceanic crust (or spreading centre) in an arc or a forearc setting (e.g. Solomon island arc picrites). The adakite–MA–NEB association also results during subduction of young, hot oceanic crust (or spreading centre). (**a**) is modified after Deschamps & Lallemand (2003).

commensurately broad pattern of subduction, where the axis of arc magmatism migrates trenchward through the accretionary prism (Şengör 1993; Şengör & Okuroğullari et al. 1991).

Lithological and structural similarities between Archaean greenstone belts and Phanerozoic orogenic systems are summarized in Tables 1 and 2. Spatial and temporal relationships between Phanerozoic plume-derived oceanic plateau and subduction-derived magmatic rocks, including calc-alkaline BADR, tholeiitic basalts, boninites, picrites, adakites, NEB, MA, and TTG have been well documented in the Caribbean accretionary orogen (Thirlwall et al. 1996; Kerr et al. 1999; White et al. 1999; Escuder Virutue et al. 2006, 2007; Jolly et al. 2007, and references therein). In addition to lithological characteristics, the structural aspects of the Caribbean orogenic system closely resemble those of Archaean Wawa greenstone belts, including tectonic juxtaposition of the fragments of oceanic plateau, island arcs, and accretionary prisms in a transpressional tectonic regime driven by oblique convergence (Burke 1988; Kerr et al. 1999; Escuder Viruete et al. 2007, and references therein). In the northern part of the Caribbean region (Greater Antilles, Dominican Republic), Escuder Viruete et al. (2006) documented an early Cretaceous (c. 116 Ma) island arc association composed dominantly of tholeiitic basalts, boninites, and rhyodacites. Tholeiitic basalts and boninites are interpreted to be derived from a sub-arc mantle wedge during subduction initiation, whereas rhyodacites originated from partial melting of thickened island arc crust. All these volcanic rocks are intruded by synvolcanic tonalites. In contrast to LREE-enriched Archaean tonalites, the early

Cretaceous tonalites display depleted to flat REE patterns with negative Nb anomalies; their origin has been attributed to melting of island arc crust. However, tonalites in the Aruba region of the southern Caribbean display trace element patterns comparable with those of Archaean counterparts (White *et al.* 1999). The Aruba tonalites appear to be derived by melting of the subducted Caribbean oceanic plateau.

In contrast to the Alpine–Himalayan-type orogen in the Phanerozoic, Archaean orogens do not represent mountain belts that formed through closure of a major ocean (e.g. Tethys; see Şengör 1987, 1990). Instead, Archaean orogens appear to have resulted from collisions of numerous intra-oceanic supra-subduction zone systems including forearcs, arcs, back-arc, fragments of oceanic plateaux (only in the Wawa and Abitibi regions), and continental blocks. In terms of rock types, style of deformation and volcanism, greenstone belts in the Abitibi–Wawa superterrane and SW Greenland are comparable with Cenozoic island arc systems in the Western Pacific. Tectonic interleaving of island arc volcanic rocks with komatiite–OPB associations in Archaean orogens requires accretionary tectonics. Similar tectonic relationships between Phanerozoic arcs and oceanic plateaux are seen in the southwestern Pacific. For example, the Ontong–Java ocean plateau has been 'captured', and jammed, in the New Ireland and Solomon arcs (Mann & Taira 2004).

It is generally accepted that Archaean continental crust grew by accretionary and magmatic processes taking place at convergent plate boundaries (Willams *et al.* 1991*a*; Taylor & McLennan 1995; Percival *et al.* 2006*a*, and references therein). Fundamentally, continental crust cannot grow and arcs cannot generate melts over millions of years without migration, subduction of water, and wedge fluxing or slab melting. Recent geochemical studies on Archaean eclogites endorse existing models that 2.5–3.5 Ga Archaean cratons were built by accretionary geodynamic processes at convergent plate boundaries (see Horodyskyj *et al.* 2007).

Uniformitarian v. non-uniformitarian models for the early Earth

Archaean orogens show both similarities and contrasts to Phanerozoic counterparts. Similarities include: (1) the patterns of deformation (e.g. thrust faulting, strike-slip faulting, recumbent folding, continental rifting, transposition); (2) nature of magmatism (e.g. tholeiitic, calc-alkaline, alkaline suites); (3) trace element geochemical signatures (i.e. LREE enrichment coupled with HFSE depletions); (4) grade of metamorphism (e.g. low- to high-grade regional metamorphism); (5) type of sedimentary rocks (e.g. siliciclastic turbidites, platform-type quartz sandstones and carbonates); (6) types of mineral deposits (e.g. volcanogenic massive sulphide deposits (VMS), orogenic Au); (7) accretionary tectonics (i.e. accretion of oceanic island arcs, continental fragments and oceanic plateaux; and (8) episodic crust formation, crustal recycling by subduction erosion, and supercontinent cycles (see Cawood *et al.* 2006; Kerrich & Polat 2006; O'Neill *et al.* 2007, and references therein). Major differences between Archaean and Phanerozoic orogens are manifested by the abundances and presence and/or absence of specific rocks types in the former orogens. For example, calc-alkaline Na-rich high-Al TTG, komatiites and komatiitic basalts, and banded iron formation are more abundant in Archaean orogens than in Phanerozoic equivalents. The anorthositic layered complexes in West Greenland are genetically associated with the metavolcanic (basaltic) amphibolites, both of which have arc-related origins. The layered complexes formed the magma chamber to the volcanic arcs (Garde 2007). Modern analogues are found in active continental margins, where comparable layered complexes were engulfed by voluminous TTG magmas (Windley & Smith 1976; Windley *et al.* 1981). Examples include the mid-Palaeozoic Black Giant anorthositic layered complex in Fiordland, New Zealand (Gibson & Ireland 1999), and the Chilas complex of layered norites in the Himalayas of Pakistan, which formed the magma chamber to the Cretaceous Kohistan island arc (Khan *et al.* 1989). Finally, blueschist-facies metamorphic rocks and intact Penrose-type ophiolites have not been documented in Archaean orogens.

Both uniformitarian and non-uniformitarian models have been proposed to argue for and against, respectively, the operation of Phanerozoic-style plate tectonics in the Archaean (Burke *et al.* 1976; Condie 1981; Kusky 1991, 2004; Williams *et al.* 1991*a*; Sleep 1992; Windley 1993; Hamilton 1998, 2003; Friend & Nutman 2005*a* and *b*; Stern 2005; Cawood *et al.* 2006; Kerrich & Polat 2006; Nutman & Friend 2007*a*). On the basis of detailed structural, lithological, seismic, geochronological, geochemical, and palaeomagnetic data, proponents of the uniformitarian models have argued that Archaean cratons were built by horizontal tectonics (see Cawood *et al.* 2006; Kerrich & Polat 2006, and references therein). In contrast, proponents of the non-uniformitarian models have contended that Archaean greenstone belts formed as autochthonous sequences in synformal basins on pre-existing continental crust (Hamilton 1998; Bédard *et al.* 2003; Bédard 2006, and references therein). According to the non-uniformitarian models, synformal basins were developed between dome-shaped TTG batholiths in response to gravity-driven vertical movements (Hamilton 1998; Bédard et al. 2003; Bédard 2006). The non-uniformitarian models, however, cannot explain the multiple phases of non-coaxial deformation, transposition, large-scale strike-slip faults, structurally juxtaposed lithotectonic assemblages and geochemical complexity in Archaean greenstone belts (Figs 7 and 12). Van Kranendonk et al. (2007) proposed that crust formation events between 3.52 and 3.24 Ga in the Pilpara Craton, Western Australia, were dominated by vertical tectonic processes related to upwelling mantle plumes. According to Van Kranendonk et al. (2007), horizontal tectonics and accretionary growth processes became dominant after 3.2 Ga. Lithological differences between the Archaean and Phanerozoic geological records can be attributed to secular cooling of the Earth through time (see Condie 2005a). Collectively, we suggest that, despite the hotter Archaean Earth, physical and chemical processes that produced the oceanic and continental crusts since the early Archaean were not fundamentally different from those that operate today (see Windley 1993).

The authors thank P. Cawood and A. Kröner for organizing this volume, and the invitation to contribute a paper. Reviewers A. Garde and H. Smithies are acknowledged for their constructive and incisive comments, which resulted in significant improvements to the paper. A. P. and R. K. acknowledge receipt of NSERC Discovery grants. Fieldwork was supported by the Bureau of Minerals and Petroleum in Nuuk and the Geological Survey of Denmark and Greenland (GEUS). A. P. thanks T. Demir for helping his family during fieldwork in Greenland. P. W. U. Appel provided some field photos.

References

APPEL, P. W. U. 1997. High bromine and low Cl/Br ratios in hydrothermally altered Archean komatiitic rocks, West Greenland. *Precambrian Research*, **82**, 177–189.

APPEL, P. W. U., FEDO, C. M., MOORBATH, S. & MYERS, J. S. 1998. Recognizable primary volcanic and sedimentary features in a low strain domain of the highly deformed, oldest known (3.7–3.8 Gyr) Greenstone belt, Isua, West Greenland. *Terra Nova*, **10**, 57–62.

APPEL, P. W. U., POLAT, A. & FREI, R. 2009. Dacitic ocelli in mafic lavas, 3.8–3.7 Isua greenstone belt, West Greenland: geochemical evidence for partial melting of oceanic crust and magma mixing. *Chemical Geology*, **258**, 105–124.

BEAKHOUSE, G. P. & DAVIS, D. W. 2005. Evolution and tectonic significance of intermediate to felsic plutonism associated with the Hemlo greenstone belt, Superior Province, Canada. *Precambrian Research*, **137**, 61–92.

BECCALUVA, L. & SERRI, G. 1988. Boninitic and low-Ti subduction-related lavas from intraoceanic arc–backarc systems and low-Ti ophiolites: A reappraisal of their petrogenesis and original tectonic setting. *Tectonophysics*, **146**, 291–315.

BÉDARD, J. H. 1999. Petrogenesis of boninites from the Betts Cove Ophiolite, Newfoundland, Canada: Identification of subducted source components. *Journal of Petrology*, **40**, 1853–1889.

BÉDARD, J. H. 2006. A catalytic delamination-driven model for coupled genesis of Archean crust and sub-continental lithospheric mantle. *Geochimica et Cosmochimica Acta*, **70**, 1188–1214.

BÉDARD, J. H., BROUILLETTE, P., MADORE, L. & BERCLAZ, A. 2003. Archean cratonization and deformation in the northern Superior Province, Canada: An evaluation of plate tectonic versus vertical tectonic models. *Precambrian Research*, **127**, 61–87.

BENOIT, M., AGUILLON-ROBLES, A. ET AL. 2002. Geochemical diversity of late Miocene volcanism in southern Baja California, Mexico: Implication of mantle and crustal sources during the opening of an asthenospheric window. *Journal of Geology*, **110**, 627–648.

BLAKE, T. S., BUICK, R., BROWN, S. J. A. & BARLEY, M. E. 2004. Geochronology of a late Archean flood basalt province in the Pilbara Craton, Australia: Constraints on basin evolution, volcanic and sedimentary accumulation, and continental drift rates. *Precambrian Research*, **133**, 143–173.

BLEEKER, W. 2002. Archean tectonics: A review, with illustrations from the Slave craton. *In*: FOWLER, M. R., EELBINGER, C. J. & HAWKEWORTH, C. J. (eds) *The Early Earth: Physical, Chemical and Biological Development*. Geological Society, London, Special Publications, **199**, 151–181.

BOILY, M. & DION, C. 2002. Geochemistry of boninite-type volcanic rocks in the Frotet–Evans greenstone belt, Opatica subprovince, Quebec: Implications for the evolution of Archean greenstone belt. *Precambrian Research*, **115**, 349–371.

BOURDON, E., EISSEN, J. P., MONZIER, M., ROBIN, C., MARTIN, H., COTTON, J. & HALL, M. L. 2002. Adakite-like lavas from Antisana volcano (Ecuador): Evidence for slab melt metasomatism beneath the Andean Northern Volcanic Zone. *Journal of Petrology*, **43**, 199–217.

BRIDGWATER, D., McGREGOR, V. R. & MYERS, J. S. 1974. A horizontal tectonic regime in the Archean of Greenland and its implications for early crustal thickening. *Precambrian Research*, **1**, 179–198.

BRIDGWATER, D., KETO, L., McGREGOR, V. R. & MYERS, J. S. 1976. Archean gneiss complex of Greenland. *In*: ESCHER, A. & WATT, W. S. (eds) *Geology of Greenland*. Geological Survey of Greenland, Copenhagen, 21–75.

BROWN, A. V. & JENNER, G. A. 1989. Geological setting, petrology and chemistry of Cambrian boninite and low-Ti tholeiitic lavas in western Tasmania. *In*: CRAWFORD, A. J. (ed.) *Boninites and Related Rocks*. Unwin Hyman, London, 233–263.

BURCHFIEL, B. C., COWAN, D. S. & DAVIS, G. A. 1992. Tectonic overview of the Cordilleran orogen in the United States. *In*: BURCHFIEL, B. C., LIPMAN, P. D. & ZOBACK, M. L. (eds) *The Cordilleran Orogeny: Conterminous United States*. The Geology of North America, **G-3**, 407–479.

BURKE, K. 1988. Tectonic evolution of the Caribbean. *Annual Review of Earth and Planetary Sciences*, **16**, 201–230.

BURKE, K., DEWEY, J. F. & KIDD, W. S. F. 1976. Dominance of horizontal movements, arc and microcontinental collisions during the later permobile regime. *In*: WINDLEY, B. F. (ed.) *The Early History of the Earth*. Wiley, London, 113–129.

CALMUS, T., AGUILLON-ROBLES, A. *ET AL*. 2003. Spatial and temporal evolution of basalts and magnesian andesites ('bajaites') from Baja California, Mexico: The role of slab melts. *Lithos*, **66**, 77–105.

CALVERT, A. J. & LUDDEN, J. N. 1999. Archean continental assembly in the southeastern Superior Province of Canada. *Tectonics*, **18**, 412–429.

CALVERT, A. J., CRUDEN, A. R. & HYNES, A. 2004. Seismic evidence for preservation of the Archean Uchi granite–greenstone belt by crustal-scale extension. *Tectonophysics*, **388**, 135–143.

CARD, K. D. 1990. A review of the Superior Province of the Canadian Shield, a product of Archean accretion. *Precambrian Research*, **48**, 99–156.

CARD, K. D. & CIESIELSKI, A. 1986. DNAG#1: Subdivision of the Superior Province of the Canadian Shield. *Geoscience Canada*, **13**, 5–13.

CASEY, J. F. & DEWEY, J. F. 1984, Initiation of subduction zones along transform and accreting plate boundaries, triple-junction evolution, and forearc spreading centres—implications for ophiolitic geology and obduction. *In*: GASS, I. G., LIPPARD, S. J. & SHELTON, A. W. (eds) *Ophiolites and Oceanic Lithosphere*. Geological Society, London, Special Publications, **13**, 269–290.

CAWOOD, A. P., KRÖNER, A. & PISAREVSKY, S. 2006. Precambrian plate tectonics: Criteria and evidence. *GSA Today*, **16**, 4–11.

CHADWICK, B. 1990. The stratigraphy of a sheet of supracrustal rocks within high-grade orthogneisses and its bearing on late Archean structure in southern West Greenland. *Journal of the Geological Society, London*, **147**, 639–652.

CHADWICK, B., VASUDEV, V. N. & HEGDE, G. V. 2000. The Dharwar craton, southern India, interpreted as the result of late Archean oblique convergence. *Precambrian Research*, **99**, 91–111.

CHEN, S. F., RIGANTI, A., WYCHE, S., GREENFIELD, J. E. & NELSON, D. R. 2003. Lithostratigraphy and tectonic evolution of contrasting greenstone successions in the central Yilgarn Craton, Western Australia. *Precambrian Research*, **127**, 249–266.

CONDIE, K. C. 1981. *Archean Greenstone Belts*. Elsevier, Amsterdam.

CONDIE, K. C. 2004. Precambrian superplume events. *In*: ERIKSSON, P. G., ALTERMANN, W., NELSON, D. R., MUELLER, W. U. & CATUNEANU, O. (eds) *The Precambrian Earth: Tempos and Events*. Elsevier, Amsterdam, 163–173.

CONDIE, K. C. 2005*a*. *Earth as an Evolving Planetary System*. Elsevier, Amsterdam.

CONDIE, K. C. 2005*b*. TTGs and adakites: Are they both slab melts? *Lithos*, **80**, 33–44.

CORFU, F. & MUIR, T. L. 1989*a*. The Hemlo–Heron Bay greenstone belt and Hemlo Au–Mo deposit, Superior Province, Ontario, Canada. 1. Sequence of igneous activity determined by zircon U–Pb geochronology. *Chemical Geology*, **79**, 183–200.

CORFU, F. & MUIR, T. L. 1989*b*. The Hemlo–Heron Bay greenstone belt and Hemlo Au–Mo deposit, Superior Province, Ontario, Canada. 2. Timing of metamorphism, alteration and Au mineralization from titanite, rutile, and monazite U–Pb geochronology. *Chemical Geology*, **79**, 201–223.

CORFU, F. & STOTT, G. M. 1998. Shebandowan greenstone belt, western Superior Province: U–Pb ages, tectonic implications, and correlations. *Geological Society of America Bulletin*, **110**, 1467–1484.

COUSENS, B., FACEY, K. & FALCK, H. 2002. Geochemistry of the late Archean Banting Group, Yellowknife greenstone belt, Slave Province, Canada: Simultaneous melting of the upper mantle and juvenile mafic crust. *Canadian Journal of Earth Sciences*, **39**, 1635–1656.

CRAWFORD, A. J., FALLOON, T. J. & GREEN, D. H. 1989. Classification, petrogenesis and tectonic setting of boninites. *In*: CRAWFORD, A. J. (ed.) *Boninites and Related Rocks*. Unwin Hyman, London, 1–49.

CROWLEY, J. L. 2002. Testing the model of late Archean terrane accretion in southern West Greenland: A comparison of the timing of geological events across the Qarliit Nunaat fault, Buksefjorden region. *Precambrian Research*, **116**, 57–80.

DAIGNEAULT, R., MUELLER, W. U. & CHOWN, E. H. 2004. Abitibi greenstone belt plate tectonics: Diachronous history of arc development, accretion and collision. *In*: ERIKSSON, P. G., ALTERMANN, W., NELSON, D. R., MUELLER, W. U. & CATUNEANU, O. (eds) *The Precambrian Earth: Tempos and Events*. Elsevier, Amsterdam, 88–103.

DANYUSHEVSKY, L. V., SOBOLEV, A. V. & FALLOON, T. J. 1995. North Tongan high-Ca boninite petrogenesis: The role of Samoan plume and subduction zone–transform fault transition. *Journal of Geodynamics*, **20**, 219–241.

DAVIS, D. W. & LIN, S. 2003. Unraveling the geological history of the Hemlo Archean gold deposit, Superior Province, Canada: A U–Pb geochronological study. *Economic Geology*, **98**, 51–67.

DAVIS, B. K. & MAIDENS, E. 2003. Archean orogen parallel extension: Evidence from the northern Eastern Goldfields Province, Yilgarn Craton. *Precambrian Research*, **127**, 229–248.

DEFANT, M. J., JACKSON, T. E. *ET AL*. 1992. The geochemistry of young volcanism throughout western Panama and southeastern Costa Rica: An overview. *Journal of the Geological Society, London*, **149**, 569–579.

DE RONDE, E. J. & DE WIT, M. J. 1994. Tectonic history of the Barberton greenstone belt, South Africa: 490 million years of Archean crustal evolution. *Tectonics*, **13**, 983–1005.

DESCHAMPS, A. & LALLEMAND, S. 2003. Geodynamic setting of Izu-Bonin-Mariana boninites. *In*: LARTER, R. D. & LEAT, P. T. (eds) *Intra-oceanic Subduction Systems: Tectonic and Magmatic Processes*.

Geological Society of London Special Publications, **219**, 163–185.

DESROCHERS, J. P., HUBERT, C., LUDDEN, J. N. & PILOTE, P. 1993. Accretion of Archean oceanic plateau fragments in the Abitibi greenstone belt, Canada. *Geology*, **21**, 451–454.

DEWEY, J. F. 1988. Extensional collapse of orogens. *Tectonics*, **7**, 1123–1139.

DEWEY, J. F. & LAMB, S. H. 1992. Active tectonics of the Andes. *Tectonophysics*, **205**, 79–95.

DE WIT, M. J. 1998. On Archean granites, greenstones, cratons, and tectonics: Does the evidence demand a verdict? *Precambrian Research*, **91**, 181–226.

DILEK, Y. & MOORES, E. M. 1999. A Tibetan model for early Tertiary western United States. *Journal of the Geological Society, London*, **156**, 929–941.

DOSTAL, J. & MUELLER, W. U. 1997. Komatiite flooding of a rifted Archean rhyolite arc complex: Geochemical signature and tectonic significance of Stoughton–Roquemaure Group, Abitibi greenstone belt, Canada. *Journal of Geology*, **105**, 545–563.

DRUMMOND, M. S., DEFANT, M. J. & KEPEZHINSKAS, P. K. 1996. Petrogenesis of slab-derived trondhjemite–tonalite–dacite/adakite magmas. *Transactions of Royal Society of Edinburgh: Earth Sciences*, **87**, 205–215.

EGGINS, S. M. 1993. Origin and differentiation of picritic arc magmas, Ambae (Aoba), Vanuatu. *Contributions to Mineralogy and Petrology*, **114**, 79–100.

ERNST, R. E., BUCHAN, K. L. & PROKOPH, A. 2004. Large igneous province record through time. *In*: ERIKSSON, P. G., ALTERMANN, W., NELSON, D. R., MUELLER, W. U. & CATUNEANU, O. (eds) *The Precambrian Earth: Tempos and Events*. Elsevier, Amsterdam, 173–180.

ESCUDER VIRUETE, J., DE NEIRA, D. A. ET AL. 2006. Magmatic relationships and ages of Caribbean island arc tholeiites, boninites, and related felsic rocks, Dominican Republic. *Lithos*, **90**, 161–186.

ESCUDER VIRUETE, J. E., CONTRERAS, F. ET AL. 2007. Magmatic relationships and ages between adakites, magnesian andesites and Nb-enriched basalt–andesites from Hispaniola: Record of a major change in the Caribbean island arc magma sources. *Lithos*, **99**, 151–177.

FENG, R. & KERRICH, R. 1992. Geodynamic evolution of the southern Abitibi and Pontiac terranes—evidence from geochemistry of granitoid magma series (2700–2630 Ma). *Canadian Journal of Earth Sciences*, **29**, 2266–2286.

FRANCIS, D., LUDDEN, J., JOHNSTONE, R. & DAWIS, W. 1999. Picrite evidence for more Fe in Archean reservoirs. *Earth and Planetary Science Letters*, **167**, 197–213.

FRIEND, C. R. L. & NUTMAN, A. P. 2005a. New pieces to the Archean jigsaw puzzle in the Nuuk region, southern West Greenland: Steps in transforming a simple insight into a complex regional tectonothermal model. *Journal of the Geological Society, London*, **162**, 147–162.

FRIEND, C. R. L. & NUTMAN, A. P. 2005b. Complex 3670–3500 Ma orogenic episode superimposed on juvenile crust accreted between 3850 and 3690, Itsaq Gneiss Complex, southern West Greenland. *Journal of Geology*, **113**, 375–397.

FRIEND, C. R. L., NUTMAN, A. P. & MCGREGOR, V. R. 1988. Late Archean terrane accretion in the Godthab region, southern West Greenland. *Nature*, **335**, 535–538.

FRIEND, C. R. L., NUTMAN, A. P., BAADSGAARD, H., KINNY, P. D. & MCGREGOR, V. R. 1996. Timing of late Archean terrane assembly, crustal thickening and granite emplacement in the Nuuk region, southern West Greenland. *Earth and Planetary Science Letters*, **142**, 353–365.

FURNES, H., DE WIT, M., STAUDIGEL, H., ROSING, M. & MUEHLENBACHS, K. 2007. A vestige of Earth's oldest ophiolite. *Science*, **315**, 1704–1707.

GAO, S., RUDNICK, R. L. ET AL. 2004. Recycling lower continental crust in the North China Craton. *Nature*, **432**, 892–897.

GARDE, A. A. 2007. A mid-Archean island arc complex in eastern Akia terrane, Godthabsfjord, southern West Greenland. *Journal of the Geological Society, London*, **164**, 565–579.

GIBSON, G. M. & IRELAND, T. R. 1999. Black Giants anorthosite, New Zealand, a Paleozoic analogue of Archean stratiform anorthosites and implications for the formation of Archean high-grade gneissic terranes. *Geology*, **27**, 131–134.

GILL, G. E. 1992. Structure and kinematics of a major tectonic contact, Michipicoten greenstone belt, Ontario. *Canadian Journal of Earth Sciences*, **29**, 2118–2132.

GOODWIN, A. M. 1996. *Principles of Precambrian Geology*. Academic Press, Toronto, Ont.

GREENOUGH, J. D., DOSTAL, J. & MALLORY-GREENOUGH, L. M. 2005. Oceanic island volcanism II: Mantle processes. *Geoscience Canada*, **32**, 77–90.

HAMILTON, W. B. 1998. Archean magmatism and deformation were not products of plate tectonics. *Precambrian Research*, **91**, 143–179.

HAMILTON, W. B. 2003. An alternative Earth. *GSA Today*, **13**, 4–12.

HANMER, S. & GREENE, D. C. 2002. A modern structural regime in the Paleoarchean (~3.64 Ga): Isua greenstone belt, southern West Greenland. *Tectonophysics*, **346**, 201–222.

HANMER, S., HAMILTON, M. A. & CROWLEY, J. L. 2002. Geochronological constraints on Paleoarchean thrust-nappe and Neoarchean accretionary tectonics in southern West Greenland. *Tectonophysics*, **350**, 255–271.

HAWKESWORTH, C. J., GALLAGHER, K., HERGT, J. M. & MCDERMOTT, F. 1993. Mantle and slab contributions in arc magmas. *Annual Review of Earth and Planetary Sciences*, **21**, 175–204.

HAWKINS, J. W. 2003. Geology of supra-subduction zones—Implications for the origin of ophiolites. *In*: DILEK, Y. & NEWCOMB, S. (eds) *Ophiolite Concept and the Evolution of Geological Thought*. Geological Society of America, Special Papers, **373**, 227–268.

HERZBERG, C. 1992. Depth and degree of melting of komatiites. *Journal of Geophysical Research*, **97**, 4521–4540.

HOFMANN, A. W. 1988. Chemical differentiation of the Earth: The relationships between mantle, continental crust, and oceanic crust. *Earth and Planetary Science Letters*, **90**, 297–314.

HOFMANN, A. W. 2004. Sampling mantle heterogeneity through oceanic basalts: Isotopes and trace elements. *In*: CARLSON, R. W. (ed.) *Treatise on Geochemistry*, **2**, 61–101. Elsevier, Amsterdam.

HOLLINGS, P. & KERRICH, R. 2000. An Archean arc basalt–Nb-enriched basalt–adakite association: The 2.7 Ga Confederation assemblage of the Birch–Uchi greenstone belt, Superior Province. *Contributions to Mineralogy and Petrology*, **139**, 208–226.

HORODYSKYJ, U. N., LEE, C. T. & DUCEA, M. N. 2007. Similarities between Archean high MgO eclogites and Phanerozoic arc eclogites cumulates and the role of arcs in Archean continent formation. *Earth and Planetary Science Letters*, **256**, 510–520.

ISHIKAWA, T., NAGAISHI, K. & UMINO, S. 2002. Boninitic volcanism in the Oman ophiolite: Implications for thermal condition during transition from spreading ridge to arc. *Geology*, **30**, 899–902.

JIANG, N., LIU, Y., ZHOU, W., YANG, J. & ZHANG, S. 2007. Derivation of Mesozoic adakitic magmas from ancient lower crust in the North China Craton. *Geochimica et Cosmochimica Acta*, **71**, 2591–2608.

JOLLY, W. T., SCHELLEKENS, J. H. & DICKIN, A. P. 2007. High-Mg andesites and related lavas from southwest Puerto Rico (Greater Antilles Island Arc): Petrogenetic links with emplacement of late Cretaceous Caribbean mantle plume. *Lithos*, **98**, 1–26.

KEAREY, P. & VINE, F. 1996. *Global Tectonics*. Blackwell Scientific, Oxford.

KELEMEN, P. B. 1995. Genesis of high Mg# andesites and the continental crust. *Contributions to Mineralogy and Petrology*, **120**, 1–19.

KELEMEN, P. B., HANGHOJ, K. & GREENE, A. R. 2004. One view of the geochemistry of subduction-related magmatic arcs, with an emphasis on primitive andesite and lower crust. *In*: RUDNICK, R. L., HOLLAND, H. D. & TUREKIAN, K. K. (eds) *Treatise on Geochemistry*, **3**. Elsevier, Amsterdam, 593–659.

KEPEZHINSKAS, P., DEFANT, M. J. & DRUMMOND, M. S. 1996. Progressive enrichment of island arc mantle by melt–peridotite interaction inferred from Kamchatka xenoliths. *Geochimica et Cosmochimica Acta*, **60**, 1217–1229.

KERR, A. C. 2004. Oceanic plateaus. *In*: RUDNICK, R. L., HOLLAND, H. D. & TUREKIAN, K. K. (eds) *Treatise on Geochemistry*, **3**. Elsevier, Amsterdam, 537–565.

KERR, A. C. & MAHONEY, J. J. 2007. Oceanic plateaus: Problematic plumes, potential paradigms. *Chemical Geology*, **241**, 332–353.

KERR, A. C., ITURRALDE-VINENT, M. A., SAUNDERS, A. D., BABBS, T. L. & TARNEY, J. 1999. A new plate tectonic model for the Caribbean: Implications from a geochemical reconnaissance of Cuban Mesozoic volcanic rocks. *Geological Society of America Bulletin*, **111**, 1581–1599.

KERRICH, R. & POLAT, A. 2006. Archean greenstone–tonalite duality: Thermochemical mantle convection models or plate tectonics in the early Earth global dynamics? *Tectonophysics*, **415**, 141–165.

KERRICH, R., WYMAN, D. A., FAN, J. & BLEEKER, W. 1998. Boninite series: Low Ti-tholeiite associations from the 2.7 Ga Abitibi greenstone belt. *Earth and Planetary Science Letters*, **164**, 303–316.

KHAN, M. A., JAN, M. Q., WINDLEY, B. F., TARNEY, J. & THIRLWALL, M. F. 1989. The Chilas mafic–ultramafic igneous complex, the root of the Kohistan island arc in Himalaya of northern Pakistan. *In*: MALINCONICO, L. L. JR & LILLIE, R. J. (eds) *Tectonics of the Western Himalayas*. Geological Society of America, Special Papers, **232**, 75–94.

KIMURA, G., LUDDEN, J. N., DESROCHERS, J.-P. & HORI, R. 1993. A model of ocean-crust accretion for the Superior Province, Canada. *Lithos*, **30**, 337–355.

KISTERS, A. F. M., STEVENS, G., DZIGGEL, A. & ARMSTRONG, R. A. 2003. Extensional detachment faulting and core complex formation in the southern Barberton granite–greenstone terrain, South Africa: Evidence for 3.2 Ga orogenic collapse. *Precambrian Research*, **127**, 355–378.

KOMIYA, T., MARUYAMA, S., MASUDA, T., NOHDA, S., HAYASHI, M. & OKAMOTO, K. 1999. Plate tectonics at 3.8–3.7 Ga: Field evidence from the Isua accretionary complex, southern west Greenland. *Journal of Geology*, **107**, 515–554.

KUSKY, T. M. 1991. Structural development of an Archean orogen, Western Point Lake, Northwest Territories. *Tectonics*, **10**, 820–840.

KUSKY, T. M. 2004. Introduction. *In*: KUSKY, T. M. (ed.) *Precambrian Ophiolites and Related Rocks*. Elsevier, Amsterdam, 1–34.

KUSKY, T. M. & KIDD, W. F. 1992. Remnants of an Archean oceanic plateau, Belingwe greenstone belt, Zimbabwe. *Geology*, **20**, 43–46.

KUSKY, T. M. & POLAT, A. 1999. Growth of granite–greenstone terranes at convergent margins, and stabilization of Archean cratons. *Tectonophysics*, **305**, 43–73.

LACROIX, S. & SAWYER, E. W. 1995. An Archean fold and thrust belt in the northwestern Abitibi greenstone belt: Structural and seismic evidence. *Canadian Journal of Earth Sciences*, **32**, 97–112.

MACPHERSON, C. G. & HALL, R. 2001. Tectonic setting of Eocene boninite magmatism in the Izu–Bonin–Mariana forearc. *Earth and Planetary Science Letters*, **186**, 215–230.

MANIKYAMBA, C. & KHANNA, T. C. 2007. Crustal growth processes as illustrated by the Neoarchean intraoceanic magmatism from Gadwal greenstone belt, Eastern Dharwar Craton, India. *Gondwana Research*, **11**, 476–491.

MANIKYAMBA, C., NAQVI, S. M., RAO, D. V. S., MOHAN, M. R., KHANNA, T. C., RAO, T. G. & REDDY, G. L. N. 2005. Boninites from the Neoarchean Gadwal Greenstone belt, Eastern Dharwar Craton, India: Implications for Archean subduction processes. *Earth and Planetary Science Letters*, **230**, 65–83.

MANIKYAMBA, C., KERRICH, R., KHANNA, T. C. & RAO, D. V. S. 2007. Geochemistry of adakites and rhyolites from the Neoarchean Gadwal greenstone belt, eastern Dharwar craton, India: Implications for source and geodynamic setting. *Canadian Journal of Earth Sciences*, **44**, 1517–1535.

MANN, P. & TAIRA, A. 2004. Global tectonic significance of the Solomon Islands and Ontong Java convergent zone. *Tectonophysics*, **389**, 137–190.

MARTIN, H. 1999. Adakitic magmas: Modern analogues of Archean granitoids. *Lithos*, **46**, 411–429.

MARTIN, H., SMITHIES, R. H., RAPP, R., MOYEN, J.-F. & CHAMPION, D. 2005. An overview of adakite, tonalite–trondhjemite–granodiorite (TTG), and sanukutoid: Relationships and some implications for crustal evolution. *Lithos*, **79**, 1–24.

MCGREGOR, V. R., FRIEND, C. R. L. & NUTMAN, A. P. 1991. The late Archaen mobile belt through Godthabsfjord, southern West Greenland: A continent–continent collision zone? *Bulletin of the Geological Society of Denmark*, **39**, 179–197.

MORRIS, P. A. 1995. Slab melting as an explanation of Quaternary volcanism and aseismicity in southwest Japan. *Geology*, **23**, 1040–1043.

MUELLER, W. U., DAIGNEAULT, R., MORTENSEN, J. K. & CHOWN, E. H. 1996. Archean terrane docking: Upper crustal collision tectonics, Abitibi greenstone belt, Quebec, Canada. *Tectonophysics*, **265**, 127–150.

MUIR, T. L. 2003. Structural evolution of the Hemlo greenstone belt in the vicinity of the world-class Hemlo gold deposit. *Canadian Journal of Earth Sciences*, **40**, 395–430.

MURPHY, J. B. 2007. Arc magmatism II: Geochemical and isotopic characteristics. *Geoscience Canada*, **34**, 7–35.

MYERS, J. S. 2001. Protoliths of the 3.7–3.8 Ga Isua greenstone belt, West Greenland. *Precambrian Research*, **105**, 129–141.

NUTMAN, A. P. 2006. Antiquity of the oceans and continents. *Elements*, **2**, 223–227.

NUTMAN, A. P. & FRIEND, C. R. L. 2007a. Adjacent terranes with ca. 2715 and 2650 Ma high-pressure metamorphic assemblages in the Nuuk region of the North Atlantic Craton, southern West Greenland: Complexities of Neoarchean collisional orogeny. *Precambrian Research*, **155**, 159–203.

NUTMAN, A. P. & FRIEND, C. R. L. 2007b. Comment 'A vestige of Earth's oldest ophiolite'. *Science*, **318**, 746c.

NUTMAN, A. P., MCGREGOR, V. R., FRIEND, C. R. L., BENNETT, V. C. & KINNY, P. D. 1996. The Itsaq Gneiss Complex of southern West Greenland; the world's most extensive record of early crustal evolution (3900–3600 Ma). *Precambrian Research*, **78**, 1–39.

NUTMAN, A. P., FRIEND, C. R. L. & BENNETT, V. C. 2002. Evidence for 3650–3600 assembly of the northern end of the Itsaq Gneiss Complex, Greenland: Implications for early Archean tectonics. *Tectonics*, doi: 10.1029/2000TC001203.

NUTMAN, A. P., FRIEND, C. R. L., BARKER, S. S. & MCGREGOR, V. R. 2004. Inventory and Paleoarchean gneiss terranes and detrital zircons in southern West Greenland. *Precambrian Research*, **135**, 281–314.

NYE, C. J. & REID, M. R. 1986. Geochemistry of primary and least fractionated lavas from Okmaok volcano, central Aleutians: Implications for arc magma genesis. *Journal of Geophysical Research*, **91**, 10271–10287.

O'NEILL, C., LENARDIC, A., MORESI, L., TORSVIK, T. H. & LEE, C. T. A. 2007. Episodic Precambrian subduction. *Earth and Planetary Science Letters*, **262**, 552–562.

PEARCE, J. A. 1982. Trace element characteristics of lavas from the destructive plate boundaries. *In*: THORPE, R. S. (ed.) *Andesites*. Wiley, Chichester, 525–548.

PEARCE, J. A. 2003. Supra-subduction zone ophiolites: The search for modern analogues. *In*: DILEK, Y. & NEWCOMB, S. (eds) *Ophiolite Concept and the Evolution of Geological Thought*. Geological Society of America, Special Papers, **373**, 269–293.

PEARCE, J. A. 2008. Geochemical fingerprinting of oceanic basalts with applications to ophiolite classification and the search for Archean oceanic crust. *Lithos*, **100**, 14–48.

PEARCE, J. A. & PEATE, D. W. 1995. Tectonic implications of the composition of volcanic arc magmas. *Annual Review of Earth and Planetary Sciences*, **23**, 251–285.

PEARCE, J. A., VAN DER LAAN, S., ARCULUS, R. J., MURTON, B. J., ISHII, T. & PEATE, D. W. 1992. Boninite and harzburgite from Leg 125 (Bonin–Mariana fore-arc): A case study of magma genesis during the initial stages of subduction. *In*: FRYER, P., PEARCE, J. A. & STOKKING, L. B. (eds) *Proceedings of the Ocean Drilling Program, Scientific Results*, **125**. Ocean Drilling Program, College Station, TX, 623–659.

PERCIVAL, J. A., STERN, R. A. & RAYNER, N. 2003. Archean adakites from the Ashuanipi complex, eastern Superior Province, Canada: Geochemistry, geochronology, and tectonic significance. *Contributions to Mineralogy and Petrology*, **145**, 265–280.

PERCIVAL, J. A., SANBORN-BARRIE, M., SKULSKI, T., STOTT, G. M., HELMSTAEDT, H. & WHITE, D. J. 2006a. Tectonic evolution of the western Superior Province from NATMAP and Lithoprobe studies. *Canadian Journal of Earth Sciences*, **43**, 1085–1117.

PERCIVAL, J. A., MCNICOLL, V. & BAILES, A. H. 2006b. Strike-slip juxtaposition of ca. 2.72 Ga juvenile arc and >2.98 Ga continental margin sequences and its implications for Archean terrane accretion, western Superior Province, Canada. *Canadian Journal of Earth Sciences*, **43**, 895–927.

POLAT, A. 2009. The geochemistry of Neoarchean (ca. 2700 Ma) tholeiitic basalts, transitional to alkaline basalts, and gabbros, Wawa Subprovince, Canada: Implications for petrogenetic and geodynamic processes. *Precambrian Research*, **168**, 83–105.

POLAT, A. & HOFMANN, A. W. 2003. Alteration and geochemical patterns in the 3.7–3.8 Ga Isua greenstone belt, West Greenland. *Precambrian Research*, **126**, 197–218.

POLAT, A. & KERRICH, R. 1999. Formation of an Archean tectonic mélange in the Schreiber–Hemlo greenstone belt, Superior province, Canada: Implications for Archean subduction–accretion process. *Tectonics*, **18**, 733–755.

POLAT, A. & KERRICH, R. 2000. Archean greenstone belt magmatism and continental growth–mantle evolution connection: Constraints from Th–U–Nb–LREE systematics of the 2.7 Ga Wawa subprovince, Superior Province, Canada. *Earth and Planetary Science Letters*, **175**, 41–54.

POLAT, A. & KERRICH, R. 2001a. Magnesian andesites, Nb-enriched basalt–andesites, and adakites from late Archean 2.7 Ga Wawa greenstone belts, Superior

Province, Canada: Implications for late Archean subduction zone petrogenetic processes. *Contributions to Mineralogy and Petrology*, **141**, 36–52.

POLAT, A. & KERRICH, R. 2001b. Geodynamic processes, continental growth, and mantle evolution recorded in late Archean greenstone belts of the southern Superior Province, Canada. *Precambrian Research*, **112**, 5–25.

POLAT, A. & KERRICH, R. 2006. Reading the geochemical fingerprints of Archean hot subduction volcanic rocks: Evidence for accretion and crustal recycling in a mobile tectonic regime. *In*: BENN, K., MARESCHAL, J. C. & CONDIE, K. C. (eds) *Archean Geodynamics and Environments*. Geophysical Monograph, American Geophysical Union, **164**, 189–213.

POLAT, A. & MÜNKER, C. 2004. Hf–Nd isotope evidence for contemporaneous subduction processes in the source of late Archean arc lavas from the Superior Province, Canada. *Chemical Geology*, **213**, 403–429.

POLAT, A., CASEY, J. F. & KERRICH, R. 1996. Geochemical characteristics of accreted material beneath the Pozanti–Karsanti ophiolite, Turkey: Intra-oceanic detachment, assembly and obduction. *Tectonophysics*, **263**, 249–276.

POLAT, A., KERRICH, R. & WYMAN, D. A. 1998. The late Archean Schreiber–Hemlo and White River–Dayohessarah greenstone belts, Superior Province: Collages of oceanic plateaus, oceanic arcs, and subduction–accretion complexes. *Tectonophysics*, **294**, 295–326.

POLAT, A., KERRICH, R. & WYMAN, D. A. 1999. Geochemical diversity in oceanic komatiites and basalts from the late Archean Wawa greenstone belts, Superior Province, Canada: Trace element and Nd isotope evidence for a heterogeneous mantle. *Precambrian Research*, **94**, 139–173.

POLAT, A., HOFMANN, A. W. & ROSING, M. 2002. Boninite-like volcanic rocks in the 3.7–3.8 Ga Isua greenstone belt, West Greenland: Geochemical evidence for intra-oceanic subduction zone processes in the Earth. *Chemical Geology*, **184**, 231–254.

POLAT, A., HERZBERG, C., MÜNKER, C., RODGERS, R., KUSKY, T., LI, J., FRYER, B. & DELANEY, J. 2006. Geochemical and petrological evidence for a suprasubduction zone origin of Neoarchean (ca. 2.55–2.50 Ga) peridotites, central orogenic belt, North China craton. *Geological Society of America Bulletin*, **118**, 771–784.

POLAT, A., APPEL, P. W. U. ET AL. 2007. Field and geochemical characteristics of the Mesoarchean (~3075 Ma) Ivisaartoq greenstone belt, southern West Greenland: Evidence for seafloor hydrothermal alteration in a supra-subduction oceanic crust. *Gondwana Research*, **11**, 69–91.

POLAT, A., FREI, R., APPEL, P. W. U., DILEK, Y., FRYER, B., ORDONEZ-CALDERON, J. C. & YANG, Z. 2008. The origin and compositions of Mesoarchean oceanic crust: Evidence from the 3075 Ma Ivisaartoq greenstone belt, SW Greenland. *Lithos*, **100**, 293–321, doi:10.1016/j.lithos.2007.06.021.

POULSEN, K. H., BORRADAILE, G. J. & KEHLENBECK, M. M. 1980. An inverted Archean succession at Rainy Lake, Ontario. *Canadian Journal of Earth Sciences*, **17**, 1358–1369.

PROUTEAU, G., SCAILLET, B., PICHAVANT, M. & MAURY, R. C. 1999. Fluid-present melting of oceanic crust in subduction zones. *Geology*, **27**, 1111–1114.

RAPP, R. P., SHIMIZU, N., NORMAN, M. D. & APPELGATE, G. S. 1999. Reaction between slab derived melts and peridotite in the mantle wedge: Experimental constraints at 3.8 GPa. *Chemical Geology*, **160**, 335–356.

RICHARDS, J. P. & KERRICH, R. 2007. Special paper: Adakite-like rocks: Their diverse origins and questionable role in metallogenesis. *Economic Geology*, **102**, 537–576.

ROGERS, G. & SAUNDERS, A. D. 1989. Magnesian andesites from Mexico, Chile, and the Aleutian islands: Implications for magmatism associated with ridge–trench subduction. *In*: CRAWFORD, A. J. (ed.) *Boninites and Related Rocks*. Unwin Hyman, London, 416–445.

ROSING, M. T., ROSE, N. M., BRIDGWATER, D. & THOMSEN, H. S. 1996. Earliest part of Earth's stratigraphic record: A reappraisal of the >3.7 Ga Isua (Greenland) supracrustal sequence. *Geology*, **24**, 43–46.

SAJONA, F. G., MAURY, R., BELLON, H., COTTON, J. & DEFANT, D. 1996. High field strength element enrichment of Pliocene–Pleistocene island arc basalts, Zamboanga Peninsula, Western Mindanao (Philippines). *Journal of Petrology*, **37**, 693–726.

SAUNDERS, A. D., NORRY, M. J. & TARNEY, J. 1991. Fluid influence on the trace element compositions of subduction zone magmas. *Philosophical Transactions of the Royal Society of London, Series A*, **335**, 377–392.

SCHUTH, S., ROHRBACH, A., MÜNKER, C., BALLHAUS, C., GARBE-SCHONBERG, D. & QOPOTO, C. 2004. Geochemical constraints on the petrogenesis of arc picrites and basalts, New Georgia Group, Solomon Islands. *Contributions to Mineralogy and Petrology*, **148**, 288–304.

ŞENGÖR, A. M. C. 1987. Tectonics of the Tethysides: Orogenic collage development in a collisional setting. *Annual Review of Earth and Planetary Sciences*, **15**, 213–244.

ŞENGÖR, A. M. C. 1990. Plate tectonics and orogenic research after 25 years: A Tethyan perspective. *Earth-Science Reviews*, **27**, 1–201.

ŞENGÖR, A. M. C. 1993. Turkic-type orogeny in the Altaids: Implications for the evolution of continental crust and methodology of regional tectonic analyses. *Transactions of the Leicester Literary and Philosophical Society*, **87**, 37–47.

ŞENGÖR, A. M. C. & NATAL'IN, B. A. 1996. Turkic-type orogeny and its role in the making of the continental crust. *Annual Review of Earth and Planetary Sciences*, **24**, 263–337.

ŞENGÖR, A. M. C. & NATAL'IN, B. A. 2004. Phanerozoic analogues of Archean basement fragments: Altaid ophiolites and ophirags. *In*: KUSKY, T. M. (ed.) *Precambrian Ophiolites and Related Rocks*. Elsevier, Amsterdam, 671–721.

ŞENGÖR, A. M. C. & OKUROĞULLARI, A. H. 1991. The rôle of accretionary wedges in the growth of continents: Asiatic examples from Argand to plate tectonics. *Eclogae Geologicae Helvetiae*, **84**, 535–597.

ŞENGÖR, A. M. C., NATAL'IN, B. A. & BURTMAN, V. S. 1993. Evolution of the Altaid tectonic collage and Paleozoic crustal growth in Eurasia. *Nature*, **364**, 299–307.

SHCHIPANSKY, A. A., SAMSONOV, A. V. *ET AL*. 2004. 2.8 Ga boninite-hosting partial subduction zone ophiolite sequences from the North Karelian greenstone belt, NE Baltic Shield, Russia. *In*: KUSKY, T. M. (ed.) *Precambrian Ophiolites and Related Rocks*. Elsevier, Amsterdam, 425–486.

SLEEP, N. H. 1992. Archean plate tectonics: What can be learned from continental geology. *Canadian Journal of Earth Sciences*, **29**, 2066–2071.

SMITHIES, R. H. 2000. The Archean tonalite–trondhjemite–granodiorite (TTG) series is not an analogue of Cenozoic adakites. *Earth and Planetary Science Letters*, **182**, 115–125.

SMITHIES, R. H., CHAMPION, D. C. & SUN, S. S. 2004. The case for Archean boninites. *Contributions to Mineralogy and Petrology*, **147**, 705–721.

SMITHIES, R. H., CHAMPION, D. C., VAN KRANENDONK, M. J., HOWARD, H. M. & HICKMAN, A. H. 2005. Modern-style subduction processes in the Mesoarchean: Geochemical evidence from the 3.12 Ga Whundo intra-oceanic arc. *Earth and Planetary Science Letters*, **231**, 221–237.

SPROULE, R. A., LESHER, C. M., AYER, J. A., THURSTON, P. C. & HERZBERG, C. T. 2002. Spatial and temporal variations in the geochemistry of komatiies and komatiitic basalts in the Abitibi greenstone belt. *Precambrian Research*, **115**, 153–186.

STEENFELT, A., GARDE, A. A. & MOYEN, J.-F. 2005. Mantle wedge involvement in the petrogenesis of Archean grey gneisses in West Greenland. *Lithos*, **79**, 207–228.

STERN, R. J. 2005. Evidence from ophiolites, blueschists, and ultrahigh-pressure metamorphic terranes that the modern episode of subduction tectonics began in Neoproterozoic time. *Geology*, **33**, 557–560.

STERN, R. J. & BLOOMER, S. H. 1992. Subduction zone infancy: Examples from the Eocene Izu–Bonin–Mariana and Jurassic California arcs. *Geological Society of America Bulletin*, **104**, 1621–1636.

STERN, C. R. & DE WIT, M. J. 1997. The Rocas Verdes 'Greenstone Belt', southernmost South America. *In*: DE WIT, M. & ASHWAL, L. (eds) *Tectonic Evolution of Greenstone Belts*. Oxford Monographs on Geology and Geophysics, **35**, 55–90.

STERN, C. R. & KILIAN, R. 1996. Role of the subducted slab, mantle wedge and continental crust in the generation of adakites from the Andean Austral Volcanic Zone. *Contributions to Mineralogy and Petrology*, **123**, 263–281.

STERN, R. J., MORRIS, J., BLOOMER, S. H. & HAWKINS, J. W. 1991. The source of the subduction component in convergent margin magmas: Trace element and radiogenic evidence from Eocene boninites, Mariana forearc. *Geochimica et Cosmochimica Acta*, **55**, 1467–1481.

STONE, W. E., CROCKET, J. H., DICKIN, A. P. & FLEET, M. F. 1995. Origin of Archean ferropicrites: Geochemical constraints from the Boston Creek flow, Abitibi greenstone belt, Ontario, Canada. *Chemical Geology*, **121**, 51–71.

STOREY, B. C. 1995. The role of mantle plumes in continental breakup: Case histories from Gondwanaland. *Nature*, **377**, 301–308.

STOREY, M., MAHONEY, J. J., KROENKE, L. W. & SAUNDERS, A. D. 1991. Are oceanic plateaus sites of komatiite formation? *Geology*, **19**, 376–379.

STOTT, G. M. 1997. The Superior Province, Canada. *In*: DE WIT, & ASHWAL, L. D. (eds) *Tectonic Evolution of Greenstone Belts*. Oxford Monographs on Geology and Geophysics, **35**, 480–507.

SUN, S. S. & MCDONOUGH, W. F. 1989. Chemical and isotopic systematics of oceanic basalts: Implications for mantle composition and processes. *In*: SAUNDERS, A. D. & NORRY, M. J. (eds) *Magmatism in the Ocean Basins*. Geological Society, London, Special Publications, **42**, 313–345.

SUN, S.-S. & NESBITT, R. W. 1978. Geochemical regularities and genetic significance of ophiolitic basalts. *Geology*, **6**, 689–693.

SVETOV, S. A., HUHMA, H., SVETOVA, A. I. & NAZAROVA, T. N. 2004. The oldest adakites of the Fennoscandian Shield. *Doklady Earth Sciences*, **397A**, 878–882.

SWAGER, C. 1997. Tectonostratigraphy of late Archean greenstone terranes in the southern Eastern Goldfields, Western Australia. *Precambrian Research*, **83**, 11–42.

TAIRA, A., PICKERING, K. T., WINDLEY, B. F. & SOH, W. 1992. Accretion of Japanese island arcs and implications for the origin of Archean greenstone belts. *Tectonics*, **11**, 1224–1244.

TATSUMI, Y. 2001. Geochemical modeling of partial melting of subducting sediments and subsequent melt–mantle interaction: Generation of high-Mg andesites in the Setouchi volcanic belt, southwest Japan. *Geology*, **29**, 323–326.

TAYLOR, S. R. & MCLENNAN, S. M. 1995. The geochemical evolution of the continental crust. *Reviews of Geophysics*, **33**, 241–265.

THIRLWALL, M. F., GRAHAM, A. M., ARCULUS, R. J., HARMON, R. S. & MACPHERSON, C. G. 1996. Resolution of the effects of crustal assimilation, sediment subduction, and fluid transport in arc magmas: Pb–Sr–Nd–O isotope geochemistry of Grenada, Lesser Antilles. *Geochimica et Cosmochimica Acta*, **60**, 4785–4810.

THORKELSON, D. J. & BREITSPRECHER, K. 2005. Partial melting of slab window margins; genesis of adakitic and non-adakitic magmas. *Lithos*, **79**, 25–41.

TOMLINSON, K. Y., HUGHES, D. J., THURSTON, P. C. & HALL, R. P. 1999. Plume magmatism and crustal growth at 2.9 to 3.0 Ga in the Steep Rock and Lumby Lake area, Western Superior Province. *Lithos*, **46**, 103–136.

UJIKE, O., GOODWIN, A. M. & SHIBATA, T. 2007. Geochemistry of Archean volcanic rocks from the Upper Keewatin assemblage (*ca*. 2.7 Ga), Lake of the Woods greenstone belt, Western Wabigoon subprovince, Superior Province, Canada. *Island Arc*, **16**, 191–208.

UMHOEFER, P. J. & SCHIARIZZA, P. 1996. Latest Cretaceous to early Tertiary dextral strike-slip faulting on the southeastern Yalakom fault system, southeastern Coast Belt. *Geological Society of America Bulletin*, **108**, 768–785.

VANDERHAEGHE, O. & TEYSSIER, C. 2001. Crustal-scale rheological transitions during late orogenic collapse. *Tectonophysics*, **335**, 211–228.

VAN KRANENDONK, M. J., SMITHIES, R. H., HICKMAN, A. H. & CHAMPION, D. C. 2007. Review: Secular tectonic evolution of Archean continental crust: Interplay between horizontal and vertical processes in the formation of the Pilbara Craton, Australia. *Terra Nova*, **19**, 1–38.

VAN THIENEN, P., VAN DEN BERG, A. P. & VLAAR, N. J. 2004. On the formation of continental silicic melts in thermochemical mantle convection models: Implications for early Earth. *Tectonophysics*, **394**, 111–124.

WAKABAYASHI, J. 1992. Nappes, tectonics of oblique plate convergence, and metamorphic evolution related to 140 million years of continuous subduction, Franciscan Complex, California. *Journal of Geology*, **100**, 19–40.

WANG, Q., WYMAN, D. A. *ET AL.* 2006. Petrogenesis of Cretaceous adakitic and shoshonitic igneous rocks in the Luzong area, Anhui Province (eastern China): Implications for geodynamics and Cu–Au mineralization. *Lithos*, **89**, 424–446.

WHALEN, J. B., PERCIVAL, J. A., MCNICOLL, V. J. & LONGSTAFFE, F. J., 2002. A mainly crustal origin for tonalitic granitoid rocks, Superior Province, Canada: Implications for late Archean Tectonomagmatic Processes. *Journal of Petrology*, **43**, 1551–1570.

WHITE, R. V., TARNEY, J., KERR, A. C., SAUNDERS, A. D., KEMPTON, P. D., PRINGLE, M. S. & KLAVER, G. T. 1999. Modification of oceanic plateau, Aruba, Dutch Caribbean: Implications for the generation of continental crust. *Lithos*, **46**, 43–68.

WILLIAMS, H. R., STOTT, G. M., THURSTON, P. C., SUTCLIFFE, R. H., BENNETT, G., EASTEN, R. M. & ARMSTRONG, D. K. 1991*a*. Tectonic evolution of Ontario: Summary and synthesis. *In*: THURSTON, P. C., WILLIAMS, H. R., SUTCLIFFE, R. H. & STOTT, G. (eds) *Geology of Ontario*. Ontario Geological Survey, Special Volume, **4**, Part 2, 1255–1332.

WILLIAMS, R. H., STOTT, G. M., HEATHER, K., MUIR, T. L. & SAGE, R. P. 1991*b*. Wawa Subprovince. *In*: THURSTON, P. C., WILLIAMS, H. R., SUTCLIFFE, H. R. & STOTT, G. M. (eds) *Geology of Ontario*. Ontario Geological Survey, Special Volume, **4**, 485–539.

WINDLEY, B. F. 1976. New tectonic models for the evolution of Archean continents and oceans. *In*: WINDLEY, B. F. (ed.) *The Early History of the Earth*. Wiley, London, 105–112.

WINDLEY, B. F. 1993. Uniformitarianism today: Plate tectonics is the key to the past. *Journal of the Geological Society, London*, **150**, 7–19.

WINDLEY, B. F. & SMITH, J. V. 1976. Archean high grade complexes and modern continental margins. *Nature*, **260**, 671–675.

WINDLEY, B. F. & GARDE, A. A. 2009. Arc-generated blocks with crustal sections in the North Atlantic craton of West Greenland: new mechanism of crustal growth in the Archean with modern analogues. *Earth Science Reviews*, **93**, 1–30.

WINDLEY, B. F., BISHOP, F. C. & SMITH, J. V. 1981. Metamorphosed layered igneous complexes in Archean granulite–gneiss belts. *Annual Review of Earth and Planetary Sciences*, **9**, 175–198.

WINDLEY, B. F., ALEXEIEV, D., XIAO, W., KRÖNER, A. & BADARCH, G. 2007. Tectonic models for accretion of the Central Asian Orogenic Belt. *Journal of the Geological Society, London*, **164**, 31–47.

WYMAN, D. A. 1999. A 2.7 Ga depleted tholeiite suite: Evidence of plume–arc interaction in the Abitibi greenstone belt, Canada. *Precambrian Research*, **97**, 27–42.

WYMAN, D. A. 2003. Upper mantle processes beneath the 2.7 Ga Abitibi belt, Canada: A trace element perspective. *Precambrian Research*, **127**, 143–165.

WYMAN, D. A., BLEEKER, W. & KERRICH, R. 1999. A 2.7 Ga komatiite, low-Ti tholeiite, arc transition, and inferred proto-arc geodynamic setting of the Kidd Creek deposit: Evidence from precise ICP MS trace element data. *In*: HANNINGTON, M. D. & BARRIE, C. T. (eds) *The Giant Kidd Creek Volcanogenic Massive Sulfide Deposit, Western Abitibi Subprovince, Canada*. Economic Geology Monograph, **10**, 511–528.

WYMAN, D. A., KERRICH, R. & POLAT, A. 2002. Assembly of Archean cratonic mantle lithosphere and crust: Plume–arc interaction in the Abitibi–Wawa subduction–accretion complex. *Precambrian Research*, **115**, 37–62.

XIE, Q., KERRICH, R. & FAN, J. 1993. HFSE/REE fractionations recorded in three komatiite–basalt sequences, Archean Abitibi greenstone belt: Implications for multiple plume sources and depths. *Geochimica et Cosmochimica Acta*, **57**, 4111–4118.

XU, J. F., SHINJO, R., DEFANT, M. J., WANG, Q. & RAPP, R. P. 2002. Origin of Mesozoic adakitic intrusive rocks in the Ningzhen area of east China: Partial melting of delaminated lower continental crust. *Geology*, **30**, 1111–1114.

YOGODZINSKI, G. M., KAY, R. W., VOLYNETS, O. N., KOLOSKOV, A. V. & KAY, S. M. 1995. Magnesian andesite in the western Aleutian Komandorsky region: Implications for slab melting processes in the mantle wedge. *Geological Society of America Bulletin*, **107**, 505–519.

ZALESKI, E. & PETERSON, V. L. 1995. Depositional setting and deformation of massive sulphide deposits, iron formation and associated alteration in the Manitouwadge greenstone belt, Superior Province, Ontario. *Economic Geology*, **90**, 2214–2261.

ZALESKI, E., VAN BRENAN, O. & PETERSON, V. L. 1999. Geological evolution of the Manitouwadge greenstone belt and Wawa–Quetico subprovince boundary, Superior Province, Ontario, constrained by U–Pb zircon dates of supracrustal and plutonic rocks. *Canadian Journal of Earth Sciences*, **36**, 945–966.

ZEGERS, T. E., KEIJZER, M., PASSCHIER, C. W. & WHITE, S. H. 1998. The Mulgandinnah Shear zone: An Archean crustal scale shear zone in the eastern Pilbara, Western Australia. *Precambrian Research*, **88**, 233–247.

Correlation of Archaean and Palaeoproterozoic units between northeastern Canada and western Greenland: constraining the pre-collisional upper plate accretionary history of the Trans-Hudson orogen

MARC R. ST-ONGE[1]*, JEROEN A. M. VAN GOOL[2,3], ADAM A. GARDE[2] & DAVID J. SCOTT[4]

[1]*Geological Survey of Canada, 601 Booth Street, Ottawa, Ontario, Canada K1A 0E8*

[2]*Geological Survey of Denmark and Greenland, Øster Voldgade 10, Copenhagen-K 1350, Denmark*

[3]*Present address: Scandinavian Highlands A/S, Hørsholm Kongevej 11, DK-2970 Hørsholm, Denmark*

[4]*Program Branch, Earth Sciences Sector, 601 Booth Street, Ottawa, Ontario, Canada K1A 0E8*

*Corresponding author (e-mail: mstonge@nrcan.gc.ca)

Abstract: Based on available tectonostratigraphic, geochronological, and structural data for northeastern Canada and western Greenland, we propose that the early, upper plate history of the Trans-Hudson orogen was characterized by a number of accretionary–tectonic events, which led to the nucleation and growth of a northern composite continent (the Churchill domain), prior to terminal collision with and indentation by the lower plate Superior craton. Between 1.96 and 1.91 Ga Palaeoproterozoic deformation and magmatism along the northern margin of the Rae craton is documented both in northeastern Canada (Ellesmere–Devon terrane) and in northern West Greenland (Etah Group–metaigneous complex). The southern margin of the craton was dominated by the accumulation of a thick continental margin sequence between c. 2.16 and 1.89 Ga, whose correlative components are recognized on Baffin Island (Piling and Hoare Bay groups) and in West Greenland (Karrat and Anap nunâ groups). Initiation of north–south convergence led to accretion of the Meta Incognita microcontinent to the southern margin of the Rae craton at c. 1.88–1.865 Ga on Baffin Island. Accretion of the Aasiaat domain (microcontinental fragment?) in West Greenland to the Rae craton resulted in formation of the Rinkian fold belt at c. 1.88 Ga. Subsequent accretion–collision of the North Atlantic craton with the southern margin of the composite Rae craton and Aasiaat domain is bracketed between c. 1.86 and 1.84 Ga (Nagssugtoqidian orogen), whereas collision of the North Atlantic craton with the eastern margin of Meta Incognita microcontinent in Labrador is constrained at c. 1.87–1.85 Ga (Torngat orogen). Accretion of the intra-oceanic Narsajuaq arc terrane of northern Quebec (no correlative in Greenland) to the southern margin of the composite Churchill domain at 1.845 Ga was followed by terminal collision between the lower plate Superior craton (no correlative in Greenland) and the composite, upper plate Churchill domain in northern and eastern Quebec at c. 1.82–1.795 Ga. Taken as a set, the accretionary–tectonic events documented in Canada and Greenland prior to collision of the lower plate Superior craton constrain the key processes of crustal accretion during the growth of northeastern Laurentia and specifically those in the upper plate Churchill domain of the Trans-Hudson orogen during the Palaeoproterozoic Era. This period of crustal amalgamation can be compared directly with that of the upper plate Asian continent prior to its collision with the lower plate Indian subcontinent in the early Eocene. In both cases, terminal continental collision was preceded by several important episodes of upper plate crustal accretion and collision, which may therefore be considered as a harbinger of collisional orogenesis and a signature of the formation of supercontinents, such as Nuna (Palaeoproterozoic Era) and Amasia (Cenozoic Era).

The correlation of Precambrian bedrock units and tectonic features in northeastern Laurentia from western Greenland to eastern Canada across Baffin Bay, Davis Strait and the Labrador Sea has been the subject of many studies, models, and reviews over the years (e.g. Bridgwater *et al.* 1973*b*, 1990; Dawes *et al.* 1982, 2000; Frisch & Dawes 1982; Korstgård *et al.* 1987; Friend *et al.* 1988; Hoffman

1990a; Van Kranendonk et al. 1993; Friend & Nutman 1994; Dawes 1997; Kerr et al. 1996, 1997; Scott 1999; Wardle et al. 2000b, 2002; Garde et al. 2002; James et al. 2002; van Gool et al. 2002, 2004; Thrane et al. 2005; Connelly et al. 2006; and references therein). The aims of this contribution are to: (1) identify and review some of the pertinent correlations; (2) integrate new tectonostratigraphic, structural, and analytical data and observations from both Canada and Greenland (in particular, new geochronological profiling of detrital zircon populations from Palaeoproterozoic supracrustal sequences); and (3) use an actualistic model to analyse and sequence the episodes of crustal accretion and growth in NE Laurentia during the Palaeoproterozoic Era.

We take a geological, field research-based approach to address the issue of bedrock correlations from Canada to Greenland by: (1) reviewing the salient Archaean and Palaeoproterozoic crustal assemblages and tectonic features on both sides of Baffin Bay–Davis Strait–Labrador Sea; and (2) considering their Proterozoic tectonic evolution and amalgamation within an Asian plate–Indian plate (Himalayan), accretion–collision context. Our premise is that if close similarities in principal crustal assemblages, as well as timing of magmatic, deformation, and metamorphic events, in western

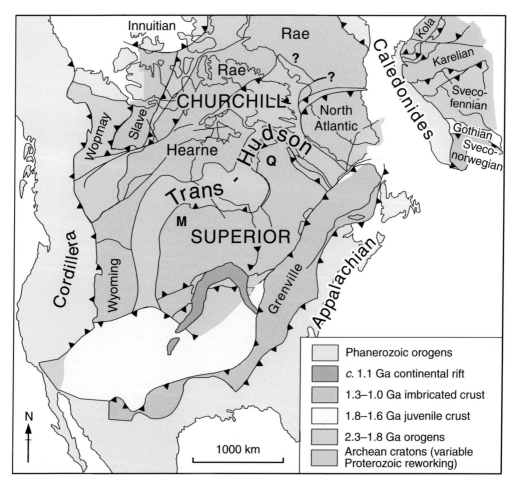

Fig. 1. Summary geological map of North America and Greenland (Laurentia), and the western Baltic Shield modified from Hoffman (1988). The map shows the extent of the Trans-Hudson orogen and the location of bounding Archaean crustal blocks and cratons. The collisional lower plate of Trans-Hudson orogen comprises the Superior craton and its associated Palaeoproterozoic supracrustal sequences. The collisional upper plate (the Churchill domain) includes the Wyoming craton, the Slave craton, the Hearne block, the Rae craton, and the North Atlantic (Nain) craton and their associated Palaeoproterozoic supracrustal sequences. M, Manitoba promontory; Q, Quebec promontory.

Greenland and northeastern Canada can be established, then there is a basis for correlation. Guided by the Mesozoic to Quaternary tectonic evolution of southeastern Asia as a modern example, we then consider the application of actualistic principles to the Palaeoproterozoic crustal history of northeastern Laurentia and document the importance of accretionary processes in the evolution of the greater Trans-Hudson orogen (Fig. 1; see below). Finally, residual differences in the accessible (exposed) Precambrian rock record of northeastern Canada and western Greenland can then be evaluated in terms of contrasting depths of erosion, primary distribution of tectonostratigraphic units, and along-strike variation within different segments of orogenic belts.

Tectonic upper plate v. lower plate context for northeastern Canada and western Greenland during the Palaeoproterozoic Era

Much of Laurentia (the Precambrian core of North America including Greenland) was assembled in the middle Palaeoproterozoic Period during a time of global amalgamation of Archaean cratons and crustal slivers, and attendant Palaeoproterozoic cover sequences (Hoffman 1988, 1989, 1990a; Zhao et al. 2002, 2004; Bleeker 2003). Within Laurentia, the Trans-Hudson orogen (Hoffman 1988, 1990b; Lewry & Collerson 1990) comprises a Himalayan-scale collisional orogenic belt that extends from the south–central part of the North American continent to its northeastern edge, where it is truncated by the younger Meso- to Neoproterozoic Grenville orogen (Fig. 1). The Trans-Hudson orogen marks the collision between a lower plate, comprising the Archaean Superior craton (and associated Palaeoproterozoic supracrustal sequences) that acted as an indentor, and an upper plate collage of Archaean crustal blocks and associated Palaeoproterozoic sequences (the Churchill domain) that includes the Wyoming craton to the west, the Slave craton, the Hearne block, and Rae craton to the north, and the North Atlantic (Nain) craton to the east (Fig. 1). Upper plate orogenic segments of the eastern Trans-Hudson orogen within the Churchill domain include the Foxe and Dorset fold belts of Baffin Island (Fig. 2), the Rinkian fold belt and Nagssugtoqidian orogen of Greenland (Fig. 3), and the Torngat orogen of Labrador (Fig. 4). The Manitoba and Quebec promontories in the west and NE (M and Q, Fig. 1) mark the corners of the indenting lower plate Superior craton.

Recently, St-Onge et al. (2006c) have argued that the tectonic record of collision and indentation of the Superior craton into the upper plate collage of cratons and terranes (Churchill domain) that formed northeastern Canada in the late Palaeoproterozoic Period (c. 1830–1785 Ma) was similar to the record of collision and indentation of India into the upper plate collage of cratons and terranes that made up central Asia in the early Eocene, beginning at c. 50.6 Ma (Rowley et al. 2004; Searle et al. 2007). Using the Asian template, the Archaean and Palaeoproterozoic geological record of western Greenland and northeastern Canada can thus be compared and sequenced (ordered) from north to south in order to document: (1) the amalgamation and growth of a northern composite upper plate (Churchill) domain mostly through terrane accretion; (2) the collision of the lower plate Superior craton with the upper plate Churchill domain; and (3) the pre- to syncollisional accretion of juvenile magmatic arcs along the southern margin of the growing Laurentian craton. Viewed in this context, and as detailed below, the collisional lower plate comprises the Superior craton and its bounding Palaeoproterozoic margin sequences (Fig. 1), which together underlie large parts of Quebec, Ontario and Manitoba. The collisional upper plate comprises the Churchill domain (Fig. 1), which in northeastern Canada and western Greenland includes (Figs 2–4) the Rae craton, the Ellesmere–Devon terrane and Inglefield mobile belt, the Piling Group–Hoare Bay Group–Karrat Group–Anap nunâ Group cratonic margin, the Foxe fold belt, the Rinkian fold belt, the Aasiaat domain, the Nagssugtoqidian orogen, the North Atlantic craton, the Torngat orogen, the Makkovik and Ketilidian orogens, the Meta Incognita microcontinent, the Dorset fold belt, and the Narsajuaq arc.

Moreover, within the composite Churchill domain, the crustal components were themselves in changing plate settings at various times during the preceding episodes of convergence and accretion that led to the systematic growth of the northern landmass prior to collision with the southern Superior craton. For example, the North Atlantic craton was in an upper plate position with respect to the Rae craton ± Aasiaat domain during the period of convergence and collision that resulted in the Rinkian fold belt and Nagssugtoqidian orogen, and the Narsajuaq arc was in a lower plate position with respect to the Meta Incognita microcontinent prior to its accretion to the composite Churchill domain, as described in more detail below.

Crustal components

An overview of the crustal framework for northeastern Canada (Ellesmere Island, Devon Island, Baffin Island, northern and central Labrador, and northern Quebec) and western Greenland is given below

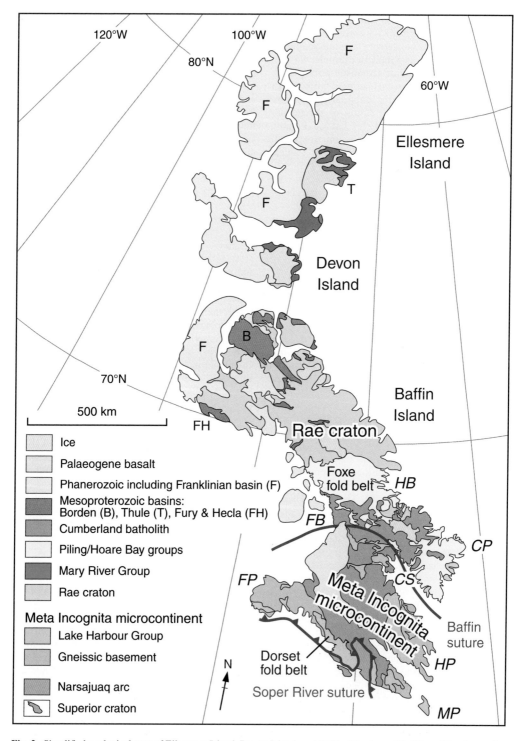

Fig. 2. Simplified geological map of Ellesmere Island, Devon Island, and Baffin Island, modified from Wheeler *et al.* (1996), showing principal tectonostratigraphic assemblages and structures discussed in the text. B, Borden basin; CP, Cumberland Peninsula; CS, Cumberland Sound; F, Franklinian basin; FB, Foxe basin; FH, Fury and Hecla basin; FP, Foxe Peninsula; HB, Home Bay; HP, Hall Peninsula; MP, Meta Incognita Peninsula; T, Thule basin.

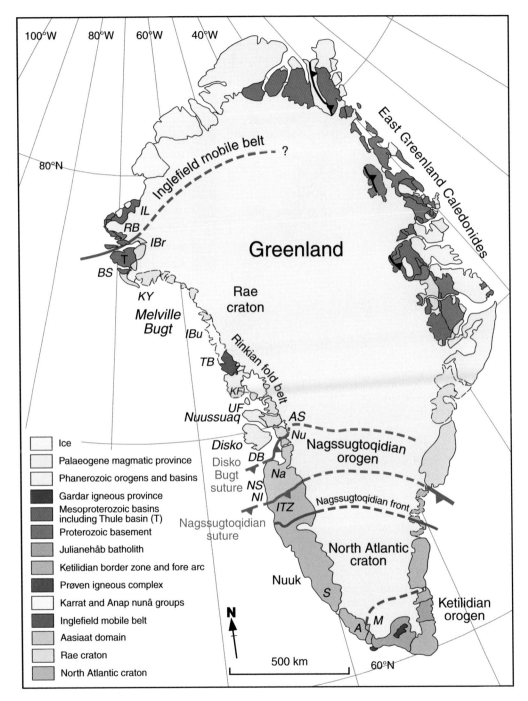

Fig. 3. Simplified geological map of Greenland modified from Escher & Pulvertaft (1995), showing principal tectonostratigraphic assemblages and structures discussed in the text. Extrapolation of geological boundaries beneath the Inland Ice (dashed lines) is constrained by the aeromagnetic data of Saltus & Gaina (2007). A, Arsuk; AS, Ataa Sund; BS, Bylot Sund; DB, Disko Bugt; IBr, Inglefield Bredning; IBu, Inussulik Bugt; IL, Inglefield Land; ITZ, Ikertôq thrust zone; KF, Karrat Fjord; KY, Kap York; M, Midternæs; Na, Naternaq; NI, Nordre Isortoq; NS, Nordre Strømfjord; Nu, Nunatarsuaq; RB, Rensselaer Bugt; S, Sermilik; T, Thule basin; TB, Tasiussaq Bugt; UF, Uummannaq Fjord.

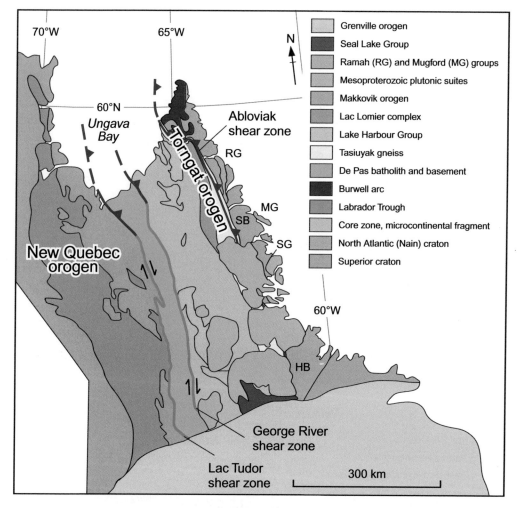

Fig. 4. Simplified geological map of Labrador modified from Wardle *et al.* (1997, 2000*a*), showing principal tectonostratigraphic assemblages and structures discussed in the text. Principal crustal boundaries are shown in red (suture) and green (strike-slip fault). HB, Hopedale block; MG, Mugford Group; RG, Ramah Group; SB, Saglek block; SG, Snyder Group.

from north to south, based on geological maps, papers, reviews, and descriptions published by A. Escher (1971), Allaart (1982), J. C. Escher (1985), Henderson & Pulvertaft (1987), Frisch (1988), Grocott & Pulvertaft (1990), Hoffman (1990*a*, *b*), Kalsbeek *et al.* (1990), Dawes (1991, 2004, 2006), Garde (1994), Escher & Pulvertaft (1995), Wardle *et al.* (1997, 2000*a*, 2002), Scott & de Kemp (1998), Garde & Steenfelt (1999*a*, *b*), Connelly & Thrane (2005), Connelly *et al.* (2000, 2006), Culshaw *et al.* (2000*b*), Dawes *et al.* (2000), Jackson (2000), Garde *et al.* (2002), James *et al.* (2002), Ketchum *et al.* (2002), St-Onge *et al.* (2002, 2006*a*–*c*), van Gool *et al.* (2002, 2004), Scott *et al.* (2003), Dawes & Garde (2004), Thrane *et al.* (2005), Thrane & Connelly (2006), and references therein. Unless otherwise specified, the geological ages quoted in the text, and summarized in Table 1, are based on conventional thermal ionization mass spectrometry (TIMS) U–Pb analyses of zircon.

Archaean Rae craton

The northern half of Baffin Island (Fig. 2) is underlain by the Archaean Rae craton (Hoffman 1988), which comprises: (1) banded granodioritic to monzogranitic orthogneiss (Fig. 5); (2) an overlying clastic sedimentary rock sequence with quartzite and banded iron formation, and dominantly mafic

Table 1. *Eastern Canada–West Greenland correlation table*

Eastern Canada		West Greenland	
Unit	Age*	Unit	Age*
Rae craton		*Rae craton*	
Felsic plutons	c. 2.73 Ga to 2658 +16/−14 Ma	Kap York metaigneous complex	c. 2.7 Ga
		Melville Bugt orthogneiss complex	c. 2.7 Ga
Prince Albert Group	2732 +8/−7 to 2718 +5/−3 Ma	Lauge Koch Kyst supracrustal complex	c. 2.7–2.6 Ga
Mary River Group	c. 2829 Ma	Supracrustal rocks on Nuussuaq	2847 ± 4 Ma
Felsic orthogneiss	c. 2.9–2.82 Ga	Torsukattak–Ataa Sund orthogneiss	2835 ± 4 to 2758 ± 2 Ma
		Inussulik Bugt orthogneiss	c. 2.82–2.57 Ga
		Thule mixed gneiss complex	c. 2.91 Ga
		Nuussuaq orthogneiss	2947 ± 23 Ma
		Nuussuaq diorite	3030 +8/−5 Ma
Ellesmere–Devon terrane		*Inglefield mobile belt*	
Plutonic rocks	1960 ± 5 to 1912 ± 2 Ma	Etah metaigneous complex	1949 ± 13 to 1915 ± 19 Ma
Supracrustal rocks		Etah Group	c. 1.98 Ga to 1949 ± 13 Ma
Piling and Hoare Bay groups		*Karrat and Anap nunâ groups*	
Longstaff Bluff Fm	≤1915 ± 8 Ma	Nûkavsak Fm	≤1946 ± 13 Ma
Astarte River Fm	≤1935 ± 25 Ma		
Bravo Lake Fm	1935 ± 25 Ma	Mafic volcanic member	≤c. 2150 Ma
Flint Lake Fm	≤2159 ± 16 Ma	Mârmorilik Fm	≤c. 2150 Ma
Dewar Lakes Fm	≤2159 ± 16 Ma	Qeqertarssuaq Fm	≤c. 1.89 Ga
		Upper Anap nunâ Group	≤c. 1.93 Ga
Cumberland batholith	1865 +4/−2 to 1848 ± 2 Ma	Lower Anap nunâ Group	
		Prøven igneous complex	1869 ± 9 Ma
Meta Incognita microcontinent		*Aasiaat domain*	
Lake Harbour Group	≤1934 ± 2 to c. 1880 Ma	Nunatarsuaq supracrustal rocks	≤c. 1.89 Ga
		Naternaq supracrustal belt	≤1904 ± 8 Ma
		Nordre Strømfjord supracrustal suite	c. 1.95–1.92 Ga
Crystalline basement	3019 ± 5 to 2784 ± 9 Ma	Orthogneiss	c. 2.87–2.75 Ga
Burwell arc	1910 ± 2 to 1869 +3/−2 Ma	Sisimiut charnockite suite	1921 ± 10 to 1873 +7/−4 Ma
		Arfersiorfik quartz diorite	1921 ± 15 to 1885 +6/−3 Ma
Nain craton†		*North Atlantic craton*†	
Ramah, Mugford and Snyder groups	≤2.0 Ga	Maligiaq supracrustal suite	≤2.1 Ga
Basaltic dykes	c. 2.2–2.0 Ga	Kangâmiut dyke swarm	2.05–2.04 Ga
(Hopedale block)		Tre Brødre terrane	c. 2.83–2.75 Ga

(*Continued*)

Table 1. *Continued*

Eastern Canada		West Greenland	
Unit	Age*	Unit	Age*
Kanairiktok plutonic suite	c. 2.89–2.82 Ga	Tasiusarsuaq terrane	c. 2.92–2.86 Ga
Supracrustal belts	c. 3.1–2.98 Ga	Kapisilik terrane	c. 3.075–2.96 Ga
Maggo Gneiss (Saglek block)	c. 3.3 Ga to 3105 +6/−9 Ma	Akia terrane	c. 3.22 Ga, 3.07–2.97 Ga
Upernavik supracrustal rocks		Færingehavn and Isukasia terranes	c. 3.8–3.5 Ga
Uivak gneisses	c. 3.7–3.3 Ga	Aasivik terrane	c. 3.78–3.55 Ga
Supracrustal (Nulliak) and plutonic (Nanok) remnants	>3.9 Ga		
Makkovik orogen		*Ketilidian orogen*	
Aillik Group	c. 1860–1807 Ma	Julianehåb batholith	1854–1795 Ma
Island Harbour Bay plutonic suite	1893 ± 2 to 1870 Ma		
Metasedimentary formation	Younger than 2013 ± 3 Ma		
Moran Lake and Post Hill groups	2178 ± 4 Ma	Vallen and Sortis groups	<2130 ± 65 Ma
Kikkertavak mafic dykes	2235 ± 2 Ma	Iggavik mafic dykes	2130 ± 65 Ma

*References for ages and age-dating methods are specified in the text.
†No correlation proposed at terrane level.

Fig. 5. Archaean granodioritic to monzogranitic orthogneiss, Rae craton, central Baffin Island. Length of hammer is 35 cm.

tholeiitic, subordinate intermediate, and minor felsic metavolcanic rocks (Mary River Group); (3) a second volcanic-dominated supracrustal sequence (Prince Albert Group) comprising tholeiitic and isotopically depleted amphibolite (including komatiite), quartzite, and calc-alkaline mafic and intermediate metavolcanic rocks; and (4) younger granodioritic to monzogranitic and rare tonalitic calc-alkaline plutonic rocks (Scott & de Kemp 1998; Jackson 2000; Jackson & Berman 2000; Bethune & Scammell 2003; Scott et al. 2003; Johns & Young 2006; St-Onge et al. 2006b; Young et al. 2007). The felsic orthogneiss is c. 2.9–2.78 Ga in age, including a 2868 +13/−12 Ma age determination and the metavolcanic rocks of the Mary River Group have been dated at c. 2829 Ma (c. 3.3–2.9 Ga Sm–Nd model ages). The Prince Albert Group yields ages that cluster between 2732 +8/−7 and 2718 +5/−3 Ma (c. 2.85–2.7 Ga Sm–Nd model ages) and the younger felsic plutons range in age between c. 2.73 Ga and 2658 +16/−14 Ma (Jackson et al. 1990; Jackson 2000; Wodicka et al. 2002b, 2007a; Bethune & Scammell 2003; Johns & Young 2006; Young et al. 2007).

In northern West Greenland, the coastal areas of Inglefield Bredning, Kap York, and Melville Bugt (Rae craton; Fig. 3) are underlain by tonalitic to granitic orthogneiss and quartzofeldspathic paragneiss of the Thule mixed gneiss complex (c. 2.91 Ga Sm–Nd model age), gabbro–tonalite–granite units of the Kap York metaigneous complex (c. 2.7 Ga Rb–Sr age), and tonalitic to granitic gneiss and granite of the Melville Bugt orthogneiss complex (c. 2.7 Ga Rb–Sr age) (Dawes 1991, 2006; and references therein). Based on lithological characteristics and associations, the Archaean, dominantly felsic, metaigneous units of the Rae craton in Greenland are correlated with the metaplutonic units of northern Baffin Island (Hoffman 1988). Along the north shore of Melville Bugt, c. 2.7–2.6 Ga (Rb–Sr and U–Pb ages; Dawes et al. 1988) quartzofeldspathic to pelitic paragneiss and schist (including ironstone), magnetite-bearing quartzite, mafic schist, amphibolite, and ultramafic rocks of the Lauge Koch Kyst supracrustal complex (Dawes 1991, 2006) are similar to the Mary River Group (Jackson 2000) and/or Prince Albert Group of northern Baffin Island (Young et al. 2007) and Melville Peninsula (Scott & de Kemp 1998), thus providing an additional basis for the correlation of units belonging to the Rae craton between West and northern West Greenland and northern Baffin Island (Figs 2 and 3).

Palaeoproterozoic northern margin of the Rae craton

The Precambrian basement of southeastern Ellesmere Island and eastern Devon Island comprises highly deformed, granulite facies, metasedimentary and metaigneous rocks of the Ellesmere–Devon terrane (Frisch 1988), which are overlain by the unmetamorphosed strata of the late Mesoproterozoic Thule basin and early Palaeozoic Franklinian basin (Fig. 2). The metasedimentary rocks of the Ellesmere–Devon terrane include quartzofeldspathic gneiss, migmatitic pelitic gneiss, marble and quartzite that are interpreted to represent a metamorphosed continental margin sequence of shale, greywacke, shallow-water carbonate and volcanogenic rocks (Frisch 1988). The plutonic rocks mainly comprise pyroxene-bearing tonalite, quartz norite and several varieties of granite, including peraluminous S-type granite. Lithological and gneissic trends are predominantly north–south on Ellesmere Island but east–west on Devon Island (Frisch 1988). Age determinations (Frisch & Hunt 1988) yield 1960 ± 5 to 1912 ± 2 Ma for the metaplutonic rocks and indicate the presence of Archaean crust on southernmost Devon Island (Fig. 2; north margin of Rae craton?).

Polydeformed and granulite facies rocks of the Inglefield mobile belt in northern West Greenland (Fig. 3; Frisch & Dawes 1982; Dawes 1988; Dawes & Garde 2004) can be divided into two main assemblages (Dawes et al. 2000; Dawes 2004); namely, high-grade Palaeoproterozoic paragneiss (Etah Group) and a polyphase igneous suite (Etah metaigneous complex) that intrudes the Etah Group on all scales. Recently, Nutman et al. (2008) described the mobile belt as also comprising a northern and a southern part, with the two separated by a late shear zone that they called the Sunrise Pynt straight belt.

The Etah Group comprises pelitic to quartzofeldspathic paragneiss and schist, as well as voluminous S-type granite, marble and calc-silicate, amphibolite, and ultramafic units. The Etah metaigneous complex is composed of intermediate to felsic orthogneiss, megacrystic monzogranite, quartz diorite, syenite, and subordinate metagabbro and magnetite-rich rocks. The available geochronological data are complex (Nutman et al. 2008), but regional subsidence and deposition of the Etah Group has been bracketed between c. 1.98 and 1.95 Ga, with the age of the Etah metaigneous complex constrained between 1949 ± 13 and 1915 ± 19 Ma by ion probe U–Pb ages on zircon. Correlations between the metasedimentary rocks of the Ellesmere–Devon terrane and the Etah Group of the Inglefield mobile belt, as well as between the metaplutonic rocks of Ellesmere and Devon Islands and those of Inglefield Land, were made by Frisch & Dawes (1982), Dawes (1988), and Dawes et al. (1988), based on similarities in lithology, metamorphic grade, and tectonic history.

Palaeoproterozoic southern margin of the Rae craton

On central Baffin Island, the southern margin of the Rae craton is unconformably overlain by the Palaeoproterozoic Piling Group (Morgan et al. 1975, 1976; Scott & de Kemp 1998; Scott et al. 2003), as well as by the stratigraphically correlative Hoare Bay Group on Cumberland Peninsula (Fig. 2; Jackson & Taylor 1972; St-Onge et al. 2006b). In the type Piling Group area of west–central Baffin Island, the stratigraphically south-facing continental margin sequence comprises (Scott et al. 2003): (1) shallow marine, continental margin clastic (Dewar Lakes Fm) and carbonate platform (Flint Lake Fm; Fig. 6) strata (younger than 2159 ± 16 Ma); (2) stratigraphically younger mafic intrusive, extrusive, and sedimentary rift units (younger than 1980 ± 11 Ma, with a feeder dyke of intermediate composition yielding an age of 1935 ± 25 Ma) (Bravo Lake Fm) and rusty-weathering sulphide schist, black shale, and sulphide-facies iron formation (Astarte River Fm) associated with foundering of the carbonate platform; and (3) foredeep turbidites younger than 1915 ± 8 Ma (Longstaff Bluff Fm). Analysed sedimentary strata are dominated by 3.61–2.72 Ga detrital zircons (lower and middle Dewar Lakes Fm), 3.02–2.16 Ga detrital zircons (upper Dewar Lakes Fm), 3.31–1.98 Ga detrital zircons (Bravo Lake Fm), and 2.95–1.91 Ga detrital zircons with a preponderance of 2.03–1.91 Ga ages (Longstaff Bluff Fm). The above geochronological constraints on the deposition of the Piling Group are based on ion probe U–Pb ages on detrital zircon in clastic rocks and igneous zircon in a volcanic dyke from Wodicka et al. (2007b). The interbedded mafic volcanic and sedimentary rift units of the Bravo Lake Formation occur at the interface between the lower clastic unit (Dewar Lakes Fm) and the overlying black shale and turbidites (Astarte River and Longstaff Bluff Fms), where the marble of the Flint Lake Formation is not present (Scott et al. 2003).

Various felsic plutonic rocks, ranging in age from $1897 +7/-4$ to $1823 +7/-4$ Ma (Wodicka et al. 2002b; Bethune & Scammell 2003) and including the northernmost components of the dominantly hypersthene-bearing $1865 +4/-2$ to 1848 ± 2 Ma Cumberland batholith (Jackson et al. 1990; Wodicka & Scott 1997; Scott & Wodicka 1998; Scott 1999), intrude the southern and western strata of the Piling and Hoare Bay groups, respectively,

Fig. 6. Flat-lying Palaeoproterozoic carbonate strata of the Flint Lake Formation, lower Piling Group, central Baffin Island. Front of helicopter for scale.

on Baffin Island. The Cumberland batholith has been interpreted as the base of a continental margin arc emplaced above a north-dipping subduction system following accretion of the Meta Incognita microcontinent to the Rae craton (see below; Thériault et al. 2001; St-Onge et al. 2006c, 2007), although this conflicts with the interpretation of the contemporaneous Prøven igneous complex in adjacent northern West Greenland (see below; Thrane et al. 2005).

Palaeoproterozoic units in central and northern West Greenland have traditionally been considered as belonging to two distinct belts, the Rinkian fold belt in the north and the Nagssugtoqidian orogen farther south (Fig. 3), largely because of their contrasting structural styles (see below; Pulvertaft 1973; Escher & Pulvertaft 1976; Escher et al. 1976b; Grocott & Pulvertaft 1990). Following a recent integration of fieldwork in 2001 to 2003 by Adam Garde and coworkers with previously published field observations, new U–Pb geochronology from the intervening region around Disko Bugt (Fig. 3), and regional tectonostructural considerations, the two belts are now considered to represent the northern and southern parts of a single, composite orogen more than 1100 km wide resulting from the collision of the Rae craton with a medial Aasiaat domain (microcontinental fragment?) of Archaean age and the North Atlantic craton to the south (van Gool et al. 2002; Garde et al. 2003, 2007; Thrane et al. 2003; Connelly & Thrane 2005; Connelly et al. 2006; Sidgren et al. 2006).

In the Inussulik Bugt to Uummannaq Fjord area of the Rinkian fold belt in northern West Greenland (Fig. 3), c. 2.86–2.57 Ga (Rb–Sr and U–Pb ages; Kalsbeek 1981, 1986; Andersen & Pulvertaft 1985) tonalitic to granodioritic gneiss of the Rae craton (Fig. 7) is overlain unconformably by the metamorphosed supracrustal units of the Palaeoproterozoic Karrat Group (Garde 1978; Escher 1985; Grocott & Pulvertaft 1990). The Karrat Group in the type Karrat Fjord area (Henderson & Pulvertaft 1967, 1987) comprises a lower sequence of shallow marine, shelf-type metasedimentary rocks, including quartzite, semipelitic to pelitic gneiss and schist (Qeqertarssuaq Fm), and marble (Mârmorilik Fm), overlain by black shale and a thick, uniform, arenaceous turbidite flysch sequence (Nûkavsak Fm; Fig. 8) that is younger than 1946 ± 13 Ma (ion probe U–Pb age of youngest detrital zircon; Kalsbeek et al. 1998). Samples of quartzite from the lower Qeqertarssuaq Formation yield detrital zircons up to 3.65 Ga (laser ablation inductively coupled plasma mass spectrometry (LA-ICP-MS) analyses, Thrane et al. 2003), whereas a sample of metagreywacke from the Nûkavsak Formation

Fig. 7. Recumbent isoclinal folds of Archaean orthogneiss (pink) and thin tectonic panels of Palaeoproterozoic supracrustal rocks (black), Uummannaq Island, Rinkian fold belt, West Greenland. Cliff is c. 1200 m high.

analysed by Kalsbeek *et al.* (1998) has yielded 3.04–1.95 Ga detrital zircons with a preponderance of 2.10–1.95 Ga ages (ion probe U–Pb ages). A maximum age for deposition of the Qeqertarssuaq and Mârmorilik formations is constrained at c. ≤ 2150 Ma (K. Thrane, pers. comm.). A mafic volcanic member comprising both pillowed lava flows and pyroclastic rocks is found at the interface between the lower clastic and marble units (Qeqertarssuaq and Mârmorilik Fms) and the overlying turbidites (Nûkavsak Fm) (Grocott & Pulvertaft 1990; Thomassen 1992). Based on the similarity in distribution and association of the sedimentary and volcanic lithologies, stratigraphic sequences, overall tectonic context, and recent detrital zircon analytical data (Kalsbeek *et al.* 1998; Thrane *et al.* 2003; Wodicka *et al.* 2007*b*), and following previous suggestions, the correlation between the Piling–Hoare Bay groups of central Baffin Island and the Karrat Group of West Greenland is retained (Figs 2 and 3; Jackson & Taylor 1972; Escher & Pulvertaft 1976; Taylor 1982; Henderson & Pulvertaft 1987).

South of Uummannaq Fjord, the peninsula of Nuussuaq (Fig. 3) is underlain by Archaean orthogneiss (2947 ± 23 Ma; LA-ICP-MS Pb–Pb age), diorite (3030 + 8/−5 Ma), a large anorthosite complex, and metavolcanic-dominated supracrustal sequences (2847 ± 4 Ma) (Garde & Steenfelt 1989, 1999*a*; with ages from Connelly *et al.* 2006). Low-grade Palaeoproterozoic cover rocks in the northern part of the peninsula have been correlated with the Karrat Group to the north (Garde & Steenfelt 1999*a*).

South of Nuussuaq, supracrustal units also correlative with the Karrat Group crop out in the Torsukattak–Ataa Sund area (Fig. 3; Anap nunâ Group of Escher & Burri 1967; Escher 1971; see updated correlation by Garde & Steenfelt 1999*a*; Higgins & Soper 1999). The Anap nunâ Group consists of platform and tidal flat sedimentary rocks that include basal, cross-bedded and ripple-marked mature quartz sandstone overlain by marble that is capped by tidal and deeper-water siltstone and fine-grained sandstone (Garde & Steenfelt 1999*a*). A sample collected near the base of the Anap nunâ Group has 3.22–1.93 Ga detrital zircons with a preponderance of 2.96–2.46 Ga ages (LA-ICP-MS Pb–Pb, Connelly *et al.* 2006). In contrast, another sample collected by the same workers from the upper part of the sequence yielded 2.94–1.89 Ga detrital zircon ages, with a dominant component

Fig. 8. Palaeoproterozoic graded metaturbidite of the Nûkavsak Fm, upper Karrat Group, Rinkian fold belt, West Greenland. Lower amphibolite facies, with abundant andalusite porphyroblasts visible in the tops of the graded beds. Coin is 2.8 cm in diameter.

between 2.10 and 1.89 Ga. The clastic and carbonaceous supracrustal strata unconformably overlie 2835 ± 4 to 2758 ± 2 Ma orthogneiss and supracrustal rocks (ion probe U–Pb ages, Nutman & Kalsbeek 1999, see also Connelly et al. 2006).

In the northern part of the southern Nagssugtoqidian orogen (Fig. 3), Palaeoproterozoic psammitic and pelitic schist and gneiss, with banded iron formation, metagreywacke, fine-grained metavolcanic rocks, and marble (Nunatarsuaq supracrustal rocks of Garde & Steenfelt 1999a; Naternaq supracrustal belt of Østergaard et al. 2002; Thrane & Connelly 2006; Nordre Strømfjord supracrustal suite of Marker et al. 1999) are dominated by 2.20–1.95 Ga detrital zircons with very little to no Archaean input (ion probe U–Pb and LA-ICP-MS Pb–Pb ages; Scott et al. 1998; Nutman et al. 1999; Connelly et al. 2006; Thrane & Connelly 2006). Deposition of the Nunatarsuaq supracrustal rocks is constrained to being in part younger than c. 1.89 Ga (Connelly et al. 2006). Accumulation of the Naternaq supracrustal belt is constrained to younger than 1904 ± 8 Ma (ion probe U–Pb age; Thrane & Connelly 2006), whereas deposition of the Nordre Strømfjord supracrustal suite is bracketed between 1.95 and 1.92 Ga (van Gool et al. 2002). These restricted (dominantly Palaeoproterozoic) age ranges for detrital zircons are in contrast to those of the Karrat Group (and correlative Anap nunâ Group) to the north. The composite lithostratigraphies of these assemblages also appear to be different. Whereas the former belts are dominated by mafic to ultramafic metavolcanic and pelitic rocks indicative of rifting (and possibly deep marine) environments (Garde & Steenfelt 1999a; Marker et al. 1999; Østergaard et al. 2002), the Karrat and Anap nunâ groups largely consist of shelf-type metasedimentary rocks deposited unconformably on Archaean basement and draped by a thick turbidite flysch sequence (Henderson & Pulvertaft 1967, 1987).

The present correlation of supracrustal units within West Greenland is critical and needs to be further tested because, in conjunction with structural and isotopic studies, it may provide the best method for determining the southern extent of the Rae craton versus the occurrence of separate cratonic domains to the south. That in turn will further constrain the location of the crustal suture (or sutures) that necessarily separate(s) the Rae craton from the North Atlantic craton to the south within the tectonic framework of the Rinkian fold belt and Nagssugtoqidian orogen (van Gool et al. 2002; Connelly et al. 2006; Garde et al. 2007).

Based on geometric, structural, and isotopic data and arguments presented by van Gool et al. (2002), Connelly & Thrane (2005), and Connelly et al. (2006), two south-dipping sutures are shown separating the Rae craton from the North Atlantic craton in Figure 3 (see also Garde et al. 2007). The northern suture, labelled 'Disko Bugt suture', separates the Anap nunâ Group and underlying Archaean basement in the north from the Nunatarsuaq supracrustal rocks and underlying basement units in the south. The southern suture roots in the Nordre Isortoq steep belt (van Gool et al. 1996, 2002) after Kalsbeek et al. (1987) and separates the North Atlantic craton in the south (see below) from a small cratonic domain to the north, which was heated but not deformed by the Nagssugtoqidian orogeny. For ease of reference and to avoid confusion with previous designations, we utilize the term 'Aasiaat domain' for the area bound by the two sutures in Figure 3. The Aasiaat domain thus comprises part of the northern and north–central Nagssugtoqidian domains of van Gool et al. (2002), the Nunatarsuaq and southern Rodebay domains of Garde & Steenfelt (1999a), and the southern domain or block of Connelly & Thrane (2005) and Connelly et al. (2006). Lastly, the recent discovery of tholeiitic metavolcanic rocks, pelitic metasedimentary rocks, chert and banded iron formation on islands north of Aasiaat (southern Disko Bugt, Fig. 3), which are lithologically similar to the Naternaq supracrustal belt but much better preserved, would corroborate the idea that the transitional area east and south of Disko Bugt, between the Rinkian fold belt and the central Nagssugtoqidian orogen, may contain one or more microcontinents, as well as relicts of intervening pieces of oceanic crust including deformed and metamorphosed mafic pillow lava and pelagic sedimentary strata (Garde et al. 2007).

In the Tasiussaq Bugt area (Fig. 3), metasedimentary rocks of the Karrat Group are intruded by K-feldspar megacrystic, hypersthene-bearing, granite of the Prøven igneous complex, which has recently been dated at 1869 ± 9 Ma (Thrane et al. 2005). A-type geochemical signatures and Sm–Nd, Lu–Hf and Rb–Sr isotopic compositions show that it is a crustal melt derived largely from Archaean continental crust (Kalsbeek 1981; Thrane et al. 2005). Thrane et al. (2005) also suggested that the Prøven igenous complex formed in response to massive underplating related to collisionally induced delamination of the overthickened upper lithosphere at an early stage of the Nagssugtoqidian–Rinkian orogenic collision.

Similarities in major element geochemistry of the constituent plutonic rock types, the stratigraphy of the encasing host-rocks, and available geochronology suggest that the Prøven igneous complex may be viewed as an eastern correlative to the Cumberland batholith of Baffin Island (Hoffman 1990a; Figs 2 and 3). However, geochemical and isotopic compositions of the complex preclude it from being interpreted as a continental margin arc, as has been suggested for the Cumberland batholith by Thériault et al. (2001) and St-Onge et al. (2006c, 2007).

Archaean to Palaeoproterozoic Meta Incognita microcontinent

Archaean and Palaeoproterozoic units of the Meta Incognita microcontinent (Fig. 2; St-Onge et al. 2000) comprise (Scott et al. 1997; St-Onge et al. 2002; Sanborn-Barrie et al. 2009): (1) 3019 ± 5 to 2784 ± 9 Ma (Scott 1998, 1999, pers. comm.; N. Rayner, pers. comm.; N. Wodicka, pers. comm.) crystalline basement exposed in the Foxe and Hall Peninsulas (Fig. 9); (2) an overlying succession of relatively homogeneous arenaceous rocks (Lona Bay sequence and Blandford Bay assemblage); (3) a heterogeneous volcanic-bearing supracrustal assemblage (Schooner Harbour sequence) of mainly basaltic composition and including lapilli tuff and variolitic units; (4) a stratigraphically north-facing 1934 ± 2 to c. 1880 Ma (Scott 1997; Scott et al. 2002), continental margin shelf and foredeep succession (Fig. 10; Lake Harbour Group); and (5) an extensive suite of $1865 +4/-2$ to 1848 ± 2 Ma (Jackson et al. 1990; Wodicka & Scott 1997; Scott & Wodicka 1998; Scott 1999) continental margin quartz diorite to monzogranitic arc plutons (southern portion of the Cumberland batholith) that intrude units (1) to (4). At this point, it remains unclear whether the Meta Incognita microcontinent was initially rifted from the Superior craton (described below) as suggested by St-Onge et al. (2000), whether it constitutes a rifted fragment of the Rae craton, or whether it represents crust that is exotic with respect to both bounding cratons.

As suggested by Jackson & Taylor (1972), supracrustal rocks of the Lake Harbour Group are correlated with similar-named supracrustal sequences on the southeastern side of Ungava Bay, based on lithological similarities (Fig. 4). St-Onge et al. (2002) utilized residual total magnetic field signatures to further strengthen the correlation and project the boundaries of the Meta Incognita microcontinent from Baffin Island to the SE shore of Ungava Bay. In this correlation, the distinctive aluminous Tasiuyak paragneiss of eastern Labrador and northeastern Quebec (Fig. 4; Wardle 1983) may represent a lateral, deeper water equivalent of

Fig. 9. Disrupted and folded Archaean tonalitic to granodioritic gneiss, with late monzogranitic seams, Meta Incognita microcontinent, Foxe Peninsula, Baffin Island. Pen is 15 cm long.

the shelf-facies Lake Harbour Group (Goulet & Ciesielski 1990). The Lac Lomier complex (Fig. 4; Ermanovics & Van Kranendonk 1998; Wardle et al. 2002) may correspond to the southern extension of the Cumberland batholith (R. Wardle 2005, pers. comm.). Detrital zircons from both the Lake Harbour Group and the Tasiuyak paragneiss show a similar Palaeoproterozoic-dominated source (2.10–1.94 Ga, Scott & Gauthier 1996; Scott 1997; Scott et al. 2002) and distinct REE and Nd-isotopic signature (Thériault et al. 2001), supporting suggestions based on lithological similarities and tectonostratigraphic context that these units may have been part of a single depositional system developed along the northeastern and eastern side of the microcontinent (see also Van Kranendonk et al. 1993; Scott 1998). Alternatively, Wardle & Van Kranendonk (1996) utilized the relatively juvenile isotopic character of the Tasiuyak paragneiss to suggest deposition within an accretionary prism environment (see also Rivers et al. 1996), associated with the Burwell arc (see below) subduction system (Van Kranendonk & Wardle 1997). More recently, Wardle et al. (2002) proposed accumulation in an oceanic arc setting in an attempt to reconcile the juvenile nature of the detrital zircon population and the Archaean input suggested by the Nd-isotopic signature of the Tasiuyak gneiss.

Archaean and Palaeoproterozoic units correlative to those of the Meta Incognita microcontinent possibly occur within central West Greenland (Aasiaat domain in Fig. 3; Scott 1999; Hollis et al. 2006b; Thrane & Connelly 2006). However, the pursuit of a definite correlation would require further field-based research, as well as geochronological and geochemical data on basement and cover units of the Hall Peninsula on eastern Baffin Island, to link central West Greenland with southern Baffin Island and the southeastern side of Ungava Bay.

Archaean North Atlantic craton

The Archaean North Atlantic craton of eastern Canada, Greenland and NW Scotland (Sutton et al. 1972; Bridgwater et al. 1973b; Bridgwater & Schiøtte 1991; Wasteneys et al. 1996) is bounded to the north and west by segments of Palaeoproterozoic orogenic belts that are tectonically related to the accretional–collisional Trans-Hudson orogen,

Fig. 10. Tonalitic gneiss (background) tectonically overlying Palaeoproterozoic marble of the middle Lake Harbour Group in the foreground, Meta Incognita microcontinent, southern Baffin Island. Cliff in background is 250 m high.

including the Nagssugtoqidian orogen and Rinkian fold belt on the north side and the Torngat orogen on the west side of the craton (Figs 3 and 4). On its south side, the craton is bounded by the Palaeoproterozoic accretionary Makkovik–Ketilidian orogens (Figs 3 and 4).

In northern Labrador, the North Atlantic craton (known in Labrador as the Nain craton or Nain Province; Stockwell 1963; Taylor 1971, 1972, 1977) comprises the isotopically distinct Saglek block in the north and Hopedale block in the south. These two crustal blocks are separated by Mesoproterozoic plutonic suites (Fig. 4), which include the Nain plutonic suite dated at 1343 ± 3 to c. 1295 Ma (Fig. 4; Simmons et al. 1986; Connelly & Ryan 1994). The two Archaean blocks in Labrador are characterized predominantly by deeply exhumed, early to late Archaean orthogneiss, greenstone belts, and late Archaean granitoid intrusions. The southern Hopedale block contains c. 3.3 Ga to 3105 +6/−9 Ma (Loveridge et al. 1987; Finn 1989; James et al. 2002), high-grade tonalite, granodiorite and granite orthogneiss (Maggo Gneiss; Ermanovics 1993), c. 3.1–2.98 Ga (James et al. 1998) supracrustal belts dominated by mafic metavolcanic rocks and local occurrences of metasedimentary rocks, and c. 2.89–2.82 Ga (Wasteneys et al. 1994, 1996) tonalitic to granitic plutonic rocks (Kanairiktok plutonic suite). North of the Mesoproterozoic plutonic suites, the Saglek block (Bridgwater et al. 1976; Bridgwater & Schiøtte 1991) comprises c. 3.7–3.3 Ga tonalitic to granodioritic metaplutonic rocks (Uivak gneisses; Schiøtte et al. 1989a; Collerson et al. 1991; Collerson & Regelous 1995; Wasteneys et al. 1996), subordinate >3.9 Ga supracrustal (Nulliak) and plutonic (Nanok) remnants or inclusions, and metasedimentary and metavolcanic rocks (Upernavik supracrustal rocks; Schiøtte et al. 1989a, b, 1992), and deformed gabbro–dolerite intrusions and dykes (Saglek dykes). It is assumed that the boundary between the Hopedale and Saglek blocks is tectonic, and that the amalgamation of the two blocks occurred at c. 2.7 Ga following distinct magmatic and tectonic histories for each block (James et al. 2002, and references therein).

In southern West Greenland, recent and current detailed studies of the Archaean magmatic and tectonic evolution in the greater Nuuk region (Fig. 3) have documented the complex and composite nature of this segment of Archaean crust (Friend et al. 1988, 1996; Nutman et al. 1989, 2005; Garde et al. 2000; Crowley 2002; Friend & Nutman 2005; Hollis et al. 2006a; Garde 2007a; van Gool et al. 2007; Hölttä et al. 2008). These studies suggest that the North Atlantic craton in Greenland comprises a still unknown number of distinct tectonostratigraphic terranes or microplates, varying in size from tens of kilometres to at least a couple of hundred kilometres, and having independent histories prior to amalgamation in the late Archaean eon at c. 2.7 Ga (e.g. Friend & Nutman 2005; overview by Garde 2003; Hollis et al. 2006a; and references therein). The northwestern Akia terrane consists of a c. 3.2 Ga core of mafic tonalitic–dioritic gneiss, surrounded by mafic supracrustal rocks including a c. 3071 ± 1 Ma disrupted oceanic island arc complex (Garde 2007a) and younger orthogneiss (3.05–2.97 Ga) metamorphosed at amphibolite- to granulite-facies conditions at c. 2.98 Ga. Several complexly interfolded terranes occur SE of the Akia terrane. Two of these, the Færingehavn and Isukasia terranes, contain at least two different groups of c. 3.8–3.5 Ga supracrustal and plutonic rocks, including the well-known Isua supracrustal belt (Nutman et al. 1996, 2000; Furnes et al. 2007), as well as c. 2.8 Ga orthogneiss. The 3.075–2.96 Ga Kapisilik terrane (Friend & Nutman 2005) is contemporaneous with the Akia terrane and may be regarded as having been rafted off the latter. The Tre Brødre terrane is dominated by uniform, c. 2.82 Ga, largely granodioritic orthogneisses with a simple crustal history (Ikkattoq gneisses; Friend et al. 1988, 1996). The southern Tasiusarsuaq terrane largely comprises c. 2.92–2.86 Ga orthogneiss with c. 2.80 Ga granulite-grade metamorphism.

Archaean terranes outside the Nuuk region are poorly defined (van Gool et al. 2004). They mainly comprise 3.0–2.7 Ga rocks (Connelly & Mengel 2000; Friend & Nutman 2001), and the small 3.78–3.55 Ga Aasivik terrane (only orthogneiss, Rosing et al. 2001) near the Nagssugtoqidian front. Windley & Garde (2009) show that the entire North Atlantic craton in SW Greenland can be described in terms of six tilted crustal blocks, which preserve relics of volcanic arcs in their upper, greenschist to amphibolite facies parts (Garde 2007; Polat et al. 2007), and arc roots with anorthosite-gabbro complexes (e.g. the Fiskenæsset complex, Windley et al. 1973; Myers 1985) in their lower, amphibolite to granulite facies parts. Windley & Garde (2009) also reinterpreted three terranes in the Kvanefjord region (Friend & Nutman 2001) as a major refolded nappe structure.

Given an unexposed, pre-drift distance of c. 300–400 km, correlations of single blocks and terranes between the Canadian and Greenlandic components of the North Atlantic craton are considered highly speculative by van Gool et al. (2004) and are referenced here only for completeness. Friend & Nutman (1994) suggested a correlation between the Akia terrane in Greenland and the Hopedale block in Labrador, based on c. 3.0 Ga metamorphism and plutonism in both, whereas an unnamed terrane south of Kangerlussuaq–Søndre Strømfjord was correlated with the Saglek block, based on a common granulite facies metamorphism at c. 2.74 Ga. More recently, James et al. (2002) highlighted the significant difference in the timing of metamorphism of the Hopedale block compared with the Akia terrane, and proposed a correlation between the Hopedale block and the Tasiusarsuaq terrane, and the Saglek block with the Færingehavn and Isukasia terranes.

The correlation of the North Atlantic craton from Labrador to southern West Greenland follows that of Bridgwater et al. (1973b, 1990), Korstgård et al. (1987), Bridgwater & Shiøtte (1991), Garde et al. (2002), James et al. (2002), and van Gool et al. (2002, 2004). It is consistent with the dating of offshore well cores by Wasteneys et al. (1996), which added additional constraints to the correlation of Precambrian bedrock units across the Labrador Sea.

Western and northern margins of the Archaean North Atlantic craton

The Torngat orogen of northwestern Labrador and northeastern Quebec, and the Nagssugtoqidian orogen of central West Greenland are Palaeoproterozoic collisional orogenic belts that developed respectively along the western and northern margins of the North Atlantic craton (Figs 3 and 4). The correlation of the two orogenic belts is based on a similarity in constituent lithotectonic units, coeval tectonic history including igneous, structural and metamorphic events, aeromagnetic data, and complementary kinematics, as proposed and documented by Bridgwater et al. (1973a, 1990), Korstgård et al. (1987), Hoffman (1990a), Van Kranendonk et al. (1993), Park (1994), Wardle & Van Kranendonk (1996), Connelly et al. (2000), van Gool et al. (2002, 2004), and Wardle et al. (2002).

In northern Labrador, basaltic dykes indicate that rifting along the western margin of the North Atlantic craton occurred at c. 2.2–2.0 Ga, and was accompanied by the contemporaneous emplacement of anorthosite–granite suites at 2.1–2.0 Ga (Connelly & Ryan 1994, 1999; Wardle & Van Kranendonk 1996; Hamilton et al. 1998). Along the eastern coast of Labrador, these rocks are unconformably overlain by Palaeoproterozoic rocks of the Ramah, Mugford, and Snyder groups (Fig. 4), each representing an upward progression from shallow- to deep-water environments (Wardle et al. 2002) and dominated by detritus of Archaean provenance in the lower part of the sequence (Scott & Gauthier 1996). To the west, the Tasiuyak gneiss (see above) comprises migmatitic pelitic and psammitic rocks, inferred to be metaturbidites, with minor mafic and ultramafic material, and as noted interpreted either as continental slope deposits off the eastern margin of the orogenic core zone to the west (i.e. deeper-water equivalent to the platformal Lake Harbour Group of Meta Incognita microcontinent) or as a distal accretionary wedge sequence that accumulated on the western margin of the North Atlantic craton.

Along the northern margin of the North Atlantic craton within the southern Nagssugtoqidian orogen, Archaean dioritic, granodioritic, tonalitic and granitic orthogneiss of the North Atlantic craton predominate, albeit in a reworked state (Fig. 11; Kalsbeek et al. 1984, 1987; Kalsbeek & Nutman 1996; Connelly & Mengel 2000; van Gool et al. 2002). The orthogneiss are generally 2.87–2.81 Ga in age, with the plutonic protoliths deformed and metamorphosed between 2.81 and 2.72 Ga, immediately after their emplacement (Kalsbeek & Nutman 1996; Connelly & Mengel 2000). The orthogneiss are cut by several sets of mafic dykes, the most voluminous of which is the north–south- to NE–SW- trending, 2.05–2.04 Ga Kangâmiut dyke swarm (Windley 1970; Escher et al. 1975, 1976a; Korstgård 1979; Bridgwater et al. 1995; Nutman et al. 1999; Connelly et al. 2000; Cadman et al. 2001; Mayborn & Lesher 2006). The Kangâmiut dyke swarm, which extends from the Archaean foreland into the centre of the orogen, becomes reworked across an abrupt transition at the southern Nagssugtoqidian front (Fig. 12). Its northern extent is abruptly terminated by the Ikertôq thrust zone (Fig. 3), a south-vergent ductile shear zone. In this zone, metasedimentary rocks with a relatively high proportion of psammite interpreted as continental-margin deposits (Maligiaq supracrustal suite of Marker et al. 1999) are interleaved with Archaean orthogneiss (Fig. 13). The clastic strata have a large proportion of Archaean detrital zircons (c. 2.85 Ga) but also contain a Palaeoproterozoic population that constrains their deposition to after 2.1 Ga (Scott et al. 1998; Marker et al. 1999; Nutman et al. 1999). The Maligiaq supracrustal suite (Fig. 13) is interpreted to document the northern reaches of the North Atlantic craton (a rift basin or continental margin, comparable with the Ramah Group in northern Labrador; van Gool et al. 2002). The suite provides an ultimate southern constraint on the position of the Aasiaat domain–North Atlantic craton suture discussed above, with the Aasiaat domain sandwiched between the two main (northern Rae and southern North Atlantic) cratons in the Disko Bugt area.

Two overlapping phases of arc magmatism preceding the Nagssugtoqidian orogen (described

Fig. 11. Nagssugtoqidian reworking of older rocks in the northern Nagssugtoqidian orogen. Palaeoproterozoic mafic dykes were emplaced into grey Archaean orthogneiss during pre-Nagssugtoqidian rifting. The dykes were subsequently rotated and deformed with their Archaean host during the Nagssugtoqidian orogeny, and subsequently cut by postkinematic pegmatites. Aasiaat region, West Greenland. Boat for scale.

Fig. 12. Deformed and boudinaged dykes of the Palaeoproterozoic Kangâmiut swarm in the southern Nagssugtoqidian orogen, within Archaean orthogneiss of the North Atlantic craton, West Greenland. Dykes in the centre of the photograph are c. 30 m wide.

below) have been documented in the core of the belt. In the Arfersiorfik fjord and Nordre Strømfjord area (Fig. 3), quartz diorite preserved as tectonic slivers and a larger body (the 1921 ± 15 to 1885 +6/−3 Ma Arfersiorfik intrusive suite of calc-alkaline affinity; Kalsbeek et al. 1987; Kalsbeek & Nutman 1996; Whitehouse et al. 1998; Connelly et al. 2000; van Gool et al. 1999, 2002; Sørensen et al. 2006) was intruded into metavolcanic and metasedimentary rocks of the Nordre Strømfjord supracrustal suite. All these rocks are tectonically interleaved with reworked basement rocks of Aasiaat domain and/or North Atlantic craton parentage in the subsequent collisional phase (Fig. 14). The Arfersiorfik intrusive suite occurs south of the proposed northern suture running through Disko Bugt and its origin has been related to subduction of oceanic crust beneath the Aasiaat domain (Garde et al. 2007).

South of the proposed southern suture (Fig. 3; van Gool et al. 2002), calc-alkaline continental arc rocks of the 1921 ± 10 to 1873 +7/−4 Ma Sisimiut intrusive suite (equivalent to the Burwell arc in Labrador) (Kalsbeek et al. 1987; Kalsbeek & Nutman 1996; Connelly et al. 2000; Campbell & Bridgwater 1996; Whitehouse et al. 1998) were emplaced into Archaean ortho- and paragneisses, consistent with the proposed location of the second south-dipping suture (van Gool et al. 2002; Garde et al. 2007) and the interpretation of the North Atlantic craton as the local overriding plate within the Nagssugtoqidian collisional orogen (see below).

South margin of the Archaean North Atlantic craton

The Makkovik orogen in central Labrador and the Ketilidian orogen along strike in South Greenland together define the southern margin of the North Atlantic craton (Figs 3 and 4). This margin was initiated as a passive continental margin at c. 2.24–2.13 Ga and became the locus of subduction, arc magmatism and juvenile crustal accretion in an overall transpressional environment within a continental margin arc setting between c. 1.89 and 1.80 Ga (see Kerr et al. 1996; Culshaw et al. 2000a, b; Garde et al. 2002; Ketchum et al. 2002; and references therein). However, in spite of long being considered along-strike segments of the same orogenic belt, the correlation of Makkovikian

Fig. 13. Panel of Palaeoproterozoic metasedimentary rocks of the Maligiaq supracrustal suite (two buff-coloured zones on either side of a grey sheet of mafic supracrustal rocks) interleaved with Archaean orthogneiss (grey) within the Ikertoq thrust zone, West Greenland. Dark layers within the Archean gneiss are deformed Kangâmiut dykes. Cliff is c. 300 m high. Photograph provided by F. Mengel.

Fig. 14. Palaeoproterozoic metasedimentary rocks of the Nordre Strømfjord supracrustal suite (brown, layered middle section of the cliff) and orthogneiss of the Arfersiorfik intrusive suite (homogeneous grey top section) in tectonic contact with Archaean gneisses (white bottom section) in the Nordre Strømfjord region, West Greenland. Cliff is c. 150 m high.

and Ketilidian events is complex (Garde et al. 2002; Ketchum et al. 2002; and references therein) and hindered by the 300–400 km wide pre-late Cretaceous drift gap between mainland Greenland and Labrador.

The Makkovik orogen appears much narrower than the Ketilidian orogen, which includes a major batholith (see below). The apparent eastward widening of the Makkovik orogen continues offshore, as documented by data from offshore wells and seismic lines (Kerr et al. 1996, 1997; Wasteneys et al. 1996; Hall et al. 2002). Overall, the belt records c. 300–600 Ma of convergent continental margin activity, including the accretion of several arc and back-arc units, as well as possibly a small Archaean terrane, punctuated by evidence of intra-accretion quiescent periods (Gower et al. 1990; Kerr 1994; Kerr et al. 1997; Culshaw et al. 2000b; Ketchum et al. 2001a, b, 2002). In contrast, supracrustal and plutonic units within the Ketilidian orogen record c. 450 Ma of relatively continuous magmatic and tectonic activity, which does not seem to involve the accretion of allochthonous crustal terranes to the southern margin of the North Atlantic craton (Chadwick & Garde 1996; Garde et al. 1998, 2002).

The onset of rifting of the southern margin of the North Atlantic craton in Labrador is marked by the 2235 ± 2 Ma age of the Kikkertavak mafic dykes (Cadman et al. 1993; Ermanovics 1993). Deposition of overlying continental margin sedimentary strata (quartzite, iron formation, shale, dolostone and greywacke) and pillowed mafic volcanic rocks (Moran Lake, Aillik and Post Hill groups; Ketchum et al. 2002; and references therein) is constrained to post-date the age of the Kikkertavak dykes, and to have begun prior to 2178 ± 4 Ma, the age of an intermediate tuff layer within the Post Hill Group (Culshaw et al. 2000b; Ketchum et al. 2001b). Overlying micaceous psammite with minor pelite and graphitic paragneiss (Metasedimentary formation; Marten 1977) has yielded both Archaean and Palaeoproterozoic detrital zircons and was deposited after 2013 ± 3 Ma, possibly in a foredeep setting (Ketchum et al. 2001b).

In South Greenland, rifting of the North Atlantic craton is poorly constrained by an Rb–Sr age of 2130 ± 65 Ma (Kalsbeek & Taylor 1985) on doleritic dykes of the Iggavik suite (Berthelsen & Henriksen 1975) west of Midternæs (Fig. 3). The mafic dykes were emplaced into Archaean quartzofeldspathic orthogneiss, with deposition of continental margin sedimentary rocks (Vallen Group; Bondesen 1970; Higgins 1970) loosely constrained by this age. Shallow marine quartz-pebble conglomerate, quartzite, dolomite, mudstone, chert and a

Fig. 15. Palaeoproterozoic arkosic metasediment displaying boudinaged ultramafic dyke, folding and abundant partial melting. Ketilidian forearc, southern East Greenland. Coin near centre is 2.8 cm in diameter.

banded iron formation characterize the lower part of the group, whereas deeper marine greywacke predominates in the upper part (Fig. 15). The Vallen Group is structurally overlain by the possibly laterally equivalent Sortis Group (Bondesen 1970), which comprises metabasaltic pillow lava, pillow breccia and sills, intercalated with minor mudstone and calcareous rock. Alternatively, the Sortis Group may represent an obducted component of a rifted basin to the south of the North Atlantic craton (Garde et al. 2002). Unpublished U–Pb ages of detrital zircons from a Vallen Group quartzite point to a late Archaean basement provenance (Garde et al. 2002). The Vallen and Sortis groups have been generally correlated with, and have apparent tectonostratigraphic equivalents to, the Moran Lake and Post Hill groups of Labrador (Wardle & Bailey 1981).

Finally, it is uncertain if the onshore part of the Makkovik orogen preserves sedimentary sequences equivalent to those found in the classic psammite and pelite zones of the Ketilidian orogen (Garde et al. 2002; and references therein). Offshore, strongly reflective components of the ECSOOT seismic line (Kerr et al. 1997; Hall et al. 2002) might correspond to a westerly continuation of the psammite–pelite zones as illustrated in figure 3 of Garde et al. (2002).

Narsajuaq island-arc terrane

South of the Meta Incognita microcontinent (described above) and in the footwall of a crustal suture exposed along the southern coast of Baffin Island (Soper River suture, Fig. 2; see below), the Narsajuaq arc terrane (Dunphy & Ludden 1998) is exposed on both sides of Hudson Strait. At c. 1845 Ma, the arc formed the leading edge of the Churchill domain (St-Onge et al. 2007). It includes forearc siliciclastic rocks (Spartan Group), a dominantly volcanic sequence (Parent Group), and a dominantly plutonic assemblage (Narsajuaq arc). The volcanic and plutonic units of the arc terrane can be grouped into two temporally and petrologically distinct suites (Dunphy & Ludden 1998). An older 1863 ± 2 to 1845 ± 2 Ma suite includes calc-alkaline layered diorite–tonalite gneiss (Fig. 16) and tholeiitic to calc-alkaline basaltic andesite to rhyolite (St-Onge et al. 1992; Machado et al. 1993; R. Parrish 1994, pers. comm.). It is interpreted as an island-arc assemblage built on Palaeoproterozoic oceanic crust (Watts Group) and a rifted sliver of Archaean continental crust (Thériault et al. 2001). Accretion of the island-arc assemblage to the composite Churchill domain across the Soper River suture is constrained to have occurred at c. 1845 Ma (St-Onge et al. 2007; see below). The

Fig. 16. Palaeoproterozoic calc-alkaline diorite (dark)–tonalite (light) gneiss complex, older suite of Narsajuaq arc, northern Quebec. Hammer is 35 cm in length.

younger, 1842 +5/−3 to 1820 +4/−3 Ma suite comprises crosscutting, gneissic to massive, monzodiorite to granite plutons (Parrish 1989; Machado et al. 1993; R. Parrish 1994, pers. comm.; Scott 1997; Scott & Wodicka 1998) and is interpreted as having been emplaced in a continental margin arc setting (Dunphy & Ludden 1998; Thériault et al. 2001) following accretion of the arc terrane to the northern Churchill domain. St-Onge et al. (2002) have correlated the younger suite of Narsajuaq arc rocks in northern Quebec and Baffin Island with the 1.84–1.81 Ga de Pas batholith south of Ungava Bay (van der Leeden et al. 1990; Dunphy & Skulski 1996; James & Dunning 2000), based on strong similarities between bedrock geology units and residual total magnetic field data, available geochronological constraints, and petrological characteristics.

Plutonic, volcanic and sedimentary units correlative with those of the Narsajuaq arc terrane are not known to occur in West and South Greenland.

Archaean Superior craton; northern and eastern margins

In northern and northeastern Quebec and in western Labrador the Archaean Superior craton (Fig. 4) predominantly comprises felsic orthogneiss and plutonic units ranging in age between 3220 +32/−23 and 2654±5 Ma (Fig. 17; Machado et al. 1989; Mortensen & Percival 1989; Parrish 1989; St-Onge et al. 1992; R. Parrish 1994, pers. comm.; Scott & St-Onge 1995; James & Dunning 2000). In the Cape Smith belt of northern Quebec (St-Onge et al. 2006a; and references therein), a suite of parautochthonous basal clastic sedimentary units, carbonatitic volcaniclastic rock, continental tholeiitic flood basalt (Fig. 18), and rhyolite of the Povungnituk Group is associated with initial Palaeoproterozoic rifting of the northern Superior craton. These units have yielded ages between 2038 +4/−2 and 1958.6 +3.1/−2.7 Ma (Parrish 1989; Machado et al. 1993). Disconformably overlying the initial-rift sedimentary and volcanic rocks is a younger succession of predominantly komatiitic to tholeiitic basalt (Chukotat Group) accumulated during renewed rifting along the northern continental margin and dated between 1887 +37/−11 and 1870 ± 4 Ma (R. Parrish 1994, pers. comm.; Wodicka et al. 2002a). The ages of the younger volcanic succession indicate that c. 150 Ma elapsed between the onset of initial continental rifting and the subsequent rifting event (St-Onge et al. 2000).

Along the eastern margin of the Superior craton within the New Quebec orogen (Fig. 4), the

Fig. 17. Archaean biotite–hornblende ± orthopyroxene granodiorite (light) with abundant mafic enclaves (dark), lower plate Superior craton, northern Quebec. Hammer is 35 cm in length.

Fig. 18. Palaeoproterozoic pillowed basalt in a continental tholeiite flow, Povungnituk Group, Cape Smith belt, northern Quebec. Hammer is 34 cm in length.

tectonostratigraphic record (Hoffman 1990b; Rohon et al. 1993; Skulski et al. 1993; Clark & Wares 2004; and references therein) includes a parautochthonous succession of: (1) basal fluvial redbeds and mildly alkalic mafic lava (Seward subgroup) that are associated with initial Palaeoproterozoic rifting of the craton; (2) marine-shelf quartzite and dolostone (Pistolet Subgroup) that mark the establishment of a platform-type continental margin; (3) interstratified black shale and turbidite (Swampy Bay Subgroup) that interfinger eastward with tholeiitic basalt, gabbro sills, and rhyolite (Bacchus Formation) and are interpreted as a foredeep sequence; (4) a regressive peritidal carbonate reef complex (Denault and Abner formations). Age constraints on deposition range between 2169 ± 2 and $2142 +4/-2$ Ma (Clark 1984; Rohon et al. 1993). The initial-rift assemblage is overlain by a succession of transgressive quartzite (Wishart Formation), which is in turn overlain by shale (Ruth Formation), banded iron formation (Sokoman Formation), coeval alkalic mafic to felsic volcanic rock, tholeiitic sills, basalt, and turbidite (Menihek Formation), all of which are intruded by layered peridotite–gabbro–diorite sills and collectively accumulated in pull-apart basins during renewed rifting along the established continental margin (Skulski et al. 1993). Ages for this package range between c. 1884 ± 1.6 and 1870 Ma (Findlay et al. 1995; Machado et al. 1997), and consequently c. 292 Ma separate the older and younger successions (see Skulski et al. 1993) and the establishment and subsequent rifting of the eastern margin of the Superior craton.

Plutonic, volcanic and sedimentary units correlative with those of Superior craton are not known to occur in West Greenland or South Greenland.

Palaeoproterozoic orogens and plate geometries in NE Laurentia

Within northeastern Laurentia, the constituent Archaean cratons and microcontinents, attendant Palaeoproterozoic cover sequences, and continental or oceanic magmatic arcs, as described above, were assembled during a period of global amalgamation between c. 1.9 and 1.8 Ga. Documentation of the geometry, age, structural evolution, magmatic context, and metamorphic framework of the intervening deformation zones and orogenic belts allows the relative upper plate v. lower plate geometry to be established in each case (Fig. 19). This in turn allows the tectonic evolution and growth of

NE Laurentia during the Palaeoproterozoic Era to be modelled as a series of cumulative accretion–collision events that can be compared directly with the tectonic growth of SE Asia during the late Mesozoic Era and Cenozoic Era (see below).

Ellesmere–Devon terrane and Inglefield mobile belt

The oldest deformation zone in northeastern Laurentia is the one that outlines the northern margin of the Rae craton in northern Canada (Ellesmere–Devon terrane) and in northern West Greenland (Inglefield mobile belt). The plutonic rocks of the Ellesmere–Devon terrane (Fig. 2) are dominated by arc-type pyroxene-bearing tonalite, quartz norite and several varieties of granite, and show evidence of penetrative to gneissic fabric development (Frisch 1988). A minimum age for the granulite-facies metamorphism was determined as $c.$ 1930 Ma (Frisch & Hunt 1988). Hoffman (1988, 1990b) utilized available ages, similarities in lithotectonic units, and the occurrence of distinctive aeromagnetic anomalies as a basis for correlation of the Ellesmere–Devon terrane with the upper plate, 2.0–1.9 Ga Taltson–Thelon magmatic zone (northwestern boundary of the Rae craton; Hoffman 1988, 1989; Thériault 1992), which is exposed south of the Queen Maud Gulf in the northwestern Canadian Shield (Wheeler et al. 1996). The inferred upper plate setting for the northern margin of the Rae craton is shown with a line marking the southern limit of upper plate arc magmatism and labelled 'Ellesmere–Inglefield belt' in Figure 19.

The tectonomagmatic history of the Inglefield mobile belt of northern West Greenland (Fig. 3) is complex, with most rock units having been repeatedly deformed under low- to medium-pressure granulite-facies conditions with several episodes of anatectic melting (Dawes 2004; Nutman et al. 2008). Coring the Inglefield belt, the Etah metaigneous complex is composed of continental margin arc-type intermediate to felsic orthogneiss, megacrystic monzogranite, quartz diorite, and syenite. At least two episodes of isoclinal folding produced map-scale structures involving Etah Group supracrustal units, with at least one episode post-dating the Etah metaigneous complex, and the major, late-kinematic Sunrise Pynt shear zone that separates the northern and southern parts of the belt. Granulite-facies metamorphism is constrained at 1923 ± 8 Ma by ion probe U–Pb ages on zircon (Dawes 2004; Nutman et al. 2008). The polydeformed, high-T character of the Inglefield mobile belt and the arc-type granitoids of the Etah metaigneous complex are interpreted as indicative of an upper plate setting for the northern margin of the Rae craton in Greenland (Fig. 19).

Foxe fold belt

Deformation along the southern margin of the Rae craton within the Foxe fold belt of central Baffin Island (Fig. 2) is characterized by early north-verging, thin-skinned imbrication and tight intrafolial isoclinal folding of the Piling Group, followed by NE–SW-trending upright folding of both Palaeoproterozoic cover and Archaean basement, and subsequent open NW–SE-trending, thick-skinned cross-folding (Scott et al. 2003). Metamorphic grade increases from greenschist facies at higher structural levels to granulite facies at lower structural levels and in proximity to the plutonic units of the Cumberland batholith. As noted, the Cumberland batholith has been interpreted as a continental margin arc emplaced above a north-dipping subduction system following accretion of the Meta Incognita microcontinent to the northern Rae craton.

The proposed Baffin suture (St-Onge et al. 2006c) separating the Archaean Rae craton and its flanking, southern Palaeoproterozoic continental margin sedimentary and volcanic sequences (Piling and Hoare Bay groups) from accreted tectonic elements to the south trends east from the Foxe Basin to the head of Cumberland Sound (Fig. 2). Closing across the proposed suture is thought to post-date the youngest dated unit in the Piling Group (1883 ± 5 Ma) and predate emplacement of the Cumberland batholith at $1865 +4/-2$ to 1848 ± 2 Ma (St-Onge et al. 2006c). However, new field-based research is required to further characterize the suture in terms of its structural context, tectonothermal history, and crustal geometry. Consequently, the proposed suture is shown without an upper or lower plate connotation in Figure 19.

Rinkian fold belt and Nagssugtoqidian orogen

Deformation in the Rinkian fold belt of northern West Greenland (Fig. 3) is characterized by first NE- and then WNW-directed ductile thrusts and fold-nappes (Fig. 20), and terminated by crustal-scale open, upright folding with broad domes and narrow cusps (Pulvertaft 1986; Grocott & Pulvertaft 1990; Escher et al. 1999; Garde et al. 2003; Lahtinen et al. 2009). Metamorphic grade increases from greenschist facies at Mârmorilik, is at lower to middle amphibolite facies over wide areas of central West Greenland, and reaches granulite facies in proximity to the Prøven igneous complex (Garde 1978; Crocott & Pulvertaft 1990). Collision across the Rinkian fold belt occurred at 1881 ± 20 Ma (Taylor & Kalsbeek 1990), based on Pb–Pb dating of an amphibolite-facies Karrat Group marble.

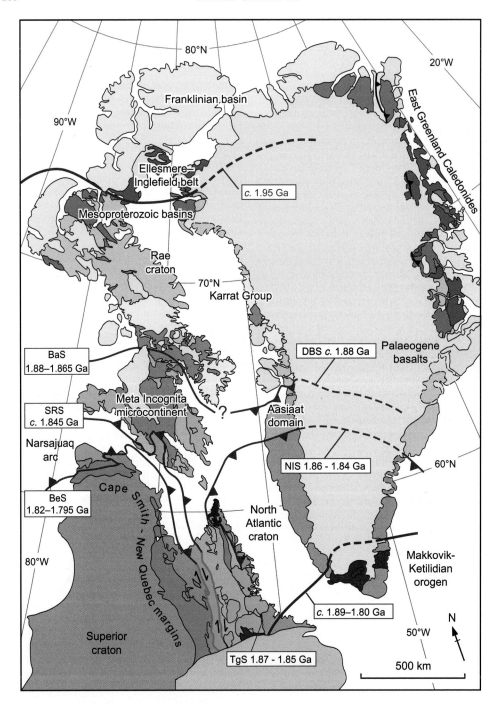

Fig. 19. Composite geological map highlighting the principal cratonic, supracrustal and tectonic entities, as well as the bounding crustal structures that can be correlated from eastern Canada to West Greenland. Greenland and Ellesmere Island, as well as Baffin Island, are shown in a pre-drift (i.e. pre-late Cretaceous) position with respect to mainland Canada, following the reconstruction of Oakey (2006, 2007, pers. comm.). It should be noted that non-rigid plate deformation (extension) on the Canadian side, which is not taken into account in the reconstruction, prevents Devon Island from being shown. Lines of latitude and longitude are pertinent to mainland Canada. Interpreted Palaeoproterozoic sutures are shown in red with known age constraints. Crustal boundaries shown in green denote a

Fig. 20. Part of Kigarsima nappe structure in the central Rinkian fold belt, West Greenland. Pale Archaean orthogneiss in the upper part of the c. 1600 m high mountain tectonically overlies Palaeoproterozoic metagreywacke of the Nûkavsak Fm, upper Karrat Group.

Fig. 19. (*Continued*) dominant strike-slip component of displacement, and magmatic arc fronts are shown in purple with their known age range. Extrapolation of geological boundaries beneath the Inland Ice (dashed lines) is constrained by the aeromagnetic data of Saltus & Gaina (2007). It should be noted that the figure documents a systematic north-to-south decrease in the age of suturing from the northern Rae craton to the lower plate Superior craton, and that with the alignment of the Palaeogene basalts on either side of Davis Strait, all other units and structures of Phanerozoic, Mesoproterozoic, Palaeoproterozoic, and Archaean age in West Greenland and northeastern Canada follow suit. Colour scheme follows that of Figs 2–4. BaS, Baffin suture; BeS, Bergeron suture; DBS, Disko Bugt suture; NIS, Nordre Isortoq steep belt; SRS, Soper River suture; TgS, Tasiuyak gneiss suture.

The earliest Palaeoproterozoic, collision-related, deformation documented within the Nagssugtoqidian orogen of central West Greenland (Fig. 3) is NW-directed thrusting and imbrication of Archaean and Palaeoproterozoic units at a scale of hundreds of metres to several kilometres (van Gool et al. 1999, 2002; and references therein). Thrust imbrication started after intrusion of the youngest arc rocks at 1873 Ma and lasted at least until $1837 + 9/-8$ Ma (Kalsbeek et al. 1987; van Gool et al. 2002). It was followed by isoclinal folding of large-scale tectonic contacts, and subsequent belt-parallel, thick-skinned, folds of basement and cover units at c. 1825 ± 1 Ma (Connelly et al. 2000) and emplacement of steep, NNE–SSW-trending pegmatites at 1837 ± 12 Ma (Thrane & Connelly 2006), both during post-collisional convergence. This phase coincides with left-lateral shearing in the Abloviak shear zone in Labrador (see below). Late strike-slip deformation in the Nordre Strømfjord shear zone and Nordre Isortoq steep belt occurred at c. 1775 Ma (Connelly et al. 2000). The granulite-facies thermal peak of metamorphism in the core of the orogen was dated at c. 1.86–1.84 Ga (Taylor & Kalsbeek 1990; Kalsbeek & Nutman 1996; Connelly et al. 2000; Willigers et al. 2001). This age interval coincides with the timing of thrusting and is interpreted as documenting the main phase of collision across the Nagssugtoqidian orogen.

As noted, the Rinkian fold belt and Nagssugtoqidian orogen are interpreted as being associated with the closure of two south-dipping Palaeoproterozoic sutures (see above; Connelly & Thrane 2005; Connelly et al. 2006; Garde et al. 2007): (1) a northern suture, the Disko Bugt suture that separates the lower plate Rae craton from the medial Aasiaat domain; and (2) a southern suture rooting in the Nordre Isortoq steep belt that separates the upper plate North Atlantic craton from the Aasiaat domain (Fig. 19).

Torngat orogen

Palaeoproterozoic crustal deformation within the Torngat orogen is interpreted to have resulted from the collision of the North Atlantic craton in the east with the Meta Incognita microcontinent in the west, locally described as the core zone, of northeastern Quebec (Fig. 19). The collision resulted in a narrow, north–south-trending, doubly vergent orogen (Rivers et al. 1996; Wardle & Van Kranendonk 1996; and references therein), with the crustal suture corresponding to the eastern margin of the Tasiuyak gneiss (Fig. 4). The collision was preceded by emplacement of the continental, calc-alkaline Burwell arc in northernmost Labrador, dated between 1910 ± 2 Ma and $1869 +3/-2$ (Scott & Machado 1995; Van Kranendonk &

Wardle 1996). The collision itself is recorded by peak granulite-facies metamorphism within the core of the orogen during west-directed thrusting within the Tasiuyak gneisses and east-directed thrusting at the North Atlantic craton margin between c. 1870 and 1845 Ma (Bertrand et al. 1993; Rivers et al. 1996; Van Kranendonk 1996; Scott 1998; Connelly 2001). Continued deformation is manifest as sinistral transpressional shear between 1844 ± 3.6 and 1822 Ma along the north–south-trending Abloviak shear zone (Bertrand et al. 1993), which developed mainly in the Tasiuyak gneiss. Younger shear deformation (related to indentation of the composite Churchill domain by the lower plate Superior craton?; see below) continued to c. 1780 Ma (Van Kranendonk & Wardle 1996, 1997).

The west-directed thrusting documented within the Tasiuyak gneiss and the location of the Burwell arc plutonic rocks suggest that the North Atlantic craton occupied an upper plate position with respect to the Meta Incognita microcontinent to the west during convergence across the Torngat orogen (Fig. 19).

Dorset fold belt

Convergence between the southern margin of Meta Incognita microcontinent and crustal domains to the south led to development of the Dorset fold belt and formation of the north-dipping Soper River suture (Fig. 3). Closing of the suture is bracketed between 1845 ± 2 Ma, the age of the youngest unit associated with the intra-oceanic phase of the accreted Narasajuaq arc (Fig. 3; Dunphy & Ludden 1998) and $1842 +5/-3$ Ma (Scott 1997), the age of the oldest plutonic unit of the continental margin arc phase of the Narsajuaq arc (St-Onge et al. 2007).

Deformation in the Dorset fold belt includes: (1) structural repetition and truncation of distinct tectonostratigraphic units yielding an overall SW-verging ramp-flat fault geometry (Scott et al. 1997); (2) associated tight to isoclinal folding during SW- to south-directed deformation (Sanborn-Barrie et al. 2009); (3) development of a penetrative granulite-facies compositional fabric; (4) formation of ribbon mylonites and transposition of cross-cutting intrusive units into parallelism in the vicinity of the Soper River suture; and (5) later open cross-folding and localized dextral transcurrent shearing. Syntectonic, granulite-facies, regional metamorphism associated with emplacement of the Cumberland batholith and closure of the Soper River suture is bracketed between c. 1849 and 1835 Ma (St-Onge et al. 2007).

The SW- to south-verging thrusting and folding documented within the Dorset fold belt and the location of the Cumberland batholith continental margin arc rocks in the northern hanging wall of the suture suggest that the Meta Incognita microcontinent occupied an upper plate position with respect to the Narsajuaq arc to the south (Fig. 19) during convergence across the Soper River suture.

Makkovik and Ketilidian orogens

Early foreland-directed thrusting and dextral transpression in the Makkovik orogen of Labrador at c. 1895–1870 Ma (Culshaw et al. 2000b) may be contemporaneous with Ketilidian thrusting of the Sortis Group and dextral transpression along the steep, ENE–WSW-trending, Kobberminebugt shear zone in South Greenland. Deformation along the Kobberminebugt shear zone is constrained by the 1848 ± 2 Ma age of crosscutting augen granite (Hamilton et al. 1999; Garde et al. 2002).

In the Makkovik orogen, early continental calc-alkaline plutonism is manifest by the c. 1893 ± 2 to 1870 Ma Island Harbour Bay plutonic suite (Ryan et al. 1983; Kerr et al. 1992; Culshaw et al. 2000a, b; Barr et al. 2001). There appears to be no direct plutonic counterpart in the Makkovik orogen to the voluminous continental margin arc of the Julianehåb batholith (Fig. 21) in South Greenland, which largely comprises 1854–1795 Ma plutons of granodiorite and granite, with minor gabbro, diorite, quartz monzodiorite, tonalite and rare quartz syenite (Chadwick et al. 1994; Chadwick & Garde 1996; Garde et al. 2002; Garde 2007b). The felsic volcanic Aillik Group dated at c. 1860–1807 Ma (Schärer et al. 1988; Ketchum 1998) is close in age to the Julianehåb batholith and, like the latter, it contains evidence of synmagmatic sinistral transpression. The Aillik Group has been interpreted as having been erupted in an extensional back-arc setting, but it might instead be a high-level equivalent of the Julianehåb batholith (Culshaw et al. 2000b).

The continental calc-alkaline plutonism (and possible related felsic volcanism) documented in both the Makkovik and Ketilidian orogens, as well as the geometry of thrusting and transpressional deformation within these orogens, points to an upper plate setting for the southern margin of the North Atlantic craton during the time interval 1.89–1.80 Ga (Fig. 19). The upper plate setting for the southern margin of the North Atlantic craton established during the middle Palaeoproterozoic appears to have been maintained during the late Palaeoproterozoic (1.71–1.60 Ga) Labradorian orogeny, which involved formation and closure of a small ocean or back-arc basin, and formation and accretion of offshore terranes (Gower 1996; and references therein).

New Quebec orogen

Within the New Quebec orogen (Fig. 4) parautochthonous sedimentary and volcanic strata along the eastern margin of the Superior craton are imbricated by thrust faults above a regional basal décollement (Wares & Goutier 1990). Fault displacement was in a west to SW direction, with thin-skinned imbrication and associated open to tight, large-scale, upright to overturned folds occurring in a piggyback sequence toward the western foreland.

A trailing fan of break-back thrusts or 'out-of-sequence' faults younging toward the eastern hinterland of the orogen re-imbricate the cover units of the Superior craton (Wares & Goutier 1990). As in the case of the Cape Smith belt (Lucas 1989; see below), out-of-sequence thrusting in the New Quebec orogen seems to be the dominant mechanism responsible for crustal shortening based on degree of imbrication and juxtaposition of domains.

Regional metamorphic grade increases from greenschist and lower amphibolite facies in the Palaeoproterozoic cover strata of the western foreland, to amphibolite and granulite facies in the Archaean basement and Palaeoproterozoic metasedimentary and metavolcanic rocks in the eastern hinterland (Perreault & Hynes 1990).

The west-verging geometry of the broad thin- to thick-skinned thrust–fold belt preserved within the New Quebec orogen and the systematic eastward increase in metamorphic grade within the tapered thrust-stack document the lower plate position of the Superior craton during its collision with the eastern portion of the composite Churchill domain (Fig. 19).

Cape Smith belt

Within the Cape Smith belt of northern Quebec (Fig. 19) parautochthonous sedimentary and volcanic strata along the northern margin of the Superior craton are imbricated by thrust faults above a regional basal décollement (Lucas 1989). Fault displacement was in a southerly direction, with thin-skinned imbrication and associated folding occurring in a piggyback sequence toward the southern foreland. Thrust deformation was initiated after 1870 ± 4 Ma, the age of the youngest unit within the parautochthonous Superior craton cover sequence.

A distinct suite of late or 'out-of-sequence' thrust faults that post-date the thin-skinned structures re-imbricate the cover units of the Superior craton (Lucas 1989). These younger south-verging structures are thick-skinned (involving both crystalline basement and Palaeoproterozoic cover

Fig. 21. Juvenile, 1792 Ma arc-type granite cut by synkinematic mafic dykes emplaced during sinistral transpression. Late-stage Julianehåb batholith, southeastern Ketilidian orogen, South Greenland. Hammer is 35 cm in length.

units) and are collisional in origin as they can be linked to terrane boundary faults within the Churchill domain (St-Onge et al. 2001). The late faults truncate the metamorphic isograds within the Cape Smith belt (see below) and thus must postdate c. 1820 +4/−3 to 1815 ± 4 Ma (Bégin 1992). They predate the age of emplacement of post-kinematic syenite plugs and syenogranite dykes at 1795 ± 2 to 1758.2 ± 1.2 Ma (St-Onge et al. 2006c).

Regional, Barrovian-facies, kyanite–sillimanite-grade, medium-pressure metamorphism is associated with early thin-skinned thrusting of cover units along the north margin of the Superior craton (Bégin 1992). Metamorphism is bracketed between 1820 +4/−3 and 1815 ± 4 Ma and is interpreted as a consequence of the relaxation of isotherms in the tectonically thickened thrust belt (St-Onge & Lucas 1991).

A south-verging tectonic boundary or crustal suture (Bergeron suture) separates the northern Superior margin strata from allochthonous crustal elements of the composite Churchill domain to the north (Fig. 19; St-Onge et al. 1999, 2001). Associated with, and sitting in the hanging wall of, the Bergeron suture are the crustal components of an obducted 1998 ± 2 Ma ophiolite (Watts Group; Parrish 1989; Scott et al. 1992, 1999), and the plutonic, volcanic and sedimentary components of the Narsajuaq arc (described above). Preservation of the ophiolite and the higher structural levels it represents within the Cape Smith belt are entirely a function of the late- to postcollisional, crustal-scale, orogen-parallel folding and orogen-perpendicular cross-folding that characterize the southern margin of the Trans-Hudson orogen in northern Quebec (Lucas & Byrne 1992). Closure of the Bergeron suture and collision of the composite Churchill domain with the Superior craton, is bracketed between 1820 +4/−3 Ma (youngest component of Narsajuaq arc) and 1795 ± 2 Ma (the age of an undeformed crosscutting syenogranite pegmatite dyke) (St-Onge et al. 2006c).

The architecture of the foreland thin- to thick-skinned thrust–fold belt, the geometry of the Bergeron suture, and the regional Barrovian metamorphism documented within the Cape Smith belt document the lower plate position of the Superior craton during its collision with the northern Churchill domain (Fig. 19).

Palaeoproterozoic tectonic evolution of NE Laurentia through multiple accretionary events

The correlation of Archaean and Palaeoproterozoic bedrock units and structures between the eastern Canadian Shield and West Greenland as presented in this paper allows the identification of an internally consistent, north-to-south sequence of accretionary and collisional tectonic events during the Palaeoproterozoic Era (Fig. 19). These tectonic events (itemized below) first resulted in the growth of the composite Churchill domain around the crustal nucleus represented by the Rae craton. Initial assembly of the upper plate Churchill domain was then followed by collision with the southern, lower plate Superior craton, which resulted in the terminal collisional phase of the Trans-Hudson orogen and significantly added to the landmass of the emerging Laurentian craton.

Based on available tectonostratigraphic, structural, and geochronological data in eastern Canada and West Greenland, the sequence of tectonic events that characterize the accretionary–collisional growth of NE Laurentia are as follows (Fig. 19):

(1) Deformation and magmatism along the northern margin of the upper plate Rae craton at c. 1.96–1.91 Ga (Inglefield mobile belt).

(2) Accumulation of a south-facing continental margin sequence along the southern margin of the Rae craton between c. 2.16 and 1.88 Ga.

(3) North–south convergence and accretion of the Meta Incognita microcontinent to the southern margin of the Rae craton across the Baffin suture between c. 1.88 and 1.87 Ga (Foxe fold belt; plate geometry during accretion undetermined); accretion of the upper plate Aasiaat domain (microcontinental fragment?) to the southern margin of the Rae craton (lower plate) across the Disko Bugt suture at c. 1.88 Ga (Rinkian fold belt).

(4) Collision of the upper plate North Atlantic craton with the southern margin of the composite Rae craton and Aasiaat domain in Greenland (lower plate) between c. 1.86 and 1.84 Ga (Nagssugtoqidian orogen), and with the eastern margin of the Meta Incognita microcontinent in western Labrador and eastern Quebec (lower plate) between c. 1.87 and 1.85 Ga (Torngat orogen).

(5) Accretion of dominantly juvenile material in an active continental margin arc setting along the southern margin of the upper plate North Atlantic craton in Labrador (Makkovik orogen) and South Greenland (Ketilidian orogen) between c. 1.89 and 1.80 Ga.

(6) Accretion of the Narsajuaq arc terrane (lower plate) to the southern margin of the composite upper plate Churchill domain at c. 1.845 Ga.

(7) Collision of the lower plate Superior craton with the composite upper plate Churchill domain between c. 1.82 and 1.795 Ga.

Discussion

St-Onge et al. (2006c) compared the structural and thermal evolution of the lower and upper collisional plates of the Trans-Hudson orogen in North America with that of the Himalaya–Karakoram–Tibetan orogen of SE Asia. That tectonic comparison, which was based on a geological transect from northern Quebec to central Baffin Island for the Trans-Hudson orogen, is strengthened and enriched by the present paper and the incorporation of observations and constraints from western Greenland and northern Labrador that further highlight

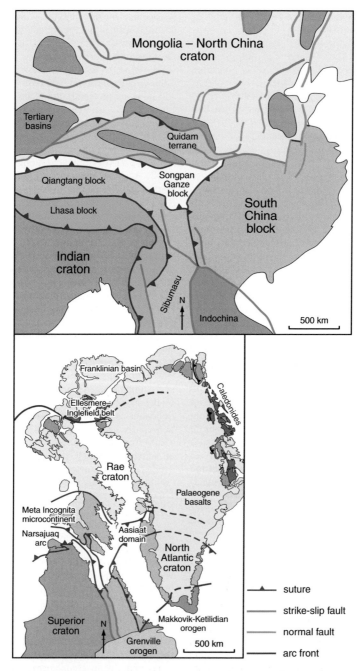

Fig. 22. Composite geological map highlighting the principal crustal elements and bounding structures in NE Laurentia and SE Asia at the same scale. Greenland, Ellesmere Island and Baffin Island are shown in a pre-drift (i.e. pre-late Cretaceous) position with respect to mainland Canada, following the reconstruction of Oakey (2006, 2007, pers. comm.). Crustal sutures are shown in red, strike-slip faults in green, normal faults in blue, and magmatic arc fronts in purple. Extrapolation of geological boundaries beneath the Inland Ice in Greenland (dashed lines) is constrained by the aeromagnetic data of Saltus & Gaina (2007). It should be noted that the figure documents similar relative tectonic positions for the Rae craton, North Atlantic craton and Superior craton in NE Laurentia, when compared with the Mongolia–North China craton, South China block and Indian craton in SE Asia. Colour scheme for NE Laurentia follows that of Figs 2–4.

the fundamental early accretionary character of the Trans-Hudson orogen (Fig. 22).

The Trans-Hudson orogen and the Himalaya–Karakoram–Tibetan orogen both record protracted, pre-terminal collision, north-to-south (present coordinates) histories of crustal accretion and growth within their respective composite upper plates (St-Onge et al. 2006c; and references therein). In northeastern Laurentia, the timing of early accretionary–collisional events in the composite upper plate Churchill domain from north to south is as follows (Fig. 19): 1.88–1.87 Ga (accretion of Meta Incognita microcontinent and closure of the Baffin suture); c. 1.88 Ga (accretion of the Aasiaat (microcontinental?) domain to the northern Rae craton and closure of the Disko Bugt suture); 1.86–1.84 Ga (collision of the North Atlantic craton with the composite Rae craton–Aasiaat domain and closure of the suture rooted in the Nordre Isortoq steep belt); 1.87–1.85 Ga (collision of the North Atlantic craton with the Meta Incognita microcontinent and closure of the suture rooted in the Tasiuyak gneiss); 1.845 Ga (accretion of Narsajuaq arc and closure of the Soper River suture); 1.82–1.795 Ga (collision with the lower plate Superior craton and closure of the Bergeron suture). Periods of calc-alkaline magmatism documented in northeastern Laurentia during the Palaeoproterozoic Era are bracketed from north to south at: 1.96–1.91 Ga (metaplutonic rocks of the Ellesmere–Devon terrane and Etah metaigneous complex), c. 1.9 Ga (southern margin of the Rae craton), 1.87–1.85 Ga (Cumberland batholith), 1.92–1.89 Ga (Arfersiorfik intrusive suite), 1.92–1.87 Ga (Sisimiut charnockite suite), 1.91–1.87 Ga (Burwell arc), 1.86–1.85 Ga (older suite of Narsajuaq arc), 1.84–1.82 Ga (younger suite of Narsajuaq arc), and 1.89–1.80 Ga (Makkovik and Ketilidian orogens), all in the composite upper plate Churchill domain.

In the Himalaya–Karakoram–Tibetan orogen, a similar north-to-south trend in the crustal accretion and growth of the Asian upper plate domain (Fig. 22; St-Onge et al. 2006c; and references therein) is indicated by the age of the crustal sutures (c. 200, 100–125 and 50 Ma) contained within the Tibetan Plateau, and by the periods of pre-collision continental margin arc magmatism dated at c. 150 Ma for the Hushe gneiss, 102–51 Ma for the Ladakh–Gangdese granites of the Trans-Himalayan batholith, and 78–54 Ma for the arc volcanic rocks of the Kohistan terrane.

Although both the Trans-Hudson orogen in the Palaeoproterozoic Era and the Himalaya–Karakoram–Tibetan orogen in the Cenozoic Era show broadly similar structural and thermal responses to the underthrusting of lower plate material (first oceanic and then continental), which is followed by the consequent tectonic thickening of continental margin units, differences in the orogenic style of these two large collisional orogens have been recognized and discussed by St-Onge et al. (2006c). These include: (1) the apparent absence of early eclogite-facies mineral assemblages in the eastern Trans-Hudson orogen (complete post-eclogite-facies retrogression?); (2) evidence that a midcrustal channel flow structure was not involved in the thermal evolution of the lower plate south of the Bergeron suture (lack of voluminous anatectic melt generation and consequent melt weakening?); (3) the attainment of granulite-facies conditions in the Churchill domain (shallower erosion levels in the Himalaya–Karakoram–Tibetan orogen?); and (4) the restricted occurrence of synorogenic molasse in the Trans-Hudson orogen (again difference in structural level of present erosion surface?).

In conclusion, it seems clear from the rock record of the Palaeoproterozoic Trans-Hudson orogen and the Cenozoic Himalaya–Karakoram–Tibetan orogen that both large continent–continent collisions followed extended periods of pre-collisional accretionary events, during which the composite collisional upper plate grew as a result of plate convergence, subduction, crustal accretion and continental margin arc magmatism. We would suggest that such a pattern of continental growth is an integral part of the global amalgamation process that eventually leads to the formation of supercontinents (Hoffman 1992) such as Nuna (Hoffman 1997; Zhao et al. 2002, 2004), Rodinia (Dalziel 1991; Hoffman 1991), or the future Amasia (Hoffman 1997).

We are grateful to the Bureau of Minerals and Petroleum (BMP), Nuuk, Greenland, and in particular H. K. Schønwandt (BMP), H. Stendal and M. N. Nielsen (Geological Survey of Denmark and Greenland) for hosting an initial workshop entitled 'Geology and Mineral Resources in Greenland and Eastern North America' in Kangerlussuaq, Greenland, 3–7 October 2005. We also thank D. James (Canada–Nunavut Geoscience Office) and H. Stendal for organizing a subsequent special session entitled 'Northeast Canada and Greenland: Geology, correlations and resource potential' at the Geological Association of Canada annual meeting in Yellowknife 2007. We are thankful to G. Oakey (Geological Survey of Canada; GSC) for providing us with the pre-drift (i.e. pre-late Cretaceous) reconstruction of Greenland, Ellesmere Island and Baffin Island with respect to mainland Canada utilized in Figure 19. T. Rivers (Memorial University), M. Van Kranendonk (Geological Survey of Western Australia), D. Wardle (Geological Survey, Newfoundland and Labrador), and N. Wodicka (GSC) are sincerely thanked for reviewing and improving early versions of this paper. This is ESS Contribution 20070581. This paper is published with permission from the Geological Survey of Denmark and Greenland.

References

ALLAART, J. H. 1982. *Geological map of Greenland, 1:500 000, Frederikshåb Isblink–Søndre Strømfjord, sheet 2*. Geological Survey of Greenland, Copenhagen.

ANDERSEN, M. C. & PULVERTAFT, T. C. R. 1985. Rb–Sr whole rock ages from reworked basement gneisses in the Umanak area, central West Greenland. *Bulletin of the Geological Society of Denmark*, **34**, 205–212.

BARR, S. M., WHITE, C. E., CULSHAW, N. G. & KETCHUM, J. W. F. 2001. Geology and tectonic setting of Paleoproterozoic granitoid suites in the Island Harbour Bay area, Makkovik Province, Labrador. *Canadian Journal of Earth Sciences*, **38**, 441–463.

BÉGIN, N. J. 1992. Contrasting mineral isograd sequences in metabasites of the Cape Smith Belt, northern Quebec, Canada: Three new bathograds for mafic rocks. *Journal of Metamorphic Geology*, **10**, 685–704.

BERTHELSEN, A. & HENRIKSEN, N. 1975. *Geological map of Greenland, 1:100 000, Ivigtut 61 V. 1 syd*. Geological Survey of Greenland, Copenhagen.

BERTRAND, J.-M., RODDICK, J. C., VAN KRANENDONK, M. J. & ERMANOVICS, I. 1993. U–Pb geochronology of deformation and metamorphism across a central transect of the Early Proterozoic Torngat Orogen, North River map area, Labrador. *Canadian Journal of Earth Sciences*, **30**, 1470–1489.

BETHUNE, K. M. & SCAMMELL, R. J. 2003. Geology, geochronology, and geochemistry of Archaean rocks in the Eqe Bay area, north–central Baffin Island, Canada: Constraints on the depositional and tectonic history of the Mary River Group of northeastern Rae Province. *Canadian Journal of Earth Sciences*, **40**, 1137–1167.

BLEEKER, W. 2003. The late Archean record: A puzzle in c. 35 pieces. *Lithos*, **71**, 99–134.

BONDESEN, E. 1970. The stratigraphy and deformation of the Precambrian rocks of the Graenseland area, South-West Greenland. *Bulletin Grønlands Geologiske Undersøgelse*, **86**, 1–210.

BRIDGWATER, D. & SCHIØTTE, L. 1991. The Archean gneiss complex of Northern Labrador: A review of current results, ideas and problems. *Bulletin of the Geological Society of Denmark*, **39**, 153–166.

BRIDGWATER, D., ESCHER, A. & WATTERSON, J. 1973a. Tectonic displacements and thermal activity in two contrasting Proterozoic mobile belts from Greenland. *Philosophical Transactions of the Royal Society of London, Series A*, **273**, 513–533.

BRIDGWATER, D., WATSON, J. & WINDLEY, B. F. 1973b. The Archaean craton of the North Atlantic region. *Philosophical Transactions of the Royal Society of London, Series A*, **273**, 493–512.

BRIDGWATER, D., COLLERSON, K. D. & MYERS, J. 1976. The development of the Archean gneiss complex of the North Atlantic region. *In*: TARLING, D. H. (ed.) *Evolution of the Earth's Crust*. Academic Press, New York, 16–69.

BRIDGWATER, D., AUSTRHEIM, H., HANSEN, B. T., MENGEL, F., PEDERSEN, S. & WINTER, J. 1990. The Proterozoic Nagssugtoqidian mobile belt of southeast Greenland: A link between the eastern Canadian and Baltic shields. *Geoscience Canada*, **17**, 305–310.

BRIDGWATER, D., MENGEL, F., FRYER, B., WAGNER, P. & HANSEN, S. 1995. Early Proterozoic mafic dykes in the North Atlantic and Baltic Cratons: Field settings and chemistry of distinctive dyke swarms. *In*: COWARD, M. P. & RIES, A. C. (eds) *Precambrian Processes*. Geological Society, London, Special Publications, **95**, 193–210.

CADMAN, A. C., HEAMAN, L., TARNEY, J., WARDLE, R. & KROGH, T. E. 1993. U–Pb geochronology and geochemical variation within two Proterozoic mafic dyke swarms, Labrador. *Canadian Journal of Earth Sciences*, **30**, 1490–1504.

CADMAN, A. C., TARNEY, J., BRIDGWATER, D., MENGEL, F., WHITEHOUSE, M. J. & WINDLEY, B. F. 2001. The petrogenesis of the Kangâmiut dyke swarm, W. Greenland. *Precambrian Research*, **105**, 183–203.

CAMPBELL, L. M. & BRIDGWATER, D. 1996. 1995 summer field investigations of the Proterozoic Sisimiut charnockite suite, Sisimiut region, West Greenland. *In*: MENGEL, F. (ed.) *Report on 1995 fieldwork in the Nagssugtoqidian Orogen, West Greenland*. Danish Lithosphere Centre, Copenhagen, 24–28.

CHADWICK, B. & GARDE, A. A. 1996. Paleoproterozoic oblique plate convergence in South Greenland: A reappraisal of the Ketilidian orogen. *In*: BREWER, T. S. (ed.) *Precambrian Crustal Evolution in the North Atlantic Region*. Geological Society, London, Special Publications, **112**, 179–196.

CHADWICK, B., ERFURT, P., FRITH, R. A., NIELSEN, T. F. D., SCHØNWANDT, H. K. & STENDAL, H. 1994. Sinistral transpression and hydrothermal activity during emplacement of the Early Proterozoic Julianehåb batholith, Ketilidian orogenic belt, South Greenland. *Rapport Grønlands Geologiske Undersøgelse*, **163**, 5–22.

CLARK, T. 1984. *Géologie de la région du lac Cambrien, Territoire du Nouveau Québec*. Ministère de l'Energie et des Ressources du Québec ET, **83-02**.

CLARK, T. & WARES, R. 2004. *Synthèse lithotectonique et métallogénique de l'Orogène du Nouveau-Québec (Fosse du Labrador)*. Ministère des Ressources naturelles et de la Faune, Québec MM, **2004-01**.

COLLERSON, K. D. & REGELOUS, M. 1995. ^{143}Nd–^{142}Nd systematics of early Archean rocks from northern Labrador and implications for crust–mantle evolution of the North Atlantic craton. *EOS Transactions, American Geophysical Union*, **76**, F687.

COLLERSON, K. D., CAMPBELL, L. M., WEAVER, B. L. & PALACZ, Z. A. 1991. Evidence for extreme mantle fractionation in early Archean ultramafic rocks from northern Labrador. *Nature*, **349**, 209–214.

CONNELLY, J. N. 2001. Constraining the timing of metamorphism: U–Pb and Sm–Nd ages from a transect across the northern Torngat Orogen, Labrador, Canada. *Journal of Geology*, **109**, 57–77.

CONNELLY, J. N. & MENGEL, F. C. 2000. Evolution of Archean components in the Nagssugtoqidian Orogen, West Greenland. *Geological Society of America Bulletin*, **112**, 747–763.

CONNELLY, J. N. & RYAN, B. 1994. Late Archean and Proterozoic events in the central Nain craton. *In*: WARDLE, R. J. & HALL, J. (eds) *Eastern Canadian*

Shield Onshore–Offshore Transect (ECSOOT). University of British Columbia, Lithoprobe Secretariat Report, **36**, 53–61.

CONNELLY, J. N. & RYAN, B. 1999. Age and tectonic implications of Paleoproterozoic granitoid intrusions within Nain Province near Nain, Labrador. *Canadian Journal of Earth Sciences*, **36**, 833–853.

CONNELLY, J. N. & THRANE, K. 2005. Rapid determination of Pb isotopes to define Precambrian allochthonous domains: An example from West Greenland. *Geology*, **33**, 953–956, doi:10.1130/G21720.1.

CONNELLY, J. N., VAN GOOL, J. A. M. & MENGEL, F. C. 2000. Temporal evolution of a deeply eroded orogen: The Nagssugtoqidian orogen, West Greenland. *Canadian Journal of Earth Sciences*, **37**, 1121–1142.

CONNELLY, J. N., THRANE, K., KRAWIEC, A. W. & GARDE, A. A. 2006. Linking the Palaeoproterozoic Nagssugtoqidian and Rinkian orogens through the Disko Bugt region of West Greenland. *Journal of the Geological Society, London*, **163**, 319–335.

CROWLEY, J. L. 2002. Testing the model of late Archaean terrane accretion in southern West Greenland: A comparison of the timing of geological events across the Qarliit Nunaat fault, Buksefjorden region. *Precambrian Research*, **116**, 57–79.

CULSHAW, N. G., BROWN, T., REYNOLDS, P. H. & KETCHUM, J. W. F. 2000a. Kanairiktok shear zone: The boundary between the Paleoproterozoic Makkovik Province and the Archean Nain Province, Labrador. *Canadian Journal of Earth Sciences*, **37**, 1245–1257.

CULSHAW, N., KETCHUM, J. & BARR, S. 2000b. Structural evolution of the Makkovik Province, Labrador, Canada: Tectonic processes during 200 Myr at a Paleoproterozoic active margin. *Tectonics*, **19**, 961–977.

DALZIEL, I. W. D. 1991. Pacific margins of Laurentia and East Antartica–Australia as a conjugate rift pair: Evidence and implications for an Eocambrian supercontinent. *Geology*, **19**, 598–601.

DAWES, P. R. 1988. Etah meta-igneous complex and the Wulff structure: Proterozoic magmatism and deformation in Inglefield Land, North-West Greenland. *Rapport Grønlands Geologiske Undersøgelse*, **139**, 1–24.

DAWES, P. R. 1991. Geological Map of Greenland, 1:500 000, Thule, sheet 5. Geological Survey of Greenland, Copenhagen.

DAWES, P. R. 1997. The Proterozoic Thule Supergroup, Greenland and Canada: History, lithostratigraphy and development. *Geology of Greenland Survey Bulletin*, **174**, 1–24.

DAWES, P. R. 2004. Explanatory notes to the geological map of Greenland, 1:500 000, Humboldt Gletscher, Sheet 6. Geological Survey of Denmark and Greenland Map Series, **1**.

DAWES, P. R. 2006. Explanatory notes to the geological map of Greenland, 1:500 000, Thule, Sheet 5. Geological Survey of Denmark and Greenland Map Series, **2**.

DAWES, P. R. & GARDE, A. A. 2004. Geological map of Greenland, 1:500 000, Humboldt Gletscher, sheet 6. Geological Survey of Greenland, Copenhagen.

DAWES, P. R., FRISCH, T. & CHRISTIE, R. L. 1982. The Proterozoic Thule Basin of Greenland and Ellesmere Island: Importance to the Nares Strait debate. *In*: DAWES, P. R. & KERR, J. W. (eds) *Nares Strait and the Drift of Greenland: A Conflict in Plate Tectonics*. Meddelelser om Grønland Geosciences, **8-1982**, 89–104.

DAWES, P. R., LARSEN, O. & KALSBEEK, F. 1988. Archean and Proterozoic crust in North-West Greenland: Evidence from Rb–Sr whole-rock age determinations. *Canadian Journal of Earth Sciences*, **25**, 1365–1373.

DAWES, P. R., FRISCH, T. *ET AL*. 2000. Kane Basin 1999: Mapping, stratigraphic studies and economic assessment of Precambrian and Lower Palaeozoic provinces in north-western Greenland. *Geology of Greenland Survey Bulletin*, **186**, 11–28.

DUNPHY, J. M. & LUDDEN, J. N. 1998. Petrological and geochemical characteristics of a Paleoproterozoic magmatic arc (Narsajuaq Terrane, Ungava Orogen, Canada) and comparisons to Superior Province granitoids. *Precambrian Research*, **91**, 109–142.

DUNPHY, J. M. & SKULSKI, T. 1996. Petrological zonation across the De Pas batholith: A tilted section through a continental arc? *In*: WARDLE, R. J. & HALL, J. (eds) *Eastern Canadian Shield Onshore–Offshore Transect (ECSOOT)*. University of British Columbia, Lithoprobe Secretariat Report, **57**, 44–58.

ERMANOVICS, I. F. 1993. *Geology of the Hopedale Block, southern Nain Province, and the adjacent Proterozoic terranes, Labrador, Newfoundland*. Geological Survey of Canada Memoir, **431**.

ERMANOVICS, I. & VAN KRANENDONK, M. J. 1998. Geology of the Archean Nain Province and Paleoproterozoic Torngat Orogen in a transect of the North River–Nutak map area, Newfoundland (Labrador) and Quebec. *Geological Survey of Canada Bulletin*, **497**, 1–156.

ESCHER, A. 1971. *Geological map of Greenland, 1:500 000, Søndre Strømfjord–Nûgssuaq, sheet 3*. Geological Survey of Greenland, Copenhagen.

ESCHER, A. & BURRI, M. 1967. Stratigraphy and structural development of the Precambrian rocks in the area north-east of Disko Bugt, West Greenland. Rapport Grønlands Geologiske Undersøgelse, **13**, 1–28.

ESCHER, A. & PULVERTAFT, T. C. R. 1976. Rinkian mobile belt of West Greenland. *In*: ESCHER, A. & WATT, W. S. (eds) *Geology of Greenland*. Geological Survey of Greenland, Copenhagen, 104–119.

ESCHER, A., ESCHER, J. C. & WATTERSON, J. 1975. The reorientation of the Kangâmiut dyke swarm, West Greenland. *Canadian Journal of Earth Sciences*, **12**, 158–173.

ESCHER, A., JACK, S. & WATTERSON, J. 1976a. Tectonics of the North Atlantic Proterozoic dyke swarm. *Philosophical Transactions of the Royal Society of London, Series A*, **280**, 529–539.

ESCHER, A., SØRENSEN, K. & ZECK, H. P. 1976b. Nagssugtoqidian mobile belt in West Greenland. *In*: ESCHER, A. & WATT, W. S. (eds) *Geology of Greenland*. Geological Survey of Greenland, Copenhagen, 76–95.

ESCHER, J. C. 1985. *Geological map of Greenland, 1:500 000, Upernavik Isfjord, sheet 4*. Geological Survey of Greenland, Copenhagen.

ESCHER, J. C. & PULVERTAFT, T. C. R. 1995. *Geological map of Greenland, 1:2 500 000*. Geological Survey of Greenland, Copenhagen.

ESCHER, J. C., RYAN, M. J. & MARKER, M. 1999. Early Proterozoic thrust tectonics east of Ataa Sund, northeast Disko Bugt, West Greenland. *Geology of Greenland Survey Bulletin*, **181**, 171–179.

FINDLAY, J. M., PARRISH, R. R., BIRKETT, T. C. & WATANABE, D. H. 1995. U–Pb ages from the Nimish Formation and Montagnais glomeroporphyritic gabbro of the central New Quebec orogen, Canada. *Canadian Journal of Earth Sciences*, **32**, 1208–1220.

FINN, G. C. 1989. Rb–Sr geochronology of the Archean Maggo gneiss from the Hopedale block, Nain Province, Labrador. *Canadian Journal of Earth Sciences*, **26**, 2512–2522.

FRIEND, C. R. L. & NUTMAN, A. P. 1994. Two Archaean granulite-facies metamorphic events in the Nuuk-Maniitsoq region, southern West Greenland: Correlation with the Saglek block, Labrador. *Journal of the Geological Society, London*, **151**, 421–424.

FRIEND, C. R. L. & NUTMAN, A. P. 2001. U–Pb zircon study of tectonically bounded blocks of 2940–2840 Ma crust with different metamorphic histories, Paamiut region, South–West Greenland: Implications for the tectonic assembly of the North Atlantic Craton. *Precambrian Research*, **105**, 143–164.

FRIEND, C. R. L. & NUTMAN, A. P. 2005. New pieces to the Archaean terrane puzzle in the Nuuk region, southern West Greenland: Steps in transforming a simple insight into a complex regional tectonothermal model. *Journal of the Geological Society, London*, **162**, 147–162.

FRIEND, C. R. L., NUTMAN, A. P. & MCGREGOR, V. R. 1988. Late Archean terrane accretion in the Godthåb region, southern West Greenland. *Nature*, **144**, 369–376.

FRIEND, C. R. L., NUTMAN, A. P., BAADSGAARD, H., KINNY, P. D. & MCGREGOR, V. R. 1996. Timing of late Archaean terrane assembly, crustal thickening and granite emplacement in the Nuuk region, southern West Greenland. *Earth and Planetary Science Letters*, **142**, 353–365.

FRISCH, T. 1988. Reconnaissance geology of the Precambrian Shield of Ellesmere, Devon and Coburg islands, Canadian Arctic Archipelago. *In*: *Radiogenic age and isotopic studies: report 2*. Geological Survey of Canada Memoir, **409**.

FRISCH, T. & DAWES, P. R. 1982. The Precambrian Shield of northernmost Baffin Bay: Correlations across Nares Strait. *In*: DAWES, P. R. & KERR, J. W. (eds) *Nares Strait and the Drift of Greenland: a Conflict in Plate Tectonics*. Meddelelser om Grønland Geosciences, **8-1982**, 79–88.

FRISCH, T. & HUNT, P. 1988. U–Pb zircon and monazite ages from the Precambrian Shield of Ellesmere and Devon islands, Arctic Archipelago. *In*: *Radiogenic age and isotopic studies: report 2*. Geological Survey of Canada Paper, **88-2**, 117–125.

FURNES, H., DE WIT, M., STAUDIGEL, H., ROSING, M. & MUEHLENBACHS, K. 2007. A vestige of Earth's oldest ophiolite. *Science*, **135**, 1704–1707, doi:10.1126/science.1139170.

GARDE, A. A. 1978. The lower Proterozoic Mârmorilik formation, east of Mârmorilik, West Greenland. *Meddelelser om Grønland*, **200(3)**.

GARDE, A. A. 1994. *Precambrian Geology between Qarajaq Isfjord and Jakobshavn Isfjord, West Greenland, scale 1:250 000*, Grønlands Geologiske Undersøgelse, Copenhagen.

GARDE, A. A. 2003. An overview of the Archaean evolution of the Nuuk region: From long-lived evolution of a sea of gneisses to episodic accretion of small continental terranes along boundary structures. *Danmarks og Grønlands Geologiske Undersøgelse Rapport*, **2003/94**, 19–44.

GARDE, A. A. 2007a. A relict island arc complex in the eastern Akia terrane, Godthåbsfjord, southern West Greenland. *Journal of the Geological Society, London*, **164**, 565–579.

GARDE, A. A. 2007b. *Geological map of Greenland, 1:500 000, Sydgrønland, sheet 2nd edn*. Geological Survey of Greenland, Copenhagen.

GARDE, A. A. & STEENFELT, A. 1989. A new anorthosite/gabbro complex at Nûgssuaq, central West Greenland. *Rapport Grønlands Geologiske Undersøgelse*, **145**, 16–20.

GARDE, A. A. & STEENFELT, A. 1999a. Precambrian geology of Nuussuaq and the area north-east of Disko Bugt, West Greenland. *Geology of Greenland Survey Bulletin*, **181**, 7–40.

GARDE, A. A. & STEENFELT, A. 1999b. Proterozoic structural overprinting of Archeaean gneisses in Nuussuaq, West Greenland. *Geology of Greenland Survey Bulletin*, **181**, 141–154.

GARDE, A. A., CHADWICK, B., GROCOTT, J., HAMILTON, M., MCCAFFREY, K. & SWAGER, C. P. 1998. An overview of the Paleoproterozoic Ketilidian orogen, south Greenland. *In*: WARDLE, R. J. & HALL, J. (eds) *Eastern Canadian Shield Onshore–Offshore Transect (ECSOOT)*. University of British Columbia, Lithoprobe Secretariat Report, **68**, 50–66.

GARDE, A. A., FRIEND, C. R. L., NUTMAN, A. P. & MARKER, M. 2000. Rapid maturation and stabilisation of middle Archaean continental crust: The Akia terrane, southern West Greenland. *Bulletin of the Geological Society of Denmark*, **47**, 1–27.

GARDE, A. A., HAMILTON, M. A., CHADWICK, B., GROCOTT, J. & MCCAFFREY, K. J. W. 2002. The Ketilidian orogen of South Greenland: Geochronology, tectonics, magmatism and forearc accretion during Palaeoproterozoic oblique convergence. *Canadian Journal of Earth Sciences*, **39**, 765–793, doi: 10.1139/E02-026.

GARDE, A. A., GROCOTT, J., THRANE, K. & CONNELLY, J. N. 2003. Reappraisal of the Rinkian fold belt in central West Greenland: Tectonic evolution during crustal shortening and linkage with the Nagssugtoqidian Orogen. *European Union of Geosciences Meeting, Geophysical Research Abstracts CD*, **5**, Abstract 09411.

GARDE, A. A., HOLLIS, J. A. & MAZUR, S. 2007. Palaeoproterozoic greenstones and pelitic schists in the northern Nagssugtoqidian orogen, West Greenland: Evidence for a second subduction zone? *Geological Association of Canada–Mineralogical Association of Canada (GAC–MAC) Program with Abstracts*, **32**, 30.

GOULET, N. & CIESIELSKI, A. 1990. The Abloviak shear zone and the NW Torngat Orogen, eastern Ungava Bay, Québec. *Geoscience Canada*, **17**, 269–272.

GOWER, C. F. 1996. The evolution of the Grenville Province in eastern Labrador. *In*: BREWER, T. S. (ed.) *Precambrian Crustal Evolution in the North Atlantic Region*. Geological Society, London, Special Publications, **112**, 197–218.

GOWER, C. F., RYAN, A. N. & RIVERS, T. 1990. Mid-Proterozoic Laurentia–Baltica: An overview of its geological evolution and a summary of the contributions made by this volume. *In*: GOWER, C. F., RIVERS, T. & RYAN, A. B. (eds) *Mid-Proterozoic Laurentia–Baltica*. Geological Association of Canada, Special Papers, **38**, 1–20.

GROCOTT, J. & PULVERTAFT, T. C. R. 1990. The Early Proterozoic Rinkian Belt of Central West Greenland. *In*: LEWRY, J. F. & STAUFFER, M. R. (eds) *The Early Proterozoic Trans-Hudson Orogen of North America*. Geological Association of Canada, Special Papers, **37**, 443–463.

HALL, J., LOUDEN, K. E., FUNCK, T. & DEEMER, S. 2002. Geophysical characteristics of the continental crust along the Lithoprobe Eastern Canadian Shield Onshore–Offshore Transect (ECSOOT): A review. *Canadian Journal of Earth Sciences*, **39**, 569–587, doi:10.1139/E02–005.

HAMILTON, M. A., RYAN, A. B., EMSLIE, R. F. & ERMANOVICS, I. 1998. Identification of Paleoproterozoic anorthositic and monzonitic rocks in the vicinity of the Mesoproterozoic Nain Plutonic Suite, Labrador: U–Pb evidence. *In: Radiogenic age and isotopic studies: report 2*. Geological Survey of Canada Paper, **1998-F**, 23–40.

HAMILTON, M. A., GARDE, A. A., CHADWICK, B., GROCOTT, J. & MCCAFFREY, K. 1999. Temporal evolution of the Palaeoproterozoic Ketilidian orogen, South Greenland: A synopsis of results from conventional and SHRIMP U–Pb geochronology. *European Union of Geosciences Journal of Conference Abstracts*, **4**, 129.

HENDERSON, G. & PULVERTAFT, T. C. R. 1967. The stratigraphy and structure of the Precambrian rocks of the Umanak area, West Greenland. *Meddelelser fra Dansk Geologisk Forening*, **17**, 1–20.

HENDERSON, G. & PULVERTAFT, T. C. R. 1987. *The lithostratigraphy and structure of a Lower Proterozoic dome and nappe complex. Descriptive text to 1:100 000 sheets Mârmorilik 71 V. 2 Syd, Nûgâtsiaq 71 V. 2 Nord and Pangnertôq 72 V. 2 Syd*, Geological Survey of Greenland, Copenhagen.

HIGGINS, A. K. 1970. The stratigraphy and structure of the Ketilidian rocks of Midternæs, South-West Greenland. *Bulletin Grønlands Geologiske Undersøgelse*, **87**, 1–96.

HIGGINS, A. K. & SOPER, N. J. 1999. The Precambrian supracrustal rocks of Nunataq, north-east Disko Bugt, West Greenland. *Geology of Greenland Survey Bulletin*, **181**, 79–86.

HOFFMAN, P. F. 1988. United Plates of America, the birth of a craton: Early Proterozoic assembly and growth of Laurentia. *Annual Review of Earth and Planetary Sciences*, **16**, 543–603.

HOFFMAN, P. F. 1989. Precambrian geology and tectonic history of North America. *In*: BALLY, A. W. & PALMER, A. R. (eds) *The Geology of North America—An Overview*. Geological Society of America, Decade of North American Geology, **A**, 447–511.

HOFFMAN, P. F. 1990a. Dynamics of the tectonic assembly of northeast Laurentia in geon 18 (1.9–1.8 Ga). *Geoscience Canada*, **17**, 222–226.

HOFFMAN, P. F. 1990b. Subdivision of the Churchill Province and extent of the Trans-Hudson Orogen. *In*: LEWRY, J. F. & STAUFFER, M. R. (eds) *The Early Proterozoic Trans-Hudson Orogen of North America*. Geological Association of Canada, Special Papers, **37**, 15–39.

HOFFMAN, P. F. 1991. Did the breakout of Laurentia turn Gondwanaland inside-out? *Science*, **252**, 1409–1412.

HOFFMAN, P. F. 1992. Supercontinents. *In*: BREKHOVSKIKH, L., TUREKIAN, K. K., EMERY, K. & TSENG, C. K. (eds) *Encyclopedia of Earth System Science*, 4. Academic Press, London, 323–328.

HOFFMAN, P. F. 1997. Tectonic genealogy of North America. *In*: VAN DER PLUIJM, B. A. & MARSHAK, S. (eds) *Earth Structures and Introduction to Structural Geology and Tectonics*. McGraw-Hill, New York, 459–464.

HOLLIS, J. A., FREI, D., VAN GOOL, J. A. M., GARDE, A. A. & PERSSON, M. 2006a. Using zircon geochronology to resolve the Archaean geology of southern West Greenland. *Geological Survey of Denmark and Greenland Bulletin*, **10**, 49–52.

HOLLIS, J. A., KEIDING, M., MØLLER STENSGAARD, B., VAN GOOL, J. A. M. & GARDE, A. A. 2006b. Evolution of Neoarcheaean supracrustal belts at the northern margin of the North Atlantic craton, West Greenland. *Geological Survey of Denmark and Greenland Bulletin*, **11**, 9–31.

HÖLTTÄ, P., BALAGANSKY, V. ET AL. 2008. Archaean of Greenland and Fennoscandia. *Episodes*, **31**, 13–19.

JACKSON, G. D. 2000. *Geology of the Clyde-Cockburn Land map area, north–central Baffin Island, Nunavut*. Geological Survey of Canada Memoir, **440**.

JACKSON, G. D. & BERMAN, R. G. 2000. Precambrian metamorphic and tectonic evolution of northern Baffin Island, Nunavut, Canada. *Canadian Mineralogist*, **38**, 399–421.

JACKSON, G. D. & TAYLOR, F. C. 1972. Correlation of major Aphebian rock units in the northeastern Canadian Shield. *Canadian Journal of Earth Sciences*, **9**, 1650–1669.

JACKSON, G. D., HUNT, P. A., LOVERIDGE, W. D. & PARRISH, R. R. 1990. Reconnaissance geochronology of Baffin Island, N. W. T. *In: Radiogenic age and isotopic studies: report 2*. Geological Survey of Canada Paper, **89-2**, 123–148.

JAMES, D. T. & DUNNING, G. R. 2000. U–Pb geochronological constraints for Paleoproterozoic evolution of the Core Zone, southeastern Churchill Province, northeastern Laurentia. *Precambrian Research*, **103**, 31–54.

JAMES, D., KAMO, S. & KROGH, T. 1998. Evolution of 3.1 and 3.0 Ga Archean greenstone belts, Hopedale Block, southern Nain Province, Labrador (Canada).

Geological Society of America, Abstracts with Programs, **30**, A395.
JAMES, D. T., KAMO, S. & KROGH, T. 2002. Evolution of 3.1 and 3.0 Ga volcanic belts and a new thermotectonic model for the Hopedale Block, North Atlantic craton (Canada). *Canadian Journal of Earth Sciences*, **39**, 687–710, doi:10.1139/E01-092.
JOHNS, S. M. & YOUNG, M. D. 2006. *Bedrock geology and economic potential of the Archean Mary River group, northern Baffin Island, Nunavut*. Geological Survey of Canada Current Research, **2006-C5**.
KALSBEEK, F. 1981. The northward extent of the Archaean basement of Greenland—a review of Rb–Sr whole rock ages. *Precambrian Research*, **14**, 203–219.
KALSBEEK, F. 1986. The tectonic framework of the Precambrian shield in Greenland: A review of new isotopic evidence. *Rapport Grønlands Geologiske Undersøgelse*, **128**, 55–64.
KALSBEEK, F. & NUTMAN, A. P. 1996. Anatomy of the Early Proterozoic Nagssugtoqidian Orogen, West Greenland, explored by reconnaissance SHRIMP U–Pb dating. *Geology*, **24**, 515–518.
KALSBEEK, F. & TAYLOR, P. N. 1985. Age and origin of early Proterozoic dykes in South–West Greenland. *Contributions to Mineralogy and Petrology*, **89**, 307–316.
KALSBEEK, F., TAYLOR, P. N. & HENRIKSEN, N. 1984. Age of rocks, structures, and metamorphism in the Nagssugtoqidian mobile belt, West Greenland—field and Pb isotope evidence. *Canadian Journal of Earth Sciences*, **21**, 1126–1131.
KALSBEEK, F., PIDGEON, R. T. & TAYLOR, P. N. 1987. Nagssugtoqidian mobile belt of West Greenland: Cryptic 1850 Ma suture between two Archaean continents—chemical and isotopic evidence. *Earth and Planetary Science Letters*, **85**, 365–385.
KALSBEEK, F. T. C., LARSEN, L. M. & BONDAM, J. 1990. *Descriptive text to 1:500 000 sheet 1, Sydgrønland*. Geological Survey of Greenland, Copenhagen.
KALSBEEK, F., PULVERTAFT, T. C. R. & NUTMAN, A. P. 1998. Geochemistry, age and origin of metagreywackes from the Palaeoproterozoic Karrat Group, Rinkian Belt, West Greenland. *Precambrian Research*, **91**, 383–399.
KERR, A. 1994. *Early Proterozoic magmatic suites of the eastern Central Mineral Belt (Makkovik Province), Labrador: Geology, geochemistry and mineral potential*. Government of Newfoundland and Labrador Geological Survey Report, **94-3**.
KERR, A., KROGH, T. E., CORFU, F., SCHÄRER, U., GANDHI, S. S. & KWOK, Y. Y. 1992. Episodic Early Proterozoic granitoid plutonism in the Makkovik Province, Labrador: U–Pb geochronological data and geological implications. *Canadian Journal of Earth Sciences*, **29**, 1166–1179.
KERR, A., RYAN, B., GOWER, C. F. & WARDLE, R. J. 1996. The Makkovik Province: Extension of the Ketilidian Mobile Belt in mainland North America. In: BREWER, T. S. (ed.) *Precambrian Crustal Evolution in the North Atlantic Region*. Geological Society, London, Special Publications, **112**, 155–177.
KERR, A., HALL, J., WARDLE, R. J., GOWER, C. F. & RYAN, B. 1997. New reflections on the structure and evolution of the Makkovikian–Ketilidian orogen in Labrador and southern Greenland. *Tectonics*, **16**, 942–965.
KETCHUM, J. W. F. 1998. The Cape Harrison metamorphic suite and Upper Aillik Group in the southeastern Makkovik Province, Labrador. In: WARDLE, R. J. & HALL, J. (eds) *Eastern Canadian Shield Onshore–Offshore Transect (ECSOOT)*. University of British Columbia, Lithoprobe Secretariat Report, **68**, 78–91.
KETCHUM, J. W. F., BARR, S. M., CULSHAW, N. G. & WHITE, C. E. 2001a. U–Pb ages of granitoid rocks in the northwestern Makkovik Province, Labrador: Evidence for 175 million years of episodic synorogenic and postorogenic plutonism. *Canadian Journal of Earth Sciences*, **38**, 359–372.
KETCHUM, J. W. F., JACKSON, S. E., CULSHAW, N. G. & BARR, S. M. 2001b. Depositional and tectonic setting of the Paleoproterozoic Lower Aillik Group, Makkovik Province, Canada: Evolution of a passive margin–foredeep sequence based on petrochemistry and U–Pb (TIMS and LAM-IC-MS) geochronology. *Precambrian Research*, **105**, 333–358.
KETCHUM, J. W. F., CULSHAW, N. & BARR, S. M. 2002. Anatomy and orogenic history of a Paleoproterozoic accretionary belt: The Makkovik Province, Labrador, Canada. *Canadian Journal of Earth Sciences*, **39**, 711–730, doi:10.1139/E01-099.
KORSTGÅRD, J. A. 1979. Metamorphism of the Kangâmiut dykes and the metamorphic and structural evolution of the southern Nagssugtoqidian boundary in the Itivdleq–Ikertôq region, West Greenland. *Rapport Grønlands Geologiske Undersøgelse*, **89**, 63–75.
KORSTGÅRD, J. A., RYAN, B. & WARDLE, R. 1987. The boundary between Proterozoic and Archean crustal blocks in central West Greenland and northern Labrador. In: PARK, R. G. & TARNEY, J. (eds) *Evolution of the Lewisian and Comparable Precambrian High Grade Terrains*. Geological Society, London, Special Publications, **27**, 247–259.
LAHTINEN, R., GARDE, A. A. & MELEZHIK, V. A. 2009. Palaeoproterozoic evolution of Fennoscandia and Greenland. *Episodes*, **31**, 20–28.
LEWRY, J. F. & COLLERSON, K. D. 1990. The Trans-Hudson Orogen; extent, subdivisions and problems. In: LEWRY, J. F. & STAUFFER, M. R. (eds) *The Early Proterozoic Trans-Hudson Orogen of North America*. Geological Association of Canada, Special Papers, **37**, 1–14.
LOVERIDGE, W. D., ERMANOVICS, I. F. & SULLIVAN, R. W. 1987. U–Pb ages on zircon from the Maggo Gneiss, the Kanairiktok Plutonic Suite and the Island Harbour Plutonic Suite, coast of Labrador, Newfoundland. In: *Radiogenic age and isotopic studies: report 2*. Geological Survey of Canada Paper, **87-2**, 59–65.
LUCAS, S. B. 1989. Structural evolution of the Cape Smith thrust belt and the role of out-of-sequence faulting in the thickening of mountain belts. *Tectonics*, **8**, 655–676.
LUCAS, S. B. & BYRNE, T. 1992. Footwall involvement during arc–continent collision, Ungava orogen, northern Canada. *Journal of the Geological Society, London*, **149**, 237–248.

MACHADO, N., GOULET, N. & GARIÉPY, C. 1989. U–Pb geochronology of reactivated Archean basement and of Hudsonian metamorphism in the northern Labrador Trough. *Canadian Journal of Earth Sciences*, **26**, 1–15.

MACHADO, N., CLARK, T., DAVID, J. & GOULET, N. 1997. U–Pb ages for magmatism and deformation in the New Quebec orogen. *Canadian Journal of Earth Sciences*, **34**, 716–723.

MACHADO, N., DAVID, J., SCOTT, D. J., LAMOTHE, D., PHILIPPE, S. & GARIÉPY, C. 1993. U–Pb geochronology of the western Cape Smith Belt, Canada. New insights on the age of initial rifting and arc magmatism. *Precambrian Research*, **63**, 211–223.

MARKER, M., WHITEHOUSE, M., SCOTT, D., STECHER, O., BRIDGWATER, D. & VAN GOOL, J. 1999. Deposition, provenance and tectonic setting for metasediments in the Palaeoproterozoic Nagssugtoqidian Orogen, West Greenland: A key for understanding crustal collision. *Terra Abstracts*, **11**, 128.

MARTEN, B. E. 1977. *The relationship between the Aillik Group and the Hopedale gneiss, Kaipokok Bay, Labrador*. PhD thesis, Memorial University of Newfoundland, St. John's.

MAYBORN, K. R. & LESHER, C. E. 2006. Origin and evolution of the Kangâmiut mafic dyke swarm, West Greenland. *Geological Survey of Denmark and Greenland Bulletin*, **11**, 61–86.

MORGAN, W. C., BOURNE, J., HERD, R. K., PICKETT, J. W. & TIPPETT, C. R. 1975. Geology of the Foxe foldbelt, Baffin Island, District of Franklin. Geological Survey of Canada Report, **75-1**, 343–347.

MORGAN, W. C., OKULITCH, A. V. & THOMPSON, P. H. 1976. *Stratigraphy, structure and metamorphism of the west half of the Foxe Fold Belt, Baffin Island*. Geological Survey of Canada Report, **76-1A**, 387–391.

MORTENSEN, J. K. & PERCIVAL, J. A. 1989. Reconnaissance U–Pb zircon and monazite geochronology of the Lac Clairambault area, Ashuanipi complex, Quebec. In: *Radiogenic age and isotopic studies: report 2*. Geological Survey of Canada Paper, **87-2**, 135–142.

MYERS, J. 1985. Stratigraphy and structure of the Fiskenæsset Complex, southern West Greenland. *Bulletin Grønlands Geologiske Undersøgelse*, **150**, 1–72.

NUTMAN, A. P. & KALSBEEK, F. 1999. SHRIMP U–Pb zircon ages for Archaean granitoid rocks, Ataa area, north-east Disko Bugt, West Greenland. *Geology of Greenland Survey Bulletin*, **181**, 49–54.

NUTMAN, A. P., FRIEND, C. R. L., BAADSGAARD, H. & MCGREGOR, V. R. 1989. Evolution and assembly of Archean gneiss terranes in the Godthabsfjord region, southern West Greenland: Structural, metamorphic and isotopic evidence. *Tectonics*, **8**, 573–589.

NUTMAN, A. P., MCGREGOR, V. R., FRIEND, C. R. L., BENNETT, V. C. & KINNY, P. D. 1996. The Itsaq Gneiss Complex of southern West Greenland: The world's most extensive record of early crustal evolution (3900–3600 Ma). *Precambrian Research*, **78**, 1–39.

NUTMAN, A. P., KALSBEEK, F., MARKER, M., VAN GOOL, J. A. M. & BRIDGWATER, D. 1999. U–Pb zircon ages of Kangâmiut dykes and detrital zircons in metasediments in the Palaeoproterozoic Nagssugtoqidian Orogen (West Greenland): Clues to the pre-collisional history of the orogen. *Precambrian Research*, **93**, 87–104.

NUTMAN, A. P., BENNETT, V. C., FRIEND, C. R. L. & MCGREGOR, V. R. 2000. The early Archaean Itsaq Gneiss Complex of southern West Greenland: The importance of field observations in interpreting age and isotopic constraints for early terrestrial evolution. *Geochimica et Cosmochimica Acta*, **64**, 3035–3060.

NUTMAN, A. P., FRIEND, C. R. L., BARKER, S. L. L. & MCGREGOR, V. R. 2005. Inventory and assessment of Paleoarchaean gneiss terrains and detrital zircons in southern West Greenland. *Precambrian Research*, **135**, 281–314.

NUTMAN, A. P., DAWES, P. R., KALSBEEK, F. & HAMILTON, M. A. 2008. Palaeoproterozoic and Archaean gneiss complexes in northern Greenland: Palaeoproterozoic terrane assembly in the High Arctic. *Precambrian Research*, **161**, 419–451.

OAKEY, G. N. 2006. *Cenozoic evolution and lithosphere dynamics of the Baffin Bay–Nares Strait region of Arctic Canada and Greenland*. PhD thesis, Vrije Universiteit Amsterdam.

OAKEY, G. N. 2007. Cenozoic tectonic framework of the Baffin Bay–Nares Strait region of Arctic Canada and Greenland. *Geological Association of Canada–Mineralogical Association of Canada (GAC–MAC) Program with Abstracts*, **32**.

ØSTERGAARD, C., GARDE, A. A., NYGAARD, J., BLOMSTERBERG, J., NIELSEN, B. M., STENDAL, H. & THOMAS, C. W. 2002. The Precambrian supracrustal rocks in the Naternaq (Lersletten) and Ikamiut areas, central West Greenland. *Geology of Greenland Survey Bulletin*, **191**, 24–32.

PARK, R. G. 1994. Early Proterozoic tectonic overview of the northern British Isles and neighbouring terrains in Laurentia and Baltica. *Precambrian Research*, **68**, 65–79.

PARRISH, R. R. 1989. U–Pb geochronology of the Cape Smith belt and Sugluk block. *Geoscience Canada*, **16**, 126–130.

PERREAULT, S. & HYNES, A. 1990. Tectonic evolution of the Kujjuaq terrane, New Quebec Orogen. *Geoscience Canada*, **17**, 238–240.

POLAT, A., FREI, R., APPEL, P. W. U., DILEK, Y., FRYER, B., ORDÓÑEZ-CALDERÓN, J. C. & YANG, Z. 2008. The origins and compositions of Mesoarchean oceanic crust: evidence from the 3075 Ma Ivisaartog greenstone belt, SW Greenland. *Lithos*, **100**, 293–321.

PULVERTAFT, T. C. R. 1973. Recumbent folding and flat-lying structure in the Precambrian of northern West Greenland. *Philosophical Transactions of the Royal Society of London, Series A*, **273**, 535–545.

PULVERTAFT, T. C. R. 1986. The development of thin thrust sheets and basement–cover sandwiches in the southern part of the Rinkian belt, Umanak district, West Greenland. *Rapport Grønlands Geologiske Undersøgelse*, **128**, 75–87.

RIVERS, T., MENGEL, F., SCOTT, D. J., CAMPBELL, L. M. & GOULET, N. 1996. Torngat Orogen—a Paleoproterozoic example of a narrow doubly vergent collisional orogen. In: BREWER, T. S. (ed.) *Precambrian Crustal Evolution in the North Atlantic Region*. Geological Society, London, Special Publications, **112**, 117–136.

ROHON, M.-L., VIALETTE, Y., CLARK, T., ROGER, G., OHNENSTETTER, D. & VIDAL, P. 1993. Aphebian mafic–ultramafic magmatism in the Labrador Trough (New Quebec): Its age and the nature of its mantle source. *Canadian Journal of Earth Sciences*, **30**, 1582–1593.

ROSING, M. T., NUTMAN, A. P. & LOFQVIST, L. 2001. A new fragment of the earliest Earth crust: The Aasivik terrain of West Greenland. *Precambrian Research*, **105**, 115–128.

ROWLEY, D. B., CURRIE, B. S., KIDD, W. S. F. & ZHU, B. 2004. India–Asia collision and Himalaya–Tibet orogenesis *sensu stricto*. *Geological Society of America, Abstracts with Programs*, **36**, 408.

RYAN, A. B., KAY, A. & ERMANOVICS, I. 1983. *The geology of the Makkovik Subprovince between Kaipokok Bay and Bay of Islands, Labrador*. Newfoundland Department of Mines and Energy Maps, **83-38** to **83-41**.

SALTUS, R. W. & GAINA, C. 2007. Circum Arctic map compilation. *EOS Transactions, American Geophysical Union*, **88**, 227, doi:10.1029/2007 EO210009.

SANBORN-BARRIE, M., ST-ONGE, M. R. & YOUNG, M. 2009. Bedrock geology of southwestern Baffin Island: Expanding the tectonostratigraphic framework with relevance to base metal mineralization. Geological Survey of Canada Paper, **2008-6**, 1–16.

SCHÄRER, U., KROGH, T. E., WARDLE, R. J., RYAN, A. B. & GANDHI, S. S. 1988. U–Pb ages of early and middle Proterozoic volcanism and metamorphism in the Makkovik Orogen, Labrador. *Canadian Journal of Earth Sciences*, **25**, 1098–1107.

SCHIØTTE, L., COMPSTON, W. & BRIDGWATER, D. 1989a. Ion probe U–Th–Pb zircon dating of polymetamorphic orthogneisses from northern Labrador, Canada. *Canadian Journal of Earth Sciences*, **26**, 1533–1556.

SCHIØTTE, L., COMPSTON, W. & BRIDGWATER, D. 1989b. U–Th–Pb ages of single zircons in Archaean supracrustals from Nain Province, Labrador, Canada. *Canadian Journal of Earth Sciences*, **26**, 2636–2644.

SCHIØTTE, L., NUTMAN, A. P. & BRIDGWATER, D. 1992. U–Pb ages of single zircons within 'Upernavik' metasedimentary rocks and regional implications for the tectonic evolution of the Archaean Nain Province, Labrador. *Canadian Journal of Earth Sciences*, **29**, 260–276.

SCOTT, D. J. 1997. Geology, U–Pb, and Pb–Pb geochronology of the Lake Harbour area, southern Baffin Island: Implications for the Paleoproterozoic tectonic evolution of north-eastern Laurentia. *Canadian Journal of Earth Sciences*, **34**, 140–155.

SCOTT, D. J. 1998. U–Pb ages of Archean crust in the southeast arm of the Rae Province, southeastern Ungava Bay, Quebec. *In: Radiogenic age and isotopic studies: report 2*. Geological Survey of Canada Paper, **1998-F**, 41–45.

SCOTT, D. J. 1999. U–Pb geochronology of the eastern Hall Peninsula, southern Baffin Island, Canada: A northern link between the Archean of West Greenland and the Paleoproterozoic Torngat orogen of northern Labrador. *Precambrian Research*, **93**, 5–26.

SCOTT, D. J. & DE KEMP, E. 1998. *Bedrock geology compilation, northern Baffin Island and northern Melville Peninsula, Northwest Territories*, scale 1:500 000. Geological Survey of Canada, Open File, **3633**.

SCOTT, D. J. & GAUTHIER, G. 1996. Comparison of TIMS (U–Pb) and Laser Ablation Microprobe ICP-MS (Pb) techniques for age determination of detrital zircons from Paleoproterozoic metasedimentary rocks from northeastern Laurentia, Canada, with tectonic implications. *Chemical Geology*, **131**, 127–142.

SCOTT, D. J. & MACHADO, N. 1995. U–Pb geochronology of the northern Torngat Orogen, Labrador, Canada: A record of Paleoproterozoic magmatism and deformation. *Precambrian Research*, **70**, 169–190.

SCOTT, D. J. & ST-ONGE, M. R. 1995. Constraints on Pb closure temperature in titanite based on rocks from the Ungava orogen, Canada: Implications for U–Pb geochronology and $P-T-t$ path determinations. *Geology*, **23**, 1123–1126.

SCOTT, D. J. & WODICKA, N. 1998. A second report on the U–Pb geochronology of southern Baffin Island. *In: Radiogenic age and isotopic studies: report 2*. Geological Survey of Canada Paper, **1998-F**, 47–57.

SCOTT, D. J., HELMSTAEDT, H. & BICKLE, M. J. 1992. Purtuniq ophiolite, Cape Smith belt, northern Quebec, Canada: A reconstructed section of Early Proterozoic oceanic crust. *Geology*, **20**, 173–176.

SCOTT, D. J., ST-ONGE, M. R., WODICKA, N. & HANMER, S. 1997. Geology of the Markham Bay–Crooks Inlet area, southern Baffin Island, Northwest Territories. *Geological Survey of Canada Paper*, **1997-C**, 157–166.

SCOTT, D. J., MARKER, M. ET AL. 1998. Age of deposition, provenance and tectonic setting of metasedimentary rocks in the Palaeoproterozoic Nagssugtoqidian Orogen, West Greenland. *In*: WARDLE, R. J. & HALL, J. (eds) *Eastern Canadian Shield Onshore–Offshore Transect (ECSOOT)*. University of British Columbia, Lithoprobe Secretariat Report, **68**, 148–149.

SCOTT, D. J., ST-ONGE, M. R., LUCAS, S. B. & HELMSTAEDT, H. 1999. The 2.00 Ga Purtuniq ophiolite, Cape Smith Belt, Canada: MORB-like crust intruded by OIB-like magmatism. *Ofioliti*, **24**, 199–215.

SCOTT, D. J., STERN, R. A., ST-ONGE, M. R. & MCMULLEN, S. M. 2002. U–Pb geochronology of detrital zircons in metasedimentary rocks from southern Baffin Island: Implications for the Paleoproterozoic tectonic evolution of Northeastern Laurentia. *Canadian Journal of Earth Sciences*, **39**, 611–623, doi:10.1139/E01–093.

SCOTT, D. J., ST-ONGE, M. R. & CORRIGAN, D. 2003. *Geology of the Archean Rae Craton and Mary River Group and the Paleoproterozoic Piling Group, central Baffin Island, Nunavut*. Geological Survey of Canada Paper, **2003-C26**.

SEARLE, M. P., ST-ONGE, M. R. & WODICKA, N. 2007. Reply to comment by Jason Ali and Jonathan C. Aitchison on 'Trans-Hudson Orogen of North America and Himalaya–Karakoram–Tibet Orogen of Asia: Structural and thermal characteristics of lower and upper plates' by M. R. St-Onge *et al. Tectonics*, **26**, TC3019, doi:10.1029/2007TC002101.

SIDGREN, A.-S., PAGE, L. & GARDE, A. A. 2006. New hornblende and muscovite ^{40}Ar/^{39}Ar cooling ages in the central Rinkian fold belt, West Greenland. *Geological Survey of Denmark and Greenland Bulletin*, **11**, 115–123.

SIMMONS, K. R., WIEBE, R. A., SNYDER, G. A. & SIMMONS, E. C. 1986. U–Pb zircon age for the Newark Island layered intrusion, Nain Anorthosite Complex, Labrador. *Geological Society of America, Abstracts with Programs*, **18**, 751.

SKULSKI, T., WARES, R. P. & SMITH, A. D. 1993. Early Proterozoic (1.88–1.87 Ga) tholeiitic magmatism in the New Quebec orogen. *Canadian Journal of Earth Sciences*, **30**, 1505–1520.

SØRENSEN, K., KORSTGÅRD, J. A., GLASSLEY, W. E. & MØLLER STENSGAARD, B. 2006. The Nordre Strømfjord shear zone and the Arfersiorfik quartz diorite in Arfersiorfik, the Nagssugtoqidian orogen, West Greenland. *Geological Survey of Denmark and Greenland Bulletin*, **11**, 145–161.

STOCKWELL, C. H. 1963. Third report on structural provinces, orogenies, and time-classification of the Canadian Precambrian Shield. *In*: LEECH, G. B., LOWDON, J. A., STOCKWELL, C. H. & WANLESS, R. K. (eds) *Age Determinations and Geological Studies (Including Isotopic Ages—Report 4)*. Geological Survey of Canada Paper, **63-17**, 125–132.

ST-ONGE, M. R. & LUCAS, S. B. 1991. Evolution of regional metamorphism in the Cape Smith Trust Belt (northern Quebec, Canada): Interaction of tectonic and thermal processes. *Journal of Metamorphic Geology*, **9**, 515–534.

ST-ONGE, M. R., LUCAS, S. B. & PARRISH, R. R. 1992. Terrane accretion in the internal zone of the Ungava orogen, northern Quebec: Part 1. Tectonostratigraphic assemblages and their tectonic implications. *Canadian Journal of Earth Sciences*, **29**, 746–764.

ST-ONGE, M. R., LUCAS, S. B., SCOTT, D. J. & WODICKA, N. 1999. Upper and lower plate juxtaposition, deformation and metamorphism during crustal convergence, Trans-Hudson Orogen (Quebec–Baffin segment), Canada. *Precambrian Research*, **93**, 27–49.

ST-ONGE, M. R., SCOTT, D. J. & LUCAS, S. B. 2000. Early partitioning of Quebec: Microcontinent formation in the Paleoproterozoic. *Geology*, **28**, 323–326.

ST-ONGE, M. R., SCOTT, D. J. & WODICKA, N. 2001. Terrane boundaries within Trans-Hudson Orogen (Quebec–Baffin segment), Canada: Changing structural and metamorphic character from foreland to hinterland. *Precambrian Research*, **107**, 75–91.

ST-ONGE, M. R., SCOTT, D. J. & WODICKA, N. 2002. Review of crustal architecture and evolution in the Ungava Peninsula–Baffin Island area: Connection to the Lithoprobe ECSOOT transect. *Canadian Journal of Earth Sciences*, **39**, 589–610, doi:10.1139/E02-022.

ST-ONGE, M. R., HENDERSON, I. & BARAGAR, W. R. A. 2006a. Geology, Cape Smith Belt and adjacent domains, Ungava Peninsula, Quebec and Nunavut, scale 1:300 000. Geological Survey of Canada Open File Map, **4930**.

ST-ONGE, M. R., JACKSON, G. D. & HENDERSON, I. 2006b. *Geology, Baffin Island (south of 70°N and east of 80°W), Nunavut, scale 1:500 000*. Geological Survey of Canada Open File Map, **4931**.

ST-ONGE, M. R., SEARLE, M. P. & WODICKA, N. 2006c. Trans-Hudson Orogen of North America and Himalaya–Karakoram–Tibetan Orogen of Asia: Structural and thermal characteristics of the lower and upper plates. *Tectonics*, **25**, TC4006, doi:10.1029/2005TC001907.

ST-ONGE, M. R., WODICKA, N. & IJEWLIW, O. 2007. Polymetamorphic evolution of the Trans-Hudson Orogen, Baffin Island, Canada: Integration of petrological, structural and geochronological data. *Journal of Petrology*, **48**, 271–302, doi:10.1093/petrology/egl060.

SUTTON, J. S., MARTEN, B. E., CLARK, A. M. S. & KNIGHT, I. 1972. Correlation of the Precambrian supracrustal rocks of coastal Labrador and southwest Greenland. *Nature*, **238**, 122–123.

TAYLOR, F. C. 1971. A revision of Precambrian structural provinces in northeastern Quebec and northern Labrador. *Canadian Journal of Earth Sciences*, **8**, 579–584.

TAYLOR, F. C. 1972. *Reconnaissance geology of a part of the Precambrian Shield, northeastern Quebec, northern Labrador*. Geological Survey of Canada Paper, **71-48**.

TAYLOR, F. C. 1977. *Geology, Hopedale, Newfoundland, scale 1:250 000*. Geological Survey of Canada Map, **1443A**.

TAYLOR, F. C. 1982. Precambrian geology of the Canadian North Atlantic borderlands. *In*: KERR, J. W. & FERGUSSON, A. J. (eds) *Geology of the North Atlantic Borderlands*. Canadian Society of Petroleum Geologists, Memoir, **7**, 11–30.

TAYLOR, P. N. & KALSBEEK, F. 1990. Dating the metamorphism of Precambrian marbles: Examples from Proterozoic mobile belts in Greenland. *Chemical Geology*, **86**, 21–28.

THÉRIAULT, R. J. 1992. Nd isotopic evolution of the Taltson Magmatic Zone, Northwest Territories, Canada: Insights into Early Proterozoic accretion along the western margin of the Churchill Province. *Journal of Geology*, **100**, 465–475.

THÉRIAULT, R. J., ST-ONGE, M. R. & SCOTT, D. J. 2001. Nd isotopic and geochemical signature of the Paleoproterozoic Trans-Hudson Orogen, southern Baffin Island, Canada: Implications for the evolution of eastern Laurentia. *Precambrian Research*, **108**, 113–138.

THOMASSEN, B. 1992. The gold and base metal potential of the Lower Proterozoic Karrat Group, West Greenland. *Rapport Grønlands Geologiske Undersøgelse*, **155**, 57–66.

THRANE, K. & CONNELLY, J. N. 2006. Zircon geochronology from the Kangaatsiaq–Qasigiannguit region, the northern part of the 1.9–1.8 Ga Nagssugtoqidian orogen, West Greenland. *Geological Survey of Denmark and Greenland Bulletin*, **11**, 87–99.

THRANE, K., CONNELLY, J. N., GARDE, A. A., GROCOTT, J. & KRAWIEC, A. W. 2003. Linking the Palaeoproterozoic Rinkian and Nagssugtoqidian belts of central West Greenland: Implications of new U–Pb and Pb–Pb zircon ages. *European Union of*

Geosciences Meeting, Geophysical Research Abstracts CD, **5**, Abstract 09275.

THRANE, K., BAKER, J., CONNELLY, J. & NUTMAN, A. 2005. Age, petrogenesis and metamorphism of the syn-collisional Prøven Igneous Complex, West Greenland. *Contributions to Mineralogy and Petrology*, **149**, 541–555, doi:10.1007/s00410-005-0660-0.

VAN DER LEEDEN, J., BÉLANGER, M., DANIS, D., GIRARD, R. & MARTELAIN, J. 1990. Lithotectonic domains in the high-grade terrain east of the Labrador Trough (Quebec). *In*: LEWRY, J. F. & STAUFFER, M. R. (eds) *The Early Proterozoic Trans-Hudson Orogen of North America*. Geological Association of Canada, Special Papers, **37**, 371–386.

VAN GOOL, J. A. M., MARKER, M., MENGEL, F. & FIELD PARTY. 1996. The Palaeoproterozoic Nagssugtoqidian Orogen in West Greenland: Current status of work by the Danish Lithosphere Centre, Report of Activities, 1995. *Bulletin Grønlands Geologiske Undersøgelse*, **172**, 88–94.

VAN GOOL, J. A. M., KRIEGSMAN, L., MARKER, M. & NICHOLS, G. T. 1999. Thrust stacking in the inner Nordre Strømfjord area, West Greenland: Significance for the tectonic evolution of the Palaeoproterozoic Nagssugtoqidian Orogen. *Precambrian Research*, **93**, 71–86.

VAN GOOL, J. A. M., CONNELLY, J. N., MARKER, M. & MENGEL, F. C. 2002. The Nagssugtoqidian Orogen of West Greenland: Tectonic evolution and regional correlations from a West Greenland perspective. *Canadian Journal of Earth Sciences*, **39**, 665–686, doi:10.1139/E02-027.

VAN GOOL, J. A. M., GARDE, A. A. & LARSEN, L. M. 2004. Correlation of Precambrian geology in Labrador and southern Greenland. Danmarks og Grønlands Geologiske Undersøgelse Rapport, **2004/29**.

VAN GOOL, J. A. M., GARDE, A. A. & HOLLIS, J. A. 2007. Crustal growth and terrane amalgamation in the Archaean North Atlantic craton of southern west Greenland. *Geological Association of Canada–Mineralogical Association of Canada (GAC–MAC) Program with Abstracts*, **32**, 84.

VAN KRANENDONK, M. J. 1996. Tectonic evolution of the Paleoproterozoic Torngat Orogen; evidence from $P-T-t-d$ paths in the North River map area, Labrador. *Tectonics*, **15**, 843–869.

VAN KRADENDONK, M. J. & WARDLE, R. J. 1996. Burwell domain of the Palaeoproterozoic Torngat orogen, northeastern Canada: Tilted cross-section of a magmatic arc caught between a rock and a hard place. *In*: BREWER, T. S. (ed.) *Precambrian Crustal Evolution in the North Atlantic Region*. Geological Society, London, Special Publications, **112**, 91–115.

VAN KRANENDONK, M. J. & WARDLE, R. J. 1997. Crustal-scale flexural slip folding during late tectonic amplification of an orogenic boundary perturbation in the Paleoproterozoic Torngat Orogen, northeastern Canada, *Canadian Journal of Earth Sciences*, **34**, 1545–1565.

VAN KRANENDONK, M. J., ST-ONGE, M. R. & HENDERSON, J. R. 1993. Paleoproterozoic tectonic assembly of northeast Laurentia through multiple indentations. *Precambrian Research*, **63**, 325–347.

WARDLE, R. J. 1983. *Nain–Churchill province cross-section, Nachvak Fiord, northern Labrador*. Newfoundland Department of Mines and Energy, Mineral Development Division Report, **83-1**, 68–90.

WARDLE, R. J. & BAILEY, D. G. 1981. Early Proterozoic sequences in Labrador. *In*: CAMPBELL, F. H. A. (ed.) *Proterozoic basins of Canada*. Geological Survey of Canada Paper, **81-10**, 331–359.

WARDLE, R. J. & VAN KRANENDONK, M. J. 1996. The Paleoproterozoic southeastern Churchill Province of Labrador–Quebec, Canada: Orogenic development as a consequence of oblique collision and indentation. *In*: BREWER, T. S. (ed.) *Precambrian Crustal Evolution in the North Atlantic Region*. Geological Society, London, Special Publications, **112**, 137–153.

WARDLE, R. J., GOWER, C. F., RYAN, B., NUNN, G. A. G., JAMES, D. T. & KERR, A. 1997. *Geological Map of Labrador, scale 1:1 million*. Government of Newfoundland and Labrador, Department of Mines and Energy, Geological Survey Map, **97-07**.

WARDLE, R. J., JAMES, D. ET AL. 2000a. *Geological map of the LITHOPROBE Eastern Canadian Shield Onshore–Offshore Transect (ECSOOT) area, Labrador and adjacent parts of Quebec, version 1.0. Lithoprobe ECSOOT synthesis project map*. Lithoprobe Secretariat, University of British Columbia, Vancouver.

WARDLE, R. J., SCOTT, D., VAN GOOL, J., GARDE, A., CULSHAW, N. & HALL, J. 2000b. An overview of the development of NE Laurentia: Nain–Superior collision and links to western Trans-Hudson Orogen. *Geological Association of Canada–Mineralogical Association of Canada (GAC–MAC) Program with Abstracts*, **25** [CD].

WARDLE, R. J., JAMES, D. T., SCOTT, D. J. & HALL, J. 2002. The southeastern Churchill Province: Synthesis of a Palaeoproterozoic transpressional orogen. *Canadian Journal of Earth Sciences*, **39**, 639–663, doi:10.1139/E02-004.

WARES, R. P. & GOUTIER, J. 1990. Deformational style in the foreland of the northern New Quebec Orogen. *Geoscience Canada*, **17**, 244–249.

WASTENEYS, H., WARDLE, R. J., KROGH, T. & ERMANOVICS, I. 1994. Preliminary U–Pb geochronology of the Ingrid Group and basement rocks in the Hopedale Block in the vicinity of the southern Nain craton–Torngat Orogen boundary. *In*: WARDLE, R. J. & HALL, J. (eds) *Eastern Canadian Shield Onshore–Offshore Transect (ECSOOT)*. University of British Columbia, Lithoprobe Secretariat Report, **45**, 206–216.

WASTENEYS, H. A., WARDLE, R. J. & KROGH, T. E. 1996. Extrapolation of tectonic boundaries across the Labrador shelf: U–Pb geochronology of well samples. *Canadian Journal of Earth Sciences*, **33**, 1308–1324.

WHEELER, J. O., HOFFMAN, P. F., CARD, K. D., DAVIDSON, A., SANDFORD, B. V., OKULITCH, A. V. & ROEST, W. R. 1996. *Geological Map of Canada, scale 1:5 000 000*. Geological Survey of Canada Map, **1860A**.

WHITEHOUSE, M. J., KALSBEEK, F. & NUTMAN, A. P. 1998. Crustal growth and crustal recycling in the Nagssugtoqidian Orogen of West Greenland: Constraints from radiogenic isotope systematics and U–Pb zircon geochronology. *Precambrian Research*, **91**, 365–381.

WILLIGERS, B. J. A., KROGSTAD, E. J. & WIJBRANS, J. R. 2001. Comparison of thermochronometers in a slowly cooled granulite terrain: Nagssugtoqidian Orogen, West Greenland. *Journal of Petrology*, **42**, 1720–1749.

WINDLEY, B. F. 1970. Primary quartz ferro-dolerite/garnet amphibolite dikes in the Sukkertoppen region of West Greenland. *Geological Journal Special Issue*, **2**, 79–92.

WINDLEY, B. F. & GARDE, A. A. 2009. Arc-generated blocks with crustal sections in the North Atlantic craton of West Greenland: crustal growth in the Archean with modern analogues. *Earth-Science Reviews*, **93**, 1–30.

WINDLEY, B. F., HERD, R. K. & BOWDEN, A. A. 1973. The Fiskenæsset complex, West Greenland. I. A preliminary study of the stratigraphy, petrology, and whole-rock chemistry from Qeqertarssuatsiaq. *Bulletin Grønlands Geologiske Undersøgelse*, **106**, 1–80.

WODICKA, N. & SCOTT, D. J. 1997. A preliminary report on the U–Pb geochronology of the Meta Incognita Peninsula, southern Baffin Island, Northwest Territories. Geological Survey of Canada Paper, **1997-C**, 167–178.

WODICKA, N., MADORE, L., LARBI, Y. & VICKER, P. 2002a. Géochronologie U–Pb de filons-couches mafiques de la Ceinture de Cape Smith et de la Fosse du Labrador. *In*: Ministère des Ressources naturelles Québec (ed.) *Séminaire d'information sur la recherche géologique*. Programme et résumés 2002, **DV 2002-10**, 48.

WODICKA, N., ST-ONGE, M. R., SCOTT, D. J. & CORRIGAN, D. 2002b. Preliminary report on the U–Pb geochronology of the northern margin of the Trans-Hudson Orogen, central Baffin Island, Nunavut. Geological Survey of Canada Paper, **2002-F7**.

WODICKA, N., WHALEN, J. B., JACKSON, G. D. & HEGNER, E. 2007a. Geochronology. *In*: ST-ONGE, M. R., FORD, A. & HENDERSON, I. (eds) *Digital geoscience atlas of Baffin Island (south of 70°N and east of 80°W), Nunavut*. Geological Survey of Canada, Open File, **5116**.

WODICKA, N., ST-ONGE, M. R., CORRIGAN, D. & SCOTT, D. J. 2007b. Depositional age and provenance of the Piling Group, central Baffin Island, Nunavut: Implications for the Paleoproterozoic tectonic development of the southern Rae margin. *Geological Association of Canada–Mineralogical Association of Canada (GAC–MAC) Program with Abstracts*, **32**, 88.

YOUNG, M., SANDEMAN, H., CREASER, R. & JAMES, D. 2007. Meso- to Neoarchean crustal growth and recycling on northern Baffin Island and correlation of Rae Province rocks across mainland Nunavut and Greenland. *Geological Association of Canada–Mineralogical Association of Canada (GAC–MAC) Program with Abstracts*, **32**, 89.

ZHAO, G., CAWOOD, P. A., WILDE, S. A. & SUN, M. 2002. Review of global 2.1–1.8 Ga orogens: Implications for a pre-Rodinia supercontinent. *Earth-Science Reviews*, **59**, 125–162.

ZHAO, G., SUN, M., WILDE, S. A. & LI, S. 2004. A Paleo-Mesoproterozoic supercontinent: Assembly, growth and breakup. *Earth-Science Reviews*, **67**, 91–123.

Palaeoproterozoic accretionary processes in Fennoscandia

RAIMO LAHTINEN[1]*, ANNAKAISA KORJA[2], MIKKO NIRONEN[1] & PEKKA HEIKKINEN[2]

[1]*Geological Survey of Finland, PO Box 96, FI-02151 Espoo, Finland*
[2]*Institute of Seismology, PO Box 68, FI-00014 University of Helsinki, Finland*
Corresponding author (e-mail: Raimo.Lahtinen@gtk.fi)

Abstract: Accretionary processes contributed to major continental growth in Fennoscandia during the Palaeoproterozoic, mainly from 2.1 to 1.8 Ga. The composite Svecofennian orogen covers $c.$ 1×10^6 km^2 and comprises the Lapland–Savo, Fennia, Svecobaltic and Nordic orogens. It is a collage of 2.1–2.0 Ga microcontinents and 2.02–1.82 Ga island arcs attached to the Archaean Karelian craton between 1.92 and 1.79 Ga. Andean-type vertical magmatic additions, especially at $c.$ 1.89 and $c.$ 1.8 Ga, were also important in the continental growth. The Palaeoproterozoic crust is the end product of accretionary growth, continental collision and orogenic collapse. Preserved accretional sections are found in areas where docking of rigid blocks has prevented further shortening. The Pirkanmaa belt represents a composite accretionary prism, and other preserved palaeosubduction zones are identified in the Gulf of Bothnia and the Baltic Sea areas. In the southern segment of the Lapland–Savo orogen collision between the Archaean continent (lower plate) and the Palaeoproterozoic arc–microcontinent assembly (upper plate) produced a special type of lateral crustal growth: the Archaean continental edge decoupled from its mantle during initial collision and overrode the arc and its mantle during continued collision.

Fennoscandia is one of the best-known Precambrian regions in the world. Lithological, petrological, geochronological, potential field, deep seismic reflection and refraction, and geoelectric data are available for many parts of Fennoscandia. Consequently, it is an important region for tracing Precambrian evolution. Archaean crust with a Palaeoproterozoic cover dominates in the east, Palaeoproterozoic Svecofennian crust in the centre, and Mesoproterozoic rocks are found in the SW (Fig. 1). The Palaeoproterozoic is the most important crust-forming era in Fennoscandia and the composite Svecofennian orogen forms the largest Palaeoproterozoic unit. It covers $c.$ 1×10^6 km^2 and extends southwards, under the Phanerozoic cover, as far as the Trans-European Suture Zone (TESZ in Fig. 1).

Hietanen (1975) was the first to propose a plate-tectonic interpretation for the Svecofennian rocks based on the comparison between the western Cordillera of North America and the Svecofennian orogen. After that study, numerous tectonic models have been presented (see Lahtinen *et al.* 2005, and references therein). Most models infer that subduction-type plate tectonics operated during 2.0–1.8 Ga in Fennoscandia.

The end phase of the classical Wilson cycle of ocean opening and closing is a continent–continent collision, a collisional orogeny. Accretionary orogeny takes place in an environment of continuing plate convergence and in most cases precedes the continent–continent collision (Cawood *et al.* 2009). Collisional orogeny involves continent-arc–continent collision that stitches two continents into one and is often followed by orogenic collapse. Accretionary orogenies increase crust vertically by magmatism and underthrusting, and laterally by arc–arc collisions and growth of accretionary wedges in the forearc (oceanic plateaux, sediments, etc.). Thus, where collisional processes predominantly rework the crust, arc magmatism and accretionary processes form new crust.

Accretionary orogens can be conceptualized with the help of two end-member types: retreating and advancing (see discussion by Cawood *et al.* 2009). Retreating orogens undergo long-term extension in response to a lower plate retreating with respect to the overriding plate and result in repeated cycles of forearc accretion; a modern example is the Japanese island arc. Lower plate retreat may also lead to back-arc basin opening, as exemplified by the Tertiary history of the western Pacific. Advancing orogens develop in an environment where the overriding plate is advancing towards the subducting plate, as exemplified by the Andes. Subduction-related voluminous calc-alkaline magmatism results in major vertical accretion, and lateral accretion occurs in extensive retro-arc fold and thrust belts.

The Palaeoproterozoic composite Svecofennian orogen is a collage of microcontinents and island

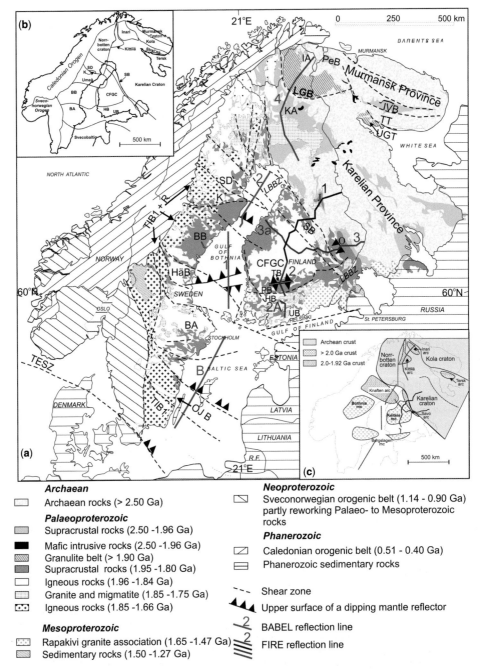

Fig. 1. Geological units of Fennoscandia. (**a**) A simplified geological map based on Koistinen *et al.* (2001) with major shear zones after Korja & Heikkinen (2005) and mantle reflections after Lahtinen *et al.* (2005). (**b**) Major Palaeoproterozoic terranes after Lahtinen *et al.* (2005). (**c**) Older than 1.92 Ga hidden and exposed microcontinental nuclei and arcs after Lahtinen *et al.* (2005). BA, Bergslagen area; BB, Bothnian basin; CFGC, Central Finland granitoid complex; HäB, Hälsingland belt; HB, Häme belt; IA, Inari area; IVB, Imandra–Varzuga belt; KA, Kittilä allochthon; LBBZ, Ladoga–Bothnian Bay zone; LGB, Lapland granulite belt; OJB, Oskarshamn–Jönköping belt; PB, Pirkanmaa belt; PeB, Pechenga belt; R, Revsund-type granite; SB, Savo belt; SD, Skellefte district; TESZ, Trans-European suture zone; TB, Tampere belt; TIB, Transscandinavian igneous belt; TT, Tersk terrane; UB, Uusimaa belt; UGT, Umba granulite terrane. J, Jormua; K, Knaften; O, Outokumpu.

arcs attached to the Archaean Karelian craton (Lahtinen et al. 2005; Korja et al. 2006a), and the Lapland–Kola orogen is a Palaeoproterozoic collision zone between the Archaean Karelian and Murmansk cratons (Daly et al. 2006). The Palaeoproterozoic orogenic history of Fennoscandia includes major accretionary stages but the present exhumed crustal level reveals continent–continent collision and collapse stage features.

We use available data on magmatic, geochemical and isotopic signatures, crustal architecture (seismic image), as well as structural and metamorphic data in Fennoscandia to probe the lithosphere in three dimensions. Based on these datasets we try to see beyond the collision and collapse stages and try to evaluate the composition and volume of accreted material and the nature of accretion processes during the Palaeoproterozoic crustal growth of Fennoscandia.

Geological outline of Fennoscandia

Fennoscandia represents the northern part of the East European craton (Gorbatschev & Bogdanova 1993). The Precambrian is exposed in the north and concealed under a thick Phanerozoic cover in the south (Fig. 1). The northeastern part of Fennoscandia is composed of Archaean and the central part of Palaeoproterozoic rocks, called the Svecofennian domain. The Archaean part has been dispersed and reassembled during the Palaeoproterozoic (Lahtinen et al. 2005; Daly et al. 2006).

The Palaeoproterozoic cover on the Archaean cratons comprises autochthonous supracrustal rocks, including rift-type basalts, deposited at and after 2.45 Ga, allochthonous series, and c. 1.95 Ga ophiolites (Melezhik & Sturt 1994; Bergman et al. 2001; Hanski & Huhma 2005; Laajoki 2005; Peltonen 2005). Continental break-up at c. 2.05 Ga has been proposed to take place at the western margin of the Karelian craton (Lahtinen et al. 2005, and references therein) and between the Karelian and Murmansk cratons in the NE (Daly et al. 2006, and references therein).

The 2.02 Ga bimodal association in the Kittilä allochthon (KA in Fig. 1a) is the oldest Palaeoproterozoic example of oceanic-affinity magmatism in Fennoscandia (island arc or oceanic plateau; Hanski & Huhma 2005). Island arc magmatism (1.98–1.96 Ga) has been recognized in the Kola region (Daly et al. 2006; TT in Fig. 1a). The oldest documented rocks in the Svecofennian domain are the 1.95 Ga Knaften granitoid and associated supracrustal rocks (Wasström 2005) in Sweden (K in Fig. 1a) and the 1.93–1.92 Ga primitive island arc rocks in the Savo belt, central Finland (SB in Fig. 1a).

Petrological, geochemical and isotopic data on magmatic rocks from some intrusive complexes or domains indicate that, in addition to juvenile crust, older 2.1–2.0 Ga lithosphere components also exist in the Svecofennian domain (Valbracht et al. 1994; Andersson 1997; Lahtinen & Huhma 1997; Rämö et al. 2001). Lahtinen et al. (2005) distinguished and delineated the following pre-1.92 Ga components in Fennoscandia (Fig. 1c): Karelian, Kola and Norrbotten Archaean cratons; Keitele, Bergslagen and Bothnia microcontinents (>2.0 Ga); Kittilä island arc and oceanic crust (c. 2.0 Ga); Savo, Knaften, Inari and Tersk island arcs (c. 1.95 Ga). The Palaeoproterozoic microcontinents (Keitele, Bergslagen and Bothnia) have no identified surface expressions, the remnants of the Knaften arc are found only at two places, and the Kittilä and Savo arcs are partly exposed at the surface.

The 1.90–1.87 Ga and 1.83–1.79 Ga age spans record the periods of most prominent Palaeoproterozoic magmatic activity in Fennoscandia. Island arc-type volcanic rocks (1.90–1.87 Ga) varying from less mature in the Skellefte district to mature in the Tampere belt, and calc-alkaline granitoids (e.g. in the Central Finland granitoid complex), are dominant in central Fennoscandia (Fig. 1a and b).

The southern part of the Svecofennian domain (Fig. 1a and b) includes the 1.90–1.88 Ga Bergslagen area (BA) and Uusimaa belt (UB), partly formed in an intra-arc basin of a mature continental arc (e.g. Kähkönen et al. 1994; Allen et al. 1996). Less-evolved island arc volcanic rocks are found in the Häme belt (HB, Fig. 1b; Kähkönen 2005). Plutonism in the area shows age peaks at 1.89–1.88, 1.87–1.85 and 1.83–1.79 Ga. These granites (partly S-type) and migmatites with granite leucosomes form a belt that extends from southeastern Finland to Sweden (Fig. 1a; Koistinen et al. 2001).

The 1.83 Ga Oskarshamn–Jönköping belt (OJB; Fig. 1) is characterized by supracrustal rocks intruded by calc-alkaline I-type granitoids (Mansfeld 1996). Otherwise, southern Sweden is dominated by granitoids of the Transscandinavian igneous belt, which continues to central Sweden (TIB in Fig. 1; Patchett et al. 1987). Three age groups of TIB have been identified: TIB 1 (1.81–1.77 Ga), TIB 2 (c. 1.7 Ga) and TIB 3 (1.68–1.65 Ga), of which TIB 1 is the most voluminous (Larson & Berglund 1992; Åhäll & Larson 2000; Gorbatschev 2004). The westernmost part of Fennoscandia (Fig. 1a and b) was strongly reworked during the Sveconorwegian–Grenvillian orogeny at 1.14–0.9 Ga (e.g. Gorbatschev & Bogdanova 1993; Bingen et al. 2005). Some of the rocks formed during the controversial Gothian orogeny (1.75–1.55 Ga; Åhäll & Larson 2000; or c. 1.6 Ga; Andersson et al. 2002).

Seismic data

Like most Precambrian shield areas, the Precambrian region in Fennoscandia comprises eroded surfaces without elevation and direct 3D information. The information on vertical direction is based on indirect geophysical measurements such as seismic reflection and refraction profiles. Surface geological and geophysical observations are correlated with the seismic reflection patterns close to the surface. The 2D seismic reflection sections image 2D projections of 3D reflective structures.

The reflections have apparent dips that depend on the true dips of the reflectors and the interception angle between the survey line and the strike of the reflector (Kukkonen et al. 2006). Depending on the view-angle the same reflector may be imaged as a near-vertical to subhorizontal reflection. The interpretation is further complicated as the structures may change dip and strike both vertically and laterally along the line; for example, crustal structures at depth may have different strikes from those observed at the surface.

The seismic data used in this paper (Fig. 1a) are from the deep reflection lines BABEL (BABEL Working Group 1990; Korja & Heikkinen 2005) and FIRE (Kukkonen et al. 2006). The BABEL lines B, 1 and 2 are marine seismic profiles in the Baltic Sea and the Gulf of Bothnia. The FIRE lines 1, 2 and 3 are land profiles in Central and Southern Finland.

The major difference between the BABEL and FIRE sections is in the reflectivity of the upper crust. In the BABEL sections the uppermost crust down to depth about 5 km looks rather transparent when compared with the FIRE sections, in which the upper crust is very reflective. This variation is caused by the differences in the acquisition environment. The air-gun source generates a strong direct pressure wave in the water column, which masks the weak reflections in marine surveys.

The seismic sections shown here are migrated normal moveout (NMO) stacks. They are displayed as instantaneous amplitude sections, averaged both horizontally and vertically and plotted as grey-scale intensities (for technical details, see Korja & Heikkinen 2005; Kukkonen et al. 2006). All the sections are plotted without normalization; that is, the amplitudes of the different areas in each section are comparable. In the sections, large-scale reflectivity changes are pronounced, whereas the details of single reflections are obscured.

Accretionary processes through time and space in Palaeoproterozoic Fennoscandia

The Palaeoproterozoic orogenic evolution of Fennoscandia can be divided into two major stages (Table 1): formation of the Lapland–Kola orogen at 1.94–1.86 Ga (Daly et al. 2006) and formation of the composite Svecofennian orogen at 1.92–1.79 Ga (Lahtinen et al. 2005; Korja et al. 2006a). The Svecofennian orogen is further divided into the Lapland–Savo, Fennia, Svecobaltic and Nordic orogens (Fig. 2).

Table 1. *Major Palaeoproterozoic orogens in Fennoscandia (see Fig. 2)*

	Age of main collision (Ga)	Type	Components	Notes
Lapland–Kola	1.93–1.91	C–C	A–A	Assembly continued until c. 1.86 Ga
Svecofennian				
Lapland–Savo; north	c. 1.92	C–C	A–A	Not well defined and includes only the Kittilä allochthon
Lapland–Savo; south	c. 1.92–1.91	C–C	A–P	Karelia–Keitele collision and docking of the Knaften arc and Bothnia microcontinent
Fennia	1.89–1.86	Acc	(A)P–P	Accretion of arcs and the Bergslagen microcontinent to a newly formed continent
Svecobaltic	1.83–1.79	C–C	(A)P–P(A)	Sarmatia–Fennoscandia collision in the east at c. 1.82 Ga and Fennoscandia–unknown continent in the SW at c. 1.80–1.79 Ga
Nordic	c. 1.82–1.79	C–C/Acc	?	Fennoscandia–Amazonia? collision or advancing accretionary orogen with retro-arc fold and thrust belts

Lapland–Kola and composite Svecofennian orogens modified after Daly et al. (2006) and Lahtinen et al. (2005), respectively. C–C, continent–continent collision; Acc, accretionary; A, Archaean; P, Palaeoproterozoic.

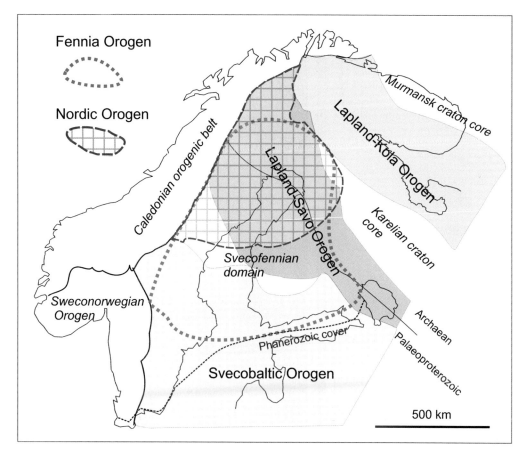

Fig. 2. Major Palaeoproterozoic orogens in Fennoscandia.

Accretion in the Lapland–Kola orogen

The Lapland–Kola orogen is a 700–800 km long and 300–400 km wide orogenic root (Fig. 2) comprising largely reworked Archaean crust. The Lapland–Kola orogen formed in a continent–continent collision between the Karelian and Murmansk cratons (Lahtinen et al. 2005; Daly et al. 2006, and references therein). Daly et al. (2006) have proposed the closure of two oceans of modest width, marked by the Pechenga–Imandra–Varzuga (PeB and IVB in Fig. 1a) and Lapland–Kola (SW boundary of LGB in Fig. 1a) sutures. Lahtinen et al. (2005) have favoured closure of only one ocean along the Pechenga–Imandra–Varzuga suture and related the Lapland and Umba granulites (Fig. 1a) to the shortening and thickening of a pre-heated, at least partly ensialic, back-arc basin.

The composite Strel'na and Inari terranes (Fig. 1b) have Archaean rocks and evolved arc-type 1.98–1.94 Ga calc-alkaline rocks (Daly et al. 2006). The Lapland granulite belt and Umba granulite terrane (Fig. 1a) are composed of sedimentary rocks intruded by 1.94–1.91 Ga arc-type calc-alkaline norites and enderbites (Daly et al. 2006; Tuisku & Huhma 2006). This calc-alkaline magmatism could be linked with an Andean-type setting with vertical addition of subduction-related magma but the amount of this addition is unknown. The 1.98–1.96 Ga Tersk terrane (Fig. 1a–c) is interpreted to consist of juvenile island arc rocks (Daly et al. 2006) and is considered as a true accreted terrane. The areal extent of the preserved juvenile arc (lateral accretion) and the Andean-type granitoids (vertical accretion) is less than 10 000 km².

Accretion in the Lapland–Savo orogen

The Lapland–Savo orogen can be divided into northern and southern segments (Table 1 and Fig. 2). The northern segment formed during the collision of the Karelian and Norrbotten Archaean cratons, and the southern segment formed during

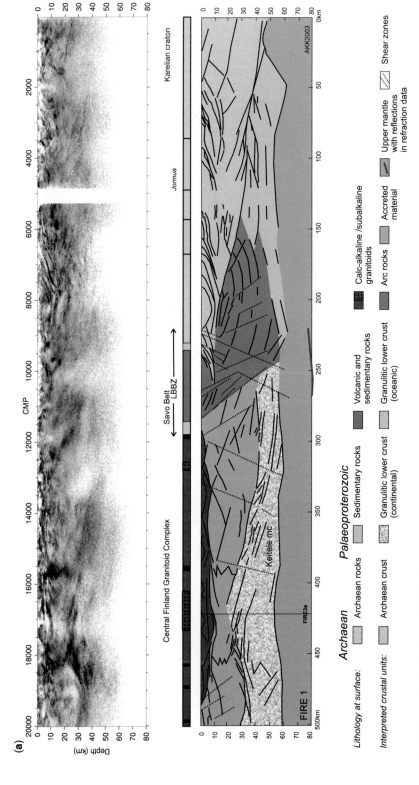

Fig. 3. Deep seismic reflection profiles FIRE 1 (a) and FIRE 3 and 3a (b) plotted as straight lines. Depth is in kilometres and distance in common mid-points (CMPs) and kilometres. No vertical exaggeration. Upper panel: an averaged instantaneous amplitude section after Kukkonen et al. (2006). Lower panel: a line drawing and a geological interpretation modified from Korja & Heikkinen (2008). Lithology at the surface is after Korsman et al. (1997) and Korja et al. (2006b). (a) FIRE 1; the cross-point of line FIRE 3a–3 is marked by a vertical line. (b) FIRE 3 and 3a; the crosspoint of line FIRE1 is marked by a vertical line.

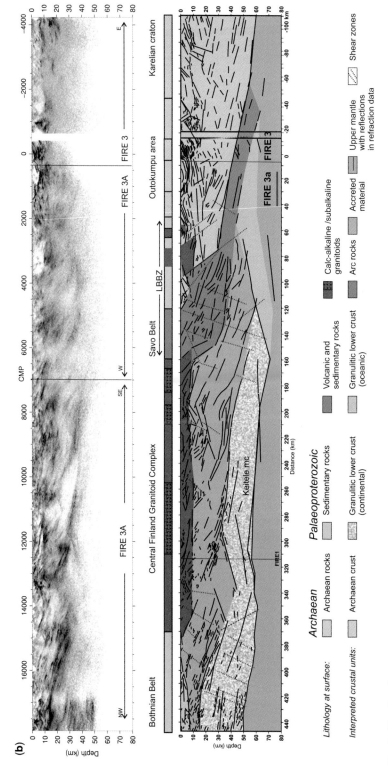

Fig. 3. (Continued).

the collision of the Karelian craton and the Keitele microcontinent (Lahtinen et al. 2005). The accreted terranes are the Kittilä allochthon (arc) in the north and the Savo belt (arc) in the south (Fig. 1a–c). The Lapland–Savo orogen includes also the docking of the Knaften arc and the Bothnia microcontinent to the western margin of the Keitele–Karelia–Norrbotten collage.

Northern segment

The 2.02 Ga felsic porphyries and coeval tholeiitic mafic volcanic rocks in the Kittilä allochthon show no indication of sialic contamination, and the geochemistry of the mafic rocks indicates oceanic island arc and oceanic plateau affinities (Hanski & Huhma 2005). The Nuttio serpentites and dunites, found at the eastern edge of the Kittilä allochthon, are cut by numerous dykes with island arc tholeite, calc-alkaline and boninitic affinities. Hanski & Huhma (2005) have interpreted this association to comprise dismembered ophiolitic rocks formed in a suprasubduction zone environment. A unit of shallow-water light rare earth element (LREE)-enriched mafic rocks with less primitive Nd isotope composition and different geochemistry (e.g. low Nb/Th) has been interpreted as crustally contaminated rocks formed possibly at a passive margin (Hanski & Huhma 2005).

Seismic reflection data indicate that the Kittilä allochthon is a 1–10 km thick unit preserved in the core of intervening orogens (Patison et al. 2006). The Kittilä allochthon has dominantly been metamorphosed in greenschist facies but in the NE there is sharp change to mid-amphibolite-facies rocks (Hölttä et al. 2007). Seismic reflection line FIRE 4 (Fig. 1a) suggests that the mid-amphibolite-facies rocks have been thrust on top of the greenschist-facies rocks from the NE along with the Lapland granulite belt (Patison et al. 2006). The available data suggest that the Kittilä allochthon is underlain and surrounded by rocks from the Karelian craton and its cover.

Our interpretation is that the Kittilä allochthon is a collage of allochthonous fragments of an island arc and parautochthonous fragments of a passive margin. The close association of oceanic plateaux and arc rocks suggests that the Kittilä island arc was of the retreating type. The Kittilä island arc was originally attached to the Norrbotten nucleus and was subsequently overthrust onto the Karelian craton during the continent–continent collision.

Southern segment

The southern segment of the Lapland–Savo orogen comprises the Central Finland granitoid complex, formed on the 2.1–2.0 Ga Keitele microcontinent, and the Savo belt adjacent to the Karelian craton (Fig. 1a–c). The geochemical and isotopic composition of plutonic rocks in the Central Finland granitoid complex indicates that the Keitele microcontinent was composed of mature crust and associated enriched subcontinental lithosphere at c. 1.9 Ga (Lahtinen 1994; Lahtinen & Huhma 1997; Nironen et al. 2000; Rämö et al. 2001). The pre-1.91 Ga evolutionary history of the Keitele microcontinent is unknown. Based on the seismic data (Kukkonen et al. 2006; Sorjonen-Ward 2006; Korja & Heikkinen 2008; Fig. 3a and b) we suggest that the Keitele microcontinent is a c. 20–30 km thick rigid unit with much wider extent before thickening.

The Savo belt, the primitive arc complex of Korsman et al. (1997), is a 350 km long and 20–50 km wide belt comprising migmatitic mica gneisses and 1.92–1.93 Ga gneissic tonalites and coeval mafic to felsic metavolcanic rocks. Based on the Sm–Nd data no Archaean component is found in these rocks (Lahtinen & Huhma 1997), and rather juvenile origin is also indicated by the island arc affinity of the metavolcanic rocks (Kousa et al. 1994; Lahtinen 1994; Kähkönen 2005). No clear suture between the Keitele microcontinent and the arc complex has been identified (Korja et al. 2006b), and thus we consider that the 1.93–1.92 Ga island arc was attached to the Keitele microcontinent prior to the 1.92–1.91 Ga collision.

The Savo belt is characterized by juvenile island arc material that continues beneath the craton edge based on the seismic interpretation (Fig. 3a and b). Mantle xenolith and xenocryst studies have identified the occurrence of Palaeoproterozoic mantle components to a depth of 110 km under the craton margin (Peltonen & Brügmann 2006). A mantle reflector at 75 km depth under the craton margin (Figs 1a and 3b) indicates an important velocity and density contrast in the upper mantle, which is interpreted as the lower boundary of the accreted oceanic crustal block and its mantle (Fig. 3b). The Archaean–Palaeoproterozoic mantle suture zone seems to be complex and interlayered. We conclude that the crust of the thinned craton margin was decoupled from its mantle, and that the crust then overthrust the main part of the Savo arc and its mantle during collision.

The strong shortening and metamorphic overprint prevents us from seeing details of the pre-collisional accretion history of the Savo arc. However, an important bimodal stage associated with voluminous VMS-type Zn–Cu mineralizations, similar to the Kuroko-type deposits in Japan, has been documented (Lahtinen 1994; Weihed et al. 2005), and some mafic volcanic rocks show rift–marginal basin or oceanic plateau geochemical affinities (Lawrie 1992; Ekdahl 1993;

Lahtinen 1994). Thus, we suggest that a retreating type accretionary setting (oceanic fragments) with intra-arc rifting (VMS mineralization) existed before the Karelian craton–Keitele microcontinent collision.

The c. 1.95 Ga Jormua ophiolite at the western edge of the Karelian craton (Fig. 1a) is one of the world's best-documented Precambrian ophiolites. The Jormua basalts are underlain by old subcontinental lithospheric mantle and the overall environment is continental (Peltonen 2005).

Knaften arc

The Lapland–Savo orogen incorporates the termination of magmatism and docking of the Knaften arc to the Keitele–Karelia–Norrbotten collage (Figs 1a–c and 2). Mid-ocean ridge basalt (MORB)- to island arc-type metavolcanic rocks found in Knaften (Fig. 1a) have a minimum age of 1.95 Ga based on the intruding granitoids (Wasström 2005, and references therein). Coeval metadacite west of the Skellefte district (Eliasson & Sträng 1998) verifies the wider occurrence of 1.96–1.95 Ga island arc magmatism. The Sm–Nd data from calc-alkaline c. 1.9 Ga plutonic rocks in the Skellefte district (Mellqvist et al. 1999, and references therein) indicate derivation mainly from a juvenile source, possibly from Knaften-type arc rocks.

The juvenile Knaften arc and related accreted material probably continues to the south under an upper crustal allochthon (Fig. 4). The allochthon (Umeå in Fig. 1b) coincides with the pre-1.9 Ga sequence outlined by Rutland et al. (2001). We suggest that the 'basement' of the Skellefte district and its continuation to the south is composed of c. 1.95 Ga juvenile oceanic material and accreted sediments (Knaften arc; Fig. 1c) with an undefined accretion history.

A well-preserved palaeosubduction zone dipping to the NE is observed in the BABEL 2 profile (BABEL working group 1990) between the Bothnia microcontinent (lower plate) and the Keitele microcontinent + Knaften arc (upper plate) (Fig. 4). As a result of docking of lower crustal rigid units (Fig. 4) the configuration of the subduction system seems to have retained much of its original shape (Korja & Heikkinen 2008).

Accretion in the Fennia orogen

The Keitele–Karelia collision (1.92–1.91 Ga) caused a change in the plate motions. After this event, convergence between the Keitele–Karelia collage and the Bergslagen microcontinent (Fig. 1b) was accommodated by northward subduction at the southern margin of the Keitele–Karelia collage (TB; Fig. 1a and b). Almost simultaneously, a southward subduction commenced. The subductions terminated when the Bergslagen microcontinent collided with the Keitele microcontinent and initiated the Fennia orogen. The major components in the Fennia orogen (Fig. 2) are the Tampere, Pirkanmaa, Häme and Uusimaa belts in Finland, and the Skellefte district, Bothnian basin, Hälsingland belt and Bergslagen area in Sweden (Fig. 1a).

Tampere and Pirkanmaa belts, and the Skellefte district and Bothnian basin

The Tampere belt (TB; Fig. 1a and b) (1.905–1.89 Ga) volcanic rocks formed in a mature continental island arc or an active continental margin-type tectonic setting (Kähkönen 1987; Lahtinen 1996). These volcanic rocks are underlain by turbiditic greywackes and mudstones, and enriched (E)-MORB- or within-plate basalt (WPB)-affinity lavas, and overlain by transitional island arc tholeiite (IAT)- or WPB-affinity mafic volcanic rocks, conglomerates and arenites (Kähkönen 2005).

The mantle-like lead data and ε_{Nd} (1.9) + 0.5 of the E-MORB- or WPB-affinity basalts (Vaasjoki & Huhma 1999) are interpreted to have been derived from the Keitele subcontinental lithospheric mantle, during a pre-1.91 Ga rift stage (Lahtinen 1996; Lahtinen & Huhma 1997). In such a scenario the bulk of the turbidites in the Tampere and Pirkanmaa belts (TB and PB, Fig. 1a) are derived from the rising Lapland–Savo orogen in the NE. Although the rift or marginal basin (Kähkönen & Nironen 1994) model for these lavas is favoured, an accreted oceanic ridge or plateau origin cannot be excluded.

Migmatites and gneisses of turbiditic origin characterize the Pirkanmaa belt (PB in Fig. 1a). These rocks are generally chemically similar to the Tampere belt rift-stage greywackes (≥ 1.91 Ga) although arc-derived forearc sediments (≤ 1.90 Ga) are also found (Lahtinen 1996). MORB- to WPB-affinity mafic and ultramafic volcanic rocks are locally found in the Pirkanmaa belt (Lahtinen 1994, 1996; Peltonen 1995).

We propose that the Pirkanmaa belt (Fig. 1a) is a composite accretionary prism including early rift-stage sediments and volcanic rocks (>1.91 Ga), forearc (1.91–1.89 Ga) sediments and 1.91–1.89 Ga accreted material. Without any clear markers it is very difficult to define the pre-collision structures but the earliest foliations (Kilpeläinen 1998) could possibly be related to deformation within the accretionary wedge. MORB-affinity basalts in the southern edge of the Pirkanmaa belt could be either oceanward representatives of the rift stage or accreted basalt–sediment slices from oceanic lithosphere.

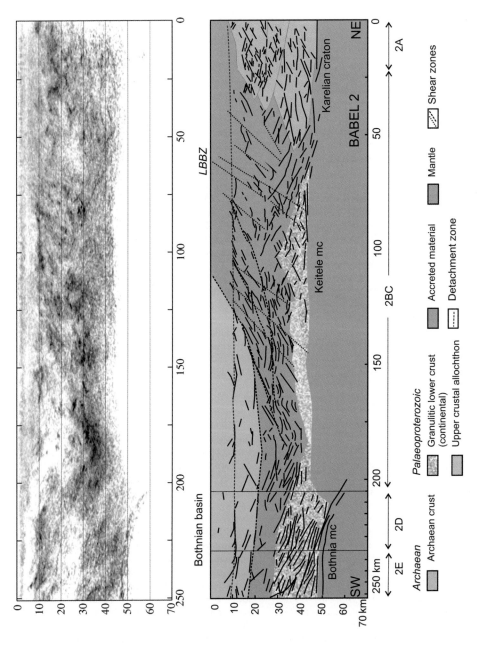

Fig. 4. Deep seismic reflection profiles BABEL 2A–E modified from Korja & Heikkinen (2005). No vertical exaggeration. The section is shown as a straight line and the major turning points at 205 km and 231 km are marked by vertical lines. Upper panel: an averaged instantaneous amplitude section; lower panel: a line drawing and a geological interpretation. mc, microcontinent.

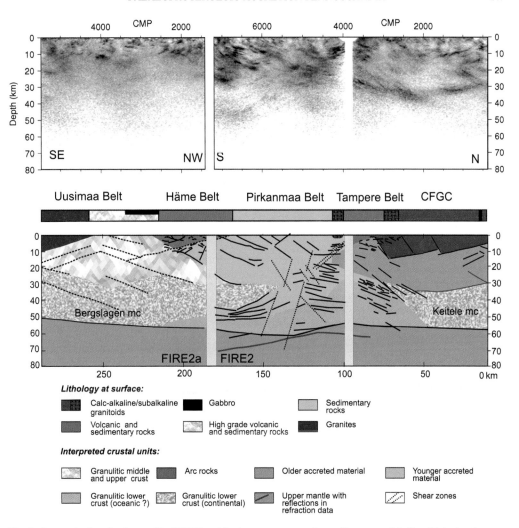

Fig. 5. Deep seismic reflection profiles FIRE 2 and 2a shown as a composite profile on a straight line. Major breaks are marked by white vertical columns. Distance is in CMPs and kilometres. No vertical exaggeration. Upper panel: an averaged instantaneous amplitude section after Kukkonen *et al.* (2006). Lower panel: a line drawing and a geological interpretation modified from Korja & Heikkinen (2008). Lithology at the surface is after Korsman *et al.* (1997) and Nironen *et al.* (2006).

The north-dipping mantle reflector (Fig. 1a), seismic reflectivity (Fig. 5) and conductivity anomaly (Hjelt *et al.* 2006) below the Tampere belt are interpreted as a north-dipping subduction system under the Keitele microcontinent. The lower crustal fragment between the Keitele and Bergslagen microcontinents (Fig. 5) could represent a part of an island arc, an oceanic plateau or a continent.

The Tampere arc is interpreted as a continental arc formed in an advancing-type accretionary setting on the Keitele microcontinent. Some accretion of oceanic crust and sediments may have also occurred but the main addition is magmatic and vertical.

Offshore to the west of the Pirkanmaa belt, a subduction system of similar geometry (a lower crustal fragment and double plunging reflectors) has been identified (Fig. 6; Korja & Heikkinen 2005). The Tampere belt curves northwestward at its western edge, where there is also a north–south-oriented shear zone along the Finnish shoreline (Fig. 1a). The Tampere belt does not have a surface continuation in Sweden and the offshore subduction system shows northward displacement relative to the Pirkanmaa belt one. The shear zone has been interpreted as part of a major transform fault system separating the two subduction systems (Lahtinen *et al.* 2005).

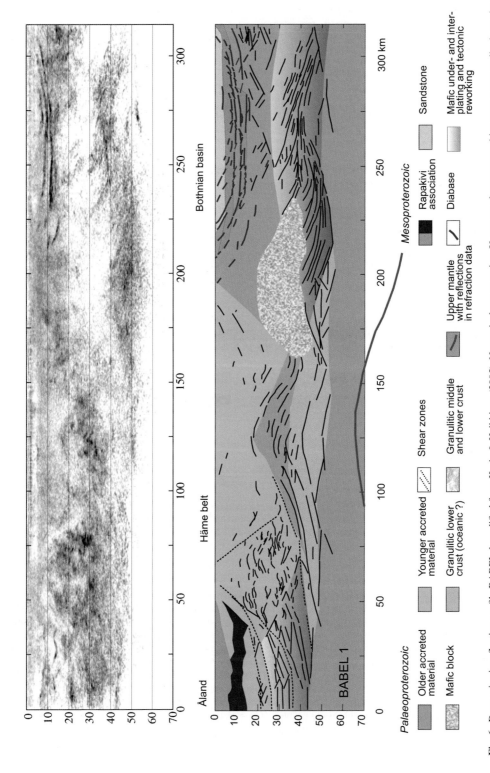

Fig. 6. Deep seismic reflection profile BABEL 1 modified from Korja & Heikkinen (2005). No vertical exaggeration. Upper panel: an averaged instantaneous amplitude section. Lower panel, a line drawing and a geological interpretation.

The 1.88–1.87 Ga volcanism in the Skellefte district is interpreted as back-arc volcanism associated with the north-directed subduction below the Gulf of Bothnia (Figs 1a and 6). The mafic block in Figure 6 is a possible example of accreted material; part of an island arc or oceanic plateau (Korja & Heikkinen 2005).

Uusimaa and Häme belts and the Hälsingland belt and Bergslagen area

The boundary between the Pirkanmaa and Häme belts is considered to be the surface extension of the boundary between two exotic terranes (Korsman *et al.* 1997; Nironen *et al.* 2006). The northern part of the Häme belt is a layer of a few kilometres thickness possibly thrust southwards during formation of the pop-up geometry of the Pirkanmaa belt (Nironen *et al.* 2006). The actual collisional boundary is buried underneath (Fig. 5). The Häme and Uusimaa belts (Fig. 1a) are partly geological and partly geographical terms, and the correlations within and between these areas are very poorly constrained.

The Häme belt metavolcanic rocks can be divided into older arc rocks and younger rift-affinity basalts (Hakkarainen 1994; Lahtinen 1996). The southernmost part of the Uusimaa belt in Finland shows isotopic evidence for evolved crust with a long crustal history (Lahtinen & Huhma 1997; Rämö *et al.* 2001) but the central part contains more juvenile arc rocks with variation from island arc (1895 Ma) to mature arc affinity (1878 Ma) (Väisänen & Mänttäri 2002). The arc stage seems to have been preceded and succeeded by mafic–ultramafic rift-related volcanism (Ehlers *et al.* 1986; Väisänen & Mänttäri 2002). Small volcanogenic massive sulphide deposits in the Uusimaa belt have been correlated with the VMS deposits in the Bergslagen area (Weihed *et al.* 2005).

The Bergslagen area is 1.90–1.87 Ga in age and the volcanic succession is dominated by calc-alkaline rhyolites. Allen *et al.* (1996) interpreted the area as an extensional intra-continental or continental margin back-arc region. The Hälsingland belt is characterized by active continental margin magmatism at 1.86–1.84 Ga, and it was later affected by migmatization at 1.82 Ga (Högdahl *et al.* 2007).

Early island arc type magmatism characterizes the Häme belt, whereas back-arc magmatism in a continental setting characterizes the Bergslagen area and parts of the southernmost Uusimaa belt (Bergslagen mc in Fig. 1c). Lahtinen *et al.* (2005) have suggested that an oceanic plate subducted southward, beneath the Bergslagen microcontinent and an attached island arc crust, at 1.90–1.88 Ga.

On the other hand, Väisänen & Mänttäri (2002) have suggested subduction to the north in the Uusimaa belt.

The close juxtaposition of continental-type (continental margin back-arc environment rocks and sedimentary carbonates), arc-type (island arc and mature arc), oceanic-affinity, cratonic (pure metasandstones) and continental-rift (sandstones and tholeiitic volcanic rocks) rocks in the Häme and Uusimaa belts and the Bergslagen area is difficult to interpret as one evolving system. It is more probable that these are crustal segments partly exotic to each other and that they have been tectonically transported to their present locations as proposed by Korja & Heikkinen (2005) and Nironen *et al.* (2006).

Accretion in the Svecobaltic orogen

A north-directed subduction was active at 1.86–1.84 Ga and caused magmatism both along the southwestern margin of the Bergslagen area and in the Hälsingland belt (Fig. 1a; Åhäll & Larson 2000; Andersson & Wikström 2001; Hermansson *et al.* 2007; Högdahl *et al.* 2007). The mantle reflector in BABEL B (Figs 1a and 7) is interpreted as the subduction zone trace for the subduction-related magmatism (Abramovitz *et al.* 1997; Korja & Heikkinen 2005). Subsequent crustal extension led to the formation of intra-orogenic clastic sedimentary basins (Bergman *et al.* 2007).

The Svecobaltic orogen (Table 1 and Fig. 2) started to form when an oblique (transpressional) collision between Sarmatia and Fennoscandia commenced in the SE. Transpressional tectonics can be observed as large-scale thrusting and margin-parallel (earth–west) shear zones in central Sweden and southern Finland (e.g. Högdahl & Sjöström 2001; Ehlers *et al.* 2004). Thickening of the pre-heated crust led to local migmatization and formation of minimum-melt granites from sediments and igneous rocks (Korja & Heikkinen 2005). No subduction-related rocks of this age are found in Finland and this favours a model of pure tectonic inversion of extended Fennian crust in continent–continent collision.

Simultaneously with the collision in the SE (1.83–1.80 Ga), a subduction regime, west of a crustal-scale shear system (Mid-Lithuanian Suture Zone; Skridlaite & Motuza 2001), was still active in the southwestern margin. The *c.* 1.83 Ga Oskarshamn–Jönköping belt (OJB; Fig. 1a), south of the Bergslagen area, shows highly positive ε_{Nd} values, indicating a juvenile island arc origin for the belt (Mansfeld & Beunk 2004; Mansfeld *et al.* 2005). A possible continuation of the Oskarshamn–Jönköping belt is found in western Lithuania (e.g. Skridlaite & Motuza 2001).

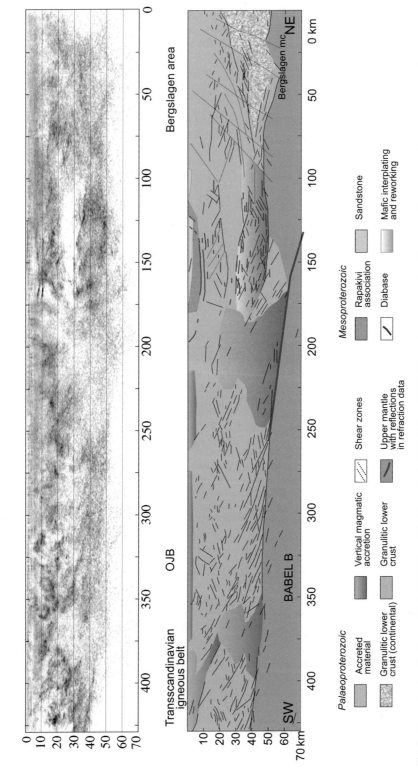

Fig. 7. Deep seismic reflection profile BABEL B modified from Korja & Heikkinen (2005). No vertical exaggeration. Upper panel: an averaged instantaneous amplitude section. Lower panel: a line drawing and a geological interpretation.

Beunk & Page (2001) interpreted metamorphism and deformation (1.82–1.80 Ga) north of the Oskarshamn–Jönköping belt to have occurred in a back-arc environment and to have been caused by an oblique accretion of a magmatic arc (OJB) onto the continental margin (Bergslagen; Figs 1a and 7). A WNW–ESE- to east–west-trending belt of c. 1.8 Ga para- and orthogneisses in the SE corner of Sweden (Gorbatschev 2004) implies that a second (continental) arc accreted to the Bergslagen microcontinent from the SW. This final collision or docking is seen as later 1.80–1.78 Ga 'cold' transpressive shortening in a north–south direction NNE of the Oskarshamn–Jönköping belt (Beunk & Page 2001).

We suggest that an Andean-type active margin retreated southwestward and caused cyclic periods of subduction-type and marginal basin-type magmatism west of the crustal-scale shear system before the final continent–continent collision at 1.80–1.79 Ga (Lahtinen et al. 2005).

Accretion in the Nordic orogen

Lahtinen et al. (2005) proposed that the Nordic orogen resulted from a continent–continent collision between the newly formed Fennoscandia and another continent, possibly Amazonia. Another possibility is to interpret the Nordic orogen as an advancing Andean-type accretionary orogen, where the inboard deformation relates to retro-arc fold and thrust belts (Table 1 and Fig. 2).

The c. 1400 km long Transscandinavian igneous belt (TIB; Patchett et al. 1987; Gorbatschev & Bogdanova 1993) is a collective term for a linear north–south-trending batholith belt, partly covered by Caledonian rocks. The WNW–ESE-trending TIB 1 (c. 1.8 Ga) belt in southern Sweden is related to the NE-directed Andean-type subduction and only the NE–SW belt of TIB 1 + R (1.81–1.77 Ga) granitoids in the north is included in the Nordic orogen (Fig. 1a–c; Lahtinen et al. 2005).

The TIB 1 granitoids intruded during extension in either a convergent margin or intracratonic setting between 1.80 and 1.78 Ga (e.g. Andersson 1997; Åhäll & Larson 2000). The 'post-orogenic' 1.80–1.78 Ga Revsund granitoids are a rather homogeneous group of predominantly A- to I-type granitoids (e.g. Claesson & Lundqvist 1995; Andersson 1997).

The intruded batholith chain, at present occupying an 800 km long and 100–200 km wide area, may be regarded as an advancing orogenic setting with Andean-type vertical magmatic accretion. The complicated and partly localized deformation seen in the north (Bergman et al. 2006; Hölttä et al. 2007) could be associated retro-arc fold and thrust belts. Thus, an advancing convergent plate-tectonic system with or without subsequent continent–continent collision is possible.

Discussion

It is a challenge to study Precambrian shield areas with their strongly reworked crust, exhumed surfaces, flat topography and the lack of direct 3D information. Accretionary stage events are overprinted by continent–continent collision and collapse events. Palaeoproterozoic collisional events in Fennoscandia have resulted in anomalous crustal and lithospheric thicknesses, up to 65 km and 250 km, respectively. The Palaeoproterozoic Svecofennian domain has an area of c. 1×10^6 km^2 of c. 50 km thick crust that results in a total volume of 50×10^6 km^3. With the help of the surface observations and few drillings (normally <500 m) we can directly observe only ≪1% of the crust and all the other interpretations are based on indirect observations.

One example of an end-product of an eroded continent–continent collision between two Archaean continents is the Lapland–Kola orogen, which is about 700–800 km long and 300–400 km wide with an area of c. 250 000 km^2. The laterally (island arc rocks) and vertically (Andean-type magmatism) accreted juvenile components cover less than 10 000 km^2 of the orogenic root at the surface. The wide and long-lived metamorphic imprint in the orogenic footwall favours the occurrence of a Himalayan-scale mountain belt (Daly et al. 2006); probably substantial amounts of juvenile material were thrust and subsequently exhumed and eroded.

Similarly, the amount of juvenile material (Kittilä allochthon) preserved in the northern segment of the Lapland–Savo orogen is less than 5000 km^2. It is impossible to estimate the original amount of juvenile addition in this case. Palaeoproterozoic metasediments are abundant in Fennoscandia and they have typically 60–70% of Palaeoproterozoic detrital zircons in the age range of 1.92–2.1 Ga (Claesson et al. 1993; Lahtinen et al. 2002).

In the southern segment of the Lapland–Savo orogen the Karelian craton collided with the Palaeoproterozoic Savo arc and the Keitele microcontinent. The more buoyant and rigid Karelian passive margin wedge overrode the young and hot island arc (Fig. 3a and b). In this case there is c. 12 000 km^2 of juvenile material exposed at the surface. In addition to material eroded during orogen exhumation, large amounts of juvenile material are preserved at depth beneath the Archaean craton margin.

Accretionary wedges can be erosional and poorly developed. The wedges have also a high potential to be thrust and subsequently eroded during collision. Thus, the preservation potential of an accretionary wedge is poor as is also the case in Fennoscandia. The Pirkanmaa belt could be a combination of a forearc and an accretionary wedge that formed at a rifted continental margin. Rather well-preserved crustal structures as well as proposed subduction traces (Fig. 5) are due to the docking of rigid fragments that prevented further shortening (Korja & Heikkinen 2008). Another example of this is seen in the preservation of the palaeosubduction zone below the BABEL 2 profile (Fig. 4).

The ultimate complexity of collision systems is seen in southern Finland, where there are rocks from oceanic, island arc to active continental margin affinity, juxtaposed with rocks of craton to continental rift affinity. The two-stage orogenic evolution, forming the Fennia and Svecobaltic orogens, partly produced similar structural successions (e.g. Skyttä et al. 2006, and references therein) and there is a need for more detailed work and dating to resolve the complexities.

Only small fragments of the accretion systems are preserved but both retreating and advancing types can be recognized in Fennoscandia. The Tersk, Kittilä, Savo and Häme arcs as well as the Knaften arc could be classified as retreating types. The Oskarshamn–Jönköping belt or arc is either retreating or advancing. These arcs show oceanic to mature but juvenile geochemical and isotopic signatures without evidence for derivation from an older evolved continental crust. Arc rocks in the Tampere, Bergslagen and TIB 1 belts show isotopic and geochemical signatures of a mature and thick crust with associated sub-continental lithospheric mantle, which favours an advancing Andean-type setting for these belts.

Early accretion stage metamorphic patterns are often overprinted by metamorphism during continent–continent collision. No data are available for the possible accretion stage metamorphism in the Lapland–Kola orogen, which is characterized by collision-related granulite-facies metamorphism (Daly et al. 2006; Tuisku & Huhma 2006; Hölttä et al. 2007).

The Kittilä arc of the Lapland–Savo orogen is characterized by greenschist-facies metamorphism (Hölttä et al. 2007). Pluton-driven (c. 1885 Ma) HT–LP amphibolite- to granulite-facies metamorphism (Hölttä 1995; Korsman et al. 1999) characterizes the Savo arc in the southern segment of the Lapland–Savo orogen. This metamorphism occurred late in the collision history and no data are available for the earlier accretion stage metamorphism.

Metamorphism in the composite accretionary prism (PB in Fig. 1a) of the Tampere belt is characterized by migmatites with early stage leucosomes poor in K-feldspar followed by increasing amounts of K-feldspar in later leucosomes (Mouri et al. 1999). HT–LP metamorphism culminated at 1.88 Ga (Mouri et al. 1999; Rutland et al. 2004) but early subhorizontal structures and associated thin veining (Kilpeläinen 1998) could be related to earlier accretion. Subduction of buoyant (young and hot) oceanic lithosphere during early stages of collision could have caused a high geothermal gradient and early migmatization (K-poor). The following collision-related contraction caused pop-up geometry, uplift and younger migmatization (K-enriched) as a result of pressure decrease.

In the Southern Finland and Bergslagen area an early (c. 1.88–1.87 Ga) metamorphic peak characterizes the Fennia orogen and a younger (1.83–1.80 Ga) metamorphic peak the Svecobaltic orogen (Andersson et al. 2006; Skyttä et al. 2006; Högdahl et al. 2007, and references therein).

Accretion and collision involves the evolution of the sub-continental lithospheric mantle. The complex crustal image of the juvenile Savo arc (former upper plate), continuing under the Karelian craton edge (former lower plate) (Fig. 3a and b), is accompanied by the complex and interlayered Archaean–Palaeoproterozoic mantle lithosphere boundary below (Sandoval et al. 2003; Bruneton et al. 2004; Yliniemi et al. 2004; Peltonen & Brügmann 2006; Plomerová et al. 2006).

Unfortunately, we do not have any mantle xenolith or xenocryst information from the Svecofennian part of Fennoscandia, and thus we do not have direct information from the Svceofennian sub-continental lithospheric mantle. We propose that the thick keel (possibly up to 250–300 km; O'Brien et al. 2005; Hjelt et al. 2006) in the central part of Fennoscandia is due to intersecting orogens (Fig. 2). The major component in the thick sub-continental lithospheric mantle under the Palaeoproterozoic Svecofennian is probably the stacked Keitele 2.0–2.1 Ga sub-continental lithospheric mantle.

Palaeoproterozoic crustal growth of Fennoscandia can be divided into the formation of 2.0–2.1 Ga microcontinents with undefined evolution history and juvenile arcs from >2.02 Ga (Kittilä) to c. 1.82 Ga (Oskarshamn–Jönköping). Andean-type vertical additions are seen as the c. 1.89 Ga magmatism at the southern rim of the Central Finland granitoid complex and as the c. 1.8 Ga magmatism in Sweden (TIB 1 and Revsund magmatism). Only few detrital zircons in the age range 2.5–2.1 Ga have been found in Fennoscandia, which also favours the main Palaeoproterozoic crustal growth episode occurring during 2.1–1.8 Ga.

Conclusions

The central part of Fennoscandia is a Precambrian cratonic area characterized by deeply exhumed and thick crust associated with a thick subcontinental lithospheric mantle. The Palaeoproterozoic crust is an end product of accretionary growth, continent collisions and orogenic collapses. Accretionary structures and configurations have often been destroyed and the preserved accretional sections are normally found in areas where docking of rigid blocks has prevented further shortening.

The Lapland–Kola orogen is a large and wide orogenic root of a mountain belt with only limited amount of juvenile material preserved. Similarly, the amount of juvenile material in the northern segment of the Lapland–Savo orogen is minor. These orogenic sections exemplify continent-continent collisions between two Archaean continents. Different collision geometry is seen between an Archaean continent and a Palaeoproterozoic arc–microcontinent collage in the southern segment of the Lapland–Savo orogen. There the arc is partly buried under the stacked Archaean continental edge and a complex and interlayered Archaean–Palaeoproterozoic mantle lithosphere boundary exists.

The Kittilä (2.02 Ga), Tersk (1.98 Ga), Knaften (1.95 Ga), Savo (1.93 Ga), Häme (1.89 Ga) and, possibly also Oskarshamn–Jönköping (1.83 Ga) arcs are classified as retreating type juvenile arcs. The Tampere, Bergslagen and TIB 1 magmatic arcs are classified as advancing type continental arcs.

The main Palaeoproterozoic crustal growth episode in Fennoscandia was during 2.1–1.8 Ga. Major lateral crustal additions were 2.0–2.1 Ga microcontinents, with undefined evolution history, and juvenile arcs from >2.02 Ga to c. 1.83 Ga. Andean-type vertical magmatic additions, especially at c. 1.89 Ga and c. 1.8 Ga, were also important in the continental growth.

We gratefully acknowledge thorough and constructive comments and suggestions of the two reviewers B. Bingen and J. S. Daly.

References

ABRAMOVITZ, T., BERTHELSEN, A. & THYBO, H. 1997. Proterozoic sutures and terranes in the southeastern Baltic Shield interpreted from BABEL seismic data. *Tectonophysics*, **270**, 259–277.

ÅHÄLL, K.-I. & LARSON, S. Å. 2000. Growth-related 1.85–1.55 Ga magmatism in the Baltic Shield; a review addressing the tectonic characteristics of Svecofennian, TIB 1-related and Gothian events. *GFF*, **122**, 193–206.

ALLEN, R. L., LUNDSTRÖM, I., RIPA, M., SIMEONOV, A. & CHRISTOFFERSON, H. 1996. Facies analysis of a 1.9 Ga, continental margin, back-arc, felsic caldera province with diverse Zn–Pb–Ag–(Cu–Au) sulfide and Fe oxide deposits, Bergslagen region, Sweden. *Economic Geology*, **91**, 979–1008.

ANDERSSON, J., MÖLLER, C. & JOHANSSON, L. 2002. Zircon geochronology of migmatite gneisses along the Mylonite Zone (S Sweden): A major Sveconorwegian terrane boundary in the Baltic shield. *Precambrian Research*, **114**, 121–147.

ANDERSSON, U. B. 1997. Petrogenesis of some Proterozoic granitoid suites and associated basic rocks in Sweden (geochemistry and isotope geology). *Sveriges geologiska undersökning, Rapporter and meddelanden*, **91**, 1–216.

ANDERSSON, U. B. & WIKSTRÖM, A. 2001. Growth-related 1.85–1.55 Ga magmatism in the Baltic Shield; a review addressing the tectonic characteristics of Svecofennian, TIB 1-related and Gothian events—A discussion. *GFF*, **123**, 55–61.

ANDERSSON, U. B., HÖGDAHL, K., SJÖSTRÖM, H. & BERGMAN, S. 2006. Multistage growth and reworking of the Palaeoproterozoic crust in the Bergslagen area, southern Sweden; evidence from U/Pb geochronology. *Geological Magazine*, **143**, 679–697.

BABEL WORKING GROUP. 1990. Evidence for Early Proterozoic plate tectonics from seismic reflection profiles in the Baltic Shield. *Nature*, **348**, 34–38.

BERGMAN, S., KÜBLER, L. & MARTINSSON, O. 2001. *Description of regional geological and geophysical maps of northern Norrbotten County (east of the Caledonian orogen)*. Sveriges Geologiska Undersökning, Ba 56.

BERGMAN, S., BILLSTRÖM, K., PERSSON, P.-O., SKIÖLD, T. & EVINS, P. 2006. U–Pb age evidence for repeated Palaeoproterozoic metamorphism and deformation near the Pajala shear zone in the northern Fennoscandian shield. *GFF*, **128**, 7–20.

BERGMAN, S., HÖGDAHL, K., NIRONEN, M., OGENHALL, E., SJÖSTRÖM, H., LUNDQVIST, L. & LAHTINEN, R. 2007. Timing of Palaeoproterozoic intra-orogenic sedimentation in the central Fennoscandian Shield; evidence from detrital zircon in metasandstone. *Precambrian Research*, doi:10.1016/j.precamres.2007.08.007.

BEUNK, F. F. & PAGE, L. M. 2001. Structural evolution of the accretional continental margin of the Palaeoproterozoic Svecofennian orogen in southern Sweden. *Tectonophysics*, **339**, 67–92.

BINGEN, B., SKÅR, Ø. ET AL. 2005. Timing of continental building in the Sveconorwegian orogen, SW Scandinavia. *Norwegian Journal of Geology*, **85**, 87–116.

BRUNETON, M., PEDERSEN, H. A. ET AL. 2004. Layered lithospheric mantle in the central Baltic Shield from surface waves and xenolith analysis. *Earth and Planetary Science Letters*, **226**, 41–52.

CAWOOD, P. A., KRÖNER, A., COLLINS, W. J., KUSKY, T. M., MOONEY, W. D. & WINDLEY, B. F. 2009. Accretionary orogens through Earth history. *In*: CAWOOD, P. A. & KRÖNER, A. (eds) *Earth Accretionary Systems in Space and Time*. Geological Society, London, Special Publications, **318**, 1–36.

CLAESSON, S. & LUNDQVIST, T. 1995. Origins and ages of Proterozoic granitoids in the Bothnian Basin, central

Sweden; isotopic and geochemical constraints. *Lithos*, **36**, 115–140.

CLAESSON, S., HUHMA, H., KINNY, P. D. & WILLIAMS, I. S. 1993. Svecofennian detrital zircon ages—implications for the Precambrian evolution of the Baltic Shield. *Precambrian Research*, **64**, 109–130.

DALY, J. S., BALAGANSKY, V. V., TIMMERMAN, M. J. & WHITEHOUSE, M. J. 2006. The Lapland–Kola orogen: Palaeoproterozoic collision and accretion of the northern Fennoscandian lithosphere. *In*: GEE, D. G. & STEPHENSON, R. A. (eds) *European Lithosphere Dynamics*. Geological Society, London, Memoirs, **32**, 561–578.

EHLERS, C., LINDROOS, A. & JAANUS-JÄRKKÄLÄ, M. 1986. Stratigraphy and geochemistry in the Proterozoic mafic volcanic rocks of the Nagu-Korpo area, SW Finland. *Precambrian Research*, **32**, 297–315.

EHLERS, C., SKIÖLD, T. & VAASJOKI, M. 2004. Timing of Svecofennian crustal growth and collisional tectonics in Åland, SW Finland. *Geological Society of Finland, Bulletin*, **76**, 63–91.

EKDAHL, E. 1993. Early Proterozoic Karelian and Svecofennian formations and the Evolution of the Raahe–Ladoga Ore Zone, based on the Pielavesi area, central Finland. *Geological Survey of Finland, Bulletin*, **373**, 1–137.

ELIASSON, T. & STRÄNG, T. 1998. Kartbladen 23H Stensele. *In*: WAHLGREN, C.-H. (ed.) *Regional berggrundsgeologisk undersökning—sammanfattning av pågående undersökningar 1997*. Sveriges Geologiska Undersökning. Rapporter och meddelanden, **97**, 55–59 [in Swedish].

GORBATSCHEV, R. 2004. The Transcandinavian Igneous Belt—introduction and background. *In*: HÖGDAHL, K., ANDERSSON, U. B. & EKLUND, O. (eds) *The Transcandinavian Igneous Belt (TIB) in Sweden: A Review of Its Character and Evolution*. Geological Survey of Finland, Special Paper, **37**, 9–15.

GORBATSCHEV, R. & BOGDANOVA, S. 1993. Frontiers in the Baltic Shield. *Precambrian Research*, **64**, 3–21.

HAKKARAINEN, G. 1994. Geology and geochemistry of the Hämeenlinna–Somero Volcanic Belt, southwestern Finland: A Paleoproterozoic island arc. *In*: NIRONEN, M. & KÄHKÖNEN, Y. (eds) *Geochemistry of Proterozoic supracrustal rocks in Finland*. IGCP Project 179 (Stratigraphic Methods as Applied to the Proterozoic Record) and IGCP Project 217 (Proterozoic Geochemistry). Geological Survey of Finland, Special Paper, **19**, 85–100.

HANSKI, E. & HUHMA, H. 2005. Central Lapland greenstone belt. *In*: LEHTINEN, M., NURMI, P. & RÄMÖ, T. (eds) *The Precambrian Bedrock of Finland—Key to the Evolution of the Fennoscandian Shield*. Elsevier, Amsterdam, 139–194.

HERMANSSON, T., STEPHENS, M. B., CORFU, F., PAGE, L. M. & ANDERSSON, J. 2007. Migratory tectonic switching, western Svecofennian orogen, central Sweden—constraints from U/Pb zircon and titanite geochronology. *Precambrian Research*, doi:10.1016/j.precamres.2007.08.008.

HIETANEN, A. 1975. Generation of potassium poor magmas in the northern Sierra Nevada and the Svecofennian in Finland. *Journal of Research, US Geological Survey*, **3**, 631–645.

HJELT, S.-E., BOCK, G., KORJA, T., KOZLOVSKAJA, E., YLINIEMI, J., BEAR & SVEKALAPKO SEISMIC TOMOGRAPHY WORKING GROUPS. 2006. Electrical conductivity and seismic velocity structures of the lithosphere beneath the Fennoscandian Shield. *In*: GEE, D. & STEPHENSON, R. (eds) *European Lithospheric Dynamics*. Geological Society, London, Memoirs, **32**, 541–559.

HÖGDAHL, K. & SJÖSTRÖM, H. 2001. Evidence for 1.82 Ga transpressive shearing in a 1.85 Ga granitoid in central Sweden: Implications for the regional evolution. *Precambrian Research*, **105**, 37–56.

HÖGDAHL, K., SJÖSTRÖM, H., ANDERSSON, U. B. & AHL, M. 2007. Continental margin magmatism and migmatisation in the west–central Fennoscandian Shield. *Lithos*, doi:10.1016/j.lithos.2007.07.019.

HÖLTTÄ, P. 1995. Contact metamorphism of the Vaaraslahti pyroxene granitoid intrusion in Pielavesi, Central Finland. *In*: HÖLTTÄ, P. (ed.) *Relationship of granitoids, structures and metamorphism at the eastern margin of the Central Finland Granitoid Complex*. Geological Survey of Finland, Bulletin, **382**, 27–79.

HÖLTTÄ, P., VÄISÄNEN, M., VÄÄNÄNEN, J. & MANNINEN, T. 2007. Paleoproterozoic metamorphism and deformation in Central Finnish Lapland. *Geological Survey of Finland, Special Paper*, **44**, 7–56.

KÄHKÖNEN, Y. 1987. Geochemistry and tectonomagmatic affinities of the metavolcanic rocks of the early Proterozoic Tampere Schist Belt, southern Finland. *Precambrian Research*, **35**, 295–311.

KÄHKÖNEN, Y. 2005. Svecofennian supracrustal rocks. *In*: LEHTINEN, M., NURMI, P. & RÄMÖ, T. (eds) *The Precambrian Bedrock of Finland—Key to the Evolution of the Fennoscandian Shield*. Elsevier, Amsterdam, 343–406.

KÄHKÖNEN, Y. & NIRONEN, M. 1994. Supracrustal rocks around the Paleoproterozoic Haveri Au–Cu deposit, southern Finland: Evolution from a spreading center to a volcanic arc environment. *In*: NIRONEN, M. & KÄHKÖNEN, Y. (eds) *Geochemistry of Proterozoic supracrustal rocks in Finland*. IGCP Project 179 (Stratigraphic Methods as Applied to the Proterozoic Record) and IGCP Project 217 (Proterozoic Geochemistry). Geological Survey of Finland, Special Paper, **19**, 141–159.

KÄHKÖNEN, Y., LAHTINEN, R. & NIRONEN, M. 1994. *Palaeoproterozoic supracrustal belts in southwestern Finland*. Geological Survey of Finland, Guide, **37**, 43–47.

KILPELÄINEN, T. 1998. Evolution and 3D modelling of structural and metamorphic patterns of the Palaeoproterozoic crust in the Tampere–Vammala area, southern Finland. *Geological Survey of Finland, Bulletin*, **397**, 1–124.

KOISTINEN, T., STEPHENS, M. B., BOGATCHEV, V., NORDGULEN, Ø., WENNERSTRÖM, M. & KORHONEN, J. (compilers) 2001. *Geological map of the Fennoscandian Shield, scale 1:2 000 000*. Geological Survey of Finland, Espoo; Geological Survey of Norway, Trondheim; Geological Survey of Sweden, Uppsala; Ministry of Natural Resources of Russia, Moscow.

KORJA, A. & HEIKKINEN, P. 2005. The accretionary Svecofennian Orogen—insight from the BABEL profiles. *Precambrian Research*, **136**, 241–268.

KORJA, A. & HEIKKINEN, P. 2008. Seismic images of Paleoproterozoic microplate boundaries in Fennoscandian Shield. *In*: CONDIE, K. & PEASE, V. (eds) *When did Plate Tectonics Begin on Planet Earth?* Geological Society of America, Special Papers, **440**, 229–248.

KORJA, A., LAHTINEN, R. & NIRONEN, M. 2006a. The Svecofennian orogen—a collage of microcontinents and island arcs. *In*: GEE, D. G. & STEPHENSON, R. A. (eds) *European Lithosphere Dynamics*. Geological Society, London, Memoirs, **32**, 561–578.

KORJA, A., LAHTINEN, R., HEIKKINEN, P., KUKKONEN, I. T. & FIRE WORKING GROUP. 2006b. A geological interpretation of the upper crust along FIRE 1. *In*: KUKKONEN, I. T. & LAHTINEN, R. (eds) *Finnish Reflection Experiment FIRE 2001–2005*. Geological Survey of Finland, Special Paper, **43**, 45–76.

KORSMAN, K., KOISTINEN, T. ET AL. 1997. *Suomen kallioperäkartta–Berggrundskarta över Finland–Bedrock map of Finland 1:1 000 000*. Geological Survey of Finland, Espoo.

KORSMAN, K., KORJA, T., PAJUNEN, M., VIRRANSALO, P. & GGT/SVEKA WORKING GROUP. 1999. The GGT/SVEKA Transect: Structure and evolution of the continental crust in the Palaeoproterozoic Svecofennian orogen in Finland. *International Geology Review*, **41**, 287–333.

KOUSA, J., MARTTILA, E. & VAASJOKI, M. 1994. Petrology, geochemistry and dating of Palaeoperoterozoic metavolcanic rocks in the Pyhäjärvi area, central Finland. *In*: NIRONEN, M. & KÄHKÖNEN, Y. (eds) *Geochemistry of Proterozoic supracrustal rocks in Finland*. IGCP Project 179 (Stratigraphic Methods as Applied to the Proterozoic Record) and IGCP Project 217 (Proterozoic Geochemistry). Geological Survey of Finland, Special Paper, **19**, 7–27.

KUKKONEN, I. T., HEIKKINEN, P., EKDAHL, E., HJELT, S.-E., YLINIEMI, J., JALKANEN, E. & FIRE WORKING GROUP. 2006. Acquisition and geophysical characteristics of reflection seismic data on FIRE transects, Fennoscandian Shield. *In*: KUKKONEN, I. T. & LAHTINEN, R. (eds) *FIRE Finnish Reflection Experiment 2001–2005*. Geological Survey of Finland, Special Paper, **43**, 13–43.

LAAJOKI, K. 2005. Karelian supracrustal rocks. *In*: LEHTINEN, M., NURMI, P. & RÄMÖ, T. (eds) *The Precambrian Bedrock of Finland—Key to the Evolution of the Fennoscandian Shield*. Elsevier, Amsterdam, 279–342.

LAHTINEN, R. 1994. Crustal evolution of the Svecofennian and Karelian domains during 2.1–1.79 Ga, with special emphasis on the geochemistry and origin of 1.93–1.91 Ga gneissic tonalites and associated supracrustal rocks in the Rautalampi area, central Finland. *Geological Survey of Finland, Bulletin*, **378**, 1–128.

LAHTINEN, R. 1996. Geochemistry of Palaeoproterozoic supracrustal and plutonic rocks in the Tampere–Hämeenlinna area, southern Finland. *Geological Survey of Finland, Bulletin*, **389**, 1–113.

LAHTINEN, R. & HUHMA, H. 1997. Isotopic and geochemical constraints on the evolution of the 1.93–1.79 Ga Svecofennian crust and mantle. *Precambrian Research*, **82**, 13–34.

LAHTINEN, R., HUHMA, H. & KOUSA, J. 2002. Contrasting source components of the Palaeoproterozoic Svecofennian metasediments: Detrital zircon U–Pb, Sm–Nd and geochemical data. *Precambrian Research*, **116**, 81–109.

LAHTINEN, R., KORJA, A. & NIRONEN, M. 2005. Palaeoproterozoic tectonic evolution of the Fennoscandian Shield. *In*: LEHTINEN, M., NURMI, P. & RÄMÖ, T. (eds) *The Precambrian Bedrock of Finland—Key to the Evolution of the Fennoscandian Shield*. Elsevier, Amsterdam, 418–532.

LARSON, S. A. & BERGLUND, J. 1992. A chronological subdivision of the Transscandinavian Igneous Belt: Three magmatic episodes? *GFF*, **114**, 459–461.

LAWRIE, K. C. 1992. Geochemical characterisation of a polyphase deformed, altered, and high grade metamorphosed volcanic terrane; implications for the tectonic setting of the Svecofennides, south–central Finland. *Precambrian Research*, **59**, 171–205.

MANSFELD, J. 1996. Geological, geochemical and geochronological evidence for a new Palaeoproterozoic terrane in southeastern Sweden. *Precambrian Research*, **77**, 91–103.

MANSFELD, J. & BEUNK, F. F. 2004. The Oskarshamn–Jönköping Belt, a well-preserved Palaeoproterozoic volcanic arc. *GFF*, **126**, 29–30.

MANSFELD, J., BEUNK, F. F. & BARLING, J. 2005. 1.83–1.82 Ga formation of a juvenile volcanic arc—implications from U–Pb and Sm–Nd analyses of the Oskarshamn–Jönköping Belt, southeastern Sweden. *GFF*, **127**, 149–157.

MELEZHIK, V. A. & STURT, B. A. 1994. General geology and evolutionary history of the early Proterozoic Polmak–Pasvik–Pechenga–Imandra/Varzuga–Ust' Ponoy Greenstone Belt in the northeastern Baltic Shield. *Earth-Science Reviews*, **36**, 205–241.

MELLQVIST, C., ÖHLANDER, B., SKIÖLD, T. & WIKSTRÖM, A. 1999. The Archaean–Proterozoic palaeoboundary in the Luleå area, northern Sweden: Field and isotope geochemical evidence for a sharp terrane boundary. *Precambrian Research*, **96**, 225–243.

MOURI, H., KORSMAN, K. & HUHMA, H. 1999. Tectonometamorphic evolution and timing of the melting processes in the Svecofennian Tonalite–Trondhjemite Migmatite Belt: An example from Luopioinen, Tampere area, southern Finland. *Geological Society of Finland, Bulletin*, **71**, 31–56.

NIRONEN, M., ELLIOTT, B. A. & RÄMÖ, O. T. 2000. 1.88–1.87 Ga post-kinematic intrusions of the Central Finland Granitoid Complex: A shift from C-type to A-type magmatism during lithospheric convergence. *Lithos*, **53**, 37–58.

NIRONEN, M., KORJA, A., HEIKKINEN, P. & FIRE WORKING GROUP. 2006. A geological interpretation of the upper crust along FIRE 2 and FIRE 2A. *In*: KUKKONEN, I. T. & LAHTINEN, R. (eds) *FIRE Finnish Reflection Experiment 2001–2005*. Geological Survey of Finland, Special Paper, **43**, 77–103.

O'BRIEN, H. E., PELTONEN, P. & VARTIAINEN, H. 2005. Kimberlites, carbonatites, and alkaline rocks. *In*: LEHTINEN, M., NURMI, P. & RÄMÖ, T. (eds) *The*

Precambrian Bedrock of Finland—Key to the evolution of the Fennoscandian Shield. Elsevier, Amsterdam, 605–644.

PATCHETT, J., GORBATSCHEV, R. & TODT, W. 1987. Origin of continental crust of 1.9–1.7 Ga age: Nd isotopes in the Svecofennian orogenic terrains of Sweden. *Precambrian Research*, **35**, 145–160.

PATISON, N. L., KORJA, A., LAHTINEN, R., OJALA, V. J. & FIRE WORKING GROUP. 2006. FIRE Seismic reflection profiles 4, 4A and 4B: Insights into the crustal structure of northern Finland from Ranua to Näätämö. *In*: KUKKONEN, I. T. & LAHTINEN, R. (eds) *FIRE Finnish Reflection Experiment 2001–2005*. Geological Survey of Finland, Special Paper, **43**, 161–223.

PELTONEN, P. 1995. Petrogenesis of ultramafic rocks in the Vammala Nickel Belt: Implications for crustal evolution of the early Proterozoic Svecofennian arc terrane. *Lithos*, **34**, 253–274.

PELTONEN, P. 2005. Ophiolites. *In*: LEHTINEN, M., NURMI, P. & RÄMÖ, T. (eds) *The Precambrian Bedrock of Finland—Key to the Evolution of the Fennoscandian Shield*. Elsevier, Amsterdam, 237–278.

PELTONEN, P. & BRÜGMANN, G. 2006. Origin of layered continental mantle (Karelian craton, Finland): Geochemical and Re–Os isotope constraints. *Lithos*, **89**, 405–423.

PLOMEROVÁ, J., BABUŠKA, V., VECSEY, L., KOZLOVSKAYA, E., RAITA, T. & SSTWG. 2006. Proterozoic–Archean boundary in the mantle lithosphere of eastern Fennoscandia as seen by seismic anisotropy. *Journal of Geodynamics*, **41**, 400–410.

RÄMÖ, O. T., VAASJOKI, M., MÄNTTÄRI, I., ELLIOTT, B. A. & NIRONEN, M. 2001. Petrogenesis of the postkinematic magmatism of the Central Finland Granitoid Complex I: Radiogenic isotope constraints and implications for crustal evolution. *Journal of Petrology*, **42**, 1971–1993.

RUTLAND, R. W. R., SKIÖLD, T. & PAGE, R. W. 2001. Age of deformation episodes in the Palaeoproterozoic domain of northern Sweden, and evidence for a pre-1.9 Ga crustal layer. *Precambrian Research*, **112**, 239–259.

RUTLAND, R. W. R., WILLIAMS, I. S. & KORSMAN, K. 2004. Pre-1.91 Ga deformation and metamorphism in the Palaeoproterozoic Vammala Migmatite Belt, southern Finland, and implications for Svecofennian tectonics. *Geological Society of Finland, Bulletin*, **76**, 93–140.

SANDOVAL, S., KISSLING, E. & ANSORGE, J. 2003. High-resolution body wave tomography beneath the SVEKALAPKO array: I. *A priori* three-dimensional crustal model and associated traveltime effects on teleseismic wave fronts. *Geophysical Journal International*, **153**, 75–87.

SKRIDLAITE, G. & MOTUZA, G. 2001. Precambrian domains in Lithuania: Evidence of terrane tectonics. *Tectonophysics*, **339**, 113–133.

SKYTTÄ, P., VÄISÄNEN, M. & MÄNTTÄRI, I. 2006. Preservation of Palaeoproterozoic early Svecofennian structures in the Orijärvi area, SW Finland—evidence for polyphase strain partitioning. *Precambrian Research*, **150**, 153–172.

SORJONEN-WARD, P. 2006. Geological and structural framework and preliminary interpretation of the FIRE 3 and FIRE 3A reflection seismic profiles, Central Finland. *In*: KUKKONEN, I. T. & LAHTINEN, R. (eds) *FIRE Finnish Reflection Experiment 2001–2005*. Geological Survey of Finland, Special Paper, **43**, 105–159.

TUISKU, P. & HUHMA, H. 2006. Evolution of migmatitic granulite complexes: Implications from Lapland Granulite Belt, Part II: Isotopic dating. *Geological Society of Finland, Bulletin*, **78**, 143–175.

VAASJOKI, M. & HUHMA, H. 1999. Lead and neodymium isotopic results from metabasalts of the Haveri Formation, southern Finland: Evidence for Palaeoproterozoic enriched mantle. *Geological Society of Finland, Bulletin*, **71**, 143–153.

VÄISÄNEN, M. & MÄNTTÄRI, I. 2002. 1.90–1.88 Ga arc and back-arc basin in the Orijärvi area, SW Finland. *Geological Society of Finland, Bulletin*, **74**, 185–214.

VALBRACHT, P. J., OEN, I. S. & BEUNK, F. F. 1994. Sm–Nd isotope systematics of 1.9–1.8-Ga granites from western Bergslagen, Sweden: Inferences on a 2.1–2.0-Ga crustal precursor. *Chemical Geology*, **112**, 21–37.

WASSTRÖM, A. 2005. Petrology of a 1.95 Ga granite–granodiorite–tonalite–trondhjemite complex and associated extrusive rocks in the Knaften area, northern Sweden. *GFF*, **127**, 67–82.

WEIHED, P., ARNDT, N. ET AL. 2005. Precambrian geodynamics and ore formation: The Fennoscandian Shield. *In*: BLUNDELL, D., ARNDT, N., COBBOLD, P. R. & HEINRICH, C. (eds) *Geodynamics and Ore Deposit Evolution in Europe*. Ore Geology Reviews, **27**, 273–322.

YLINIEMI, J., KOZLOVSKAYA, E., HJELT, S.-E., KOMMINAHO, K. & USHAKOV, A. 2004. Structure of the crust and uppermost mantle beneath southern Finland revealed by analysis of local events registered by the SVEKALAPKO seismic array. *Tectonophysics*, **394**, 41–67.

The underestimated Proterozoic component of the Canadian Cordillera accretionary margin

D. B. SNYDER[1]*, M. PILKINGTON[1], R. M. CLOWES[2] & F. A. COOK[3]

[1] *Geological Survey of Canada, 615 Booth Street, Ottawa, ON K1A 0E9, Canada*

[2] *LITHOPROBE and Department of Earth and Ocean Sciences, University of British Columbia, Vancouver, BC V6T 1Z4, Canada*

[3] *Department of Geology and Geophysics, University of Calgary, Calgary, AB T2N 1N4, Canada*

Corresponding author (e-mail: dsnyder@NRCan.gc.ca)

Abstract: Analysis of several types of seismic and potential field geophysical data consistently indicate that the majority of the crust underlying the Canadian Cordillera and much of western Canada was originally Proterozoic sedimentary rocks shed off the Canadian Shield into rift or basin structures between 1.84 and 0.54 Ga. These variably metamorphosed strata were primarily quartz- and limestone-rich sediments and thus have distinctive geophysical signatures because of their lower density, lower magnetization, and lower Poisson's ratio compared with more mafic rocks. The sediments formed a prograding wedge that has a distinctive, internally reflective, seismic stratigraphy. In the east, these Proterozoic sedimentary rocks thicken at a 'hinge line' defined by the margin of the pre-1.84 Ga crystalline basement of the Canadian Shield; previous work mapped this hinge line locally using deep reflection profiles and regionally using distinctive gravity gradients. Here we assemble previously published results of several geophysical methods to define the overall shape of the wedge along the margin and westward to where it pinches out at the modern Moho beneath the crustal collage of exotic and suspect terranes accreted onto North America during the Mesozoic. The volume of crust occupied by this wedge limits the thickness of most accreted terranes to several kilometres and suggests that deeper portions of the accreted blocks detached or underthrust the wedge during accretion and are no longer contiguous to crust exposed at the surface. This type Cenozoic accretionary orogen thus spent most of its prior geological history as a passive or extensional margin punctuated by only a few, brief convergent or accretionary events.

The Canadian Cordillera is one of the principal regions from which the hypothesis of exotic, accreted, or suspect terranes developed a few decades ago (e.g. Monger *et al.* 1972). The recognition of a vast volume of Proterozoic strata of largely North American affinity underlying this accretionary orogen (Young *et al.* 1979; Cook *et al.* 2005, and references therein), leaving scant space for the suspect terranes, has required an important modification to this fundamental plate-tectonic hypothesis. Relatively quartz-rich and non-magnetic Proterozoic strata, deposited in at least three distinct periods between 1.85 and 0.54 Ga, form a reflective tectono-depositional prism or wedge that has a volume greater than 1×10^6 km^3, extends over 1000 km in strike length, and makes up most of the crust of the Canadian Cordillera. This sedimentary prism starts where Proterozoic strata thicken sharply at the so-called continental hinge line located at the edge of the Canadian Shield (Cook *et al.* 2005). Crystalline continental and transitional basement rocks are inferred beneath these strata. Further south in North America, the equivalent continental hinge line lies near the Wasatch Front in Utah (Burchfiel & Davis 1975); this location coincides with the eastern limit of the Basin and Range and thus later extensional structures partly obscure the geometry of the Proterozoic feature.

The part of the Canadian Cordillera near 60°N latitude formed on the western margin of the Canadian Shield that includes the Archaean Slave craton and the Wopmay orogen (Fig. 1). The 1.84 Ga terminal contraction phase of the *c*. 2.10–1.84 Ga Wopmay orogen included accretion of *c*. 1.85 Ga crystalline basement rocks, probably composed largely of volcanic arcs and exotic continental fragments. Through subduction and internal wedging, these Fort Simpson and Nahanni terranes accreted onto the western margin of North America that had grown westward via the 1.90–1.88 Ga Calderian orogeny during which the Hottah arc terrane converged with the Slave craton margin (Hoffman 1980; Hildebrand *et al.* 1987; Cook *et al.* 1999). The *c*. 1.8 Ga margin of North America is today buried beneath Phanerozoic and Proterozoic rocks

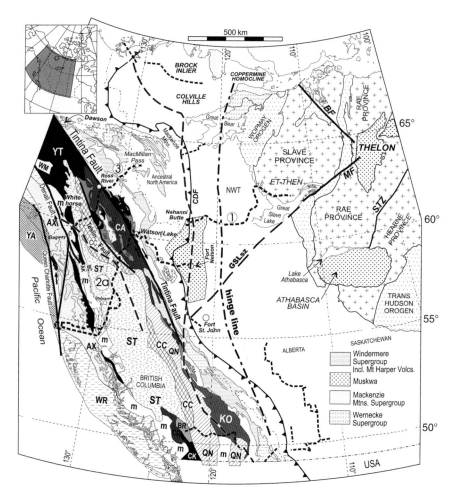

Fig. 1. Map of the Canadian Cordillera showing major elements of the western Canadian Shield, major accreted terranes, known inliers of exposed Proterozoic strata, and the location of LITHOPROBE SNorCLE (labelled 1, 2a, 3) and other relevant seismic reflection profiles (short-dash lines). The continuous line with barbs labelled CDF marks the frontal thrust of the Cordillera Deformation Front. The long-dash lines mark limits of where Proterozoic strata are here interpreted to occupy much of the crust; the dash–dot line to the east marks the continental 'hinge line' where these sedimentary rocks begin to thicken significantly (Clark & Cook 1992). Four main periods of sedimentation are indicated: the 0.8–0.54 Ga Windermere, the 1.08–0.78 Ga Mackenzie Mountains and 1.84–1.71 Ga Wernecke Supergroups, and the less well-known 1.815–1.5 Ga Muskwa assemblage (Cook & MacLean 1995; Ross et al. 2001; Thorkelson et al. 2005). The Muskwa unit near Fort Nelson is inferred from drill-holes (Ross et al. 2001). GSLsz, Great Slave Lake strike-slip fault zone. Suspect terranes: AX, Alexander; BR, Bridge River; CA, Cassier; CC, Cache Creek; CD, Cadwallader; CK, Chilliwack; KO, Kootenay; QN, Quesnellia; SM, Slide Mountain; ST, Stikinia; YA, Yukatat; YT, Yukon–Tanana; WR, Wrangelia; m, undifferentiated metamorphic terranes.

of the Western Canada Sedimentary basin and adjacent Cordillera and is thus detectable primarily using geophysical methods. Eight deep seismic reflection profiles and correlatable isostatic gravity anomalies locate a series of west-facing crustal-scale monoclines with 15–20 km of relief over >1500 km of strike (MacLean & Cook 2004; Cook et al. 2005). This monocline, or hinge line formed by westward-thickening Proterozoic sedimentary rocks, is best documented near Fort Simpson, NWT (near Nahanni Butte in Fig. 1), where the relatively low-density sedimentary rocks compose the Fort Simpson Basin and the clear linear feature in isostatic gravity anomaly maps associated with the basin margin was named the Fort Simpson trend.

The Proterozoic prism thins away from the Canadian Shield and disappears beneath a collage of accreted continental fragments, plutons and batholiths and Cenozoic accretionary rocks associated with continuing convergence along the Pacific margin. Cordilleran suspect terranes apparently were wedged apart during accretion, with the upper crust overriding this metamorphosed sedimentary prism during a complex interplay of thrusting and strike-slip displacements 190–170 Ma ago (Colpron & Nelson 2006). Thus, the Cordilleran suspect terranes, excepting Stikinia, have shallow roots only a few kilometres deep, with no pre accretion deep crust or mantle attached. The lower crust and any attached mantle presumably were underthrust or subducted. These terranes appear exotic and diverse at the surface, but when the entire crust and lithosphere are considered, they represent only a very minor rock volume of the Cordilleran accretionary margin.

Here we synthesize a number of independent geophysical observations, most published previously, that collectively provide an estimate of the total volume of this sedimentary prism, whose overall extent and geometry were previously undefined. It is the great time-stratigraphic extent of 1.710–0.540 Ga that this >25 km rock-stratigraphic thickness implies that is especially significant to studies of accretionary orogens. It implies that the Canadian Cordillera was a very long-lived (1.8 Ga) continental margin, but nearly two-thirds of that history was characterized by passive subsidence, possibly active rifting, bracketed by classic arc-terrane accretion periods at 1.88–1.85 Ga and 0.19 Ga–present.

Proterozoic stratigraphy

The various Proterozoic rock units within the part of the Canadian Cordillera discussed here remain incompletely identified in their extent and depositional history (e.g. Thorkelson et al. 2005). In a few locations, composite stratigraphic sections as great as 27 km thick have been constructed, but seismic stratigraphic sections of the entire crust as interpreted from LITHOPROBE SNorCLE deep seismic reflection profiles cannot be comprehensively tied directly to outcrop or wells. Opportunistic outcrop correlations allow a few reflective layers to be matched with Proterozoic sedimentary rock units that include the 0.78–0.54 Ga Windermere, the 1.08–0.78 Mackenzie Mountains, the 1.84–1.71 Ga Wernecke Supergroups, and the lesser known 1.735–0.775 Ga Muskwa assemblage (Cook & MacLean 1995; Ross et al. 2001; MacLean & Cook 2004; Thorkelson et al. 2005; Fig. 2). Young et al. (1979) originally divided the supracrustal strata recognized at that time into unconformity bound sequences A, B and C. Subsequent mapping in Proterozoic inliers, structural culminations that are mainly contractional in origin, revealed more details in the Yukon portion of the Cordillera. For example, Thorkelson et al. (2005) have recently reported at least seven unconformities and inferred greater age differences within each original sequence than occur between them, thus making the A–C nomenclature less useful.

A rock stratigraphic column of the Yukon section shows about 22 km of Proterozoic sedimentary rocks; the correlative time stratigraphic column indicates that most of these sediments, as in most basins, were deposited in a relatively small fraction of the 1 Ga represented (Fig. 2). The inferred palaeogeography of northwestern Laurentia during this period was low subaerial terrain, possibly similar to the upland or foothills setting postulated for the Torridonian succession of the North Atlantic (Dalziel & Soper 2001; Rainbird et al. 2001). Denudation of significant amounts of topography is documented at two unconformities in the Yukon (Eisbacher 1978; Thorkelson et al. 2005). Horizontal contractional deformation that has been associated with a number of orogenies, such as the Racklan orogeny, implies an active convergent margin. Overall, the Palaeo- and Meso-Proterozoic geological history of the Yukon section is that of a long-lived (c. 1.2 Ga) passive margin punctuated by a relatively small number of horizontal shortening, erosional, intrusive, or hydrothermal events of types documented on other passive or accretionary continental margins in non-orogenic settings (e.g. Stagg et al. 1999).

The base of the Wernecke Supergroup is nowhere exposed in the Yukon, so that the preserved c. 13 km section (former Sequence A) is assumed to overlie highly attenuated continental crust (Fig. 2; MacLean & Cook 2004; Thorkelson et al. 2005). Its depositional history is represented by two clastic-to-carbonate grand cycles (sensu Aitken 1978) followed by shallow-water to locally emergent conditions during a period of protracted subsidence and localized normal faulting (Delaney 1981; Thorkelson 2000). The Wernecke Supergroup may be correlative with parts of the Hornby Bay Group to the east (Kerans et al. 1981; Aitken & McMechan 1992; MacLean & Cook 2004) and with the Muskwa assemblage in northern British Columbia (Long & Pratt 2000; Ross et al. 2001), but not with the 1.47–1.37 Ga Belt–Purcell Supergroup further south (Anderson & Davis 1995).

The Wernecke Supergroup was then deformed and metamorphosed, uplifted and partially exhumed, intruded by magma and covered by volcanic rocks before it was altered by hydrothermal fluids that accompanied formation of the Wernecke breccias at 1.60 Ga (Thorkelson et al. 2005). The much redefined >1.60 Ga Racklan orogeny caused greenschist metamorphic conditions, sets of east- and south-directed tight folds, and locally substantial

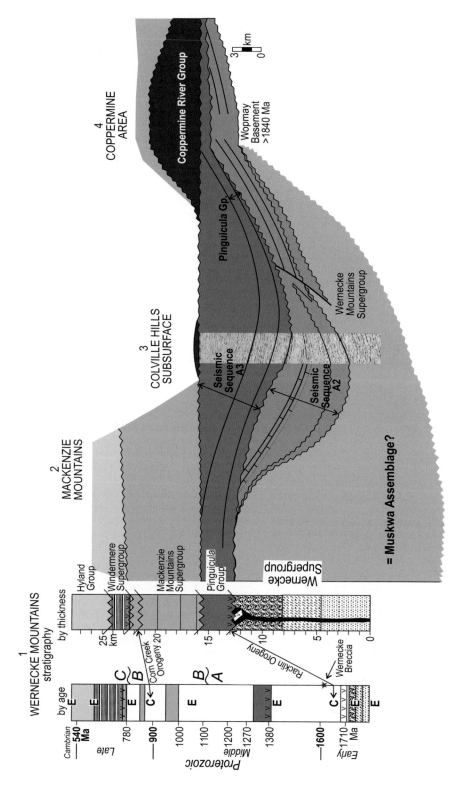

Fig. 2. Time (left) and rock (right) stratigraphic columns for the Proterozoic rocks in the Wernecke Mountains (1) of the Canadian Cordillera (modified from Thorkelson et al. 2005). It should be noted that 28 km of section represents less than half of the 1.3 Ga geological record preserved in these sections; C and E indicate compressive and extensional events, respectively. Less detailed stratigraphic columns are shown for areas in the Mackenzie Mountains (2), Colville Hills (3) and near Coppermine (4) as correlated via seismic sequences interpreted from an extensive grid of shallow seismic reflection profiles (modified from MacLean & Cook 2004). The strong lateral variations between stratigraphic units and 'hinge line' located near Coppermine should be noted. Areas are indicated in Figure 1.

erosion. It is regarded temporally equivalent to the c. 1.66 Ga Forward orogeny as documented 300 km further east (Cook & MacLean 1995; Thorkelson et al. 2005). Mantle-derived magmas of dioritic to gabbroic composition intruded the Wernecke Supergroup to form the Bonnet Plume River Intrusions at c. 1.71 Ga. These enriched continental tholeiites probably originated in an extensional, perhaps rift, setting (Thorkelson et al. 2001).

The scant 1.60–1.38 Ga rock record in the Yukon suggests that northwestern Laurentia was largely emergent and tectonically quiescent between Wernecke brecciation and intrusion of sills and renewed sedimentation at 1.38 Ga, but further east, the 4–5 km section of Dismal Lakes Group north of Great Bear Lake and other marine carbonates in the Athabasca, Elu and Thelon basins were deposited at this time. This quiescence is also in marked contrast to igneous and tectonic activity further south, in SW Canada and beyond, activity that included formation of the Belt–Purcell basin (Anderson & Davis 1995). In the Yukon, the 3.5 km thick Pinguicula Group was deposited in a southward-deepening basin coeval with marine volcanism and emplacement of mafic sills at 1.38 Ga (Abbot 1997). Mackenzie Mountains Supergroup and correlative strata formed throughout much of northwestern Laurentia in the late Meso- and early Neoproterozoic (Aitken & McMechan 1992; Rainbird et al. 1997). The base of the sequence is a succession of mainly marine siltstone overlain by cycles of fluvial to shallow marine sandstones. This, in turn, is overlain by a sequence typified by shallow marine platform carbonates with intervening evaporites signifying restricted basin conditions. The Mackenzie Mountains Supergroup is capped by the 0.779 Ga Little Dal basalts (Jefferson & Parrish 1989).

Eisbacher (1978, 1981) recognized that the Mackenzie Mountains Supergroup was locally involved in contractional deformation prior to deposition of the Windermere Supergroup in the Yukon. Thorkelson (2000) provided additional evidence for his Corn Creek orogeny in the form of overturned, west-verging thrust faults and folds, all local to the dextral Snake River fault. Another observation of horizontal shortening within Mackenzie Mountains Supergroup rocks are two post-1.27 Ga (post Coppermine basalt) contractions in the Coppermine River group (Hildebrand & Baragar 1991). More regionally, the sub-Windermere unconformity is associated with uplift, extensional tectonism and widespread bevelling of exhumed land surface, although the Coates Lake Group was deposited at this time (Jefferson & Parrish 1989).

Collectively, the thick sedimentary successions represented by the Wernecke and Mackenzie Mountains Supergroups and Muskwa assemblage are interpreted as either a clastic apron deposited along an episodically subsiding Proterozoic continental margin of Laurentia or a series of extensional events forming deep intracontinental basins. In both scenarios, sedimentation was punctuated by local orogenesis, magmatism or hydrothermal activity. Locally, the depositional environment is described as varying from prograding delta to shallow margin to deep-water turbidites (Aitken & McMechan 1992). The sediment budget may be similar to that of the present-day Ganges–Brahmaputra system, where about one-third of the total sediment load that originates in the interior of the continent is stored in the Bengal fan (Goodbred & Kuehl 1999). The other two-thirds are floodplain and deltaic components of the same sedimentary units, and equivalent components in the Cordillera may be partly represented by the Proterozoic Athabasca, Hornby Bay and Thelon basins located further east on the Canadian Shield (Fig. 1; Fraser et al. 1970). More specifically, future detrital geochronology studies may confirm that the Wernecke Supergroup, which was possibly deposited over a period as short as a few million years to as long as 130 Ma, represents synorogenic distal deposits of the Wopmay or perhaps Trans-Hudson orogens (e.g. Rainbird et al. 1997; Ross et al. 2001; Thorkelson et al. 2005). The central point here is that modern sedimentary units have similar thicknesses, volumes and geometries to Proterozoic ones described here. Subsequent tectonism and metamorphism of the Cordilleran rocks makes detailed, one-to-one comparison of seismic characteristics impossible.

Stratigraphic studies of these Proterozoic sedimentary units document thicknesses of at least 7–21 km (Figs 2 and 3), but correlations with the seismic reflectors indicate a total sequence as thick as 25–30 km at the ancient continental slope or hinge line (Clark & Cook 1992; Cook et al. 1999, 2005). Within the eastern half of the collective Proterozoic sedimentary prism we estimate that 0–5 km of Windermere Supergroup strata overlie a combined 5–25 km sequence of Muskwa and Wernecke strata and crystalline basement layers or tectonically inserted thrust slices. These dimensions make the wedge comparable in scale with the modern Indus fan, but the Cordilleran wedge today represents a compound feature deposited over a period up to 10 times longer and much deformed by horizontal contraction.

Geophysical evidence for a Proterozoic prism

The nature of Proterozoic strata within the Canadian Cordillera is known primarily from exposures within a few inliers located in largely inaccessible

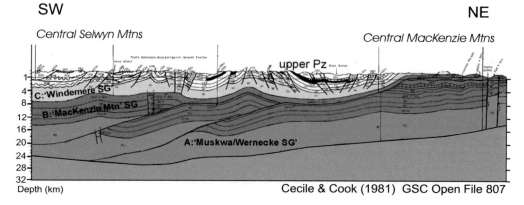

Fig. 3. Crustal section from the Mackenzie Mountains (Fig. 1) based on down-dip projection of mapped units and illustrating that westward dipping thrust sheets of Proterozoic stata make up at least 24 km of the crust at this location (from Cecile & Cook 1981). The SNorCLE transect line 3 lies about 50 km to the south, and downdip.

mountain terrain and a few drill-holes; nevertheless, these strata were inferred to represent over 25 km of crust in most locations. Limited geological accessibility means that geophysical methods are required to estimate better the 3D extent of these rocks. A number of geophysical datasets are currently available. Here we focus on deep seismic reflection profiles, coincident large-offset refraction surveys, and aeromagnetic maps.

Evidence from seismic reflection profiling

In 1999–2000, LITHOPROBE acquired nearly 1900 km of deep seismic reflection profiles in two transects of the North American Cordillera in British Columbia and the Yukon (Fig. 1; Cook et al. 2004). These profiles completed the SNorCLE transect of western North America that began with a 1996 survey from Yellowknife to Nahanni Butte in the Northwest Territories (Fig. 1; Cook et al. 1999). The combined SNorCLE profiles take advantage of the regional north–south distribution of Proterozoic and younger rocks in northwestern Canada to address continental evolution from the Early Archaean margin of the Slave Province to the modern active accretionary orogen in the area studied by the ACCRETE project (Fig. 1; Morozov et al. 2001). When processed, the data produced two continuous deep seismic reflection cross-sections traversing most of the Cordilleran mountain belt and penetrating to depths as great as 100 km (Cook & Erdmer 2005).

When viewed as a whole, the SNorCLE seismic reflection sections are most notable for the triangular, or wedged-shaped, areas where reflections are more densely spaced or of higher amplitude than elsewhere (Fig. 4; Cook et al. 2004). The continuity of the profiles allows single crustal reflective layers that make up these wedges to be mapped from the near surface along the eastern margin of the mountains, where they tie with outcrop (e.g. Young et al. 1979; Clark & Cook 1992), to 30 km depth beneath the western Cordillera (Figs 3–5). Similar up-dip projections and outcrop correlations in continental interior (shield) settings globally suggest that the vast majority of such prominent basement reflectors are mafic intrusions into the felsic upper crust, stratigraphic contacts or major shear zones (Snyder & Hobbs 1999, and references therein). Reflectors recorded by SNorCLE are interpreted as primarily stratigraphic contacts where west-dipping and deformational shear zones where east-dipping (Cook et al. 2004). Impedance contrasts associated with strong reflectors presumably became enhanced by granulite-facies metamorphism and generation of granitic melts associated with the lower crust. Any such melts must have percolated through residual rocks at sufficiently low volumes so as to preserve reflector continuity observed today.

Synthetic seismogram and stratigraphic correlations

In accretionary orogen settings such as the Canadian Cordillera, correlations of a series of subparallel reflections with stratigraphic contacts at the surface are often possible because thrust faults expose oblique sections. The Muskwa anticlinorium hosts a particularly favourable structure (Fig. 5). Reflectors projecting to the surface at this exposed section can be traced without breaks or offsets to depths of 20–25 km, and are thus assumed to represent those same strata at depth (Cook et al. 2004). The detailed stratigraphic section (Bell 1968) was converted to an impedance series using laboratory measurements of relevant rocks, geophysical well logs and appropriate seismic wavelets to

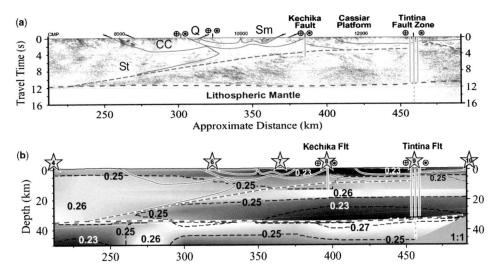

Fig. 4. (a) Crustal section from SNorCLE line 2 looking north showing reflections that define the sedimentary prism (below blue dashed line and above red dashed line representing the Moho). (b) Crustal section from coincident line showing Poisson's ratios as calculated from co-registered P- and S-wave velocity models derived from SNorRE refraction survey data (Clowes et al. 2005). The leftmost third of the crustal section is occupied by the Stikina accreted terrane with a consistent average Poisson's ratio of 0.26 indicating mafic rocks. Suspect terranes: CC, Cache Creek; Q, Quesnellia; St, Stikinia; Sm, Slide Mountain.

generate a predictive synthetic seismogram for the LITHOPROBE Muskwa section (Cook & Siegel 2006). This synthetic seismic trace matches moderately well the nearby SNorCLE seismic section located within a few kilometres of the rock section (three examples are shown in Fig. 5).

The largely unbroken continuation of this seismic signature to 25 km depths implies that Muskwa strata also continue to those lower crustal depths along dipping reflectors and inferred thrusts. Further west, particularly near the toe of the prism (300–380 km in Fig. 4), most reflectors are east-dipping and therefore interpreted as post-depositional imbricate thrusts that thickened the prism, most probably during accretion of the overlying terranes (Cook et al. 2004).

Bulk seismic wave speeds

Long-offset seismic refraction data were acquired in addition to seismic reflection profiles along SNorCLE lines in the Northwest Territories and the Yukon (Fig. 1; Fernandez-Viejo et al. 2005, and references therein). Seismic P- and S-wave phase arrivals recorded on these data were modelled to produce crustal sections of bulk P- and S-wave speeds with resolution of several kilometres. These two parameters can be combined into a ratio, traditionally known as Poisson's ratio, which is often used to estimate rock types at depth (Christensen 1996; Musacchio et al. 2004). Ratios less than 0.24 are typical of quartz-rich rocks because pure quartz has a ratio of 0.08 and other major minerals such as feldspar have higher ratios. The western part and eastern upper crust are characterized by ratios of 0.26, whereas the eastern lower crust has values of 0.23 (Fig. 4). When averaged over tens of kilometres of crust this contrast in Poisson's ratio suggests a significantly higher quartz content in parts of the lower crust that coincide with the reflective wedge described above.

The current high heat flow throughout the Cordillera (Lewis et al. 2003) and the great depth of burial of both wedge and underlying rocks indicate that basement and Proterozoic rocks within the lower half of the wedge are now metamorphosed into gneiss and granulites (but see Evenchick et al. 1984, for discussion on crystalline basement rocks). Proterozoic sedimentary rocks could retain modal compositions, primary layering and large-scale structures during this metamorphism if large-percentage melts did not accumulate locally. High-level Cretaceous granites common in the Canadian Cordillera probably represent melts 'sweated out' of the Proterozoic sedimentary layers in the lower crust; although related melt migration could reduce bulk silica content of the crust, its presumed pervasive nature apparently did not greatly disrupt the primary 'layered' reflectivity.

Magnetically quiet crust

The magnetic field over the Canadian Cordillera is characterized by complex, short-wavelength

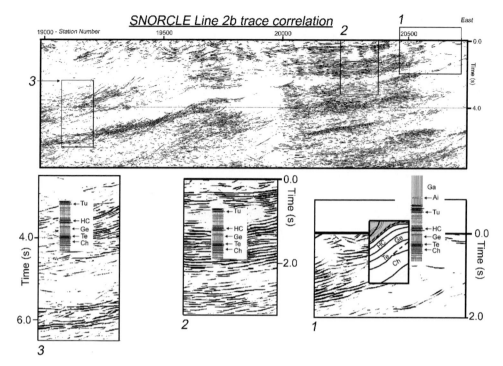

Fig. 5. Detail of seismic reflection section near the Muskwa anticlinorium (adapted from Cook & Siegel 2006 © 2008 NRC Canada or its licensors. Reproduced with permission). The three highlighted panels illustrate that the westward dipping thrust sheets bring Muskwa assemblage strata to the surface near Fort Nelson, along the Cordilleran frontal thrust (Fig. 1). These reflectors can be traced continuously into the lower crust (Cook et al. 2004). The synthetic seismograms superimposed on the section where the strata crop out were constructed using physical property measurements and logs from the exposed Muskwa section (see Bell 1968; Cook & Siegel 2006, for formation names); the amplitudes on the synthetic traces match reflective and non-reflective sequences observed on the SNorCLE seismic section near the outcrop and at 10–12 km depths.

anomalies so that its field at wavelengths >100 km is subdued and featureless; the large regional extent of non-magnetic crust can be confirmed by modelling of satellite magnetometer data (Pilkington et al. 2006). Spectral analysis of the magnetic field in areas located within the Cordillera reveals magnetic sources at consistently shallow (<5 km) depths (Fig. 6). The slope of the power spectrum is proportional to the average depth to the causative magnetic sources (Spector & Grant 1970). Intrusive, metamorphic and volcanic rocks can be readily associated with these sources, geographically and via their magnetic properties, within accreted terranes. Similar spectral analysis of the magnetic field over areas near and east of the hinge line, and thus largely underlain by pre-1.8 Ga basement rocks, reveals magnetic sources distributed over a much larger depth range, to >20 km depth (Fig. 6c). Power levels for wavelength components <40 km (0.025 cycles km^{-1}) are stronger using study areas located within the Cordillera whereas the opposite is true for study areas within the Canadian Shield.

Both low-level (300 m altitude) and high-altitude magnetic field analysis suggest that the Cordilleran crust comprises a shallow, thin, magnetic upper crustal source layer corresponding to igneous and metamorphic rocks of accreted terranes underlain by a weakly or non-magnetic crust. Magnetic signatures of continental crust depend largely on the distribution of magnetite minerals from the surface down to the Curie isotherm. Crystalline igneous and metamorphic rocks with tens of per cent mafic minerals are typically the main magnetic source contributors. In contrast, sedimentary rocks with several per cent iron detrital mineral grains are relatively non-magnetic, even at greenschist metamorphic facies.

Extent of the prism

The several independent geophysical studies collectively summarized here take advantage of several distinct physical properties of relatively quartz-rich, clastic sedimentary rocks to support

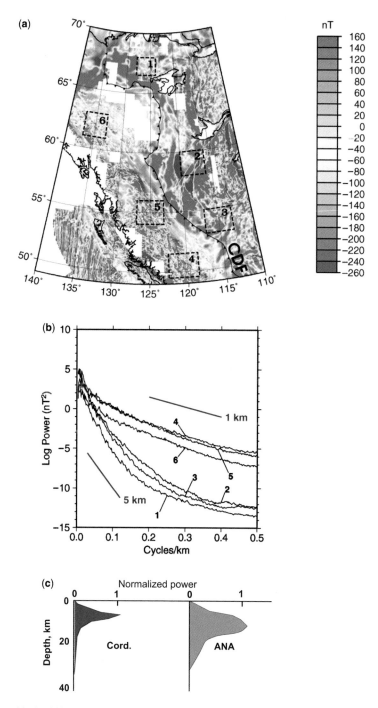

Fig. 6. (a) Low-altitude (300 m mean terrain clearance) aeromagnetic anomaly map of the Canadian Cordillera showing six areas (box outlines) used in this analysis (modified from Pilkington et al. 2006). (b) Spectral analysis of the aeromagnetic data illustrating shallow (several kilometres) crustal source of magnetism within the Cordillera (boxes 4–6) as compared with 5–10 km source depths for ancestral (pre-Proterozoic) North American crust (boxes 1–3). Line segments labelled 1 km and 5 km indicate slope of spectra expected for source bodies located at that depth. (c) Schematic depth–distribution plots for source anomalies as indicated by analysis in (b). ANA, ancestral North America.

the interpretation that Proterozoic metaturbidites and similar metasedimentary rocks volumetrically make up much of the crust of the Canadian Cordillera. Synthetic seismic trace modelling of distinctive reflector sequences within a generally reflective prism establishes that Muskwa-age strata reach modern Moho depths in at least one location near 60°N latitude. Combinations of bulk P- and S-wave speeds (Poisson's ratios), averaged over crustal volumes of several kilometres, are distinctively characteristic of quartz-rich rocks; Poisson's ratios <0.25 are unusual globally in the lower crust. Magnetic source bodies inferred to be largely restricted to the uppermost 5 km further indicate lack of mafic minerals at greater depths in the crust in proximity to the seismic surveys but equally so throughout the Cordillera in Canada and possibly beyond in Alaska and the contiguous USA.

The continental hinge line along the western margin of the Canadian Shield, as primarily delineated by continuous gradient trends in isostatic gravity anomalies (Cook et al. 2005), provides one general eastern delimiter to crust predominantly composed of Proterozoic sedimentary rocks (Fig. 1). The western limit is less obvious. In places it coincides with a few major accreted blocks such as the Stikinia terrane, but elsewhere the keels or bases of accreted terranes have not been clearly defined geophysically, geochemically or geologically (e.g. Cook et al. 2004).

If the minimum amount of the proposed 450–800 km of right-hand strike-slip displacement along the Tintina–Northern Rocky Mountain trench fault (Gabrielse 1985) is restored, the presumed >900 km reconstructed strike length and uniformity of this reflective wedge along the continental margin appear more clearly. The reflective prism represents an inferred $(1-10) \times 10^6$ km^3 volume of layered sedimentary rock deposited on the attenuated North American margin between 1.85 and 0.54 Ga.

The continuity of this reflective wedge, both along strike between the two SNorCLE profiles and across its width, is an obvious and significant characteristic. An important factor is that this segment of the North American Cordillera has undergone only limited amounts of extension (i.e. <10%) in contrast to the considerable extension associated with the southern Canadian Cordillera (Cook & Varsek 1994) or the western USA (e.g. Allmendinger et al. 1987). As a result, single reflection packets observed in the northern Cordillera were not stretched or dismembered beyond recognition. However, because of sparse relevant data we cannot at present assess the full northern or southern extent of the reflective wedge (Figs 1 and 7).

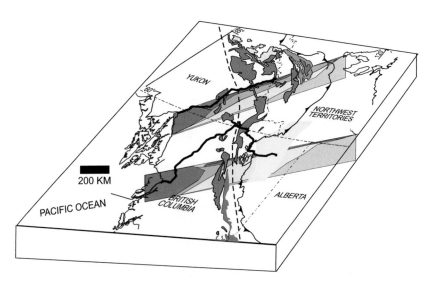

Fig. 7. Three-dimensional perspective view of Proterozoic sedimentary sequences exposed in western Canada and crustal cross-sections derived from the seismic profiles. Proterozoic outcrop locations are shown with respect to the seismic reflection profiles (bold continuous lines) and major tectonic features such as the Tintina fault zone (dashed line), Cordilleran frontal thrust (barbed line) and Pacific coastline. Brown shaded units are Proterozoic Windermere Supergroup; green units are Mackenzie Mountains Supergroup; yellow shaded units are Muskwa assemblage strata; pink units mark the exotic terrane Stikinia in cross-section. The yellow shaded area straddling the NWT and British Columbia border is subcrop of Muskwa assemblage as delimited by drill-holes. Dark grey is Laurentia, mainly crystalline basement.

Thin suspect terranes

The Canadian Cordillera is best known for its Neoproterozoic and Phanerozoic history, in which western Laurentia transformed from a rifted or passive margin to a complex accretionary orogen. None of the numerous exotic (sometimes called suspect) terranes that represent most of the surface rocks along the northwestern margin of North America (Monger et al. 1972; Wheeler & McFeely 1991; Colpron & Nelson 2006) are observed to underlie the Proterozoic sedimentary prism that crops out only in a few outliers. These terranes, some exotic to North America, some perhaps originally marginal to it, apparently were thrust on top of the tapered prism of Proterozoic and early Palaeozoic rock layers and nowhere wedged into it. Some terranes, such as Yukon–Tanana, Cache Creek and Quesnellia, were detached from their pre-accretion roots and thrust 200–400 km onto the margin of North America as thin (<5 km) crustal flakes (Figs 1 and 4; Cook et al. 2004). Other terranes, such as Stikinia, docked above and outside the leading edge of the layered wedge and may make up most of the crust there today (Fig. 4).

Most previous workers inferred that these suspect terranes do not occupy the entire crust (e.g. Gabrielse 1985) nor fully represent the deeper structure of the orogen (e.g. Oldow et al. 1990). A recent heavy-mineral survey in northwestern-most British Columbia sought the local source rocks hosting micro-diamonds and associated garnets and pyroxenes (Canil et al. 2005). No primary source was identified, but an uplift event, dated at 190 Ma, probably exhumed mantle peridotite and eclogite, which are preserved within coarse pebble conglomerates in forearc basin deposits of the Laberge Group and Whitehorse trough (Johnson et al. 1996; Dusel-Bacon et al. 2002; Mackenzie et al. 2005). Exhumation of mantle rocks from depths approaching the diamond stability field (100 km) suggests very active tectonism, probably associated with dismemberment of lithospheric-scale terranes during accretion to North America.

The recognition that most of these terranes, as well as tectonic slices of North American Precambrian strata, collectively form a hanging-wall block above a gently west-dipping detachment surface at the top of the Proterozoic wedge helps to constrain the order and timing of their accretion to the North American continent (Colpron & Nelson 2006). For example, if terranes occupy only the uppermost 10 km of crust, it becomes mechanically improbable that the various suspect terranes were stacked on top of or overrode one another. The simplest order of terrane accretion is the order in which they appear at the surface today. The interpreted thinness of the terranes also de-emphasizes their volumetric importance in the overall architecture of the Cordillera. As a corollary, the recognition of the thin flakes of suspect terranes emphasizes the volumetric importance of missing mantle lithosphere (Canil et al. 2005), material detached from the suspect terranes during their obduction that has been recycled into the mantle, melted, and potentially reintroduced into the crust as magma.

Origin and more general implications

Accretionary orogens form at sites of continuing subduction of oceanic lithosphere and consist of accretionary wedges containing material transferred from the downgoing plate, recrystallized rock from melt products, and detritus from the upper plate. Most investigators agree that the major site of juvenile continental crust production globally is convergent plate margins associated with accretionary orogens. Juvenile crust is produced in arc systems from magmas derived primarily from the mantle wedge and upper plate crust; it also is added to continents as exotic or suspect terranes by collision and at least partial accretion of oceanic arcs, oceanic plateaus and oceanic crust (ophiolites). Some outstanding questions related to the Canadian Cordillera accretionary orogen are the following. (1) In which tectonic setting was continental crust produced *in situ* and has the setting varied with time? (2) What is the lifetime of this accretionary orogen and what was the lifespan of component terranes and the margin prior to convergence? (3) What was the net addition rate of juvenile crust in this accretionary orogen and has it spiked with time?

The accretion of exotic and arc terranes to the Canadian Cordillera at 190–170 Ma represents a type accretionary margin. Geophysical evidence now demonstrates that these terranes were mostly accreted by thin-skinned thrusting of crustal flakes onto a prism composed of metasedimentary Proterozoic rocks and that pre-accretion lower crust and mantle lithosphere of the terranes are no longer recognized and presumably subducted into the mantle. The major question addressed here, however, is the setting of this accretion and the influence of the rock types forming the upper plate lithosphere of Laurentia.

Throughgoing seismic reflectors and related bulk properties indicate that almost the entire crust of the upper plate is composed of rocks that were once Proterozoic sediments shed off the northwestern continental shield of Laurentia. Reflector geometries indicate that this margin had >25 km of relief or subsidence and that single reflectors are generally listric. Although interpretational sections are drawn with metasedimentary rocks projecting into the lower crust almost to the present-day

Moho, the Moho location at the time of Proterozoic sediment deposition and its subsequent deformation is not known. Modification of the Moho's depth is controlled by post-deformational thermal events (e.g. Hynes & Snyder 1995) of which there were probably many since deposition of the oldest (Wernecke Supergroup) rocks. As such, the 'sub-Wernecke' basement could be (1) highly attenuated continental crust contiguous with the Fort Simpson terrane, (2) oceanic crust now in eclogite facies or (3) juvenile transitional crust as typically found in wide rift basins. In any case, its present-day seismic properties make it largely indistinguishable from mantle (e.g. Fig. 4). It and its underlying mantle have sufficient strength or viscosity to buttress the present-day North America plate margin against convergence of the Kula–Pacific oceanic plate, but high regional geotherms suggest dynamic support, high degrees of melting and generally low viscosities (Hyndman et al. 2005).

Speculation on the Proterozoic palaeogeography of NW Laurentia varies from connections to another cratonic landmass such as Australia (Moores 1991), Siberia (Condie & Rosen 1994; Sears & Price 2003) or South China (Li et al. 1995) to no connection at all, with open ocean offshore Yukon (Delaney 1981). The Wernecke and Muskwa sequences are too thick and fine-grained to represent an intercontinental basin setting, whereas the Pinguicula, Belt and Mackenzie Mountains sequences are good interior basin candidates with thick platformal carbonates, thick terrestrial sandstones and thinner intercalated shallow marine deposits. The Wernecke and Muskwa sequences could represent a newly rifted margin or a long-lived passively subsiding margin (Thorkelson et al. 2005). Each of these options might be viable along different segments of the Laurentia margin.

Possible settings for the >1.60 Ga Racklan orogeny include (Thorkelson et al. 2001) either deformation resulting from far-field stresses generated along the southern margin of Laurentia, or foreland deformation associated with a now-rifted collisional orogen on the northwest margin of Laurentia. The more enigmatic post-1.27 Ga Corn Creek orogen structures may be local effects of strike-slip faulting when viewed regionally. Strong, wide spread horizontal shortening and uplift typically associated with 'collisional' or accretionary plate margins is lacking at this time along the NW margin of Laurentia. Because chances of interaction with other continental blocks or plates during the Proterozoic are high, either this margin was protected within a continental margin embayment (e.g. Sears & Price 2003) or continental interactions were unusually few and probably oblique.

Although inferred from previous stratigraphic studies, the vast volume and lateral extent of Proterozoic sedimentary deposits along the western margin of North America requires rethinking of some long-held ideas. Are these deposits orogenic or rift related? Although the genesis is different, the final structure of a slowly filling rift or coalescing alluvial fans cannot be distinguished if only one margin is preserved (see Ross et al. 2001; Thorkelson et al. 2001, 2005, for further discussion). Stratigraphic studies indicate that a long-lived, slowly subsiding rift margin is the most probable setting: an open margin or very wide rift basin was fed by at least four major sedimentation pulses that were widely spaced in time over about 1 Ga. Potential sources for these pulses include interior orogens such as the Wopmay and Trans-Hudson orogen during the Palaeoproterozoic assembly of Laurentia (Nuna or Columbia) and the Grenvillian during the Neoproterozoic assembly of Rodinia (Rainbird et al. 1997; Ross et al. 2001; Thorkelson et al. 2005).

A long-lived, largely passive margin setting remains the simplest one that explains current observations. This Proterozoic sedimentary prism was subjected to several periods of contractional deformations: some Proterozoic, possibly Devonian, and certainly Mesozoic–Tertiary. It would be improbable to recognize onlap, faulting and fold structures characteristic of any original active rift setting on seismic reflection data from such a heterogeneously shortened and thickened crustal volume. Not all characteristics of modern major sedimentary fans along continental margins are analogous with the Proterozoic sedimentary sequences described here. Nevertheless, the volumes, thicknesses, accumulation rates and overall geometries of the Wernecke and Bengal sedimentary fans are comparable given the known information, especially if the Proterozoic Wopmay and Trans-Hudson orogens are deemed comparable with the modern Himalayas.

Geochemical analysis of mantle xenoliths entrained in Cretaceous to Recent volcanic eruptions indicates that the North American plate ends somewhere between the Tintina and Teslin zones (Fig. 1; Creaser et al. 1997; Abraham et al. 2001). This implies that North American mantle (and attenuated crust) underlies the Proterozoic sedimentary prism. Thus reflectors associated with Proterozoic sedimentary sequences, such as the Muskwa assemblage, indicate that these sediments prograded largely onto attenuated (rifted?) continental crust that episodically subsided because of variable cooling and loading rates. Overall, this crust subsided sufficiently to reach depths near 25 km while accommodating subsequent sedimentary unit deposition. Underlying crystalline rocks metamorphosed into rocks with seismic characteristics of mantle.

Heat flow studies throughout the Cordillera indicate that the average heat flow is remarkably high, about 105 mW m^{-2}, and results in estimates of high crustal temperatures (Lewis *et al.* 2003). If similar temperatures existed during the orogenic evolution of the Cordillera, only the upper crust would have been strong, allowing for mid- to lower crustal detachment within and above the reflective prism. The high temperatures would also produce melts of generally granitic composition such as the numerous Cretaceous granites observed in the Cordillera.

Magnetotelluric results suggest a laterally varied mantle and crust (Ledo *et al.* 2002), in contrast to the horizontally stratified wedge of reflectors described here. These observations are not incompatible. Highly attenuated North American continental mantle lithosphere within the upper plate is juxtaposed complexly with delaminated, underthrust lithosphere of more primitive, oceanic affinity as is observed for crustal rocks of the northern Cordillera (e.g. Colpron & Nelson 2006). These primitive mantle blocks would represent the deeper parts of the accreted exotic terranes.

Partitioning of strain and horizontal shortening within the lithosphere of Canadian Cordillera and the effect of the large Proterozoic sedimentary prism on processes of plate convergence in the Mesozoic accretionary orogens remain unquantified. Recent estimates from the Andes (the Southern Cordillera) suggests that 58% of the westward drift of South America during the last 35 Ma was accommodated by trench rollback of the Nazca Plate, 37% by tectonic shortening of the South American plate and 5% by subduction erosion (Sobolev & Babeyko 2005). The relative convergence of North America and subducting plates from 400 to 250 Ma when the current collage of Canadian Cordilleran terranes accreted onto North America was similar; however, the large portion of felsic crust and attenuated underlying mantle inferred here would greatly increase the resulting horizontal shortening of the overriding plate when compared with modern Andean analogues (Sobolev & Babeyko 2005). The apparent shortening of North America would be reduced by 'flake tectonic' accretion of offshore terranes (e.g. Colpron & Nelson 2006).

The relatively shallow décollement between the prism and the overlying suspect terranes revisits concepts of thin-skinned thrusting and the compressive strength of the footwall material. Although much of the wedge of Proterozoic continental lithosphere is sedimentary in origin, it is now metamorphosed and deformed into gneiss and therefore acts mechanically as granitic basement. Deeper parts of the wedge also incorporated blocks of mantle lithosphere from accreting terranes. The geometry described resembles that of thin-skinned thrust zones in many locations worldwide; for example, the Appalachians, Caledonides and Urals. Footwall sedimentary prisms of the volume described here are not known in those accretionary or collisional orogens, but perhaps are just not yet recognized.

Summary answers to the questions posed at the beginning of this section are as follows:

(1) Much of the continental crust of the northern Canadian Cordillera began as erosional products from the shield interior deposited in a passive margin setting; exotic upper crustal flakes were thrust on top of this crust in a classic accretionary margin setting.

(2) The passive margin existed since at least 1.85 Ga; component exotic terranes formed offshore during the Palaeozoic and early Cenozoic; the current accretionary margin was assembled from 190 Ma until today.

(3) Proterozoic addition of sedimentary rocks to the margin occurred in three or four pulses of largely undetermined length; exotic terranes accreted primarily during the Jurassic.

This paper is LITHOPROBE Contribution 1474; Geological Survey of Canada Contribution 20070227.

References

ABBOT, J. G. 1997. *Geology of the upper Hart River area, eastern Ogilvie Mountains, Yukon Territory (116A/10, 116A/11).* Exploration and Geological Services Division, Yukon Region, Indian and Northern Affairs Canada, Bulletin, **9**.

ABRAHAM, A. C., FRANCIS, D. & POLVE, M. 2001. Recent alkaline basalts as probes of the lithospheric roots of the northern Canadian Cordillera. *Chemical Geology*, **175**, 361–386.

AITKEN, J. D. 1978. Revised models for depositional grand cycles, Cambrian of the southern Rocky Mountains, Canada. *Bulletin of the Canadian Society of Petroleum Geology*, **26**, 515–542.

AITKEN, J. D. & MCMECHAN, M. E. 1992. Middle Proterozoic assemblages. *In*: GABRIELSE, H. & YORATH, C. J. (eds) *Geology of Cordilleran Orogen in Canada*. Geological Survey of Canada, Geology of Canada, **4**, 97–124.

ALLMENDINGER, R. W., HAUGE, T. ET AL. 1987. Overview of the COCORP 40°N Transect, Western United States: The fabric of an orogenic belt. *Geological Society of America Bulletin*, **98**, 308–319.

ANDERSON, H. E. & DAVIS, D. W. 1995. U–Pb geochronology of the Moyie sills, Purcell Supergroup, southeastern British Columbia: Implications for the Mesoproterozoic geological history of the Purcell (Belt) basin. *Canadian Journal of Earth Sciences*, **32**, 1180–1193.

BELL, R. T. 1968. *Proterozoic stratigraphy of northeastern British Columbia*. Geological Survey of Canada, Paper, **67–68**, 1–75.

BURCHFIEL, B. C. & DAVIS, G. A. 1975. Nature and controls of Cordilleran orogenesis, western United States: Extensions of an earlier synthesis. *American Journal of Science (Rodgers Volume)*, **275-A**, 363–396.

CANIL, D., MIHALYNUK, M. G., MACKENZIE, J. M., JOHNSTON, S. T. & GRANT, B. 2005. Diamonds in the Atlin–Nakina region, British Columbia: Insights from heavy minerals in stream sediments. *Canadian Journal of Earth Sciences*, **42**, 2161–2171.

CECILE, M. P. & COOK, D. G. 1981. *Structural cross-section northern Selwyn and Mackenzie mountains, Yukon and District of Mackenzie*. Geological Survey of Canada, Open File, **807**.

CHRISTENSEN, N. I. 1996. Poisson's ratio and crustal seismology. *Journal of Geophysical Research*, **101**, 3139–3156.

CLARK, E. A. & COOK, F. A. 1992. Crustal-scale ramp in a middle Proterozoic orogen, Northwest Territories, Canada. *Canadian Journal of Earth Sciences*, **29**, 142–157.

CLOWES, R. M., HAMMER, P. T. C., FERNANDEZ-VIEJO, G. & WELFORD, K. 2005. Lithospheric structure in northwestern Canada from Lithoprobe seismic refraction and related studies: a synthesis. *Canadian Journal of Earth Sciences*, **42**, 1277–1293.

COLPRON, M. & NELSON, J. L. 2006. *Paleozoic evolution and metallogeny of pericratonic terranes at the ancient Pacific margin of North America, Canadian and Alaskan Cordillera*. Geological Association of Canada, Special Paper, **45**.

CONDIE, K. C. & ROSEN, O. M. 1994. Laurentia-Siberia. *Geology*, **22**, 168–170.

COOK, D. G. & MACLEAN, B. C. 1995. The intracratonic Forward Orogeny, and implications for regional correlations, Northwest Territories, Canada. *Canadian Journal of Earth Sciences*, **32**, 1991–2008.

COOK, F. A. & ERDMER, P. 2005. An 1800 km cross section of the lithosphere through the northwestern North American plate: Lessons from 4.0 billion years of Earth's history. *Canadian Journal of Earth Sciences*, **42**, 1295–1311.

COOK, F. A. & SIEGEL, S. M. 2006. From Proterozoic strata to a synthesized seismic reflection trace: Implications for regional seismic reflection patterns in northwestern Canada. *Canadian Journal of Earth Sciences*, **43**, 1639–1651.

COOK, F. A. & VARSEK, J. L. 1994. Orogen-scale décollements. *Reviews of Geophysics*, **32**, 37–60.

COOK, F. A., VAN DER VELDEN, A. J., HALL, K. & ROBERTS, B. 1999. Frozen subduction in Canada's Northwest Territories: LITHOPROBE deep lithospheric reflection profiling of the western Canadian Shield. *Tectonics*, **18**, 1–24.

COOK, F. A., CLOWES, R. M., SNYDER, D. B., VAN DER VELDEN, A. J., HALL, K., ERDMER, P. & EVENCHICK, C. A. 2004. Precambrian crust and lithosphere beneath the Northern Canadian Cordillera discovered by LITHOPROBE seismic reflection profiling. *Tectonics*, **23**, TC2010, doi:10.1029/2002TC001412.

COOK, F. A., HALL, K. & LYNN, C. E. 2005. The edge of northwestern North America at ~1.8 Ga. *Canadian Journal of Earth Sciences*, **42**, 9831–9997.

CREASER, R. A., ERDMER, P., STEVENS, R. A. & GRANT, S. L. 1997. Tectonic affinity of the Nisutlin and Anvil assemblage strata from the Teslin Tectonic Zone, Northern Canadian Cordillera: Constraints from neodymium isotope and geochemistry evidence. *Tectonics*, **16**, 107–121.

DALZIEL, I. W. D. & SOPER, N. J. 2001. Neoproterozoic extension on the Scottish promontory of Laurentia; paleogeographic and tectonic implications. *Journal of Geology*, **109**, 299–317.

DELANEY, G. D. 1981. The mid-Proterozoic Wernecke Supergroup, Wernecke Mountains, Yukon Territory. *In*: CAMPBELL, F. H. A. (ed.) *Proterozoic Basins of Canada*. Geological Survey of Canada Paper, **81–10**, 1–23.

DUSEL-BACON, C., LANPHERE, M. A., SHARP, W. D., LAYER, P. W. & HANSEN, V. L. 2002. Mesozoic thermal history and timing of structural events for the Yukon–Tanana Upland, east–central Alaska: $^{40}Ar/^{39}Ar$ data from metamorphic and plutonic rocks. *Canadian Journal of Earth Sciences*, **39**, 1013–1051.

EISBACHER, G. H. 1978. Two major Proterozoic unconformities, Northern Cordillera. *In: Current Research, Part A*. Geological Survey of Canada Paper, **78-1A**, 53–58.

EISBACHER, G. H. 1981. *Sedimentary tectonics and glacial record in the Windermere Supergroup, Mackenzie Mountains, northwestern Canada*. Geological Survey of Canada Paper, **80–27**.

EVENCHICK, C. A., PARRISH, R. R. & GABRIELSE, H. 1984. Precambrian gneiss and late Proterozoic sedimentation in north–central British Columbia. *Geology*, **12**, 233–237.

FERNANDEZ-VIEJO, G., CLOWES, R. M. & WELFORD, J. K. 2005. Constraints on the composition of the crust and uppermost mantle in northwestern Canada: Vp/Vs variations along LITHOPROBE's SNorCLE transect. *Canadian Journal of Earth Sciences*, **42**, 1205–1222.

FRASER, J. A., DONALDSON, J. A., FAHRIG, W. F. & TREMBLAY, L. P. 1970. Helikian basins and geosynclines of the northwestern Canadian Shield. *In*: BAER, A. J. (ed.) *Symposium on Basins and Geosynclines of the Canadian Shield*. Geological Survey of Canada Paper, **70–40**, 213–238.

GABRIELSE, H. 1985. Major dextral transcurrent displacements along the northern Rocky Mountain Trench and related lineaments in north central British Columbia. *Geological Society of America Bulletin*, **96**, 1–14.

GOODBRED, S. L. & KUEHL, S. A. 1999. Holocene and modern sediment budgets for the Ganges–Brahmaputra river system: Evidence for highstand dispersal to flood-plain, shelf, and deep-sea depocenters. *Geology*, **27**, 559–562.

HILDEBRAND, R. S. & BARANGAR, W. R. A. 1991. On folds and thrusts affecting the Coppermine River Group, northwest Canada. *Canadian Journal of Earth Sciences*, **28**, 523–531.

HILDEBRAND, R. S., HOFFMAN, P. F. & BOWRING, S. 1987. Tectono-magmatic evolution of the 1.9 Ga Great Bear magmatic zone, Wopmay orogen, northwest Canada. *Journal of Volcanology and Geothermal Research*, **32**, 99–118.

HOFFMAN, P. F. 1980. Wopmay Orogen: A Wilson cycle of Early Proterozoic age in the northwest of the Canadian Shield. *In*: STRANGWAY, D. W. (ed.) *The Continental Crust and its Mineral Deposits*. Geological Association of Canada Special Paper, **20**, 523–549.

HYNDMAN, R. D., FLUECK, P., MAZZOTTI, S., LEWIS, T. J., RISTAU, J. & LEONARD, L. 2005. Current tectonics of the northern Canadian Cordillera, Canada. *Canadian Journal of Earth Sciences*, **42**, 1117–1136.

HYNES, A. & SNYDER, D. 1995. Deep-crustal mineral assemblages and potential for crustal rocks below the

Moho in the Scottish Caledonides. *Geophysical Journal International*, **123**, 323–339.

JEFFERSON, C. W. & PARRISH, R. R. 1989. Late Proterozoic stratigraphy, U–Pb zircon ages, and rift tectonics, Mackenzie Mountains, northwestern Canada. *Canadian Journal of Earth Sciences*, **26**, 1784–1801.

JOHNSON, S. T., MORTENSEN, J. K. & ERDMER, P. 1996. Igneous and meta-igneous age constraints for the Aishihik metamorphic suite, southwest Yukon. *Canadian Journal of Earth Sciences*, **33**, 1543–1555.

KERANS, C., ROSS, G. M., DONALDSON, J. A. & GELDSETZER, H. J. 1981. Tectonism and depositional history of the Helikian Hornby Bay and Dismal Lakes Groups, District of Mackenzie. *In*: CAMPBELL, F. H. A. (ed.) *Proterozoic Basins of Canada*. Geological Survey of Canada Paper, **81–10**, 157–182.

LEDO, J., JONES, A. G., FERGUSON, I. J. & WENNBERG, G. 2002. SNORCLE Corridor 3 MT experiment. *In*: COOK, F. & ERDMER, P. (eds) LITHOPROBE Secretariat, University of British Columbia, Vancouver, Lithoprobe Report, **82**, 21–23.

LEWIS, T., HYNDMAN, R. & FLUECK, P. 2003. Heat flow, heat generation, and crustal temperatures in the northern Canadian Cordillera: Thermal control of tectonics. *Journal of Geophysical Research*, **108**, doi:10.1029/2002JB002090.

LI, Z.-X., ZHANG, L. & POWELL, C. 1995. South China in Rodinia: Part of the missing link between Australia–east Antarctica and Laurentia? *Geology*, **23**, 407–410.

LONG, D. G. F. & PRATT, B. R. 2000. Mesoproterozoic evolution of northeastern British Columbia. *In*: COOK, F. & ERDMER, P. (eds) *Slave–Northern Cordillera Lithospheric Evolution (SNorCLE) Transect and Cordilleran Tectonics Workshop Meeting, Calgary, AB, 25–27 February, 2000*. LITHOPROBE Secretariat, University of British Columbia, Vancouver, LITHOPROBE Report, **72**, 91–92.

MACKENZIE, J. M., CANIL, D., JOHNSTON, S. T., ENGLISH, J. E., MIHALYNUK, M. G. & GRANT, B. 2005. First evidence for ultrahigh pressure garnet peridotite in the North America Cordillera. *Geology*, **33**, 105–108.

MACLEAN, B. C. & COOK, D. G. 2004. Revisions to the Paleoproterozoic Sequence A, based on reflection seismic data across the western plains of the Northwest Territories, Canada. *Precambrian Research*, **129**, 271–289.

MONGER, J. W. H., GABRIELSE, H. & SOUTHER, J. A. 1972. Evolution of the Canadian Cordillera: A plate tectonic model. *American Journal of Science*, **272**, 577–602.

MOORES, E. M. 1991. Southwest U.S.–east Antarctic (SWEAT) connection: A hypothesis. *Geology*, **19**, 425–428.

MOROZOV, I. B., SMITHSON, S. B., CHEN, J. & HOLLISTER, L. S. 2001. Generation of new continental crust and terrane accretion in SE Alaska and western British Columbia; constraints from P- and S-wave wide-angle seismic data (ACCRETE). *Tectonophysics*, **341**, 49–67.

MUSACCHIO, G., WHITE, D. J., ASUDEH, I. & THOMSON, C. J. 2004. Lithospheric structure and composition of the Archaean western Superior Province from seismic refraction/wide-angle reflection and gravity modeling. *Journal of Geophysical Research*, **109**, B03304, doi:10.1029/2003JB002427.

OLDOW, J. S., BALLY, A. W. & AVE LALLEMENT, H. G. 1990. Transpression, orogenic float, and lithospheric balance. *Geology*, **18**, 991–994.

PILKINGTON, M., SNYDER, D. B. & HEMANT, K. 2006. Weakly magnetic crust in the Canadian Cordillera. *Earth and Planetary Science Letters*, **248**, 461–470.

RAINBIRD, R. H., MCNICOLL, V. J., THERIAULT, R. J., HEAMAN, L. M., ABBOT, J. G., LONG, D. G. F. & THORKELSON, D. J. 1997. Pan-continental river system draining Grenville orogen recorded by U–Pb and Sm–Nd geochronology of Neoproterozoic quartzites and mudrocks, northwestern Canada. *Journal of Geology*, **105**, 1–17.

RAINBIRD, R. H., HAMILTON, M. A. & YOUNG, G. M. 2001. Detrital zircon geochronology and provenance of the Torridonian, NW Scotland. *Journal of the Geological Society, London*, **158**, 15–27.

ROSS, G. M., VILLENEUVE, M. E. & THERIAULT, R. J. 2001. Isotopic provenance of the lower Muskwa assemblage (Mesoproterozoic, Rocky Mountains, British Columbia): New clues to correlation and source areas. *Precambrian Research*, **111**, 57–77.

SEARS, J. W. & PRICE, R. A. 2003. Tightening the Siberian connection to western Laurentia. *Geological Society of America Bulletin*, **115**, 943–953.

SNYDER, D. B. & HOBBS, R. H. 1999. *The BIRPS Atlas II: A Second Decade of Deep Seismic Reflection Profiling*. Geological Society, London.

SOBOLEV, S. V. & BABEYKO, A. Y. 2005. What drives orogeny in the Andes? *Geology*, **33**, 617–620.

SPECTOR, A. & GRANT, F. S. 1970. Statistical methods for interpreting aeromagnetic data. *Geophysics*, **35**, 293–302.

STAGG, H. M. J., WILCOX, J. B. ET AL. 1999. Architecture and evolution of the Australian continental margin. *AGSO Journal of Australian Geology and Geophysics*, **17**, 17–33.

THORKELSON, D. J. 2000. *Geology and mineral occurrences of the Slats Creek, Fairchild Lake and 'Dolores Creek' areas, Wernecke Mountains, Yukon Territory (106D/16, 106C/13, 106C/14)*. Exploration and Geological Services Division, Yukon Region, Indian and Northern Affairs Canada, Bulletin, **10**.

THORKELSON, D. J., MORTENSEN, J. K., DAVIDSON, G. J., CREASER, R. A., PEREZ, W. A. & ABBOTT, J. G. 2001. Early Mesoproterozoic intrusive breccias in Yukon, Canada: The role of hydrothermal systems in reconstructions of North America and Australia. *Precambrian Research*, **111**, 31–55.

THORKELSON, D. J., ABBOTT, J. G., MORTENSEN, J. K., CREASER, R. A., VILLENEUVE, M. E., MCNICOLL, V. J. & LAYER, P. W. 2005. Early and Middle Proterozoic evolution of Yukon, Canada. *Canadian Journal of Earth Sciences*, **42**, 1045–1071.

WHEELER, J. O. & MCFEELY, P. 1991. *Tectonic assemblage map of the Canadian Cordillera and adjacent parts of the United States of America, scale 1:2 000 000*. Geological Survey of Canada Map, **1712A**.

YOUNG, G. M., JEFFERSON, C. M., LONG, D. G., DELANEY, G. D. & YEO, C. M. 1979. Middle and late Proterozoic evolution of the northern Canadian Cordillera and Shield. *Geology*, **7**, 125–128.

A Palaeozoic Northwest Passage: incursion of Caledonian, Baltican and Siberian terranes into eastern Panthalassa, and the early evolution of the North American Cordillera

MAURICE COLPRON[1]* & JOANNE L. NELSON[2]

[1]*Yukon Geological Survey, PO Box 2703 (K-10), Whitehorse, Yukon, Canada Y1A 2C6*

[2]*British Columbia Geological Survey, PO Box 9333 Stn Prov Gov't, Victoria, British Columbia, Canada V8W 9N3*

Corresponding author (e-mail: maurice.colpron@gov.yk.ca)

Abstract: Palaeozoic to early Mesozoic terranes of the North American Cordillera mostly originated from three distinct regions in Palaeozoic time: the western peri-Laurentian margin, western (Asian) Panthalassa, and the northern Caledonides–Siberia. A review of geological history, fossil and provenance data for the Caledonian–Siberian terranes suggests that they probably occupied an intermediate position between northern Baltica, northeastern Laurentia and Siberia, in proximity to the northern Caledonides, in early Palaeozoic time. Dispersion of these terranes and their westward incursion into eastern Panthalassa are interpreted to result from development of a Caribbean- or Scotia-style subduction system between northern Laurentia and Siberia in mid-Palaeozoic time, termed here the Northwest Passage. Westward propagation of a narrow subduction zone coupled with a global change in plate motion, related to the collision of Gondwana with Laurentia–Baltica, are proposed to have led to initiation of subduction along the western passive margin of Laurentia and development of the peri-Laurentian terranes as a set of rifted continental fragments, superimposed arcs and marginal ocean basin(s) in mid- to late Palaeozoic time. Diachronous orogenic activity from Late Silurian in Arctic Canada, to Early Devonian in north Yukon and adjacent Alaska, Middle Devonian in southeastern British Columbia, and Late Devonian–Early Mississippian in the western USA records progressive development of the Northwest Passage and southward propagation of subduction along western Laurentia.

The North American Cordillera is regarded as one of the type accretionary orogens on Earth, where growth occurred as the result of progressive addition of terranes, crustal elements that preserve a geological record distinct from their neighbours, to the western margin of Laurentia beginning in late Palaeozoic time (Coney *et al.* 1980). The western margin of Laurentia originated during late Neoproterozoic breakup of Rodinia, and passive continental margin deposition prevailed through the early Palaeozoic (Price 1994). In mid-Palaeozoic time, it was converted to an active plate boundary and subduction was initiated, as indicated by first occurrences of arc and back-arc magmatism in Devonian strata of the distal continental margin and the accreted terranes (Rubin *et al.* 1990; Monger & Nokleberg 1996). This convergent margin geodynamic setting, which has prevailed along the western margin of North America up to the present time, provided the environment for generation, dispersion and accretion of the Cordilleran terranes.

Early terrane analysis of the North American Cordillera suggested that allochthonous terranes were of uncertain palaeogeographical origins and that the Cordilleran orogen represented a collage of disparate crustal fragments (Helwig 1974; Coney *et al.* 1980). Since then, detailed mapping coupled with application of analytical tools (Nd, Hf, Sr isotopes; geochemistry; U–Pb geochronology, particularly of detrital zircon suites; palaeomagnetism) and the fossil evidence have greatly improved our knowledge of the geological history and geodynamic affinities of the accreted terranes (Fig. 1). It is now recognized that a group of terranes that generally occupy the core of the orogen (Yukon–Tanana, Quesnellia, Stikinia and related terranes) were generated along the western margin of Laurentia as a series of rifted continental fragments, superimposed arcs and marginal ocean basin(s) in mid-Palaeozoic to early Mesozoic time (Fig. 1; Nelson & Colpron 2007; Nelson *et al.* 2006; Colpron *et al.* 2007). These terranes enclose oceanic rocks with Palaeozoic faunal elements of Tethyan (Asian) affinity that were incorporated during the early Mesozoic development of the Cordilleran orogen (Mihalynuk *et al.* 1994).

In contrast, terranes that generally occupy more outboard positions in the orogen (the Arctic realm of Colpron *et al.* (2007)) have been recognized by

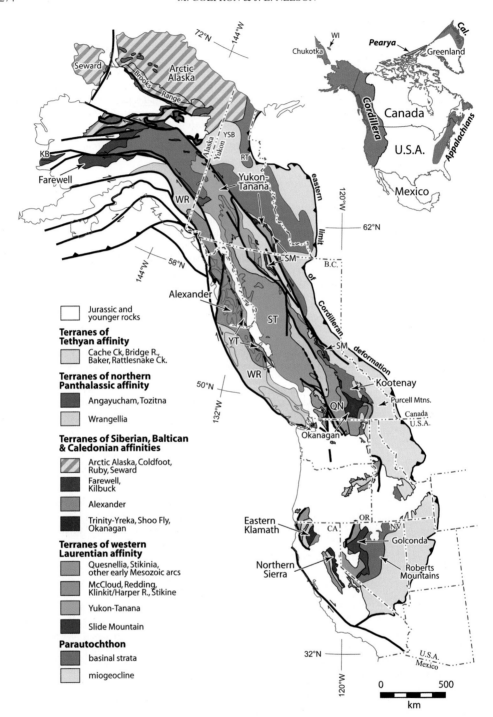

Fig. 1. Palaeozoic to early Mesozoic terranes of the North American Cordillera. Terranes are grouped according to faunal affinity and/or source region in early Palaeozoic time. Terrane and geological abbreviations: KB, Kilbuck; QN, Quesnellia; RT, Richardson trough; SM, Slide Mountain; ST, Stikinia; YSB, Yukon Stable Block; YT, Yukon–Tanana terrane in the Coast Mountains; WR, Wrangellia. Inset shows location of the Cordilleran orogen in western North America with respect to Chukotka and Wrangel Island (WI), Pearya in the Arctic Islands, the Greenland Caledonides (Cal.) and the Appalachians along the east coast.

a growing consensus as manifesting Palaeozoic and older affinities with northern Baltica (Alexander; Bazard et al. 1995; Gehrels et al. 1996; Soja & Antoshkina 1997; Antoshkina & Soja 2006; Soja & Krutikov 2008), Siberia (e.g. Farewell; Blodgett et al. 2002; Dumoulin et al. 2002; Bradley et al. 2003, 2007; Fryda & Blodgett 2008), the northern Caledonides (part of Arctic Alaska; Nilsen 1981; Moore et al. 1994), or a combination of these end-members and northern Laurentia (e.g. Arctic Alaska, Eastern Klamath, Northern Sierra and others; Dumoulin et al. 2002; Wright & Wyld 2006; Lindsley-Griffin et al. 2008; Fig. 1). Equally important, shared faunas, igneous and deformational events, and similar detrital zircon populations suggest that all of these terranes developed in some proximity to each other, and therefore constitute elements of a single, albeit complex tectonic system.

Here, we summarize the evidence that constrains the tectonic evolution and probable sites of origin of Palaeozoic–early Mesozoic terranes of the North American Cordillera, with particular emphasis on the convergence of published opinion as to the probable palaeo-Arctic origins of the outboard terranes. We then present a hypothesis that accounts for the incursion of crustal fragments of inferred Siberian, Baltican and Caledonian affinities into eastern Panthalassa, the late Palaeozoic World Ocean (Scotese 2002), and at the same time provides a mechanism for the mid-Palaeozoic initiation of subduction along the western passive margin of Laurentia. Our model calls for the development of a Caribbean- or Scotia-style subduction system along the northern margin of Laurentia in mid-Palaeozoic time. A similar concept was proposed earlier by Wright & Wyld (2006), who suggested that a mid-Palaeozoic Caribbean-style system, which developed between southern Laurentia and Gondwana, allowed south to north migration of the Alexander and parts of the Eastern Klamath and Northern Sierra terranes from Iapetus into Panthalassa. Our review of the geological evidence (below) leads us to conclude that these terranes have much stronger affinities with northern Caledonian, northern Baltican and Siberian source regions than with Appalachian Iapetan terranes. Our model also provides an explanation for the enigmatic mid-Palaeozoic compressional (transpressional) deformational events documented along the northern and western continental margin of Laurentia.

Our discussion is focused on terranes with inferred Caledonian–Baltican–Siberian affinities and their Palaeozoic interactions with northern and western Laurentia. Terranes of Tethyan (western Panthalassan) affinity in Palaeozoic time were introduced to the Cordilleran arena only in the early Mesozoic and will not be considered any further here. Aspects of their evolution have been presented by Mihalynuk et al. (1994, 2004) and English & Johnston (2005).

Our review of the tectonic evolution of the North American Cordillera is primarily focused on its pre-Mesozoic history, prior to the Jurassic–Palaeocene orogenesis that led to development of the modern Cordilleran mountain belt (Fig. 1). Recent reviews of the Mesozoic to early Cenozoic evolution of the Cordillera have been presented by Monger & Nokleberg (1996), Monger & Price (2002) and Evenchick et al. (2007).

Peri-Laurentian terranes

The inner belt of terranes of the Canadian Cordillera, the pericratonic terranes of Wheeler et al. (1991; Kootenay, Yukon–Tanana and others, Fig. 1), have faunal affinities and provenance ties with western Laurentia, but their mid-Palaeozoic to early Mesozoic geological record differs from that of the continental margin (Fig. 2). They probably formed a mid- to late Palaeozoic set of island-arc and rifted crustal fragments that were separated from the Laurentian continental margin by a marginal ocean basin, remnants of which are preserved in the Slide Mountain terrane (Fig. 1; Nelson et al. 2006; Colpron et al. 2007). Development of these terranes began with the onset of subduction along western Laurentia in mid-Palaeozoic time (Rubin et al. 1990). Early indications of this convergent plate setting are recorded in local Middle Devonian shortening of continental margin strata in southeastern British Columbia (Purcell Mountains, Fig. 1; Root 2001) and in the Late Devonian to Early Mississippian Antler orogeny of the southwestern USA (Roberts Mountains allochthons, Fig. 1; Johnson & Pendergast 1981). In contrast, Late Devonian–Early Mississippian continental margin strata in the northern Cordillera record extension and localized alkaline to calc-alkaline bimodal volcanism and a distal influx of detritus from the Ellesmerian orogen to the north (Gordey et al. 1987; Smith et al. 1993).

The earliest record of arc magmatism in the peri-Laurentian realm occurred in late Middle Devonian time (c. 387 Ma) in the Yukon–Tanana terrane of coastal British Columbia and southeastern Alaska (Nelson et al. 2006) and the western Kootenay terrane of southern British Columbia (Schiarizza & Preto 1987; Fig. 1). By Late Devonian–earliest Mississippian time, arc magmatism was widespread in Yukon–Tanana and locally significant on the distal Laurentian margin (Nelson et al. 2006; Piercey et al. 2006).

The western Kootenay terrane (Eagle Bay assemblage; Schiarizza & Preto 1987; Paradis et al. 2006) includes Late Devonian to earliest

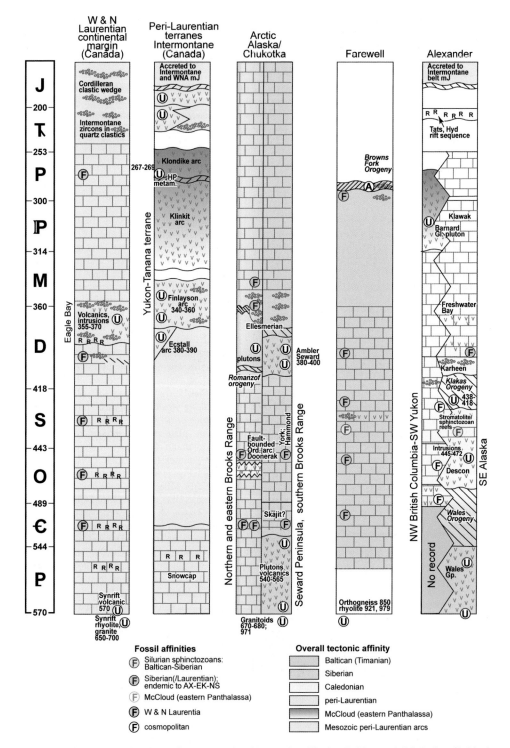

Fig. 2. Composite stratigraphic columns for terranes of peri-Laurentian, Siberian, Baltican and Caledonian affinities in western and northern North America (compiled from sources cited in the text). AMQ, Antelope Mountain Quartzite; AX, Alexander terrane; EK, Eastern Klamath terrane; NS, Northern Sierra terrane.

Mississippian calc-alkaline plutons and coeval bimodal volcanic rocks that intrude and overlie a Neoproterozoic–early Palaeozoic sequence of predominantly clastic metasedimentary rocks and rift-related alkalic to tholeiitic metavolcanic rocks, including an archaeocyathid-bearing marble. This Neoproterozoic–early Palaeozoic sequence is correlated in part with the lower Palaeozoic Lardeau Group east of the Monashee complex (Fig. 3), which conformably overlies Ediacaran

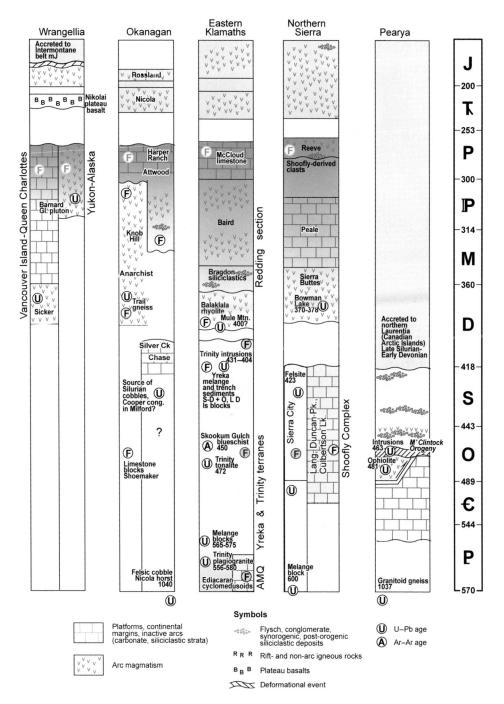

Fig. 2. (*Continued*).

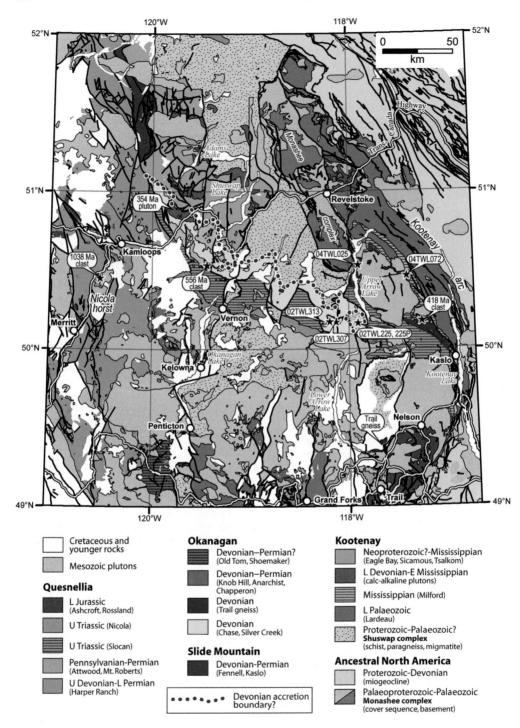

Fig. 3. Geological map of the Okanagan–Kootenay region of southern British Columbia (modified after Massey et al. 2005; Thompson et al. 2006). Detrital zircon samples reported by Lemieux et al. (2007) and other significant features of the Okanagan subterrane and southern Quesnellia are located by stars. Dotted red line shows location of a possible Devonian accretion boundary.

to Lower Cambrian quartzite and marble of the distal Laurentian miogeocline (Logan & Colpron 2006). Limited detrital zircon data from the Lardeau Group and underlying continental margin strata are dominated by c. 1.8 Ga grains that suggest a common source (recycling?) in Precambrian domains of Alberta and match the miogeocline reference data for southern British Columbia (Fig. 4c and f; Smith & Gehrels 1991; Gehrels et al. 1995). The evolved isotopic composition (εNd_{360} values −6.5 to −6.8) and common inheritance in zircons from calc-alkaline rocks of the Kootenay terrane suggest a continental arc setting (Paradis et al. 2006). Depositional ties with the miogeocline in the eastern Kootenay terrane and its occurrence inboard of Devonian–Permian oceanic rocks of the Slide Mountain terrane strongly suggest that the Kootenay terrane is parautochthonous with respect to the Laurentian continental margin.

A similar intepretation is proposed for basinal, clastic metasedimentary rocks and voluminous felsic magmatism of Late Devonian–earliest Mississippian age in east–central Alaska (Fig. 1; Dusel-Bacon et al. 2006). Magmatism ended at c. 354 Ma along the entire parautochthonous continental margin, presumably as the Slide Mountain ocean widened and the axis of magmatism and the Yukon–Tanana terrane migrated away to the west (Nelson et al. 2006; Colpron et al. 2007). Parautochthonous rocks of east–central Alaska and the Kootenay terrane therefore preserve the remnant, inboard arc that was isolated behind the Slide Mountain back-arc basin in Mississippian time.

The Yukon–Tanana terrane lies outboard (west) of the oceanic Slide Mountain terrane in the northern Cordillera (Fig. 1; Mortensen 1992). It consists of a metasedimentary basement (Snowcap assemblage) overlain by up to three unconformity-bounded Upper Devonian to Permian volcanic arc sequences (Finlayson, Klinkit, Klondike; Fig. 2) that are coeval with oceanic chert, argillite and mafic volcanic rocks of the Slide Mountain back-arc terrane (Colpron et al. 2006). The Snowcap assemblage is pre-Late Devonian in age and comprises varying amounts of quartzite, pelite, psammite, marble and calc-silicate, and minor mafic metavolcanic and meta-intrusive rocks (Colpron et al. 2006). Its lithological, geochemical and isotopic compositions suggest that the Snowcap assemblage represents a distal portion of the continental margin that was rifted off western Laurentia in mid-Palaeozoic time and subsequently formed the nucleus upon which magmatic arcs of the Finlayson, Klinkit and Klondike assemblages were deposited (Nelson et al. 2006; Piercey & Colpron 2006). Detrital zircons from Upper Devonian–Lower Mississippian sandstone of the Yukon–Tanana terrane are dominated by grains of

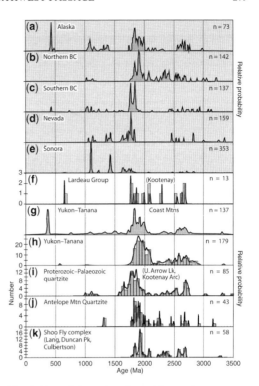

Fig. 4. Detrital zircon data for peri-Laurentian terranes compared with reference data of Gehrels et al. (1995) and Gehrels & Ross (1998) for the North American miogeocline (**a**–**e**; after Gehrels 2000); (**f**) Broadview Formation, Lardeau Group (Kootenay terrane), southern British Columbia (after Smith & Gehrels 1991); (**g**) Yukon–Tanana terrane from the Coast Mountains of southeastern Alaska (after Gehrels & Kapp 1998); (**h**) Yukon–Tanana terrane of northern British Columbia and Yukon (composite of three samples; after Devine et al. 2006; Nelson & Gehrels 2007; M. Colpron, unpubl. data); (**i**) Proterozoic–lower Palaeozoic schist from Upper Arrow Lake region, southern British Columbia and basal Milford quartzite of the Kootenay Arc region (Fig. 3; composite of samples 02TWL225P, 02TWL307 and 04TWL072 of Lemieux et al. 2007); (**j**) Antelope Mountain Quartzite, Eastern Klamath terrane (after Wallin et al. 2000); (**k**) Lang, Duncan Peak and Culbertson Lake allochthons of the Shoo Fly Complex, Northern Sierra terrane (after Harding et al. 2000). Reference data are shown with shaded background in this and other detrital zircon plots.

1.8–2.1 Ga and 2.5–2.7 Ga, with minor populations at c. 1.3–1.4 Ga (Fig. 4g and h; Gehrels & Kapp 1998; Devine et al. 2006; Nelson & Gehrels 2007) that match well with the miogeoclinal reference for northern British Columbia (Gehrels et al. 1995; Gehrels & Ross 1998). Rocks of the Snowcap assemblage appear to have been deformed and metamorphosed prior to deposition of the overlying Upper Devonian to Permian strata. The cause of this deformational event remains enigmatic. Berman et al. (2007) interpreted a Late Devonian low-pressure metamorphic and deformational event in the western Yukon–Tanana terrane to be related to arc plutonism above an east-dipping subduction zone.

Arc and back-arc magmatism began in Late Devonian time (c. 365 Ma) in most of Yukon–Tanana terrane and was widespread in the terrane by Early Mississippian time (c. 355–345 Ma), forming large calc-alkaline plutons intruding the Snowcap basement and extensive, coeval volcanic successions of the Finlayson assemblage (Colpron et al. 2006; Piercey et al. 2006). Back-arc volcanism occurred in association with basinal, carbonaceous phyllite and syngenetic sulphide deposits that have highly radiogenic Pb isotopic compositions, similar to those of broadly coeval syngenetic sulphide occurrences in basinal continental margin strata to the east; a lead isotopic signature that is unique to the Late Devonian of northwestern Laurentia (Godwin & Sinclair 1982; Mortensen et al. 2006; Nelson et al. 2006). The wide range of Nd isotopic compositions (εNd values of -5.1 to $+7.0$), widespread felsic magmatism and the ubiquitous Palaeoproterozoic–Archaean inheritance in zircons all suggest a continental arc setting.

By Middle Mississippian time (c. 340 Ma), the Yukon–Tanana arc had reached a mature stage, as indicated by the predominance of andesite, basaltic andesite and volcaniclastic rocks, and locally thick carbonate of the Klinkit assemblage (mid-Mississippian to Lower Permian; Colpron et al. 2006). Primitive geochemical and isotopic compositions (εNd $= +6.7$ to $+7.4$) indicate limited interaction with evolved crustal material (Simard et al. 2003). The Klinkit assemblage is correlated on the basis of stratigraphy and geochemistry with late Palaeozoic sequences of Quesnellia (Simard et al. 2003; Nelson & Friedman 2004). These locally include limestone with giant Parafusulina, a southern Laurentia endemic fauna that also occurs in the McCloud terranes, which are interpreted as fragments of late Palaeozoic arcs that developed some distance west of the continent (Stevens 1995).

The youngest arc succession in the Yukon–Tanana terrane, the Middle to Late Permian Klondike assemblage (Colpron et al. 2006), is dominated by felsic magmatism (Mortensen 1990; Piercey et al. 2006). It is paired with a belt of eclogite and blueschist (U–Pb zircon ages of 267–269 Ma) that lies along the eastern edge of the Yukon–Tanana terrane (Erdmer et al. 1998), separating it from the oceanic Slide Mountain terrane. Arc magmatism of the Klondike assemblage is interpreted to record west-dipping subduction of the Slide Mountain back-arc oceanic lithosphere beneath the Yukon–Tanana arc terrane in mid- to Late Permian time (Nelson et al. 2006). The cause of this mid-Permian reversal in subduction polarity and the fate of the earlier east-dipping subduction of Panthalassa oceanic lithosphere remain enigmatic. By Triassic time, generally fine-grained siliciclastic rocks that contain metamorphic detritus and zircons derived from the Yukon–Tanana terrane were deposited unconformably on the Yukon–Tanana and Slide Mountain terranes, as well as miogeoclinal strata to the east (Beranek & Mortensen 2007), suggesting that the Slide Mountain ocean had closed. Significant crustal thickening in the Yukon–Tanana terrane is indicated by c. 239 Ma amphibolite-facies metamorphism (c. 9 kbar and 600 °C) in western Yukon (Berman et al. 2007). Subsequent early Mesozoic arc magmatism of Quesnellia and Stikinia was developed in part on top of the Palaeozoic successions of Yukon–Tanana and above renewed east-dipping subduction of Panthalassa oceanic lithosphere beneath western Laurentia (Colpron et al. 2007).

In summary, the peri-Laurentian terranes show early linkages to and relationships with the western margin of Laurentia, such as the following:

(1) There are stratigraphic similarities between basement siliciclastic–carbonate–basalt sequences in the Yukon–Tanana terrane and Neoproterozoic–early Palaeozoic units on the autochthonous continental margin.

(2) Detrital zircon populations with marked peaks in the 1.8–2.0 Ga range, and distributions into the Archaean and in some cases minor grains in the range 1.0–1.3 Ga; the latter are considered to represent very distal Grenvillian detritus (Rainbird et al. 1997; Fig. 4).

(3) Continental arc magmatism commenced in Middle to Late Devonian time, accompanied by widespread evidence of extension and local compressional tectonics.

(4) There is complete lack of evidence for subduction and/or magmatic arc development in the Neoproterozoic to early Palaeozoic. In addition, Neoproterozoic arc development adjacent to western Laurentia would have been unlikely, as the rifting events responsible for creating an open, ocean-facing margin were not yet complete (e.g. Colpron et al. 2002, and references therein). A few Neoproterozoic (700–540 Ma) rift-related igneous bodies have been identified; they may be the source of small detrital zircon populations in this age range that are present in miogeoclinal samples.

Terranes of Siberian, Baltican and Caledonian affinities

The Palaeozoic parts of the outboard terranes of the Cordillera (Arctic realm of Colpron et al. 2007) do not share these characteristics. They present tectonic histories, detrital zircon populations and fossil assemblages that differ profoundly from western Laurentia and the peri-Laurentian terranes of the Cordillera. In general, their characteristics are more compatible with an Appalachian–Caledonian origin (Wright & Wyld 2006). Recent studies of these terranes strongly favour linkages with northern elements of the Caledonian orogen in northeastern Laurentia and Baltica as the closest match for some, and Siberia for others. The main lines of evidence include the following:

(1) affinities of early Palaeozoic macro- and microfossils with those in Siberia, and in some cases NE Laurentia–Baltica;

(2) for terranes of inferred Baltican and Caledonian affinities, detrital zircon signatures reflecting a heterogeneous basement with multiple sources between 1.0 and 2.0 Ga, including significant populations in the 1.49–1.61 Ga North American magmatic gap (Van Schmus et al. 1993), and only minor Archaean source terranes;

(3) evidence of Grenvillian magmatism, both direct and reflected in robust detrital zircon populations;

(4) Late Neoproterozoic magmatism and arc development (700–540 Ma);

(5) Ordovician–Silurian arc development.

In the following sections, the inferred Siberian, Baltican and Caledonian terranes, now incorporated in the Cordillera, are described in present geographical order from north to south, with a focus on the characteristics that identify their palaeogeographical affinities, and also their early relationships to each other.

Arctic Alaska–Seward–Chukotka

A sinuous, composite pericratonic terrane extends from far northern Yukon, through the Arctic Alaska region (Brooks Range and North Slope); with correlatives in the Seward Peninsula and the Chukotka Peninsula and Wrangel Island of northeastern Russia (Figs 1 and 5). Its pre-Devonian geological record and early Palaeozoic faunal affinities are distinct from those of northern Laurentia (Fig. 2). Comprehensive faunal analysis led Dumoulin et al. (2002) to propose that it developed as an isolated crustal fragment originally located between the Siberian and Laurentian cratons. Metamorphosed basement units include orthogneiss and metavolcanic units of c. 970 and c. 750–540 Ma in the southern Brooks Range (Hammond and Coldfoot subterranes; Amato et al. 2006; McClelland et al. 2006), c. 680–670 Ma and c. 560–540 Ma on the Seward Peninsula (Patrick & McClelland 1995; Amato et al. 2006), and c. 700–630 Ma on Wrangel Island (Cecile et al. 1991). Igneous suites of these ages are rare on the northwestern Laurentian margin, except for a few Neoproterozoic (700–570 Ma) rift-related bodies. On the other hand, magmatic ages in the ranges of 980–900 Ma and 700–600 Ma are more common along the eastern margin of Laurentia and in Barentsia (Svalbard), which led Patrick & McClelland (1995) to propose that Arctic Alaska was originally positioned between Siberia, Barentsia and Greenland (NE Laurentia) within Rodinia. However, calc-alkaline magmatism in the c. 560–540 Ma range is not known in either Barentsia or Laurentia but is widespread in the late Neoproterozoic Timanide orogen of eastern Baltica (Gee et al. 2006). Finally, Moore et al. (2007) reported significant populations of non-western Laurentian detrital zircons (e.g. 1.0–1.2 and 1.5–1.7 Ga; 475–600 Ma) in Proterozoic and younger sandstones from all but far eastern Arctic Alaska, which they interpreted as derived from non-Laurentian sources related to the northern Caledonides.

On Wrangel Island, 540 Ma and older igneous rocks are unconformably overlain by a lower Palaeozoic platformal cover sequence (Kos'ko et al. 1993; Amato et al. 2006). On the Seward Peninsula, Ediacaran and older metamorphic and igneous rocks are presumably overlain by less-metamorphosed strata, including immature sandstone with detrital zircons as young as c. 540 Ma, and fossiliferous Lower Ordovician carbonate (Amato et al. 2006). In most of Arctic Alaska, the early Palaeozoic was a period of tectonic quiescence, characterized by platformal to basinal deposition. These strata contain both macrofaunal and microfaunal assemblages with Siberian and Siberian–Laurentian affinities (Blodgett et al. 2002; Dumoulin et al. 2002). Megafossils with Siberian affinities include Middle Cambrian trilobites in the central Brooks Range, Ordovician trilobites from the Seward Peninsula, and Late Ordovician brachiopods and gastropods from the Seward Peninsula and western and eastern Brooks Range. Characteristic Laurentian forms include Early and Late Cambrian trilobites from the eastern Brooks Range and Late Ordovician corals, stromatoporoids and brachiopods from the Seward Peninsula and central Brooks Range. Faunal components with Siberian affinities, including Ordovician conodonts, decrease markedly from west to east across northern Alaska (Dumoulin et al. 2002). Lane (2007) pointed out that Proterozoic to Silurian strata in the far eastern part of the terrane show lithological linkages to autochthonous Laurentian elements such as the Richardson trough and Yukon Stable Block (Fig. 1).

There are several occurrences of lower Palaeozoic oceanic and arc-related assemblages in the Arctic Alaska terrane (Moore et al. 1994). In the Mt. Doonerak fenster, a structural window in the central Brooks Range, a metasedimentary assemblage occupying a high structural level consists of Middle Cambrian limestone containing trilobites of Siberian affinity, along with Ordovician and Silurian basinal strata. A structurally lower volcanic assemblage has a supra-subduction zone geochemical signature and has yielded a K–Ar age of c. 470 Ma (Dutro et al. 1976). The palaeogeographical and palaeotectonic setting of the Ordovician arc rocks is uncertain. They were probably incorporated into the Arctic Alaska terrane in Devonian time, as the Palaeozoic rocks in the Mt. Doonerak fenster are unconformably overlain by Lower Mississippian clastic rocks of the Endicott Group, which blankets the terrane as a whole. Lower Palaeozoic rocks of oceanic character are also recognized in the Romanzof Mountains in the eastern Brooks Range in a disrupted assemblage that may structurally overlie miogeoclinal facies (Moore et al. 1994). The deformed oceanic rocks are truncated by an angular unconformity, which is overlain by Middle Devonian chert-rich sandstone. The presence of these oceanic assemblages, along with the Proterozoic to Cambrian magmatic suites in the southern Brooks Range and strong Siberian faunal affinities in the western Brooks Range all support the concept of Moore et al. (1994) that much of the Arctic Alaska terrane evolved apart from western Laurentia in the early Palaeozoic, prior to Devonian time.

The Romanzof orogeny is a late Early to earliest Middle Devonian (c. 400 Ma) event that caused shortening in the eastern Arctic Alaska terrane, in the Romanzof and British–Barn Mountains near the Alaska–Yukon border (Fig. 5; Lane 2007). Thrust fault displacement was to the NE and east, and deformation was followed by widespread Late Devonian granitic plutonism (375–362 Ma). The lack of Early to Middle Devonian deformation further to the south and the southward progradation of Early Devonian turbidites suggest that the responsible collision took place in what is now the Beaufort Sea (Fig. 5). Lane (2007) hypothesized that it involved accretion to northern Laurentia of a single continent-scale terrane, which included Pearya in the Arctic Islands, a fragment that was accreted in Late Silurian time (Fig. 5). Moore et al. (2007) also considered that Devonian deformation in Arctic Alaska represents a northern element of the Caledonian deformational system that probably once linked up with Caledonian structures in the Canadian Arctic Islands and adjacent continental margin region. A tectonic highland persisted north of present-day Arctic Alaska from Early Mississippian to Triassic time, based on successive onlaps and northward-coarsening of siliciclastic strata (Moore et al. 1994). Ediacaran–Cambrian (600–500 Ma) and Ordovician (490–445 Ma) zircons in northerly derived Triassic units in the eastern Arctic Alaska terrane and the autochthonous Sverdrup Basin of Arctic Canada (Miller et al. 2006) may have sourced this now-submerged terrane.

A two-fold belt of mid-Devonian plutonic and volcanic bodies occurs along the southern fringe of the Arctic Alaska terrane (Fig. 5). The more

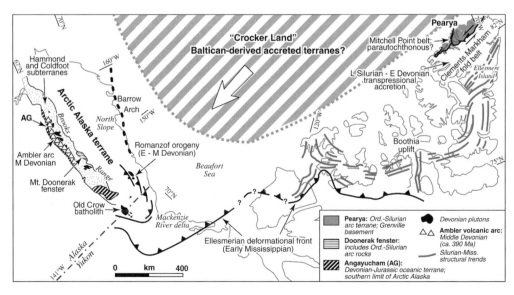

Fig. 5. Distribution of the main mid-Palaeozoic tectonic elements of northern Canada and adjacent Alaska (after Moore et al. 1994; Lane 2007).

northerly Hammond terrane lacks volcanic rocks, but is intruded by an east–west-trending belt of I-type and crustally derived granitoids with discordant U–Pb zircon ages between 366 and 402 Ma (Moore et al. 1994). The Ambler sequence in the Coldfoot subterrane to the south (Fig. 5) is a Middle Devonian (378–386 Ma, U–Pb zircon; McClelland et al. 2006) pericratonic arc, intruded by comagmatic plutons. Arc polarity was probably southward above a north-dipping (in present-day coordinates) subduction zone that consumed oceanic lithosphere now represented by the Angayucham ophiolitic terrane (Figs 1 and 5). Limited detrital zircon data from a quartzite in the Coldfoot subterrane yielded primarily Late Devonian grains (360–370 Ma) with scattered Proterozoic and Archaean grains at c. 1.3, 1.8–2.0, 2.3.–2.5 and 2.8 Ga (Fig. 6b; Moore et al. 1997a). Ambler arc activity commenced about 5–10 Ma after the Romanzof orogeny had finished. Initiation of this arc along the present southern margin of the Arctic Alaska crustal fragment could have represented a subduction polarity reversal following Romanzof terrane collision along its present NE (Fig. 5).

In the Mackenzie Delta region of northern Yukon, the Early Mississippian Ellesmerian orogeny is expressed as a southerly to southeasterly vergent fold and thrust belt that apparently terminates near the eastern limit of rocks assigned to the Arctic Alaska terrane, and east of rocks affected by the older Romanzof orogeny (Fig. 5; Lane 2007). The opposing vergences, differing ages and distinct geographical distribution of Ellesmerian and Romanzof structures suggests that they represent two discrete tectonic events formed as a result of mid-Palaeozoic terrane interactions in the Arctic region (Lane 2007).

A sub-Mississippian unconformity is present throughout much of Arctic Alaska, overlain by northward-transgressive and coarsening quartz-

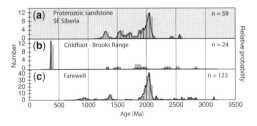

Fig. 6. Detrital zircon data for terranes of Siberian affinities compared with data from Proterozoic sandstone of the eastern Siberian platform (**a**) after Khudoley et al. 2001; (**b**) Marion Schist of the Coldfoot terrane, Brooks Range, Alaska (Moore et al. 1997); (**c**) Farewell terrane quartzites, central Alaska (composite of three samples; data from Bradley et al. 2007).

and chert-rich marine and non-marine clastic rocks of the Endicott Group (Moore et al. 1994). Prominent redbeds in this sequence more closely resemble the thick synorogenic and postorogenic Devonian redbeds of the Franklinian, Caledonian, and Appalachian orogens rather than the abundant Upper Devonian and Lower Mississippian flysch of the adjacent Cordilleran orogen to the south (Nilsen 1981). The tectonic environment in northwestern Arctic Alaska at this time was one of mild extension characterized by several broad-scale uplifts and basins developed on a south-facing continental margin. An exception to this setting is shown by the Nuka Formation, which lies within the highest allochthon emplaced during the north-vergent Jurassic–Cretaceous Brookian orogeny, and which therefore restores south of all other elements of Arctic Alaska, including the Hammond and Coldfoot terranes. The Nuka Formation is a coarse, Mississippian–Pennsylvanian(?) arkose with subangular potassium feldspar grains indicative of a local granitic source, and 2.0–2.1 Ga detrital zircons (Moore et al. 1994, 1997b). It indicates the presence of a Palaeoproterozoic basement block to the south (present coordinates), similar in age to basement rocks of the Kilbuck terrane (2.07–2.04 Ga; Box et al. 1990; Fig. 1).

Farewell terrane

The Farewell terrane, subdivided into the Nixon Fork, Dillinger, and parts of the Mystic terrane, comprises Proterozoic basement overlain by Palaeozoic shelf and slope strata and a Permian clastic wedge (Fig. 2; Bradley et al. 2003). The oldest rocks in the Farewell terrane include c. 1200 Ma metasedimentary rocks, 980–920 Ma rhyolites and c. 850 Ma orthogneiss (Bradley et al. 2003, 2007; McClelland et al. 2006). Detrital zircons show a prominent peak at c. 2050 Ma, and minor peaks at c. 1375 and 950 Ma, which correspond to basement ages cited above (Fig. 6c; Bradley et al. 2007). Parallels for the 2050 Ma peak can be found in the Nuka Formation in the Brooks Range (Moore et al. 2007), in the Kilbuck terrane of western Alaska (Box et al. 1990) and in Proterozoic sandstones from the southeastern Siberian platform (Fig. 6a; Box et al. 1990; Khudoley et al. 2001; Bradley et al. 2007). The 980 Ma rhyolite is coeval with an igneous body in the southern Brooks Range (McClelland et al. 2006). Detrital grains with ages between 2030 and 890 Ma are reported from Seward Peninsula (Amato 2004), a limited dataset that nevertheless suggests similarities to the Farewell terrane.

The overlying carbonate and siliciclastic strata were deposited on an Ediacaran to Devonian continental platform and slope like that of Arctic Alaska–Chukotka; minor Silurian tuffs suggest nearby arc activity (Fig. 2; Bradley et al. 2007). Dumoulin et al. (2002) pointed out that the overall pattern of lithofacies correlates closely with that in Arctic Alaska, both consisting of Ediacaran ooid-rich dolostones, Middle Cambrian outer shelf deposits, and Ordovician to Devonian platform and basin facies. Early Palaeozoic faunas, similar to those of Arctic Alaska, show the influence of both Siberian and Laurentian provinces (Blodgett et al. 2002; Dumoulin et al. 2002). Middle Cambrian trilobites are of Siberian aspect, like those in the central Brooks Range. Identical species of Ordovician brachiopods (*Tsherkidium*) and gastropods are found in both terranes; Ordovician conodont faunas with mixed Siberian–Laurentian affinities characterize both as well (Dumoulin et al. 2002).

Silurian stromatolite–sphinctozoan reefs in the Farewell terrane resemble those in the Ural Mountains, as well as those in the Alexander terrane (Soja & Antoshkina 1997; Antoshkina & Soja 2006). Combined with the other faunal evidence, these data favour an early Palaeozoic palaeogeography in which the Farewell and Arctic Alaska crustal fragments were proximal to each other, in a position between Laurentia and Siberia.

The Early Permian (c. 285 Ma) Browns Fork orogeny is a distinctive event of deformation, metamorphism and deposition of a clastic wedge in the Farewell terrane. It has no correlatives in the nearby Arctic Alaska–Chukotka terrane. Instead, Bradley et al. (2003) have linked this event to Uralian orogenesis related to Permian collision between Baltica, Siberia and Taimyr. This constraint, along with the evidence that Arctic Alaska was interacting with the northwestern margin of Laurentia in Devonian time, suggests that at this time Farewell and Arctic Alaska had separated, with Farewell remaining linked to eastern Siberia and the Uralian orogen until the end of the Palaeozoic.

Alexander terrane

The Alexander terrane occupies a broad belt in the western Cordillera, in close spatial association with Wrangellia (Fig. 1). Together they make up the Insular superterrane. The geology of the Alexander terrane is best known from the detailed work that has been done on Prince of Wales and nearby islands in southeastern Alaska (Gehrels & Saleeby 1987; Gehrels 1990; Gehrels et al. 1996). The remaining parts, at least 1000 km long, is mainly described in publications and maps released by the Canadian and US geological surveys; modern high-resolution geochronology, geochemistry and isotopic work is very sparse. Alexander is probably a composite terrane. Stratigraphic successions in various parts of it indicate widely differing, coeval tectonic environments, ranging from the well-known succession of Neoproterozoic to lower Palaeozoic arc-related rocks preserved in southeastern Alaska (Gehrels & Saleeby 1987; Gehrels 1990; Gehrels et al. 1996) to a thick Proterozoic–lower Palaeozoic continental platform sequence in the northern part of the terrane near the Tatshenshini River in northwestern British Columbia (Fig. 2; Mihalynuk et al. 1993).

The oldest known rocks in the southern Alexander terrane belong to the Ediacaran, arc-related Wales Group (U–Pb ages of c. 595 and c. 554 Ma; Gehrels et al. 1996). These strata were deformed in the pre-Early Ordovician Wales orogeny. Arc magmatism resumed in Early Ordovician–Early Silurian time, represented by the Descon Formation. This was followed by clastic and carbonate sedimentation and pluton emplacement later in the Silurian. The Middle Silurian to Early Devonian Klakas orogeny was marked by thrust imbrication, metamorphism, ductile deformation, and deposition of the Lower Devonian Karheen Formation clastic wedge, a redbed unit that has been compared with the Old Red Sandstone of northern Europe (Bazard et al. 1995; Gehrels et al. 1996).

The combination of Ediacaran arc magmatism followed by late Neoproterozoic–early Palaeozoic orogenesis has no match in western Laurentia. It resembles tectonic events recorded at that time along the Pacific margin of Gondwana (Cawood & Buchan 2007) or the Timanide orogen of eastern Baltica (Gee et al. 2006). The Wales orogeny could also be coeval with the earliest phase of the Taconic orogeny (latest Cambrian) or the Penobscotian orogeny (Early Ordovician) in the northern Appalachians (van Staal et al. 1998; van Staal 2007), or the Finnmarkian phase of the Caledonian orogeny in Baltica (Late Cambrian–Early Ordovician; McKerrow et al. 2000). Timing of Ordovician–Silurian arc building and Silurian–Devonian orogenesis resembles events that are preserved in the Appalachians and Caledonides (Bazard et al. 1995). Detrital zircon ages from the Karheen Formation show a strong population between 400 and 500 Ma, a scatter of Proterozoic ages between 1.0 and 2.1 Ga, and a few Archaean grains, a pattern very unlike northwestern Laurentia (Fig. 7c; compare with Fig. 4a and b). Instead, they show a strong Grenvillian influence and zircons that fall within the 1.49–1.61 Ga North American magmatic gap of Van Schmus et al. (1993), which suggest a possible connection with Baltica (Fig. 7a and c; Bazard et al. 1995; Gehrels et al. 1996; Grove et al. 2008) or northeastern Laurentia (Fig. 7b; Greenland Caledonides; Cawood et al. 2007, and references therein). The 400–500 Ma

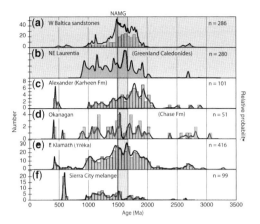

Fig. 7. Detrital zircon data for terranes of Caledonian and Baltican affinities compared with data from (**a**) Neoproterozoic–lower Palaeozoic sandstone from western Baltica (data from Knudsen et al. 1997; Åhäll et al. 1998; de Haas et al. 1999; Bingen et al. 2005) and (**b**) Neoproterozoic metasedimentary rocks from the Greenland Caledonides (northeastern Laurentia; after Cawood et al. 2007); (**c**) Karheen Formation, Alexander terrane (after Gehrels et al. 1996; Grove et al. 2008); (**d**) Chase Formation, Okanagan terrane, southern British Columbia (composite of samples 02TWL225, 02TWL313 and 04TWL025 of Lemieux et al. 2007); (**e**) Yreka terrane, Eastern Klamaths (after Grove et al. 2008); (**f**) Sierra City mélange, Northern Sierra terrane (after Grove et al. 2008). NAMG indicates the 1.49–1.61 Ga North American magmatic gap (Van Schmus et al. 1993).

population matches well the ages of intrusive and metamorphic events in the northern Appalachians and Caledonides, and resembles detrital mica $^{40}Ar/^{39}Ar$ ages from the Old Red Sandstone (Sherlock et al. 2002). Silurian stromatolite–sphinctozoan reef faunas resemble those in the Ural Mountains, as well as the Farewell terrane (Soja & Antoshkina 1997; Antonishka & Soja 2006). Early Devonian (Pragian–early Emsian) rugose corals from Alexander terrane show strong similarities with those of Siberia, Omulevka and Baltica (Pedder 2006). An Early Devonian palaeopole, integrated with the geological record of igneous and orogenic events, fossil affinities and detrital zircon signatures, led Bazard et al. (1995) to favour a mid-Palaeozoic location for the Alexander terrane near eastern Siberia and Baltica.

Devonian volcanic-derived strata and hypabyssal igneous rocks occur in a number of localities in the southern Alexander terrane, indicating that magmatic arc activity was re-established there after the Klakas orogeny. They occur, from south to north, on Prince of Wales Island (Eberlein & Churkin 1970; Gehrels & Saleeby 1987), Kupreanof Island (Muffler 1967), and Chigagof and Baranof Islands (Loney 1964), and in the Glacier Bay area (Brew & Ford 1985).

Palaeozoic strata in the northern Alexander terrane of northwestern British Columbia are generally platformal, dominated by thick shallow-water carbonate sequences with lesser clastic strata and, particularly in the Cambrian section, basalt flows and sills (Mihalynuk et al. 1993). Fossil data from this section are sparse and generally non diagnostic of palaeogeographical affinity, and no detrital zircon studies have been carried out. The lowest unit comprises Cambro-Ordovician siliciclastic rocks, typically fine sandstone–siltstone couplets that display intricate cross-laminations and topsets. Abundant basalt flows and sills are interbedded with this unit. It is overlain by a thick unit of Ordovician–Silurian pure and impure carbonates deposited in shallow, subtidal marine environment. Shallowing-upwards megacycles are noted in some units. The lack of bioturbation and skeletal debris could be due to hypersaline conditions. Intraclasts are abundant in some units. Sedimentologically, these carbonates are comparable with those of the Farewell terrane as described by Dumoulin et al. (2002), a possible connection that it is hoped will be addressed in future studies. A minor argillite–chert facies contains early Middle Ordovician graptolites, assigned to the Pacific province; however, the almost cosmopolitan distribution of these faunas does not lend itself to robust palaeogeographical assignment (Norford & Mihalynuk 1994). Silurian siliciclastic strata mark a transition from shallow-water carbonate platform to deltaic and possibly non-marine depositional environments (Mihalynuk et al. 1993). Sandstones and coarse conglomerates are present in some areas, with clasts of chert, argillite and pyritiferous volcanic(?) rocks. They locally contain abundant fern fossils. This clastic influx could reflect tectonism related to the Klakas orogeny in the southern part of the terrane. Late Devonian (Frasnian) and Mississippian (Viséan) rugose corals in the Alexander terrane resemble those of the western Canada sedimentary basin, thus suggesting proximity of the terrane to western Laurentia at that time (Pedder 2006).

At the northern end of the Alexander terrane in eastern Alaska, a Pennsylvanian intrusion, the c. 309 Ma Barnard Glacier pluton, cuts rocks of the oldest unit of the Kaskawulsh Group, a lower to mid-Palaeozoic metamorphic unit that forms part of the terrane and continues southeastwards into the Kluane Ranges of western Yukon (Gardner et al. 1988). The pluton also intrudes Pennsylvanian arc-related, and probably comagmatic volcanic strata of the Station Creek Formation, which locally is the oldest unit in Wrangellia. These relationships indicate that the late Palaeozoic arc festoon that underlies northern Wrangellia either developed upon, or at least was anchored to, older Alexander basement, perhaps in the way that the eastern end of the modern Aleutian arc laps onto the Alaska Peninsula. The latter option may be more plausible, as elsewhere in the Alexander terrane Mississippian to Pennsylvanian carbonate and fine-grained siliciclastic strata indicate platformal conditions without arc activity. In any event, as pointed out by Gardner et al. (1988), the Barnard Glacier pluton and its contact relationships are taken to signify that the Alexander terrane and Wrangellia were contiguous by Pennsylvanian time.

Wrangellia

Wrangellia is a long-lived, multi-episodic, Devonian and younger arc terrane. It extends over 2500 km, from southern Vancouver Island to south–central Alaska (Fig. 1). There are significant variations in Palaeozoic stratigraphy along its length, particularly between its exposures north and south of the Alexander terrane (Fig. 2). Its defining characteristic is the widespread and voluminous Carnian (Upper Triassic) Nikolai-Karmutsen basalts, which have been interpreted as the product of oceanic plateau volcanism (Lassiter et al. 1995).

Pennsylvanian–Permian arc deposits are widespread as the oldest unit of northern Wrangellia in eastern Alaska and Yukon (Skolai arc; Nokleberg et al. 2000); whereas in the Talkeetna Mountains of southern Alaska and southern Wrangellia (Queen Charlotte and Vancouver Islands), Upper Mississippian to Permian strata are typically thin and non-volcanogenic (Yorath et al. 1999; Schmidt & Rogers 2007), comparable with sections in most of the Alexander terrane.

The oldest rocks in southern Wrangellia on Vancouver Island are the Middle(?) and Upper Devonian Sicker Group, an intra-oceanic arc sequence (Yorath et al. 1999). The lowest unit is the Duck Lake Formation, which comprises basalts of tholeiitic to calc-alkaline affinity. It is overlain by pyroxene–feldspar-porphyritic basalts and basaltic andesites, and a heterogeneous Upper Devonian sequence of mafic to felsic volcanic and volcaniclastic rocks. The overlying, mainly non-volcanic Buttle Lake Group is made up of a sedimentary sequence dominantly comprising epiclastic rocks and bioclastic limestone of Mississippian to Early Permian age. Within it, the Fourth Lake Formation comprises mostly thin-bedded, often cherty sedimentary rocks with minor massive and pillowed non-arc basalt flows. The basalts are slightly undersaturated olivine tholeiites or transitional basalts, with somewhat enriched incompatible trace-element contents akin to ocean-island tholeiite or enriched ocean-floor basalt. Pennsylvanian–Permian crinoidal calcarenite, chert and argillite of the Mount Mark Formation conformably overlie the Fourth Lake Formation.

The upper Palaeozoic Buttle Lake Group was deposited in a cool temperate setting based on the presence and dominance of crinoids, bryozoans, sponges, and temperate-water conodont faunas; warm-water taxa such as fusulinaceans are rare or absent (Katvala & Henderson 2002). Statistical analysis of marine macrofaunas (brachiopods, corals and fusulinids) led Belasky et al. (2002) to an Early Permian reconstruction of Wrangellia (and presumably the Alexander terrane, to which it was then attached) north of the peri-Laurentian terranes Quesnellia and Stikinia, and perhaps a few thousand kilometres west of the northern Laurentian continental margin. If their reconstruction is correct then, by late Palaeozoic time, the Alexander terrane had been transported out of the Arctic realm and into northeastern Panthalassa.

The Insular terranes apparently remained isolated from the western peri-Laurentian terranes until mid-Jurassic time. The earliest indications of their approach are in post-Triassic, pre-Late Jurassic structures that juxtapose the Alexander terrane with the Yukon–Tanana terrane in southeastern Alaska (Gehrels 2001, 2002). There, a regional, low-angle ductile fault system is crosscut by Late Jurassic (162–139 Ma) dykes (Saleeby 2000). Middle Jurassic (c. 177–168 Ma) volcanic rocks and Upper Jurassic–Lower Cretaceous strata of the Gravina belt unconformably overlie both terranes (Gehrels 2001). Farther north, the inboard margin of the Alexander terrane was deformed by a wide dextral shear zone of mid-Jurassic age (McClelland & Gehrels 1990). In the central Coast Mountains of British Columbia, intrusion of granitoids, ductile deformation, and metamorphism occurred at c. 160–155 Ma. These events were interpreted to reflect collision between the Insular and Intermontane (peri-Laurentian) terranes (van der Heyden 1992).

Okanagan

The inner terranes of southern British Columbia include the probably parautochthonous Kootenay terrane, discussed above, the Slide Mountain terrane, and the southern part of Quesnellia (Fig. 3). Both southern and northern Quesnellia contain Mississippian to Permian arc and related strata that contain faunas of McCloud affinity. In southern British Columbia, these include the Harper Ranch and Attwood groups (Fig. 2) and the Mt. Roberts Formation, which correlate with the Lay Range assemblage of central British Columbia (Beatty et al. 2006). The Devonian and older units of southern Quesnellia differ significantly from coeval units in the northern part of the terrane, where the peri-Laurentian Yukon–Tanana terrane forms its basement (Nelson & Friedman 2004). These older units of southern Quesnellia form a roughly east–west-trending belt: the Trail gneiss complex, the Knob Hill complex and Anarchist group between Grand Forks and Penticton, the Old Tom–Shoemaker assemblage south of Penticton, and the Chapperon Group near Kelowna (Fig. 3). Most of these were included in the Okanagan subterrane of Quesnellia (Monger et al. 1991). To the north, between Shuswap and Upper Arrow lakes (Fig. 3), the Chase and Silver Creek formations form a NW–SE-striking belt bordering the western Kootenay terrane and underlying late Paleozoic and younger rocks of Quesnellia (Thompson et al. 2006). Recently, Thompson et al. (2006) have interpreted these terranes as an extension of the western Laurentian margin. However, many aspects of these early, proto-Quesnellian elements are at odds with such an interpretation, and some show closer similarities to the 'outboard' terranes. What follows is a comparatively more detailed treatment than for other terranes, given that the idea of an exotic origin for the pre-Mississippian rocks of southern Quesnellia is evaluated here for the first time.

Farthest east, the Trail gneiss complex comprises paragneiss and orthogneiss that are at least in part c. 372 Ma (Simony et al. 2006; Fig. 3). Orthogneisses of the Trail complex have a primitive geochemical and isotopic character (εNd values of +4.7 to +5.6 and T_{DM} model ages of 880–1050 Ma; Simony et al. 2006). The lack of evidence of any continental influence in these rocks contrasts strongly with the more evolved, pericratonic arcs that characterize the basement to Quesnellia in northern British Columbia and Yukon (Yukon–Tanana terrane, Fig. 1; Nelson & Friedman 2004). Simony et al. (2006) related these rocks to a Late Devonian intra-oceanic arc.

West of Trail, between Grand Forks and Penticton, the Okanagan subterrane includes chert, greenstone and ultramafic rocks of the Knob Hill complex, and argillite–phyllite, chert, carbonate and greenstone of the Kobau and Anarchist groups (Massey 2007, and references therein). Greenstones of the Knob Hill complex have compositions varying from normal mid-ocean ridge basalt (N-MORB), to enriched (E)-MORB and island arc tholeiites (Dostal et al. 2001; Massey 2007) with a juvenile εNd value of +7.2 (Ghosh 1995). The Knob Hill complex has yielded a Late Devonian U–Pb date (c. 370 Ma; N. Massey, pers. comm.), coeval with the Trail gneiss complex, and Late Devonian and Pennsylvanian–earliest Permian conodont assemblages (Massey 2007, and unpubl. data). No age constraints are available for the Anarchist and correlative units, but Massey (2007) suggested that these argillite-dominated rocks are more complexly deformed lateral equivalents of the Knob Hill complex. Collectively they represent a primitive arc to back-arc assemblage.

Near Kelowna, undated basalts of the Chapperon Group are apparently depositionally overlain by Devonian to Permian strata of the Harper Ranch Group (Thompson et al. 2006), which comprises a volcaniclastic and carbonate succession correlated with the McCloud belt (Beatty et al. 2006). Little is known about the Chapperon Group depositional setting; Thompson et al. (2006; R. Creaser, pers. comm.) reported primitive εNd values that support an association with the Knob Hill complex to the south.

At the western end of this belt of Palaeozoic rocks, south of Penticton (Fig. 3), the Old Tom–Shoemaker assemblage comprises structurally intermixed greenstone, silicified tuff, minor limestone and chert breccia (Monger et al. 1991). Limestone blocks in argillite matrix have yielded conodonts of Ordovician age and enigmatic faunal affinity (Pohler et al. 1989). Radiolarian chert interbedded with greenstone is in part latest Devonian to Carboniferous in age. Parts of this assemblage may represent a subduction complex (Monger et al. 1991).

The Nicola horst is a basement high exposed in central Quesnellia south of Kamloops. A conglomerate within the horst contains metaplutonic clasts of Grenvillian age (c. 1038 Ma), along with Mesozoic detrital grains (Erdmer et al. 2002). The presence of a locally derived Grenvillian clast (as opposed to minor populations of detrital grains) suggests that basement of this part of Quesnellia was markedly different from either the adjacent autochthonous continental margin (Villeneuve et al. 1993) or the Yukon–Tanana terrane, with their northwestern Laurentian Archaean–Palaeoproterozoic provenance (Fig. 4).

To the north of the primitive rocks of the Okanagan subterrane described above, the Devonian Chase and Silver Creek formations are a regionally extensive platformal succession of calcareous quartzite, psammitic schist, pelite and minor marble (the 'Okanagan high' of Thompson et al. 2006), which forms a roughly WNW–ESE-trending belt along the southern margin of the Kootenay terrane (Figs 2 and 3). It is juxtaposed to the north with variable rock successions. To the west, it is in uncertain contact relationship with Neoproterozoic–Mississippian metasedimentary and metavolcanic rocks of the Sicamous and Tsalkom formations and Eagle Bay assemblage (Fig. 3). To the east, it is in sharp and locally highly strained contact with amphibolite-grade schist and paragneiss of the Shuswap complex (Fig. 3; probable Laurentian continental margin strata that were exhumed from mid-crustal level in Eocene time; Parrish *et al.* 1988). Thompson *et al.* (2006) interpreted these contacts as a transposed unconformity. The relationship between the platformal succession of the Chase and Silver Creek formations and the Okanagan subterrane to the south is uncertain (Fig. 3).

Detrital zircons from the schist underlying the Chase Formation near Upper Arrow Lake (Unit 1 of Lemieux *et al.* 2007) are predominantly in the range of 1.6–2.0 Ga, with a peak at c. 1.8 Ga, and minor populations at c. 2.5 and 2.7 Ga, and two grains of Grenvillian age (1.0–1.3 Ga; Fig. 8a). These ages match well with the Precambrian basement provinces of southern Alberta and the miogeoclinal detrital zircon reference for southern British Columbia (Fig. 4c and i; Villeneuve *et al.* 1993; Gehrels *et al.* 1995; Gehrels & Ross 1998), and thus confirm the continental margin setting for the amphibolite-grade rocks of the Shuswap metamorphic complex (Lemieux *et al.* 2007).

In contrast, the detrital zircon data presented by Lemieux *et al.* (2007, fig. 4) for the Devonian Chase Formation show a near-continuous spread of ages between 0.85 and 2.7 Ga, which they interpreted to indicate a mixing between adjacent Laurentian sources to the east and younger Precambrian crust to the west. In their original treatment of the data, Lemieux *et al.* (2007, fig. 4) presented only a composite age probability density plot for four samples of the Chase Formation. When considered individually, samples attributed to the Chase Formation exhibit significant differences in their distribution of detrital zircon ages, rather than a mixing of ages (Fig. 8b–e). The easternmost sample (04TWL072; Fig. 3), a calcareous quartzite occurring near the base of the Mississippian Milford Group in the Kootenay Arc region (Thompson *et al.* 2006), shows a distribution of detrital zircon ages that is strikingly similar to that of Lemieux's Unit 1 (compare Fig. 8a and b) but very different from other Chase samples collected near Upper Arrow Lake to the west (Figs 3 and 8c–e). The Kootenay Arc sample is most reasonably interpreted as having the same source as Lemieux's Unit 1 in the Precambrian basement domains of southern Alberta to the east (Fig. 4c and i).

The two samples of Chase Formation collected SW of Upper Arrow Lake (02TWL225 and 02TWL313; Fig. 3) have similar distributions of detrital zircon ages of 0.8–0.9, 1.0–1.2, 1.3–1.4, 1.45–1.55 and 1.65–2.05 Ga, with a few Archaean grains (Fig. 8d and e). Sample 04TWL025, collected along the shore of Upper Arrow Lake (Fig. 3), contains fewer concordant analyses ($n = 14$, Fig. 8c) but displays some similarities to other Chase samples collected in the Upper Arrow Lake area (Fig. 8d and e). Its main distinction is in the presence of two Early Devonian grains (c. 403 and 412 Ma) and one Neoproterozoic grain (c. 561 Ma) (Lemieux *et al.* 2007); ages that match respectively two detrital zircon ages and a granitic cobble from the Chase Formation and related conglomerate NW of Vernon (Erdmer *et al.* 2001; Thompson *et al.* 2006).

All three Chase samples from the Upper Arrow Lake area have zircons in the 1.49–1.61 Ga range, the North American magmatic gap of Van Schmus *et al.* (1993), and a number of Grenvillian grains (1.0–1.3 Ga), and thus are probably derived from an exotic source (Fig. 8). Taken together, their detrital signatures compare remarkably well with those in other terranes of Caledonian and Baltican affinities discussed in this paper (Alexander, Yreka–Trinity, Sierra City; Fig. 7). We propose that at least part of the basement of southern Quesnellia, including the source of the Nicola horst Grenvillian clast, is of exotic, Caledonian affinity. These crustal fragments were probably first accreted to the western margin of Laurentia in mid-Palaeozoic time. A Middle Devonian episode of compressional deformation is documented in the Purcell Mountains of southeastern British Columbia (Root 2001). An Early Mississippian (354 Ma) pluton intrudes the contact between the Silver Creek Formation and rocks of the Eagle Bay assemblage west of Shuswap Lake (Fig. 3), thus suggesting that the platformal succession of the Chase and Silver Creek formations was juxtaposed with the western Kootenay terrane (i.e. distal continental margin) before development of the Late Devonian arc that characterizes

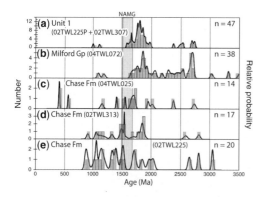

Fig. 8. Detrital zircon age-spectra for single samples reported by Lemieux *et al.* (2007). (**a**) Composite of two samples of Proterozoic–lower Palaeozoic schist (Unit 1). Single samples show identical age distribution. (**b**) Quartzite near the base of Milford Group, Kootenay Arc. (**c–e**) Chase Formation in the Upper Arrow Lake region. Sample locations are shown in Figure 3. NAMG indicates the 1.49–1.61 Ga North American magmatic gap (Van Schmus *et al.* 1993).

the western Kootenay terrane. Occurrence of a Late Silurian (c. 418 Ma) granitic cobble in a conglomerate from the Milford Group (Kootenay–Slide Mountain terranes, Fig. 3; Roback et al. 1994) and detrital zircons in the Mt. Roberts Formation of southern Quesnellia (Roback & Walker 1995) suggest that both western Laurentia and the accreted 'Okanagan high' were available as source regions in late Palaeozoic time.

Eastern Klamath terranes

Pre-Devonian basement rocks in the Eastern Klamath Mountains of southern Oregon and northern California comprise two terranes, the Trinity and Yreka (Fig. 2; summarized by Lindsley-Griffin et al. 2006, 2008; Wright & Wyld 2006). The Trinity terrane is a composite Ediacaran (c. 565–570 Ma) and Cambro-Ordovician(?) ophiolitic complex dominated by ultramafic and mafic rocks, including ductilely deformed mantle tectonites (Lindsley-Griffin et al. 2008). It is crosscut by post-tectonic Ordovician plagiogranites, and also by a suite of Silurian to Early Devonian supra-subduction zone plutons with related basalt flows (c. 435–412 Ma). The Yreka terrane structurally overlies the Trinity ophiolite on a SE-vergent fault (Fig. 9). It consists of numerous related and unrelated tectonic slivers and mélanges, and was interpreted by Lindsley-Griffin et al. (2008) as a forearc complex related to Ordovician and Siluro-Devonian subduction. It includes Late Ordovician blueschist slivers (c. 454 Ma; Grove et al. 2008). Mélange units near its structural base carry Ediacaran tonalite blocks (c. 560–570 Ma). Most of the terrane consists of imbricated Upper Silurian to Lower Devonian sedimentary strata and sediment-matrix mélange units with mixed siliciclastic and volcaniclastic provenance. They are interpreted as forearc and/or trench sedimentary deposits. In contrast, the structurally highest unit in the terrane, the Antelope Mountain Quartzite, is a shallow-water pericontinental siliciclastic unit that has been identified as Neoproterozoic (Ediacaran) based on the presence of cyclomedusoids (Lindsley-Griffin et al. 2006). Waggoner (1999) defined three global biogeographical groups for Ediacaran faunas. The Antelope Mountain cyclomedusoids most closely resemble those of the 'White Sea assemblage', which includes faunas of Baltica, northwestern Laurentia, Siberia and also Australia (Lindsley–Griffin et al. 2008). The Trinity and Yreka terranes were juxtaposed, and internally deformed and imbricated in Early to Middle(?) Devonian time; this is modelled as a west-facing subduction complex (present coordinates) related to the onset of Devonian arc activity in the Eastern Klamath terranes (Fig. 9; Lindsley-Griffin et al. 2006, 2008; Wright & Wyld 2006).

Detrital zircon populations in the Yreka blueschist and in crustally derived Lower Devonian units show broad, multipeaked distributions from 1.0 to 2.0 Ga, including significant peaks in the range 1.49–1.61 Ga (Fig. 7e; Grove et al. 2008). These are similar to populations in the Alexander terrane and Neoproterozoic–lower Palaeozoic sandstone from western Baltica and NE Laurentia (Fig. 7a–c; Cawood et al. 2007; Grove et al. 2008), but contrast sharply with detrital zircon references for western Laurentia (Fig. 4a–e). Other arc-derived clastic units contain mostly early Palaeozoic zircons, with predominant ages from 476 to 381 Ma, and a few 550–560 Ma grains that match those of tonalite blocks in the mélange.

The Antelope Mountain Quartzite is generally considered an integral part of the Yreka terrane,

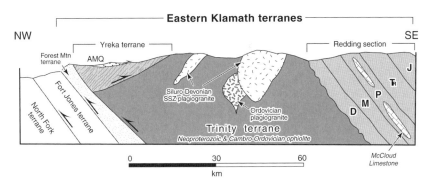

Fig. 9. Schematic cross-section of the Eastern Klamath terranes (after N. Lindsley-Griffin, pers. comm.). AMQ, Antelope Mountain Quartzite. The Forest Mountain terrane is an oceanic ophiolite, metamorphosed at c. 400 Ma, during accretion to the eastern Klamath nucleus. Fort Jones and North Fork terranes are late Palaeozoic to early Mesozoic oceanic terranes accreted in early Mesozoic time. Yreka and Trinity terranes and the Redding section are described in the text.

which developed in some proximity to the rest of the terrane throughout Neoproterozoic to mid-Palaeozoic time (Lindsley-Griffin et al. 2008, and references therein). This possibility is not refuted by the available evidence. However, its unique character raises questions. In Ediacaran time, it resided in a continental platform setting, completely different from the Trinity ophiolite. Its detrital zircon population is unlike that of any other unit in the Yreka terrane; it is dominated by Archaean and Palaeoproterozoic grains with a marked peak between 1.78 and 1.95 Ma in a pattern that closely resembles that of the northwestern Laurentian miogeocline (Fig. 4b and j; Wallin et al. 2000; compare with the Yreka pattern in Fig. 7e). The presence of significant feldspar contents and coarse, relatively immature siliciclastic detritus in the unit precludes large-scale sediment transport as a means of introducing a northerly zircon signature into the terrane; instead, Wallin et al. (2000) favoured large-scale sinistral strike-slip motion along the continental margin. Three grains out of the 46 in the Antelope Mountain sample are c. 1.3 Ga, which corresponds to minor populations in autochthonous Cambrian strata of Alaska (Fig. 4a; Gehrels et al. 1995, 1999).

Macrofossils from Ordovician and Silurian limestone blocks in the Yreka mélanges comprise both endemic species and species linked to faunas of Laurentia, Siberia, Baltica, Kazakhstan, Scotland, Australia and China (Lindsley-Griffin et al. 2008). Late Ordovician sphinctozoan sponges are part of a rare fauna that is known only in Australia, the Farewell terrane, autochthonous rocks of northern Yukon to east–central Alaska (Potter et al. 1990b), the Montgomery limestone in the Shoo Fly Complex of the western Sierra Nevada (see below), and in the Alexander terrane on Prince of Wales Island (Rigby et al. 2005). The combination of highly endemic and shared faunas is suggestive of an intra-oceanic setting, and Lindsley-Griffin et al. (2008) favoured an Ordovician–Silurian location for the Yreka terrane as part of a chain of islands in the seaway between northern Laurentia, Baltica, Kazakhstan and Siberia.

Devonian fossils from mixed siliciclastic–volcaniclastic units, and clasts and matrix in mélange units of the Yreka terrane, include corals, brachiopods and conodonts. Like older Yreka terrane faunas, some are very restricted in occurrence whereas others have more widespread correlatives. Brachiopods in the Gregg Ranch Complex, one of the mélange units, are similar to those in Nevada and northern Canada; some conodonts and corals are found only in the Yreka terrane and in Nevada (Lindsley-Griffin et al. 2008). Compared with older faunas, these exhibit a greater connection to western Laurentia, as opposed to northern Laurentia–Baltica–Siberia. A Middle Devonian palaeopole from the Redding section (Fig. 9) places the combined Eastern Klamath terranes at 31° either north or south latitude (Mankinen et al. 2002); combined with faunal linkages the terrane probably lay near northwestern Laurentia at that time (Lindsley-Griffin et al. 2008, fig. 11). This pole, however, is based on only two sites, and should be considered preliminary.

In summary, the Yreka and Trinity terranes (with the exception of the Antelope Mountain Quartzite) comprise Ediacaran ophiolite, Ordovician blueschist, Silurian–Devonian supra-subduction zone ophiolite, Silurian to Lower Devonian mélange units and mixed siliciclastic–volcaniclastic marine sedimentary strata. Except for the Ediacaran oceanic crust, these characteristics and the detrital zircon signature of the Yreka terrane suggest linkages between the Eastern Klamath and Alexander terranes (Wright & Wyld 2006), and also with terranes of the northern Caledonian orogen (Grove et al. 2008). The favoured palaeogeographical position for these terranes during Neoproterozoic to Silurian time was offshore near the Baltican end of the Caledonides (Lindsley-Griffin et al. 2008). Lindsley-Griffin et al. considered a 'southern' location between Laurentia and Gondwana in Silurian time unlikely, because of the lack of faunal similarities between the Yreka terrane and Gondwana.

The Antelope Mountain Quartzite, unlike the rest of the Yreka terrane, has a detrital zircon signature that is compatible with that of northwestern Laurentia. By Early Devonian time, it was structurally incorporated into a forearc setting along with the other units of the Yreka terrane. It is possible that this was the result of an early encounter of the allochthonous Yreka and Trinity terranes and the northern Laurentian margin, in which the Antelope Mountain pericratonic sliver was transferred to the mobile plate. It is interesting to note that imbrication of the Yreka terrane is roughly coeval with the Romanzof orogeny in eastern Arctic Alaska.

The long-lived sequence of arc-related strata of the Early Devonian to Jurassic Redding section (Irwin 1981; Redding subterrane of Lindsley-Griffin et al. 2008) of the Eastern Klamath terrane was presumably built on Yreka–Trinity basement after its Silurian–Early Devonian amalgamation (Potter et al. 1990b; Gehrels & Miller 2000; Lindsley-Griffin et al. 2008). The oldest rocks in the section are greenstones and rhyolites; coeval, c. 390 Ma intrusions cut both the Yreka and Trinity terranes. The Redding section contains a Lower Mississippian siliciclastic unit, the Bragdon Formation, in which detrital zircon populations reflect Precambrian basement and Neoproterozoic and early Palaeozoic arc sources in the Trinity terrane (Gehrels & Miller 2000). Also within this section is the Permian McCloud Limestone, which contains the defining fusulinid genera of the McCloud faunal belt, described by Miller (1988) as the dispersed remnants of a northeastern Pacific fringing arc. Differences between the McCloud and western Laurentian faunas suggest that in Early to mid-Permian time, the various elements of this belt probably lay 2000–3000 km west of the continental margin and at somewhat more southerly latitudes than at present (Belasky et al. 2002). The peri-Laurentian terranes of western Canada (Stikinia, Quesnellia and Yukon–Tanana) also contain faunas of McCloud affinity (Stevens & Rycerski 1989; Stevens 1995; Nelson et al. 2006). In the Early Permian reconstruction of Belasky et al. (2002), the Eastern Klamath terrane lay south of Stikinia–Quesnellia, and somewhat south of its present location. Therefore, if the Eastern Klamath terrane was already interacting with distal parts of northwestern Laurentia in Early Devonian time, it had travelled over 3000 km southwards by the Permian. This would require sinistral motion with respect to western Laurentia that averaged slightly more than 2 cm a^{-1} for 130 Ma.

Northern Sierra terrane: the Shoo Fly Complex

The lower Palaeozoic Shoo Fly Complex forms the basement for Late Devonian to Permian arc and related strata of the Northern Sierra terrane (Fig. 2; Girty et al. 1990). It is composed of four allochthons (Fig. 10). The three structurally lower allochthons, the Lang sequence, Duncan Peak allochthon, and Culbertson Lake allochthon, contain siliciclastic and basinal strata. Alkalic 'intraplate' basalts occur within the Culbertson Lake allochthon. Conodonts of Middle to Late Ordovician age are present in the structurally lowest Lang sequence (Harwood et al. 1988). The other two allochthons are also thought to be early Palaeozoic in age, based on poorly preserved radiolaria, and tentative stratigraphic linkages with the Lang sequence. Detrital zircon signatures of the terrigenous sedimentary strata show consistency between the three allochthons (Harding et al. 2000). The dominant peaks are Archaean (2.55–2.70 Ga) and Palaeoproterozoic (1.80–2.10 and 2.20–2.45 Ga), consistent with the northwestern Laurentian margin in British Columbia and Yukon (Fig. 4b and k). There are no Mesoproterozoic, Neoproterozoic or Palaeozoic grains present. Although the alkalic basalts in the Culbertston Lake allochthon have been interpreted as seamounts by some workers, it is important to note that basalts of similar chemistry are widespread within lower Palaeozoic parautochthonous basinal sequences throughout the northern Cordillera (Goodfellow et al. 1995), where they are associated with periodic episodes of extension and second-order basin development.

The Sierra City mélange is the highest and also the most easterly unit within the Shoo Fly Complex (Fig. 10). It differs radically from the underlying allochthons in all aspects. It contains blocks of ophiolitic affinity (serpentinite, gabbro, plagiogranite, basalt), and of sedimentary origin (chert, limestone, sandstone) within a sheared matrix of slate, chert and sandstone; it is interpreted as a combination of tectonic mélange and olistostrome (Schweickert et al. 1984). A plagiogranite block in the mélange has yielded a Neoproterozoic (c. 600 Ma) U–Pb zircon age, and a felsic body (either a tuff or a dyke) is Silurian (c. 423 Ma; Saleeby et al. 1987; Saleeby 1990). Detrital zircon populations are predominantly Ediacaran in age (550–600 Ma), corresponding to the age of the single dated igneous block, and show a scatter of ages between 0.95–1.55 Ga and 1.65–1.95 Ga (Fig. 7f; Harding et al. 2000; Grove et al. 2008).

Upper Ordovician limestone blocks in olistostromes within the Sierra City mélange contain brachiopods, conodonts, rugose and tabulate corals, and sphinctozoan sponges (Potter et al. 1990a).

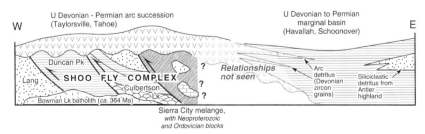

Fig. 10. Schematic cross-section of the Northern Sierra terrane and Golconda allochthon. Shoo Fly relationships after Wright & Wyld (2006). Late Palaeozoic strata and relationships conceptually after Harwood & Murchey (1990) and Miller & Harwood (1990).

Most of the brachiopod, coral and sponge genera are also present within the Yreka terrane mélanges. In particular, the sponges include *Amblysiphonolella* sp., *Corymbospongia adnata*, and *Girtyocoelia* sp., which the Montgomery Limestone shares with the Yreka terrane (Potter *et al.* 1990*b*) as well as with an Ordovician limestone block in the Alexander terrane (Rigby *et al.* 2005). *Cystothalamiella* sp., *Girtycoelia epiporata* and *Ribyetia obconica* occur in the Montgomery Limestone and in limestones of the Farewell terrane (Potter *et al.* 1990*a*). Brachiopod faunas are similar to those in the Yreka terrane and also the autochthon of northern Yukon. A rugosan coral, *Grewingkia penobscotensis*, is known only from the Montgomery Limestone, the Yreka terrane, and northern Maine (eastern Laurentia; Potter *et al.* 1990*a*). Other faunas are less provincial.

The detrital zircon signature of the Sierra City mélange, including abundant Neoproterozoic grains, Grenvillian zircons and grains with ages within the 1.49–1.61 Ga North American magmatic gap, as well as detritus that is clearly derived from an ophiolitic source favour a relationship to the other Caledonian terranes of the Cordillera, such as Yreka–Trinity and Alexander (Fig. 7). Faunal connections between the Sierra City mélange and other Arctic terranes such as Alexander and Farewell, as well as autochthonous sites in far northern Yukon, are consistent with an Arctic location in the early Palaeozoic.

The assembly of the Shoo Fly Complex took place after the Late Silurian, based on the age of its youngest rocks, but prior to intrusion of the post-tectonic Late Devonian Bowman Lake batholith (*c.* 364 Ma; Miller & Harwood 1990). This event involved the tectonic juxtaposition of the lower allochthons, with their northwestern Laurentian margin detrital zircon affinities, with the inferred Caledonian Sierra City mélange. There are clear parallels between this event and the proposed Siluro-Devonian imbrication of the northwestern Laurentian Antelope Mountain Quartzite with the exotic rocks of the Yreka and Trinity terranes (Fig. 9). In both cases, initial collision of transported Caledonian crustal fragments would have been with the northernmost part of the western Laurentian margin. In the Shoo Fly Complex, the most exotic component now lies structurally inboard of less far-travelled rocks. However, if after Silurian–Devonian time the amalgamated fragment has been transported thousands of kilometres to the south, an accompanying rotation up to 180° is unsurprising.

The deformed Shoo Fly Complex is unconformably overlain by the Upper Devonian to Upper Permian arc-related Taylorsville and Tahoe sequences, which have a coarse clastic basal unit of Famennian age, the Grizzly Formation (Miller & Harwood 1990). The Late Devonian initiation of arc magmatism in the Northern Sierra terrane is younger than the Middle Devonian or older onset of arc activity in the Eastern Klamaths. In general, differences between these two terranes are probably a result of somewhat different positions within an evolving arc system (Miller & Harwood 1990). An important shared element is that McCloud faunas occur within the Permian of the Northern Sierras (Miller 1988). The Northern Sierra arc is modelled as facing west, away from the continent and towards Panthalassa. East of it, the Havallah and Schoonover sequences of the Golconda allochthon record the corresponding Devonian to Permian back-arc marginal basin, which received detritus both from the continental margin and from the Northern Sierra arc (Fig. 10; Miller *et al.* 1984; Harwood & Murchey 1990). Early Mississippian and Early Permian influxes of arc-derived sediment into the Havallah sequence correspond to discrete magmatic episodes in the Northern Sierra terrane. Lower Mississippian chert–quartz-rich siliciclastic rocks in the Havallah basin were derived from erosion of the Antler orogenic belt on the continent margin to the east (Miller *et al.* 1984; Harwood & Murchey 1990). Detrital zircon populations reflect derivation from the Devonian–Mississippian arc (338–358 Ma), and from recycled Palaeoproterozoic and Archaean sources similar to those in the Roberts Mountains allochthon (Riley *et al.* 2000).

The inferred relationships between the Northern Sierra terrane, Golconda basin and Roberts Mountains allochthon suggest that by Early Mississippian time, the Northern Sierra terrane had arrived close to its present position near the southwestern Laurentian margin. Early Mississippian emplacement of the Roberts Mountains allochthon may have been related to the arrival of this crustal fragment by sinistral transpression along the margin. Interestingly, extension within the Havallah basin began in Late Devonian to Early Mississippian time, coeval with initial development of the Northern Sierra arc (Harwood & Murchey 1990). They probably represent a conventional SW Pacific-style extensional arc–back-arc pair. The Antler collisional belt may have acted as a pinning point for the arc festoon, comparable with the role and position of the recently extinct collision zone in northern Taiwan, which is overprinted by the younger, west-migrating Okinawa back-arc trough behind the Ryukyu arc (Letouzey & Kimura 1986).

This proposed scenario would require passage of the Shoo Fly Complex (and perhaps also the Yreka–Trinity terranes, with which it was probably affiliated) from a site of collision with northwestern Laurentia to a location nearer southwestern Laurentia as Devonian in age, a maximum possible interval of 58 Ma and probably less (time scale of Gradstein et al. 2004). In this case average motion would need to be about 5 cm a^{-1}, comparable with rates of advance of short modern arc segments such as the Scotia and New Hebrides arcs (Schellart et al. 2007).

Palaeogeographical implications

The evidence summarized above clearly points to common, non-western Laurentian origins for a number of western Cordilleran terranes, as has been previously argued by many of the workers we cite. Much of the geological history (Ordovician–Silurian arcs, Cambrian–Ordovician and Silurian–Devonian deformational events) and detrital zircons from the Alexander terrane suggest early Palaeozoic interactions with the Appalachian–Caledonian orogen and a source region dominated by Mesoproterozoic and late Palaeoproterozoic igneous rocks, including significant contributions from Grenvillian (1.0–1.3 Ga) and 1.49–1.61 Ga sources (Figs 2 and 7; Wright & Wyld 2006; Grove et al. 2008). The occurrence of Ediacaran arc-related rocks at the base of the Alexander terrane contrasts, however, with the Neoproterozoic history of either eastern Laurentia or western Baltica, each of which was characterized by rifting and development of a passive continental margin at that time (e.g. Cawood et al. 2001). Wright & Wyld (2006) proposed an early affinity with Avalonia and other peri-Gondwanan terranes. But when considering Silurian–Devonian palaeomagnetic, faunal and detrital zircon data together, a position near northern Baltica, adjacent to the north end of the Caledonides, seems to provide a better fit (Fig. 11; Bazard et al. 1995; Soja & Antoshkina 1997; Pedder 2006; Soja & Krutikov 2008).

Neoproterozoic arc magmatism (c. 600–550 Ma) and tectonism characterize the Timanide orogen, which extends along eastern Baltica from the southern Urals northward to the Barents–Kara Sea region, where it is locally overprinted by the Caledonian deformation front (Gee & Pease 2004; Gee et al. 2006). The North Kara terrane is inferred to underlie much of the Barents–Kara Shelf and

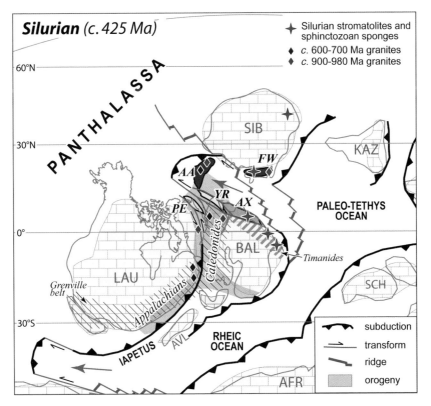

Fig. 11. Palaeogeographical setting of Cordilleran exotic terranes in Silurian time. Continental reconstructions in this and subsequent figures are modified after Scotese (2002). Continent abbreviations: AFR, Africa; AVL, Avalonia; BAL, Baltica; KAZ, Kazakhstania; LAU, Laurentia; SCH, South China; SIB, Siberia. Terrane abbreviations: AA, Arctic Alaska; AX, Alexander; FW, Farewell; PE, Pearya; YR, Yreka (including Trinity, and parts of Shoo Fly and Okanagan).

northern Taimyr. It comprises local evidence of reworked Mesoproterozoic (Grenvillian) basement, a Neoproterozoic succession of turbidite and shale with c. 555 Ma detrital zircons, evidence for latest Cambrian–earliest Ordovician deformation, and a lower Palaeozoic succession that includes a Lower Devonian Old Red Sandstone unit with a western, Caledonian source (Gee et al. 2006, and references therein). The mixed Timanian–Caledonian affinity of the North Kara terrane may be an analogue for some of the key characteristics of the Alexander terrane (Fig. 11).

Detrital zircon populations, occurrences of Ediacaran and Ordovician arc-related blocks in mélange, Upper Ordovician limestone blocks with distinctive sphinctozoan sponges, and Ordovician blueschists in the Yreka–Trinity terranes and the Sierra City mélange, ally them with the Alexander terrane (Fig. 7; Wright & Wyld 2006; Grove et al. 2008), and thus suggest similar origins near the northern Caledonides for at least these parts of the Eastern Klamath and Northern Sierra terranes (Fig. 11). Limited data from the Okanagan terrane of southern British Columbia, including detrital zircons from the Chase Formation, Grenvillian and Ediacaran granitoid cobbles, and enigmatic Ordovician limestone blocks from the Shoemaker assemblage, also support a possible association with the Alexander, Yreka–Trinity and Sierra City terranes in early Palaeozoic time (Figs 2 and 7). Other parts of the Eastern Klamath (Antelope Mountain Quartzite), Northern Sierra (Lang, Duncan Peak and Culbertson Lake allochthons) and Okanagan terranes may have originated in northwestern Laurentia (Fig. 4), in which case they would have been juxtaposed with Caledonian elements in mid-Palaeozoic time (see below).

The Farewell terrane has overall stronger Palaeozoic lithological, faunal and provenance ties to Siberia (Blodgett et al. 2002; Dumoulin et al. 2002; Bradley et al. 2007); however, it also shares distinctive Silurian stromatolite–sphinctozoan reefs with the Alexander terrane, and its early Neoproterozoic igneous rocks are more akin to northeastern Laurentia and Barentsia than Siberia (Fig. 11). The Arctic Alaska terrane shares this igneous suite and Devonian and older lithological character with the Farewell terrane. Its Palaeozoic faunal affinities are transitional between Siberian and Laurentian provinces, therefore suggesting that Arctic Alaska may have been situated between Siberia and Laurentia in early Palaeozoic time (Fig. 11; Dumoulin et al. 2002). Arc magmatism of Ediacaran age on the Seward Peninsula and Ordovician age in the Doonerak fenster parallels that of the Alexander terrane (Fig. 2).

In summary, most of the outer Palaeozoic terranes of the North American Cordillera appear to have originated from the same general region near the northern end of the Caledonides, in a position intermediate between Baltica, Siberia and northeastern Laurentia in early Palaeozoic time (Fig. 11). This implies that they have travelled thousands of kilometres around northern Laurentia during the Palaeozoic, before finding their final resting place in the Cordillera in Mesozoic time.

Pearya: an Arctic connection

If the exotic, Caledonian and Siberian terranes of the Cordillera migrated around northern Laurentia in Palaeozoic time, then the geological record of Arctic Canada may have recorded this history. The early Palaeozoic record of northern Laurentia is to a large extent identical to that of its western continental margin, with passive margin deposition along the Arctic platform, Franklinian shelf, and adjacent slope and basin (Fig. 2; Trettin et al. 1991a). Pearya is the only exotic, accreted terrane exposed in Arctic Canada along the northern Laurentian margin (Figs 1 and 5), although related crustal fragments may also form part of the Arctic Ocean sea floor. Pearya lies on the northern side of Ellesmere and Axel Heiberg Islands, juxtaposed against lower Palaeozoic miogeoclinal strata and Silurian–Lower Devonian flysch deposits of the Clements Markham fold belt (Trettin 1991; Trettin et al. 1991a). Pearya is a composite terrane (Fig. 2). The oldest rocks in it, Succession I of Trettin (1991), are schists and gneisses of Grenvillian age (c. 1060–1040 Ma), based on limited Rb–Sr and U–Pb dating. The crystalline rocks are overlain by Succession II, a Neoproterozoic to Lower Ordovician rift-related and passive margin sequence. The Lower to Middle Ordovician, suprasubduction zone related Maskell Inlet assemblage was juxtaposed with the continental margin rocks during the Middle Ordovician M'Clintock orogeny (Trettin 1991). This tectonic event is age-equivalent to the Taconic orogeny in the peri-Laurentian terranes of the Canadian Appalachians, and similarly juxtaposes intra-oceanic arc and Grenvillian continent-margin crustal blocks (van Staal 2007). The characteristic features of Pearya, in particular the evidence for Early Ordovician subduction and Middle Ordovician tectonism, are strikingly similar to those of the southwestern terranes of Spitsbergen (western Svalbard), thus suggesting a derivation of Pearya from the northern end of the Caledonian orogen (Trettin 1991; Gee & Teben'kov 2004). The Middle Ordovician to Upper Silurian strata of Succession IV represent a successor basin and arc sequence developed across the M'Clintock orogen. Late Ordovician faunas have diagnostic elements in common with faunas from Siberia, northern Greenland and the Arctic platform (Trettin 1991).

The initial approach of Pearya to the present Arctic Islands is suggested by quartzite and marble clasts in coarse Upper Silurian conglomerates of the Danish River Formation in central Ellesmere Island (Trettin *et al.* 1991*b*). Its actual emplacement is probably marked by the sub-Middle Devonian unconformity that overlies strata as young as Silurian (and possibly Lower Devonian) in the Clements Markham fold belt (Trettin *et al.* 1991*b*). A *c.* 390 Ma post-tectonic pluton cuts rocks of the terrane. The southern boundary of Pearya is a high-angle, sinistral fault, which curves northward near its western end into a thrust (Fig. 5). Along its northwestern extent, Pearya has been dissected into a set of slivers, possibly structurally interleaved with parautochthonous strata. Trettin *et al.* (1991*b*) favoured a model of westward transpressional emplacement. Sinistral transcurrent motion dominated in Silurian time, with Pearya acting as an indentor or indentors into the continental margin (Trettin *et al.* 1991*b*; see their fig. 12B.2). This was succeeded by a more widespread shortening in latest Devonian and Early Mississippian time: the Ellesmerian orogeny, which produced a broad, south-verging fold-and-thrust belt in the Canadian Arctic region.

At present there are no detrital zircon data available from Pearya, and study of its basement rocks is fairly limited. Although remote, this terrane is a crucial target for further studies, in that it solely represents the most proximal of the possible Caledonian crustal blocks that were located between Laurentia and Siberia (Fig. 11).

Geodynamic model

The geological record of the Alexander and parts of the Eastern Klamath, Northern Sierra and Okanagan terranes indicates decreasing affinity with the Caledonides through mid-Palaeozoic time, and for some terranes (with the exception of Alexander) a firm position in eastern Panthalassa by late Palaeozoic time as basement terranes to the McCloud arcs (Fig. 2). Dispersion of these terranes and their westward travel around northern Laurentia began as the Iapetus and Rheic oceans closed and the Appalachian–Caledonian orogen developed along eastern Laurentia and western Baltica in mid-Palaeozoic time (Fig. 11). We postulate that upper mantle flow out of the shrinking Iapetus–Rheic oceans opened a mid-Palaeozoic 'gateway' between Laurentia and Siberia, termed here the Northwest Passage, similar to the Miocene to recent development of the Scotia Sea through Drake Passage between South America and Antarctica (Pearce *et al.* 2001). Initial rifting and rapid westward migration of a narrow subduction zone (Schellart *et al.* 2007) led to dispersion of the crustal fragments that once lay between Baltica, Siberia and northeastern Laurentia (Fig. 11). The southern boundary of the Northwest Passage developed as a sinistral transpressive zone along which Pearya, the least displaced of these terranes, was emplaced along the northern Laurentian margin in Late Silurian–Early Devonian time (Fig. 12). This sinistral transpressive zone was probably kinematically linked with Silurian to Devonian sinistral transpression that characterized the late Caledonian deformation of Svalbard and NE Greenland (eastern Laurentia; Figs 11 and 12; Gee & Page 1994; Gee & Teben'kov 2004).

The early record of a Scotia-style arc is probably preserved in the Early to Middle Devonian magmatism of the Ambler district in the southern Brooks Range and the Seward Peninsula. By late Early Devonian time, the Arctic Alaska terrane was accreted to the northwestern margin of Laurentia during the Romanzof orogeny (Lane 2007). We speculate that the Caledonian-derived elements of the Eastern Klamath, Northern Sierra and perhaps Okanagan terranes also came into contact with their northwestern Laurentian counterparts (Figs 9 and 10) during the Romanzof orogeny and later migrated southward along western Laurentia as composite terranes (see also Wallin *et al.* 2000). This southward transport of terranes most probably occurred along a sinistral transform fault system that developed along the western edge of Laurentia in Middle Devonian time (Fig. 13). Progressively younger deformation in continental margin strata, from the late Early Devonian Romanzof orogeny in northern Yukon and Alaska (Lane 2007), to early Middle Devonian (Eifelian) folding and faulting in the Purcell Mountains of southeastern British Columbia (Root 2001), and the Late Devonian to Early Mississippian Antler orogeny of the SW USA (Johnson & Pendergast 1981), may record the southward propagation of this transpressional fault system. The pre-Late Devonian deformation recorded in the Snowcap assemblage of the Yukon–Tanana terrane could be related to this event. The Okanagan terrane appears to have been emplaced against the western Kootenay terrane by Late Devonian time (Thompson *et al.* 2006), an event that could relate to Middle Devonian deformation in the Purcell Mountains to the east. Also, arrival of the Northern Sierra terrane before intrusion of the 364 Ma Bowman Lake batholith could have triggered the emplacement of the Roberts Mountains allochthons during the Antler orogeny.

The question of how Devonian subduction was initiated along the entire western margin of Laurentia in Devonian time remains an enduring problem. The expected strength of old oceanic lithosphere at a passive margin is considered to be too great to be

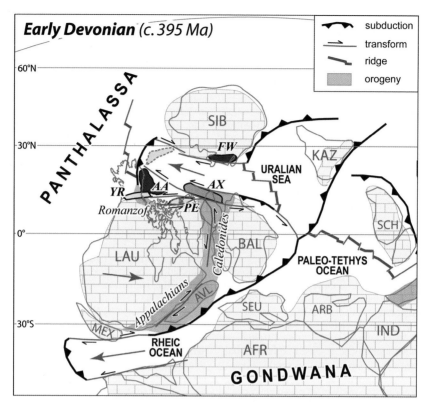

Fig. 12. Early Devonian palaeogeography and development of the Northwest Passage between Laurentia–Baltica and Siberia. Light grey terrane with dashed outline represents possible additional crustal fragments now submerged in the Arctic Ocean. ARB, Arabia; IND, India; MEX, Mexico; SEU, southern Europe. Other abbreviations as in Figure 11.

overcome by either gravity or compression, unless it has been previously weakened (Stern 2004). Propagation of a sinistral transform fault that apparently nucleated out of the Northwest Passage in Devonian time could have provided the weakness along which the oceanic lithosphere collapsed and subduction propagated southward (Figs 13 and 14). Onset of subduction and its southward propagation is recorded by magmatism of 380–400 Ma in the Arctic Alaska terrane, 380–390 Ma in the Yukon–Tanana terrane of the Coast Mountains (which probably restored near present-day Alaska in Palaeozoic time; Mihalynuk et al. 1994), 360–370 Ma in the parautochthonous continental margin of eastern Alaska and Yukon, and c. 360 Ma along the entire margin of western Laurentia (Nelson et al. 2006). These events were probably the result of a global plate reorganization that followed the Middle Devonian Acadian orogeny in the Appalachians and continued with the Carboniferous collision of Gondwana (Fig. 14). A narrow subduction zone that propagated westward through the Northwest Passage in Silurian–Devonian time could have provided the seed point from which subduction was initiated along western Laurentia (Figs 11–14).

Along northern Laurentia, this change in plate motion led to a collision with an enigmatic crustal block, the mythical Crocker Land of Arctic explorers, and development of the Late Devonian–Early Mississippian Ellesmerian orogeny as Laurentia apparently tracked north during collision with Gondwana (Fig. 14). Crocker Land was possibly one of the Caledonian crustal fragments associated with the Alexander and other terranes (Fig. 5). It apparently supplied sediments intermittently to the Sverdrup basin to the south until mid-Mesozoic time (Davies & Nassichuk 1991) and was probably removed during Jurassic–Cretaceous opening of the Arctic Ocean. Future provenance studies will provide more information about the nature and origins of Crocker Land (Omma et al. 2007).

Shortly after initiation of subduction along western Laurentia, slab rollback is thought to have caused extension in the back-arc region, which led to rifting of parts of the distal continental margin, such as the Snowcap assemblage (basement to

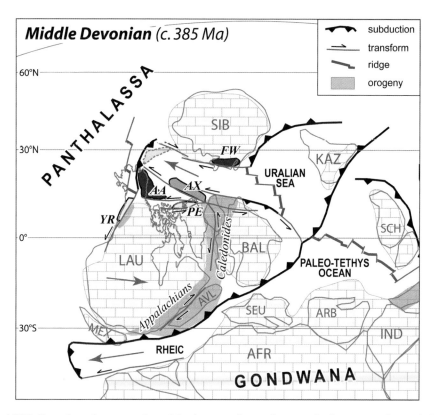

Fig. 13. Middle Devonian palaeogeography and development of a transform margin along western Laurentia. Light grey terrane with dashed outline represents possible additional crustal fragments now submerged in the Arctic ocean. Abbreviations as in Figures 11 and 12.

Yukon–Tanana) and possibly the Shoo Fly Complex (basement to the Northern Sierra terrane; Fig. 14). This rifting culminated in opening of the Slide Mountain ocean in Early Mississippian time and hence migration of the late Palaeozoic peri-Laurentian arcs away from the continental margin (Figs 14 and 15). The Yukon–Tanana terrane shares Late Devonian to earliest Mississippian (370–355 Ma) magmatism with the Laurentian margin, but younger Carboniferous to Permian arc magmatism is unique to the terrane (Nelson et al. 2006). The western Kootenay terrane (Eagle Bay assemblage; Fig. 3) of southern British Columbia appears to be a portion of the Late Devonian–earliest Mississippian remnant arc that remained stranded on the Laurentian margin after opening of the Slide Mountain ocean. The exact distribution of the Slide Mountain terrane in southern British Columbia has been obscured by the penetrative Mesozoic deformation and severe early Cenozoic extension that affected this region. However, the occurrence of upper Palaeozoic arc sequences with McCloud faunal elements overlying parts of the Okanagan terrane (Harper Ranch and Attwood groups, Mt. Roberts Formation; Figs 2 and 3) requires a more southerly palaeolatitude in Early Permian time (Belasky et al. 2002). The Havallah and Schoonover basinal sequences in the Golconda allochthon are probably the southern extension of the Slide Mountain ocean (Figs 1 and 10; Miller et al. 1984; Harwood & Murchey 1990).

The Slide Mountain ocean apparently reached its maximum width in Early Permian time (Fig. 15; Nelson et al. 2006). Differences between the McCloud and western Laurentian faunas, based on statistical analysis, suggest that the McCloud belt probably lay 2000–3000 km west of the continental margin (Belasky et al. 2002), providing a maximum estimate for the width of the Slide Mountain ocean. By Middle Permian time (c. 270 Ma), subduction polarity was reversed and the Slide Mountain lithosphere was being subducted beneath the Yukon–Tanana and related terranes. This is recorded in the belt of high-pressure rocks that lies along the eastern edge of the Yukon–Tanana terrane and Middle to Late Permian magmatic rocks of the Klondike arc (Fig. 2). By Triassic time, the Slide Mountain ocean had closed and the

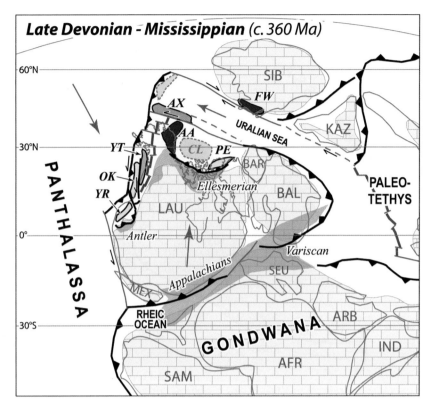

Fig. 14. Late Devonian to Early Mississippian palaeogeography. Development of the Antler and Ellesmerian orogens, initiation of subduction along western Laurentia and onset of rifting in the back-arc region. BAR, Barentsia; CL, 'Crocker Land'; OK, Okanagan terrane; SAM, South America; YT, Yukon–Tanana terrane. Other abbreviations as in Figures 11 and 12.

late Palaeozoic peri-Laurentian arcs were accreted to western Laurentia, by then a part of Pangaea, during the Sonoma orogeny (Fig. 16; Dickinson 2004). Triassic synorogenic clastic rocks overlying the Yukon–Tanana and Slide Mountain terranes, as well as the Laurentian continental margin, and amphibolite-facies metamorphism in the Yukon–Tanana terrane provide records of the Sonoman event in the northern Cordillera (Beranek & Mortensen 2007; Berman et al. 2007).

The Alexander terrane is inferred to have migrated out of the Northwest Passage during Carboniferous time (Fig. 14). By Pennsylvanian time, it had joined Wrangellia in northern Panthalassa, where they apparently evolved in an isolated intraoceanic setting until their Middle Jurassic accretion (Figs 15 and 16).

The Farewell terrane is thought to have originated from the northern margin of the Northwest Passage, where it originally evolved as part of the Siberian Platform until at least Early Permian time, when it was deformed during the c. 285 Ma Browns Fork orogeny, an event related to development of the Uralian and Taimyr fold belts (Figs 11–15; Bradley et al. 2003). Details of its Mesozoic history are sparse. The Farewell terrane may have been expelled from its site of origin during or following Uralian tectonism (Figs 15 and 16).

By Middle to Late Triassic time, east-dipping subduction was re-established along the entire western margin of Laurentia (now part of Pangaea; Fig. 16), giving rise to voluminous Triassic–Jurassic arc magmatism of Stikinia, Quesnellia and related terranes of the western USA, which were in part built upon Palaeozoic basements of the Yukon–Tanana, Okanagan, Eastern Klamath and Northern Sierra terranes (Fig. 2). This more stable, wide-slab geometry (Schellart et al. 2007) apparently persisted more or less in its original form along western North America until at least early Cenozoic time.

Convergence between the North American plate and the various oceanic plates that succeeded Panthalassa (e.g. Farallon, Kula and Pacific) began with the Jurassic opening of the North Atlantic

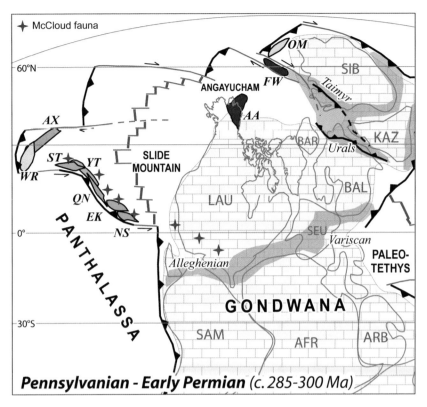

Fig. 15. Pennsylvanian to Early Permian palaeogeography. By this time, the Slide Mountain ocean had reached its maximum width and volcanic arcs of the McCloud belt (late Palaeozoic sequences of Stikinia (ST), Quesnellia (QN), Eastern Klamaths (EK) and Northern Sierra (NS) terranes) were developing on top of pericratonic mid-Palaeozoic and older fragments of Yukon–Tanana (YT), Yreka–Trinity and Shoo Fly terranes. Onset of Uralian tectonism along northern Baltica, Kazakhstania and Siberia, and inferred expulsion of the Farewell terrane (FW) from the Siberian margin. OM, Omulevka ridge; WR, Wrangellia. Other abbreviations as in Figures 11–14.

Ocean and the westward drift of North America over its western subduction zone (Monger & Price 2002).

It is possible that another Caribbean- or Scotia-style subduction system developed between southern Laurentia and Gondwana (Wright & Wyld 2006) and coexisted with the Northwest Passage in Silurian–Devonian time (Figs 11–13), much like the modern Caribbean and Scotia systems developed at either end of South America. However, our review of the geological evidence leads us to conclude that Cordilleran terranes of inferred Caledonian and Siberian affinities were more probably introduced into eastern Panthalassa via the Northwest Passage rather than its southern equivalent.

Conclusions

Palaeozoic to early Mesozoic terranes of the North American Cordillera are proposed to have originated from three major regions in Palaeozoic time: (1) the western peri-Laurentian margin; (2) western Panthalassa in proximity to the Palaeo-Tethys realm; (3) in proximity to the northern Caledonides, occupying an intermediate position between NE Laurentia, Baltica and Siberia (Figs 1 and 11). Dispersion of the Caledonian–Siberian terranes and their westward migration into eastern Panthalassa is interpreted to result from development of a Caribbean- or Scotia-type subduction system between northern Laurentia–Baltica and Siberia in mid-Palaeozoic time: the Northwest Passage. This system was probably driven by upper mantle outflow from the closing Iapetus–Rheic oceans along eastern Laurentia, as Pangaea was being amalgamated (Figs 11–14). The rapid westward migration of a narrow subduction zone through the Northwest Passage entrained Caledonian and Siberian terranes into eastern Panthalassa and provided a seed point for propagation of subduction along western Laurentia in

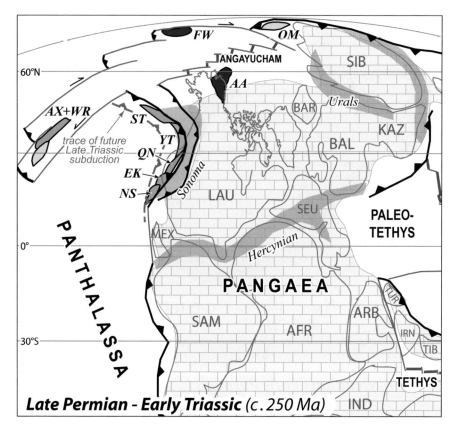

Fig. 16. Late Permian to Early Triassic palaeogeography. By then, all Caledonian, Baltican and Siberian terranes now found in the North American Cordillera had entered Panthalassa. The Slide Mountain ocean was closing and the late Palaeozoic peri-Laurentian arcs were accreted to western Laurentia, by then a part of Pangaea. IRN, Iran; TIB, Tibet; TUR, Turkey. Other abbreviations as in Figures 11–15.

Late Devonian–Early Mississippian time. Subduction along western Laurentia is inferred to have been initiated as a result of a global plate reorganization related to Devonian convergence in the Appalachian orogen of eastern Laurentia and Carboniferous collision with Gondwana. By early Mesozoic time, this subduction system had evolved to a stable, wide-slab geometry that persisted along western North America at least until early Cenozoic time.

This paper, as indeed our entire learning experience in the Cordillera, has been shaped by the observations and inferences of many others. As well as citing their papers, we highlight here some principal influences on our grasp of the Caledonian–Siberian terranes: P. Belasky, D. Bradley, J. Dumoulin, G. Gehrels, D. Harwood, E. Katvala, L. Lane, N. Lindsley-Griffin, W. McClelland, M. Mihalynuk, E. Miller, M. Miller, J. Monger, T. Moore, B. Murchey, and C. Soja; it is hoped that their ideas appear here in forms that are true to the originals. In particular, we wish to acknowledge recent conversations with J. Wright, whose innovative solution to some problems of Cordilleran terrane origin was the spark for the present endeavour. Thanks to P. Cawood for encouraging us to submit a western Laurentian story to this volume. We are grateful to N. Lindsley-Griffin for her careful corrections to the section on the eastern Klamaths. Formal reviews by D. Bradley, G. Gehrels and C. van Staal have helped to clarify our arguments. This is Yukon Geological Survey Contribution YGS2008-001.

References

ÅHÄLL, K.-I., CORNELL, D. H. & ARMSTRONG, R. 1998. Ion probe zircon dating of three metasedimentary units bordering the Oslo Rift: New constraints for early Mesoproterozoic growth of the Baltic Shield. *Precambrian Research*, **87**, 117–134.

AMATO, J. 2004. Crystalline basement ages, detrital zircon ages, and metamorphic ages from Seward Peninsula: Implications for Proterozoic and Cambrian–Ordovician paleogeographic reconstructions of the Arctic–Alaska terrane. *Geological Society of America, Abstracts with Programs*, **36**, 22.

AMATO, J., TORO, J., MILLER, E. & GEHRELS, G. E. 2006. Late Proterozoic magmatism in Alaska and its implications for paleogeographic reconstructions of the Arctic Alaska–Chukotka plate. *Geological Society of America, Abstracts with Programs*, Cordillera Section, Anchorage, Alaska, May, 2006, **38**, 13.

ANTOSHKINA, A. I. & SOJA, C. M. 2006. Late Silurian reconstruction indicated by migration of reef biota between Alaska, Baltica (Urals), and Siberia (Salair). *GFF-online*, **128**, 75–78.

BAZARD, D. R., BUTLER, R. F., GEHRELS, G. E. & SOJA, C. M. 1995. Early Devonian paleomagnetic data from the Lower Devonian Karheen Formation suggest Laurentia–Baltica connection for the Alexander terrane. *Geology*, **23**, 707–710.

BEATTY, T. W., ORCHARD, M. J. & MUSTARD, P. S. 2006. Geology and tectonic history of the Quesnel terrane in the area of Kamloops, British Columbia. *In*: COLPRON, M. & NELSON, J. L. (eds) *Paleozoic Evolution and Metallogeny of Pericratonic Terranes at the Ancient Pacific Margin of North America, Canadian and Alaskan Cordillera*. Geological Association of Canada, Special Papers, **45**, 483–504.

BELASKY, P., STEVENS, C. H. & HANGER, R. A. 2002. Early Permian location of western North American terranes based on brachiopod, fusulinid and coral biogeography. *Palaeogeography, Palaeoclimatology, Palaeoecology*, **179**, 245–266.

BERANEK, L. P. & MORTENSEN, J. K. 2007. Investigating a Triassic overlap assemblage in Yukon: On-going field studies and preliminary detrital-zircon age data. *In*: EMOND, D. S., LEWIS, L. L. & WESTON, L. H. (eds) *Yukon Exploration and Geology 2006*. Yukon Geological Survey, Whitehorse, 83–92.

BERMAN, R. G., RYAN, J. J., GORDEY, S. P. & VILLENEUVE, M. 2007. Permian to Cretaceous polymetamorphic evolution of the Stewart River region, Yukon–Tanana terrane, Yukon, Canada: P–T evolution linked with *in situ* SHRIMP monazite geochronology. *Journal of Metamorphic Geology*, **25**, 803–827.

BINGEN, B., GRIFFIN, W. L., TORSVIK, T. H. & SAEED, A. 2005. Timing of Late Neoproterozoic glaciation on Baltica constrained by detrital zircon geochronology in the Hedmark Group, south-east Norway. *Terra Nova*, **17**, 250–258.

BLODGETT, R. B., ROHR, D. M. & BOUCOT, A. J. 2002. Paleozoic links among some Alaskan accreted terranes and Siberia based on megafossils. *In*: MILLER, E. L., GRANTZ, A. & KLEMPERER, S. (eds) *Tectonic Evolution of the Bering Shelf–Chukchi Sea–Arctic Margin and Adjacent Land Masses*. Geological Society of America, Special Papers, **360**, 273–290.

BOX, S. E., MOLL-STALCUP, E. J., WOODEN, J. L. & BRADSHAW, J. Y. 1990. Kilbuck terrane: Oldest rocks in Alaska. *Geology*, **18**, 1219–1222.

BRADLEY, D. C., DUMOULIN, J. *ET AL*. 2003. Late Paleozoic orogeny in Alaska's Farewell terrane. *Tectonophysics*, **372**, 23–40.

BRADLEY, D. C., MCCLELLAND, W. C. *ET AL*. 2007. Detrital zircon geochronology of some Neoproterozoic to Triassic rocks in interior Alaska. *In*: RIDGWAY, K. D., TROP, J. M., GLEN, J. M. G. & O'NEILL, J. M. (eds) *Tectonic Growth of a Collisional Continental Margin: Crustal Evolution of Southern Alaska*. Geological Society of America, Special Papers, **431**, 155–189.

BREW, D. A. & FORD, A. B. 1985. *Preliminary reconnaissance geologic map of the Juneau, Taku River, Atlin and part of Skagway 1:250,000 quadrangles, southeastern Alaska*. US Geological Survey, Open File, **85–395**.

CAWOOD, P. A. & BUCHAN, C. 2007. Linking accretionary orogenesis with supercontinent assembly. *Earth-Science Reviews*, **82**, 217–256.

CAWOOD, P. A., MCCAUSLAND, P. J. A. & DUNNING, G. R. 2001. Opening Iapetus: Constraints from the Laurentian margin in Newfoundland. *Geological Society of America Bulletin*, **113**, 443–453.

CAWOOD, P. A., NEMCHIN, A. A., STRACHAN, R., PRAVE, A. R. & KRABBENDAM, M. 2007. Sedimentary basin and detrital zircon record along East Laurentia and Baltica during assembly and breakup of Rodinia. *Journal of the Geological Society, London*, **164**, 257–275.

CECILE, M. P., HARRISON, J. C., KOS'KO, M. K. & PARRISH, R. R. 1991. Precambrian U–Pb ages of igneous rocks, Wrangel complex, Wrangel Island, USSR. *Canadian Journal of Earth Sciences*, **28**, 1340–1348.

COLPRON, M., LOGAN, J. M. & MORTENSEN, J. K. 2002. U–Pb zircon age constraint for late Neoproterozoic rifting and initiation of the lower Paleozoic passive margin of western Laurentia. *Canadian Journal of Earth Sciences*, **39**, 133–143.

COLPRON, M., NELSON, J. L. & MURPHY, D. C. 2006. A tectonostratigraphic framework for the pericratonic terranes of the northern Cordillera. *In*: COLPRON, M. & NELSON, J. L. (eds) *Paleozoic Evolution and Metallogeny of Pericratonic Terranes at the Ancient Pacific Margin of North America, Canadian and Alaskan Cordillera*. Geological Association of Canada, Special Papers, **45**, 1–23.

COLPRON, M., NELSON, J. L. & MURPHY, D. C. 2007. Northern Cordilleran terranes and their interactions through time. *GSA Today*, **17**, 4–10.

CONEY, P. J., JONES, D. L. & MONGER, J. W. H. 1980. Cordilleran suspect terranes. *Nature*, **288**, 329–333.

DAVIES, G. R. & NASSICHUK, W. W. 1991. Carboniferous and Permian history of the Sverdrup Basin, Arctic Islands. *In*: TRETTIN, H. P. (ed.) *Geology of the Innuitian Orogen and Arctic Platform of Canada and Greenland*. Geological Survey of Canada, Geology of Canada, **3**, 345–367.

DE HAAS, G. J. L. M., ANDERSEN, T. & VESTIN, J. 1999. Detrital zircon geochronology: New evidence for an old model for accretion of the southwest Baltic Shield. *Journal of Geology*, **107**, 569–586.

DEVINE, F., CARR, S. D., MURPHY, D. C., DAVIS, W. J., SMITH, S. & VILLENEUVE, M. E. 2006. Geochronological and geochemical constraints on the origin of the Klatsa metamorphic complex: Implications for Early Mississippian high-pressure metamorphism within Yukon–Tanana terrane. *In*: COLPRON, M. & NELSON, J. L. (eds) *Paleozoic Evolution and Metallogeny of Pericratonic Terranes at the Ancient Pacific Margin of North America, Canadian and Alaskan*

Cordillera. Geological Association of Canada, Special Papers, **45**, 107–130.

DICKINSON, W. R. 2004. Evolution of the North American Cordillera. *Annual Review of Earth and Planetary Sciences*, **32**, 12–45.

DOSTAL, J., CHURCH, B. N. & HOY, T. 2001. Geological and geochemical evidence for variable magmatism and tectonics in the southern Canadian Cordillera: Paleozoic to Jurassic suites, Greenwood, southern British Columbia. *Canadian Journal of Earth Sciences*, **38**, 75–90.

DUMOULIN, J. A., HARRIS, A. G., GAGIEV, M., BRADLEY, D. C. & REPETSKI, J. E. 2002. Lithostratigraphic, conodont, and other faunal links between lower Paleozoic strata in northern and central Alaska and northeastern Russia. *In*: MILLER, E. L., GRANTZ, A. & KLEMPERER, S. (eds) *Tectonic Evolution of the Bering Shelf–Chukchi Sea–Arctic Margin and Adjacent Land Masses*. Geological Society of America, Special Papers, **360**, 291–312.

DUSEL-BACON, C., HOPKINS, M. J., MORTENSEN, J. K., DASHEVSKY, S. S., BRESSLER, J. R. & DAY, W. C. 2006. Paleozoic tectonic and metallogenic evolution of the pericratonic rocks of east–central Alaska and adjacent Yukon. *In*: COLPRON, M. & NELSON, J. L. (eds) *Paleozoic Evolution and Metallogeny of Pericratonic Terranes at the Ancient Pacific Margin of North America, Canadian and Alaskan Cordillera*. Geological Association of Canada, Special Papers, **45**, 25–74.

DUTRO, J. T. JR, BROSGÉ, W. P., LANPHERE, M. A. & REISER, H. N. 1976. Geologic significance of Doonerak structural high, central Brooks Range, Alaska. *AAPG Bulletin*, **60**, 952–961.

EBERLEIN, G. D. & CHURKIN, M. JR. 1970. *Paleozoic stratigraphy in the northwest coastal area of Prince of Wales Island*. US Geological Survey, Bulletin, **1284**.

ENGLISH, J. M. & JOHNSTON, S. T. 2005. Collisional orogenesis in the northern Canadian Cordillera; implications for Cordilleran crustal structure, ophiolite emplacement, continental growth, and the terrane hypothesis. *Earth and Planetary Science Letters*, **232**, 333–344.

ERDMER, P., GHENT, E. D., ARCHIBALD, D. A. & STOUT, M. Z. 1998. Paleozoic and Mesozoic high-pressure metamorphism at the margin of ancestral North America in central Yukon. *Geological Society of America Bulletin*, **110**, 615–629.

ERDMER, P., HEAMAN, L. M., CREASER, R. A., THOMPSON, R. I. & DAUGHTRY, K. L. 2001. Eocambrian granite clasts in southern British Columbia shed light on Cordilleran hinterland crust. *Canadian Journal of Earth Sciences*, **38**, 1007–1016.

ERDMER, P., MOORE, J. M., HEAMAN, L., THOMPSON, R. I., DAUGHTRY, K. L. & CREASER, R. A. 2002. Extending the ancient margin outboard in the Canadian Cordillera: Record of Proterozoic crust and Paleocene regional metamorphism in the Nicola horst, southern British Columbia. *Canadian Journal of Earth Sciences*, **39**, 1605–1623.

EVENCHICK, C. A., MCMECHAN, M. E., MCNICOLL, V. J. & CARR, S. D. 2007. A synthesis of the Jurassic–Cretaceous tectonic evolution of the central and southeastern Canadian Cordillera: Exploring links across the orogen. *In*: SEARS, J. W., HARMS, T. A. & EVENCHICK, C. A. (eds) *Whence the Mountains? Inquiries into the Evolution of Orogenic Systems: A Volume in Honor of Raymond A. Price*. Geological Society of America, Special Papers, **433**, 117–145.

FRYDA, J. & BLODGETT, R. B. 2008. Paleobiogeographic affinities of Emsian (late Early Devonian) gastropods from Farewell terrane (west–central Alaska). *In*: BLODGETT, R. B. & STANLEY, G. D. JR (eds) *The Terrane Puzzle: New Perspectives on Paleontology and Stratigraphy from the North American Cordillera*. Geological Society of America, Special Papers, **442**, 107–120.

GARDNER, M. C., BERGMAN, S. C. ET AL. 1988. Pennsylvanian pluton stitching of Wrangellia and the Alexander terrane, Wrangell Mountains, Alaska. *Geology*, **16**, 967–971.

GEE, D. G. & PAGE, L. M. 1994. Caledonian terrane assembly on Svalbard: New evidence from $^{40}Ar/^{39}Ar$ dating in Ny Friesland. *American Journal of Science*, **294**, 1166–1186.

GEE, D. G. & PEASE, V. 2004. The Neoproterozoic Timanide Orogen of eastern Baltica: Introduction. *In*: GEE, D. G. & PEASE, V. (eds) *The Neoproterozoic Timanide Orogen of Eastern Baltica*. Geological Society, London, Memoirs, **30**, 1–3.

GEE, D. G. & TEBEN'KOV, A. M. 2004. Svalbard: A fragment of the Laurentian margin. *In*: GEE, D. G. & PEASE, V. (eds) *The Neoproterozoic Timanide Orogen of Eastern Baltica*. Geological Society, London, Memoirs, **30**, 191–206.

GEE, D. G., BOGOLEPOVA, O. K. & LORENZ, H. 2006. The Timanide, Caledonide and Uralide orogens in the Eurasian high Arctic, and relationships to the palaeo-continents Laurentia, Baltica and Siberia. *In*: GEE, D. G. & STEPHENSON, R. A. (eds) *European Lithosphere Dynamics*. Geological Society, London, Memoirs, **32**, 507–520.

GEHRELS, G. E. 1990. Late Proterozoic–Cambrian metamorphic basement of the Alexander terrane on Long and Dall Islands, southeast Alaska. *Geological Society of America Bulletin*, **102**, 760–767.

GEHRELS, G. E. 2000. Introduction to the detrital zircon studies of Paleozoic and Triassic strata in western Nevada and Northern California. *In*: SOREGHAN, M. J. & GEHRELS, G. E. (eds) *Paleozoic and Triassic paleogeography and tectonics of western Nevada and northern California*. Geological Society of America, Special Papers, **347**, 1–17.

GEHRELS, G. E. 2001. Geology of the Chatham Sound region, southeast Alaska and coastal British Columbia. *Canadian Journal of Earth Sciences*, **38**, 1579–1599.

GEHRELS, G. E. 2002. Detrital zircon geochronology of the Taku terrane, southeast Alaska. *Canadian Journal of Earth Sciences*, **39**, 921–931.

GEHRELS, G. E. & KAPP, P. A. 1998. Detrital zircon geochronology and regional correlation of metasedimentary rocks in the Coast Mountains, southeastern Alaska. *Canadian Journal of Earth Sciences*, **35**, 269–279.

GEHRELS, G. E. & MILLER, M. M. 2000. Detrital zircon geochronologic study of upper Paleozoic strata in the eastern Klamath terrane, northern California. *In*: SOREGHAN, M. J. & GEHRELS, G. E. (eds) *Paleozoic and Triassic Paleogeography and Tectonics of western Nevada and northern California*. Geological Society of America, Special Papers, **347**, 99–107.

GEHRELS, G. E. & ROSS, G. M. 1998. Detrital zircon geochronology of Neoproterozoic to Permian miogeoclinal strata in British Columbia and Alberta. *Canadian Journal of Earth Sciences*, **35**, 1380–1401.

GEHRELS, G. E. & SALEEBY, J. B. 1987. Geologic framework, tectonic evolution and displacement history of the Alexander terrane. *Tectonics*, **6**, 151–174.

GEHRELS, G. E., DICKINSON, W. R., ROSS, G. M., STEWART, J. H. & HOWELL, D. G. 1995. Detrital zircon reference for Cambrian to Triassic miogeoclinal strata of western North America. *Geology*, **23**, 831–834.

GEHRELS, G. E., BUTLER, R. F. & BAZARD, D. R. 1996. Detrital zircon geochronology of the Alexander terrane, southeastern Alaska. *Geological Society of America Bulletin*, **108**, 722–734.

GEHRELS, G. E., JOHNSSON, M. J. & HOWELL, D. G. 1999. Detrital zircon geochronology of the Adams Argillite and Nation River Formation, east–central Alaska, U.S.A. *Journal of Sedimentary Research*, **69**, 135–144.

GHOSH, D. K. 1995. U–Pb geochronology of Jurassic to early Tertiary granitic intrusives from the Nelson–Castlegar area, southeastern British Columbia, Canada. *Canadian Journal of Earth Sciences*, **32**, 1668–1680.

GIRTY, G. H., GESTER, K. C. & TURNER, J. B. 1990. Pre-Late Devonian geochemical, stratigraphic, sedimentologic and structural patterns, Shoo Fly Complex, northern Sierra Nevada, California. *In*: HARWOOD, D. S. & MILLER, M. M. (eds) *Paleozoic and early Mesozoic paleogeographic relations, Sierra Nevada, Klamath Mountains, and related terranes*. Geological Society of America, Special Papers, **255**, 43–56.

GODWIN, C. J. & SINCLAIR, A. J. 1982. Average lead isotope growth curves for shale-hosted zinc–lead deposits, Canadian Cordillera. *Economic Geology*, **77**, 675–690.

GOODFELLOW, W. D., CECILE, M. P. & LEYBOURNE, M. I. 1995. Geochemistry, petrogenesis, and tectonic setting of lower Paleozoic alkalic and potassic volcanic rocks, Northern Canadian Cordillera Miogeocline. *Canadian Journal of Earth Sciences*, **32**, 1236–1254.

GORDEY, S. P., ABBOTT, J. G., TEMPELMAN-KLUIT, D. J. & GABRIELSE, H. 1987. 'Antler' clastics in the Canadian Cordillera. *Geology*, **15**, 103–107.

GRADSTEIN, F. M., OGG, J. G. & SMITH, A. G. 2004. *A Geologic Time Scale 2004*. Cambridge University Press, Cambridge.

GROVE, M., GEHRELS, G. E., COTKIN, S. J., WRIGHT, J. E. & ZOU, H. 2008. Non-Laurentian cratonal provenance of Late Ordovician eastern Klamath blueschists and a link to the Alexander terrane. *In*: WRIGHT, J. E. & SHERVAIS, J. W. (eds) *Ophiolites, Arcs, and Batholiths*. Geological Society of America, Special Papers, **438**, 223–250.

HARDING, J. P., GEHRELS, G. E., HARWOOD, D. S. & GIRTY, G. H. 2000. Detrital zircon geochronology of the Shoo Fly Complex, northern Sierra terrane, northeastern California. *In*: SOREGHAN, M. J. & GEHRELS, G. E. (eds) *Paleozoic and Triassic Paleogeography and Tectonics of western Nevada and northern California*. Geological Society of America, Special Papers, **347**, 43–55.

HARWOOD, D. S. & MURCHEY, B. L. 1990. Biostratigraphic, tectonic and paelogeographic ties between upper Paleozoic volcanic and basinal rocks in the northern Sierra terrane, California, and the Havallah sequence, Nevada. *In*: HARWOOD, D. S. & MILLER, M. M. (eds) *Paleozoic and early Mesozoic paleogeographic relations, Sierra Nevada, Klamath Mountains, and related terranes*. Geological Society of America, Special Papers, **255**, 157–174.

HARWOOD, D. S., JAYKO, A. S., HARRIS, A. G., SILBERLING, N. J. & STEVENS, C. H. 1988. Permo-Triassic rocks slivered between the Shoo Fly Complex and the Feather River peridotite belt, northern Sierra Nevada, California. *Geological Society of America, Abstracts with Programs*, **20**, 167–168.

HELWIG, J. 1974. Eugeosynclinal basement and a collage concept of orogenic belts. *In*: DOTT, R. H. JR. & SHAVER, R. H. (eds) *Modern and Ancient Geosynclinal Sedimentation*. Society of Economic Paleontologists and Mineralogists, Special Publications, **19**, 359–380.

IRWIN, W. P. 1981. Tectonic accretion of the Klamath Mountains. *In*: ERNST, G. G. (ed.) *The Geotectonic Development of California—Rubey Volume I*. Prentice–Hall, Englewood Cliffs, NJ, 29–49.

JOHNSON, J. G. & PENDERGAST, A. 1981. Timing and mode of emplacement of the Roberts Mountains allochthon, Antler orogeny. *Geological Society of America Bulletin*, **92**, 648–658.

KATVALA, E. C. & HENDERSON, C. M. 2002. Conodont sequence biostratigraphy and paleogeography of the Pennsylvanian–Permian Mount Mark and Fourth Lake formations, southern Vancouver Island. *In*: HILLS, L. V., HENDERSON, C. M. & BAMBER, E. W. (eds) *Carboniferous and Permian of the World*. Canadian Society of Petroleum Geologists, Memoirs, **19**, 461–478.

KHUDOLEY, A. K., RAINBIRD, R. H. ET AL. 2001. Sedimentary evolution of the Riphean–Vendian basin of southeastern Siberia. *Precambrian Research*, **111**, 129–163.

KNUDSEN, T.-L., ANDERSEN, T., WHITEHOUSE, M. J. & VESTIN, J. 1997. Detrital zircon ages from southern Norway—implications for the Proterozoic evolution of the southwestern Baltic Shield. *Contributions to Mineralogy and Petrology*, **130**, 47–58.

KOS'KO, M., CECILE, M., HARRISON, J., GANELIN, V., KANDOSHKO, N. & LOPATIN, B. 1993. *Geology of Wrangel Island, between Chukchi and eastern Siberian seas, northeastern Russia*. Geological Survey of Canada, Bulletin, **461**.

LANE, L. S. 2007. Devonian–Carboniferous paleogeography and orogenesis, northern Yukon and adjacent Arctic Alaska. *Canadian Journal of Earth Sciences*, **44**, 679–694.

LASSITER, J. C., DEPAOLO, D. J. & MAHONEY, J. J. 1995. Geochemistry of the Wrangellia flood basalt province: Implications for the role of continental and oceanic lithosphere in flood basalt genesis. *Journal of Petrology*, **36**, 983–1009.

LEMIEUX, Y., THOMPSON, R. I., ERDMER, P., SIMONETTI, A. & CREASER, R. A. 2007. Detrital zircon geochronology and provenance of Late Proterozoic and mid-Paleozoic successions outboard of the miogeocline, southeastern Canadian Cordillera. *Canadian Journal of Earth Sciences*, **44**, 1675–1693.

LETOUZEY, J. & KIMURA, M. 1986. The Okinawa trough: Genesis of a back arc basin along a continental margin. *Tectonophysics*, **125**, 209–230.

LINDSLEY-GRIFFIN, N., GRIFFIN, J. R., FARMER, J. D., SIVERS, E. A., BRUCKNO, B. & TOZER, M. K. 2006. Ediacaran cyclomedusoids and the paleogeographic setting of the Neoproterozoic–early Paleozoic Yreka and Trinity terranes, eastern Klamath Mountains, California. *In*: SNOKE, A. W. & BARNES, C. G. (eds) *Geological Studies in the Klamath Mountains Province, California and Oregon*. Geological Society of America, Special Papers, **410**, 411–432.

LINDSLEY-GRIFFIN, N., GRIFFIN, J. R. & FARMER, J. D. 2008. Paleogeographic significance of Ediacaran cyclomedusoids within the Antelope Mountain Quartzite, Yreka terrane, eastern Klamath Mountains, California. *In*: BLODGETT, R. B. & STANLEY, G. D., JR (eds) *The Terrane Puzzle: New Perspectives on Paleontology and Stratigraphy from the North American Cordillera*. Geological Society of America, Special Papers, **442**, 1–37.

LOGAN, J. M. & COLPRON, M. 2006. Stratigraphy, geochemistry, syngenetic sulphide occurrences and tectonic setting of the lower Paleozoic Lardeau Group, northern Selkirk Mountains, British Columbia. *In*: COLPRON, M. & NELSON, J. L. (eds) *Paleozoic Evolution and Metallogeny of Pericratonic Terranes at the Ancient Pacific Margin of North America, Canadian and Alaskan Cordillera*. Geological Association of Canada, Special Papers, **45**, 361–382.

LONEY, R. A. 1964. *Stratigraphy and petrography of the Pybus–Gambier area, Admiralty Island, Alaska*. US Geological Survey, Bulletin, **1178**.

MANKINEN, E. A., LINDSLEY-GRIFFIN, N. & GRIFFIN, J. R. 2002. Concordant paleolatitudes for Neoproterozoic ophiolitic rocks of the Trinity Complex, Klamath Mountains, California. *Journal of Geophysical Research*, **94**, 10444–10472.

MASSEY, N. W. D. 2007. Boundary project: Rock Creek area (NTS 082/02W, 03E), southern British Columbia. *In: Geological Fieldwork 2006*. British Columbia Ministry of Energy, Mines and Petroleum Resources, Paper, **2007-1**, 117–128.

MASSEY, N. W. D., MACINTYRE, D. G., DESJARDINS, P. J. & COONEY, R. T. 2005. *Digital geology map of British Columbia: Whole Province*. British Columbia Ministry of Energy, Mines and Petroleum Resources, Geological Survey Branch, GeoFile, **2005-1**.

MCCLELLAND, W. C. & GEHRELS, G. E. 1990. The Duncan Canal shear zone: A right-lateral shear zone of Jurassic age along the inboard margin of the Alexander terrane. *Geological Society of America Bulletin*, **102**, 1378–1392.

MCCLELLAND, W. C., SCHMIDT, J. M. & TILL, A. B. 2006. New SHRIMP ages from Devonian felsic volcanic and Proterozoic plutonic rocks of the southern Brooks Range, Alaska. *Geological Society of America, Abstracts with Programs*, **38**, 12.

MCKERROW, W. S., MAC NIOCAILL, C. & DEWEY, J. F. 2000. The Caledonian orogeny redefined. *Journal of the Geological Society, London*, **157**, 1149–1154.

MIHALYNUK, M. G., SMITH, M. T., MACINTYRE, D. G. & DESCHÊNES, M. 1993. Tashenshini Project Part B: Stratigraphic and magmatic setting of mineral occurrences. *In: Geological Fieldwork 1992*. British Columbia Ministry of Energy, Mines and Petroleum Resources, Paper, **1993-1**, 189–228.

MIHALYNUK, M. G., NELSON, J. & DIAKOW, L. J. 1994. Cache Creek terrane entrapment: Oroclinal paradox within the Canadian Cordillera. *Tectonics*, **13**, 575–595.

MIHALYNUK, M. G., ERDMER, P., GHENT, E. D., CORDEY, F., ARCHIBALD, D. A., FRIEDMAN, R. M. & JOHANNSON, G. G. 2004. Coherent French Range blueschist: Subduction to exhumation in <2.5 m.y.? *Geological Society of America Bulletin*, **116**, 910–922.

MILLER, E. L., HOLDSWORTH, B. K., WHITEFORD, W. B. & RODGERS, D. 1984. Stratigraphy and structure of the Schoonover sequence, northeastern Nevada: Implications for Paleozoic plate-margin tectonics. *Geological Society of America Bulletin*, **95**, 1063–1076.

MILLER, E. L., TORO, J. ET AL. 2006. New insights into Arctic paleogeography and tectonics from U–Pb detrital zircon geochronology. *Tectonics*, **25**, **TC3013** 1–19.

MILLER, M. M. 1988. Displaced remnants of a northeast Pacific fringing arc: Upper Paleozoic terranes of Permian McCloud faunal affinity, western U.S. *Tectonics*, **6**, 807–830.

MILLER, M. M. & HARWOOD, D. S. 1990. Paleogeographic setting of upper Paleozoic rocks in the northern Sierra and eastern Klamath terranes. *In*: HARWOOD, D. S. & MILLER, M. M. (eds) *Paleozoic and early Mesozoic paleogeographic relations: Sierra Nevada, Klamath Mountains, and related terranes*. Geological Society of America, Special Papers, **255**, 175–192.

MONGER, J. W. H. & NOKLEBERG, W. J. 1996. Evolution of the northern North American Cordillera: Generation, fragmentation, displacement and accretion of successive North American plate-margin arcs. *In*: COYNER, A. R. & FAHEY, P. L. (eds) *Geology and Ore Deposits of the American Cordillera*. Geological Society of Nevada, Symposium Proceedings, **III**, 1133–1152.

MONGER, J. W. H. & PRICE, R. A. 2002. The Canadian Cordillera: Geology and tectonic evolution. *Canadian Society of Exploration Geophysicists Recorder*, **27**, 17–36.

MONGER, J. W. H., WHEELER, J. O. ET AL. 1991. Part B. Cordilleran terranes, Upper Devonian to Middle Jurassic assemblages. *In*: GABRIELSE, H. & YORATH, C. J. (eds) *Geology of the Cordilleran orogen in Canada*. Geological Survey of Canada, Geology of Canada. **4**, 281–327.

MOORE, T. E., WALLACE, W. K., BIRD, K. J., KARL, S. M., MULL, C. G. & DILLON, J. T. 1994. Geology of northern Alaska. *In*: PLAFKER, G. & BERG, H. C. (eds) *The Geology of Alaska*. Geological Society of America, The Geology of North America, **G-1**, 49–140.

MOORE, T. E., ALEINIKOFF, J. N. & HARRIS, A. G. 1997a. Stratigraphic and structural implications of conodont and detrital zircon U–Pb ages from metamorphic rocks of the Coldfoot terrane, Brooks Range, Alaska. *Journal of Geophysical Research*, **102**, 20797–20820.

MOORE, T. E., HEMMING, S. & SHARP, W. D. 1997*b*. Provenance of the Carboniferous Nuka Formation, Brooks Range, Alaska: A multicomponent isotope provenance study with implications for age of cryptic crystalline basement. *In*: DUMOULIN, J. A. & GRAY, J. E. (eds) *Geological Studies in Alaska by the US Geological Survey, 1995*. US Geological Survey, Professional Papers, **1574**, 173–194.

MOORE, T. E., POTTER, C. J., O'SULLIVAN, P. B. & ALEINIKOFF, J. N. 2007. Evidence from detrital zircon U–Pb analysis for suturing of pre-Mississippian terranes in Arctic Alaska. *EOS Transanctions, American Geophysical Union, Fall Meeting Supplement*, **88**, Abstract T13D-1570.

MORTENSEN, J. K. 1990. Geology and U–Pb geochronology of the Klondike District, west–central Yukon. *Canadian Journal of Earth Sciences*, **27**, 903–914.

MORTENSEN, J. K. 1992. Pre-Mid-Mesozoic tectonic evolution of the Yukon–Tanana terrane, Yukon and Alaska, *Tectonics*, **11**, 836–853.

MORTENSEN, J. K., DUSEL-BACON, C., HUNT, J. A. & GABITES, J. 2006. Lead isotopic constraints on the metallogeny of middle and late Paleozoic syngenetic base metal occurrences in the Yukon–Tanana and Slide Mountain/Seventymile terranes and adjacent portions of the North American miogeocline. *In*: COLPRON, M. & NELSON, J. L. (eds) *Paleozoic Evolution and Metallogeny of Pericratonic Terranes at the Ancient Pacific Margin of North America, Canadian and Alaskan Cordillera*. Geological Association of Canada, Special Papers, **45**, 261–279.

MUFFLER, L. J. P. 1967. *Stratigraphy of Keku Islets and neighboring parts of Kuiu and Kupreanof Islands, southeastern Alaska*. US Geological Survey, Bulletin, **1241-C**.

NELSON, J. L. & COLPRON, M. 2007. Tectonics and metallogeny of the Canadian and Alaskan Cordillera, 1.8 Ga to present. *In*: GOODFELLOW, W. D. (eds) *Mineral Deposits of Canada: A Synthesis of Major Deposit Types, District Metallogeny, the Evolution of Geological Provinces, and Exploration Methods*. Mineral Deposit Division, Geological Association of Canada, Special Publications, **5**, 755–791.

NELSON, J. L. & FRIEDMAN, R. M. 2004. Superimposed Quesnel (late Paleozoic–Jurassic) and Yukon–Tanana (Devonian–Mississippian) arc assemblages, Cassiar Mountains, northern British Columbia: Field, U–Pb and igneous petrochemical evidence. *Canadian Journal of Earth Sciences*, **41**, 1201–1235.

NELSON, J. L. & GEHRELS, G. E. 2007. Detrital zircon geochronology and provenance of the southeastern Yukon–Tanana terrane. *Canadian Journal of Earth Sciences*, **44**, 297–316.

NELSON, J. L., COLPRON, M., PIERCEY, S. J., DUSEL-BACON, C., MURPHY, D. C. & ROOTS, C. F. 2006. Paleozoic tectonic and metallogenic evolution of the pericratonic terranes in Yukon, northern British Columbia and eastern Alaska. *In*: COLPRON, M. & NELSON, J. L. (eds) *Paleozoic Evolution and Metallogeny of Pericratonic Terranes at the Ancient Pacific Margin of North America, Canadian and Alaskan Cordillera*. Geological Association of Canada, Special Papers, **45**, 323–360.

NILSEN, T. H. 1981. Upper Devonian and Lower Mississippian redbeds, Brooks Range, Alaska. *In*: MIALL, A. D. (ed.) *Sedimentation and Tectonics in Alluvial Basins*. Geological Association of Canada, Special Papers, **23**, 187–219.

NOKLEBERG, W. J., PARFENOV, L. M. ET AL. 2000. *Phanerozoic tectonic evolution of the circum-North Pacific*. US Geological Survey, Professional Papers, **1626**.

NORFORD, B. S. & MIHALYNUK, M. G. 1994. Evidence of the Pacific faunal province in the northern Alexander terrane, recognition of two Middle Ordovician graptolite zones in northern British Columbia. *Canadian Journal of Earth Sciences*, **31**, 1389–1396.

OMMA, J. E., SCOTT, R. A., PEASE, V., EMBRY, A. & MORTON, A. 2007. Sediment provenance of northerly-derived sediment in the Sverdrup Basin, Arctic Canada. *In*: BREKKE, H., HENRIKSEN, S. & HAUKDAL, G. (eds) *Abstract and Proceedings*. Geological Society of Norway, Trondheim, **2**, 115.

PARADIS, S., BAILEY, S. L., CREASER, R. A., PIERCEY, S. J. & SCHIARIZZA, P. 2006. Paleozoic magmatism and syngenetic massive sulphide deposits of the Eagle Bay assemblage, Kootenay terrane, southern British Columbia. *In*: COLPRON, M. & NELSON, J. L. (eds) *Paleozoic Evolution and Metallogeny of Pericratonic Terranes at the Ancient Pacific Margin of North America, Canadian and Alaskan Cordillera*. Geological Association of Canada, Special Papers, **45**, 383–414.

PARRISH, R. R., CARR, S. D. & PARKINSON, D. L. 1988. Eocene extensional tectonics and geochronology of the southern Omineca Belt, British Columbia and Washington, *Tectonics*, **7**, 181–212.

PATRICK, B. E. & McCLELLAND, W. C. 1995. Late Proterozoic granitic magmatism on Seward Peninsula and a Barentian origin for Arctic Alaska–Chukotka. *Geology*, **23**, 81–84.

PEARCE, J. A., LEAT, P. T., BARKER, P. F. & MILLAR, I. L. 2001. Geochemical tracing of Pacific-to-Atlantic upper-mantle flow through the Drake Passage. *Nature*, **410**, 457–461.

PEDDER, A. E. H. 2006. Zoogeographic data from studies of Paleozoic corals of the Alexander terrane, southeastern Alaska and British Columbia. *In*: HAGGART, J. W., ENKIN, R. J. & MONGER, J. W. H. (eds) *Paleogeography of the North American Cordillera: Evidence For and Against Large-Scale Displacements*. Geological Association of Canada, Special Papers, **46**, 29–57.

PIERCEY, S. J. & COLPRON, M. 2006. Geochemistry, Nd–Hf isotopic composition and detrital zircon geochronology of the Snowcap assemblage, Yukon: Insights into the provenance and composition of the basement to Yukon–Tanana terrane. *Geological Society of America, Abstracts with Programs*, **385**, 5.

PIERCEY, S. J., NELSON, J. L., COLPRON, M., DUSEL-BACON, C., ROOTS, C. F. & SIMARD, R.-L. 2006. Paleozoic magmatism and crustal recycling along the ancient Pacific margin of North America, northern Cordillera. *In*: COLPRON, M. & NELSON, J. L. (eds) *Paleozoic Evolution and Metallogeny of Pericratonic*

Terranes at the Ancient Pacific Margin of North America, Canadian and Alaskan Cordillera. Geological Association of Canada, Special Papers, **45**, 281–322.

POHLER, S. M., ORCHARD, M. J. & TEMPELMAN-KLUIT, D. J. 1989. Ordovician conodonts identify the oldest sediments in the Intermontane Belt, Olalla, south–central British Columbia. In: Current Research. Geological Survey of Canada, Paper, **89-1E**, 61–67.

POTTER, A. W., WARTKINS, R., BOUCOT, A. J., FLORY, R. A. & RIGBY, J. K. 1990a. Biogeography of the Upper Ordovician Montgomery Limestone, Shoo Fly Complex, northern Sierra Nevada, California. In: HARWOOD, D. S. & MILLER, M. M. (eds) *Paleozoic and early Mesozoic paleogeographic relations, Sierra Nevada, Klamath Mountains, and related terranes*. Geological Society of America, Special Papers, **255**, 33–42.

POTTER, A. W., BOUCOT, A. J. ET AL. 1990b. Early Paleozoic stratigraphic, paleogeographic and biogeographic relations of the eastern Klamath belt, northern California. In: HARWOOD, D. S. & MILLER, M. M. (eds) *Paleozoic and early Mesozoic paleogeographic relations, Sierra Nevada, Klamath Mountains, and related terranes*. Geological Society of America, Special Papers, **255**, 57–74.

PRICE, R. A. 1994. Cordilleran tectonics and the evolution of the western Canada sedimentary basin. In: MOSSOP, G. & SHESTIN, I. (eds) *Geologic Atlas of the Western Canada Sedimentary Basin*. Alberta Research Council and Canadian Society of Petroleum Geologists, Edmonton, 13–24.

RAINBIRD, R. H., MCNICOLL, V. J., THÉRIAULT, R. J., HEAMAN, L. M., ABBOTT, J. G., LONG, D. G. F. & THORKELSON, D. J. 1997. Pan-continental river system draining the Grenville orogen recorded by U–Pb and Sm–Nd geochronology of Neoproterozoic quartzarenites and mudrocks, northwestern Canada. *Journal of Geology*, **105**, 1–17.

RIGBY, J. K., KARL, S. M., BLODGETT, R. B. & BAICHTAL, J. F. 2005. Ordovician 'sphinctozoan' sponges from Prince of Wales Island, southeastern Alaska. *Journal of Paleontology*, **79**, 862–870.

RILEY, B. C. D., SNYDER, W. S. & GEHRELS, G. E. 2000. U–Pb detrital zircon geochronology of the Golconda allochthon, Nevada. In: SOREGHAN, M. J. & GEHRELS, G. E. (eds) *Paleozoic and Triassic Paleogeography and Tectonics of western Nevada and northern California*. Geological Society of America, Special Papers, **347**, 65–75.

ROBACK, R. C. & WALKER, N. W. 1995. Provenance, detrital zircon U–Pb geochronometry, and tectonic significance of Permian to Lower Triassic sandstone in southeastern Quesnellia, British Columbia and Washington. *Geological Society of America Bulletin*, **107**, 665–675.

ROBACK, R. C., SEVIGNY, J. H. & WALKER, N. W. 1994. Tectonic setting of the Slide Mountain terrane, southern British Columbia. *Tectonics*, **13**, 1242–1258.

ROOT, K. G. 2001. Devonian Antler fold and thrust belt and foreland basin development in the southern Canadian Cordillera: Implications for the Western Canada Sedimentary Basin. *Bulletin of Canadian Petroleum Geology*, **49**, 7–36.

RUBIN, C. M., MILLER, M. M. & SMITH, G. M. 1990. Tectonic development of Cordilleran mid-Paleozoic volcano-plutonic complexes; Evidence for convergent margin tectonism. In: HARWOOD, D. S. & MILLER, M. M. (eds) *Paleozoic and early Mesozoic paleogeographic relations; Sierra Nevada, Klamath Mountains, and related terranes*. Geological Society of America, Special Papers, **255**, 1–16.

SALEEBY, J., HANNAH, J. L. & VARGA, R. J. 1987. Isotopic age constraints on middle Paleozoic deformation in the northern Sierra Nevada, California. *Geology*, **15**, 757–760.

SALEEBY, J. B. 1990. Geochronological and tectonostratigraphic framework of Sierran and Klamath ophiolitic assemblages. In: HARWOOD, D. S. & MILLER, M. M. (eds) *Paleozoic and early Mesozoic paleogeographic relations, Sierra Nevada, Klamath Mountains, and related terranes*. Geological Society of America, Special Papers, **255**, 93–114.

SALEEBY, J. B. 2000. Geochronologic investigations along the Alexander–Taku terrane boundary, southern Revillagigedo Island to Cape Fox areas, southeast Alaska. In: STOWELL, H. H. & MCCLELLAND, W. C. (eds) *Tectonics of the Coast Mountains in southeast Alaska and British Columbia*. Geological Society of America, Special Papers, **343**, 107–143.

SCHELLART, W. P., FREEMAN, J., STEGMAN, D. R. & MAY, D. 2007. Evolution and diversity of subduction zones controlled by slab width. *Nature*, **446**, 308–311.

SCHIARIZZA, P. & PRETO, V. A. 1987. *Geology of the Adams Plateau–Clearwater–Vavenby area*. British Columbia Ministry of Energy, Mines and Petroleum Resources, Paper, **88**.

SCHMIDT, J. M. & ROGERS, R. K. 2007. Metallogeny of the Nikolai large igneous province (LIP) in southern Alaska and its influence on the mineral potential of the Talkeetna Mountains. In: RIDGWAY, K. D., TROP, J. M., GLEN, J. M. G. & O'NEILL, J. M. (eds) *Tectonic Growth of a Collisional Continental Margin: Crustal Evolution of Southern Alaska*. Geological Society of America, Special Papers, **431**, 623–648.

SCHWEICKERT, R. A., HARWOOD, D. S., GIRTY, G. H. & HANSON, R. F. 1984. *Tectonic development of the Northern Sierra terrane: An accreted late Paleozoic island arc and its basement*. Geological Society of America, Guidebook.

SCOTESE, C. R. 2002. Paleomap project. World Wide Web Address: http://www.scotese.com/earth.htm

SHERLOCK, S. C., JONES, K. A. & KELLEY, S. P. 2002. Fingerprinting polyorogenic detritus using $^{40}Ar/^{39}Ar$ ultraviolet laser microprobe. *Geology*. **30**, 515–518.

SIMARD, R.-L., DOSTAL, J. & ROOTS, C. F. 2003. Development of late Paleozoic volcanic arcs in the Canadian Cordillera: An example from the Klinkit Group, northern British Columbia and southern Yukon. *Canadian Journal of Earth Sciences*, **40**, 907–924.

SIMONY, P. S., SEVIGNY, J. H., MORTENSEN, J. K. & ROBACK, R. C. 2006. Age and origin of the Trail Gneiss Complex: Basement to Quesnel terrane near Trail, southeastern British Columbia. In: COLPRON, M. & NELSON, J. L. (eds) *Paleozoic evolution and metallogeny of pericratonic terranes at the ancient Pacific margin of North America, Canadian and Alaskan Cordillera*. Geological Association of Canada, Special Papers, **45**, 505–515.

SMITH, M. T. & GEHRELS, G. E. 1991. Detrital zircon geochronology of Upper Proterozoic to Lower Paleozoic continental margin strata of the Kootenay arc: Implications for the Early Paleozoic tectonic development of the eastern Canadian Cordillera. *Canadian Journal of Earth Sciences*, **28**, 1271–1284.

SMITH, M. T., DICKINSON, W. R. & GEHRELS, G. E. 1993. Contractional nature of Devonian–Mississippian Antler tectonism along the North American continental margin. *Geology*, **21**, 21–24.

SOJA, C. M. & ANTOSHKINA, A. I. 1997. Coeval development of Silurian stromatolite reefs in Alaska and the Ural Mountains: Implications for paleogeography of the Alexander terrane. *Geology*, **25**, 539–542.

SOJA, C. M. & KRUTIKOV, L. 2008. Provenance, depositional setting, and tectonic implications of Silurian polymictic conglomerate in Alaska's Alexander terrane. *In*: BLODGETT, R. B. & STANLEY, G. D. JR. (eds) *The Terrane Puzzle: New Perspectives on Paleontology and Stratigraphy from the North American Cordillera*. Geological Society of America, Special Papers, **442**, 63–75.

STERN, R. J. 2004. Subduction intitiation: Spontaneous and induced. *Earth and Planetary Science Letters*, **226**, 275–292.

STEVENS, C. H. 1995. A giant parafusulina from east–central Alaska with comparisons to all giant fusulinids in western North America. *Journal of Paleontology*, **69**, 802–812.

STEVENS, C. H. & RYCERSKI, B. 1989. Early Permian colonial rugose corals from the Stikine River area, British Columbia, Canada. *Journal of Paleontology*, **63**, 158–181.

THOMPSON, R. I., GLOMBICK, P., ERDMER, P., HEAMAN, L. M., LEMIEUX, Y. & DAUGHTRY, K. L. 2006. Evolution of the ancestral Pacific margin, southern Canadian Cordillera: Insights from new geologic maps. *In*: COLPRON, M. & NELSON, J. L. (eds) *Paleozoic Evolution and Metallogeny of Pericratonic Terranes at the Ancient Pacific Margin of North America, Canadian and Alaska Cordillera*. Geological Association of Canada, Special Papers, **45**, 433–482.

TRETTIN, H. P. 1991. The Proterozoic to Late Silurian record of Pearya. *In*: TRETTIN, H. P. (ed.) *Geology of the Innuitian Orogen and Arctic Platform of Canada and Greenland*. Geological Survey of Canada, Geology of Canada, **3**, 239–261.

TRETTIN, H. P., MAYR, U., LONG, G. D. F. & PACKARD, J. J. 1991a. Cambrian to Early Devonian basin development, sedimentation and volcanism, Arctic Islands. *In*: TRETTIN, H. P. (eds) *Geology of the Innuitian Orogen and Arctic Platform of Canada and Greenland*. Geological Survey of Canada, Geology of Canada, **3**, 165–238.

TRETTIN, H. P., OKULITCH, A. V. ET AL. 1991b. Silurian–Early Carboniferous deformational phases and associated metamorphism and plutonism, Arctic Islands. *In*: TRETTIN, H. P. (ed.) *Geology of the Innuitian Orogen and Arctic Platform of Canada and Greenland*. Geological Survey of Canada, Geology of Canada, **3**, 293–341.

VAN DER HEYDEN, P. 1992. A Middle Jurassic to Early Tertiary Andean–Sierran arc model for the Coast Belt of British Columbia. *Tectonics*, **11**, 82–97.

VAN SCHMUS, W. R., BICKFORD, M. E. ET AL. 1993. Transcontinental Proterozoic provinces. *In*: REED, J. C. JR, BICKFORD, M. E., HOUSTON, R. S., LINK, P. K., RANKIN, D. W., SIMS, P. K. & VAN SCHMUS, W. R. (eds) *Precambrian Conterminous US*. Geological Society of America, The Geology of North America, **C-2**, 171–334.

VAN STAAL, C. R. 2007. Pre-Carboniferous tectonic evolution and metallogeny of the Canadian Appalachians. *In*: GOODFELLOW, W. D. (ed.) *Mineral Deposits of Canada: A Synthesis of Major Deposit Types, District Metallogeny, the Evolution of Geological Provinces and Exploration Methods*. Geological Association of Canada, Mineral Deposits Division, Special Publications, **5**, 793–818.

VAN STAAL, C. R., DEWEY, J. F., MAC NIOCAILL, C. & MCKERROW, W. S. 1998. The Cambrian–Silurian tectonic evolution of the northern Appalachians and British Caledonides: History of a complex, west and southwest Pacific-type segment of Iapetus. *In*: BLUNDELL, D. J. & SCOTT, A. C. (eds) *Lyell: the Past is the Key to the Present*. Geological Society, London, Special Publications, **143**, 199–242.

VILLENEUVE, M. E., ROSS, G. M., THERIAULT, R. J., MILES, W., PARRISH, R. R. & BROOME, J. 1993. *Tectonic subdivision and U–Pb geochronology of the crystalline basement of the Alberta basin, western Canada*. Geological Survey of Canada, Bulletin, **447**.

WAGGONER, B. 1999. Biogeographic analyses of the Ediacaran biota: A conflict with paleotectonic reconstructions. *Paleobiology*, **25**, 440–458.

WALLIN, E. T., NOTO, R. C. & GEHRELS, G. E. 2000. Provenance of the Antelope Mountain Quartzite, Yreka terrane, California: Evidence for large-scale late Paleozoic sinistral displacement along the North American Cordilleran margin and implications for the mid-Paleozoic fringing arc model. *In*: SOREGHAN, M. J. & GEHRELS, G. E. (eds) *Paleozoic and Triassic Paleogeography and Tectonics of Western Nevada and Northern California*. Geological Society of America, Special Papers, **347**, 119–132.

WHEELER, J. O., BROOKFIELD, A. J., GABRIELSE, H., MONGER, J. W. H., TIPPER, H. W. & WOODSWORTH, G. J. 1991. *Terrane map of the Canadian Cordillera, 1:2 000 000*. Geological Survey of Canada, Map **1713A**.

WRIGHT, J. E. & WYLD, S. J. 2006. Gondwanan, Iapetan, Cordilleran interactions: A geodynamic model for the Paleozoic tectonic evolution of the North American Cordillera. *In*: HAGGART, J. W., ENKIN, R. J. & MONGER, J. W. H. (eds) *Paleogeography of the North American Cordillera: Evidence For and Against Large-Scale Displacements*. Geological Association of Canada, Special Papers, **46**, 377–408.

YORATH, C. J., SUTHERLAND-BROWN, A. & MASSEY, N. W. D. 1999. *Lithoprobe, southern Vancouver Island, British Columbia: Geology*. Geological Survey of Canada, Bulletin, **498**.

Arc imbrication during thick-skinned collision within the northern Cordilleran accretionary orogen, Yukon, Canada

A. M. TIZZARD[1], S. T. JOHNSTON[1]* & L. M. HEAMAN[2]

[1]*School of Earth and Ocean Sciences, University of Victoria, PO Box 3055 STN CSC, Victoria, British Columbia, Canada V8W 3P6*

[2]*Department of Earth and Atmospheric Sciences, University of Alberta, Edmonton, Alberta, Canada T6G 2E3*

Corresponding author (e-mail: stj@uvic.ca)

Abstract: We present the results of geological mapping and geochronological studies of the Tally Ho shear zone (THSZ) and adjacent rocks. The shear zone crops out near the west margin of Stikinia, an oceanic arc and the largest of the accreted terranes within the Cordilleran orogen of western North America. The hanging wall of the largely flat-lying shear zone consists of coarsely crystalline leucogabbro and cumulate pyroxenite interpreted as the lower crustal and possibly lithospheric mantle roots of a magmatic arc. Rocks in the footwall consist of volcanic and volcano-sedimentary sequences of the Lewes River Arc, a Late Triassic magmatic arc characteristic of Stikinia. Because the shear zone places lower crustal plutonic rocks over a supracrustal sequence, we interpret it as a crustal-scale thrust fault. Kinematic indicators imply top-to-the-east displacement across the shear zone. The geometry of folds of the shear zone is consistent with deformation in response to displacement over ramps in deeper-seated thrust faults kinematically linked to the THSZ. Crystallization of the hanging-wall leucogabbro at 208 ± 4.3 Ma provides a maximum age constraint for deformation, whereas a post-kinematic granitoid pluton that plugs the shear zone and that crystallized at about 173 Ma provides a lower age limit. The THSZ is, therefore, coeval with: (1) a series of latest Triassic–Early Jurassic shear and fault zones that characterize the length of the west margin of Stikinia; (2) the termination of isotopically juvenile arc magmatism of the Lewes River Arc; (3) crustal loading of Stikinia giving rise to a foreland basin that rapidly filled with westerly derived orogenic molasse that includes clasts of ultrahigh-pressure metamorphic rocks; and (4) juxtaposition of Stikinia against continental crust of the Nisling Assemblage of the Yukon–Tanana terrane to the west. These constraints are consistent with a model of deformation in response to the entry of the continental Nisling Assemblage into the trench of the west-facing Lewes River Arc, terminating subduction and imbricating the arc along a series of east-verging thrust faults, including the THSZ.

The Canadian Cordillera orogen, consisting of a mosaic of oceanic and pericratonic terranes, has developed in the absence of any terminal continental collision and is considered to be a classic accretionary orogen. As such, the Cordillera provides us with an opportunity to determine the processes involved in the development of accretionary orogens, and to better understand the role of crustal-scale structures therein. Delineating and distinguishing between terrane bounding faults and structures internal to a terrane can, however, be complicated. It is rarely clear where one terrane ends and another begins, especially where similar oceanic or arc terranes have been juxtaposed. Post-accretionary dispersion of terranes by margin-parallel translations can 'shuffle the deck', and syn- to post-accretionary magmatism commonly obscures both structure and stratigraphy. None the less, understanding how strain is partitioned between and within accreting crustal blocks provides a first-order constraint on the tectonic processes responsible for accretionary orogens.

In the northern Canadian Cordillera the Tally Ho shear zone (THSZ) crops out near the northwestern margin of Stikinia, a large accreted oceanic arc terrane (Fig. 1) (Hart & Radloff 1990; Tizzard & Johnston 2005). The THSZ is a 40 km long, NW–SE-striking package of highly strained rocks just north of the Yukon–British Columbia border. West of the THSZ is the pericratonic Nisling Assemblage of the Yukon–Tanana terrane (Fig. 2; Mortensen 1992). To the north the THSZ is covered by Neogene basalt flows, and to the south it is truncated by younger intrusive rocks. The tectonic location of the THSZ near the western margin of an accreted terrane (Stikinia) offers an example of

From: CAWOOD, P. A. & KRÖNER, A. (eds) *Earth Accretionary Systems in Space and Time.*
The Geological Society, London, Special Publications, **318**, 309–327.
DOI: 10.1144/SP318.11 0305-8719/09/$15.00 © The Geological Society of London 2009.

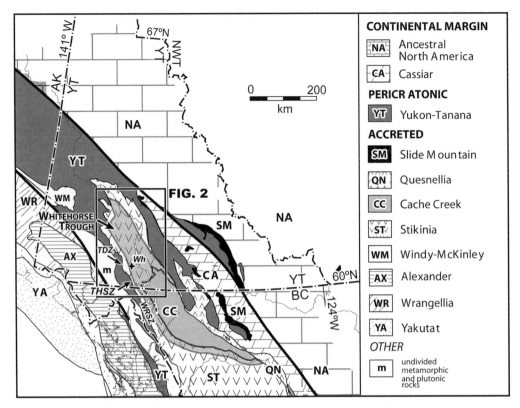

Fig. 1. Simplified terrane map of the northern Canadian Cordillera (modified after Colpron *et al.* 2006). THSZ, Tally Ho shear zone; TDZ, Takhini deformation zone; WRSZ, Wann River shear zone; Wh, Whitehorse.

a terrane-marginal structure and may therefore provide insight into the lithospheric-scale geometry of terranes and the nature of deformation along their bounding structures. Characterizing the structural style of the THSZ and applying geochronological constraints on the timing of deformation should aid in developing an accretionary history for part of the northern Canadian Cordillera. For example, is the THSZ the structure along which the oceanic Stikinia and continental Nisling Assemblage were juxtaposed? Alternatively, is the THSZ a secondary structure developed during collision, or is it a post-collisional or reactivated feature? Answering these questions may aid in improving our general understanding of terrane bounding structures.

This paper reports the results of detailed geological mapping and U–Pb geochronology of the THSZ, and aims to: (1) establish the geometry and kinematics of the high-strain zone; (2) constrain the timing of deformation; (3) compare the THSZ with other structures along the northwestern margin of Stikinia; and (4) examine the significance of the THSZ in a regional context. We first describe Stikinia and the Nisling Assemblage and introduce the main structures along and adjacent to the boundary between these terranes. We then describe the lithologies and structures within and adjacent to the THSZ. Lastly we provide geochronological constraints on the deformation in the THSZ. A tectonic model for the evolution of the western margin of Stikinia is presented.

Regional geology

Stikinia

In Yukon, western Stikinia consists of the Upper Palaeozoic Takhini assemblage and the Upper Triassic to Lower Jurassic Lewes River and Laberge groups of the Whitehorse Trough (Fig. 3; Hart 1997). The Takhini assemblage consists of variably deformed and metamorphosed mafic volcanic rocks and minor felsic and sedimentary rocks. The Lewes River Group consists of a lower volcanic succession (the Povoas formation) and an upper sedimentary succession (the Aksala formation). In northern British Columbia the Stuhini formation is correlative with the Povoas formation and together they form the Lewes River Arc (Hart 1997). The Whitehorse Trough is a Late Triassic to Middle Jurassic

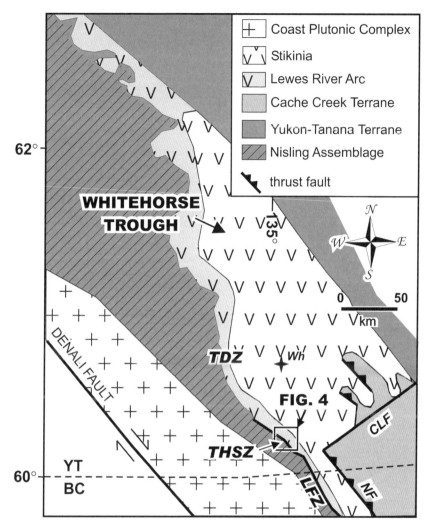

Fig. 2. Simplified tectonic map of central southern Yukon showing location of mapping carried out in this study. THSZ, Tally Ho shear zone; TDZ, Takhini deformation zone; LFZ, Llewellyn fault zone; NF, Nahlin fault; CLF, Crag Lake fault; Wh, Whitehorse. Modified from Hart (1997).

arc-marginal basin (Eisbacher 1981) in which volcanic and sedimentary rock of the Aksala Formation and Laberge Group accumulated. Based on detailed stratigraphy of the Laberge Group, Dickie & Hein (1995) interpreted the Whitehorse Trough as an east-facing (present-day coordinates) forearc basin that developed above a west-dipping subduction zone. As we demonstrate below, the trough is a composite basin with a much more complicated history than previously envisaged.

The NW margin of Stikinia is characterized by a series of high-strain, shear and fault zones. These include, from north to south, the Takhini deformation zone, the THSZ (the subject of this paper), the Llewellyn fault zone, and the Wann River shear zone (Figs 1 and 2). The Takhini deformation zone lies 60 km along strike to the NW of the THSZ and consists of variably strained schist and phyllite with minor tuff, rhyolite, marble and greenstone of the Mississippian Takhini assemblage. The deformation zone is intruded by unstrained granitic rocks of the Jurassic and Cretaceous Little River (183 ± 2 Ma) and Annie Ned batholiths, respectively (Hart 1997). The Llewellyn fault zone is a NW–SE-striking brittle to brittle–ductile dextral strike-slip to dip-slip fault system that parallels the THSZ in southern Yukon and extends southeastwards into northern British Columbia (Hart & Radloff 1990; Fig. 2). The Llewellyn fault zone in southern Yukon records dextral movement and is

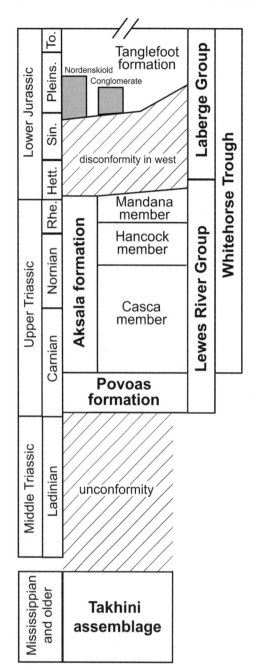

Fig. 3. Stratigraphy of the western margin of Stikinia in southern Yukon. Names discussed in text are in bold. Modified from Hart (1997).

located along the eastern margin of the THSZ. The fault zone affects rocks as young as Eocene, although it may be in part a reactivation of an older fault zone (Mihalynuk & Rouse 1988). The Wann River shear zone is subvertical to subhorizontal, varies in strike from east–west to north–south, places Nisling Assemblage schist over Stikinia, and was active between 185 Ma (the crystallization age of a deformed pluton adjacent to the shear zone) and c. 170 Ma (U–Pb cooling ages on titanite and rutile) (Currie & Parrish 1993).

Nisling Assemblage

Rocks of the Nisling Assemblage occur as pendants and blocks within the Coast Plutonic Complex, west of the THSZ and Takhini assemblage (Fig. 2; Hart & Radloff 1990; Hart 1997). The Nisling Assemblage consists of regionally deformed and metamorphosed Palaeozoic mica–quartz–feldspar schist, quartzite, marble and orthogneiss, and is interpreted as part of the Yukon–Tanana terrane, a pericratonic continental assemblage (Gehrels et al. 1990; Hart & Radloff 1990; Mortensen 1992). A U–Pb zircon age determination of meta-igneous rocks sampled along-strike 100 km to the north within the correlative Aishihik metamorphic suite provides a Mississippian age constraint for the Nisling Assemblage (Johnston et al. 1996).

Lithological units

The map area is divisible into three structural sequences: a high-strain central sequence (the THSZ) separates little strained rocks of an overlying hanging-wall sheet from an underlying footwall sequence. A layered sequence of variably deformed gabbro and pyroxenite forms the hanging wall of the THSZ. The THSZ consists of mylonitic to schistose rocks, metabasalt, and minor marble. Rock units in the footwall are variably deformed and include augite-phenocrystic basalt, volcaniclastic, and sedimentary rocks. Megacrystic granite, granodiorite, quartz diorite and numerous sub-volcanic rhyolite dykes are in intrusive or fault contact with the hanging-wall, THSZ and footwall rocks (Figs 4–6).

Hanging-wall lithological units

Leucogabbro. The leucogabbro unit consists of leucogabbro with rare lenses of pyroxenite (Fig. 5). The leucogabbro is medium- to very coarse-grained and consists of plagioclase, amphibole, clinopyroxene and quartz with minor accessory magnetite. The occurrence of chlorite and amphibole replacing pyroxene suggests that the gabbro has been metamorphosed to at least greenschist facies. Chlorite and sericite alteration is common. The leucogabbro is weakly to well-foliated and weakly to non-lineated. The foliation planes are variably anastamosing and are defined by the

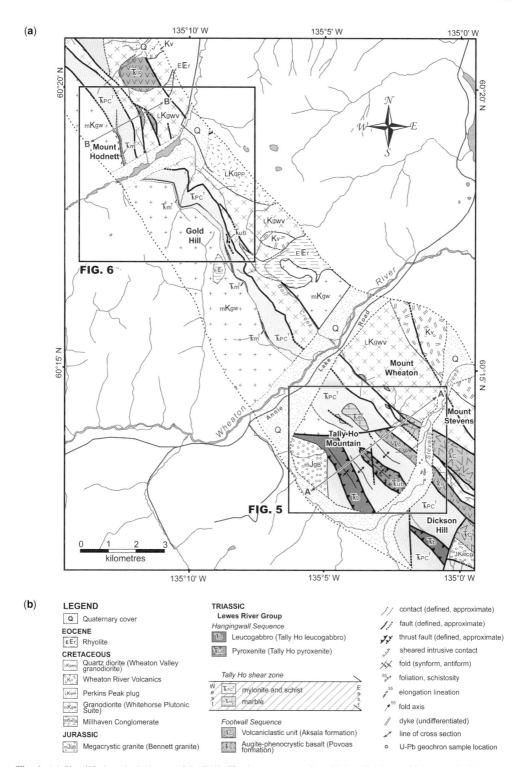

Fig. 4. (a) Simplified geological map of the Tally Ho shear zone, southern Yukon Territory. (b) Legend for Figures 4–6. Names in parenthesis refer to unit names of Hart & Radloff (1990).

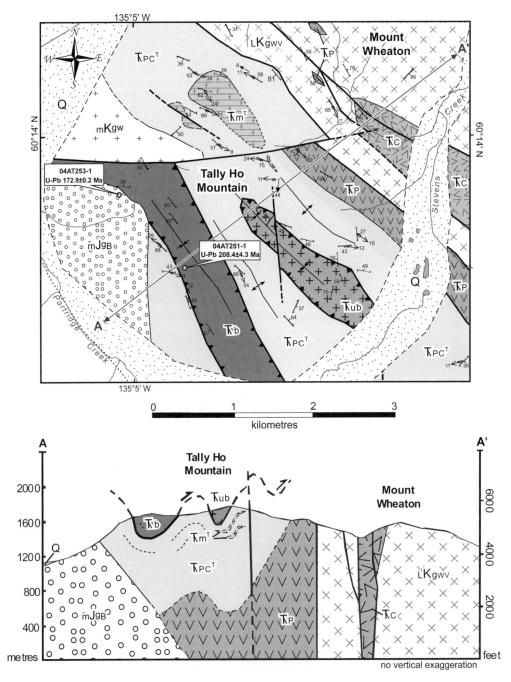

Fig. 5. Geological map and cross-section of the Tally Ho shear zone showing locations of U–Pb geochron samples in this study. Legend as in Figure 4b.

preferential orientation of amphibole and pyroxene. Locally, elongate pyroxene–amphibole aggregates that lie in the foliation plane define a stretching lineation.

Lenses of pyroxenite up to 1 m long have irregular and diffuse boundaries (Fig. 7a), are interpreted to be the result of magmatic processes, and therefore provide a temporal link between the leucogabbro

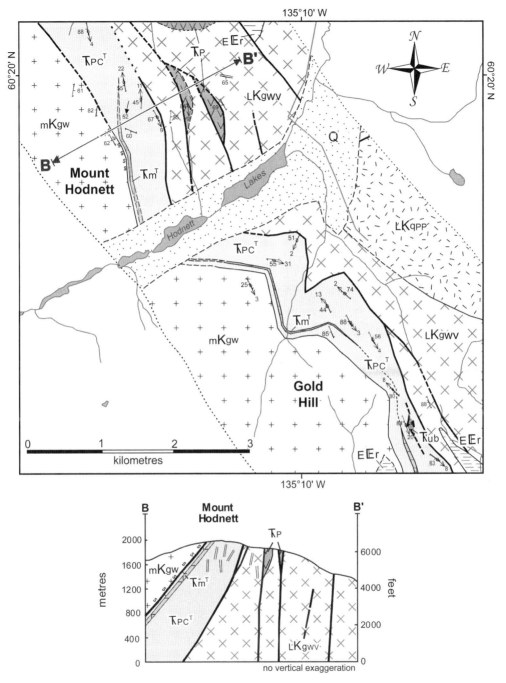

Fig. 6. Geological map and cross-section of the Mount Hodnett and Gold Hill areas, Tally Ho shear zone. Legend as in Figure 4b. Modified from Tizzard & Johnston (2005).

and pyroxenite units (described below). A sharp contact exists between the leucogabbro and a megacrystic granite, with xenoliths of gabbro found in the megacrystic granite (Fig. 7b). The gabbro is, therefore, older than and has been intruded by the megacrystic granite. A contact with augite-porphyroclastic schist of the THSZ structural unit is covered. The contact is, however, interpreted as

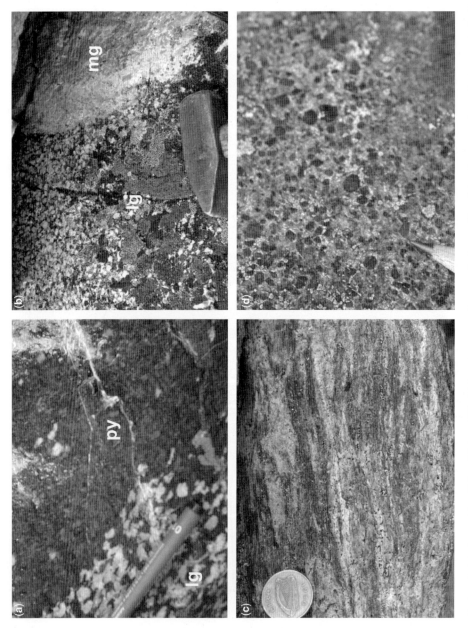

Fig. 7. (a) Pyroxenite lens (py) in leucogabbro (lg) on Dickson Hill; (b) intrusive contact between the leucogabbro (lg) and megacrystic granite (mg) on Tally Ho Mountain; (c) sheared pyroxenite near contact with THSZ on Tally Ho Mountain; (d) augite-phenocrystic basalt adjacent to the Tally Ho shear zone on Dickson Hill.

gradational; within 50 m of the contact the leucogabbro is characterized by a weak foliation that increases in intensity toward the contact with the THSZ. The increasing strain toward the contact implies that the leucogabbro was caught up in the deformation that gave rise to the THSZ. The leucogabbro unit in the map area is lithologically similar to and continuous with the Tally Ho leucogabbro (Hart & Radloff 1990). A U–Pb zircon age date of 215 ± 1 Ma for the Tally Ho leucogabbro indicates crystallization in the Late Triassic (Hart et al. 1995).

Pyroxenite. The pyroxenite unit consists of black to dark green pyroxenite and crops out along a blocky ridgeline on top of Tally Ho Mountain and near Gold Hill (Fig. 4). The pyroxenite is medium- to coarse-grained and dominantly consists of clinopyroxene with minor amounts of orthopyroxene, olivine and opaque minerals. Serpentine-rich layers 5–10 cm thick in the pyroxenite are interpreted as primary olivine-rich cumulate in which serpentine has subsequently replaced olivine. The presence of serpentine and chlorite indicates that the pyroxenite has been metamorphosed to at least greenschist facies. The pyroxenite is massive to well-foliated with foliation planes defined by the preferential orientation of pyroxene and chlorite. Lineations are poorly developed. To the west, the contact of the pyroxenite with the leucogabbro is covered. Lenses of pyroxenite in the leucogabbro are inferred to indicate that the units are coeval. To the east the pyroxenite is increasingly strained and forms a gradational contact with mylonite of the THSZ structural unit (Fig. 7c). The transition zone from highly strained pyroxenite to rocks of the THSZ varies in width from 1 to 50 cm.

The leucogabbro and pyroxenite are interpreted as being a coeval, cumulate-style magmatic body based on the presence of large pyroxenite lenses in the leucogabbro unit. Cumulate processes in the pyroxenite are also indicated by serpentinized cumulate olivine layers. The coarse crystallinity and petrology of the pyroxenite and leucogabbro units are consist with their having originated within the deep crust and potentially lithospheric mantle part of an arc system (Hamilton 1988); hence the pyroxenite and leucogabbro units are interpreted as part of the roots of a volcanic arc, probably the Lewes River.

Footwall lithological units

Augite-phenocrystic basalt. East of the THSZ is a massive augite-phenocrystic to aphyric basalt unit (Fig. 4). The basalt consists of clinopyroxene, plagioclase and minor orthopyroxene and amphibole. Augite phenocrysts are up to 10 mm in diameter and the aphanitic matrix is variably epidotized, choloritized and sericitized (Fig. 7d).

A granodiorite pluton intrudes the basalt in the northern map area. Elsewhere, brittle faults form the contact between the basalt and granodiorite. To the west, the basalt becomes progressively strained over several hundred metres, forming a gradational contact with metabasaltic mylonite and schist of the THSZ. The augite-phenocrystic basalt unit is lithologically similar to and locally contiguous with basalt mapped as the Upper Triassic Povoas formation (Hart & Radloff 1990).

Volcaniclastic unit. The volcaniclastic unit includes tuff, agglomerate, greywacke, pebble conglomerate and minor shale. The dominant lithology is fine- to coarse-grained tuff that contains both crystal and lithic fragments. Crystal fragments of plagioclase and pyroxene are common; lithic fragments are angular to sub-angular, up to cobble size, and are dominantly fine-grained and mafic in composition. Epidote, chlorite and sericite are common alteration minerals, and characterize the volcaniclastic rocks adjacent to NW–SE-striking brittle faults. The volcaniclastic unit is well-bedded to moderately well-foliated with foliation planes defined by an anastomosing cleavage. On Mount Wheaton, xenoliths that are of a similar composition to the volcaniclastic unit are found in quartz diorite (Fig. 5). The volcaniclastic unit is therefore interpreted to be older than and has been intruded by quartz diorite. Elsewhere younger brittle faults juxtapose the volcaniclastic unit against quartz diorite and schist of the THSZ.

The volcaniclastic unit is lithologically similar to and locally continuous with the Upper Triassic Casca member of the Aksala formation (Hart & Radloff 1990; Fig. 3). The volcaniclastic and augite-porphyritic basalt units in the footwall of the THSZ are therefore interpreted as a part of the supracrustal sequence of Stikinia.

Tally Ho Shear Zone

The THSZ is a structural unit (Figs 8 and 9) that separates the hanging-wall Tally Ho pyroxenite and leucogabbro units from the footwall augite-porphyroclastic basalt and volcaniclastic units. The THSZ is composed of mylonite and augite-porphyroclastic schist with minor phyllite, foliated metabasalt and marble. The mylonite is laminated, very fine- to fine-grained and consists of quartz, amphibole, feldspar and minor clinopyroxene, biotite and opaque minerals. The augite-porphyroclastic schist is fine- to medium-grained and consists of clinopyroxene, plagioclase and minor amphibole and opaque minerals. Augite-porphyroclasts are up to 8 mm in diameter (Fig. 8a). The mylonite and the augite-porphyroclastic schist are moderately epidotized,

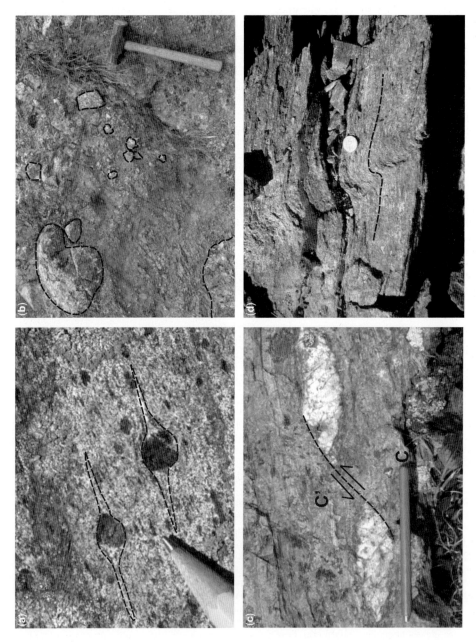

Fig. 8. (a) Rotated augite porphyroclasts in augite schist on Dickson Hill indicating a sinistral and top-to-the-east vorticity; (b) zone of fault gouge on the south face of Mount Hodnett containing clasts of the granodiorite unit; larger clasts are outlined; (c) C′ extensional shears in mylonite on Tally Ho Mountain indicating top-down and top-to-the-east movement; (d) east-vergent crenulations in mylonite on Gold Hill.

sericitized and chloritized. The abundance of chlorite in the fine-grained matrix of the mylonite and schist is interpreted to indicate that the THSZ structural unit has been metamorphosed to at least greenschist facies. To the west, the THSZ structural unit grades up-section into strained and foliated pyroxenite and leucogabbro that in turn decrease in strain away from the mylonite and augite-porphyroclastic schist over a distance of 0.5–50 m. The THSZ decreases in strain over 50 m to the east where it grades into augite-phenocrystic basalt and volcaniclastic rocks down-section into the footwall units. Locally THSZ mylonite is faulted against granodiorite along younger steeply dipping brittle faults (Fig. 4).

Younger intrusions

Megacrystic granite unit. Four main intrusive units that cross-cut and therefore post-date the THSZ are recognized and include, from west to east (Fig. 4): a megacrystic granite, a granodiorite, a quartz diorite, and a sub-volcanic rhyolite. The megacrystic granite is leucocratic, coarse-grained, and massive, and consists of quartz, plagioclase, orthoclase and minor biotite, hornblende and magnetite; alkali-feldspar megacrysts are up to 2 cm long. Sericite and chlorite alteration is locally evident, but not pervasive. Xenoliths in the megacrystic granite of a medium-grained gabbro and a sharp contact with the leucogabbro unit indicate that the megacrystic granite is younger than and intruded the leucogabbro (Fig. 7b). West of the map area the megacrystic granite intrudes rocks of the Nisling Assemblage of the Yukon–Tanana terrane (Hart & Radloff 1990). The megacrystic granite unit is lithologically similar to and locally continuous with the Bennett granite (Hart & Pelletier 1989).

Granodiorite unit. Medium- to coarse-grained leucocratic granodiorite consists of quartz, plagioclase and minor orthoclase and biotite 'books' up to 8 mm in diameter. Epidote, chlorite and sericite alteration is locally developed, although not pervasively. A foliation in the granodiorite, defined by an anastomosing cleavage, is present adjacent to and is associated with a fault that juxtaposes the granodiorite against the THSZ on Mount Hodnett (Figs 5 and 8b). Fine-grained and foliated xenoliths up to 50 cm in width that are texturally similar to and interpreted as mylonitic rocks of the THSZ are found within the granodiorite in the Gold Hill area. The granodiorite unit is lithologically similar to and locally continuous with granodiorite of the Whitehorse plutonic suite (Hart & Radloff 1990). A U–Pb zircon age date of 119 Ma from Whitehorse plutonic suite granodiorite (Doherty & Hart 1988) is interpreted as the age of crystallization of the granodiorite unit.

Quartz diorite unit. The quartz diorite unit underlies much of the eastern portion of the map area (Fig. 4). Leucocratic quartz diorite is the most common lithology in this unit; granodiorite is a minor phase. The quartz diorite is coarse-grained and massive and consists of quartz, plagioclase and minor orthoclase, hornblende and, less commonly, biotite. Chlorite and sericite alteration is common, and is pervasive adjacent to brittle faults. Locally, the quartz diorite unit contains fine-grained, mafic xenoliths. A foliation defined by the weak alignment of hornblende and biotite is commonly developed adjacent to NW–SE-striking brittle faults. To the west, the quartz diorite unit is in fault contact with the THSZ and augite-phenocrystic basalt unit. The quartz diorite unit in the map area is lithologically similar to and locally continuous with the Wheaton Valley granodiorite (Hart & Radloff 1990). The quartz diorite unit is interpreted to be a Late Cretaceous intrusion based on a U–Pb zircon date of 77.1 ± 0.7 Ma (Hart 1995).

Rhyolite unit. Numerous rhyolite dykes and sills cross-cut all lithologies (Fig. 4). The dykes and sills, previously mapped by Hart & Radloff (1990), range in composition from feldspar-porphyritic rhyolite to aphanitic rhyolite and are 0.1 m to several metres in width. In the Gold Hill area, feldspar-porphyritic rhyolite forms a map-scale sill that plugs a NW–SE-striking fault. In the Mount Hodnett area rhyolite dykes are cut by a NW–SE-trending fault. K–Ar age determinations have been interpreted as indicating a Palaeocene to Early Eocene age for the dykes (Hart 1995; Breitsprecher & Mortensen 2004).

Geochronology

One sample of each of the leucogabbro and megacrystic granite units were collected in the southwestern map area for U–Pb zircon geochronology to constrain the age of the THSZ (Fig. 5). The leucogabbro forms the hanging wall of, and passes into mylonites of the THSZ along a gradational contact. It is therefore inferred to have crystallized prior to the formation of the THSZ. The massive unstrained megacrystic granite intrudes the leucogabbro and mylonite of the THSZ and is inferred to post-date the deformation that gave rise to the THSZ. U–Pb geochronology can therefore provide us with the maximum age of formation of the THSZ (age of crystallization of the leucogabbro) and a minimum age (age of crystallization of the granite).

Methods

U–Pb geochron samples of the leucogabbro and megacrystic granite were processed at the Radiogenic Isotope Facility at the University of Alberta. The leucogabbro sample was very coarse-grained and weakly foliated. The megacrystic granite sample was massive, coarse-grained and alkali-feldspar megacrystic. Zircon was isolated from each 15–25 kg sample using conventional Wilfley table, heavy-liquid and magnetic separation techniques (Heaman & Parrish 1991). Fractions for analysis were differentiated by grouping zircons with similar sizes, shapes, colour and clarity. Geochemical procedures followed that of Heaman et al. (2002). Concordia diagrams were generated using Isoplot 3.00 (Ludwig 1998) and all errors are given at the 2σ level. The mean square of weighted derivatives (MSWD) for the discordia is provided as a measure of the quality of fit of the regression line between zircon fractions. Results for the leucogabbro and megacrystic granite analyses are presented in Table 1.

Leucogabbro

Three fractions of zircons were isolated from the leucogabbro. Each analysed fraction consisted of 3–17 tan-coloured zircon fragments that had no obvious zoning or inclusions. The data points lie below the concordia curve between 215 and 221 Ma and form a linear array (Fig. 10a). A three-point regression line through the three fractions yielded an upper intercept age of 520 ± 140 Ma and a lower intercept age of 208.4 ± 4.3 Ma (MSWD = 0.034). The lower intercept of the discordia probably represents the age of crystallization of the leucogabbro. A Cambrian upper intercept of the discordia suggests some component of inherited zircon in the leucogabbro.

Megacrystic granite

Three fractions of zircon were isolated and analysed from the megacrystic granite. Each fraction consisted of 19–50 colourless prismatic zircons with crystal aspect ratios of 3:1. The zircons had no obvious zoning although colourless bubble-shaped inclusions were present in a small proportion of the grains. The data points lie below the concordia curve between 173 and 183 Ma and do not form a linear array (Fig. 10b). The discordance in fraction 2 may be due in part to a component of inheritance in the granite; this fraction was characterized by an elevated level of total common lead (Table 1). The combination of only a few analysed fractions, potential inheritance, and an excess in total common lead in fraction 2 prevents precise determination of the age of crystallization of the megacrystic granite. Fraction 3 was concordant yielding a $^{206}Pb/^{238}U$ age of 172.9 ± 0.2 Ma that is comparable with a two-point regression through fractions 2 (least concordant) and 3 (most concordant) that yielded a lower intercept of 172.8 ± 0.5 Ma. The same regression yielded an upper intercept of 1809 ± 100 Ma. The combination of the $^{206}Pb/^{238}U$ age of fraction 3 and the lower intercept of the fractions 3 and 2 discordia line probably provides the age range in which the granite crystallized (172.9 ± 0.2 to 172.8 ± 0.5 Ma).

Structure

THSZ

The THSZ consists of flaggy mylonite with a strong planar fabric, blastomylonite and schist. A well-developed lineation is common. The mylonitic fabric is cut by brittle and brittle–ductile faults and shear zones. The strong flaggy foliation is referred to as S_1 because it is the oldest observable planar fabric within the THSZ. S_1 is defined by preferential alignment of minerals, including amphiboles, by a planar parting along which the rock readily breaks, by a finely developed lamination, and by compositional laminae, including quartz–feldspar layers. The S_1 fabric parallels the margins of the THSZ, striking NW–SE (Fig. 9a). The well-developed lineation is defined by the alignment of elongate aggregates of amphibole and is interpreted as a stretching lineation (L_1). L_1 plunges shallowly to the NW and SE (Fig. 9b).

Kinematic indicators in the THSZ include rotated porphyroclasts, C–C′ extensional shear bands and rare S–C fabric (Fig. 8). On Tally Ho Mountain and Dickson Hill the mylonites yield top-to-the-SE ($n = 6$) and top to-the-east ($n = 7$) kinematic indicators. A small number of kinematic indicators on Tally Ho Mountain and Dickson Hill ($n = 2$), observed along steeply dipping portions of the THSZ, yield sinistral indicators. Mount Hodnett is similarly dominated by top to-the-SE kinematic indicators ($n = 5$); however, a small number of sinistral and dextral kinematic indicators were observed on steeply dipping portions of the THSZ ($n = 2$ and $n = 1$, respectively). C–C′ extensional shear bands were the most commonly observed kinematic indicator in the THSZ ($n = 14$) followed in frequency by rotated augite porphyroclasts ($n = 10$; Fig. 8c). The SE–NW trend of the L_1 stretching lineation and the dominance of top-to-the-east and top-to-the-SE kinematic indicators suggest that the THSZ mylonites developed during and provide a record of top-to-the-east to -SE shear. Formation of the THSZ is limited to have occurred between about 208 Ma, the age of the hanging-wall leucogabbro, and

Table 1. *U–Pb zircon results for Yukon samples 04AT251-1 and 04AT253-1*

Description	Weight (mg)	U (ppm)	Th (ppm)	Pb (ppm)	Th/U	TCPb (pg)	$^{206}Pb/^{204}Pb$	$^{206}Pb/^{238}U$	$^{207}Pb/^{235}U$	$^{207}Pb/^{206}Pb$	Model ages (Ma) $^{206}Pb/^{238}U$	Model ages (Ma) $^{207}Pb/^{235}U$	Model ages (Ma) $^{207}Pb/^{206}Pb$	%Disc.
04AT251-1 (leucogabbro)														
1z, tan fragments NM@1.2A (10)	24.6	2280.7	747.3	77.3	0.328	7	16640	0.03390 ± 4	0.23683 ± 30	0.05066 ± 2	214.9 ± 0.3	215.3 ± 0.2	225.5 ± 0.9	4.8
2z, tan fragments NM@1.2A (17)	34.8	901.1	353.1	31.7	0.392	9	7420	0.03445 ± 4	0.24137 ± 33	0.05085 ± 3	218.2 ± 0.3	219.6 ± 0.3	233.8 ± 1.5	6.8
3z, tan fragments NM@1.2A (3)	16.1	1299.3	357.9	44.3	0.275	3	15404	0.03460 ± 5	0.24288 ± 32	0.05092 ± 3	219.2 ± 0.3	220.8 ± 0.3	237.1 ± 1.3	7.7
04AT253-1 (megacrystic granite)														
1z, larger colourless 3:1 prisms NM@1.8A (19)	54.1	479.3	169.3	13.3	0.353	7	6643	0.02749 ± 3	0.18833 ± 24	0.04969 ± 3	174.8 ± 0.2	175.2 ± 0.2	180.3 ± 1.5	3.1
2z, smaller colourless 3:1 prisms NM@1.8A (50)	26.7	597.8	125.9	18.0	0.211	20	1418	0.02855 ± 4	0.20765 ± 43	0.05274 ± 9	181.5 ± 0.2	191.6 ± 0.4	317.7 ± 4.0	43.5
3z, large colourless prisms NM@1.8A (20)	44.0	489.4	176.3	13.4	0.360	4	8362	0.02718 ± 3	0.18571 ± 22	0.04956 ± 3	172.9 ± 0.2	173.0 ± 0.2	174.3 ± 1.3	0.8

Th concentration calculated from amount of ^{208}Pb and $^{207}Pb/^{206}Pb$ age with a slight correction for Pb loss in some discordant fractions. All errors quoted in this table are reported at 1σ. Atomic ratios were corrected for fractionation and initial common Pb (Stacey & Kramers 1975). Number in parentheses corresponds to the number of grains analysed. z, zircon; IF, initial Frantz. NM, non-magnetic (at indicated angle of tilt on Frantz Isodynamic Separator).

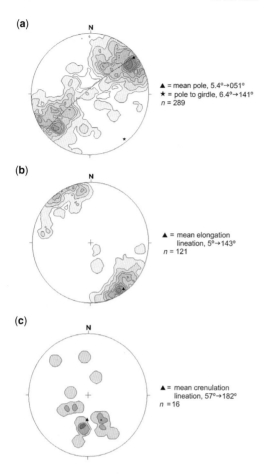

Fig. 9. (a) Stereographic plot of contoured poles to foliation plane in the THSZ. (b) Stereographic contour plot of elongation lineations in the THSZ. (c) Stereographic contour plot of crenulation lineations in the THSZ.

173 Ma, the age of the cross-cutting, post-tectonic megacrystic granite.

The THSZ, including its contacts with the overlying hanging-wall leucogabbro and pyroxenite and underlying footwall volcanic and volcaniclastic units, is folded. Folding is indicated by change in the orientation of the S_1 mylonitic fabric, by metre- to centimetre-scale parasitic folds of S_1, and by the presence of a crenulation lineation and cleavage. Folds of the S_1 mylonitic fabric are referred to as F_2. Crenulations of the S_1 fabric are asymmetric with steeply west-dipping axial planes (S_2), and short steep east limbs, indicating eastward vergence (Fig. 8d). These parasitic folds are congruent to the interpreted shape of map-scale folds of the THSZ and its hanging- and footwall sequences (Fig. 5). The majority of crenulation lineations (L_2) trend south, consistent with east-vergent kinematic indicators and the asymmetry of folds on Tally Ho Mountain and Dickson Hill (Fig. 9c). The eastward vergence of the THSZ and of the F_2 folds affecting the THSZ may indicate that the two phases of deformation are kinematically linked. F_2 folding post-dates the formation of the S_1 mylonitic fabric. The folds are, however, truncated by and older than the 173 Ma megacrystic granite (Fig. 5). Folding did not, therefore, significantly post-date formation of the THSZ, further strengthening the suggested kinematic link between mylonitization and subsequent folding.

The THSZ separates structurally overlying coarse-grained ultramafic and mafic crystalline rocks from an underlying sequence of volcanic and volcaniclastic rocks. Strain decreases progressively structurally up- and down-section from mylonitic and schistose rocks into undeformed footwall and hanging-wall lithological units, respectively. The mylonite is of similar composition to and is inferred to consist of highly strained rocks of the footwall volcaniclastic unit and locally of the hanging-wall pyroxenite and leucogabbro units. Because the THSZ has an overall flat, sheet-like geometry (albeit subsequently folded), and places a lower crustal intrusive sequence inferred to be part of an arc root over top of a supracrustal arc sequence of volcanic and volcaniclastic rocks, we interpret the shear zone as a crustal-scale, intra-arc thrust fault. The presence of coarsely crystalline ultramafic rocks carried in the THSZ hanging wall indicates that the thrust fault rooted into the lower crust and potentially the lithospheric mantle of the arc. The distribution of hanging-wall units west of the footwall units, the eastward extension of the footwall units into the more easterly Whitehorse Trough, and the overall eastward vergence of the THSZ, all imply that the THSZ rooted to the west, and climbed up-section to the east. The east vergence of F_2 folds, and the close temporal relationship between THSZ mylonites and the subsequent F_2 folds, implies a genetic relationship. We interpret the map-scale F_2 folds as fault-bend folds that developed during movement of the THSZ over ramps in structurally deeper-seated, kinematically linked thrust faults, a commonly observed relationship in fold and thrust belts.

Brittle deformation

Steeply dipping, brittle to brittle–ductile fault zones strike parallel to and cross-cut the mylonitic THSZ, and are characterized by fault gouge, tension gashes, Riedel shears, slickenlines and rare S–C fabric. Tension gashes and Riedel shears on Mount Wheaton indicate sinistral ($n = 2$) and dextral ($n = 2$) slip. Brittle faults commonly

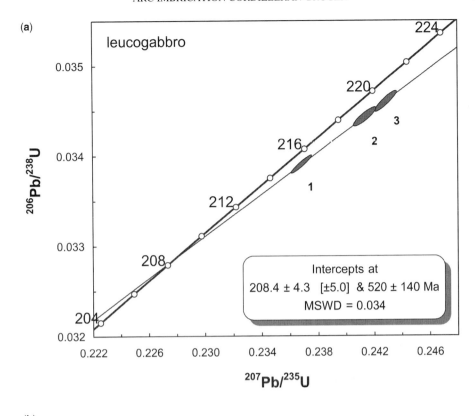
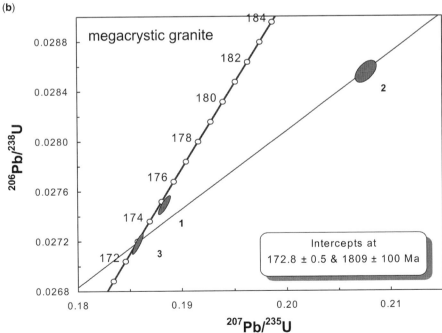

Fig. 10. (a) U–Pb concordia diagram of three zircon fractions from leucogabbro (04AT251-1). (b) U–Pb concordia diagram of three zircon fractions from megacrystic granite (04AT253-1).

contain gouge, strike NW–SE and dip subvertically (Fig. 8b). Although slickenline orientations along the fault zones are variable, strike-slip to oblique strike-slip displacements are most common. Dextral kinematic indicators were the most common sense of motion observed ($n = 33$; sinistral $n = 2$). Direct measurement of the amount of displacement along the brittle faults is difficult because of the lack of offset markers.

The brittle faults have a strike and dip similar to and are locally continuous with faults mapped as part of the Llewellyn fault zone (Hart & Radloff 1990). The steeply dipping brittle to brittle–ductile faults that overprint and parallel the THSZ are therefore interpreted as part of the Llewellyn fault zone. Rocks as young as early Eocene are affected by the brittle deformation. However, near Gold Hill, an undeformed Eocene rhyolite intrusion truncates and therefore post-dates movement along the Llewellyn fault zone (Fig. 4; Hart & Radloff 1990).

Discussion

The THSZ places lower crust and potentially mantle of a Late Triassic arc onto the supracrustal volcanic and volcaniclastic rocks of the Late Triassic Lewes River Arc of Stikinia. The THSZ is, therefore, an intra-arc fault developed within and responsible for imbrication of Stikinia; it is not a terrane-bounding structure marking the primary contact between Stikinia and the more westerly Nisling Assemblage. Kinematic indicators, including S–C fabric, C′ extensional shears, rotated porphyroclasts, and asymmetric parasitic folds and crenulations indicate that the subhorizontal shear zone was characterized by top-to-the-east shear. The geometry of folds of the THSZ is consistent with their interpretation as fault-bend folds developed above ramps in deeper-seated shear zones kinematically linked to the THSZ. Hence the THSZ appears to be one of a series of faults forming an east-verging thrust belt.

The well-developed foliation in the leucogabbro and, to a lesser extent, the pyroxenite is typically defined by the parallel alignment of coarsely crystalline, unstrained elongate amphibole and pyroxene grains. The lack of internal strain of single grains indicates that the foliation formed in response to flow within a deep crustal (high-P, high-T) setting. Rapid exhumation of these rocks to shallow crustal levels resulted in retrogressive greenschist metamorphism and overprinted the high-T ductile foliation with brittle–ductile shear zones and brittle fractures.

The age of the THSZ is tightly constrained. Leucogabbro carried in the hanging wall of the THSZ crystallized at 208.4 ± 4.3 Ma, providing a maximum age constraint on shear zone development. A minimum age constraint is provided by the megacrystic granite that intrudes into and plugs the THSZ and that crystallized at 173 Ma. Formation of the THSZ is, therefore, constrained to the latest Late Triassic to Early Jurassic.

The age, geometry and intra-Stikinian nature of the THSZ indicate that it is one of a series of Early Jurassic shear zones that characterize the Early Jurassic west margin of Stikinia. Timing constraints for the more northerly Takhini deformation zone, although few, are permissive of an Early Jurassic age. North and west of the Takhini deformation zone, in the Aishihik Lake region, little metamorphosed basalts of the Upper Triassic Lewes River Group are in fault contact with upper amphibolite-facies schist and quartzite of the Aishihik metamorphic suite, which is here referred to as the Nisling Assemblage. The fault is plugged by and older than the 186 Ma Aishihik batholith and 185 Ma granitic intrusions of the Long Lake plutonic suite. The Early Jurassic Wann River shear zone to the south is, of the documented Stikinian shear zones, perhaps the most similar to the THSZ, given its top-to-the-east sense of shear. Together, these coeval fault and shear zones accommodated both intra-Stikinia strain as well as juxtaposition of isotopically juvenile arc crust of Stikinia against the continental Nisling Assemblage. Younger brittle–ductile shear zones, like the Llewellyn, are probably the product of reactivation of this network of crustal-scale Early Jurassic faults that extend along and characterize the west margin of Stikinia.

Formation of the THSZ was coincident with the latest Triassic cessation of the isotopically juvenile Lewes River Arc magmatism and the formation of a major unconformity within the Whitehorse Trough, and immediately predated deposition of coarse clastic sediments of the Lower Jurassic Laberge Group (Johannson 1994; Hart 1997; Fig. 3). It is, therefore, likely that deformation coincided with and was related to the termination of subduction beneath the Lewes River Arc. The termination of magmatism in arcs is commonly linked to collision with buoyant, unsubductable lithosphere along an arcs bounding trench. The development of coeval shear zones along the length of the western margin of Stikinia is consistent with a model of collision of Stikinia with continental crust to the west.

The deposition of orogenic clastic sediments across Stikinia in the Early Jurassic, including boulder conglomerates of the Laberge Group, provides further support for involvement of Stikinia in a major Early Jurassic orogenic event. In northwestern British Columbia, Pleinsbachian greywacke of the Laberge Group is locally characterized by clasts of eclogite, peridotite and ultrahigh-pressure metamorphic rocks (MacKenzie et al. 2005). These sedimentary rocks provide a record of uplift and exhumation of crustal rocks that had been buried to depths of c. 100 km and constitute the only known vestiges of ultrahigh-pressure rocks in the entire Cordilleran orogen of western North America (MacKenzie et al. 2005; Canil et al. 2006). The rapid exhumation, erosion and subsequent deposition of ultrahigh-pressure minerals immediately post-dates the large-scale change in sedimentation in the Whitehorse Trough, consistent with the cessation of the Lewes River Arc being related to a latest Triassic–Early Jurassic thick-skinned collisional event.

Juxtaposition of isotopically juvenile Triassic arc rocks of Stikinia against the isotopically evolved continental Nisling Assemblage of the Yukon–Tanana terrane occurred in the Early Jurassic (Johnston et al. 1996). Hence it seems likely that it was the entry of the buoyant and unsubductable continental crust of the Nisling Assemblage into a subduction zone that dipped east beneath the west margin of Stikinia that resulted in deformation and the cessation of arc magmatism. Evidence for deep tectonic burial of the Nisling Assemblage, consistent with its entry into a subduction zone in the Late Triassic–Early Jurassic, includes synkinematic, high-pressure (pressures of 8–12 kbar)–high-temperature regional metamorphism in the Early Jurassic (Johnston & Erdmer 1995a), and the coeval development of an east-dipping axial planar foliation (Johnston & Erdmer 1995b). The Aiskihik batholith, which intrudes the Nisling Assemblage, crystallized at a depth of c. 30 km at 186.0 ± 2.8 Ma (Johnston et al. 1996). The high-level (intruded at depths of <6 km) Long Lake plutonic suite intruded the Aishihik batholith and Nisling Assemblage at $185.6 + 2.0/-2.4$ Ma, indicating rapid exhumation of the Aishihik batholith and previously deeply buried Nisling Assemblage (Johnston et al. 1996).

Collision of Nisling with Stikinia implies east-dipping subduction beneath the west margin of Stikinia (present-day coordinates). In this model, the Late Triassic Lewes River Arc, including the cumulate leucogabbro–pyroxenite that forms the hanging wall of the THSZ, faced west and developed in response to the consumption of oceanic lithosphere that was continuous with continental crust represented by the Nisling Assemblage (Fig. 11). Contrary to previous interpretations of the Whitehorse Trough as a forearc basin (Dickie & Hein 1995), in our model the trough forms the back-arc located east of the west-facing Lewes River Arc in the Late Triassic. Stratigraphic similarities between Stikinia and the Nisling Assemblage (Jackson et al. 1991; McClelland et al. 1992) may indicate that Stikinia and Nisling

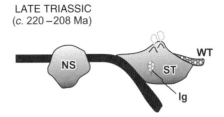
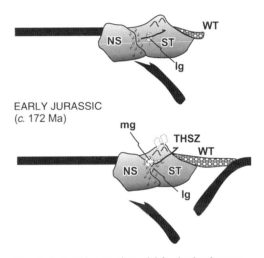

Fig. 11 Schematic tectonic model for the development of the Tally Ho shear zone along the western margin of Stikinia. NS, Nisling Assemblage; ST, Stikinia; WT, Whitehorse Trough; THSZ, Tally Ho shear zone; lg, leucogabbro and pyroxenite units; mg, megacrystic granite unit.

originated as part of one terrane that subsequently rifted apart.

Entry of the buoyant continental Nisling Assemblage into the trench bounding Stikinia to the west terminated subduction (Fig. 11). The collision drove structurally deep-seated portions of the Stikinia arc crust up-section to the east, giving rise to an east-verging intra-arc fold and thrust belt that included the THSZ. Thickening of the arc crust loaded the upper plate lithosphere and gave rise to a foreland basin east of the arc, within which the Laberge Group orogenic clastic sediments were deposited. The Early Jurassic Whitehorse Trough was, therefore, neither a forearc nor a back-arc; it was a foreland basin. Confusion of the nature of the Whitehorse Trough in part results from the misinterpretation of the foreland basin Laberge Group strata as the result of normal arc-marginal basin sedimentation. The presence of clasts of ultrahigh-pressure metamorphic rocks within the foreland basin indicates that collision resulted in the exhumation of rocks from depths of >100 km, probably along faults related to and kinematically linked to the THSZ. The post-collisional resumption of arc magmatism, including the 185 Ma Long Lake plutonic suite to the north and the 173 Ma Bennett granite in the study area, implies that collision was not terminal and may indicate that the Nisling terrane was a narrow ribbon continent. The alkalic nature of some of the post-tectonic magmatism may, however, be better explained as a result of breakoff of the slab attached to the Nisling Assemblage.

Our model of the tectonic setting that gave rise to the THSZ rests in part on our interpretation of the pyroxenite and leucogabbro that forms the hanging wall to the shear zone as an arc root, rather than an ophiolite (slice of oceanic lithosphere). Future geochemical and isotopic studies of the magmatic hanging-wall sequence will, therefore, constitute a major test of our model. Furthermore, the complicated composite evolution of the Whitehorse Trough predicted by our model is testable by constraining subsidence rates through time by performing a back-stripping analysis of the strata that make up the Laberge and Lewes River groups.

Conclusions

The northern Canadian Cordillera is at least in part the product of thick-skinned collisions between a juvenile oceanic arc (Stikinia) and a pericratonic terrane (the Nisling Assemblage). Entry of the buoyant Nisling Assemblage into the Stikine subduction zone resulted in the cessation of magmatism in the Lewes River Arc and the imbrication of the arc along crustal-scale east-vergent thrust faults such as the THSZ. The collision also resulted in the rapid exhumation of the Nisling Assemblage and Aishihik batholith to the west and deposition of ultrahigh-pressure rocks into the Whitehorse Trough to the east. Models of the Cordillera as an eastward amalgamation of thin-skinned crustal flakes that rode over North American continental basement (Snyder et al. 2002; Cook et al. 2004) are inconsistent with the tectonic model presented here. Collision of the Nisling Assemblage with Stikinia in the Early Jurassic demonstrates that thick-skinned accretion played a significant role in crustal growth along the ancient North American continental margin.

This research was supported by NSERC Discovery grants to STJ and LMH, and by a grant from the Yukon Geological Survey to STJ. Derek Turner provided able field assistance; geologists of the Yukon Geological Survey provided mentorship and advice. A review by Brendan Murphy and comments by the editor, Peter Cawood, greatly improved the manuscript.

References

BREITSPRECHER, K. & MORTENSEN, J. K. 2004. *Yukon Age 2004: A database of isotopic age determinations for rock units from Yukon Territory, Canadian Geochronology Knowledgebase*. Yukon Geological Survey, Whitehorse [CD-ROM].

CANIL, D., MIHALYNUK, M. & CHARNELL, C. 2006. Sedimentary record for exhumation of ultrahigh pressure (UHP) rocks in the northern Cordillera, British Columbia. *Geological Society of America Bulletin*, **118**, 1171–1184.

COLPRON, M., NELSON, J. L. & MURPHY, D. C. 2006. A tectonostratigraphic framework for the pericratonic terranes of the northern Canadian Cordillera. *In*: COLPRON, M. & NELSON, J. L. (eds) *Paleozoic Evolution and Metallogeny of Pericratonic Terranes at the Ancient Pacific Margin of North America, Canadian and Alaskan Cordillera*. Geological Association of Canada, St. John's, NFLD, Special Paper, **45**, 1–23.

COOK, F. A., CLOWES, R. M., SNYDER, D. B., VAN DER VELDEN, A. J., HALL, K. W., ERDMER, P. & EVENCHICK, C. A. 2004. Precambrian crust beneath the Mesozoic northern Canadian Cordillera discovered by Lithoprobe seismic reflection profiling. *Tectonics*, **23**, TC2010.

CURRIE, L. & PARRISH, R. R. 1993. Jurassic accretion of Nisling terrane along the western margin of Stikinia, Coast Mountains, northwestern British Columbia. *Geology*, **21**, 235–238.

DICKIE, J. R. & HEIN, F. J. 1995. Conglomeratic fan deltas and submarine fans of the Jurassic Laberge Group, Whitehorse Trough, Yukon Territory, Canada: Fore-arc sedimentation and unroofing of a volcanic island arc complex. *Sedimentary Geology*, **98**, 263–292.

DOHERTY, R. A. & HART, C. J. R. 1988. *Preliminary Geology of Fenwick Creek (105D/3) and Alligator Lake (105D/6) map areas, Open File 1988-2*. Exploration and Geological Services Division, Indian and Northern Affairs Canada, Yukon Region, 1–84.

EISBACHER, G. H. 1981. Late Mesozoic–Paleogene Bowser Basin molasse and Cordilleran tectonics, western Canada. *In*: MIALL, A. D. (ed.) *Sedimentation and Tectonics in Alluvial Basins*. Geological Association of Canada, Toronto, Ont., 125–151.

GEHRELS, G. E., MCCLELLAND, W. C., SAMSON, S. D., PATCHETT, P. J. & JACKSON, J. L. 1990. Ancient continental margin assemblage in the northern coast Mountains, southeast Alaska and northwest Canada. *Geology*, **18**, 208–211.

HAMILTON, W. B. 1988. Plate tectonics and island arcs. *Geological Society of America Bulletin*, **100**, 1503–1527.

HART, C. J. R. 1995. *Magmatic and tectonic evolution of the Intermontane Superterrane and Coast Plutonic Complex in southern Yukon Territory*. MSc thesis, University of British Columbia, Vancouver.

HART, C. J. R. 1997. *A transect across northern Stikinia: Geology of the northern Whitehorse map area, southern Yukon Territory (105D/13–16)*. Exploration and Geological Services Division, Indian and Northern Affairs Canada, Bulletin, **8**, 1–112.

HART, C. J. R. & PELLETIER, K. S. 1989. *Geology of Carcross (105D/2) and part of Robinson (105D/7) map areas, Open File 1989-1*. Exploration and Geological Services Division, Indian and Northern Affairs Canada, Yukon Region, 1–92.

HART, C. J. R. & RADLOFF, J. K. 1990. *Geology of Whitehorse, Alligator Lake, Fenwick Creek, Carcross and part of Robinson map areas (105D/11, 6, 3, 2, & 7), Open File 1990-4*. Indian and Northern Affairs Canada, Indian and Northern Affairs Canada, 1–113.

HART, C. J. R., DICKIE, J. R., GHOSH, D. K. & ARMSTRONG, R. L. 1995. Provenance constraints for Whitehorse Trough conglomerate: U–Pb zircon dates and initial Sr ratios of granitic clasts in Jurassic Laberge Group, Yukon Territory. *In*: MILLER, D. M. & BUSBY, C. (eds) *Jurassic Magmatism and Tectonics of the North American Cordillera*. Geological Society of America, Special Paper, **299**, 47–63.

HEAMAN, L. & PARRISH, R. R. 1991. U–Pb geochronology of accessory minerals. *In*: HEAMAN, L. & LUDDEN, J. N. (eds) *Applications of Radiogenic Isotope Systems to Problems in Geology; Short Course Handbook*. Mineralogical Association of Canada, Saskatoon, Sask., 59–102.

HEAMAN, L. M., ERDMER, P. & OWEN, J. V. 2002. U–Pb geochronologic constraints on the crustal evolution of the Long Range Inlier, Newfoundland. *Canadian Journal of Earth Sciences*, **39**, 845–865.

JACKSON, J. L., GEHRELS, G. E., PATCHETT, J. P. & MIHALYNUK, M. G. 1991. Stratigraphic and isotopic link between the northern Stikine Terrane and an ancient continental margin assemblage, Canadian Cordillera. *Geology*, **19**, 1177–1180.

JOHANNSON, G. G. 1994. *Provenance constraints on Early Jurassic evolution of the northern Stikinian arc: Laberge Group, Whitehorse Trough, northwestern British Columbia*. MSc thesis, University of British Columbia, Vancouver.

JOHNSTON, S. T. & ERDMER, P. 1995a. Hot-side-up aureole in southwest Yukon and limits on terrane assembly of the northern Canadian Cordillera. *Geology*, **23**, 419–422.

JOHNSTON, S. T. & ERDMER, P. 1995b. Magmatic flow and emplacement foliations in the Early Jurassic Aishihik Batholith, southwest Yukon: Implications for northern Stikinia. *In*: MILLER, D. M. & BUSBY, C. (eds) *Jurassic Magmatism and Tectonics of the North American Cordillera*. Geological Society of America, Special Paper, **299**, 65–82.

JOHNSTON, S. T., MORTENSEN, J. K. & ERDMER, P. 1996. Igneous and metaigneous age constraints for the Aishihik metamorphic suite, southwest Yukon. *Canadian Journal of Earth Sciences*, **33**, 1543–1555.

LUDWIG, K. R. 1998. *Isoplot/Ex (version 3.00): A geochronological toolkit for Microsoft Excel*. Berkeley Geochronology Centre, Special Publications, **1**, 1–43.

MACKENZIE, J. M., CANIL, D., JOHNSTON, S. T., ENGLISH, J., MIHALYNUK, M. G. & GRANT, B. 2005. First evidence for ultrahigh-pressure garnet peridotite in the North American Cordillera. *Geology*, **33**, 105–108.

MCCLELLAND, W. C., GEHRELS, G. E., SAMSON, S. D. & PATCHETT, P. J. 1992. Structural and geochronologic relations along the western flank of the Coast Mountains batholith: Stikine River to Cape Fanshaw, central southeastern Alaska. *Journal of Structural Geology*, **14**, 475–489.

MIHALYNUK, M. G. & ROUSE, J. N. 1988. *Preliminary Geology of the Tutshi Lake Area, Northwestern British Columbia (104M/15), Geological Fieldwork 1987: A Summary of Field Activities and Current Research, Paper 1988*. Province of British Columbia, Ministry of Energy, Mines and Petroleum Resources, Mineral Resources Division, Geological Survey Branch, Victoria BC, 217–232.

MORTENSEN, J. K. 1992. Pre-mid-Mesozoic tectonic evolution of the Yukon–Tanana Terrane, Yukon and Alaska. *Tectonics*, **11**, 836–853.

SNYDER, D. B., CLOWES, R. M., COOK, F. A., ERDMER, P., EVENCHICK, C. A., VAN DER VELDEN, A. J. & HALL, K. W. 2002. Proterozoic prism arrests suspect terranes: Insights into the ancient Cordilleran margin from seismic reflection data. *GSA Today*, **12**, 4–10.

STACEY, J. S. & KRAMERS, J. D. 1975. Approximation of terrestrial lead isotope evolution by a 2-stage model. *Earth and Planetary Science Letters*, **26**, 207–221.

TIZZARD, A. & JOHNSTON, S. T. 2005. Structural evolution of the Tally Ho shear zone (NTS 105D), southern Yukon. *In*: EMOND, D. S., LEWIS, L. L. & BRADSHAW, G. D. (eds) *Yukon Exploration and Geology 2004*. Yukon Geological Survey, Whitehorse, 237–246.

Palaeozoic Lachlan orogen, Australia; accretion and construction of continental crust in a marginal ocean setting: isotopic evidence from Cambrian metavolcanic rocks

DAVID A. FOSTER[1]*, DAVID R. GRAY[2], CATHERINE SPAGGIARI[3], GEORGE KAMENOV[1] & FRANK P. BIERLEIN[4]

[1]*Department of Geological Sciences, University of Florida, Gainesville, FL 32611-2120, USA*
[2]*School of Earth Sciences, University of Melbourne, Melbourne, Vic. 3010, Australia*
[3]*Geological Survey of Western Australia, 100 Plain Street, East Perth, WA 6004, Australia*
[4]*University of Western Australia, 35 Stirling Highway, Crawley, WA 6009, Australia*
*Corresponding author (e-mail: dafoster@ufl.edu)

Abstract: The Lachlan orogen developed as a classic accretionary orogen in an oceanic setting between the palaeo-Pacific subduction zone and the Australian craton. Direct evidence for the composition of the lower crust and the basement to the thick Palaeozoic turbidite fan of the Lachlan orogen is limited. Exposures of Cambrian metavolcanic rocks and geophysical data suggest that most of the basement is the mafic oceanic crust along with possible small fragments of older continental crust. The trace element compositions of Cambrian metavolcanic rocks in the western and central Lachlan orogen are similar to those of volcanic rocks formed in modern back-arc and forearc settings. Pb, Nd and Sr isotopic data from these Cambrian rocks suggest a supra-subduction zone setting with little or no influence of continental crust other than subducted sediment.

Accretionary orogens form from the addition of material to continents from plate margin processes at convergent or transpressive plate boundaries, and are one of the most important 'factories' for generating, recycling and maturing continental crust (e.g. Condie 2007). Palaeozoic–Mesozoic accretionary margins extended over 20 000 km around the margin of Gondwana from the present northern Andes to eastern Australia (Fig. 1a; e.g. Foster & Gray 2000; Cawood 2005). Accretionary processes along the Gondwanan margin added some 20–30% to the area of some continents including Australia, and are still continuing at the edge of the Australian plate in New Zealand (Gray *et al.* 2007).

The Tasmanides (Fig. 1b) of Australia show eastward younging in accretion of submarine fans, fragments of ocean crust, volcanic arcs and forearc basins from the Delamerian–Ross orogen to New Zealand (e.g. Foster & Gray 2000; Cawood 2005; Gray *et al.* 2007). The Lachlan orogen (Fig. 2), which is the central belt within the Australian Tasmanides, is an exceptionally well-preserved accretionary orogen. This belt formed largely in a deep-water oceanic setting from structural thickening and accretion of continental detritus and juvenile igneous components, without the incorporation of extensive blocks of older continental crust, and has great potential for improving our understanding of continental growth and recycling processes (Foster & Gray 2000; Collins 2002*a*; Glen 2005). The Lachlan orogen is widely applicable as a template for circum-Pacific accretionary orogens typified by marginal basins and large volumes of turbidite.

In this paper we review the tectonic history of the Lachlan orogen and present trace element and isotopic data from Cambrian metavolcanic rocks, which form the basement for the orogen. The focus of the geochemical investigation is on the Middle Cambrian calc-alkaline Jamieson–Licola volcanic rocks that formed within the proto-Lachlan marginal ocean basin. The data obtained are used to infer the nature of the lower crust and define the crustal growth and recycling processes in the Lachlan orogen.

The Lachlan orogen

Geological framework

Continental crust of eastern Australia formed along the margin of the supercontinent of Gondwana during the Palaeozoic and early Mesozoic as a result of accretion of oceanic crust, recycled continent-derived turbidite, and volcanic arcs.

From: CAWOOD, P. A. & KRÖNER, A. (eds) *Earth Accretionary Systems in Space and Time.*
The Geological Society, London, Special Publications, **318**, 329–349.
DOI: 10.1144/SP318.12 0305-8719/09/$15.00 © The Geological Society of London 2009.

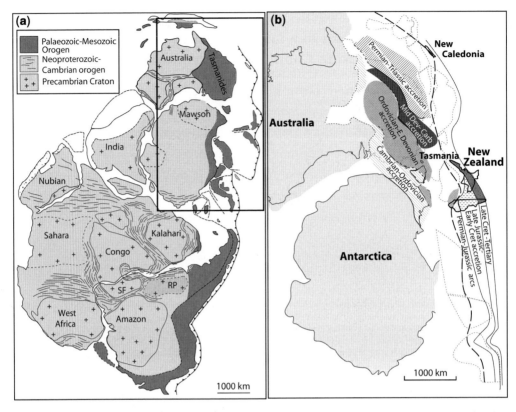

Fig. 1. (a) Map of Gondwana showing the location of the Tasmanides within the Palaeozoic–Mesozoic orogenic belts along the margin of Gondwana. (b) Reconstruction of Australia and Antarctica along with the timing of accretion for the orogenic belts on the Palaeopacific margin (modified after Gray et al. 2007). SF, Sao Francisco; RP, Rio de la Plata.

Eastern Australia is composed of distinct orogenic belts collectively referred to as the Tasmanides (Fig. 2). Accretion occurred with an eastward younging of peak deformation of Middle Cambrian, Ordovician–Devonian and Permian–Triassic age in the respective belts. Outboard of the Tasman orogen, the continental landmass of New Zealand records continuous sedimentation and accretionary prism development above the subduction system from Permian to Late Cretaceous times (Bradshaw 1989; Mortimer 2004).

The Palaeozoic Lachlan orogen (Fig. 2) is a composite orogen dominated by Cambrian to Ordovician turbidites that formed a large submarine fan system comparable in dimension with the Bengal fan (Fergusson & Coney 1992a). Rocks of the Western and Central Lachlan orogen are mainly early Palaeozoic quartz-rich sandstone and black shale turbidites, which are laterally extensive over the 800 km width, and have thicknesses upwards of 10 km. The Eastern Lachlan orogen consists of andesite, volcaniclastic rocks, and limestone, as well as quartz-rich turbidite and extensive black shale (VandenBerg & Stewart 1992; Glen et al. 2007). Age populations of detrital zircon (U–Pb) and muscovite ($^{40}Ar/^{39}Ar$) indicate that the turbidites were derived from the Delamerian–Ross and Pan-African orogenic belts throughout Gondwana (e.g. Turner et al. 1996; Foster et al. 1998; Veevers 2000; Squire et al. 2005). The Lachlan turbidite fan accumulated on Cambrian back-arc and forearc crust (Fig. 3a), consisting of predominantly low-K to arc tholeiite basalts and gabbros, high-Mg, low-Ti boninites, and calc-alkaline rocks (Crawford & Keays 1987). The orogen developed by accretion of the oceanic sequences accompanied by marked Late Ordovician–Devonian structural thickening (c. 300%) and shortening (c. 75%) to form c. 35–40 km thick crust (Coney et al. 1990; Foster et al. 1999; Fergusson 2003). Orogeny included widespread magmatism, which chemically and thermally matured the crustal section.

The Lachlan orogen comprises three thrust systems that constitute the western, central and eastern parts, respectively (Fig. 2; Gray & Foster 1997). The orogen has similar lithotectonic

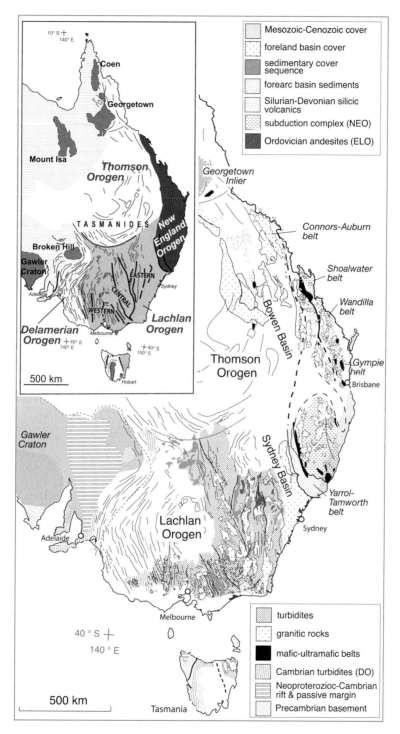

Fig. 2. Geological map of eastern Australia highlighting the main rock units of the Lachlan orogen and the Tasmanides (modified from Gray & Foster 2004). The inset map shows the orogenic belts of the Tasmanides and the three subprovinces (Western, Central and Eastern) of the Lachlan orogen. NEO, New England orogen; ELO, Eastern Lachlan orogen; DO, Delamerian orogen.

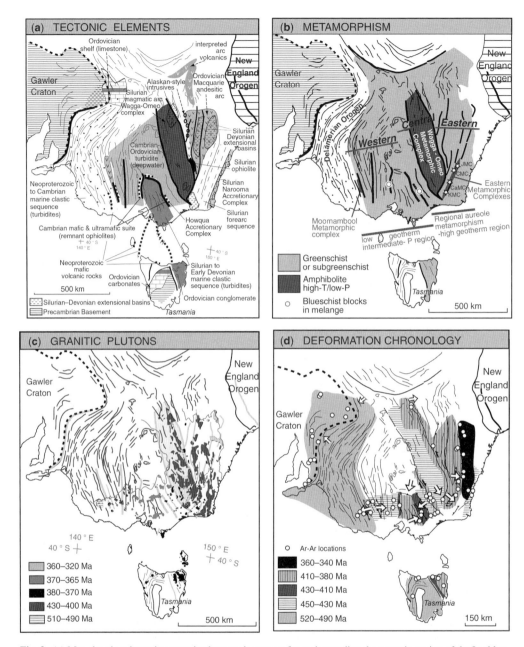

Fig. 3. (**a**) Map showing the major tectonic elements important for understanding the tectonic setting of the Lachlan orogen. The bold lines show locations of major faults and the fine lines show the orientation of the major structural grain. (**b**) Map showing the average grade of metamorphism. (**c**) Map showing the age and location of major granitic plutons. (**d**) Map showing the timing of deformational and metamorphic events across the Lachlan orogen. The arrows show tectonic vergence. JMC, Jerangle Metamorphic Complex; CMC, Cooma Metamorphic Complex; CaMC, Camblong Metamorphic Complex; KMC, Kuark Metamorphic Complex.

assemblages, general structural style and average level of exposure and metamorphism along the entire >1000 km exposed length and across the 700–800 km width of the belt (Fig. 3; Powell 1983; Coney et al. 1990; Fergusson & Coney 1992b; Gray & Foster 1997, 2004; Foster & Gray 2000; Glen 2005). The Western Lachlan is an east-vergent thrust system with zones of NW–SE- and north–south-trending structures. The Central Lachlan is dominated by NW–SE-trending structures and consists of a SW-vergent thrust belt linked to an extensive high-T/low-P metamorphic complex. The Eastern Lachlan has a north–south structural grain with east-directed faults, with an older, subduction–accretion complex in the southeastern part (Narooma complex; Fig. 3a).

The structure of the Western Lachlan is characterized by chevron folds cut by a series of linked thrust fault systems (Gray & Foster 1998; Korsch et al. 2008; Figs 4 and 5). Major faults in the Western Lachlan are spaced about 100–120 km apart, and are polydeformed zones characterized by transposition layering, crenulation cleavages, high, non-coaxial strains, intense mica fabrics, and isoclinal folds (e.g. Gray & Willman 1991; Gray & Foster 1998). The large fault zones contain dismembered Cambrian ophiolite slivers, as well as relicts of serpentinite and mud-matrix mélange incorporating blueschist knockers (Spaggiari et al. 2003a, 2004).

The Eastern Lachlan orogen is made up of several major, west-dipping fault zones that penetrate to the base of the crust (Glen et al. 2002; Gray et al. 2006b). These faults are crustal-scale imbrications of the Ordovician Macquarie Volcanic Arc and Silurian–Early Devonian platform and deep basin sedimentary rocks. The major faults are spaced 10–15 km apart and have continuous 100–150 km length segments in profile. These major west-dipping faults evolved from Silurian–Devonian normal faults that were reactivated as thrusts during the Silurian–late Early Devonian and the Carboniferous (Glen 1992; Glen et al. 1992, 2002). As a consequence, former extensional basins (e.g. Cowra and Hill End basins) have west-directed thrust faults defining their western margins.

Relicts of accretionary complexes are exposed in the southwestern part of the Central Lachlan (Howqua) and the southeastern part of the Eastern Lachlan (Narooma) (Fig. 3a). In the Narooma Accretionary Complex large-scale imbrication is associated with chaotic block-in-matrix mélange, broken formation along high-strain zones, early bedding-parallel cleavage, recumbent folds in turbidites, and structural complexity in cherts. This succession has been interpreted as the outer-arc slope and imbricated zone of an accretionary wedge that was part of a Late Ordovician–Silurian subduction zone (e.g. Powell 1983; Miller & Gray 1997; Offler et al. 1998b; Fergusson & Frikken 2003). In the Howqua Accretionary Complex an imbricated and chevron-folded, Late Ordovician–Silurian turbidite succession contains mud-matrix and serpentinite-matrix mélanges in the frontal fault zone (Spaggiari et al. 2002a, 2003b, 2004) as well as mud-matrix mélange along major faults within the wedge (Watson & Gray 2001) and a possible detached seamount (Spaggiari et al. 2004). These have been interpreted as part of a Late Ordovician–Silurian subduction accretionary wedge (e.g. Foster & Gray 2000; Fergusson 2003).

Deformation and metamorphism started in all of the subprovinces of the Lachlan orogen between c. 455 and 440 Ma (Fig. 3d; e.g. Foster et al. 1999; Foster & Gray 2007). Metamorphism in the high-temperature–low-pressure Wagga–Omeo Metamorphic Complex and in the eastern metamorphic complexes (Fig. 3b and d) occurred at c. 430 Ma (Maas et al. 2001; Williams 2001). The interval c. 420–410 Ma was characterized by fault reactivation in the Western Lachlan, strike-slip motion on the large boundary faults and mylonitic shear zones of the Wagga–Omeo Metamorphic Complex, and significant exhumation (e.g. Foster et al. 1999). The eastern part of the Western Lachlan and the Central Lachlan provinces underwent significant deformation between c. 400 and 385 Ma when they collided. The inland parts of the Eastern Lachlan are dominated by c. 400–380 Ma deformation with a central and northeastern region of c. 380–360 Ma deformation (Glen et al. 1992; Foster et al. 1999; Fig. 3d). Widespread Silurian extension and basin formation predated this phase of shortening in the Macquarie Arc. These events also overprinted the earlier fabrics in the Eastern Lachlan such as the c. 445–440 Ma fabrics in the Narooma Accretionary Complex. Although very widespread, intense Carboniferous deformation is focused in the northern part of the Lachlan orogen in the Eastern subprovince (Powell 1983; Glen 1992).

Silurian–Devonian granitoids make up about 20% of the outcrop area of the Lachlan orogen and up to 36% in the Eastern and Central subprovinces (Fig. 3c; e.g. Chappell et al. 1988; Gray & Foster 2004). Most of the plutons were emplaced at pressures ≤ 2 kbar within low-grade upper crustal rocks. Most plutons are post-tectonic and unmetamorphosed, although the older intrusions, such as those within the Wagga–Omeo Metamorphic Complex and Kosciusko Batholith, are significantly foliated and were emplaced synkinematically at mid-crustal depths (e.g. Hine et al. 1978; Morand & Gray 1991). Felsic volcanic sequences are also widespread, and form c. 15% of outcrop area in the Eastern Lachlan, large post-tectonic caldera complexes in the Western Lachlan, and other fields associated with shallow-level plutons. Older volcanic provinces (≥ 440 Ma) include the basaltic–andesitic Ordovician Macquarie Volcanic Arc in the Eastern Lachlan orogen (e.g. Glen et al. 2007).

The nature of the lower crust beneath the Lachlan turbidite succession is a subject of

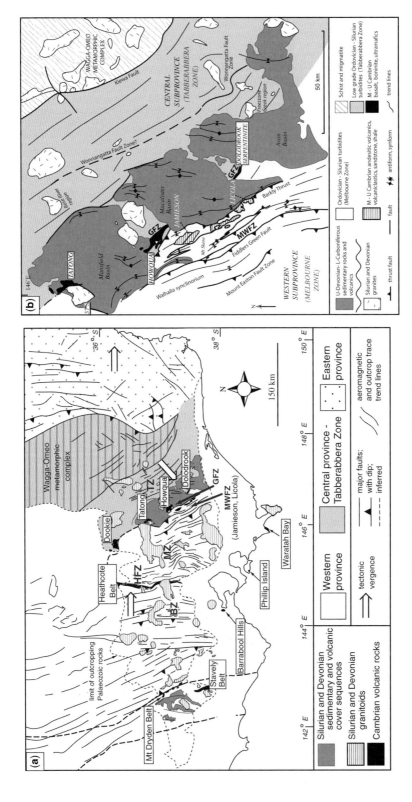

Fig. 4. (a) Geological and structural form map of the southern Lachlan orogen highlighting the locations of Cambrian metavolcanic rocks within the Mount Wellington Belt comprise the exposures within the Governor Fault Zone (GFZ) and the Mount Wellington Fault Zone (MWFZ). BZ, Bendgo Zone; MZ, Melbourne Zone; HFZ, Heathcote Fault Zone. (b) More detailed geological map of the boundary zone between the Western and Central subprovinces of the Lachlan orogen. Exposures of the Jamieson–Licola assemblage are within the Mount Wellington Fault Zone (modified from Spaggiari *et al.* 2004).

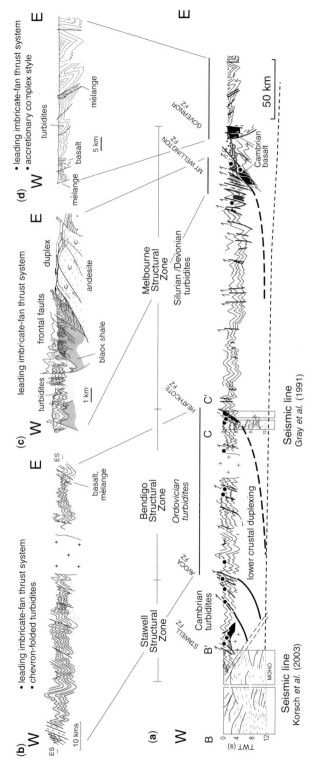

Fig. 5. West to east structural cross-section of the western Lachlan orogen (modified from Gray & Foster 1998; Gray et al. 2006b).

continued research. In the Western Lachlan Cambrian mafic volcanic rocks of oceanic affinities underlie the quartz-rich turbidite succession (Crawford & Keays 1987), whereas in the Eastern Lachlan the oldest rocks observed are Ordovician volcanic rocks and a Late Cambrian or Early Ordovician chert–turbidite–mafic volcanic assemblage (VandenBerg & Stewart 1992; Glen et al. 2007). Chappell et al. (1988) suggested that trace element and isotopic data (McCulloch & Chappell 1982) from the Silurian–Devonian granitoids indicated that basement terranes comprising attenuated Proterozoic crust existed beneath the turbidites (see Anderson et al. 1998; Handler & Bennett 2001; Maas et al. 2001). Whole-rock Nd- and Sr-isotopic data (Gray 1984; Keay et al. 1997; Soesoo et al. 1997; Collins 1998; Rossiter 2003), Hf- and O-isotopic compositions of igneous zircons (Kemp et al. 2005, 2007), similar inherited zircon age populations from S- and I-type granites and the Ordovician turbidite detrital zircon populations (Keay et al. 1997; Williams 2001; Kemp et al. 2005, 2007), however, indicate that the granitoids are derived from mixtures of the Palaeozoic turbidite, Cambrian oceanic basement and juvenile mantle-derived mafic magmas (e.g. Collins 1998; Kemp et al. 2007). Re–Os isotopic data from mantle xenoliths in basalts from the southwestern part of the Lachlan orogen give Proterozoic models ages for the lithospheric mantle (McBride et al. 1996; Handler & Bennett 2001). The xenolith locations are within the boundary zone between the Lachlan and Delamerian orogens, and the Delamerian orogen sits on Precambrian continental crust so that these data do not require Precambrian lower crust for the Lachlan orogen.

Seismic reflection profiles in the northeastern Lachlan (Pinchin 1980; Korsch et al. 1986, 1993, 1997; Leven et al. 1992; Glen et al. 1994, 2002; Finlayson et al. 2002), and the southern Lachlan (Gray et al. 1991; Korsch et al. 2002, 2008) have provided an image of the crustal structure (Gray & Foster 2004; Gray et al. 2006b). These studies show that most major fault zones dip to the west, that steeply dipping faults at the present erosion surface tend to decrease in dip with depth, and that regions between major faults sometimes show complex intersecting networks of both east- and west-dipping faults and shear zones. The reflection studies in the Eastern Lachlan image the structure of the dismembered Ordovician volcanic arc crust (Glen et al. 2007). Isostatic considerations require a dense (>2.9 g cm^{-3}) lower crust (e.g. O'Halloran & Rey 1999) matching lower crustal P-wave velocities of >6 km s^{-1} (Finlayson et al. 1979, 1980, 2002; Gibson et al. 1981), which are consistent with a mafic oceanic basement for much of the Lachlan orogen.

Many models for the tectonic evolution of the Lachlan orogen suggest development in an intraplate setting involving a single marginal basin floored by attenuated Precambrian continental crust located between a subduction zone and the Australian continent (e.g. Powell 1983; Fergusson & Coney 1992b; Li & Powell 2001; Willman et al. 2002; Squire & Miller 2003; Braun & Pauselli 2004). More complex scenarios for the evolution of the Lachlan orogen involve multiple subduction systems and microplates in a marginal oceanic setting behind the major, long-lived subduction system (Gray & Foster 1997, 2004; Soesoo et al. 1997; Collins 1998; Foster & Gray 2000; Fergusson 2003). The multiple subduction zone–microplate system implies that the basement for the Lachlan orogen is mainly oceanic and not continental, which is consistent with a suggestion made by Crook 40 years ago (Crook 1969). A geodynamic setting involving marginal basin closure by subduction to form the Western and Central subprovinces of the Lachlan orogen, which are within the core of the orogen (Gray & Foster 2004) is suggested by: (1) the presence of dismembered ophiolite slivers along some major fault zones (Spaggiari et al. 2003a, 2004); (2) the lower temperature, intermediate-pressure metamorphic conditions preserved in metasandstone and slate sequences of the Western Lachlan and external part of the Central Lachlan (Offler et al. 1998a; Spaggiari et al. 2003b); (3) the presence of broken formation in the Central and Eastern Lachlan (Miller & Gray 1997; Watson & Gray 2001); and (4) the presence of serpentinite-matrix mélange incorporating blueschist blocks similar to those in the Franciscan Complex of California (Spaggiari et al. 2002a, b, 2003a).

Cambrian metavolcanic rocks of the Western and Central Lachlan orogen

Narrow fault-bounded belts of Cambrian ultramafic to andesitic metavolcanic rocks are exposed in the southwestern Lachlan orogen (Crawford 1988). Boninites and low-Ti andesites intruded by arc tholeiites provide evidence for Cambrian back-arc and forearc basin along with island arc settings and indicate a supra-subduction zone origin for the metavolcanic rocks (Crawford et al. 1984; Nelson et al. 1984; Crawford & Cameron 1985; Crawford & Keays 1987). These belts have, therefore, been interpreted to be faulted slivers of the upper crust of a marginal oceanic basin (Fergusson 2003; Spaggiari et al. 2003a, 2004; Foster et al. 2005). The two largest exposures of metavolcanic rocks are in the Heathcote Belt and Mount Wellington Belt (Fig. 4; Crawford 1988; Spaggiari et al. 2004).

The Mount Wellington Belt includes exposed segments of Middle Cambrian, supra-subduction zone ophiolitic rocks at Dookie, Tatong, Howqua, and Dolodrook, which are within the Governor Fault Zone (Fig. 4a and b). The adjacent Mount Wellington Fault Zone contains fault slices of calk-alkaline andesitic rocks of the Jamieson–Licola assemblage that were thrust over the metavolcanic sequences in the Governor Fault Zone (Crawford 1988; Spaggiari et al. 2002b, 2003b; Fig. 5).

The Dookie segment (Fig. 4) is the most northern exposure, and is dominated by tholeiitic basalt, dolerite and gabbro (Crawford 1988). The mafic rocks were metamorphosed at prehnite–pumpellyite to lower greenschist facies and show significant hydrothermal alteration. Epidote, calcite, quartz ($\pm$ axinite) veins are widespread, and barite veins occur locally. U–Pb zircon data from a hornblende gabbro in the Dookie segment yielded an age of 501 ± 0.7 Ma (Spaggiari et al. 2003a).

The Howqua segment (Fig. 4) represents the most complete section of ophiolitic rocks in the Mount Wellington Belt (Crawford 1988; Spaggiari et al. 2002b). The sequence consists of an imbricated section of tholeiitic pillow basalt, hyaloclasite, volcaniclastic rocks, dolerite, and gabbro, in fault contact with and underlain by mafic and ultramafic boninitic lavas and intrusions, underlain by mélange. The mélange contains large fault slivers of both tholeiitic and boninitic rocks, blocks of these rocks metamorphosed to blueschist facies in talc schist matrix, and slivers of Ordovician phyllite, slate and minor sandstone (Spaggiari et al. 2002a, b, 2003a, c, 2004). The structurally highest, tholeiitic sequence was metamorphosed to prehnite–pumpellyite facies. Metamorphic pressure and temperature, as well as deformation intensity, increase down sequence into the mélange. The metamorphic pattern is interpreted as having formed by accretionary processes during underplating of the mélange, accompanied by duplexing of the upper sequences (Spaggiari et al. 2002a, 2004). Part of the tholeiitic pillow basalt sequence is conformably overlain by and interbedded with chert and silicified black shale, which, in turn, is conformably overlain by turbidites. Earliest Ordovician (Lancefieldian, La2, c. 490 Ma) graptolites occur within the upper part of the chert sequence and Latest Cambrian (Datsonian) conodonts occur approximately midway (VandenBerg & Stewart 1992). These fauna provide a minimum age of 491 Ma for the basalts, suggesting that the basalts are approximately the same age as the Dookie gabbro (c. 501 Ma). Tholeiitic rocks show mid-ocean ridge basalt (MORB) to arc tholeiite characteristics and typical Fe-enrichment trends (Crawford & Keays 1987). Older boninitic lavas are also present and most are very mafic (Crawford 1988).

Calk-alkaline rocks of the Jamieson–Licola assemblage are exposed within the Mount Wellington Fault Zone structurally above the Howqua segment (Fig. 4). The sequence is best exposed in the Jamieson River and Licola regions and consists of andesitic to rhyodacitic lavas and minor pyroclastic deposits, voluminous volcanic breccias with interbedded volcanogenic sandstone, siltstone, lenses of black shale, and occasional limestone olistoliths (VandenBerg et al. 1995). The lavas are locally pillowed or columnar jointed. The volcanic rocks are strongly deformed, except for some sections of massive lava, and metamorphosed to pumpellyite–actinolite and greenschist facies. Zircons from greenschist-facies andesite lava in the Licola region gave a U–Pb age of 500 ± 8 Ma (Spaggiari et al. 2003a). This age is in accord with Earliest Ordovician (Lancefieldian, La2, c. 490 Ma) graptolites that occur in a thin, black, pyritic shale lens in fault contact with underlying volcaniclastic rocks (VandenBerg et al. 1995).

The Cambrian calc-alkaline igneous rocks in the Mount Wellington Belt were erupted through the basement of the Lachlan orogen and predate the Palaeozoic turbidite blanket. Cayley et al. (2002) argued that this belt of andesitic to rhyodacitic rocks was erupted through Precambrian continental crust of the Selwyn block because of the presence of intermediate to felsic compositions (see also Scheibner & Veevers 2000; VandenBerg et al. 2000). Crawford et al. (2003) noted that the compositions of these rocks were very similar to that of the Mount Read volcanic assemblage of Tasmania (Crawford & Berry 1992). Crawford & Berry (1992) interpreted the Mount Read volcanic rocks to be a post-collisional assemblage associated with extension. The volcanic arc-like characteristics of the andesitic rocks, however, are also similar to 'mature-stage', supra-subduction zone rocks described by Shervais (2001).

In the next section we present trace element and whole-rock isotopic data from the metaigneous rocks in the Mount Wellington Belt along with a smaller number of samples from the Heathcote Belt in Tables 1 and 2, and Figures 6–9. The trace element, rare earth element (REE) and isotopic data for the metavolcanic rocks of the Jamieson–Licola assemblage and related metavolcanic rocks (Tables 1 and 2) supplement published results (e.g. Nelson et al. 1984; Crawford 1988; VandenBerg et al. 1995). These data have significant implications for the Cambrian tectonic setting of the Lachlan marginal basin and the nature of the lower crust, because they are not contaminated by continent-derived turbidites such as the Ordovician–Devonian igneous rocks of the Lachlan orogen.

Table 1. *Elemental and isotopic data from Cambrian metavolcanic rocks in the Western Lachlan orogen*

Exposure area/fault slice		Dookie				Jamieson					
Sample number:	CS0040	CS0045	CS0046	CS9953	CS0054	HED5-TRAY69	HED5-TRAY70	320b	328a	1064a
Rock type:	gabbro	basalt	basalt	andesite	andesite	basalt	andesite	rhyolite	andesite	andesite
Major elements (wt%)										
SiO_2	49.2	48.0	48.1	54.1	54.4	49.2	57.4	73.4	54.1	56.4
Al_2O_3	16.0	17.0	16.1	16.2	14.4	14.9	14.8	10.8	16.6	19.6
TiO_2	1.0	0.4	1.2	0.4	0.3	0.3	0.4	0.4	0.4	0.3
Fe_2O_3	11.9	12.8	13.2	8.1	10.0	7.7	7.6	7.3	7.4	6.6
MgO	7.6	7.3	7.1	4.8	7.4	5.7	5.5	2.0	4.6	3.2
MnO	0.2	0.3	0.2	0.1	0.1	0.1	0.1	0.0	0.1	0.1
CaO	8.2	7.5	9.1	11.4	7.0	3.3	4.0	0.4	8.2	8.1
Na_2O	4.0	3.9	3.2	0.1	2.8	1.5	3.1	0.0	3.5	3.8
K_2O	0.1	0.2	0.1	0.1	1.0	1.8	0.9	2.2	1.0	1.1
P_2O_5	0.1	0.1	0.1	0.1	0.2	0.1	0.1	0.5	0.2	0.2
LOI	2.6	2.1	1.9	4.5	2.4	6.2	6.4	2.9	2.1	2.0
Total	100.8	99.6	100.5	99.8	99.8	99.7	100.4	99.9	100.4	100.7
Trace elements (ppm)										
Sc	43.0	44.1	42.1	24.4	28.0	22.8	23.0	14.0	21.0	23.2
V	238.1	301.6	319.8	196.1	201.9	196.2	186.4	71.7	210.9	206.5
Cr	15.0	276.5	170.1	192.5	292.6	69.7	103.9	110.1	172.4	48.8
Co	47.0	49.4	46.9	21.9	32.7	17.8	21.1	10.4	21.6	13.9
Ni	67.2	107.3	90.6	27.7	46.1	13.1	16.0	35.7	38.8	18.4
Cu	199.6	159.8	173.4	132.8	123.0	170.2	196.1	69.8	138.2	57.6
Zn	70.2	91.6	87.1	59.0	68.3	130.9	79.9	76.7	73.6	42.7
Ga	11.6	15.5	16.6	20.4	14.9	15.1	15.6	14.0	18.1	17.7
Rb	1.0	3.6	2.2	1.6	16.6	30.4	14.9	45.6	15.9	27.2
Sr	68.6	124.9	274.1	185.2	389.3	110.6	116.8	14.4	626.2	418.9
Y	17.3	26.3	26.8	9.6	14.7	9.2	9.5	41.7	16.4	12.9
Zr	43.7	63.1	67.4	68.7	71.1	83.8	94.7	85.7	107.7	74.2
Nb	2.7	3.9	4.2	2.9	2.8	3.7	4.1	8.4	6.1	4.4
Ba	32.2	5060.5	567.6	39.0	1606.0	657.8	313.0	1516.9	513.7	665.9
La	2.2	3.7	3.8	7.9	11.8	7.4	11.4	19.5	32.6	13.6
Ce	6.1	9.5	10.2	16.1	23.6	15.2	22.6	31.7	43.3	24.6
Pr	1.0	1.5	1.6	2.1	2.9	1.8	2.6	4.4	6.9	2.9
Nd	5.1	8.0	8.4	9.2	13.5	7.8	11.1	20.0	29.6	12.7
Sm	1.7	2.6	2.7	2.0	2.9	1.7	2.1	4.4	5.2	2.6
Eu	0.6	2.2	1.0	0.6	1.1	0.5	0.6	1.4	1.3	0.8
Gd	2.1	3.5	3.5	1.6	2.6	1.7	1.8	5.0	3.9	2.4
Tb	0.4	0.6	0.7	0.3	0.4	0.3	0.3	0.9	0.5	0.4
Dy	2.8	4.1	4.2	1.5	2.3	1.6	1.6	5.5	3.1	2.1
Ho	0.6	0.9	0.9	0.3	0.5	0.3	0.3	1.2	0.6	0.4
Er	1.6	2.4	2.4	0.8	1.2	0.9	0.9	3.4	1.5	1.1
Tm	0.3	0.4	0.4	0.1	0.2	0.1	0.1	0.6	0.2	0.2
Yb	1.7	2.6	2.6	0.8	1.2	1.1	1.0	3.8	1.5	1.1
Lu	0.3	0.4	0.4	0.1	0.2	0.2	0.2	0.6	0.2	0.2
Hf	1.2	1.8	1.9	1.9	2.0	2.4	2.7	2.4	3.2	2.1
Ta	0.2	0.3	0.3	0.2	0.2	0.3	0.3	0.6	0.7	0.4
Pb	0.7	1.1	0.7	7.9	5.7	5.8	3.9	7.9	14.4	9.8
Th	0.2	0.3	0.3	3.5	3.6	4.5	5.5	8.9	10.1	4.7
U	0.1	0.1	0.1	0.9	1.1	1.5	1.7	2.5	2.6	1.4
Isotopes										
Sr (ppm)	65.6	110.3	250.8	185	383.1	101.7	107.1		584.5	380.1
Rb (ppm)		3.3		nd	nd	nd	17.2		20.2	35.1
Sm (ppm)	1.2	1.8	1.6	1.7	2.7	1	1.3		nd	1.6
Nd (ppm)	3.6	5.4	4.9	7.5	12.3	4.4	6.7		23.1	17.5
$^{143}Nd/^{144}Nd$ (corrected)	0.512927	0.51291	0.512915	0.512654	0.512637	0.512649	0.512594		0.512584	0.512628
ε_{Nd} (500 Ma)	4.8	5	5.2	4.1	4	4	4.2		4.4	4.1
$^{87}Sr/^{86}Sr$	0.7048	0.7068	0.7049	0.7044	0.7055	0.7103	0.7079		0.7049	0.7055
$^{87}Sr/^{86}Sr$ (500 Ma)	0.7045	0.7062	0.7048	0.7042	0.7046	0.7048	0.7053		0.7044	0.7042
$^{206}Pb/^{204}Pb$ (500 Ma)		17.879		17.895	17.875	17.502		17.352		
$^{207}Pb/^{204}Pb$ (500 Ma)		15.549		15.528	15.553	15.509		15.547		
$^{208}Pb/^{204}Pb$ (500 Ma)		37.688		37.717	37.702	37.355		37.23		

(Continued)

Table 1. *Continued*

	Licola			Waratah Bay	Heathcote		Howqua		Phillip Island
	CS9761 andesite	C9762 andesite	CS9857 andesite	CS9736 basalt	CS9707 boninite	CS9708 andesite	CS9741 boninite	CS9751 basalt	PI basalt
	61.5	57.6	57.9	48.3	56.4	64.6	52.4	50.0	47.7
	14.7	18.7	13.5	13.8	9.7	12.7	1.8	13.1	19.4
	0.4	0.3	0.4	1.5	0.3	0.3	0.1	1.9	1.0
	6.1	5.7	7.7	14.2	9.8	6.3	11.7	14.8	10.3
	4.3	3.7	7.0	7.4	10.1	6.1	22.8	6.5	9.1
	0.1	0.1	0.1	0.2	0.2	0.1	0.2	0.2	0.2
	5.5	7.9	8.0	8.1	6.7	3.8	6.8	7.5	7.7
	5.3	3.1	1.8	3.5	4.1	4.0	0.3	3.2	4.2
	1.1	0.9	0.8	0.5	0.1	1.8	0.0	0.1	0.2
	0.3	0.2	0.2	0.1	0.1	0.1	0.0	0.2	0.1
	1.4	2.5	3.0	2.6	4.1	2.2	4.0	3.1	1.3
	100.6	100.7	100.4	100.3	101.6	101.4	100.1	100.6	101.0
	16.1	15.2	26.9	40.4	28.6	15.5	18.8	37.5	41
	139.8	147.5	180.8	306.2	211.5	99.9	76.4	323.6	215
	167.5	152.3	259.5	143.9	876.0	364.9	4104.3	80.6	471
	24.2	15.9	23.6	45.6	43.5	34.8	77.1	44.8	54
	39.5	35.8	47.5	77.1	208.5	70.1	567.3	48.5	144
	66.9	87.4	150.6	161.2	29.1	32.5	17.7	153.4	83
	57.3	52.0	59.6	85.2	54.5	48.3	61.4	102.7	63
	12.6	16.6	13.6	15.3	7.5	11.0	1.2	16.7	12
	48.5	37.5	26.1	6.5	3.3	19.7	0.8	1.8	3
	350.0	986.9	596.6	198.2	49.9	125.0	49.8	156.8	146
	13.5	12.4	13.5	28.9	7.7	8.9	0.7	32.7	20
	115.8	111.5	109.2	84.8	28.0	88.1	14.3	109.9	49
	7.9	7.3	6.4	5.4	0.6	4.9	0.4	8.3	5.1
	797.0	690.7	773.2	170.6	171.1	347.2	16.0	207.3	51
	48.6	43.6	20.2	4.6	3.8	7.5	1.0	6.6	3.52
	85.5	76.6	38.5	11.7	8.1	16.0	2.3	16.5	8.75
	8.7	7.9	4.4	1.8	1.1	2.0	0.3	2.4	1.25
	34.2	31.2	18.7	9.6	5.3	8.4	0.8	12.8	5.98
	5.2	4.8	3.6	3.0	1.3	1.8	0.1	3.8	1.80
	1.2	1.2	1.0	1.0	0.4	0.5	0.0	1.3	0.62
	4.3	4.1	2.6	4.0	1.3	1.7	0.1	4.8	2.58
	0.6	0.5	0.4	0.7	0.2	0.3	0.0	0.9	0.48
	2.4	2.2	2.3	4.6	1.2	1.5	0.1	5.4	3.13
	0.4	0.4	0.4	1.0	0.3	0.3	0.0	1.1	0.69
	1.3	1.2	1.2	2.7	0.7	0.8	0.1	3.0	1.90
	0.2	0.2	0.2	0.4	0.1	0.1	0.0	0.5	0.30
	1.2	1.1	1.2	2.9	0.8	0.9	0.1	3.3	2.13
	0.2	0.2	0.2	0.5	0.1	0.1	0.0	0.5	0.34
	3.4	3.2	3.2	2.3	0.7	2.4	0.2	3.0	1.30
	0.6	0.6	0.5	0.4	0.1	0.4	0.0	0.5	0.33
	14.7	40.5	8.2	0.6	1.4	6.1	0.9	1.1	2.12
	25.3	22.7	9.9	0.5	0.9	1.8	0.5	0.7	0.47
	4.8	5.6	3.0	0.2	0.7	0.6	0.2	0.2	0.12
	325.7	989.9	548.3	184.9	47.9	146.2	47.9	146.2	139
	202.9	70.9	nd	nd	3.3	26.6	0.7	nd	nd
	3.6	nd	3.4	2.4	0.9	1.1	nd	2.7	1.1
	22.9	21.6	17.5	7.6	3.8	4.9	nd	8.8	3.6
	0.512313	0.512322	0.512564	0.512887	0.51277	0.512557		0.512867	0.512827
	0.1	0.2	3.5	5.1	5.5	2.4		5.2	4.2
	0.7083	0.7065	0.7056	0.7066	0.7088	0.7078	0.7087	0.7055	0.7065
	0.7055	0.7058	0.7047	0.7059	0.7074	0.7045		0.7053	0.7060
	18.333	18.222	17.892		18.482			18.732	18.586
	15.588	15.583	15.535		15.534			15.545	15.616
	37.282	37.961	37.568		38.025			38.551	38.357

Major elements analysed by X-ray fluorescence. Trace elements analysed by inductively coupled plasma mass spectrometry (ICP-MS) using an Element-2 at the University of Florida. Sr isotopic ratios measured by thermalionization mass spectrometry using a Micromass Sector 54; Nd and Pb isotopic ratios measured by multi-collector ICP-MS using a Nu-Plasma instrument from spiked solutions at the University of Florida. LOI, loss on ignition.

Table 2. Locations of Cambrian metavolcanic rocks

Sample number	Rock type	Fault slice	Location	Map coordinates
		Dookie		
CS0040	Gabbro		Kellows Road Quarry	378000E, 5979000N
CS0045	Basalt		Mt. Major, south of communication towers	383800E, 5975400N
CS0046	Basalt		Mt. Major, south of communication towers	384100E, 5975420N
		Jamieson		
CS9953	Andesite		Jamieson River, Wren's Flat	444205E, 5866760N
CS0054	Andesite		South of Mt. Sunday Track, near eastern margin	446705E, 5864740N
TRAY69	Basalt		Hill 800 (drill core)	444936E, 5868966N
TRAY70	Andesite		Hill 800 (drill core)	444936E, 5868966N
320b	Rhyolite		Handford Creek Track, rockfall	443243E, 5863869N
328a	Andesite		Prickle Spur Track	446270E, 5867470N
1064a	Andesite		Jamieson River (Silver Mine Spur Track)	
		Licola		
CS9761	Andesite		Jamieson–Heyfield Road, Wallaby Creek	462365E, 5839120N
CS9862	Andesite		Jamieson–Heyfield Road, Wallaby Creek	463200E, 5835660N
CS9857	Andesite		Jamieson–Heyfield Road, junction of Violet Hill Track	458675E, 5840765N
		Waratah Bay		
CS9736	Basalt		Point Grinder	410000E, 5693750N
		Heathcote		
CS9707	Boninite		Sheoak Gully, Shuran's Lane property	294250E, 5917650N
CS9708	Andesite		Lady's Pass, Colbinabbin turnoff	295600E, 5923000N
		Howqua		
CS9741	Boninite		Cold Creek, trail south of Howqua Track	437800E, 5885570N
CS9751	Basalt		Howqua River, west of Noonan's Hut	443100E, 5883950N
		Phillip Island		
PI	Basalt		Near Watt Point, south coast of Phillip Island	

Map coordinates are from the Australian Map Grid, Zone 55.

Geochemistry of the Cambrian volcanic rocks

Normalized trace element plots show enrichment in large ion lithophile elements (LILE) and strong depletion in high field strength elements (HSFE) for the andesitic rocks (Fig. 6). REE trends (Fig. 7) of the andesitic rocks show light REE (LREE) enrichment and flat heavy REE (HREE) patterns typical of oceanic island arcs lacking deep melting processes in the presence of garnet. The one rhyolitic rock is more enriched in HREE. Eu anomalies are not exhibited by the samples so that plagioclase fraction was limited. All of the basaltic rocks show HSFE compositions similar to MORB, but with arc-type enrichment in LILE consistent with oceanic back-arc basin igneous rocks. Tholeiitic basalts show flat REE patterns, which are also typical of oceanic back-arc basins. The Howqua boninite is extremely depleted in REE. Nelson et al. (1984) reported similar trace element data for these rock types.

Initial $^{87}Sr/^{86}Sr$ (500 Ma) ratios for all of the samples analysed range from about 0.704 to 0.708 (Fig. 8). There is significant evidence for seawater alteration and mobilization of Sr isotopes in some samples so that some do not record primary values (see Nelson et al. 1984). ε_{Nd} (500 Ma) values for the andesitic rocks are all positive and range from 0.1 to 4.2, and for the basaltic rocks are in the range of c. 4–8. The basaltic rocks approach the depleted mantle values for 500 Ma. The boninites also fall into this range. Nelson et al. (1984) reported one boninite sample with an initial ε_{Nd} of −9, which is far less radiogenic than anything we measured and is probably an artefact of very low Nd. Common Pb isotopic ratios for almost all of the samples, including the Jamieson–Licola andesites and rhyolite, lie above the curve for Bulk Earth and below that for average continental crust (Fig. 9).

Taken together, the elemental and isotopic data are consistent with previous interpretations of Nelson et al. (1984) and Crawford (1988) that the Cambrian metavolcanic rocks in the Western

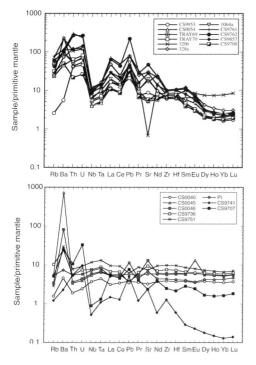

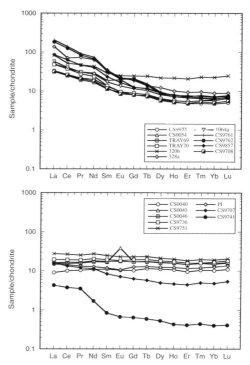

Fig. 6. Trace element variation diagrams for Cambrian metavolcanic rocks from the Lachlan orogen. The top plot includes the andesite samples and the bottom plot shows the samples of basalt and boninite. The sample numbers correspond to those in Table 1.

Fig. 7. Rare earth element variation diagrams for Cambrian metavolcanic rocks from the southern Lachlan orogen. The top plot includes the andesitic samples and the bottom plot shows the samples of associated basalt and boninite. The sample numbers correspond to those in Table 1.

and Central Lachlan were formed in a western Pacific-style oceanic marginal basin off the Australian continent. The older boninitic rocks may record early rifting of the forearc of a Delamerian subduction–arc system (Foster *et al.* 2005). The younger tholeiitic rocks resemble basalts erupted during spreading of the proto-Lachlan back-arc basin, whereas the calc-alkaline rocks of the Jamieson–Licola assemblage were probably erupted on the back-arc basin crust in an ocean volcanic arc setting. LILE enrichment and strong depletion in HFSE in these rocks are consistent with derivation from metasomatized mantle above a subduction zone. The isotopic data are also consistent with an oceanic arc and back-arc setting. The spread in ε_{Nd} values to values approaching zero as well as the Pb isotopic data are consistent with sediment contamination in a subduction environment. There is no requirement in the geochemical data that Precambrian crust underlies the Jamieson–Licola volcanic rocks or any other part of the Western or Central Lachlan. The results support the interpretation that the basement for the Lachlan orogen is largely oceanic, and that no

volumetrically significant Precambrian continental blocks contributed to the geochemical signatures of these volcanic rocks. Geochemical and isotopic data from the younger Ordovician Macquarie Arc andesites and basalts also indicate construction

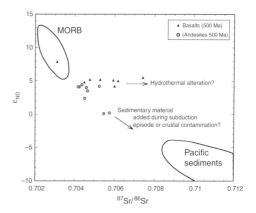

Fig. 8. Plot of ε_{Nd} v. $^{87}Sr/^{86}Sr_i$ for the Cambrian volcanic rocks. The values are corrected to 500 Ma.

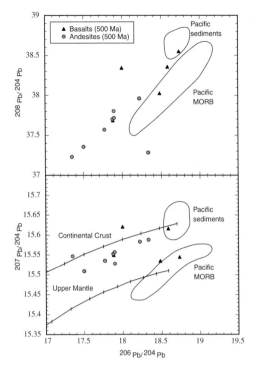

Fig. 9. Plot of common Pb isotopic data for the Cambrian metavolcanic rocks.

on Cambrian oceanic crust (reviewed by Glen et al. 2007).

It is possible, however, that the intermediate Pb isotopic values and less radiogenic Nd isotopic values of some samples reflect minor contamination of older continental material. If Proterozoic lower crust is present under the Jamieson–Licola volcanic rocks then the Cambrian volcanic rocks were not significantly contaminated by it, or the Precambrian component was relatively young because it has not imparted an obvious signature on the isotopic data. Direct correlation between the Jamieson–Licola volcanic rocks and those in the Mount Read Belt of Tasmania (e.g. Cayley et al. 2002; Crawford et al. 2003), are possible only if Precambrian crust of the Tasmanian microcontinent was much thinner under central Victoria. The results suggest that the Jamieson–Licola volcanic rocks developed as a volcanic arc above the Lachlan back-arc basin crust at the same time that the Mount Read volcanic rocks were erupted on the Tasmanian microcontinent, and the Mount Stavely volcanic rocks (Fig. 4) were erupted on the edge of the former Delamerian orogen in western Victoria.

Discussion

Accretionary orogens are major sites of continental growth, reconstruction, and world-class mineralization. They have been active over much of Earth history, and the modern circum-Pacific orogenic belts and marginal basin–arc systems show typical examples of many of the variations in processes and stages of development. Ancient examples include: Phanerozoic orogens such as the Tasmanides, Altides, and early history of the Appalachian orogen; Proterozoic belts such as the Yavapai and Central Plains orogens of North America, and the late Palaeoproterozoic to Mesoproterozoic orogens of central and western Australia; and possibly even some Archaean gneiss–granite–greenstone belts. The variations in geodynamic processes within accretionary orogens is very wide, with some consisting primarily of recycled and reworked older

---

Fig. 10. (*Continued*) between c. 490 and 470 Ma with detritus derived from the Delamerian–Ross orogen. At c. 485 Ma calc-alkaline volcanism in the Macquarie Island Arc initiated as a result of subduction of the palaeo-Pacific (Glen et al. 2007). The Macquarie Arc developed more than 2000 km off the Gondwanan margin based on palinspastic reconstructions (Gray et al. 2006a) At c. 460 Ma subduction initiated along both sides of the Lachlan marginal basin (Soesoo et al. 1997; Foster & Gray 2000). Oceanic thrust systems were active in both the eastern and western parts of the basin from c. 455 to 439 Ma (Foster et al. 1999). Oblique convergence at c. 410 Ma resulted in strike-slip motion on the shear zones bounding the Wagga–Omeo Metamorphic Belt (WOMB), and thrusting along the southern margin (Foster et al. 1999). Widespread post-orogenic magmatism in the western part of the Western Lachlan orogen at c. 400 Ma was followed by final closure of the marginal basin (Melbourne Zone) at c. 390 Ma (Foster et al. 1999). The collision between the Western Lachlan accretionary-style thrust belt and the Central Lachlan accretionary complex and magmatic arc was accompanied by localized strong to intense north–south folding and regional, meridional crenulation cleavage development in the Central Lachlan orogen (Morand & Gray 1999). In the Eastern Lachlan orogen periods of shortening separated longer intervals of extension and extension-related volcanism above the outer subduction zone between c. 440 and 420 Ma, along with high-T metamorphism and syndeformational granitic magmatism (e.g. Cooma Complex; Fergusson & VandenBerg 1990; Zen 1995; Collins 2002b). Closure of Early Devonian extensional basins formed within the older Macquarie Arc occurred between 410 and 390 Ma (Cobar Basin: Glen et al. 1992) and again between c. 360 and 340 Ma (Hill End Basin: Foster et al. 1999). Post-orogenic magmatism in the eastern part of the Western Lachlan (central Victorian magmatic province) occurred at c. 370–360 Ma, and was followed by post-orogenic magmatism in the Eastern Lachlan orogen between 360 and 340 Ma.

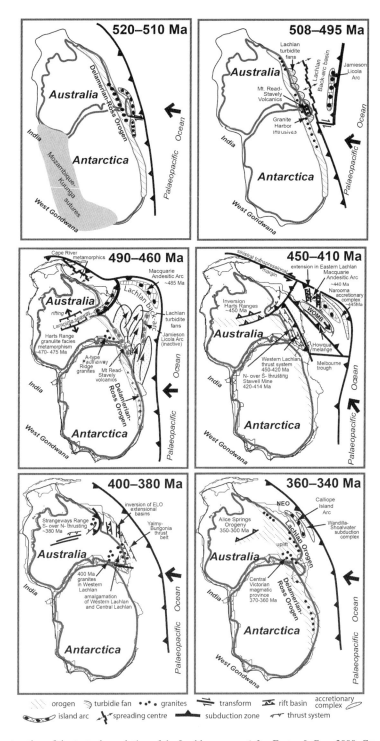

Fig. 10. Reconstruction of the tectonic evolution of the Lachlan orogen (after Foster & Gray 2000; Gray & Foster 2004). The Lachlan orogenic cycle started after collapse of the Delamerian–Ross orogen and subduction rollback opened the proto-Lachlan back-arc basin between c. 505 and 495 Ma (Foster et al. 2005). The Jamieson–Licola volcanic rocks were erupted during this time. Extensive turbidite fan deposition took place in the developing basin

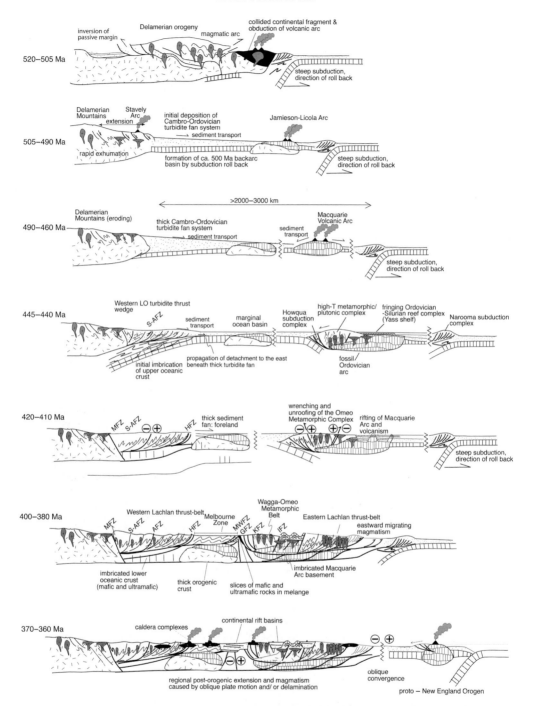

Fig. 11. Schematic cross-sections depicting the events and major tectonic elements that accreted to Australia to form the Lachlan orogen in Palaeozoic time (see Fig. 10 caption for a summary of key events). Abbreviations of major faults: AFZ, Avoca fault zone; GFZ, Governor fault zone; HFZ, Heathcote fault zone; KFZ, Kiewa fault zone; IFZ, Indi fault zone; MFZ, Moyston fault zone; MWFZ, Mt Wellington fault zone; SAFZ, Stawell–Ararat fault zone.

continental crust whereas others are constructed of almost entirely new continental crust composed of juvenile contributions from the mantle and structural thickening of oceanic materials. The most primitive accretionary orogens begin with oceanic elements combined with recycled continental detritus and many eventually become cratonized through thermal and chemical maturation processes. The Lachlan orogen is one of the type examples of a primitive accretionary orogen that began as a back-arc or marginal oceanic basin covered with a thick turbidite blanket, which was eventually closed by subduction processes (Fig. 10). Episodes of voluminous granitoid magmatism, driven by subduction and extension, eventually thermally and chemically matured the crust leading to cratonization. The Lachlan style of accretion fits many continental growth and recycling models going back through geological time to the Archaean (e.g. Condie 1982; Hamilton 1988; Coney 1992; Royden 1993; Sengor & Natal'in 1996; Foster & Gray 2000; Collins 2002a, b; Ingersoll et al. 2003; Percival et al. 2004). It is, however, critical to realize that long-term accretion in convergent settings does not typically occur simply through continuous subduction–accretion over hundreds of millions of years, but through a complex series of extensional events, some leading to marginal basins thousands of kilometres wide, interspersed with basin closure via subduction and shortening (e.g. Foster & Gray 2000; Collins 2002b; Gray & Foster 2004; Bickford & Hill 2007).

Most structural, geochemical, and geophysical data from the Lachlan and New England orogens suggest that they were constructed primarily on oceanic basement (reviewed by Gray & Foster 2004; Glen 2005). Whenever basement rocks to the Palaeozoic turbidite successions are exposed in the Lachlan orogen they consist of fault-bounded slices of Cambrian (and late Neoproterozoic?) mafic and ultramafic igneous rocks (Crawford 1988; Spaggiari et al. 2002a, b, 2003a–c, 2004; Crawford et al. 2003). There may be small intervening blocks of thin continental basement involved in accretion, but they are subordinate to the overall oceanic basin setting. Ordovician island arc volcanic rocks form the dominant basement of the Eastern Lachlan orogen, but these were probably constructed on oceanic crust (Glen et al. 2002, 2007). The extrapolation of aeromagnetic anomalies from Tasmania, granitoid generation models, and sedimentary facies models for the early Palaeozoic turbidites and black shales have led several workers to conclude that a larger block of Proterozoic continental crust underlies the central–southern Lachlan orogen (Scheibner & Veevers 2000; VandenBerg et al. 2000; Cayley et al. 2002; Clemens 2003). The new and previously published elemental and isotopic data from the Cambrian volcanic rocks that would have been erupted through this basement do not require the presence of Precambrian continental lower crust in this region, but do not totally rule it out. A structurally homogeneous and more felsic lower crust in this part of the Lachlan orogen could alternatively be the buried roots of a more extensive Cambrian Jamieson–Licola Arc assemblage.

In our interpretation, Western and Central Lachlan orogen accretion occurred by thickening and imbrication of extensive and thick submarine fans and the underlying marginal basic crust, in a Molucca Sea-style double-divergent subduction system (e.g. Hall et al. 1995). In this scenario subduction zones were initiated on the opposing sides of a former Cambrian–Ordovician back-arc basin (Fig. 11). Thrusting in the western Lachlan orogen resulted in marked shortening (c. 65–70%) by chevron folding, cleavage development and thrust imbrication of the turbidites. Subduction along the eastern side of the marginal basin, represented by the Howqua Accretionary Complex, produced large elongate composite batholiths within a shear zone-bounded, NW–SE-trending magmatic arc, as well as the associated high-temperature–low-pressure metamorphism. The Eastern Lachlan orogen is dominated by crust composed of the fragmented Ordovician Macquarie Arc, which was periodically rifted and then shortened above an outboard subduction zone. Mafic to felsic magmatism throughout the Lachlan orogen was generated by a variety of processes over time including subduction, extension, crustal thickening, and post-orogenic extension (Figs 10 and 11). The primitive nature of the initial tectonic elements of the Lachlan orogen also explains why this belt hosts a diverse range of rich mineral deposit types (e.g. Bierlein et al. 2002, 2006).

This research was funded by grants from the Australian Research Council and National Science Foundation. C. Fergusson and R. Korsch are thanked for providing helpful reviews of the manuscript.

References

ANDERSON, J. A. C., PRICE, R. C. & FLEMING, P. D. 1998. Implications for tectonic evolution and granite petrogenesis in the southern Lachlan Fold Belt, Australia. *Geology*, **26**, 119–122.

BICKFORD, M. E. & HILL, B. M. 2007. Does the arc accretion model adequately explain the Palaeoproterozoic evolution of southern Laurentia?: An expanded interpretation. *Geology*, **35**, 167–170.

BIERLEIN, F. P., GRAY, D. R. & FOSTER, D. A. 2002. Metallogenic relationships to tectonic evolution—the Lachlan Orogen, Australia. *Earth and Planetary Science Letters*, **202**, 1–13.

BIERLEIN, F. P., GROVES, D. I., GOLDFARB, R. J. & DUBÉ, B. 2006. Lithospheric controls on the formation of provinces hosting giant orogenic gold deposits. *Mineralium Deposita*, **40**, 874–887.

BRADSHAW, J. D. 1989. Cretaceous geotectonic patterns in the New Zealand region. *Tectonics*, **8**, 803–820.

BRAUN, J. & PAUSELLI, C. 2004. Tectonic evolution of the Lachlan Fold Belt, southeastern Australia: Constraints from coupled numerical models of crustal deformation and surface erosion driven by subduction of the underlying mantle. *Physics of the Earth and Planetary Interiors*, **141**, 281–301.

CAWOOD, P. A. 2005. Terra Australis Orogen: Rodinia breakup and development of the Pacific and Iapetus margins of Gondwana during the Neoproterozoic and Palaeozoic. *Earth-Science Reviews*, **69**, 249–279.

CAYLEY, R. A., TAYLOR, D. H., VANDENBERG, A. H. M. & MOORE, D. H. 2002. Proterozoic–early Palaeozoic rocks and the Tyennan Orogeny in central Victoria: The Selwyn block and its tectonic implications. *Australian Journal of Earth Sciences*, **49**, 225–254.

CHAPPELL, B. W., WHITE, A. J. R. & HINE, R. 1988. Granite provinces and basement terranes in the Lachlan Fold Belt, southeastern Australia. *Australian Journal of Earth Sciences*, **35**, 505–521.

CLEMENS, J. D. 2003. S-type granitic magmas—petrogenetic issues, models and evidence. *Earth-Science Reviews*, **61**, 1–18.

COLLINS, W. J. 1998. Evaluation of petrogenetic models for Lachlan fold belt granitoids: Implications for crustal architecture and tectonic models. *Australian Journal of Earth Sciences*, **45**, 483–500.

COLLINS, W. J. 2002a. Hot orogens, tectonic switching, and creation of continental crust. *Geology*, **30**, 535–538.

COLLINS, W. J. 2002b. Nature of extensional accretionary orogens. *Tectonics*, **21**, 1024–1036.

CONDIE, K. C. 1982. Plate tectonics model for Proterozoic continental accretion in the southwestern United States. *Geology*, **10**, 37–42.

CONDIE, K. C. 2007. Accretionary orogens in space and time. *In*: HATCHER, R. D. JR, CARLSON, M. P., MCBRIDE, J. H. & MARTINEZ CATALAN, J. R. (eds) *The 4D Framework of Continental Crust*. Geological Society of America, Memoirs, **200**, 145–158.

CONEY, P. J. 1992. The Lachlan belt of eastern Australia and Circum-Pacific tectonic evolution. *Tectonophysics*, **214**, 1–25.

CONEY, P. J., EDWARDS, A., HINE, R., MORRISON, F. & WINDRUM, D. 1990. The regional tectonics of the Tasman Orogenic system, eastern Australia. *Journal of Structural Geology*, **125**, 19–43.

CRAWFORD, A. J. 1988. Cambrian. *In*: DOUGLAS, J. G. & FERGUSON, J. A. (eds) *Geology of Victoria*. Geological Society of Australia, Victorian Division, Melbourne, 37–621.

CRAWFORD, A. J. & BERRY, R. F. 1992. Tectonic implications of Late Proterozoic–Early Palaeozoic igneous rock associations in western Tasmania. *Tectonophysics*, **214**, 37–56.

CRAWFORD, A. J. & CAMERON, W. E. 1985. Petrology and geochemistry of Cambrian boninites and low-Ti andesites from Heathcote, Victoria. *Contributions to Mineralogy and Petrology*, **91**, 93–104.

CRAWFORD, A. J. & KEAYS, R. R. 1987. Petrogenesis of Victorian Cambrian tholeiites and implications for the origin of associated boninites. *Journal of Petrology*, **28**, 1075–1109.

CRAWFORD, A. J., CAMERON, W. E. & KEAYS, R. R. 1984. The association boninite low-Ti andesite–tholeiite in the Heathcote greenstone belt, Victoria; ensimatic setting for the early Lachlan fold belt. *Australian Journal of Earth Sciences*, **31**, 161–175.

CRAWFORD, A. J., CAYLEY, R. A. ET AL. 2003. Neoproterozoic and Cambrian continental rifting, continent–arc collision and post-collisional magmatism. *In*: BIRCH, W. D. (ed.) *Geology of Victoria*. Geological Society of Australia Special Publications, **23**, 73–92.

CROOK, K. A. W. 1969. Contrasts between Atlantic and Pacific geosynclines. *Earth and Planetary Science Letters*, **5**, 429–438.

FERGUSSON, C. L. 2003. Ordovician–Silurian accretion tectonics of the Lachlan Fold Belt, southeastern Australia. *Australian Journal of Earth Sciences*, **50**, 475–490.

FERGUSSON, C. L. & CONEY, P. J. 1992a. Implications of a Bengal Fan-type deposit in the Palaeozoic Lachlan fold belt of southeastern Australia. *Geology*, **20**, 1047–1049.

FERGUSSON, C. L. & CONEY, P. J. 1992b. Convergence and intraplate deformation in the Lachlan Fold Belt of southeastern Australia. *Tectonophysics*, **214**, 417–439.

FERGUSSON, C. L. & FRIKKEN, P. 2003. Diapirism and structural thickening in an Early Palaeozoic subduction complex, southeastern New South Wales, Australia. *Journal of Structural Geology*, **25**, 43–58.

FERGUSSON, C. L. & VANDENBERG, A. H. M. 1990. Middle Paleozoic thrusting in the eastern Lachlan Fold Belt, southeastern Australia. *Journal of Structural Geology*, **12**, 577–589.

FINLAYSON, D. M., PROEDEHL, C. & COLLINS, C. D. N. 1979. Explosion seismic profiles and implications for crustal evolution in southeastern Australia. *Bureau of Mineral Resources Journal of Australian Geology and Geophysics*, **4**, 243–252.

FINLAYSON, D. M., COLLINS, C. D. N. & DENHAM, D. 1980. Crustal structure under the Lachlan Fold belt, southeastern Australia. *Physics of the Earth and Planetary Interiors*, **21**, 32–342.

FINLAYSON, D. M., KORSCH, R. J., GLEN, R. A., LEVEN, J. H. & JOHNSTONE, D. W. 2002. Seismic imaging and crustal architecture across the Lachlan Transverse Zone, a possible early cross-cutting feature of eastern Australia. *Australian Journal of Earth Sciences*, **49**, 311–321.

FOSTER, D. A. & GRAY, D. R. 2000. The structure and evolution of the Lachlan Fold Belt (Orogen) of eastern Australia. *Annual Review of Earth and Planetary Sciences*, **28**, 47–80.

FOSTER, D. A. & GRAY, D. R. 2007. Strain rate in Palaeozoic thrust sheets, the western Lachlan Orogen, Australia: Strain analysis and fabric geochronology. *In*: SEARS, J. W., HARMS, T. & EVENCHICK, C. A. (eds) *Whence the Mountains? Enquiries into the Evolution of Orogenic Systems: A Volume in Honor of Raymond Price*. Geological Society of America, Special Papers, **433**, 349–368.

FOSTER, D. A., GRAY, D. R., KWAK, T. A. P. & BUCHER, M. 1998. Chronology and tectonic framework of turbidite-hosted gold deposits in the western Lachlan Fold Belt, Victoria: ^{40}Ar–^{39}Ar results. *Ore Geology Reviews*, **13**, 229–250.

FOSTER, D. A., GRAY, D. R. & BUCHER, M. 1999. Chronology of deformation within the turbidite-dominated Lachlan orogen: Implications for the tectonic evolution of eastern Australia and Gondwana. *Tectonics*, **18**, 452–485.

FOSTER, D. A., GRAY, D. R. & SPAGGIARI, C. V. 2005. Timing of subduction and exhumation along the Cambrian East Gondwana margin, and the formation of Palaeozoic back-arc basins. *Geological Society of America Bulletin*, **117**, 105–116.

GIBSON, G., WESSON, V. & CUTHBERTSON, R. 1981. Seismicity of Victoria to 1980. *Journal of the Geological Society of Australia*, **28**, 341–356.

GLEN, R. A. 1992. Thrust, extensional and strike-slip tectonics in an evolving Palaeozoic orogen—a structural synthesis of the Lachlan Orogen of southeastern Australia. *Tectonophysics*, **214**, 341–380.

GLEN, R. A. 2005. The Tasmanides of eastern Australia. In: VAUGHAN, A. P. H., LEAT, P. T. & PANKHURST, R. J. (eds) *Terrane Processes at the Margins of Gondwana*. Geological Society, London, Special Publications, **246**, 23–96.

GLEN, R. A., DALLMEYER, R. D. & BLACK, L. P. 1992. Isotopic dating of basin inversion—the Palaeozoic Cobar Basin, Lachlan Orogen, Australia. *Tectonophysics*, **214**, 249–268.

GLEN, R. A., DRUMMOND, B. J., GOLEBY, B. R., PALMER, D. & WAKE-DYSTER, K. D. 1994. Structure of the Cobar basin, New South Wales, based on seismic reflection profiling. *Australian Journal of Earth Sciences*, **41**, 341–352.

GLEN, R. A., KORSCH, R. J. ET AL. 2002. Crustal structure of the Ordovician Macquarie arc, eastern Lachlan Orogen, based on seismic-reflection profiling. *Australian Journal of Earth Sciences*, **49**, 323–348.

GLEN, R. A., CRAWFORD, A. J., PERCIVAL, I. G. & BARRON, L. M. 2007. Early Ordovician development of the Macquarie Arc, Lachlan Orogen, New South Wales. *Australian Journal of Earth Sciences*, **54**, 167–179.

GRAY, C. M. 1984. An isotopic mixing model for the origin of granitic rocks in southeastern Australia. *Earth and Planetary Science Letters*, **70**, 47–60.

GRAY, D. R. & FOSTER, D. A. 1997. Orogenic concepts—application and definition: Lachlan fold belt, eastern Australia. *American Journal of Science*, **297**, 859–891.

GRAY, D. R. & FOSTER, D. A. 1998. Character and kinematics of faults within the turbidite-dominated Lachlan Orogen: Implications for tectonic evolution of eastern Australia. *Journal of Structural Geology*, **20**, 1691–1720.

GRAY, D. R. & FOSTER, D. A. 2004. Tectonic evolution of the Lachlan Orogen, southeast Australia: Historical review, data synthesis and modern perspectives. *Australian Journal of Earth Sciences*, **51**, 773–817.

GRAY, D. R. & WILLMAN, C. E. 1991. Thrust-related strain gradients and thrusting mechanisms in a chevron-folded sequence, southeastern Australia. *Journal of Structural Geology*, **13**, 691–710.

GRAY, D. R., WILSON, C. J. L. & BARTON, T. J. 1991. Intracrustal detachments and implications for crustal evolution within the Lachlan fold belt, southeastern Australia. *Geology*, **19**, 574–577.

GRAY, D. R., WILLMAN, C. E. & FOSTER, D. A. 2006a. Crust restoration for the western Lachlan Orogen using the strain-reversal, area-balancing technique: Implications for crustal components and original thicknesses. *Australian Journal of Earth Sciences*, **53**, 329–341.

GRAY, D. R., FOSTER, D. A., KORSCH, R. J. & SPAGGIARI, C. V. 2006b. Structural style and crustal architecture of the Tasmanides of eastern Australia, example of a composite accretionary orogen. In: MAZZOLI, S. & BUTLER, B. (eds) *Styles of Continental Compression*. Geological Society of America, Special Papers, **414**, 119–232.

GRAY, D. R., FOSTER, D. A., MAAS, R., SPAGGIARI, C. V., GREGORY, R. T., GOSCOMBE, B. D. & HOFFMANN, K. H. 2007. Continental growth and recycling by accretion of deformed turbidite fans and remnant ocean basins: Examples from Neoproterozoic and Phanerozoic orogens. In: HATCHER, R. D. JR, CARLSON, M. P., MCBRIDE, J. H. & MARTINEZ CATALAN, J. R. (eds) *The 4D Framework of Continental Crust*. Geological Society of America, Memoirs, **200**, 63–92.

HALL, R., ALI, J. R., ANDERSON, C. D. & BAKER, S. J. 1995. Origin and motion history of the Philippine Sea Plate. *Tectonophyics*, **251**, 229–250.

HAMILTON, W. B. 1988. Plate tectonics and island arcs. *Geological Society of America Bulletin*, **100**, 1503–1527.

HANDLER, M. R. & BENNETT, V. C. 2001. Constraining continental structure by integrating Os isotopic ages of lithospheric mantle with geophysical and crustal data; an example from southeastern Australia. *Tectonics*, **20**, 177–188.

HINE, R., WILLIAMS, I. S., CHAPPELL, B. W. & WHITE, A. J. R. 1978. Contrasts between I- and S-type granitoids of the Kosciusko Batholith. *Journal of the Geological Society of Australia*, **25**, 219–234.

INGERSOLL, R. V., DICKINSON, W. R. & GRAHAM, S. A. 2003. Remnant-ocean submarine fans: Largest sedimentary systems on Earth. In: CHAN, M. A. & ARCHER, A. W. (eds) *Extreme Depositional Environments: Mega End Members in Geologic Time*. Geological Society of America, Special Papers, **370**, 191–208.

KEAY, S., COLLINS, W. J. & MCCULLOCH, M. T. 1997. A three-component Sr–Nd isotopic mixing model for granitoid genesis, Lachlan fold belt, eastern Australia. *Geology*, **25**, 307–310.

KEMP, A. I. S., WORMALD, R. J., WHITEHOUSE, M. J. & PRICE, R. C. 2005. Hf isotopes in zircon reveal contrasting sources and crystallization histories for alkaline to peralkaline granites of Temora, southeastern Australia. *Geology*, **33**, 797–800.

KEMP, A. I. S., HAWKESWORTH, C. J. ET AL. 2007. Magmatic and crustal differentiation history of granitic rocks from Hf–O isotopes in zircon. *Science*, **315**, 980–983.

KORSCH, R. J., LINDSAY, J. F., O'BRIEN, P. E., SEXTON, M. J. & WAKE-DYSTER, K. D. 1986. Deep crustal seismic reflection profiling, New England Orogen, eastern Australia: Telescoping of the crust and a hidden, deep, layered sedimentary sequence. *Geology*, **14**, 982–985.

KORSCH, R. J., WAKE-DYSTER, K. D. & JOHNSTONE, D. W. 1993. The Gunndeah Basin–New England Orogen deep seismic reflection profile: Implications for New England tectonics. *In*: FLOOD, P. G. & AITCHISON, J. C. (eds) *New England Orogen, Eastern Australia*. University of New England, Armidale, 85–100.

KORSCH, R. J., WAKE-DYSTER, K. D. & JOHNSTONE, D. W. 1997. Crustal architecture of the New England Orogen based on deep crustal seismic reflection profiling. *In*: ASHLEY, P. M. & FLOOD, P. G. (eds) *Tectonics and Metallogenesis of the New England Orogen, Alan H. Voisey Memorial Volume*. Geological Society of Australia Special Publications, **19**, 29–51.

KORSCH, R. J., BARTON, T. J., GRAY, D. R., OWEN, A. J. & FOSTER, D. A. 2002. Geological interpretation of a deep seismic-reflection transect across the boundary between the Delamerian and Lachlan Orogens, in the vicinity of the Grampians, western Victoria. *Australian Journal of Earth Sciences*, **49**, 1057–1075.

KORSCH, R. J., MOORE, D. H. ET AL. 2008. Crustal architecture of Central Victoria: results from the 2006 deep crustal reflection seismic survey. *Geological Society of Australia Abstracts*, **89**, 155.

LEVEN, J. H., STUART-SMITH, P. G., MUSGRAVE, R. J., RICKARD, M. J. & CROOK, K. A. W. 1992. A geophysical transect across the Tumut Synclinorial Zone, N.S.W. *Tectonophysics*, **214**, 239–248.

LI, Z. X. & POWELL, C. MCA. 2001. An outline of the palaeogeographic evolution of the Australasian region since the beginning of the Neoproterozoic. *Earth-Science Reviews*, **53**, 237–277.

MAAS, R., NICHOLLS, I. A., GREIG, A. & NEMCHIN, A. 2001. U–Pb zircon studies of mid-crustal metasedimentary enclaves from the S-type Deddick Granodiorite, Lachlan Fold Belt, SE Australia. *Journal of Petrology*, **42**, 1429–1448.

MCBRIDE, J. S., LAMBERT, D. D., GREIG, A. & NICHOLLS, I. A. 1996. Multistage evolution of the Australian subcontinental mantle: Re–Os isotopic constraints from the Victorian mantle xenoliths. *Geology*, **24**, 631–634.

MCCULLOCH, M. T. & CHAPPELL, B. W. 1982. Nd isotopic characteristics of S- and I-type granites. *Earth and Planetary Science Letters*, **58**, 51–64.

MILLER, J. M MCL. & GRAY, D. R. 1997. Subduction related deformation and the Narooma anticlinorium, eastern Lachlan Orogen. *Australian Journal of Earth Sciences*, **44**, 237–251.

MORAND, V. & GRAY, D. R. 1991. Major fault zones related to the Omeo metamorphic complex, northeastern Victoria. *Australian Journal of Earth Sciences*, **38**, 203–221.

MORTIMER, N. 2004. New Zealand's geological foundations. *Gondwana Research*, **7**, 261–272.

NELSON, D. R., CRAWFORD, A. J. & MCCULLOCH, M. T. 1984. Nd–Sr isotopic and geochemical systematics in Cambrian boninites and tholeiites from Victoria, Australia. *Contributions to Mineralogy and Petrology*, **88**, 164–172.

OFFLER, R., MCKNIGHT, S. & MORAND, V. 1998a. Tectonothermal history of the western Lachlan Fold Belt, Australia: Insights from white mica studies. *Journal of Metamorphic Geology*, **16**, 531–540.

OFFLER, R., MILLER, J. M MCL., GRAY, D. R., FOSTER, D. A. & BALE, R. 1998b. Crystallinity and b_0 spacing of K-white micas in a Palaeozoic accretionary complex, eastern Australia: Metamorphism, palaeogeotherms, and structural style of an underplated sequence. *Journal of Geology*, **106**, 495–509.

O'HALLORAN, G. J. & REY, P. 1999. Isostatic constraints on the central Victorian lower crust: Implications for the tectonic evolution of the Lachlan Fold Belt. *Australian Journal of Earth Sciences*, **46**, 633–639.

PERCIVAL, J. A., BLEEKER, W., COOK, E. A., RIVERS, T., ROSS, G. & VAN STAAL, C. 2004. PanLITHOPROBE Workshop IV: Intra-orogen correlations and comparative Orogenic anatomy. *Geoscience Canada*, **31**, 23–39.

PINCHIN, J. 1980. Intracrustal seismic reflections from the Lachlan Fold Belt near Canberra. *Bureau of Mineral Resources Journal of Australian Geology and Geophysics*, **5**, 305–309.

POWELL, C. MCA. 1983. Tectonic relationship between the late Ordovician and Late Silurian palaeogeographies of southeastern Australia. *Journal of the Geological Society of Australia*, **30**, 353–373.

ROSSITER, A. G. 2003. Granitic rocks of the Lachlan Fold Belt in Victoria. *In*: BIRCH, W. D. (ed.) *Geology of Victoria*. Geological Society of Australia Special Publication, **23**, 217–237.

ROYDEN, L. H. 1993. The tectonic expression of slab pull at continental convergent boundaries. *Tectonics*, **12**, 303–325.

SCHEIBNER, E. & VEEVERS, J. 2000. Tasman fold belt system. *In*: VEEVERS, J. J. (ed.) *Billion-year Earth History of Australia and Neighbors in Gondwanaland*. GEMOC Press, Sydney, 154–234.

SENGOR, C. A. M. & NATAL'IN, B. A. 1996. Turkic-type orogeny and its role in the making of the continental crust. *Annual Review of Earth and Planetary Sciences*, **24**, 263–337.

SHERVAIS, J. W. 2001. Birth, death, and resurrection: The life cycle of suprasubduction zone ophiolites. *Geochemistry, Geophysics, Geosystems*, **2**, 2000GC000080.

SOESOO, A., BONS, P. D., GRAY, D. R. & FOSTER, D. A. 1997. Divergent double subduction: Tectonic and petrologic consequences. *Geology*, **25**, 755–758.

SPAGGIARI, C. V., GRAY, D. R. & FOSTER, D. A. 2002a. Blueschist metamorphism during accretion in the Lachlan Orogen, southeastern Australia. *Journal of Metamorphic Geology*, **20**, 711–726.

SPAGGIARI, C. V., GRAY, D. R., FOSTER, D. A. & FANNING, C. M. 2002b. Occurrence and significance of blueschists in the southern Lachlan Orogen. *Australian Journal of Earth Sciences*, **49**, 255–269.

SPAGGIARI, C. V., GRAY, D. R. & FOSTER, D. A. 2003a. Tethyan and Cordilleran-Type Ophiolites of Eastern Australia: Implications for the evolution of the Tasmanides. *In*: DILEK, Y. & ROBINSON, P. T. (eds) *Ophiolites in Earth's History*. Geological Society, London, Special Publications, **218**, 517–539.

SPAGGIARI, C. V., GRAY, D. R., FOSTER, D. A. & MCKNIGHT, S. 2003b. Evolution of the boundary between the western and central Lachlan Orogen: Implications for Tasmanide tectonics. *Australian Journal of Earth Sciences*, **50**, 725–749.

SPAGGIARI, C. V., GRAY, D. R. & FOSTER, D. A. 2003c. Formation and emplacement of the Dolodrook serpentinite body, Lachlan Orogen, southeastern Australia. *Australian Journal of Earth Sciences*, **50**, 709–723.

SPAGGIARI, C. V., GRAY, D. R. & FOSTER, D. A, 2004. Ophiolite accretion in the Lachlan Orogen, southeastern Australia. *Journal of Structural Geology*, **26**, 87–112.

SQUIRE, R. & MILLER, J. MC. 2003. Synchronous compression and extension in East Gondwana: Tectonics controls on world-class gold deposits at 440 Ma. *Geology*, **31**, 1073–1076.

SQUIRE, R. J., CAMPBELL, I. H., ALLEN, C. M. & WILSON, C. J. L. 2005. Did the Transgondwana supermountain trigger the explosive radiation of animals on Earth? *Earth and Planetary Science Letters*, **250**, 116–133.

TURNER, S. P., FODEN, J., KELLEY, S. P., VANDENBERG, A. H. M., SANDIFORD, M. & FLÖTTMANN, T. 1996. Source of the Lachlan fold belt flysch linked to convective removal of the lithospheric mantle and rapid exhumation of the Delamerian–Ross fold belt. *Geology*, **24**, 941–944.

VANDENBERG, A. H. M. & STEWART, I. R. 1992. Ordovician terranes of the southeastern Lachlan Fold Belt: Stratigraphy, structure and palaeogeographic reconstruction. *Tectonophysics*, **214**, 159–176.

VANDENBERG, A. H. M., WILLMAN, C. E., HENDRICKX, M., BUSH, B. D. & SANDS, B. C. 1995. *The geology and prospectivity of the 1993 Mount Wellington Airborne survey area*. Geological Survey of Victoria, Victorian Initiative for Minerals and Petroleum Report, **2**.

VANDENBERG, A. H. M., WILLMAN, C. E. ET AL. 2000. *The Tasman Fold Belt System in Victoria*. Geological Survey of Victoria Special Publication.

VEEVERS, J. J. 2000. *Billion-year Earth History of Australia and Neighbours in Gondwanaland*. GEMOC Press, Sydney.

WATSON, J. M. & GRAY, D. R. 2001. Character, extent and significance of broken formation for the Tabberabbera Zone, central Lachlan Orogen. *Australian Journal of Earth Sciences*, **48**, 943–954.

WILLIAMS, I. S. 2001. Response of detrital zircon and monazite, and their U–Pb isotopic systems, to regional metamorphism and host-rock partial melting, Cooma Complex, southeastern Australia. *Australian Journal of Earth Sciences*, **48**, 557–580.

WILLMAN, C. E., VANDENBERG, A. H. M. & MORAND, V. J. 2002. Evolution of the southeastern Lachlan Fold Belt in Victoria. *Australian Journal of Earth Sciences*, **49**, 271–289.

ZEN, E.-A. 1995. Crustal magma generation and low-pressure high-temperature regional metamorphism in an extensional environment: Possible application to the Lachlan Fold Belt, Australia. *American Journal of Science*, **295**, 851–874.

The Eurasian SE Asian margin as a modern example of an accretionary orogen

ROBERT HALL

SE Asia Research Group, Department of Geology, Royal Holloway University of London, Egham TW20 0EX, UK (e-mail: robert.hall@es.rhul.ac.uk)

Abstract: The Eurasian margin in SE Asia is a geologically complex region situated at the edge of the Sundaland continent, and is mainly within Indonesia. The external margins of Sundaland are tectonically active zones characterized by intense seismicity and volcanic activity. The region is an obvious modern analogue for older orogens, with a continental core reassembled from blocks rifted from Gondwana, and surrounded by subduction zones for much of the Mesozoic and Cenozoic. It is a mountain belt in the process of formation, and contains many features typically associated with older Pacific margin orogens: there is active subduction, transfer of material at subduction and strike-slip boundaries, collision of oceanic plate buoyant features, arcs and continents, and abundant magmatism. The orogenic belt surrounds Sundaland and stretches from Sumatra into eastern Indonesia and the Philippines. The orogen changes character and width from west to east. Its development can be tectonically described only in terms of several small plates and it includes several suture zones. The western part of the orogenic belt, where the Indian plate is subducted beneath continental crust, is a relatively narrow single suture. Further east the orogenic belt includes multiple sutures and is up to 2000 km wide; there is less continental crust and more arc and ophiolitic crust, and there are several marginal oceanic basins. The orogen has grown to its present size during the Mesozoic and Cenozoic as a result of subduction. Continental growth has occurred in an episodic way, related primarily to arrival of continental fragments at subduction margins, after which subduction resumed in new locations. There have been subordinate contributions from ophiolite accretion, and arc magmatism. Relatively small amounts of material have been accreted during subduction from the downgoing plate. In eastern Indonesia the wide plate boundary zone includes continental fragments and several arcs, but the arcs are most vulnerable to destruction and disappearance. Rollback in the Banda region has produced major extension within the collision zone, but future contraction will eliminate most of the evidence for it, leaving a collage of continental fragments, similar to the older parts of Sundaland.

The Eurasian SE Asian margin (Fig. 1) surrounds Sundaland, which is the continental core of SE Asia (van Bemmelen 1949; Katili 1975: Hamilton 1979; Hutchison 1989) and forms the southern part of the Eurasian plate. Sundaland is bordered to the west, south and east by tectonically active regions characterized by intense seismicity and volcanic activity (Fig. 2). This tectonically active zone is effectively a mountain belt in the process of formation, and contains many of the features typically thought to be associated with accretionary orogens: there is active subduction, transfer of material at plate boundaries, examples of collision with buoyant features on oceanic plates, arcs and continents, and abundant magmatism.

The present orogenic belt is situated at the junction of three major plates: the Eurasian, Indian–Australian and Pacific–Philippine Sea plates (Fig. 3). It surrounds Sundaland and stretches from Sumatra to the Philippines via eastern Indonesia. It changes character and width from west to east and is composed of different segments or sutures with different character. Gordon (1998) suggested that regions of deforming lithosphere, which he called diffuse plate boundaries, are in this region typically about 200 km wide except in east Indonesia where the deforming lithosphere may be as wide as 1000 km; McCaffrey (1996) reached a similar conclusion. The boundary between the Eurasian and Indian plates in western Indonesia is a relatively narrow zone of deformation at the active subduction margin extending from Sumatra to Java, which is now a continent–ocean boundary. East of Java and Borneo, the region is a wide and complex suture zone up to 2000 km wide, and even today can be tectonically described only in terms of several small plates and multiple subduction zones (Hamilton 1979; Hall & Wilson 2000; Hall 2002). In the east there is some continental crust but much more arc and ophiolitic crust compared with western Indonesia.

The SE Asian accretionary orogen has grown to its present size during the Mesozoic and Cenozoic (Fig. 3) and is therefore a useful comparison for older and longer-lived accretionary orogens, in terms of assessing the processes that formed them,

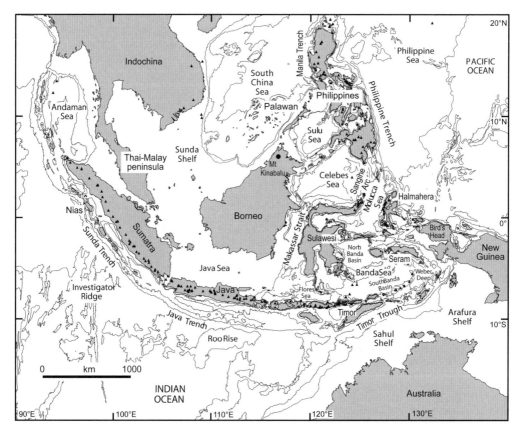

Fig. 1. Geography of SE Asia and surrounding regions. ▲ symbols, volcanoes from the Smithsonian Institution, Global Volcanism Program (Siebert & Simkin 2002); bathymetry is simplified from the GEBCO (2003) digital atlas. Bathymetric contours are at 200, 1000, 3000 and 5000 m.

the roles of different tectonic processes in crustal growth, and the amount of crustal growth. To make this assessment for the SE Asian accretionary orogen the principal tectonic elements of the orogenic belt are first identified and a brief history of each of the sutures is given. Then the growth of the orogen is considered, and its possible future development, leading to some conclusions of relevance for older orogenic belts.

Major structural elements of the region

At present the contrast between intense tectonic activity, manifested by seismicity and volcanism, at the margins of Sundaland and the apparent stability of its interior is striking. To describe the orogenic belt and discuss its Cenozoic development the region is here subdivided into a Sundaland continental core, surrounded by a number of different segments or sutures (Hall & Wilson 2000), each with different characteristics (Fig. 3). To the south and SW there is subduction of oceanic crust of the Indian plate at the Sunda and Java trenches, below the Sunda Arc (Fig. 2), and there is a relatively narrow plate boundary zone, here named the Sunda Suture, at the ocean–continent boundary that extends from Sumatra through Java into west Sulawesi. To the north of the Sundaland core is the Borneo Suture, which is situated at the political border of Indonesia with Malaysia and extends into the Philippines. There was active subduction at this boundary during the Eocene and Oligocene, and continent–continent collision in the Early Miocene. To the east of Sundaland is a complex region of continent–continent collision in the Sulawesi Suture, arc–continent collision in the Banda Suture, and arc–arc collision in the Molucca Suture. Hall & Wilson (2000) identified a Sorong Suture, which is characterized by strike-slip faulting and is here grouped with the Molucca Suture. These will arguably all be part of a single orogenic belt, which will remain after Australia–Eurasia collision is completed.

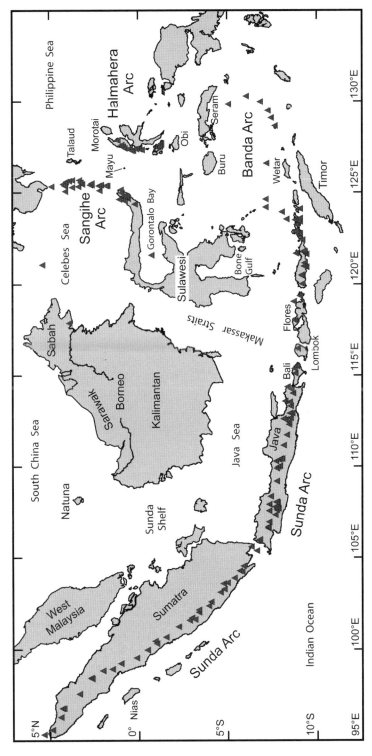

Fig. 2. The major volcanic arcs of the SE Asian margin between Sumatra and the southern Philippines.

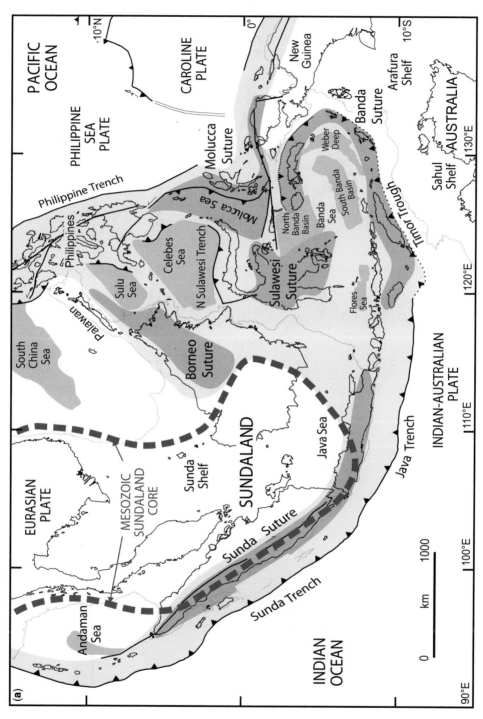

Fig. 3. (a) Principal tectonic features of the SE Asian accretionary orogen. The various suture zones within the orogenic belt are shown in orange. The area commonly considered to form the Mesozoic Sundaland continental core is outlined; the southern boundary is usually drawn at the limits of Cretaceous continental crust inferred by Hamilton (1979) shown in Figure 4. (b) The late Mesozoic and Cenozoic growth of the SE Asian accretionary orogen. It is suggested here that Sundaland grew in the Cretaceous by the addition of two main fragments: SW Borneo and east Java–west Sulawesi. In the Early Miocene new continental crust was added to Sundaland by collisions in Borneo and east Indonesia.

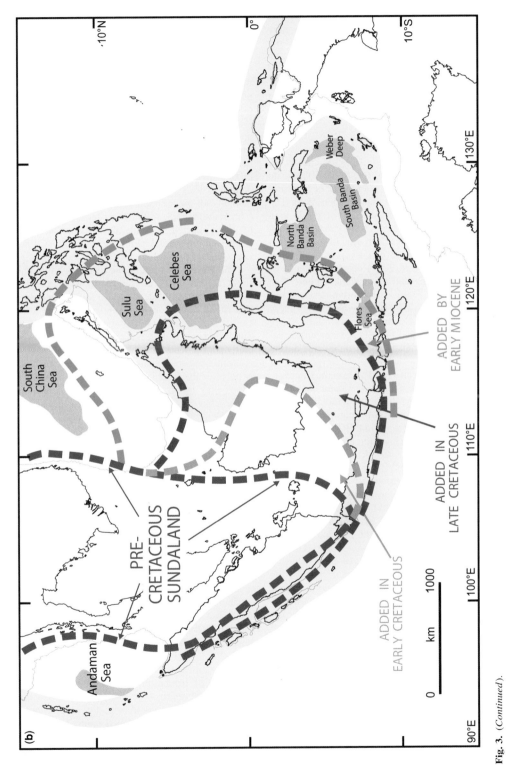

Fig. 3. (*Continued*).

Sundaland core

Sundaland comprises Indochina, the Thai–Malay Peninsula, Sumatra, Java, Borneo, and the shallow marine Sunda Shelf (Fig. 1), and formed an exposed landmass during the Pleistocene. Most of the shelf is flat and extremely shallow, with water depths considerably less than 200 m. The interior of Sundaland is largely devoid of seismicity and volcanic activity, in marked contrast to its margins. Seismicity (Cardwell & Isacks 1978; Engdahl et al. 1998) and global positioning system (GPS) measurements (Rangin et al. 1999; Michel et al. 2001; Bock et al. 2003; Simons et al. 2007) indicate that a SE Asian or Sunda microplate is currently moving slowly relative to the Eurasian plate. The Sunda Shelf is widely regarded as a stable area (e.g. Geyh et al. 1979; Tjia 1992) and sea-level data from the region have been used in construction of global eustatic sea-level curves (e.g. Fleming et al. 1998; Bird et al. 2007) despite evidence of very young faulting and vertical movements (e.g. Bird et al. 2006). Sundaland is often described as a shield or craton (e.g. Ben-Avraham & Emery 1973; Gobbett & Hutchison 1973; Tjia 1996; Barber et al. 2005) and there is a widespread misconception that Sundaland was a stable area during the Cenozoic. Some workers wrongly consider Sundaland to have been a rigid plate in the Cenozoic (e.g. Davies 1984; Replumaz & Tapponnier 2003; Replumaz et al. 2004) that rotated clockwise as a block during the last 8–10 Ma (e.g. Rangin et al. 1999) or over a longer period (Replumaz et al. 2004). However, the character of the Sundaland deep crust and mantle is very different from nearby continental regions and from cratons (Hall & Morley 2004; Hyndman et al. 2005; Currie & Hyndman 2006). Unlike the well-known shields or cratons (e.g. Baltic, Canadian, African, Australian) Sundaland is not underlain by a thick cold lithosphere that was stabilized in the Precambrian. The region has been very far from stable (Hall & Morley 2004) and it cannot have behaved as a single rigid block during most of the Cenozoic; the considerable evidence for a heterogeneous pattern of subsidence and elevation requires a much more complex tectonic model for the region with significant internal deformation of Sundaland (Hall 1996, 2002; Hall et al. 2008).

During the Late Palaeozoic and Mesozoic Sundaland grew by the addition of continental fragments rifted from Gondwana (Metcalfe 1990, 1996). There was amalgamation of continental blocks during the Triassic Indosinian orogeny, and during the Permo-Triassic stages of suturing there was extensive granite magmatism from Thailand to Malaya, associated with subduction preceding collision, and with post-collisional thickening of the continental crust (Hutchison 1989, 1996). Sundaland was part of Eurasia by the Early Mesozoic and since then has been largely emergent or submerged to very shallow depths. The boundary of the Sundaland continental core is commonly drawn to include part of west Borneo (Fig. 3) and the southern boundary follows the limits of Cretaceous continental crust interpreted by Hamilton (1979) from offshore drilling in the Java Sea (Fig. 4). In west Borneo the oldest rocks known are Palaeozoic metamorphic rocks intruded by Mesozoic granites (Tate 1996, 2001) in the Schwaner Mountains, and there are Devonian limestones as river boulders in east Kalimantan (Sugiaman & Andria 1999); detrital zircons derived from west Borneo indicate a Proterozoic continental basement (van Hattum 2005; van Hattum et al. 2006) similar to that inferred beneath the Malay peninsula from isotopic studies (Liew & McCulloch 1985; Liew & Page 1985).

The Mesozoic stratigraphic record is limited but suggests that much of Sundaland was emergent. Mesozoic terrestrial deposits are found throughout Sundaland whereas marine deposits are rare. Some continental fragments may have been added to Sundaland during the Cretaceous, separated by generally poorly exposed sutures that include Mesozoic ophiolitic and arc igneous rocks. The Luconia block is north of the Kuching zone of Hutchison (2005) which may mark a subduction margin continuing south from East Asia at which ophiolitic, island arc and microcontinental crustal fragments were added during the Mesozoic from Asia as there are rocks with Cathaysian affinities (Williams et al. 1988, 1989; Hutchison 2005). SW Borneo (Metcalfe 1990) may be another continental block that arrived in the Early Cretaceous, with a suture between the older part of Sundaland that runs south from the Natuna Islands (Fig. 2) where there are ophiolites. The Cretaceous granites in the Schwaner Mountains (Williams et al. 1988) and western Sarawak (Tate 2001) have been interpreted as the product of Andean-type magmatism but they are far from any possible subduction zones to the west and south. Granite magmatism might be explained as the product of flat-slab subduction from the west or south, but the positions of the Cretaceous arcs of southern Sundaland are well known in Sumatra, Java and SE Borneo (Hamilton 1979; Parkinson et al. 1998; Barber et al. 2005) and are outboard of the granites. If SE Sundaland in Borneo was rotated by 90° from its present position, as suggested by palaeomagnetism (e.g. Fuller et al. 1999) it is possible the granites could be part of an east-facing Andean margin that continued south from South China. Cretaceous granites are known in North China but it is still debated if they were products of a subduction margin (e.g. Lin & Wang 2006;

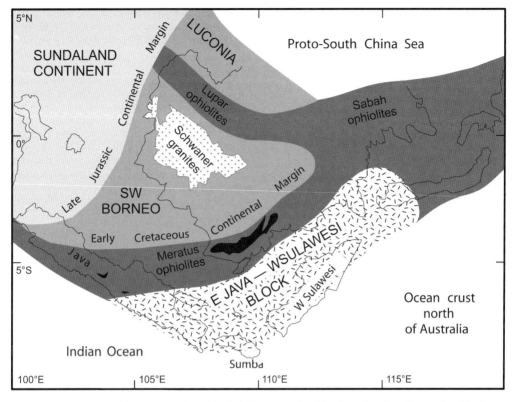

Fig. 4. Inferred position of the Late Jurassic and the Early Cretaceous Sundaland margin, adapted in part from Hamilton (1979), Parkinson et al. (1998) and Smyth et al. (2007). A suture between pre-Cretaceous Sundaland and the SW Borneo block is assumed to be present beneath the Sunda Shelf. Its position is uncertain. The extent and boundaries of the Cathaysian Luconia block are also uncertain. The youngest Cretaceous suture runs from Java and through the Meratus Mountains of SE Borneo, and to the south and east of this is a continental fragment derived from western Australia (Smyth et al. 2007).

Li & Li 2007; Yang et al. 2007). In South China acid magmatism ceased at the beginning of the Cretaceous (Sewell et al. 2000), and a younger Cretaceous granite belt equivalent to that in North China either does not exist or is offshore where there is currently no evidence for its existence. It is therefore very speculative if an Andean margin continued for at least another 2000 km further south into Borneo from North China, or even if such an Andean margin existed in the west Pacific. A simpler explanation is that the granites are products of collisional thickening. Further north in Sundaland Cretaceous granites have been interpreted as the products of continental collision (Barley et al. 2003). It is suggested here that Cretaceous granites of Borneo and the Sunda Shelf represent post-collisional magmatism following Cretaceous accretion of the SW Borneo continental fragment or the East Java–West Sulawesi continental fragment (see below).

The Sundaland Cretaceous active margin (Fig. 4) is interpreted to have run the length of Sumatra into west Java and then turned NE into SE Borneo (Hamilton 1979) as suggested by the distribution of Cretaceous high-pressure–low-temperature subduction-related metamorphism (Miyazaki et al. 1998; Parkinson et al. 1998). West Sulawesi and east Java are now also known to be underlain in part by Archaean continental crust, and geochemistry and zircon dating indicates derivation of this crust from the Australian margin (Bergman et al. 1996; Elburg et al. 2003; van Leeuwen et al. 2007; Smyth et al. 2007) and probable collision in the Cretaceous. Subduction ceased beneath Java in the Late Cretaceous following this collision (Smyth et al. 2007).

Little is known of the Late Cretaceous and Palaeocene history because of the paucity of sedimentary rocks of this age on land or offshore. By the beginning of the Cenozoic there was probably a passive margin on the south side of Sundaland, and possibly active margins, with uncertain polarity, to the west and east. During the Cenozoic, subduction around Sundaland, from Borneo to the Sunda Arc, via Sulawesi, the Banda Arc and the Moluccas, added to the core that had been constructed in the Mesozoic. The Cenozoic histories of the various sutures within the orogenic belt are summarized below; more detailed explanations, with an animated reconstructions, have been given by Hall (2002).

The Sunda Suture

The Cenozoic Sunda Arc formed on the southern side of the continental region formed by amalgamation of Gondwana fragments in the Mesozoic and is the product of subduction of the Indian–Australian plate beneath Sundaland. Subduction began in the early Cenozoic.

The Sunda Arc has been active since the Early Cenozoic, although opinions differ on the timing of subduction initiation (e.g. Hamilton 1988; Hall 2002; Crow 2005; Garwin et al. 2005; Whittaker et al. 2007). Typically, plate-tectonic reconstructions (e.g. Hall 1996, 2002; Metcalfe 1996; Barber et al. 2005) have assumed subduction at the Sunda Trench before 45 Ma, but there is little direct evidence to support this assumption. From Sumatra to Sulawesi the southern part of Sundaland was probably entirely emergent during the Late Cretaceous and Early Cenozoic and there was widespread erosion; throughout the region the oldest Cenozoic rocks rest unconformably on Cretaceous or older rocks. In Java, there is no evidence to support the interpretation of Whittaker et al. (2007) of subduction beneath the Sunda Arc during the Late Cretaceous and Early Cenozoic (Smyth et al. 2007), although there may have been some volcanic activity in south Sumatra (Crow 2005), Sulawesi (van Leeuwen 1981) and Sumba (Burollet & Salle 1981a; Abdullah et al. 2000). In west Sulawesi, the oldest Cenozoic sedimentary rocks rest in places on poorly dated volcanic rocks that may be Cretaceous or Palaeocene (Calvert & Hall 2007). It is suggested here that collision of continental fragments at the Sunda Trench in the Late Cretaceous terminated subduction until the Eocene.

From the Middle Eocene there was widespread extension and basin formation. In Sumatra the Palaeogene arc was in a similar position to older subduction arcs (McCourt et al. 1996; Crow 2005) but in Java it formed well to the south of the youngest Cretaceous active margin, which extended from west Java, through central Java, to the Meratus Mountains of SE Borneo (Hamilton 1979; Parkinson et al. 1998). Volcanic activity began from Sumatra to Sulawesi after about 45 Ma, when Australia began to move northwards relatively rapidly as the rate of Australia–Antarctica separation increased. In Sumatra volcanic activity became widespread from the Middle Eocene (Crow 2005). Recent work in Java suggests that subduction resumed in the Middle Eocene, forming a volcanic arc that ran the length of Java (Hall et al. 2007; Smyth et al. 2007). Further east, rifting had begun in the Makassar Straits by the Middle Eocene and there was calc-alkaline magmatism in parts of west Sulawesi, interpreted to be subduction-related (Polvé et al. 1997) and representing the continuation of the Sunda Arc from Sumatra and Java. The geological evidence from Sumatra, Java and Sulawesi suggests that there was a passive margin at the southern edge of Sundaland during the Late Cretaceous and Early Cenozoic, and subduction resumed only when Australia began to move north in the Eocene.

Since the Eocene there has been continuous subduction of ocean lithosphere beneath the Sunda Arc. During the Eocene and Oligocene, from Sumatra to Sulawesi, abundant volcanic activity accompanied northward subduction of the Indian–Australian plate. In the Java–Sulawesi sector of the Sunda Arc volcanism greatly diminished during the Early and Middle Miocene although northward subduction continued. This diminution in volcanic activity is not evident in Sumatra although there may have been a period of reduced volcanic activity in the Late Miocene (Crow 2005). The decline in magmatism between Java and Sulawesi, despite continued subduction, is interpreted as the result of subduction hinge movement preventing replenishment of the mantle wedge (Macpherson & Hall 1999, 2002). Following Australian continental collision in east Indonesia (Fig. 5), the Sunda subduction hinge advanced northwards as Borneo rotated counter-clockwise (Fuller et al. 1999). At the end of the Middle Miocene at about 10 Ma volcanic activity resumed in the Java–Sulawesi sector of the Sunda Arc, after the termination of Borneo–Java rotation. In Java the new arc formed in a position north of the Palaeogene arc. Since the Late Miocene there has been thrusting and contractional deformation in Sumatra and Java, which may be related to arrival of buoyant features at the trench or increased coupling between the overriding and downgoing plates, or both.

Borneo Suture

Northern Borneo records a collision between the extended passive continental margin of South China and an active margin on the northern side of Sundaland (Rangin et al. 1990; Tan & Lamy 1990; Tonkgul 1991; Hazebroek & Tan 1993; Hutchison et al. 2000; Hutchison 2005). Collision began in the Early Miocene after Eocene–Oligocene subduction of the proto-South China Sea (Fig. 5). East of Borneo, the continent–continent collision suture

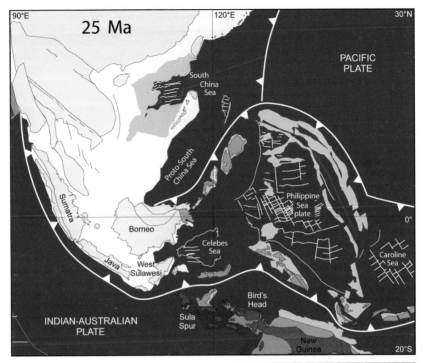

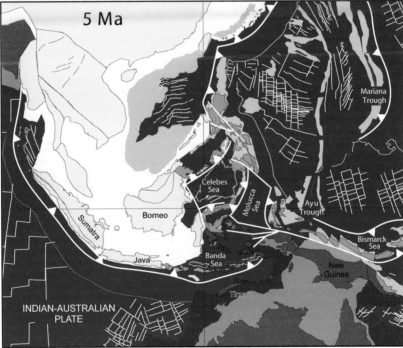

Fig. 5. Reconstructions of SE Asia at 25 and 5 Ma modified from Hall (2002). At 25 Ma collision was about to begin at the Borneo Suture as the Proto-South China Sea was eliminated by subduction beneath north Borneo, and at the Sulawesi Suture as the Sula Spur collided with the Sunda Arc in east Sulawesi. At 5 Ma collision between the Sangihe and Halmahera Arcs in the Molucca Suture was about to begin, as the Molucca Sea was eliminated by subduction. Rollback of the subduction hinge south of Sulawesi into the Banda embayment caused extension within the orogen and the development of new young oceanic crust in the North and South Banda basins. Collision between the Banda volcanic arc and the Australian continent was about to begin in East Timor.

passes into an arc–continent collisional suture between the Neogene Cagayan arc and the Palawan sector of the South China rifted margin.

In north and east Borneo the basement is predominantly ophiolitic and mainly of Cretaceous age. In north Borneo possible older crust is indicated by K–Ar dating of igneous and metamorphic rocks (Reinhard & Wenk 1951; Dhonau & Hutchison 1966; Koopmans 1967; Leong 1974), which could be deformed ophiolitic rocks. The ophiolitic rocks are intruded by diorites and granites, which resemble plutonic rocks of a Mesozoic intra-oceanic arc (Omang & Barber 1996; Hall & Wilson 2000). The ophiolite–arc complex was emplaced before the Eocene (Omang & Barber 1996).

At the beginning of the Cenozoic the Borneo part of eastern Sundaland was separated from South China by the proto-South China Sea. In north Borneo deep marine sediments of the Upper Cretaceous–Eocene Rajang Group form much of the Central Borneo Mountains and Crocker Ranges interpreted as an active (Haile 1969, 1974; Hutchison 1973, 1989, 1996; Hamilton 1979; Holloway 1982; Williams et al. 1988, 1989) or passive margin (Moss 1998). This depositional phase was terminated by uplift and deformation in the Sarawak orogeny (Hutchison 1996). As explained above, passive margins around Sundaland before 45 Ma is the interpretation favoured here, implying that this orogeny marks the initiation of subduction of the proto-South China Sea beneath northern Borneo, rather then a collisional event. From the Eocene there was an accretionary margin in north Borneo with subduction towards the SE. The Rajang Group is unconformably overlain by Eocene to Lower Miocene Crocker turbidites (Hutchison 1996). The Crocker Fan (Crevello 2001; van Hattum et al. 2006) is the largest volume of Palaeogene sediment in a single basin in SE Asia with a thickness that may locally exceed 10 km (Collenette 1958; Hutchison 1996). Crocker Fan deposition terminated in the Early Miocene as the thinned passive margin of South China underthrust north Borneo during the Sabah orogeny (Hutchison 1996) forming the major Top Crocker Unconformity (TCU) of van Hattum et al. 2006).

After the collision there was a significant change in the character of sedimentation around Borneo. Large amounts of clastic sediments began to pour into the deep basins to the north and east of the island and major delta systems formed, which prograded rapidly away from the island. Since the Early Miocene collision Borneo has continued to rise and shed vast amounts of sediment into the deep basins in and around the island, filling pre-existing accommodation space in the South China Sea, Celebes Sea and Makassar Straits margins (Hall & Nichols 2002; Hall & Morley 2004), and new accommodation space related to Sulu Sea extension and flexural loading, supplied from erosion of the central Borneo highlands to the south and west. In north Borneo marginal and shallow marine sediments were deposited unconformably upon the older accretionary complex rocks on land. Offshore the sequences above the TCU include shelf and delta deposits, turbidites and debris flows (Sandal 1996; Petronas 1999; McGilvery & Cook 2003). Since the end of the Miocene there has been emergence, uplift and exhumation of NE Borneo.

There was little Cenozoic plutonic magmatism but north Borneo includes the highest mountain in SE Asia, the 4100 m granite peak of Mt Kinabalu (Fig. 1). The cause of Kinabalu melting is uncertain. A geochemical study (Vogt & Flower 1989) argued that the composite body could be the product of multiple intrusions with separate sources, or remelting of underplated basic lower crust that produced and mixed with silicic melts, modified by fractional crystallization. Subduction of the proto-South China Sea was suggested as the cause of melting. However, it is now known that melting post-dated the end of subduction by more than 10 Ma. K–Ar dating has yielded a range of ages to 14 Ma (Jacobson 1970; Bellon & Rangin 1991; Swauger et al. 2000) but thermochronological work currently in progress at Royal Holloway University of London (M. A. Cottam & C. Sperber, pers. comm.) indicates a crystallization age of 7–8 Ma. Chiang (2002) suggested fluid-absent melting of arc-like lower crust as a result of collisional thickening after subduction ceased. After melting the body was emplaced at 3–10 km in the crust (Vogt & Flower 1989). Reconnaissance fission-track data indicate 4–8 km of Late Miocene denudation throughout the Crocker Range (Hutchison et al. 2000) Thermochronological work in progress indicates Pliocene exhumation of Kinabalu and young elevation of the mountain to several kilometres above sea level (C. Sperber, pers. comm.).

The rapid exhumation on land was accompanied offshore by shelf edge progradation and by folding and thrusting, including inversion of growth structures (Morley et al. 2003), and by repeated failure of the shelf producing deep-water debris flows, which continues to the present (McGilvery & Cook 2003). The cause of deformation is uncertain. Geochemistry of Sabah volcanic rocks (Chiang 2002) indicates a change in mantle source character of melts in the Pliocene. The similarity in timing of the end Miocene emergence, the deformation, and the regional change in magmatic character, suggests a deep cause such as lithospheric delamination. The melting and rise of Kinabalu could be due to loss of a lithospheric root or breakoff of a subducted slab (van de Zedde & Wortel 2001). Mantle tomography (Bijwaard et al. 1998) shows a poorly resolved high-velocity anomaly beneath north Borneo, which could be a detached root or slab. There is no evidence for young plate convergence at the north Borneo continental margin, and the deformation and exhumation are interpreted here as a gravity-driven response to loss of a lithospheric root.

Eastern Indonesia: Sulawesi Suture

Sulawesi contains a complex record of multiple tectonic events at the Cenozoic eastern Sundaland margin that are still poorly understood; they include rifting, arc activity, ophiolite emplacement, microcontinental collisions and strike-slip faulting.

At the beginning of the Cenozoic Borneo and west Sulawesi formed the eastern part of Sundaland (Katili 1978; Hamilton 1979) following the Cretaceous collisions of Gondwana fragments. The basement in these areas includes Mesozoic continental metamorphic rocks and ophiolites. West Sulawesi then separated from east Borneo by rifting that began in the Middle Eocene (Situmorang 1982a, b; Hall 1996; Moss et al. 1997; Guntoro 1999; Calvert & Hall 2007) and led to formation of the Makassar Straits. The cause of rifting is not known; Hall (1996) and Fraser & Ichram (2000) proposed that rifting developed as the spreading centre of the Celebes Sea propagated into the Makassar Straits, suggesting a link to marginal basin formation in the west Pacific. It may also be a consequence of back-arc extension of the Sunda margin associated with development of the Cenozoic Sunda Arc.

Debate continues about whether the crust beneath the Makassar Straits is oceanic (e.g. Cloke et al. 1999; Guntoro 1999) or continental (e.g. Burollet & Salle 1981b; Situmorang 1982a, b). Hamilton (1979) showed a spreading centre extending all the way down the straits. However, recent studies based on new seismic data suggest attenuated continental rather than oceanic crust in both the deeper north and shallower south Makassar Straits (Nur'Aini et al. 2005; Puspita et al. 2005). After rifting the Sulawesi side of the Makassar Straits was a passive margin with Eocene and Oligocene shelf carbonate and mudstone deposition during the Eocene and Oligocene, and an elevated but low continental area to the east. Further east was a boundary with the northward-moving Australian plate, which was a strike-slip and subduction boundary, now largely destroyed by subsequent deformation. In SW Sulawesi there was a Palaeogene arc that represents the eastern continuation of the Sunda Arc.

During the Early Miocene a promontory of the Australian margin, the Sula Spur, collided with the eastern end of the Sunda Arc, causing ophiolite emplacement in SE Sulawesi (Fig. 5). The ophiolites of east, SE and SW Sulawesi represent a forearc, subduction zone and oceanic area between Sundaland and the Australian plate (Silver et al. 1978, 1983a; Mubroto et al. 1994; Wijbrans et al. 1994; Monnier et al. 1995; Parkinson 1998a, b). Since the Early Miocene other fragments of continental crust have been sliced from the Bird's Head microcontinent and transported west along the Sorong fault system to collide with Sulawesi.

Sulawesi is far from adequately understood and has a complex history still to be completely unravelled. The deep Gorontalo Bay and Bone Gulf (Fig. 2) are enigmatic features between the arms of Sulawesi, which may be the product of Mio-Pliocene extension. Recent work (van Leeuwen & Muhardjo 2005; van Leeuwen et al. 2007) in north and west Sulawesi suggests Miocene core complex metamorphism, and there are rapidly exhumed upper mantle and lower crustal rocks (Helmers et al. 1990) in the Palu–Koro region, where there are also recently exhumed young granites (Priadi et al. 1999; Bellier et al. 2006) and active faulting (Sukamto 1973; Tjia 1973; Stevens et al. 1999; Beaudoin et al. 2003).

Geological mapping, palaeomagnetic investigations and GPS observations indicate a complex Neogene history of deformation in Sulawesi including extension, block rotations, and strike-slip faulting accompanying uplift and exhumation (e.g. Silver et al. 1983b; Surmont et al. 1994; Vigny et al. 2002). During the Pliocene the character of sedimentation across the whole of western, central and eastern Sulawesi changed significantly. Compressional deformation began at the beginning of the Pliocene (Fig. 5), partly as result of the collision of the Banggai–Sula Block and east Sulawesi, causing contraction and uplift throughout Sulawesi, and in particular the rise of mountains in west Sulawesi (Calvert 2000; Calvert & Hall 2007). The compressional episode formed a fold–thrust belt in western Sulawesi which has since propagated westwards into the north Makassar Straits (Puspita et al. 2005). Convergence continues in northern Sulawesi at the present day. There is subduction of the Celebes Sea to the south beneath the north arm of Sulawesi, and subduction on the opposite side of the north arm at its eastern end of the Molucca Sea towards the west. However, there is no volcanic activity as a result of subduction of the Celebes Sea, and the few volcanoes at the east end of the north arm represent the last stages of volcanism caused by subduction of the Molucca Sea.

Eastern Indonesia: Banda Suture

The Banda Arc is the horseshoe-shaped arc that today extends east from Flores to Buru, passing through Timor and Seram, and includes both an outer non-volcanic arc and an inner volcanic arc. It is an unusual region of major extension that developed within the collision zone after Australia–Sunda Arc collision began. It formed by subduction of an oceanic embayment within the northward-moving Australian plate (Hamilton 1979; Hall 1996, 2002; Charlton 2000) as a result of subduction hinge rollback following collision.

After the Eocene the Sunda Arc can be traced east through Sulawesi into the Pacific arcs of the east Philippines and Halmahera. Remnants of the Palaeogene Sunda Arc are found in the uppermost nappes of Timor and in other Banda islands where there are Cenozoic volcanic rocks and ophiolitic rocks formed above a north-dipping subduction zone. The first collision of Australian crust with SE Asia began in the Early Miocene (Fig. 5). This caused the long Sunda–east Philippines–Halmahera subduction system to separate into two parts. West of Sulawesi, northward subduction continued beneath the Sunda Arc, but to the east subduction ceased, and the Australia–Philippine Sea plate boundary became a strike-slip system (Hall 2002). Between the two, Australia–SE Asia convergence formed a mountain belt in east Sulawesi.

By the late Middle Miocene, at about 12 Ma, convergence in east Sulawesi could no longer be accommodated by orogenic contraction. At this time the oldest oceanic lithosphere in the Indian Ocean, of Jurassic age, arrived at the eastern end of the Java Trench. This area of old crust north of the NW Shelf of Australia formed an embayment in the Australian margin, the proto-Banda Sea. A tear developed at the northern edge of the embayment, which propagated east. Because of its age and thickness, the Jurassic ocean lithosphere fell away rapidly, causing the Banda subduction hinge to roll back very rapidly to the south and east, inducing massive extension in the overlying plate. In western Sulawesi this induced extensional magmatism, which began at about 11 Ma (Polvé et al. 1997). As the hinge rolled back into the Banda embayment (Fig. 5) it led to formation of the Neogene Banda Volcanic Arc and opening of the North Banda Sea, the Flores Sea and later the South Banda Sea (Hinschberger et al. 2001, 2003, 2005).

The amount of rollback was limited by the size of the embayment south of the Bird's Head microcontinent. Rollback resulted in collision at about 3 Ma of the Banda Volcanic Arc with the Australian margin in the region of Timor (Audley-Charles 2004). After collision, convergence and volcanic activity ceased in the Timor sector although volcanic activity continued to the west and east. New plate boundaries developed north of the arc between Flores and Wetar and to the north of the South Banda Sea associated with subduction polarity reversal (Price & Audley-Charles 1987). The subduction hinge has now reached the continent–ocean boundary at the margins of the embayment. The large region of extension is now contracting. At the southern margins of the Bird's Head microcontinent there has been significant shortening and intra-continental subduction within the last 3 Ma south of the Seram Trough (Stevens et al. 2002; Pairault et al. 2003).

East Indonesia: Molucca Suture

In the northern part of eastern Indonesia the Halmahera and Sangihe arcs (Fig. 2) are the only arcs in the world currently colliding. The Sangihe Arc is arguably on the outermost edge of Sundaland and can be traced from Sulawesi northwards into the Philippines; the deepest parts are built on oceanic crust. The modern Halmahera Arc is constructed on older arcs, of which the oldest known is an intra-oceanic arc formed in the Pacific in the Mesozoic (Hall et al. 1988, 1995b) presumably built on older oceanic crust. Both of the currently active arcs formed during the Neogene. The Molucca Suture differs from other parts of the Sundaland margins in the absence of continental crust. There are a few slivers of continental crust caught up in the Sorong fault zone, at its southern end, but they are very small.

The Sangihe Arc is constructed on Eocene oceanic crust (Evans et al. 1983) and formed near the Sundaland margin in the Early Cenozoic (Hall 2002). Before the Eocene the location of the Halmahera Arc is not well known; at 45 Ma it was at equatorial latitudes (Hall et al. 1995a) far out in the western Pacific on the southern margin of the Philippine Sea plate. Between 45 and 25 Ma the Philippines–Halmahera Arc developed above a north-dipping subduction zone where there was subduction of Indian–Australian lithosphere north of Australia as Australia moved north (Hall 1996, 2002; Hall & Spakman 2002).

At about 25 Ma (Fig. 5) there was arc–continent collision between the east Philippines–Halmahera Arc and northern Australian margin in New Guinea that terminated northward subduction of oceanic lithosphere north of Australia (Hall et al. 1995a, b). The Philippine Sea plate rotated clockwise and moved northwards after the collision. A major strike-slip boundary developed in northern New Guinea and arc terranes were translated westwards within the left-lateral Sorong fault zone. At the western end of the fault system there was subduction beneath the Sangihe Arc and collisions in Sulawesi of fragments sliced from New Guinea.

Initiation of east-directed Halmahera subduction probably resulted from locking of strands of the left-lateral Sorong fault zone at the southern edge of the Molucca Sea as a result of collisions in Sulawesi. The present-day Molucca Sea double subduction system was initiated at about 15 Ma and the oldest Neogene volcanic rocks in the Halmahera Arc have ages of about 11 Ma (Baker & Malaihollo 1996). Since 11 Ma the Molucca Sea has been eliminated by subduction at both its eastern and western sides (Fig. 5). The two arcs first came into contact at about 3 Ma and began to build the central Molucca Sea accretionary complex as the two forearcs collided. Convergence did not cease as the two forearcs came into contact; instead, there was repeated shortening on the Halmahera side with thrusting of both forearc and back-arc towards the active volcanic arc (Hall 2000; Hall & Smyth 2008). At the same time, parts of the Halmahera Arc have disappeared by overthrusting of the Sangihe–north Sulawesi Arc in the Molucca Sea. The central Molucca Sea mélange wedge and ophiolites represent the forearc basin and basement of the Sangihe Arc that will soon have completely overridden the Halmahera Arc (Hall 2000).

Growth of the accretionary orogen

If it is assumed that mountain building will cease only after complete continent–continent collision, then the SE Asian orogenic belt has a long future. South of Sumatra and Java there is no northward-moving continent, and in the eastern part of the region Australia continues to move north despite peripheral continent–continent collision between Sundaland and westernmost Australian margin that began about 25 Ma ago. The orogenic belt is almost entirely a result of subduction at the Sundaland margins. However, conventional subduction processes have not caused the orogenic belt to grow significantly. There has been little growth at accretionary wedges or by magmatism. There have been no major additions by collisions of plateaux or other topographic features on oceanic plates. The late Mesozoic and Cenozoic growth of the orogen predominantly reflects the arrival of continental crust at the subduction trench, which was then added to the orogenic belt, terminating subduction for a period, and causing the site of subduction to move. It is suggested here that during the Cretaceous first the SW Borneo fragment was added, followed by the east Java–Sulawesi fragment (Fig. 3). In the Cenozoic part of the South China margin was added in north Borneo (Fig. 3), at about the same time that Australian continental crust first collided with the Sundaland margin in eastern Indonesia (Fig. 5). Complex deformation in eastern Indonesia continues to the present day.

Accretion

In the west the accretionary orogen is simply a single-sided arc–trench system with an arc–trench gap up to 350 km wide and a total width of orogenic deformation of about 400 km (Simandjuntak & Barber 1996). This Sunda Suture is the narrowest part of the orogenic belt (Fig. 3). At the northwestern end of Sumatra mud-dominated sediment at the distal end of the Bengal Fan (Stow et al. 1989; Curray 1994) is arriving at the Sunda Trench. Nias, at the north end of the Sumatra forearc, is commonly cited as the classic example of a forearc high elevated by accretion, based on the work of Moore & Karig (1980). However, later mapping of Nias and other forearc islands does not support this model (Samuel 1994; Samuel et al. 1995, 1997; Samuel & Harbury 1996). The ophiolites are not slices accreted during the Cenozoic but Cretaceous basement that was in place by the Eocene. They unconformably overlain by thin Eocene marine rocks, several kilometres of Oligocene to Lower Miocene deep marine clastic sedimentary rocks, and 1–2 km of Lower to Upper Miocene shallow marine clastic sedimentary rocks with a few limestones and some tuff layers. The sedimentary rocks are not material scraped off the downgoing plate but represent the fill of a forearc basin, later inverted, in which most material was carried from Sumatra and included recycled Sundaland continental crust and volcanic material. A large part of the material that has been incorporated in the Sunda Suture is reworked from Sumatra rather than being new crust added to the continental margin.

Further SE, to the south of south Sumatra and Java, the sedimentary cover on the subducting plate is much thinner yet the arc–trench gap is just as wide. Near the trench there is a young accretionary complex (Masson et al. 1990) constructed against an older backstop interpreted to be an older accretionary complex (Kopp et al. 2002, 2006; Schlüter et al. 2002) but this is only about 50 km wide. Schlüter et al. (2002) suggested there was a progressive growth of the arc–trench gap south of Sumatra since the Palaeocene but the age control on this development is poor. In the region where the Roo Rise is arriving at the trench there is evidence of subduction erosion (Masson et al. 1990; Kopp et al. 2006). There seems to have been relatively little growth of the forearc region by the addition of material from the downgoing plate and the width of material added is less than 200 km (Hall & Smyth 2008).

Accretion of oceanic plateaux and seamounts

It is difficult to identify any major oceanic plateaux or seamounts in the Sundaland margins anywhere in Indonesia, although such features are known from the older sutures in Indochina and SW China (Metcalfe 1996, 1998). This implies that they have generally disappeared during subduction, leaving only ephemeral traces of their return to the mantle in the form of deformation, debris flows and slumped units, which may or may not be preserved. South of Java and Sumatra such features arriving at the trench today (Fig. 1) cause deformation but seem to subduct without difficulty. South of Java the Roo Rise is contributing to subduction erosion (Kopp et al. 2006), the trench shallows where the Roo Rise has arrived at the trench, and there are embayments associated with seamount subduction (Masson et al. 1990). The subduction of the Investigator Ridge beneath Sumatra has been linked to elevation of the forearc (McCann & Habermann 1989; Malod & Kemal 1996), although this has been disputed (Matson & Moore 1992), to activity in the arc and possibly to the eruptions of Toba (Chesner 1998; Page et al. 1979) but the ridge is being subducted relatively smoothly and without accretion.

Accretion during arc collision

In the central Molucca Sea the collision between the Sangihe and Halmahera arcs has formed a substantial accretionary complex as the two forearcs have come into contact. However, it is far from clear if this will ultimately result in addition of a large volume of arc crust to the Sundaland margin.

As the Molucca Sea was eliminated from south to north in the last 3 Ma the Halmahera Arc failed repeatedly with thrusting in different directions at different stages in the collision (Hall 2000; Hall & Smyth 2008). First the back-arc was thrust over the volcanic arc, and later the forearc was thrust towards the volcanic arc. In south Halmahera the back-arc region was thrust onto the forearc, in places entirely eliminating the Neogene arc. At the southern end of the Halmahera Arc the arc was thrust onto the forearc (Ali & Hall 1995; Ali et al. 2001). After west-vergent thrusting volcanism in the Halmahera Arc resumed between Bacan and north Halmahera. At the south and north ends of the arc volcanism ceased. In the northern Molucca Sea the Sangihe forearc was then thrust east onto the Halmahera forearc and arc. In the region between Morotai and the Snellius Ridge parts of the Neogene Halmahera Arc and forearc have now disappeared (Hall 2000). Further south this east-vergent thrusting carried the Halmahera forearc onto the flanks of the active Halmahera Arc, and pre-Neogene rocks of the Halmahera forearc basement are now exposed in islands of the Bacan group and off the coast of NW Halmahera.

In the northern Molucca Sea almost the whole of the forearc and volcanic arc have disappeared (Hall 2000; Widiwijayanti et al. 2004). The two arcs are still converging, and contracting in the process. One arc (Sangihe) is overriding the other (Halmahera) and parts of the Halmahera arc system (meaning the entire region between the trench and back-arc region) have disappeared by thrusting (Hall 2000; Hall & Smyth 2008). Little is known about the deep structure of the collision zone and it is not clear if the overthrusting of one arc by the other is causing an increase in crustal thickness, or if the Halmahera Arc is being subducted into the mantle, in which case it will disappear completely.

At the present there is a large area of young arc crust in eastern Indonesia, and certainly in the Philippines, but the probability of this crust remaining after termination of orogenic activity is small. The collision of the two arcs in the Molucca Suture would be expected to add new crust to the Sundaland margin, but as the convergence has proceeded the width of the arc has progressively decreased. In areas where there has been addition of arcs to the Sundaland margins in the past, such as the Cretaceous Woyla Arc of north Sumatra, or in the Meratus Mountains of SE Borneo, the width of the preserved arc is small. Where continent–continent collision has been achieved in the SE Asian orogenic belt the area of arc crust is small compared with that of continental crust.

Magmatism

The amount of new crust added by magmatism is very difficult to assess. Volcanic rocks are common in all the arc regions of Indonesia and the Philippines, large areas of plutonic rocks are common in Sumatra (McCourt et al. 1996), and smaller plutons are known from Borneo, Sulawesi and Java. The Schwaner Mountains granites of SW Borneo represent a significant volume of melt, although still small compared with the width of the orogen, but if they are post-collisional melts, as suggested above, they are not new crust, but simply melts formed from existing crust. In the young Cenozoic arcs, such as the Lesser Sunda Islands, the Banda Volcanic Arc, the north arm of Sulawesi, the Sangihe and Halmahera arcs, only the upper levels of the arcs are seen. However, in all the arcs the width of the magmatic zone is rather narrow, typically less than 100 km. Unlike areas such as South America (Pitcher et al. 1985; Kay et al. 2005; Haschke et al. 2006), where there has been long-term magmatism that intrudes and extrudes across a wide zone, the positions of magmatic arcs in Indonesia appear to have been relatively stable over time. Even in Sumatra with its long record of igneous activity that extends back to the Late Palaeozoic, the present volcanic arc is built upon older plutonic belts and the total width of the zone of magmatism is of the order of 100–200 km (McCourt et al. 1996; Cobbing 2005). In many of the younger arcs the position of the volcanoes has remained fixed, although there have been abrupt movements of volcanic activity to new positions in some arcs such as Java and Halmahera (Hall & Smyth 2008); none the less, even in these arcs the shift was only of the order of 50 km. Thus, if it is assumed that all the modern arcs are underlain by plutons of similar width that will be exposed in the future, and that all the arc crust is preserved and not lost by subduction in the process of collision, the total width of new crust added by magmatism is probably no more than 20% of the entire width of the orogenic belt, and in many parts of Indonesia is much less.

Continental fragments

The central segment of the accretionary orogen is closest to completion (Fig. 3). Considering a NW–SE profile from the South China Sea to Australia there is little or no oceanic crust remaining to subduct and there is continental crust, mainly already thickened, from Indochina to Australia. As Australia moves north there will be some potential for further contraction but the orogenic belt is currently between 2000 and 2500 km wide and is unlikely to become much narrower. This sector crosses three Cenozoic sutures (Borneo, Sulawesi and Banda) and at least two Mesozoic sutures (the north Borneo Lupar Line and the SE Borneo Meratus Mountains). The sutures include arc and ophiolitic rocks but most of the width of the orogen is made up of continental crustal blocks. These have separated from, or represent, the extended South China margin, or have separated from the western Australian margin. This sector encapsulates the history of the accretionary orogen, and represents its likely future: the orogen has grown principally by the addition of continental fragments to the Sundaland core. The growth has been incremental, and as convergence between Australia and Eurasia has continued the subduction zones at the Sundaland margin have moved outboard of the most recently added fragments.

Future development of orogen

In western Indonesia the orogen will continue to develop as the product of ocean–continent convergence for many tens of millions of years unless continent–continent collision between Australia and SE Asia causes subduction to cease, or there are major changes in plate motions as a result of other causes. However, judging from its Cenozoic history there is no reason to expect significant increase in size of the orogen. There are no continental fragments in the Indian Ocean that will arrive on the Sunda margin and areas of elevated crust are currently subducting without difficulty. There will be addition of new crust by magmatic activity and possibly some accretion, although this could be partly offset by subduction erosion.

To the east, the development of the orogen is far from complete and is likely to be very complex. Between South China and Australia there are several marginal basins floored by oceanic crust (Fig. 3a and b): the South China Sea, the Sulu Sea, the Celebes Sea, and the North and South Banda basins. Several of these are beginning to disappear by subduction at their margins (Sulu, Celebes and South Banda) and as Australia moves north it might be expected that each of these oceans will disappear leaving a wide orogenic belt composed of continental fragments, arc and ophiolitic crust. However, the Neogene history of the Sulawesi and Banda sutures suggests caution in reaching this conclusion. The arrival at the Sunda Trench of the oceanic Banda embayment led to rapid rollback and induced extension in Sulawesi, and formation of new ocean crust and arcs in the Banda Sea. It is possible to imagine the Banda scenario repeated for the Celebes Sea. For example, the present north Sulawesi trench could migrate to the north, causing renewed extension of Sulawesi. This could lead to formation of small new ocean basins of similar size to the Banda basins within the orogenic belt. These, in turn, would be subducted but their subduction could lead to renewed hinge rollback, extension and formation of new marginal basins. It may appear that this is merely delaying the inevitable completion of collision as Australia becomes sutured to East Asia, but the Mediterranean illustrates that hinge rollback in small ocean basins can be a very persistent process (e.g. Negredo et al. 1997; Wortel & Spakman 2000; Rosenbaum & Lister 2004). The Banda Arc is a small-scale example of Gondwana reassembly that has formed Sundaland. Continental fragments now found in the outer arc are the result of extension following Early Miocene collision of the Sula Spur in Sulawesi. As the Banda basins close by subduction these continental fragments are being restored approximately to their Early Miocene position in the orogenic belt.

At the eastern end of the orogen there is essentially a complex intra-oceanic arc extending from eastern Indonesia into the Philippines. There is currently subduction in several different directions, and ocean crust on both sides of the intra-oceanic arc in marginal basins of the Eurasian margin (South China, Sulu and Celebes seas) and the Philippine Sea. It is unclear if this arc will end up as part of the Eurasian margin, or if most of it will continue to move around the Pacific by related subduction and strike-slip faulting.

Conclusions

It is impossible to say if SE Asia is a typical accretionary orogen. In the 100 Ma of its Late Cretaceous and Cenozoic history there have been numerous events, and many of them would not be recognized in a Palaeozoic or Precambrian accretionary orogen simply because of the resolution of most dating methods. Furthermore, much of the evidence has been removed or degraded by subsequent events. Arcs and ocean basins are disappearing or have disappeared with little or no trace.

Each sector of the orogenic belt has its own story and the whole orogenic belt is very complex. It should not therefore be a surprise to find that there are different interpretations of, or difficulties in understanding, older orogenic belts. The SE Asian accretionary orogen is ultimately a mosaic of different orogens or sutures. Fragmentation by strike-slip faulting has not been discussed here but adds difficulty to reassembling the mosaic. In SE Asia there has been no simple story of addition of large pieces of crust by collision, and then termination of orogenic development. Subduction has resumed and new collisions have followed. Occam's razor is a valuable approach to interpreting our data but everything we know about the history of the SE Asian orogenic belt tells us that it will produce an oversimplified history of older accretionary orogenic belts.

There has been considerable growth of the orogen, but little material has been added by subduction accretion, or by accretion of oceanic plateaux or seamounts. Arcs do not seem to have contributed much additional crust in terms of width, and plutonic rocks are not likely to constitute more than about 10% of the entire orogen width in its widest parts. Intra-orogen ocean basins can completely disappear, and are also difficult to identify in the orogenic belt because they are commonly short-lived, are small, and leave little trace such as remnants of their sedimentary cover or arc volcanism as they subduct. Most of the growth has been by addition of continental fragments. After collisions subduction has resumed elsewhere.

In the area of most obvious continent–continent collision, in the Sulawesi and Banda region, the evolution has been exceptionally complex, partly because of existence of an oceanic embayment within the Australian continental margin. This led to subduction hinge rollback, extension, and arc–continent collision. The orogenic belt may be an almost self-perpetuating system as long as major parts of it remain at the edge of major ocean basins. The development of SE Asia brings to mind Hutton's (1795) famous adage 'no vestige of a beginning, no prospect of an end'.

However, there is one special feature without which the SE Asian orogenic belt would certainly be much smaller. The orogen has grown mainly by addition of continental fragments, and almost all of these have come from Australia. The continental fragments have formed by rifting of continental fragments from the Australian margin, and by slicing of continental fragments from Australia; for example, in the Sorong fault zone. Of these the most important process has been the rifting of narrow slivers from Australia. This process has been repeated numerous times since the Palaeozoic (e.g. Metcalfe 1996). Without this peculiar nibbling away of small slices of the Australian passive continental margin there would be a much smaller Sundaland and SE Asian accretionary orogen, which leads to a final question: Has there been any continental growth at all or is SE Asia merely one stage in the reassembly of Gondwana in a different region? Observations from SE Asia favour the latter.

I am grateful to the University of London Central Research Fund, the Natural Environment Research Council, and the Royal Society, but especially the consortium of oil companies that has supported our projects in SE Asia for many years. I thank colleagues in Indonesia at the Pusat Survei Geologi Bandung, Lemigas, Indonesian Institute of Sciences, and Institut Teknologi Bandung, and many colleagues, friends and students in the UK, Europe and SE Asia, particularly H. Smyth and M. van Hattum, for help and discussion. I thank A. Barber for discussions about the origins of Sundaland fragments and particularly his suggestion of a SW Borneo Argoland. I thank I. Metcalfe and J. Ali for helpful comments on the manuscript.

References

ABDULLAH, C. I., RAMPNOUX, J.-P., BELLON, H., MAURY, R. C. & SOERIA-ATMADJA, R. 2000. The evolution of Sumba Island (Indonesia) revisited in the light of new data on the geochronology and geochemistry of the magmatic rocks. *Journal of Asian Earth Sciences*, **18**, 533–546.

ALI, J. R. & HALL, R. 1995. Evolution of the boundary between the Philippine Sea Plate and Australia: Palaeomagnetic evidence from eastern Indonesia. *Tectonophysics*, **251**, 251–275.

ALI, J. R., HALL, R. & BAKER, S. J. 2001. Palaeomagnetic data from a Mesozoic Philippine Sea Plate ophiolite on Obi Island, Eastern Indonesia. *Journal of Asian Earth Sciences*, **19**, 535–546.

AUDLEY-CHARLES, M. G. 2004. Ocean trench blocked and obliterated by Banda forearc collision with Australian proximal continental slope. *Tectonophysics*, **389**, 65–79.

BAKER, S. & MALAIHOLLO, J. 1996. Dating of Neogene igneous rocks in the Halmahera region: Arc initiation and development. *In*: HALL, R. & BLUNDELL, D. J. (eds) *Tectonic Evolution of SE Asia*. Geological Society, London, Special Publications, **106**, 499–509.

BARBER, A. J., CROW, M. J. & MILSOM, J. S. (eds) 2005. *Sumatra: Geology, Resources and Tectonic Evolution*. Geological Society, London, Memoirs, **31**.

BARLEY, M. E., PICKARD, A. L., ZAW, K., RAK, P. & DOYLE, M. G. 2003. Jurassic to Miocene magmatism and metamorphism in the Mogok metamorphic belt and the India–Eurasia collision in Myanmar. *Tectonics*, **22**, doi:10.1029/2002TC001398.

BEAUDOUIN, T., BELLIER, O. & SEBRIER, M. 2003. Present-day stress and deformation fields within the Sulawesi Island area (Indonesia): Geodynamic implications. *Bulletin de la Société Géologique de France*, **174**, 305–317.

BELLIER, O., SEBRIER, M., SEWARD, D., BEAUDOUIN, T., VILLENEUVE, M. & PUTRANTO, E. 2006. Fission track and fault kinematics analyses for new insight into the Late Cenozoic tectonic regime changes in West–Central Sulawesi (Indonesia). *Tectonophysics*, **413**, 201–220.

BELLON, H. & RANGIN, C. 1991. Geochemistry and isotopic dating of the Cenozoic volcanic arc sequences around the Celebes and Sulu Seas. *In*: SILVER, E. A., RANGIN, C. & VON BREYMANN, M. (eds) *Proceedings of the Ocean Drilling Program, Scientific Results*, **124**. Ocean Drilling Program, College Station, TX, 321–338.

BEN-AVRAHAM, Z. & EMERY, K. O. 1973. Structural framework of Sunda Shelf. *AAPG Bulletin*, **57**, 2323–2366.

BERGMAN, S. C., COFFIELD, D. Q., TALBOT, J. P. & GARRARD, R. J. 1996. Tertiary tectonic and magmatic evolution of western Sulawesi and the Makassar Strait, Indonesia: Evidence for a Miocene continent–continent collision. *In*: HALL, R. & BLUNDELL, D. J. (eds) *Tectonic Evolution of SE Asia*. Geological Society, London, Special Publications, **106**, 391–430.

BIJWAARD, H., SPAKMAN, W. & ENGDAHL, E. R. 1998. Closing the gap between regional and global travel time tomography. *Journal of Geophysical Research*, **103**, 30055–30078.

BIRD, M. I., PANG, W. C. & LAMBECK, K. 2006. The age and origin of the Straits of Singapore. *Palaeogeography, Palaeoclimatology, Palaeoecology*, **241**, 531–538.

BIRD, M. I., FIFIELD, L. K., TEH, T. S., CHANG, C. H., SHIRLAW, N. & LAMBECK, K. 2007. An inflection in the rate of early mid-Holocene eustatic sea-level rise: A new sea-level curve from Singapore. *Estuarine, Coastal and Shelf Science*, **71**, 523–536.

BOCK, Y., PRAWIRODIRDJO, L. ET AL. 2003. Crustal motion in Indonesia from Global Positioning System measurements. *Journal of Geophysical Research*, **108**, doi:10.1029/2001JB000324.

BUROLLET, P. F. & SALLE, C. 1981a. A contribution to the geological study of Sumba (Indonesia). *Indonesian Petroleum Association, Proceedings 10th Annual Convention*, 331–344.

BUROLLET, P. F. & SALLE, C. 1981b. Seismic reflection profiles in the Makassar Strait. *In*: BARBER, A. J. & WIRYOSUJONO, S. (eds) *The Geology and Tectonics of Eastern Indonesia*. Geological Research and Development Centre Bandung, Special Publication, **2**, 273–276.

CALVERT, S. J. 2000. *The Cenozoic geology of the Lariang and Karama regions, western Sulawesi, Indonesia*. PhD thesis, University of London.

CALVERT, S. J. & HALL, R. 2007. Cenozoic Evolution of the Lariang and Karama regions, North Makassar Basin, western Sulawesi, Indonesia. *Petroleum Geoscience*, **13**, 353–368.

CARDWELL, R. K. & ISACKS, B. L. 1978. Geometry of the subducted lithosphere beneath the Banda Sea in eastern Indonesia from seismicity and fault plane solutions. *Journal of Geophysical Research*, **83**, 2825–2838.

CHARLTON, T. R. 2000. Tertiary evolution of the Eastern Indonesia Collision Complex. *Journal of Asian Earth Sciences*, **18**, 603–631.

CHESNER, C. A. 1998. Petrogenesis of the Toba Tuffs, Sumatra, Indonesia. *Journal of Petrology*, **39**, 397–438.

CHIANG, K. K. 2002. *Geochemistry of the Cenozoic igneous rocks of Borneo and tectonic implications*. PhD thesis, University of London.

CLOKE, I. R., MILSOM, J. & BLUNDELL, D. J. B. 1999. Implications of gravity data from east Kalimantan and the Makassar Straits: A solution to the origin of the Makassar Straits? *Journal of Asian Earth Sciences*, **17**, 61–78.

COBBING, E. J. 2005. Granites. *In*: BARBER, A. J., CROW, M. J. & MILSOM, J. S. (eds) *Sumatra: Geology, Resources and Tectonic Evolution*. Geological Society, London, Memoirs, **31**, 54–62.

COLLENETTE, P. 1958. *The geology and mineral resources of the Jesselton–Kinabalu area, North Borneo*. Malaysia Geological Survey Borneo Region, Memoir, **6**.

CREVELLO, P. D. 2001. The great Crocker submarine fan: A world-class foredeep turbidite system. *Indonesian Petroleum Association, Proceedings 28th Annual Convention*, 378–407.

CROW, M. J. 2005. Tertiary volcanicity. *In*: BARBER, A. J., CROW, M. J. & MILSOM, J. S. (eds) *Sumatra: Geology, Resources and Tectonic Evolution*. Geological Society, London, Memoirs, **31**, 98–119.

CURRAY, J. R. 1994. Sediment volume and mass beneath the Bay of Bengal. *Earth and Planetary Science Letters*, **125**, 371–383.

CURRIE, C. A. & HYNDMAN, R. D. 2006. The thermal structure of subduction zone back arcs. *Journal of Geophysical Research*, **111**, B08404, doi:10.1029/2005JB004024.

DAVIES, P. R. 1984. Tertiary structural evolution and related hydrocarbon occurrences, North Sumatra basin. *Indonesian Petroleum Association, Proceedings 13th Annual Convention*, 19–50.

DHONAU, T. J. & HUTCHISON, C. S. 1966. The Darvel Bay area, east Sabah, Malaysia. *Malaysia Geological Survey Borneo Region, Annual Report for 1965*, 141–160.

ELBURG, M., VAN LEEUWEN, T., FODEN, J. & MUHARDJO. 2003. Spatial and temporal isotopic domains of contrasting igneous suites in western and northern Sulawesi, Indonesia. *Chemical Geology*, **199**, 243–276.

ENGDAHL, E. R., VAN DER HILST, R. & BULAND, R. 1998. Global teleseismic earthquake relocation with improved travel times and procedures for depth determination. *Bulletin of the Seismological Society of America*, **88**, 722–743.

EVANS, C. A., HAWKINS, J. W. & MOORE, G. F. 1983. Petrology and geochemistry of ophiolitic and associated volcanic rocks on the Talaud Islands, Molucca Sea Collision Zone, northeast Indonesia. *In*: HILDE, T. W. C. & UYEDA, S. (eds) *Geodynamics of the Western Pacific–Indonesian Region*. American Geophysical Union and Geological Society of America, Geodynamic Series, **11**, 159–172.

FLEMING, K., JOHNSTON, P., ZWARTZ, D., YOKOYAMA, Y., LAMBECK, K. & CHAPPELL, J. 1998. Refining the eustatic sea-level curve since the last glacial maximum using far- and intermediate-field sites. *Earth and Planetary Science Letters*, **163**, 327–342.

FRASER, T. & ICHRAM, L. A. 2000. Significance of the Celebes Sea spreading centre to the Paleogene petroleum systems of the SE Sunda margin, Central Indonesia. *Indonesian Petroleum Association, Proceedings 27th Annual Convention*, 431–441.

FULLER, M., ALI, J. R., MOSS, S. J., FROST, G. M., RICHTER, B. & MAHFI, A. 1999. Paleomagnetism of Borneo. *Journal of Asian Earth Sciences*, **17**, 3–24.

GARWIN, S., HALL, R. & WATANABE, Y. 2005. Tectonic setting, geology and gold and copper mineralization in Cenozoic magmatic arcs of Southeast Asia and the West Pacific. *Economic Geology*, **100**, 891–930.

GEBCO. 2003. *IHO–UNESCO, General Bathymetric Chart of the Oceans, Digital Edition 2003*. World Wide Web Address: www.ngdc.noaa.gov/mgg/gebco

GEYH, M. A., KUDRASS, H. R. & STREIF, H. 1979. Sea-level changes during the late Pleistocene and Holocene in the Straits of Malacca. *Nature*, **278**, 441–443.

GOBBETT, D. J. & HUTCHISON, C. S. (eds) 1973. *Geology of the Malay Peninsula (West Malaysia and Singapore)*, Wiley–Interscience, New York.

GORDON, R. G. 1998. The plate tectonic approximation: Plate nonrigidity, diffuse plate boundaries, and global plate reconstructions. *Annual Review of Earth and Planetary Sciences*, **26**, 615–642.

GUNTORO, A. 1999. The formation of the Makassar Strait and the separation between SE Kalimantan and SW Sulawesi. *Journal of Asian Earth Sciences*, **17**, 79–98.

HAILE, N. S. 1969. Geosynclinal theory and the organizational pattern of the North-west Borneo geosyncline. *Quarterly Journal of the Geological Society of London*, **124**, 171–194.

HAILE, N. S. 1974. Borneo. *In*: SPENCER, A. M. (ed.) *Mesozoic–Cenozoic Orogenic Belts*. Geological Society, London, Special Publications, **4**, 333–347.

HALL, R. 1996. Reconstructing Cenozoic SE Asia. *In*: HALL, R. & BLUNDELL, D. J. (eds) *Tectonic Evolution of SE Asia*. Geological Society, London, Special Publications, **106**, 153–184.

HALL, R. 2000. Neogene history of collision in the Halmahera region, Indonesia. *Indonesian Petroleum Association, Proceedings 27th Annual Convention*, 487–493.

HALL, R. 2002. Cenozoic geological and plate tectonic evolution of SE Asia and the SW Pacific: Computer-based reconstructions, model and animations. *Journal of Asian Earth Sciences*, **20**, 353–434.

HALL, R. & MORLEY, C. K. 2004. Sundaland Basins. *In*: CLIFT, P., WANG, P., KUHNT, W. & HAYES, D. E. (eds) *Continent–Ocean Interactions within the East Asian Marginal Seas*. American Geophysical Union, Geophysical Monograph, **149**, 55–85.

HALL, R. & NICHOLS, G. J. 2002. Cenozoic sedimentation and tectonics in Borneo: Climatic influences on orogenesis. *In*: JONES, S. J. & FROSTICK, L. (eds) *Sediment Flux to Basins: Causes, Controls and Consequences*. Geological Society, London, Special Publications, **191**, 5–22.

HALL, R. & SMYTH, H. R. 2008. Cenozoic arc processes in Indonesia: Identification of the key influences on the stratigraphic record in active volcanic arcs. *In*: DRAUT, A. E., CLIFT, P. D. & SCHOLL, D. W. (eds) *Formation and Applications of the Sedimentary Record in Arc Collision Zones*. Geological Society of America Special Papers, **436**, 27–54.

HALL, R. & SPAKMAN, W. 2002. Subducted slabs beneath the eastern Indonesia–Tonga region: Insights from tomography. *Earth and Planetary Science Letters*, **201**, 321–336.

HALL, R. & WILSON, M. E. J. 2000. Neogene sutures in eastern Indonesia. *Journal of Asian Earth Sciences*, **18**, 787–814.

HALL, R., AUDLEY-CHARLES, M. G., BANNER, F. T., HIDAYAT, S. & TOBING, S. L. 1988. Basement rocks of the Halmahera region, eastern Indonesia: A Late Cretaceous–Early Tertiary arc and fore-arc. *Journal of the Geological Society, London*, **145**, 65–84.

HALL, R., ALI, J. R. & ANDERSON, C. D. 1995a. Cenozoic motion of the Philippine Sea Plate—paleomagnetic evidence from eastern Indonesia. *Tectonics*, **14**, 1117–1132.

HALL, R., ALI, J. R., ANDERSON, C. D. & BAKER, S. J. 1995b. Origin and motion history of the Philippine Sea Plate. *Tectonophysics*, **251**, 229–250.

HALL, R., CLEMENTS, B., SMYTH, H. R. & COTTAM, M. A. 2007. A new interpretation of Java's structure. *Indonesian Petroleum Association, Proceedings 31st Annual Convention*, 11–33.

HALL, R., VAN HATTUM, M. W. A. & SPAKMAN, W. 2008. Impact of India–Asia collision on SE Asia: The record in Borneo. *Tectonophysics*, **451**, 366–389.

HAMILTON, W. 1979. *Tectonics of the Indonesian region*. US Geological Survey, Professional Papers, **1078**.

HAMILTON, W. B. 1988. Plate tectonics and island arcs. *Geological Society of America Bulletin*, **100**, 1503–1527.

HASCHKE, M., GÜNTHER, A., MELNICK, D., ECHTLER, H., REUTTER, K.-J., SCHEUBER, E. & ONCKEN, O. 2006. Central and Southern Andean tectonic evolution inferred from arc magmatism. *In*: ONCKEN, O., CHONG, G., FRANZ, G. ET AL. (eds) *The Andes—Active Subduction Orogeny*. Springer, Berlin, 337–353.

HAZEBROEK, H. P. & TAN, D. N. K. 1993. Tertiary tectonic evolution of the NW Sabah continental margin.

Bulletin of the Geological Society of Malaysia, **33**, 195–210.

HELMERS, H., MAASKANT, P. & HARTEL, T. H. D. 1990. Garnet peridotite and associated high grade rocks from Sulawesi, Indonesia. *Lithos*, **25**, 171–188.

HINSCHBERGER, F., MALOD, J. A., DYMENT, J., HONTHAAS, C., REHAULT, J. P. & BURHANUDDIN, S. 2001. Magnetic lineations constraints for the back-arc opening of the Late Neogene South Banda Basin (eastern Indonesia). *Tectonophysics*, **333**, 47–59.

HINSCHBERGER, F., MALOD, J. A., REHAULT, J. P. & BURHANUDDIN, S. 2003. Apport de la bathymétrie et de la géomorphologie à la géodynamique des mers de l'Est-indonésien. *Bulletin de la Société Géologique de France*, **174**, 545–560.

HINSCHBERGER, F., MALOD, J.-A., REHAULT, J.-P., VILLENEUVE, M., ROYER, J.-Y. & BURHANUDDIN, S. 2005. Late Cenozoic geodynamic evolution of eastern Indonesia. *Tectonophysics*, **404**, 91–118.

HOLLOWAY, N. H. 1982. The stratigraphic and tectonic evolution of Reed Bank, North Palawan and Mindoro to the Asian mainland and its significance in the evolution of the South China Sea. *AAPG Bulletin*, **66**, 1357–1383.

HUTCHISON, C. S. 1973. Tectonic evolution of Sundaland: A Phanerozoic synthesis. *Bulletin of the Geological Society of Malaysia*, **6**, 61–86.

HUTCHISON, C. S. 1989. *Geological Evolution of South-East Asia*. Oxford Monographs on Geology and Geophysics, **13**.

HUTCHISON, C. S. 1996. The 'Rajang Accretionary Prism' and 'Lupar Line' Problem of Borneo. *In*: HALL, R. & BLUNDELL, D. J. (eds) *Tectonic Evolution of SE Asia*. Geological Society, London, Special Publications, **106**, 247–261.

HUTCHISON, C. S. 2005. *Geology of North-West Borneo*. Elsevier, Amsterdam.

HUTCHISON, C. S., BERGMAN, S. C., SWAUGER, D. A. & GRAVES, J. E. 2000. A Miocene collisional belt in north Borneo: Uplift mechanism and isostatic adjustment quantified by thermochronology. *Journal of the Geological Society, London*, **157**, 783–793.

HUTTON, J. 1795. *Theory of the Earth, Volume I*. World Wide Web Address: http://www.gutenberg.org/etext/12861

HYNDMAN, R. D., CURRIE, C. A. & MAZZOTTI, S. 2005. Subduction zone backarcs, mobile belts, and orogenic heat. *GSA Today*, **15**, 4–9.

JACOBSON, G. 1970. Gunung Kinabalu area, Sabah, Malaysia. Malaysia Geological Survey, Report, **8**.

KATILI, J. A. 1975. Volcanism and plate tectonics in the Indonesian island arcs. *Tectonophysics*, **26**, 165–188.

KATILI, J. A. 1978. Past and present geotectonic position of Sulawesi, Indonesia. *Tectonophysics*, **45**, 289–322.

KAY, S. M., GODOY, E. & KURTZ, A. 2005. Episodic arc migration, crustal thickening, subduction erosion, and magmatism in the south-central Andes. *Geological Society of America Bulletin*, **117**, 67–88.

KOOPMANS, B. N. 1967. Deformation of the metamorphic rocks and the Chert–Spilite Formation in the southern part of the Darvel Bay area, Sabah. *Malaysia Geological Survey Borneo Region Bulletin*, **8**, 14–24.

KOPP, H., KLAESCHEN, D., FLUEH, E. R., BIALAS, J. & REICHERT, C. 2002. Crustal structure of the Java margin from seismic wide-angle and multichannel reflection data. *Journal of Geophysical Research*, **107**, 2034, doi:10.1029/2000JB000095.

KOPP, H., FLUEH, E. R., PETERSEN, C. J., WEINREBE, W., WITTWER, A. & SCIENTISTS, M. 2006. The Java margin revisited: Evidence for subduction erosion off Java. *Earth and Planetary Science Letters*, **242**, 130–142.

LEONG, K. M. 1974. *The geology and mineral resources of the Upper Segama valley and Darvel Bay, Sabah*. Malaysia Geological Survey Borneo Region Memoir, **4**.

LI, Z.-X. & LI, X.-H. 2007. Formation of the 1300-km-wide intracontinental orogen and postorogenic magmatic province in Mesozoic South China: A flat-slab subduction model. *Geology*, **35**, 179–182.

LIEW, T. C. & MCCULLOCH, M. T. 1985. Genesis of granitoid batholiths of Peninsular Malaysia and implications for models of crustal evolution: Evidence from Nd–Sr isotopic and U–Pb zircon study. *Geochimica et Cosmochimica Acta*, **49**, 587–600.

LIEW, T. C. & PAGE, R. W. 1985. U–Pb zircon dating of granitoid plutons from the West Coast of Peninsular Malaysia. *Journal of the Geological Society, London*, **142**, 515–526.

LIN, W. & WANG, Q. 2006. Late Mesozoic extensional tectonics in the North China block: A crustal response to subcontinental mantle removal? *Bulletin de la Société Géologique de France*, **177**, 287–297.

MACPHERSON, C. G. & HALL, R. 1999. Tectonic controls of geochemical evolution in arc magmatism of SE Asia. *Australian Institute of Mining and Metallurgy, Proceedings 4th PACRIM Congress, Bali, Indonesia*, 359–368.

MACPHERSON, C. G. & HALL, R. 2002. Timing and tectonic controls in the evolving orogen of SE Asia and the western Pacific and some implications for ore generation. *In*: BLUNDELL, D. J., NEUBAUER, F. & VON QUADT, A. (eds) *The Timing and Location of Major Ore Deposits in an Evolving Orogen*. Geological Society, London, Special Publications, **204**, 49–67.

MALOD, J. & KEMAL, B. M. 1996. The Sumatra Margin: Oblique subduction and lateral displacement of the accretionary prism. *In*: HALL, R. & BLUNDELL, D. J. (eds) *Tectonic Evolution of SE Asia*. Geological Society, London, Special Publications, **106**, 19–28.

MASSON, D. G., PARSON, L. M., MILSOM, J., NICHOLS, G. J., SIKUMBANG, N., DWIWANTO, B. & KALLAGHER, H. 1990. Subduction of seamounts at the Java Trench; a view with long range side-scan sonar. *Tectonophysics*, **185**, 51–65.

MATSON, R. G. & MOORE, G. F. 1992. Structural influences on Neogene subsidence in the central Sumatra forearc basin. *In*: WATKINS, J. S., FENG, Z. Q. & MCMILLEN, K. J. (eds) *Geology and Geophysics of Continental Margins*. American Association of Petroleum Geologists, Memoirs, **53**, 157–181.

MCCAFFREY, R. 1996. Slip partitioning at convergent plate boundaries of SE Asia. *In*: HALL, R. & BLUNDELL, D. J. (eds) *Tectonic Evolution of SE Asia*. Geological Society, London, Special Publications, **106**, 3–18.

MCCANN, W. R. & HABERMANN, R. E. 1989. Morphologic and geologic effects of the subduction of bathymetric highs. *Pure and Applied Geophysics*, **129**, 41–69.

MCCOURT, W. J., CROW, M. J., COBBING, E. J. & AMIN, T. C. 1996. Mesozoic and Cenozoic plutonic evolution of SE Asia: Evidence from Sumatra, Indonesia. *In*: HALL, R. & BLUNDELL, D. J. (eds) *Tectonic Evolution of SE Asia*. Geological Society, London, Special Publications, **106**, 321–335.

MCGILVERY, T. A. & COOK, D. L. 2003. The influence of local gradients on accommodation space and linked depositional elements across a stepped slope profile, offshore Brunei. *In*: ROBERTS, H. R., ROSEN, N. C., FILLON, R. F. & ANDERSON, J. B. (eds) *Shelf Margin Deltas and Linked Down Slope Petroleum Systems*. Gulf Coast Section SEPM, 387–419.

METCALFE, I. 1990. Allochthonous terrane processes in Southeast Asia. *Philosophical Transactions of the Royal Society of London, Series A*, **331**, 625–640.

METCALFE, I. 1996. Pre-Cretaceous evolution of SE Asian terranes. *In*: HALL, R. & BLUNDELL, D. J. (eds) *Tectonic Evolution of SE Asia*. Geological Society, London, Special Publications, **106**, 97–122.

METCALFE, I. 1998. Palaeozoic and Mesozoic geological evolution of the SE Asian region: Multidisciplinary constraints and implications for biogeography. *In*: HALL, R. & HOLLOWAY, J. D. (eds) *Biogeography and Geological Evolution of SE Asia*. Backhuys, Leiden, 25–41.

MICHEL, G. W., YU, Y. Q. ET AL. 2001. Crustal motion and block behaviour in SE Asia from GPS measurements. *Earth and Planetary Science Letters*, **187**, 239–244.

MIYAZAKI, K., SOPAHELUWAKAN, J., ZULKARNAIN, I. & WAKITA, K. 1998. A jadeite–quartz–glaucophane rock from Karangsambung, central Java, Indonesia. *Island Arc*, **7**, 223–230.

MONNIER, C., GIRARDEAU, J., MAURY, R. & COTTEN, J. 1995. Back-arc basin origin for the East Sulawesi ophiolite (eastern Indonesia). *Geology*, **23**, 851–854.

MOORE, G. F. & KARIG, D. E. 1980. Structural geology of Nias island, Indonesia: Implications for subduction zone tectonics. *American Journal of Science*, **280**, 193–223.

MORLEY, C. K., BACK, S., VAN RENSBERGEN, P., CREVELLO, P. & LAMBIASE, J. J. 2003. Characteristics of repeated, detached, Miocene–Pliocene tectonic inversion events, in a large delta province on an active margin, Brunei Darussalam, Borneo. *Journal of Structural Geology*, **25**, 1147–1169.

MOSS, S. J. 1998. Embaluh group turbidites in Kalimantan: Evolution of a remnant oceanic basin in Borneo during the Late Cretaceous to Palaeogene. *Journal of the Geological Society, London*, **155**, 509–524.

MOSS, S. J., CHAMBERS, J., CLOKE, I., CARTER, A., SATRIA, D., ALI, J. R. & BAKER, S. 1997. New observations on the sedimentary and tectonic evolution of the Tertiary Kutai Basin, East Kalimantan. *In*: FRASER, A. J., MATTHEWS, S. J. & MURPHY, R. W. (eds) *Petroleum Geology of Southeast Asia*. Geological Society, London, Special Publications, **126**, 395–416.

MUBROTO, B., BRIDEN, J. C., MCCLELLAND, E. & HALL, R. 1994. Palaeomagnetism of the Balantak ophiolite, Sulawesi. *Earth and Planetary Science Letters*, **125**, 193–209.

NEGREDO, A. M., SABADINI, R. & GIUNCHI, C. 1997. Interplay between subduction and continental convergence: A three-dimensional dynamic model for the Central Mediterranean. *Geophysical Journal International*, **131**, f9–f13.

NUR'AINI, S., HALL, R. & ELDERS, C. F. 2005. Basement architecture and sedimentary fills of the North Makassar Straits basin. *Indonesian Petroleum Association, Proceedings 30th Annual Convention*, 483–497.

OMANG, S. A. K. & BARBER, A. J. 1996. Origin and tectonic significance of the metamorphic rocks associated with the Darvel Bay Ophiolite, Sabah, Malaysia. *In*: HALL, R. & BLUNDELL, D. J. (eds) *Tectonic Evolution of SE Asia*. Geological Society, London, Special Publications, **106**, 263–279.

PAGE, B. N. G., BENNETT, J. D., CAMERON, N. R., BRIDGE, D. M., JEFFREY, D. H., KEATS, W. & THAIB, J. 1979. A review of the main structural and magmatic features of northern Sumatra. *Journal of the Geological Society, London*, **136**, 569–579.

PAIRAULT, A. A., HALL, R. & ELDERS, C. F. 2003. Structural styles and tectonic evolution of the Seram Trough, Indonesia. *Marine and Petroleum Geology*, **20**, 1141–1160.

PARKINSON, C. 1998a. Emplacement of the East Sulawesi Ophiolite: Evidence from subophiolite metamorphic rocks. *Journal of Asian Earth Sciences*, **16**, 13–28.

PARKINSON, C. 1998b. An outline of the petrology, structure and age of the Pompangeo Schist Complex of central Sulawesi, Indonesia. *Island Arc*, **7**, 231–245.

PARKINSON, C. D., MIYAZAKI, K., WAKITA, K., BARBER, A. J. & CARSWELL, D. A. 1998. An overview and tectonic synthesis of the pre-Tertiary very-high-pressure metamorphic and associated rocks of Java, Sulawesi and Kalimantan, Indonesia. *Island Arc*, **7**, 184–200.

PETRONAS. 1999. *The Petroleum Geology and Resources of Malaysia*. Petronas, Kuala Lumpur.

PITCHER, W. S., ATHERTON, M. P., COBBING, E. J. & BECKINSALE, R. D. 1985. *Magmatism at a Plate Edge. The Peruvian Andes*. Blackie, Glasgow.

POLVÉ, M., MAURY, R. C., BELLON, H. ET AL. 1997. Magmatic evolution of Sulawesi (Indonesia): Constraints on the Cenozoic geodynamic history of the Sundaland active margin. *Tectonophysics*, **272**, 69–92.

PRIADI, B., SUCIPTA, I. G. B. E., UTOYO, H., SOPAHELUWAKAN, J. & SUDARSONO, 1999. Distribution of Neogene granitoid along the Palu–Koro fault zone, Central Sulawesi. *In*: DARMAN, H. & SIDI, F. H. (eds) *Tectonics and Sedimentation of Indonesia (Abstracts Volume)*. Indonesian Sedimentologists Forum Special Publication, 72–74.

PRICE, N. J. & AUDLEY-CHARLES, M. G. 1987. Tectonic collision processes after plate rupture. *Tectonophysics*, **140**, 121–129.

PUSPITA, S. D., HALL, R. & ELDERS, C. F. 2005. Structural styles of the offshore West Sulawesi fold belt, North Makassar Straits, Indonesia. *Indonesian Petroleum Association, Proceedings 30th Annual Convention*, 519–542.

RANGIN, C., BELLON, H., BENARD, F., LETOUZEY, J., MÜLLER, C. & TAHIR, S. 1990. Neogene arc–continent collision in Sabah, N. Borneo (Malaysia). *Tectonophysics*, **183**, 305–319.

RANGIN, C., LE PICHON, X. ET AL. 1999. Plate convergence measured by GPS across the Sundaland/Philippine sea plate deformed boundary: The Philippines and eastern Indonesia. *Geophysical Journal International*, **139**, 296–316.

REINHARDT, M. & WENK, E. 1951. *Geology of the colony of North Borneo*. Bulletin of the Geological Survey Department of the British Territories in Borneo, **1**.

REPLUMAZ, A. & TAPPONNIER, P. 2003. Reconstruction of the deformed collision zone between India and Asia by backward motion of lithospheric blocks. *Journal of Geophysical Research*, **108**, 2285, doi:10.1029/2001JB000661.

REPLUMAZ, A., KARASON, H., VAN DER HILST, R. D., BESSE, J. & TAPPONNIER, P. 2004. 4-D evolution of SE Asia's mantle from geological reconstructions and seismic tomography. *Earth and Planetary Science Letters*, **221**, 103–115.

ROSENBAUM, G. & LISTER, G. S. 2004. Neogene and Quaternary rollback evolution of the Tyrrhenian Sea, the Apennines, and the Sicilian Maghrebides. *Tectonics*, **23**, TC1013, doi:10.1029/2003TC001518.

SAMUEL, M. A. 1994. *The structural and stratigraphic evolution of islands at the active margin of the Sumatran Forearc, Indonesia*. PhD thesis, University of London.

SAMUEL, M. A. & HARBURY, N. A. 1996. The Mentawai fault zone and deformation of the Sumatran forearc in the Nias area. *In*: HALL, R. & BLUNDELL, D. J. (eds) *Tectonic Evolution of SE Asia*. Geological Society, London, Special Publications, **106**, 337–351.

SAMUEL, M. A., HARBURY, N. A., JONES, M. E. & MATTHEWS, S. J. 1995. Inversion of an outer-arc ridge: the Sumatran Forearc, Indonesia. *In*: BUCHANAN, J. G. & BUCHANAN, P. G. (eds) *Basin Inversion*. Geological Society, London, Special Publications, **88**, 473–492.

SAMUEL, M. A., HARBURY, N. A., BAKRI, A., BANNER, F. T. & HARTONO, L. 1997. A new stratigraphy for the islands of the Sumatran Forearc, Indonesia. *Journal of Asian Earth Sciences*, **15**, 339–380.

SANDAL, S. T. (ed.) 1996. *The Geology and Hydrocarbon Resources of Negara Brunei Darussalam*. Syabas, Bandar Seri Begawan, Brunei Darussalam.

SCHLÜTER, H. U., GAEDICKE, C. ET AL. 2002. Tectonic features of the southern Sumatra–western Java forearc of Indonesia. *Tectonics*, **21**, 1047, doi:10.1029/2001TC901048.

SEWELL, R. J., CAMPBELL, S. D. G., FLETCHER, C. J. N., LAI, K. W. & KIRK, P. A. 2000. *The Pre-Quaternary Geology of Hong Kong*. Hong Kong Geological Survey, Geological Memoirs.

SIEBERT, L. & SIMKIN, T. 2002. Volcanoes of the World: An Illustrated Catalog of Holocene Volcanoes and their Eruptions. *In*: *Smithsonian Institution, Global Volcanism Program Digital Information Series, GVP-3*. World Wide Web Address: http://www.volcano.si.edu/world/

SILVER, E. A., MCCAFFREY, R. & JOYODIWIRYO, Y. 1978. Gravity results and emplacement geometry of the Sulawesi ultramafic belt, Indonesia. *Geology*, **6**, 527–531.

SILVER, E. A., MCCAFFREY, R., JOYODIWIRYO, Y. & STEVENS, S. 1983a. Ophiolite emplacement by collision between the Sula Platform and the Sulawesi island arc, Indonesia. *Journal of Geophysical Research*, **88**, 9419–9435.

SILVER, E. A., MCCAFFREY, R. & SMITH, R. B. 1983b. Collision, rotation, and the initiation of subduction in the evolution of Sulawesi, Indonesia. *Journal of Geophysical Research*, **88**, 9407–9418.

SIMANDJUNTAK, T. O. & BARBER, A. J. 1996. Contrasting tectonic styles in the Neogene orogenic belts of Indonesia. *In*: HALL, R. & BLUNDELL, D. J. (eds) *Tectonic Evolution of SE Asia*. Geological Society, London, Special Publications, **106**, 185–201.

SIMONS, W. J. F., SOCQUET, A. ET AL. 2007. A decade of GPS in Southeast Asia: Resolving Sundaland motion and boundaries. *Journal of Geophysical Research*, **112**, B06420, doi:10.1029/2005JB003868.

SITUMORANG, B. 1982a. *The formation and evolution of the Makassar Basin, Indonesia*. PhD thesis, University of London.

SITUMORANG, B. 1982b. Formation, evolution, and hydrocarbon prospects of the Makassar Basin, Indonesia. *In*: WATSON, S. T. (ed.) *Transactions of the 3rd Circum Pacific Energy and Mineral Resources Conference, Honolulu, Hawaii*, 227–232.

SMYTH, H. R., HAMILTON, P. J., HALL, R. & KINNY, P. D. 2007. The deep crust beneath island arcs: Inherited zircons reveal a Gondwana continental fragment beneath East Java, Indonesia. *Earth and Planetary Science Letters*, **258**, 269–282.

STEVENS, C., MCCAFFREY, R. ET AL. 1999. Rapid rotations about a vertical axis in a collisional setting revealed by the Palu fault, Sulawesi, Indonesia. *Geophysical Research Letters*, **26**, 2677–2680.

STEVENS, C. W., MCCAFFREY, R., BOCK, Y., GENRICH, J. F., PUBELLIER, M. & SUBARYA, C. 2002. Evidence for block rotations and basal shear in the world's fastest slipping continental shear zone in NW New Guinea. *In*: STEIN, S. & FREYMEULLER, J. T. (eds) *Plate Boundary Zones*. American Geophysical Union, Geodynamic Series, **30**, 87–99.

STOW, D. A. V., COCHRAN, J. R. & ODP LEG 116 SHIPBOARD SCIENTIFIC PARTY. 1989. The Bengal Fan: Some preliminary results from ODP drilling. *Geo-Marine Letters*, **9**, 1–10.

SUGIAMAN, F. & ANDRIA, L. 1999. Devonian carbonate of Telen River, East Kalimantan. *Berita Sedimentologi*, **10**, 18–19.

SUKAMTO, R. 1973. *Reconnaissance geologic map of Palu Area, Sulawesi—scale 1:250 000*. Geological Survey of Indonesia, Directorate of Mineral Resources, Geological Research and Development Centre, Bandung, Open File.

SURMONT, J., LAJ, C., KISSAL, C., RANGIN, C., BELLON, H. & PRIADI, B. 1994. New paleomagnetic constraints on the Cenozoic tectonic evolution of the North Arm of Sulawesi, Indonesia. *Earth and Planetary Science Letters*, **121**, 629–638.

SWAUGER, D. A., HUTCHISON, C. S., BERGMAN, S. C. & GRAVES, J. E. 2000. Age and emplacement of the Mount Kinabalu pluton. *Bulletin of the Geological Society of Malaysia*, **44**, 159–163.

TAN, D. N. K. & LAMY, J. M. 1990. Tectonic evolution of the NW Sabah continental margin since the late Eocene. *Bulletin of the Geological Society of Malaysia*, **27**, 241–260.

TATE, R. B. 1996. *The geological evolution of Borneo Island*. MSc thesis, University of Malaya, Kuala Lumpur, Malaysia.

TATE, R. B. 2001. *The Geology of Borneo Island*. Geological Society of Malaysia, Kuala Lumpur, Malaysia.

TJIA, H. D. 1973. Palu–Koro fault zone, Sulawesi. *Berita Direktorat Geologi, Geosurvey Newsletter*, **5**, 1–3.

TJIA, H. D. 1992. Holocene sea-level changes in the Malay–Thai Peninsula, a tectonically stable environment. *Bulletin of the Geological Society of Malaysia*, **31**, 157–176.

TJIA, H. D. 1996. Sea-level changes in the tectonically stable Malay–Thai peninsula. *Quaternary International*, **31**, 95–101.

TONGKUL, F. 1991. Tectonic evolution of Sabah, Malaysia. *Journal of Southeast Asian Earth Sciences*, **6**, 395–406.

VAN BEMMELEN, R. W. 1949. *The Geology of Indonesia*. Government Printing Office, Nijhoff, The Hague.

VAN DE ZEDDE, D. M. A. & WORTEL, M. R. J. 2001. Shallow slab detachment as a transient source of heat at mid-lithospheric depths. *Tectonics*, **20**, 868–882.

VAN HATTUM, M. W. A. 2005. *Provenance of Cenozoic sedimentary rocks of northern Borneo*. PhD thesis, University of London.

VAN HATTUM, M. W. A., HALL, R., PICKARD, A. L. & NICHOLS, G. J. 2006. SE Asian sediments not from Asia: Provenance and geochronology of North Borneo sandstones. *Geology*, **34**, 589–592.

VAN LEEUWEN, T. M. 1981. The geology of southwest Sulawesi with special reference to the Biru area. *In*: BARBER, A. J. & WIRYOSUJONO, S. (eds) *The Geology and Tectonics of Eastern Indonesia*. Geological Research and Development Centre, Bandung, Special Publication, **2**, 277–304.

VAN LEEUWEN, T. M. & MUHARDJO, 2005. Stratigraphy and tectonic setting of the Cretaceous and Paleogene volcanic–sedimentary successions in northwest Sulawesi, Indonesia: Implications for the Cenozoic evolution of Western and Northern Sulawesi. *Journal of Asian Earth Sciences*, **25**, 481–511.

VAN LEEUWEN, T. M., ALLEN, C. M., KADARUSMAN, A., ELBURG, M., PALIN, J. M., MUHARDJO, & SUWIJANTO. 2007. Petrologic, isotopic, and radiometric age constraints on the origin and tectonic history of the Malino Metamorphic Complex, NW Sulawesi, Indonesia. *Journal of Asian Earth Sciences*, **29**, 751–777.

VIGNY, C., PERFETTINI, H. *ET AL*. 2002. Migration of seismicity and earthquake interactions monitored by GPS in SE Asia triple junction: Sulawesi, Indonesia. *Journal of Geophysical Research*, **107**, 2231, doi:10.1029/2001JB000377.

VOGT, E. T. & FLOWER, M. F. J. 1989. Genesis of the Kinabalu (Sabah) granitoid at a subduction–collision junction. *Contributions to Mineralogy and Petrology*, **103**, 493–509.

WHITTAKER, J. M., MÜLLER, R. D., SDROLIAS, M. & HEINE, C. 2007. Sunda–Java trench kinematics, slab window formation and overriding plate deformation since the Cretaceous. *Earth and Planetary Science Letters*, **255**, 445–457.

WIDIWIJAYANTI, C., TIBERI, C., DEPLUS, C., DIAMENT, M., MIKHAILOV, V. & LOUAT, R. 2004. Geodynamic evolution of the northern Molucca Sea area (Eastern Indonesia) constrained by 3-D gravity field inversion. *Tectonophysics*, **386**, 203–222.

WIJBRANS, J. R., HELMERS, H. & SOPAHELUWAKAN, J. 1994. The age and thermal evolution of blueschists from South-East Sulawesi, Indonesia: The case of slowly cooled phengites. *Mineralogical Magazine*, **58A**, 975–976.

WILLIAMS, P. R., JOHNSTON, C. R., ALMOND, R. A. & SIMAMORA, W. H. 1988. Late Cretaceous to Early Tertiary structural elements of West Kalimantan. *Tectonophysics*, **148**, 279–298.

WILLIAMS, P. R., SUPRIATNA, S., JOHNSTON, C. R., ALMOND, R. A. & SIMAMORA, W. H. 1989. A Late Cretaceous to Early Tertiary accretionary complex in West Kalimantan. *Bulletin, Geological Research and Development Centre, Bandung*, **13**, 9–29.

WORTEL, M. J. R. & SPAKMAN, W. 2000. Subduction and slab detachment in the Mediterranean–Carpathian region. *Science*, **290**, 1910–1917.

YANG, J.-H., WU, F.-Y., CHUNG, S.-L., LO, C.-H., WILDE, S. A. & DAVIS, G. A. 2007. Rapid exhumation and cooling of the Liaonan metamorphic core complex: Inferences from $^{40}Ar/^{39}Ar$ thermochronology and implications for Late Mesozoic extension in the eastern North China Craton. *Geological Society of America Bulletin*, **119**, 1405–1414.

Evolution from an oblique subduction back-arc mobile belt to a highly oblique collisional margin: the Cenozoic tectonic development of Thailand and eastern Myanmar

C. K. MORLEY

PTTEP, 555 Vibhavadi-Rangsit Road, Chatuchak, Bangkok 10900, Thailand
(e-mail: Chrissmorley@gmail.com)

Abstract: Previous tectonic models (escape tectonics, topographic ooze) for SE Asia have considered that Himalayan–Tibetan processes were dominant and imposed on cool, rigid SE Asian crust. However, present-day geothermal gradients, metamorphic mineral assemblages, structural style and igneous intrusions all point to east Myanmar and Thailand having hot, ductile crust during Cenozoic–Recent times. North to NE subduction beneath SE Asia during the Mesozoic–Cenozoic resulted in development of hot, thickened crust in the Thailand–Myanmar region in a back-arc mobile belt setting. This setting changed during the Eocene–Recent to highly oblique collision as India coupled with the west Burma block. The characteristics of the orogenic belt include: (1) a hot and weak former back-arc area about 200–300 km wide (Shan Plateau) heavily intruded by I-type and S-type granites during the Mesozoic and Palaeogene; (2) high modern geothermal gradients (3–7 °C per 100 m) and heat fl ow (70–100 mW m^{-2}); (3) widespread Eocene–Pliocene basaltic volcanism; (4) Late Cretaceous–earliest Cenozoic and Eocene–Oligocene high-temperature–low-pressure metamorphism; (5) c. 47–29 Ma peak metamorphism in the Mogok metamorphic belt followed by c. 30–23 Ma magmatism and exhumation of the belt between the Late Oligocene and early Miocene; (6) a broad zone of Eocene–Oligocene sinistral transpression in the Shan Plateau, later reactivated by Oligocene–Recent dextral transtension; (7) diachronous extensional collapse during the Cenozoic, involving both high-angle normal fault and low-angle normal fault (LANF) bounded basins; (8) progressive collapse of thickened, ductile crust from south (Eocene) to north (Late Oligocene) in the wake of India moving northwards; and (9) the present-day influence on the stress system by both the Himalayan orogenic belt and the Sumatra–Andaman subduction zone.

Thailand and Myanmar lie in a tectonically active region, south of the Himalayan orogenic belt, and east of the Sumatra–Andaman Trench (Fig. 1). The general Cenozoic tectonic history of the area is well established (Lee & Lawver 1995; Hall 2002): NE-directed subduction of the Indian plate oceanic crust under Sumatra and Myanmar lasted from the Mesozoic to the early Cenozoic. The arrival of the Indian continental crust in the vicinity of Myanmar during the Eocene terminated the role of the subduction zone as a sharply defi ned plate boundary. India coupled with western Myanmar and detached the western part of Myanmar (known as the west Burma block) from the continental core of SE Asia called Sundaland. During the Oligocene this coupling became manifest as a north–south-trending belt of dextral strike-slip or dextral transtension some 300 km east of the trench, that developed between the west Burma block and Sundaland as India dragged the west Burma block progressively north with respect to Sundaland (Curray 2005).

Until recently eastern Myanmar and Thailand were viewed as having a relatively passive involvement in the development of the Himalayan orogenic belt. Eastward to southward extrusion or escape of rigid crustal blocks along large Cenozoic strike-slip faults emanating from the eastern margin of the Himalayan–Tibetan orogen was the main model for Cenozoic deformation of the region (e.g. Tapponnier *et al.* 1986; Lacassin *et al.* 1997). In Thailand escape tectonics was proposed to be manifest as strike-slip faults and the development of associated rift basins, which largely represented the sum of Cenozoic tectonic activity (Tapponnier *et al.* 1986; Polachan *et al.* 1991). Little contractional deformation, magma generation or metamorphism was suspected. However, a wide range of new geological and geophysical data has been acquired in the last 15 years that suggests that considerably more varied deformation has affected NW Sundaland during the Cenozoic than that offered by the escape tectonics model.

Some of the important types of recent research over the last 15 years are outlined below. Progressive building of radiometric age databases and improvements in techniques by a number of research groups have begun to shed considerable new light

From: CAWOOD, P. A. & KRÖNER, A. (eds) *Earth Accretionary Systems in Space and Time*.
The Geological Society, London, Special Publications, **318**, 373–403.
DOI: 10.1144/SP318.14   0305-8719/09/$15.00 © The Geological Society of London 2009.

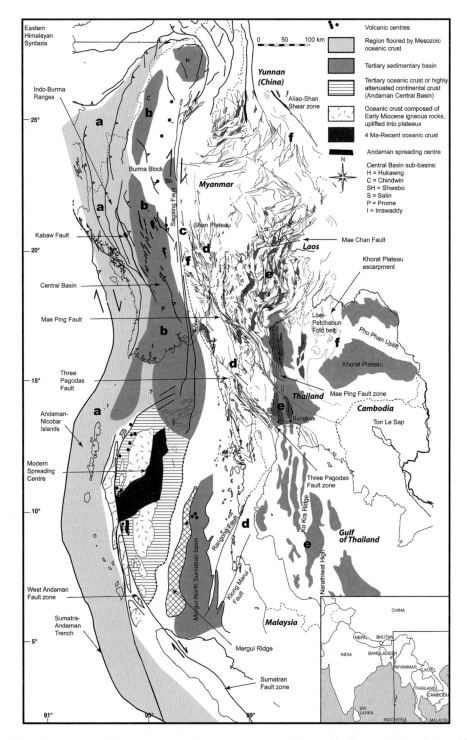

Fig. 1. Location map for the Myanmar–western Thailand region. Compiled from Pivnik *et al.* (1998), Morley (2004) and Curray (2005). (**a**) Sumatran-Andaman trench, Andaman Islands and Indo-Burma Ranges (developed on Mesozoic oceanic crust); (**b**) Central Basin; (**c**) Sagaing Fault-Sumatran Fault strike-slip system; (**d**) Shan Plateau; (**e**) Eocene-Miocene rift basins onshore and offshore Thailand; (**f**) Palaeogene folds and thrusts.

on the type, geometry and timing of deformation, and metamorphism, experienced by the region during the Cenozoic (e.g. Charusiri et al. 1993; Macdonald et al. 1993; Dunning et al. 1995; Ahrendt et al. 1997; Lacassin et al. 1997; Bertrand et al. 1999, 2001; Upton 1999; Barr et al. 2002; Barley et al. 2003; Searle et al. 2007). Seismic reflection and well data have provided considerable insights into the styles of synrift and post-rift basin deformation and history (e.g. Polachan et al. 1994; Jardine 1997; Morley 2002; Chantraprasert 2003; Uttamo et al. 2003; Morley et al. 2004, 2007a, b). Fieldwork and satellite data have more clearly revealed Cenozoic structural timing and styles (e.g. Mitchell 1989; Lacassin et al. 1997; Rhodes et al. 1997, 2000, 2005; Morley et al. 2000, 2001; Mitchell et al. 2002; Morley 2004). New marine geophysical data have considerably improved understanding of the Andaman Sea region (Nielsen et al. 2004; see review by Curray 2005; Fig. 1). Global positioning system (GPS) and earthquake data have defined the modern plate geometries and strain rates (e.g. Chamot-Rooke et al. 1999; Vigny et al. 2003, 2005). This body of data has only just reached the stage where the interrelationships of mantle, crustal and surface processes across a wide region can be integrated in the context of a unifying orogenic model. Many aspects still require considerable further study. In particular, the remote region that straddles eastern Myanmar and western Thailand, called the Shan Plateau (Fig. 1), remains poorly described: the thickness of the crust is not established in any detail, radiometric age dating of igneous and metamorphic events is sparse, and even basic geological maps are unreliable in detail. Clearly, much more work is required and there are significant limitations to the available data. However, there is considerable information that can be reviewed and synthesized in this paper, which indicates that the region experienced very dynamic tectonics during the Cenozoic.

This paper reviews the evidence for Cenozoic orogenic events in Myanmar and Thailand and describes some of the unusual characteristics of this region of NE Sundaland. Building on the regional geology of SE Asia discussed by Hall & Morley (2004), this paper proposes that the prolonged subduction history of the region strongly influenced the way deformation has progressed in eastern Myanmar and Thailand during the Cenozoic. The tectonic evolution of the region is described in two main sections, corresponding to the key tectonic stages. The first stage represents the subduction-dominated deformation and magmatism from the Mesozoic to the Early Cenozoic. The second stage marks the highly oblique collisional orogenic event when eastern India coupled with western Myanmar and subsequently developed a transform margin through eastern Myanmar. The main focus of the section is to describe the resulting deformation seen in NW Sundaland (i.e. eastern Myanmar and Thailand). A review of the Central Basin in Myanmar follows, because of its anomalous deformation history with respect to regions to the east and west. The discussion section addresses what plate-scale scenarios best explain the metamorphic, igneous and structural characteristics of the region described here. Before the evolution of the region can be described a broad outline of the geographical distribution of the main tectonic elements is necessary, and is given below.

Cenozoic tectonic elements of Myanmar and Thailand

Thailand and Myanmar form a region of Cenozoic deformation that extends c. 900 km in an east–west direction and c. 1800 km north–south. The region can be divided into six main Cenozoic tectonic provinces that trend approximately north–south (Fig. 1). Passing from west to east these provinces are: the Sumatra–Andaman Trench, Central Basin, Sagaing Fault zone, Shan Plateau, Eocene–Miocene rifts, and a region of Palaeogene folds and thrusts. A brief introduction to the provinces is given below:

(1) *The Sumatra–Andaman trench and overlying Andaman Islands–Indo-Burman ranges accretionary prism complex.* Oceanic crust has been subducted beneath the trench from the Mesozoic to Recent. Convergence became increasingly oblique as the overlying plate rotated clockwise during the Cenozoic (e.g. Lee & Lawver 1995; Curray 2005). The Indo-Burman ranges contain Cretaceous olistostrome mélanges, with blocks of gabbro, pillow basalt, serpentinite, banded chert, limestone and schist interpreted as having developed as a result of north- to NE-directed subduction during the Mesozoic–Cenozoic (Mitchell 1981).

(2) *The Central Basin.* In the south of Myanmar the basin forms a broad, flat-lying region of persistent and rapid subsidence and fluvial to shallow marine sedimentation from the Late Cretaceous to Recent, in a forearc location (Pivnik et al. 1998). Passing northward the topography becomes more hilly in places, as a result of inversion of the basin. Remnant arc-related volcanism affects the eastern part of the basin.

(3) *The Sagaing Fault–Sumatran Fault (SFSF) strike-slip system.* This is the main province that accommodates the northward motion of the west Burma block with respect to Sundaland. The SFSF system is a series of dextral transform fault zones that meet in the Andaman Sea at a large releasing bend that forms a back-arc spreading centre

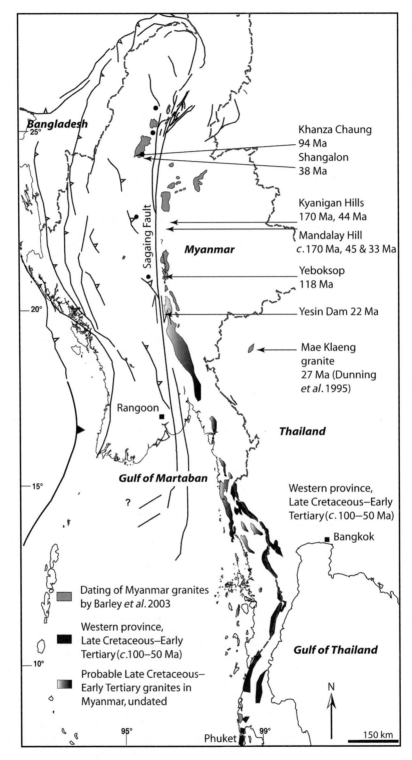

Fig. 2. Distribution of probable and dated Cretaceous–Cenozoic granite in Myanmar and Thailand. Compiled from Charusiri *et al.* (1993), Schwartz *et al.* (1995), Puttahpiban (2002) and Barley *et al.* (2003).

(Curray et al. 1978; Curray 2005). The Sagaing Fault is one of the largest and most active strike-slip faults in the world, with modern dextral slip rates of the order of c. 2.4 cm a^{-1} (Vigny et al. 2003; Curray 2005).

(4) *The Shan Plateau.* This is an uplifted region straddling eastern Myanmar and western Thailand, composed predominantly of Palaeozoic sedimentary and metasedimentary rocks extensively intruded by Mesozoic and early Cenozoic granites. In central Thailand the limit of the plateau is sharply defined by the low-lying topography of the Cenozoic post-rift basins (see (5) below). Typical maximum elevations in the plateau area are c. 1500 m. Around the western margin of the plateau are two important north–south-striking features, the Sagaing Fault and the Shan Scarp. The Shan Scarp appears to mark a Mesozoic terrane boundary between the Shan–Thai block to the east and the western Burma terrane (Mitchell et al. 2002). The Mogok metamorphic belt and the Carboniferous pebbly mudstones (Mergui Group) lie within the western Burma terrane. The Shan Scarp was later reactivated during the Cenozoic as the important (dextral strike-slip) Paunglaung Fault zone (Mitchell et al. 2002; Morley 2004). About 60–70 km west of the Shan Scarp lies the Sagaing Fault. The west Burma block came into existence during the Cenozoic as a result of development of the Sagaing Fault as a transform margin-defining fault. Extensive systems of NW–SW and north–south Cenozoic strike-slip faults also affect the Shan Plateau (Morley 2004).

(5) *Eocene–Miocene rift basins of onshore and offshore Thailand.* Post-rift subsidence has made the region of Cenozoic rift basins flat-lying and low. Offshore this region forms the Gulf of Thailand. Onshore the central plains are 450 km long and up to 125 km wide, and range in elevation from sea level to +50 m. Only in northern Thailand is the transition from the Shan Plateau less distinct and rift-related topography is still present. Over 40 isolated intermontane basins lie at elevations between 200 and 500 m. The basins are flanked by high hills with elevations up to 1500 m.

(6) *Palaeogene folds and thrusts.* This region includes surface folds seen on the eastern side of the Thailand rift basins, notably those that affect the Mesozoic clastic rocks of the Khorat Plateau (Cooper et al. 1989; Booth 1998). Palaeogene folds and thrusts are also seen, on seismic reflection data and in some wells, to underlie several onshore and offshore rift basins (e.g. Morley et al. 2004, 2007a).

Some of the provinces geographically overlap as a result of different timing of structural events. For example, the Palaeogene folds and thrusts were probably once present across the Shan Plateau, but folded Mesozoic rocks have been mostly removed by erosion or covered by later rift basins.

Mesozoic–Early Cenozoic geological evolution of the region

Mesozoic–Early Cenozoic plate setting of western Indochina

A number of major terranes rifted off Pangaea during the Palaeozoic and were conveyed northward as the Palaeo-Tethys was subducted beneath the proto-Asian margin (Metcalfe 1998). The collision of these terranes with the proto-Asian margin during the Permo-Triassic marks the Indosinian orogeny. Typically in Thailand and Myanmar the oldest peak metamorphic ages found in basement are Indosinian (c. 260–200 Ma; Ahrendt et al. 1997; Hansen et al. 2002; Barley et al. 2003). The most outboard of the terranes was the Shan–Thai (Sibumasu) block. Renewed rifting from northern Australia–Antarctica in the Late Triassic–early Jurassic defined the most western to southwestern terranes found in Sumatra (Sikuleh terrane) and Myanmar (western Burma terrane; Metcalfe 1998, 2002). Northward subduction and closure of the Meso-Tethys resulted in accretion of several terranes such as the Lhasa, Burma and Sikuleh blocks onto the Eurasian margin at different times during the Cretaceous (Metcalfe 1998). Finally, India rifted off the northern Antarctica–Australian margin during the early Cretaceous, resulting in closure of the Ceno-Tethys ocean, and creation of the Indian Ocean.

The transition of the SE Asian margin from post-Indosinian orogeny passive margin to Andean-type margin is marked by intrusion of Jurassic I-type granites and granitoids (c. 170 Ma; Searle et al. 1999; Barley et al. 2003). This margin extended westward well beyond SE Asia to at least the Karakorum region (Searle et al. 1999). The accretionary prism-related geology of the Indo-Burma ranges supports an Andean-type margin origin (Mitchell 1981). Cretaceous and early Cenozoic Andean-type granitoids have also been been identified in eastern Myanmar (Cobbing et al. 1986; Darbyshire & Swainbank 1988; Putthapiban 1992; Mitchell 1993; Barley et al. 2003).

In Thailand interpretation of the origin of the Late Cretaceous–Early Cenozoic (c. 90–50 Ma; Charusiri et al. 1993) granites and granitoids differs from that in Myanmar. S-type two-mica, tin-bearing granites predominate in volume over I-type granitoids (hornblende- and biotite-bearing tonalite and granodiorite). Late Cretaceous S-type granite forms bodies up to 120 km × 15 km, whereas I-type granite forms plutons about 5 km × 5 km

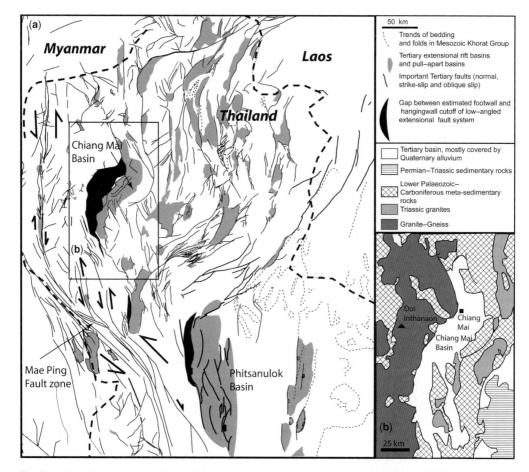

Fig. 3. (a) Location map for the Chiang Mai basin, northern Thailand, mostly based on Morley (2007). (b) Detailed location map of the highlands west of the Chiang Mai basin (modified from 1:500 000 geological map of Thailand, and Dunning et al. (1995)).

(Hutchison 1996). On the eastern margin of the Shan Plateau near Chiang Mai in Thailand (Fig. 3) the Doi Inthanon gneiss records a thermal event in the Late Cretaceous, as indicated by two U–Pb monazite ages of 72 ± 1 Ma and 84 ± 2 Ma (Macdonald et al. 1993; Dunning et al. 1995).

Origin of the Late Cretaceous–Early Cenozoic orogenic events

Charusiri et al. (1993) and Schwartz et al. (1995) suggested that the Late Cretaceous–Early Cenozoic S-type granites are related to crustal thickening caused by collision of the west Burma block with Sundaland. If there was a Late Cretaceous collision between the west Burma block and Sundaland, evidence for the suture is scant, with the Sagaing Fault lying approximately along the suture. The only candidate for ophiolite remnants are the rocks of the poorly dated jadeite belt, whose suggested ages range between Late Cretaceous and Eocene (e.g. Hutchison 1996). However, if the jadeite belt rocks are simply parts of the ophiolites present to the west and east, then their age could be Late Jurassic (Mitchell 1981, 1993). The jadeite belt lies along fault strands at the northern end of the Sagaing Fault zone.

Although the generalities of the terrane accretion story of Metcalfe (1998, 2002) outlined above describe the tectonic setting of SE Asia well, in detail the origin and timing of accretion of some of the terranes is disputed. Most critical for the Shan Plateau area is the origin of the Sikuleh terrane. Barber et al. (2005) proposed that the Sikuleh terrane (including west Sumatra and western Myanmar) is of Cathaysia (Indochina) affinity. This terrane was detached from Indochina and moved to a position outboard

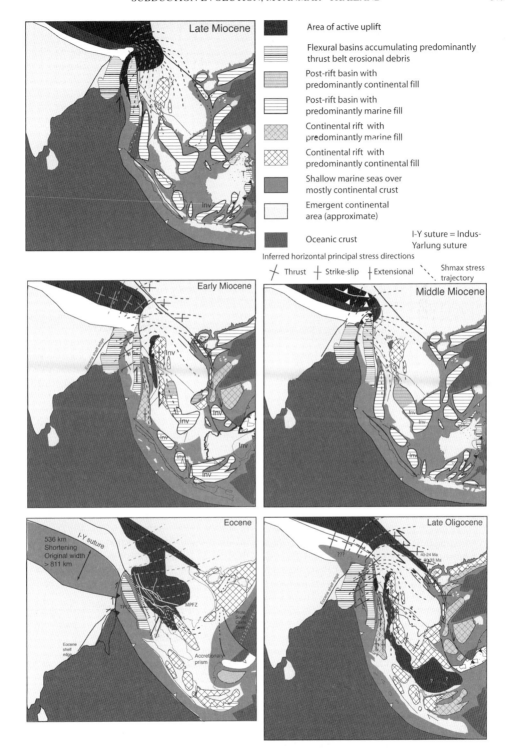

Fig. 4. Regional restoration of plates, and associated basins. Reconstruction based largely on Hall (2002), Hall & Morley (2004) and Morley (2004) with palaeostress orientations modified from Huchon et al. (1994). INV, basin undergoing inversion; MPFZ, Mae Ping fracture zone; T, Tarakan Basin; TPFZ, Three Pagodas fracture zone.

of the Sibumasu terrane along a transcurrent fault during the Triassic. Subsequently, an island arc collided with the west Sumatran block in the mid-Cretaceous. This scenario implies that late Cretaceous collision of the west Burma block with the Shan–Thai block could not have occurred as a result of closure of the Meso-Tethys. In an alternative scenario, Mitchell et al. (2007) suggested that early Cretaceous collision of an east-facing island arc with the Shan Plateau (including the west Burma block) was followed by early Cenozoic metamorphism in the Mogok metamorphic belt as a result of a reversal of tectonic polarity following arc–Shan Plateau collision.

Although there remains considerable variety in the current terrane accretion scenarios it seems reasonable to state that, despite its previous popularity, Late Cretaceous terrane collision in Myanmar is an unlikely cause of the Late Cretaceous–Early Cenozoic orogenic event. Accretion appears to have occurred considerably earlier in all current scenarios. One subduction-related event that could cause deformation, and a considerable thermal event, is subduction of a spreading centre (e.g. Thorkelsen & Taylor 1989; Schoonmaker & Kidd 2006; Whittaker et al. 2007). Hafkensheid (2004) has used restoration of estimated volumes of subducted material based on seismic tomography to test various plate-tectonic scenarios for the Tethyan realm. The best-fit model for the India–Eurasia collision was found to be that of Stampfli & Borel (2004). This model has a small back-arc ocean (Spontang Ocean) lying between the Indian plate oceanic crust and Eurasia. The reconstructions show that oblique subduction of the Spontang Ocean spreading centre during the Late Cretaceous–Early Cenozoic in the Myanmar–Sumatra region is a distinct possibility.

Another potential effect of subduction on continental deformation has been recently highlighted by Watkinson et al. (2009), who noted that the Sunda Trench between 90° and 100°E was the site of a dextral transform zone that accommodated differential motion between the fast (21 cm a^{-1}) northward moving India plate to the west and the slow or stalled Australia plate to the east. They proposed that ductile dextral deformation present along the NE–SW-striking Khlong Marui and Ranong faults in the Thai Peninsula (Fig. 1) was related to this differential motion. The timing of the dextral deformation is loosely constrained between 72 and 56 Ma intrusive events (Watkinson et al. 2009).

It is not the purpose of this paper to try and resolve the problems of interpreting sparse information on the Early Cretaceous–Early Cenozoic history of the region. The important key elements for the Cenozoic evolution of NW Sundaland are that Late Cretaceous–Early Cenozoic granite and granitoid emplacement occurred whether there was a Late Cretaceous microplate collision or subduction-related processes along an Andean-type margin. In either case, the Shan–Thai block (Shan Plateau area) is thought to have developed hot, thickened crust by the early Cenozoic (e.g. Charusiri et al. 1993; Mitchell 1993; Schwartz et al. 1995; Barley et al. 2003; Searle et al. 2007). This region of hot, weak crust became an important site of deformation when India tangentially collided and coupled with the Burma block during the Palaeogene (Fig. 4).

Coupling of India with western Myanmar

Regional tectonic evolution of the Andaman Sea region

During the early Cenozoic the western margin of what became the Andaman Sea region was probably an oblique convergence subduction zone–island arc system similar to present-day Sumatra to the south (Hall 2002; Curray 2005). The relative movement rate and direction of India with respect to Sundaland changed considerably during the Cenozoic (Curray 2005; Fig. 5). These changes affected the spreading rate in the Andaman Sea and whether deformation in Myanmar and Thailand was transpressional or transtensional. The key points from Curray (2005) are listed below, and illustrated for the Oligocene–Recent in Figure 6.

(1) *44–32 Ma.* Convergence following collision of India with Asia at about 59 Ma was initially about 100 mm a^{-1}; by 44 Ma this rate had slowed to about 60 mm a^{-1}.

(2) *32–23 Ma.* Continental extension was initiated in the North Sumatra–Mergui Basin (Polachan & Racey 1994; Andreason et al. 1997). Approximately 60 km of continental extension occurred in a 310° to 270° direction, at a rate of about 7 mm a^{-1} (Curray 2005). Accompanying the onset

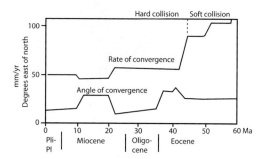

Fig. 5. Changes in direction and velocity of India–Eurasia plate convergence during the Cenozoic. Redrawn from Curray (2005).

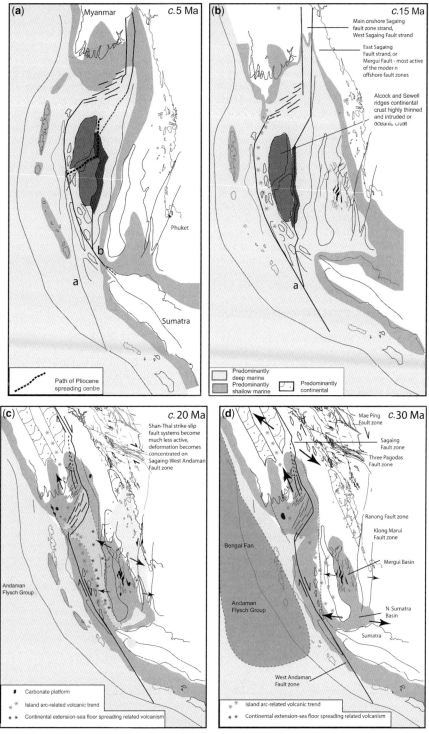

Fig. 6. Oligocene–Recent structural and tectonic evolution of the Andaman Sea area. Mostly based on data and maps of Hall (2002), Morley (2004) and Curray (2005). It should be noted that although the clockwise rotation of Thailand and Myanmar is show following Curray (2005), it is also possible that less rotation has occurred (e.g. Hall 2002).

of extensional and strike-slip deformation within the back-arc area there was decreasing activity along the Andaman–Nicobar part of the subduction zone. Recently there has been only one historically recorded volcanic eruption, and the subducted slab dips more steeply than in the Sumatran sector (Curray 2005; Shapiro et al. 2008). The onset of extension in the Mergui–north Sumatra basin marks the time when the margin evolved from an oblique Sumatra-like island arc margin to a more complex margin as the plate boundary. Accommodation of the differential motion between India and Sundaland broadened eastward from the Sumatra–Andaman subduction zone to more distributed dextral deformation within Myanmar and Thailand, particularly along the Sagaing Fault zone and Shan Plateau (Figs 4 and 6).

(3) *23–15 Ma*. The convergence rate of India with Eurasia dropped to less than 50 mm a^{-1}. The extension rate of the Andaman back-arc area was about 15 mm a^{-1}.

(4) *15–Ma*. The amount of extension along the NNE–SSW-striking Miocene spreading centre was c. 100 km in a 335° direction. The average opening rate was c. 9 mm a^{-1}.

(5) *4 Ma–present*. The present-day convergence rate of India with Asia is about 35 mm a^{-1} (Vigny et al. 2003). The present-day spreading centre, with dated magnetic anomalies, was initiated at c. 4 Ma, with an initial spreading rate of c. 16 mm a^{-1}, which increased to 38 mm a^{-1}, around 2.0–2.5 Ma (Kamesh Raju et al. 2004; Curray 2005; Fig. 5). The spreading direction was about 335°. Both Kamesh Raju et al. (2004) and Curray (2005) reached a similar conclusion that the Central Andaman Basin opened about 118 km in c. 4 Ma. The recent rate of dextral motion along the Sagaing Fault zone is of the order of 17–20 mm a^{-1}, suggesting that a further c. 15–18 mm a^{-1} is accommodated along other structures further west (Vigny et al. 2003; Sahu et al. 2006). Recent studies imaging shallow faults (Nielsen et al. 2004) and earthquakes coupled with modelling of global positioning system (GPS)-derived motions (Sahu et al. 2006) have identified major dextral faults in the Indo-Burma ranges and accretionary prism area that accommodated the dextral motion not taken up by the Sagaing–West Andaman fault system.

The Andaman Sea region records mostly extensional and strike-slip events in the trailing wake of the coupled India–Burma block region (Fig. 6). Towards the south the strike-slip faults, particularly the West Andaman Fault (Figs 1 and 6) connect the broad region of deformation in the Andaman Sea back to the more discrete plate boundary of Sumatra. Onshore in Myanmar and Thailand more complex orogenic effects related to the same plate-scale events are recorded.

Geological evidence for India coupling with the Burma block

In a highly simplified version of the structural style described by Curray (2005) the Indo-Burma Ranges–Andaman accretionary prism area can be characterized as an imbricate stack of predominantly eastward-dipping fault slices and folds (Fig. 7). Cretaceous ophiolites and older deep-sea sedimentary rocks generally lie at the top and on the eastward side of the prism. Progressively younger Neogene sedimentary rocks are found on the western side, and towards the base of the stack as well as draping the stack.

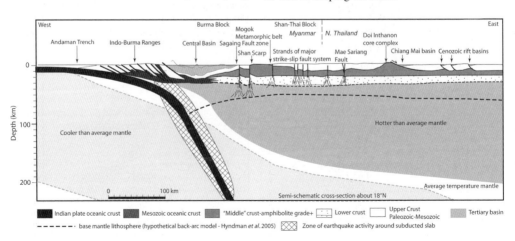

Fig. 7. A semi-schematic but true-scale cross-section across southern Myanmar and northern Thailand illustrating some of the main large-scale tectonic features of the area. The surface geology used is taken from c. 18°N. Mantle temperature distribution, and subduction zone geometry from Shapiro et al. (2008), their line 'A'.

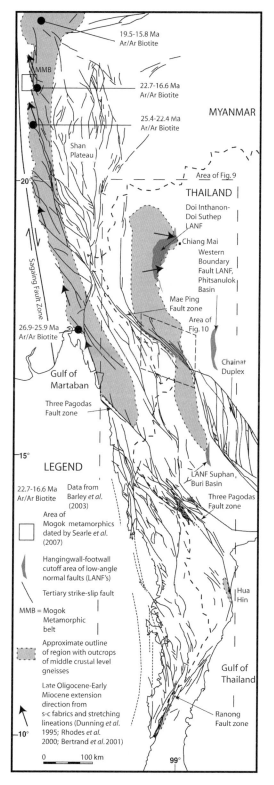

Onshore, the Kabaw Fault separates the Late Cretaceous–Cenozoic clastic deposits of the Central Basin to the east from the Indo-Burma ranges to the west (Fig. 1). Although the Kabaw Fault has a reverse sense of motion, it probably also accommodates a considerable amount of dextral motion (Magung 1987, 1989; Zaw 1989). Offshore along approximately the same trend are the major dextral Diligent and West Andaman strike-slip fault trends (Curray 2005; Fig. 1).

Evidence for India coupling with Myanmar in the early Cenozoic has been found in a number of places. The Indo-Burman ranges were uplifted into a major subaerial mountain range during the Oligocene (Brunnschweiler 1974). Recent radiometric age dating has begun to reveal significant details about the response of the Mogok metamorphic belt to the coupling of the west Burma block with the Indian plate. Barley et al. (2003) reported c. 43–47 Ma rims to Jurassic zircons in orthogneiss and syenite interpreted as recrystallization during Eocene high-grade metamorphism. The magmatic age of a syntectonic syenite was determined as 30.9 ± 0.7 Ma and that of a late tectonic syenogranite as 22.6 ± 0.4 Ma (Barley et al. 2003). U–Pb and U–Pb–Th geochronology was conducted on the metamorphic belt by Searle et al. (2007). They combined their results with those of Barley et al. (2003) to conclude that a protracted period of sillimanite + muscovite grade, high-temperature metamorphism (c. 606–680 °C, 4.4–4.9 kbar), lasted from around 43 to 29 Ma and intermittent crustal melting occurred between 45.5 and 23 Ma (see Fig. 8 for location).

Although data are more restricted in Thailand there is evidence for similar magmatic events at around 45 Ma and younger in Thailand from granite–gneiss outcrops east (Chonburi) and west (Hua Hin) of Bangkok, as well as from the western highlands (Lan Sang, Doi Inthanon; Dunning et al. 1995; Charusiri 1989; S. Meffre, pers. comm.).

Shan Plateau: a Palaeogene sinistral transpressional strike-slip orogen

The most obvious structural features on satellite images of western Thailand and eastern Myanmar

Fig. 8. Regional map of the Shan Plateau area. Fault patterns based on Landsat and digital elevation model (DEM) interpretation (revised from earlier interpretation by Morley 2004); the fault traces are inferred to be those associated with Cenozoic strike-slip faulting. The early sense of motion (Early Cenozoic to about 30 Ma) on these faults was sinistral; later fault movements were predominantly dextral. Biotite $^{40}Ar/^{39}Ar$ cooling ages along the Mogok metamorphic belt are from Bertrand et al. (1999, 2001).

are the networks of sharp, curvilinear fault traces (Le Dain et al. 1984; Morley 2004; Fig. 9). These faults include major strike-slip faults (Mae Ping, Three Pagodas, Ranong faults; e.g. Tapponnier et al. 1986; Lacassin et al. 1997; Morley 2004, 2007) and occur in a north–south-trending belt, up to 250 km wide (east–west) and 1000 km long (north–south).

The best information on the timing and sense of motion of the strike-slip fault network comes from the Lan Sang National Park area of the NW–SE-striking Mae Ping Fault zone. Lacassin et al. (1993, 1997) determined a left-lateral sense of motion from an exhumed mid-crustal level, c. 5 km wide mylonitic shear zone (Figs 10 and 11). $^{40}Ar/^{39}Ar$ biotite cooling ages ranged between 33 and 30 Ma for the Mae Ping Fault zone and c. 36 and 33 Ma for the Three Pagodas Fault zone (Lacassin et al. 1997). The cooling ages were interpreted by Lacassin et al. (1997) as documenting the last increments of ductile sinistral deformation. At a restraining bend along the Mae Ping Fault zone the Umphang gneisses were sampled for zircon and apatite fission-track (AFT) dating (Upton

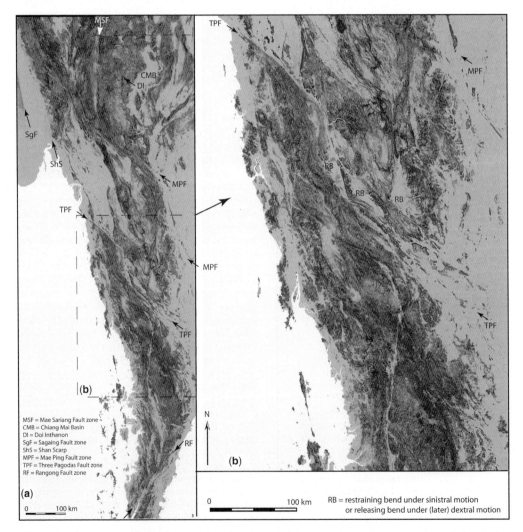

Fig. 9. Dip map of topography showing how clearly Cenozoic strike-slip faults show up as sharp linear features in the landscape. An interpretation of the fault patterns and location for the map are given in Figure 8. The strike-slip faults tend to form networks of north–south- and NW–SE-striking faults. The dense network of faults over a broad area suggests more of a transpressional fault style, compared with sharp, block-bounding strike-slip characteristic of escape tectonics.

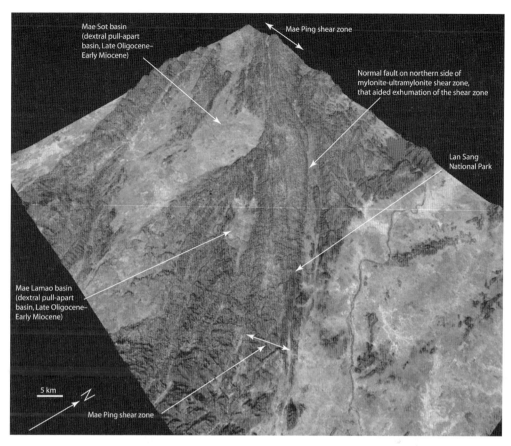

Fig. 10. A 3D perspective of a satellite image draped over a 90 m DEM with a view to the NW along the Mae Ping fault zone. The linear ridge formed by the fault zone comprises sheared middle-crustal level rocks, about 5 km wide, within which are zones of mylonites and ultramylonites of hundreds of metres to c. 1 km wide (Lacassin et al. 1997); the shear zone is best exposed in Lan Sang National Park. The shear zones are indicative of sinistral motion (Fig. 12), and associated $^{39}Ar/^{40}Ar$ biotite cooling ages range between 33 and 30 Ma (Lacassin et al. 1997). Exhumation of the shear zone appears to have been aided by the presence of a normal fault on the northern side of the shear zone (Lacassin et al. 1997). The geometry of the Mae Sot Basin indicates that the Mae Ping fault zone reversed its sense of motion during the Late Oligocene–Early Miocene, undergoing (minor, probably <10 km) dextral motion. (See Fig. 8 for location.)

1999; Fig. 10). The results of the dating indicated exhumation between c. 50 and 40 Ma, which Morley et al. (2007b) interpreted as representing the timing of early sinistral transpressional faulting.

The geometry of the broad fault network is very different from the long, single straight strike-slip faults that accommodated rigid block movement in the region (Sagaing Fault zone, Fig. 1, Red River Fault zone; e.g. Leloup et al. 1995). The sinistral sense of motion on NW–SE-oriented fault strands requires an approximately east–west maximum horizontal stress direction (Huchon et al. 1994; Figs 4 and 12). On the basis of fault geometry, Morley (2004, 2007) inferred that the region of strike-slip faulting was characteristic of a transpressional orogenic belt related to India–Burma coupling.

If there once was a more classic fold and thrust belt associated with the transpressional orogenic belt, it is not immediately obvious from the mapped geology of Thailand. With the exception of the Late Triassic–Cretaceous Khorat Group, the remainder of the rocks cropping out in Thailand were those previously caught up in the Indosinian orogeny, then intruded by Triassic and Cretaceous granites. Hence the highly heterogeneous nature of the upper crust means that a regular fold and thrust belt exploiting a simple layer cake stratigraphy did not develop. Instead, the early Cenozoic deformation that is detectable in the Palaeozoic rocks is the transpressional sinistral system (Figs 8, 9 and 12).

The thick, extensive Late Triassic–Cretaceous Khorat Group sedimentary rocks preserved in the

Fig. 11. Mid-crustal level Oligocene deformation: shear zones in the Mae Ping fault zone at Lan Sang National Park (see Fig. 10 for location). (**a**) High-strain zone within paragneiss, boudinaged more resistant units within marble; (**b**) typical steeply dipping mylonitic fabric (striking NW–SE) with some cross-cutting leucocratic veins that have been rotated to a low angle with respect to the shear zone; (**c**) downward view onto flat-lying surface showing δ structure in boudinaged leucocratic vein indicative of sinistral displacement.

Khorat Plateau of eastern Thailand display a number of broad folds (Cooper et al. 1989; Kozar et al. 1992; Booth 1998; Fig. 12). The Khorat Group is up to 4000 m thick (Racey et al. 1996) and thins only gradually towards the western margin of the plateau. Isolated remnants of Khorat Group are found scattered in synclines in central Thailand, southern peninsular Thailand and the Gulf of Thailand (Fig. 12). Beneath the Late Oligocene–Miocene sediments of the Phitsanulok basin the folded and imbricated Khorat Group is visible on seismic reflection data (Morley et al. 2007a; Fig. 12). Hence widespread folding and thrusting seems to have occurred but uplift and erosion of the Khorat Group across Thailand removed much of the evidence. AFT dating of the western edge of the Khorat Plateau (Fig. 12) indicates that the samples experienced near total annealing between 70 and 50 Ma (Upton 1999). The age range is later than the c. 110 Ma age for the upper part of the sequence (Racey et al. 1996), hence there is the possibility that an orogenic-related thermal event resulted in the annealing, not simple burial. Samples left the partial annealing zone between 49 and 11 Ma, with those nearest the strongly deformed areas uplifted between c. 49 and 30 Ma (Upton 1999). This dating suggests that the timing of folding in the Khorat Plateau was coincident with sinistral displacement on the sinistral transpressional zone to the west.

Oligocene–Recent dextral transtensional strike-slip system: unroofing of the orogen

After 30 Ma the strike-slip faults of the Shan Plateau reversed their sense of motion, with the dominant NW–SE to north–south trend undergoing dextral motion. Evidence for dextral motion comes from fault shear-sense indicators (Bertrand et al. 1999; Morley et al. 2007b), pull-apart basins developed along large-scale releasing bends as seen on geological maps and satellite images (e.g. Lacassin et al. 1998; Morley 2001, 2007; Figs 9 and 10), and from earthquake focal mechanisms (Bott et al. 1997). Morley (2007) contended that the late dextral strike-slip motion was largely confined to the Shan Plateau area and died out eastward, so that the north–south-trending Oligocene–Miocene rift basins east of the plateau (Fig. 1) are largely extensional in origin. The extensional basins appear to overlie the more external parts of the older sinistral transpressional belt, and may be associated with orogenic collapse.

The sinistral motion is inferred to be a result of strain partitioning and transpression caused by India–west Burma coupling, whereas the change to dextral motion is related to the subsequent northward movement of the coupled India–west Burma block. At the southern part of the system the Andaman Sea underwent extension–transtension during the Oligocene, probably commencing around 32 Ma (Polachan et al. 1994). To the north

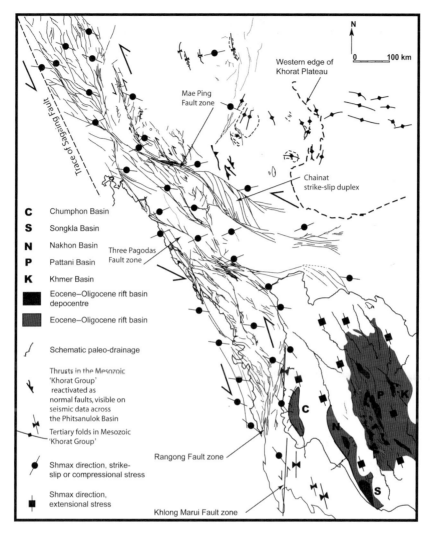

Fig. 12. Illustration of the Eocene–Early Oligocene deformation that affects Thailand and eastern Myanmar. The orientation of the structures has been rotated 20° anticlockwise from the present day to reflect Eocene orientations (e.g. Curray 2005). Strike-slip fault patterns from interpretation of surface geology and satellite geology (revised from Morley 2004, 2007). Offshore the rift basins may have developed on an earlier Cenozoic transpressional belt. Rift basin configuration from Morley & Westaway (2006). Palaeostress orientations are based on Andersonian assumptions of S_{hmax} direction to faults, with north–south- to NW–SE-trending faults on the western side of the map having sinistral displacement (Lacassin et al. 1997). Fold patterns in the Khorat area have a range of orientations that probably does not directly reflect palaeostress directions, but instead represents the influence of pre-existing fabrics, particularly inverted Triassic rift basins.

the Mogok metamorphic belt displays NNW–SSE extensional stretching lineations on low-angle faults, and mica cooling ages young from south (c. 25–27 Ma) to north (22.7–16.6 Ma; Bertrand et al. 1999). NNW–SSE-oriented transtension is also indicated from data in the Central Basin to the west (Pivnik et al. 1998).

The peak metamorphic ages between 37 and 29 Ma obtained by Searle et al. (2007) came from the same region of the Mogok gneisses where Bertrand et al. (1999) determined 22.7–16.6 Ma biotite cooling ages (Fig. 8). Hence the transition from transpression, and peak metamorphism, to transtensional exhumation in one area probably occurred in about 7 Ma. However, along the entire system dextral transtension in the south probably occurred at the same time as sinistral transpression occurred further north, which may suggest a clockwise rotational pivoting of the Indian continent. The northward motion of India appears to have caused progressively younger transtensional activation of the Sagaing–Mergui Fault zone passing northward (Bertrand et al. 1999).

For the Mogok metamorphic belt Torres et al. (1997) reported a similar cooling history to Bertrand et al. (1999), supplemented with AFT dating that yielded central ages between 14.6 and 18.7 Ma. These Late Oligocene–early Miocene cooling ages fit well with the duration of the large pull-apart basins formed during strike-slip deformation in Thailand; in particular, the youngest ages of the Mae Sot, Mae Lamao (Fig. 10) and Nong-y-Plong (Three Pagodas Fault zone, Fig. 1) basins are Early Miocene. The timing of these strike-slip related basins differs from the larger, extensional basins of the central plains to the east, which continued to develop during the Middle and Late Miocene (Fig. 4; Gibling et al. 1985; Ratanasthien 1989, 2002; Watanasak 1990; Morley et al. 2001).

Today north–south dextral motion is no longer so widely distributed, but instead has become focused on the linear Sagaing Fault zone (Bertrand et al. 1999; Vigny et al. 2005) and the Indo-Burma ranges and Andaman–Nicobar Islands (Nielsen et al. 2004; Sahu et al. 2006). The eastern side of the Shan Plateau has only a low level of recorded earthquake activity (Bott et al. 1997).

Oligocene–Miocene central Thailand extensional province

Central Thailand is occupied by a string of large and small Oligocene–Miocene rift basins that lie immediately east of the Shan Plateau (Fig. 1). These basins display a northward transition to a more strike-slip dominated province passing into Laos (Fig. 1; Lacassin et al. 1998; Morley 2007). The basins have also undergone episodic inversion during the Miocene, indicating fluctuations in the stress state with time (e.g. Morley et al. 2000, 2001, 2007b; Morley 2007). Four unusual characteristics of the extensional province are discussed below: (1) episodes of basin inversion; (2) post-rift basin thickness in the eastern Gulf of Thailand; (3) the diachroneity of rift basin onset and termination; and (4) the presence of Oligocene–Middle Miocene low-angle normal faults (LANFs).

Basin inversion. Worldwide many rift basins are characterized by episodes of inversion; with the exception of long-lived basins there is usually only one significant phase, such as the post-rift inversion found in the central Sumatran basin (e.g. Pertamina BPPKA 1996). In the rifts of Thailand inversion is highly variable; some basins do not exhibit evidence for any inversion, whereas others display multiple events and alternate periods of extension with phases of inversion over a period of 15–20 Ma (Morley et al. 2001, 2007; Morley 2007). Inversion is generally poorly developed in the Gulf of Thailand, and becomes more important northward, particularly from the Phitsanulok Basin. Seismic reflection data from the southern Phitsanulok Basin display a strong inversion event during the early Miocene. At least four Miocene inversion events were identified from coal mines in the Li Basin, and many of the rift basins in northern and central Thailand terminated in the Late Miocene or early Pliocene with an inversion event (Morley et al. 2001). The episodic inversion appears to require considerable changes in the stress state during the Miocene, Figure 13 illustrates some of the likely permutations of the stress field during the late Cenozoic. Probably most of the time the rifts developed under extension (Fig. 13c) but for short periods were subject to stresses appropriate for inversion (Fig. 13b). Figure 13d shows the present-day stress state, which may be appropriate as far back as the latest Miocene. The current stress state is largely determined from earthquake focal mechanisms, but is supplemented by borehole breakout data in some areas (Morley 2007).

Post-rift basin thickness in the Gulf of Thailand. As reviewed by Morley & Westaway (2006), the Miocene–Recent post-rift thickness of the Pattani and Malay basins in the Gulf of Thailand exceeds 6–7 km in places. The post-rift basins are synformal in cross-section and subsidence is clearly not strongly fault controlled. The problem of understanding the rapid post-rift subsidence was discussed in detail by Morley & Westaway (2006). They concluded that conventional post-rift thermal subsidence models annot explain such subsidence, and consequently proposed that sediment loading initiated flow of hot lower crust away from the basin, thereby creating accommodation space (Fig. 14). The basin centre heat flows of around 100 mW m^{-2} and geothermal gradients of 6 °C per 100 m (Pigott & Sattayarak 1993) indicate that hot, ductile crust is present beneath the basin. For ductile flow away from the basin to operate, a pressure head at the top of the ductile crust had to be established. The pressure head is achieved in the model by erosion causing the brittle–ductile transition zone under the sediment source area (Shan Plateau) to be uplifted higher than the transition zone under the sedimentary basin. The sediment flux moved to the SE, in the opposite direction to the flow of ductile crust (Fig. 14b–d; Morley & Westaway 2006).

The schematic evolution of the Pattani Basin – Shan Plateau source area shown in Figure 14 is explained below. During the initial stage (Fig. 14a)

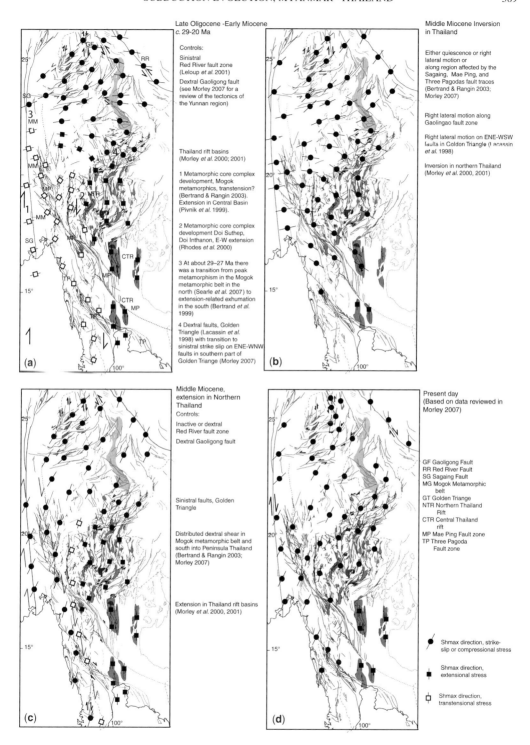

Fig. 13. Illustrations of the likely range of stresses that have affected the Yunnan–Myanmar–Laos–northern Thailand region from the Late Oligocene onwards. Stresses are inferred from Andersonian assumptions of S_{hmax} direction to faults, palaeostress studies (Morley et al. 2000; Bertrand & Rangin 2003; Morley 2007) and modern stresses (see review by Morley 2007).

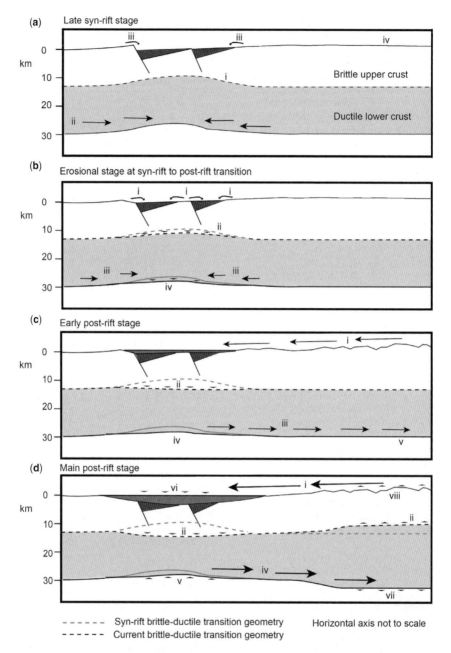

Fig. 14. Schematic summary of the Morley & Westaway (2006) model for the evolution, from synrift to post rift conditions, of the sedimentary basins in the Gulf of Thailand. Inferred senses of lower crustal flow are indicated at each stage. (**a**) Late synrift stage (Early Oligocene). (**b**) Erosional stage at the synrift to post-rift transition (Late Oligocene). (**c**) Early post-rift stage (?Early Miocene). (**d**) Main post-rift stage (post-(?)Early Miocene).

moderate crustal extension ($b \sim 1.3$) occurred, accompanied by sedimentation at a low rate (estimated as typically c. 1 km in c. 10 Ma, or c. 0.1 $mm^{-1} a^{-1}$). The sediment was mainly of local provenance (iii, Fig. 14a) (Lockhart et al. 1997; Jardine 1997). When extension ended during the late Oligocene the lithosphere cooled and the base of the brittle layer thus adjusted to progressively deeper levels (ii, Fig. 14b). The sense of any contemporaneous lower crustal flow was probably still inward, but weak. During the Early Miocene the post-rift basin was increasingly supplied by sediments derived from the Shan Plateau. Return lower crustal flow from under the basin towards the sediment source area began (iii, Fig. 14c), as a result of both sediment loading in the basin and erosion from the sediment source area. Following a change of climate and tectonics (including extension and dextal transtension), erosion in the sediment source area greatly increased (i, Fig. 14d) and resulted in much faster sedimentation rates in the basin. The faster sedimentation caused the base of the brittle upper crust to advect downwards beneath the basin (ii, Fig. 14d), and the faster erosion caused upward advection beneath the sediment source (iii). These changes increased the lateral pressure gradient and consequently increased lower crustal flow (iv, Fig. 14d).

Diachroneity of rift onset and termination. The onset of rifting in the Gulf of Thailand tends to young from east (Eocene–Early Oligocene) to west (Late Oligocene) except in the southern gulf where the Chumphon, Songkla and Nakhon basins all appear to have a Palaeogene history (Fig. 12; Morley et al. 2001; Hall & Morley 2004; Morley & Westaway 2006). With possibly one or two exceptions, the oldest age for rift basin initiation onshore in central and northern Thailand is Late Oligocene. Younging of the synrift to post-rift transition is from east (Late Oligocene) to west (Middle Miocene–Late Miocene) in the Gulf of Thailand, and south (Late Miocene) to north (Pliocene) passing onshore. Some basins in northern Thailand are bounded by faults that have remained active until the present day (Fenton et al. 1997, 2003). The northward younging of the synrift to post-rift transition suggests a link between rifting and the northward passage of India.

Low-angle normal faults. The Cenozoic rifts of Thailand are bounded by high-angle normal faults, and form classic half-graben type geometries (e.g. Flint et al. 1988; O'Leary & Hill 1989; Pradidtan & Dook 1992; Jardine et al. 1997; Lockhart et al. 1997; Morley et al. 2001; Chantraprasert 2003). Extension generally appears to be low ($b < 1.3$) within the basins (e.g. Watcharanantakul & Morley 2000). However, on the eastern edge of the Shan Plateau, two putative metamorphic core complexes (Doi Inthanon, Doi Suthep) have been identified adjacent to the Chiang Mai Basin (Fig. 3; Macdonald et al. 1993; Dunning et al. 1995; Rhodes et al. 2000, 2005; Barr et al. 2002). Other low-angle rift-bounding faults have been identified from seismic reflection data across the Suphan Buri and Phitsanulok basins, in the Mergui Basin and in the Gulf of Thailand (Chumphon, Songkla, north Malay basins) but the footwalls to the faults display Palaeozoic sedimentary, metasedimetary and igneous rocks, not mid-crustal level amphibolites. The large LANFs seen on seismic reflection data (Fig. 15) display heaves >2 km (more typically between 5 and 15 km), and dips less than 30°, typically about 20°. Based on published data, and unpublished seismic lines across the Chiang Mai Basin, the evolution of the Doi Inthanon core complex is presented in Figure 15 and is discussed below.

The temperature scale for Figure 16 is derived from Macdonald et al. (1993), who determined high-temperature (600–700 °C), low-pressure (280 ± 60 MPa) conditions of formation from mineral assemblages for the orthogneiss and paragneiss in the Doi Inthanon area. These conditions indicate a high geothermal gradient of about 5 °C per 100 m in the upper crust (Fig. 17). The presence of numerous hot springs in the area today and geothermal gradients of up to 7 °C per 100 m in some sedimentary basins (e.g. Fang Basin, Pradidtan & Tongtaow 1984), demonstrate that high temperatures are still present in northern Thailand.

One indication of high temperatures lasting into the Late Oligocene–Early Miocene is the grade of coal in Mae Cham mine. The coal mine lies in a very small basin, which has only about 200 m of section preserved, yet the grade of the coal (vitrinite reflectance value (Ro) = 0.65%) is considerably higher than those of the other Cenozoic coals mined in the region (Ro ~ 0.40–0.45%) (Morley et al. 2001). The high grade of the coal probably indicates both relatively high geothermal gradient and that the basin was much deeper (probably at least 1.5 km deep) and more extensive than it is today. One possible scenario is that the Mae Cham Basin was part of the Chiang Mai Basin left behind in the footwall of the LANF (Fig. 16d and e).

The stages of the Doi Inthanon LANF development are outlined below (see Fig. 16). They provide a summary of the key Cenozoic orogenic events that affected the eastern Shan Plateau.

85–70 Ma. This period represents the time of peak metamorphism (Dunning et al. 1995), marked by the emplacement of S-type granites (Charusiri et al. 1993) and possible transpressional thickening of the crust (Charusiri et al. 1993; Morley 2004; Watkinson et al. 2009). Evidence for transpressional or contractional structures related to this event are not well established in the Western Highlands. However, to preserve the extensive detachment geometry between the amphibolite gneisses and the overlying Palaeozoic rocks would seem to require that earlier deformation did not extensively warp or offset the contact. Hence the simplest solution is to have the Palaeozoic section thrust along a décollement above the paragneiss.

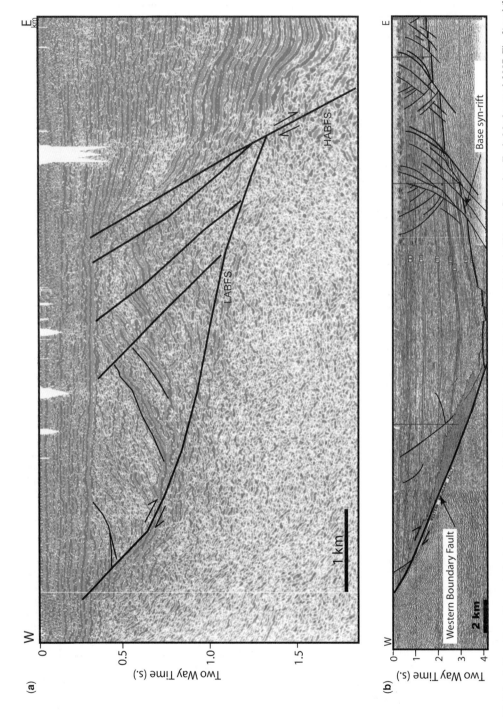

Fig. 15. Seismic lines illustrating east-dipping low-angle normal faults (LANFs) bounding (**a**) the Phitsanulok Basin (Western Boundary Fault, Morley *et al.* 2007: Fig. 3) and (**b**) the Suphan Buri Basin. The Phitsanulok Basin LANF has maximum heave of around 10–15 km; the Suphan Buri LANF has about 5 km maximum heave. (See Fig. 8 for locations.)

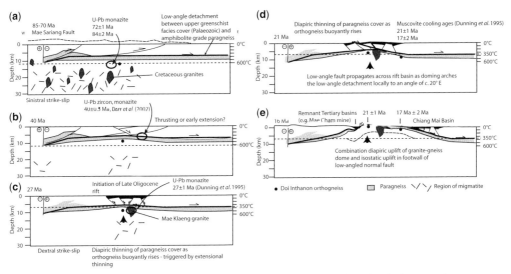

Fig. 16. Regional evolution of the Chiang Mai Basin and Western Highlands illustrating the key tectonic events affecting the eastern Shan Plateau during the Late Cretaceous–Miocene. Based on data of Macdonald et al. (1993), Dunning et al. (1995), Upton (1999), Morley et al. (2001) and Barr et al. (2002), and unpublished seismic reflection data.

40–27 Ma. Monazite and zircon grains from mylonitic gneiss in the Doi Suthep area yielded $^{235}U-^{207}Pb$ ages of 40 ± 0.5 Ma (Barr et al. 2002), which Barr et al. interpreted to define the upper age limit for mylonitization. The c. 27 Ma Mae Klaeng granite intrudes across a mylonitic fabric and is thought to constrain the upper age of mylonitization (Barr et al. 2002). Extensional collapse of the transpressional belt along the mylonitic detachment sometime between c. 40 and 27 Ma is a

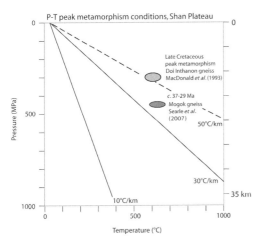

Fig. 17. Two estimates of pressure and temperature from metamorphic assemblages on the eastern and western side of the Shan Plateau. The data suggest prolonged high temperatures in the Shan Plateau area consistent with a back-arc model (Figs 7 and 18).

possible scenario (Barr et al. 2002). However, no sedimentary basins of this age are documented in the area, so the Eocene extensional event must be regarded as poorly established. One other possibility is that the detachment records a top-to-the-east compressional décollement around the brittle–ductile transition; the timing coincides with development of peak metamorphism (c. 40–29 Ma) in the Mogok metamorphic belt on the western side of the Shan Plateau (Searle et al. 2007) and sinistral transpression along the Mae Ping and Three Pagodas fault zones of Thailand and Myanmar (Lacassin et al. 1997; Morley 2004). An Eocene–early Oligocene transpressional thickening event followed by extensional collapse in the early Miocene is the scenario favoured here.

27 Ma. Late Oligocene–early Miocene rifts are widespread onshore in Thailand (e.g. Morley et al. 2001). From the seismic reflection data across the Chiang Mai Basin, the low-angle fault appears to have truncated an older high-angle rift system. Hence a high-angle rift system is shown for the Late Oligocene–Early Miocene, together with development of a gneiss dome. The Himalayan orogen provides models for contractional deformation triggering anatexis and migmatitic gneiss dome development as a result of large-scale release of fluids by overthrusting of low-grade, fluid-rich sedimentary rocks (Le Fort 1986; Le Fort et al. 1987), or by shear heating along major thrusts (Harrison et al. 1997). Once gneiss domes are initiated, a positive feedback between decompression and partial melting helps fuel further buoyant uplift and more melting (Teyssier & Whitney

2002). The partial melting of crust can help initiate subsequent orogenic collapse and crustal thinning. Hence, although it can be argued that purely extensional processes drive gneiss dome development, in orogenic belts it may more often be the case that gneiss domes are initiated during contraction, and enhancement of the process occurs subsequently during extension (e.g. Lee *et al.* 2004); the latter is the preferred evolution for the northern Thailand core complexes (Fig. 16).

21–16 Ma. During the Early Miocene the low-angle detachment developed and sliced through the existing Late Oligocene–Early Miocene high-angle fault rift basin, transporting it up to 35 km to the east. The Doi Inthanon gneisses were exhumed from temperatures of about 350 °C to about 50 °C; that is, from depths of $c. \geq 8$ km to $c.$ 1 km (Macdonald *et al.* 1993; Dunning *et al.* 1995). Exhumation was probably driven both by diapiric uplift of the granite gneiss dome and by isostatic footwall uplift caused by low-angle fault motion. Late Early Miocene cooling ages limited between muscovite ^{40}Ar–^{39}Ar cooling ages of 21 Ma and AFT ages as young as 14 ± 1 Ma (Dunning *et al.* 1995; Upton *et al.* 1997; Upton 1999; Barr *et al.* 2002) indicate the duration of LANF and core complex development. Two-dimensional seismic reflection data suggest that a high-angle fault-bounded rift basin of probable Oligocene age was truncated by the younger, early Miocene LANF activity.

Central Basin

The Central Basin in Myanmar (Fig. 7) accommodated a tremendous thickness (up to about 12 km) of Palaeogene sedimentary rock, in particular Eocene–Oligocene fluvio-deltaic to shallow marine deposits (Pivnik *et al.* 1998), yet these sediments appear to have been deposited in a broad synformal basin, with little internal deformation. Adjacent areas to the west (Indo-Burma ranges) and the east (Shan Plateau) were undergoing considerable uplift and deformation during the Eocene–Oligocene, whereas the intervening Central Basin shows virtually no compressional deformation until the Miocene (Pivnik *et al.* 1998). The Central Basin was apparently a region of relatively strong crust between two areas of much weaker crust. The western part of the Central Basin has accumulated a much greater thickness of sediment than the eastern half (Fig. 7), which can be explained by the western basin being floored by strong, dense oceanic crust, passing eastward into less dense continental crust (Pivnik *et al.* 1998).

The Central Basin has long been recognized as a $c.$ 1000 km long Cenozoic palaeo-gulf that was filled primarily by a north to south prograding major delta system (e.g. Chhibber 1934; Pivnik *et al.* 1998). For much of the Eocene and Oligocene the delta remained in the northern half of the basin, with the boundary between freshwater and marine conditions fluctuating between about 20° and 23°N. Then from the late Oligocene onward dominantly alluvial plain redbeds (Irrawadian) prograded southward, reaching the Salin sub-basin at about 20°N (Fig. 1) during the Pliocene. Hence it appears that at least until the latest Miocene the Central Basin accommodated the great majority of the sediment transported by the Irrawaddy River.

In the Salin sub-basin the Miocene section shows expansion into active normal faults (Pivnik *et al.* 1998). Then sometime in the Late Miocene–Early Pliocene a tectonic reorganization occurred, and the basin was affected by inversion, accompanied by folding and thrusting (Pivnik *et al.* 1998). Deformation in the Salin sub-basin matches the change in deformation style along the Sagaing Fault proposed by Bertrand & Rangin (2003), who described a switch from transtensional strike-slip deformation to transpressional strike-slip deformation around the Miocene–Pliocene boundary. Inversion effectively halted large-scale basin subsidence and forced rapid progradation of the Irrawaddy Delta southward into the Gulf of Martaban.

Information on crustal temperature from Myanmar is not well established; however, in the Gulf of Martaban geothermal gradients from hydrocarbon exploration wells are in the range of 3–3.5 °C per 100 m; onshore the Central Basin is thought to have been cool in the past because of the depth of maturation of hydrocarbon source rocks, and geothermal gradients from the onshore fields are around 2 °C per 100 m. Hence the available data suggests that the Central Basin was indeed a stronger, cooler region of crust than the Shan Plateau area during the Cenozoic. Probably only after the Shan Plateau area had effectively strain-hardened during the Miocene did deformation start to extensively affect the Central Basin.

Discussion

The aim of this discussion is to suggest a plate-scale scenario that explains the metamorphic, igneous and structural characteristics of the study region. The discussion considers a number of themes.

Application of the back-arc mobile belt model to eastern Myanmar and Thailand

In the 1990s Thailand and the Shan Plateau of Myanmar were largely viewed as one of the peripheral regions to the Himalayas that underwent escape tectonics, where rigid continental blocks were squeezed out laterally away from the orogenic belt by tens to hundreds of kilometres along large bounding strike-slip faults (Le Dain et al. 1984; Tapponnier et al. 1986; Polachan et al. 1991; Lacassin et al. 1993, 1997; Replumaz & Tapponnier 2003). The escape tectonic model required the rigid blocks to be composed of cold crust, and little deformed internally during the Cenozoic. The deep sedimentary basins of Thailand could also be explained by the strike-slip setting (Tapponnier et al. 1986; Polachan et al. 1991). More recently, modern regional displacement patterns based on GPS data have been used to constrain a numerical model of the Himalayan orogen that treats the crust as a continuously deforming solid (not a rigid plate) under the influence of gravity (England & Molnar 1997a, b, 2005). This model predicts little impact of Himalayan deformation on Indochina. The view that little of orogenic significance affected eastern Myanmar and Thailand during the Cenozoic is implicit in the topographic ooze model of Clark & Royden (2000). This model explained the decreasing topography passing from the Tibetan Plateau into northern Myanmar and Thailand as being created by a wedge of ductile lower crust that flowed southwards from Tibet in the Late Miocene. This focus on imposing Tibetan Plateau-related processes on the region is understandable, but ignores the role played by the other plate-scale feature of the region: the Sumatra–Andaman subduction zone. Here it is contended that the structural development of NW Sundaland is best understood in the context of a back-arc mobile belt related to the evolution of the Andaman Sea section of the Java–Sumatra–Andaman arc, plus the effects of highly oblique continent–continent collision, rather than ad hoc models related to only the India–Eurasia collision.

Hyndman et al. (2005) and Currie & Hyndman (2006) reviewed and described the characteristics of back-arc mobile belts as follows: most mobile belts arise in a back-arc situation as a result of shallow asthenosphere convection, facilitated by water derived from the underlying slab driving vigorous flow rates (Fig. 18). Heat flow is typically of the order of 70–90 mW m^{-2}. Former back-arc regions are likely to remain hot, weak areas where orogenic shortening is concentrated. The high temperatures also aid widespread orogenic plutonism and ductile crustal deformation, in particular facilitating detachments within the lithosphere and lower crustal flow. High surface elevations may also be associated with little or no crustal thickening, and instead rely on dynamic mantle support. Mobile belts can be persistently weak regions for long periods (hundreds of millions of years) of geological time (Hyndman et al. 2005).

NW Sundaland fits many of the back-arc mobile belt criteria above: it has been the site of subduction and terrane accretion since the Permo-Triassic Indosinian orogeny began at around 260 Ma (e.g. Metcalfe 1998). Two pronounced phases of tin-bearing, two-mica, granite intrusion occurred during the Triassic and Late Cretaceous–Palaeogene, which appear to link with collisional events or an Andean margin setting (Charusiri et al. 1993; Barley et al. 2003).

Many of Thailand's sedimentary basins have high present-day geothermal gradients (3–7 °C per 100 m), and high heat flows. For example, the Pattani Basin in the Gulf of Thailand has heat flows in the range of 70–100 mW m^{-2} (Pigott & Sattayarak 1993) despite being at the post-rift stage of development for the last 25 Ma (Morley & Westaway 2006; Fig. 14). U–Pb dating of monazites and zircons and analysis of mineral assemblages from the Shan Plateau have revealed high palaeo-temperatures of c. 4–6 °C per 100 m in the upper crust during the Eocene–Oligocene in the Mogok metamorphic belt (Searle et al. 2007) and Late Cretaceous around Doi Inthanon (Macdonald et al. 1993; Dunning et al. 1995; Figs 8 and 17). Seismic tomography indicates that the upper mantle is anomalously hot under Indochina (Fig. 7), and SE Asia as a whole, in a broad region overlying the subduction zones that ring SE Asia (Hall & Morley 2004; Shapiro et al. 2008). The Shan Plateau forms a high topographic region with many peaks at around 1500 m. Regional seismic tomography suggests that crustal thickness in northern Thailand and most of Myanmar, including the Shan Plateau, ranges between 40 and 50 km (Engdahl & Ritzwoller 2001).

Other manifestations of the high geothermal gradient are Cenozoic basic mantle-derived igneous activity, small basins with mature oil source rocks (Fang Basin) or relatively high-grade coal (Mae Chaem Basin), extensive occurrences of hot springs, super-deep post-rift Miocene sedimentary basins that may be indicative of lower crustal flow (Morley & Westaway 2006), and LANFs and putative metamorphic core complexes on the western and eastern side of the Shan Plateau and LANFs in many of the rift basins (Figs 7, 8 and 15).

Hyndman et al. (2005) noted that 'the high temperatures of current mobile belts are also indicated by widespread sporadic Cenozoic basaltic volcanism'. This feature is seen in Thailand, where despite being 500 km east of the active volcanic arc Cenozoic basic igneous activity occurs extensively (Putthapiban 2002). Although most of the surface volcanic rocks are of Miocene–Pliocene age (Barr & Macdonald 1981), some basic intrusions encountered in wells in the Gulf of Thailand are as old as Eocene. A number of seismic lines across rift

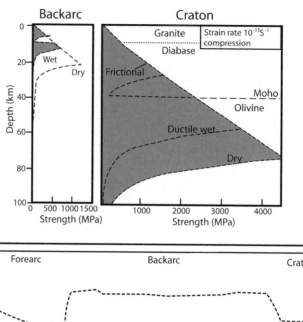

Fig. 18. Model for back-arc mobile belt modified from Hyndman *et al.* (2005).

basins clearly show the presence of igneous intrusions and flows. Wells have penetrated basic Cenozoic igneous bodies in the Suphan Buri, Kamphaeng Saeng, Phitsanulok and Phetchabun basins (Remus *et al.* 1993). The geochemistry of the volcanic rocks indicates that they were derived from the mantle with little crustal contamination (Barr & Macdonald 1981).

The Shan Plateau is a cryptic orogenic belt in many ways. There is no obvious Cenozoic fold and thrust belt, although remnants can be found (Fig. 12), and with the exception of the 2500 m high Doi Inthanon (caused by LANF activity) the height of the plateau is limited to about 1500 m. Therefore it is not surprising that the escape tectonics and topographic ooze models have been imposed upon it. However, the characteristics of tectonics affecting the Shan Plateau seem to owe much to the subduction history. The low height of the plateau is characteristic of the back-arc mobile belt model (Hyndman *et al.* 2005), the crust has been weakened thermally from below by hot mantle, and internally by the heat from high-radioactivity I-type and S-type granitoid intrusions, both mechanisms being ultimately related to subduction.

Characteristics of the transition from back-arc to oblique collisional deformation

During the Eocene Thailand and Myanmar underwent the transition from a margin dominated by subduction-related processes to one where a highly oblique collision with India focused deformation in the Shan–Thai block. It might be expected that the linkage with India would have focused deformation to a much greater extent in western Myanmar than appears to be the case. A fold and thrust belt did continue to develop in the old accretionary prism area of the Indo-Burma ranges, yet immediately to the east the Central Basin continued to accumulate sediment apparently undisturbed by compressional deformation. The surface of the basin remained at approximately sea level during the Oligocene and received fluvio-deltaic sediments in a broad synformal trough (Pivnik et al. 1998; Fig. 7). This is not a geometry typical of Andean-type margins. The location of the sinistral transpressional belt in the Shan Plateau suggests that the lithosphere was much weaker in comparison with the Central Basin. The key explanation for this strength difference lies in the much lower temperatures typically found in the forearc area compared with the back-arc (Hyndman et al. 2005; Fig. 18).

Origin of the distributed deformation accommodating India–Indochina relative motion

During the Oligocene the Sumatra–Andaman Trench ceased to be the primary site accommodating the differential motion of India and Indochina, and the zone broadened eastwards by about 300 km to the Shan Plateau (Figs 4 and 13). The reason for the shift may lie in (1) the presence of hot weak crust in eastern Myanmar and western Thailand, and (2) the abandoned, north–south-striking Indian plate subduction zone hanging down in the mantle below Myanmar, which would have acted as a giant sheet anchor exerting drag against the overall northward motion of the lithosphere. If the mantle on both sides of the slab were part of the same plate moving in the same direction less drag would occur than when the slab formed the interface between two plates undergoing predominantly strike-slip relative motions. An eastward relocation of the region that accommodated relative motion between India and Indochina would have had the effect of keeping the slab within one plate and hence may explain the location of the dextral strike-slip zone in the Shan Plateau.

The present-day intraplate position of the subduction zone is supported by the seismicity study of Guzman-Speziale & Ni (1996), which suggested the absence of interplate earthquakes in western Myanmar. They favoured interpreting the earthquake data in terms of the overriding plate being coupled with India, a conclusion supported in studies by Radha Krishna & Sanu (2000) and Rao & Kalpna (2005).

From about 40 Ma western Indochina has rotated clockwise, and the Andaman segment of the subduction zone has rotated from a roughly NW–SE strike to a north–south strike (e.g. Lee & Lawver 1995; Curray 2005; Fig. 4), hence the sheet anchor effect would have become increasingly important with time as the slab became more oblique to the northward motion of India. The (local) jump or broadening of the plate boundary was manifest during the Oligocene when the Mogok metamorphic complex became unroofed on the western side of the Shan Plateau (Bertrand et al. 1999). The transtensional component of the dextral strike-slip motion and western ranges uplift and erosion had the effect of exhuming the hottest parts of the Shan Plateau. Exhumation of the Shan Plateau provided a large sediment source for basins in the Gulf of Thailand (Morley & Westaway 2006) and Gulf of Martaban in the Early and Middle Miocene. A consequence of exhumation was cooling and effectively strain-hardening of the Shan Plateau, which may have facilitated concentration of deformation onto the Sagaing Fault zone and Indo-Burma ranges from the Middle Miocene onwards.

Extension in Thailand

Crustal thickening of the Shan–Thai block led to anatexis in the Doi Inthanon area and the intrusion of the Late Oligocene Mae Klaeng granite migmatites within a gneiss dome. However, the development of the metamorphic core complexes west of Chiang Mai from c. 21–15 Ma does not appear to coincide with the unroofing of the adjacent segment of Mogok metamorphic belt to the west, which dates predominantly from c. 27–25 Ma (but is possibly as young as 22.4 Ma) according to data of Bertrand et al. (2001; Fig. 8). Nor are the low-angle fault-related stretching directions (NNW–SSE and east–west) in the two provinces similar (Fig. 8). Although there appears to be no direct kinematic linkage between the two provinces, they may mark different stages in the northward passage of India. The NNW–SSE unroofing of the Mogok metamorphic belt directly records the northward movement (Bertrand & Rangin 2003), whereas stress relaxation in the Shan Plateau, and gravity spreading in the trailing wake of the Indian plate, is marked by development of the Doi Suthep and Doi Inthanon metamorphic core complexes, and Late Oligocene–Miocene extension in the western

Gulf of Thailand and onshore. This broad extensional province with mixed high-angle and low-angle extensional faults shifted westward and northward from the focus of earlier (Eocene–Oligocene) extension.

Extensive rifting affected SE Asia during the Cenozoic, and most of the basins had passed into post-rift subsidence by the end of the Oligocene (Hall & Morley 2004; Fig. 4). The continuation of rifting in Thailand during the Miocene is regionally anomalous. This observation, coupled with the progressive northward younging of the onset of post-rift subsidence in Thailand, strongly suggests that the northward passage of India had a strong influence on the continuation of rifting (e.g. Huchon et al. 1994). However, another influence on the strain pattern in Thailand is the Sumatra–Andaman subduction zone, as indicated by the pattern and magnitude of co-seismic deformation resulting from the 2004 Sumatra–Andaman earthquake. Southern peninsular Thailand (Phuket), was displaced to the WSW c. 27 cm (40–45 cm including post-seismic motions), Bangkok c. 8 cm, and Chiang Mai c. 2.6 cm (Vigny et al. 2005). How such differential movements translate to strain within Thailand is less certain, but GPS stations south, east and NE of Thailand showed considerably less displacement than those in Thailand (Vigny et al. 2005). Assuming a mega-earthquake every 500 years, and differential motion was focused on the Thailand rifts over a period of 25 Ma, then Phuket-magnitude displacements could account for c. 13–20 km WSW displacement relative to Vietnam, whereas displacement in Chiang Mai would be c. 1.3 km. Consequently, it seems reasonable to speculate that a considerable portion of the Late Cenozoic extension in the Gulf of Thailand could be related to events associated with the subduction zone, not only the Himalayan collision.

Conclusions

The application of the Himalayan-dominated tectonic models of escape tectonics (Tapponnier et al. 1986) and topographic ooze (Clark & Royden 2000) to Thailand and Myanmar involves an assumption of relatively cold crust in the region, prior to the imposition of Himalayan-dominated orogenic effects. However, the eastern Myanmar–western Thailand orogenic belt discussed in this paper displays a range of evidence that indicates it has been a high-temperature but low-relief orogenic belt, characterized by ductile lower crust and episodic deformation during much of the Cenozoic. The orogenic belt characteristics are atypical for Cordilleran-type, Andean-type or Himalayan-type orogenic belts, but fit well with a back-arc mobile belt setting (Hyndman et al. 2005; Currie & Hyndman 2006) modified by the effects of highly oblique continent–continent collision. Some key characteristics are as follows:

(1) The region of most intense deformation (eastern Myanmar–western Thailand orogenic belt) begins about 300–350 km inland from the subduction zone. A low-relief, weakly deformed sedimentary basin (Central Basin) separates the accretionary prism compressional–transpressional region (Indo-Burma ranges) from the main 'orogenic' belt. This is explained by the Central Basin occupying a region of stronger crust (oceanic crust with relatively low lithospheric temperatures) whereas the Myanmar–western Thailand orogenic belt was a hot and weak former back-arc area heavily intruded by I-type and S-type granites during the Triassic, Cretaceous and Palaeogene. High-temperature, low-pressure metamorphic or thermal events occurred during the Late Cretaceous–earliest Cenozoic and the Eocene–Oligocene (Barr et al. 2002; Barley et al. 2003; Searle et al. 2007).

(2) A clearly defined foreland fold and thrust belt is absent. A broad belt of sinistral transpression and later dextral transtension forms the clearest evidence of deformation in the Myanmar–western Thailand orogenic belt. However, folds and thrusts formed during the Palaeogene are extensively exposed in the Khorat Plateau, and glimpsed elsewhere in the highly eroded Khorat Group remnants seen in a few outcrops between rift basins, and as patchy remnants in seismic reflection data beneath basins.

(3) Extensional collapse of thickened continental crust is suggested by the widepread development of low-angle normal faults with and without putative metamorphic core complexes on the eastern and western margins of the orogenic belt and concomitant high-angle rifting in Thailand. On the eastern margin of the Shan Plateau extension was approximately perpendicular (east–west) to the north–south trend of the orogenic belt, whereas on the western margin it was subparallel (NNW–SSE) to the relative movement direction of India with respect to Sundaland.

(4) Exhumation of the eastern side of the present-day Shan Plateau area during the Early Miocene produced a great flux of sediment into the Gulf of Thailand, which during the post-rift stage loaded the Pattani Basin and caused the lower crust to flow away from under the post-rift basin to accommodate the load (Morley & Westaway 2006).

(5) A possible sheet-anchor effect from the steeply dipping subducted oceanic slab beneath Myanmar resulted from increasingly oblique convergence of India and Sundaland. The effect could have influenced the broad accommodation of India–Sundaland relative motion along strike-slip

zones in the upper crust distributed between the Andaman Trench and the Shan Plateau.

(6) The origin of the forces giving rise to Late Cenozoic extension and episodic inversion in the Thailand rift basins appears to be complex. Local buoyancy forces may play a role, and the Himalayan collision has strongly influenced stress patterns (e.g. Huchon *et al.* 1994), but events along the Sumatra–Andaman subduction zone also appear to be important, as indicated by GPS-defined displacements in Thailand associated with the 2004 Sumatra–Andaman earthquake (Vigny *et al.* 2005).

I would like to thank P. Charusiri, R. Hall, B. Rhodes, A. Mitchell, C. Hutchison and M. Searle for many helpful discussions about the geology of Thailand and Myanmar over the years. PTTEP has provided access to a considerable amount of industry data on the region that aided this review. S. Barr, P. Charusiri and S. Johnson are thanked for helpful reviews that considerably improved the manuscript.

References

AHRENDT, H., HANSEN, B. T., LUMJUAN, A., MICKEIN, A. & WEMMER, K. 1997. Tectonometamorphic evolution of NW-Thailand deduced from U/Pb- Sm/Nd- and K/Ar- isotope investigations. *In*: DHEERADILOK, P., HINTHONG, C. *ET AL.* (eds) *Proceedings of the International Conference on the Stratigraphy and Tectonic Evolution of Southeast Asia and the South Pacific*. Department of Mineral Resources, Bangkok, 314–319.

ANDREASON, M. W., MUDFORD, B. & ST ONGE, J. E. 1997. Geologic evolution and petroleum system of the Thailand Andaman Sea Basin. *In*: Indonesian Petroleum Association (ed.) *Proceedings of the Petroleum Systems of SE Asia and Australasia Conference*, Jakarta, 337–345.

BARBER, A. J., CROW, M. J. & MILSOM, J. S. (eds) 2005. *Sumatra: Geology, Resources and Tectonics*. Geological Society, London, Memoirs, **31**.

BARLEY, M. E., PICKARD, A. L., KHIN, Z., RAK, P. & DOYLE, M. G. 2003. Jurassic to Miocene magmatism and metamorphism in the Mogok metamorphic belt and the India–Eurasian collision in Myanmar. *Tectonics*, **22**, 4-1–4-10, doi:10.1029/2002TC001298.

BARR, S. M. & MACDONALD, A. S. 1981. Geochemistry and geochronology of late Cenozoic basalts of Southeast Asia. *Geological Society of America Bulletin*, **92**, 1069–1142.

BARR, S. M., MACDONALD, A. S., MILLER, B. V., REYNOLDS, P. H., RHODES, B. P. & YOKART, B. 2002. New U–Pb and Ar/Ar ages from the Doi Inthanon and Doi Suthep metamorphic core complexes, Northwestern Thailand. *In*: MANTAJIT, N. & POTISAT, S. (eds) *Proceedings of the Symposium on the Geology of Thailand*. Department of Mineral Resources, Bangkok, 284–308.

BERTRAND, G. & RANGIN, C. 2003. Tectonics of the western margin of the Shan Plateau (central Myanmar): Implications for the India–Indochina oblique convergence since the Oligocene. *Journal of Asian Earth Sciences*, **21**, 1139–1157.

BERTRAND, G., THEIN, M. *ET AL.* 1999. Cenozoic metamorphism along the Shan scarp (Myanmar): Evidence for ductile shear along the Sagaing fault or the northward migration of the eastern Himalayan syntaxis? *Geophysical Research Letters*, **26**, 915–918.

BERTRAND, G., RANGIN, C., MALUSKI, H., BELLON, H. & GIAC SCIENTIFIC PARTY 2001. Diachronous cooling along the Mogok Metamorphic belt (Shan scarp, Myanmar): The trace of the northward migration of the Indian syntaxis. *Journal of Asian Earth Sciences*, **19**, 649–659.

BOOTH, J. E. 1998. The Khorat Plateau of NE Thailand—exploration history and hydrocarbon potential. *Seapex Exploration Conference Proceedings*. Seapex, Singapore, 169–202.

BOTT, J., WONG, I., PRACHUAB, S., WECHBUNTHUNG, B., HINTHONG, C. & SURAPIROME, S. 1997. Contemporary seismicity in northern Thailand and its tectonic implications. *In*: DHEERADILOK, P., HINTHONG, C. *ET AL.* (eds) *Proceedings of the International Conference on the stratigraphy and Tectonic Evolution of Southeast Asia and the South Pacific*. Department of Mineral Resources, Bangkok, 453–464.

BRUNNSCHWEILER, R. O. 1974. Indoburman ranges. *In*: SPENCER, A. M. (ed.) *Mesozoic–Cenozoic Orogenic Belts*. Geological Society, London, Special Publications, **4**, 279–299.

CHAMOT-ROOKE, N., LE PICHON, X., RANGIN, C., HUCHON, P., PUBELLIER, M., VIGNY, C. & WALPERSDORF, A. 1999. Sundaland motion in a global reference frame detected from GEODYSSEA GPS measurements: Implications for relative motions at the boundaries with the Australo-Indian plates and the south China block. *In*: WILSON, P. & MICHEL, G. W. (eds) *The Geodynamics of S and SE Asia [GEODYSSEA] Project, Final Report of the GEODYSSEA Project to the EC*. GeoForschungsZentrum, Potsdam, 39–74.

CHANTRAPRASERT, S. 2003. Oblique zones in Cenozoic extensional fault systems, Southern Pattani Basins. *In*: RATANASATHIEN, B., RIEB, S. L. & CHANTRAPRASERT, S. (eds) *8th International Congress on Pacific Neogene Stratigraphy*. Chiang Mai, February 2003, 332–348.

CHARUSIRI, P. 1989. *Lithophile metallogenetic epochs of Thailand: a geological and geochronological investigation*. Ph.D. Thesis, Queen's University, Kingston, Canada.

CHARUSIRI, P., CLARK, A. H., FARRAR, E., ARCHIBALD, D. & CHARUSIRI, B. 1993. Granite belts in Thailand: Evidence from the $^{40}Ar/^{39}Ar$ geochronological and geological synthesis. *Journal of Southeast Asian Earth Sciences*, **8**, 127–136.

CHHIBER, H. L. 1934. *The Geology of Burma*. Macmillan, London.

CLARK, M. K. & ROYDEN, L. H. 2000. Topographic ooze: Building the eastern margin of Tibet by lowercrustal flow. *Geology*, **28**, 703–706.

COBBING, E. J., MALLICK, D. I. J., PITFIELD, P. E. J. & TEOH, L. H. 1986. The granites of the Southeast Asian Tin Belt. *Journal of the Geological Society, London*, **143**, 537–550.

COOPER, M. A., HERBERT, R. & HILL, G. S. 1989. The structural evolution of Triassic intermontane basins in northeastern Thailand. In: THANASUTHIPITAK, T. & OUNCHANUM, P. (eds) *Proceedings of the International Symposium on Intermontane basins: Geology and Resources*. Chiang Mai University, Chiang Mai, 231–242.

CURRAY, J. R. 2005. Tectonics and history of the Andaman Sea region. *Journal of Asian Earth Sciences*, **25**, 187–232.

CURRAY, J. R., MOORE, D. G., LAWVER, L. A., EMMEL, F. J. & RAITT, R. W. 1978. Tectonics of the Andaman Sea and Burma. In: WATKINS, J., MONTADERT, L. & DICKENSON, P. W. (eds) *Geological and Geophysical Investigations of Continental Margins*. American Association of Petroleum Geologists, Memoirs, **29**, 189–198.

CURRIE, C. A. & HYNDMAN, R. D. 2006. The thermal structure of subduction zone back arcs. *Journal of Geophysical Research*, **111**, B08404, doi:10.1029/2005JB004108.

DARBYSHIRE, D. P. F. & SWAINBANK, I. G. 1988. Southeast Asia granite project—geochronology of a selection of granites from Burma. Natural Environment Research Council, London Isotope Geology Centre Report, **88/6**.

DUNNING, G. R., MACDONALD, A. S. & BARR, S. M. 1995. Zircon and monazite U–Pb dating of the Doi Inthanon core complex, northern Thailand: Implications for extension within the Indosinian Orogen. *Tectonophysics*, **251**, 197–213.

ENGDAHL, E. R. & RITZWOLLER, M. H. 2001. Crust and upper mantle P- and S-wave delay times at Eurasian seismic stations. *Physics of the Earth and Planetary Interiors*, **123**, 205–219.

ENGLAND, P. C. & MOLNAR, P. 1997a. Active deformation of Asia: From kinematics to dynamics. *Science*, **278**, 647–650.

ENGLAND, P. C. & MOLNAR, P. 1997b. The field of crustal velocity in Asia, calculated from Quaternary rates of slip on faults. *Geophysical Journal International*, **131**, 551–582.

ENGLAND, P. C. & MOLNAR, P. 2005. Late Quaternary to decadal velocity fields in Asia. *Journal of Geophysical Research*, **110**, doi:10.1029/2004JB003541.

FENTON, C. H., CHARUSIRI, P. & WOOD, S. H. 2003. Recent paleoseismic investigations in Northern Thailand. *Annals of Geophysics*, **46**, 957–981.

FENTON, C. H., CHARUSIRI, P., HINTHONG, C., LUMJUAN, A. & MANGKONKARN, B. 1997. Late Quaternary faulting in northern Thailand. In: DHEERADILOK, P., HINTHONG, C. ET AL. (eds) *Proceedings of the International Conference on the Stratigraphy and Tectonic Evolution of Southeast Asia and the South Pacific*. Department of Mineral Resources, Bangkok, 436–452.

FLINT, S., STEWART, D. J., HYDE, T., GEVERS, C. A., DUBRULE, O. R. F. & VAN RIESSEN, E. D. 1988. Aspects of reservoir geology and production behaviour of Sirikit Oil Field, Thailand: An integrated study using well and 3-D seismic data. *AAPG Bulletin*, **72**, 1254–1268.

GIBLING, M. R., TANTISUKRIT, C., UTTAMO, W., THANASUTHIPITAK, T. & HARALUCK, M. 1985. Oil shale sedimentology and geochemistry in Cenozoic Mae Sot Basin, Thailand. *AAPG Bulletin*, **69**, 767–780.

GUZMAN-SPEZIALE, M. & NI, J. F. 1996. Seismicity and active tectonics of the western Sunda Arc. In: YIN, A. & HARRISON, T. M. (eds) *The Tectonic Evolution of Asia*. Cambridge University Press, New York, 63–84.

HAFKENSCHEID, E. 2004. *Subduction of the Tethys Oceans reconstructed from plate kinematics and mantle tomography*. PhD thesis, Utrecht University.

HALL, R. 2002. Cenozoic geological and plate tectonic evolution of SE Asia and the SW Pacific: Computer-based reconstructions and animations. *Journal of Asian Earth Sciences*, **20**, 353–434.

HALL, R. & MORLEY, C. K. 2004. Sundaland Basins. In: CLIFT, P., WANG, P., KUHNT, W. & HAYES, D. (eds) *Continental–Ocean Interactions within East Asian Marginal Seas*. American Geophysical Union, Geophysical Monograph, **149**, 55–87.

HANSEN, B. T., PAPE, B., WEMMER, K., ASSAVAPATCHARA, S. & SALYAPONGSE, S. 2002. Still searching for the Pre-Cambrian basement: New isotope age determinations from crystalline rocks in SE and S Thailand. In: MANTAJIT, N. & POTISAT, S. (eds) *Symposium on Geology of Thailand*. Department of Mineral Resources, Bangkok, 22–23.

HARRISON, T. M., LOVERA, O. M. & GROVE, M. 1997. New insights into the origin of two contrasting Himalayan granite belts. *Geology*, **25**, 899–902.

HUCHON, P., LE PICHON, X. & RANGIN, C. 1994. Indo-China Peninsula and the collision of India and Eurasia. *Geology*, **22**, 27–30.

HUTCHISON, C. S. 1996. *Geological Evolution of South-East Asia*. Geological Society of Malaysia, Kuala Lumpur.

HYNDMAN, R. D., CURRIE, C. A. & MAZZOTTI, S. P. 2005. Subduction zone backarcs, mobile belts and orogenic heat. *GSA Today*, **15**, 4–10.

JARDINE, E. 1997. Dual petroleum systems governing the prolific Pattani basin, offshore Thailand. In: Indonesian Petroleum Association (ed.) *Proceedings of the Petroleum Systems of SE Asia and Australasia Conference*, Jakarta, 351–363.

KAMESH RAJU, K. A., RAMPRASAD, T., RAO, P. S., RAMALINGESWARA RAO, B. & VARGHESE, J. 2004. New insights into the tectonic evolution of the Andaman basin, northeast Indian Ocean. *Earth and Planetary Science Letters*, **221**, 145–162.

KOZAR, M. G., CRANDALL, G. F. & HALL, S. E. 1992. Integrated structural and stratigraphic study of the Khorat Basin, Rat Buri Limestone (Permian), Thailand. In: PIANCHAROEN, C. (ed.) *Proceedings of a National Conference on Geological Resources of Thailand: Potential for Future Development*. Department of Mineral Resources, Bangkok, 682–736.

LACASSIN, R., LELOUP, P. H. & TAPPONNIER, P. 1993. Bounds on strain in large Tertiary shear zones of SE Asia from boudinage restoration. *Journal of Structural Geology*, **15**, 677–692.

LACASSIN, R., HINTHONG, C. ET AL. 1997. Tertiary diachronic extrusion and deformation of western Indochina: Structure and $^{40}Ar/^{39}Ar$ evidence from NW Thailand. *Journal of Geophysical Research*, **102**, 10013–10037.

LACASSIN, R., REPLUMAZ, A. & LELOUP, H. P. 1998. Hairpin river loops and slip-sense inversion on southeast Asian strike-slip faults. *Geology*, **26**, 703–706.

LE DAIN, A. Y., TAPPONNIER, P. & MOLNAR, P. 1984. Active faulting and tectonics of Burma and surrounding regions. *Journal of Geophysical Research*, **89**, 453–472.

LEE, J., HACKER, B. & WANG, Y. 2004. Evolution of North Himalayan gneiss domes: structural and metamorphic studies, in Mabja Dome, southern Tibet. *Journal of Structural Geology*, **26**, 2297–2316.

LEE, T.-Y. & LAWVER, L. A. 1995. Cenozoic plate reconstructions of Southeast Asia. *Tectonophysics*, **251**, 85–138.

LE FORT, P. 1986. Metamorphism and magmatism during the Himalayan orogeny. *In*: COWARD, M. & RIES, A. C. (eds) *Collision Tectonics*. Geological Society, London, Special Publications, **19**, 159–172.

LE FORT, P., CUNEY, M., DENIEL, C., FRANCE-LANORD, C., SHEPPARD, S. M. F., UPRETI, B. N. & VIDAL, P. 1987. Crustal generation of the Himalayan leucogranites. *Tectonophysics*, **134**, 39–57.

LELOUP, P. H., LASASSIN, R. ET AL. 1995. The Ailao Shan–Red River shear zone (Yunnan, China), Tertiary transform boundary of Indochina. *Tectonophysics*, **251**, 3–84.

LOCKHART, B. E., CHINOROJE, O., ENOMOTO, C. B. & HOLLOMON, G. A. 1997. Early Tertiary deposition in the southern Pattani Trough, Gulf of Thailand. *In*: DHEERADILOK, P., HINTHONG, C. ET AL. (eds) *Proceedings of the International Conference on the Stratigraphy and Tectonic Evolution of Southeast Asia and the South Pacific*. Department of Mineral Resources, Bangkok, 476–489.

MACDONALD, A. S., BARR, S. M., DUNNING, G. R. & YAOWANOIYOTHIN, W. 1993. The Doi Inthanon metamorphic core complex in NW Thailand: Age and tectonic significance. *Journal of Southeast Asian Earth Sciences*, **8**, 117–126.

MAGUNG, H. 1987. Transcurrent movements in Burma–Andaman Sea region. *Geology*, **13**, 941–961.

MAGUNG, H. 1989. Comments and Reply on 'Transcurrent movements in Burma–Andaman Sea region'. *Geology*, **15**, 911–912.

METCALFE, I. 1998. Palaeozoic and Mesozoic geological evolution of the SE Asian region: Multidisciplinary constraints and implications for biogeography. *In*: HALL, R. & HOLLOWAY, J. D. (eds) *Biogeography and Geological Evolution in SE Asia*. Backhuys, Leiden, 25–41.

METCALFE, I. 2002. Permian tectonic framework and paleography of SE Asia. *Journal of Asian Earth Sciences*, **20**, 551–566.

MITCHELL, A. H. G. 1981. Phanerozoic plate boundaries in mainland SE Asia, the Himalayas and Tibet. *Journal of the Geological Society, London*, **138**, 109–122.

MITCHELL, A. H. G. 1989. The Shan Plateau and western Burma: Mesozoic–Cenozoic plate boundaries and correlations with Tibet. *In*: SENGOR, A. M. C. (ed.) *Tectonic Evolution of the Tethyan Region*. Kluwer, Dordrecht, 567–583.

MITCHELL, A. G. H. 1993. Cretaceous–Cenozoic tectonic events in the Western Myanmar (Burma)–Assam region. *Journal of the Geological Society, London*, **150**, 1089–1102.

MITCHELL, A. G. H., HIANG, T. & HTAY, N. 2002. Mesozoic orogenies along the Mandalay–Yangon margin of the Shan Plateau. *In*: MANTAJIT, N. & POTISAT, S. (eds) *Proceedings of the Symposium on the Geology of Thailand*. Department of Mineral Resources, Bangkok, 136–149.

MITCHELL, A. G. H., HTAY, M. T., HTUN, K. M., WIN, M. N., OO, T. & HIANG, T. 2007. Rock relationship in the Mogok metamorphic belt, Tatkon to Mandalay, central Myanmar. *Journal of Asian Earth Sciences*, **29**, 891–910.

MORLEY, C. K. 2001. Combined escape tectonics and subduction rollback–back arc extension: A model for the evolution of Cenozoic rift basins in Thailand, Malaysia and Laos. *Journal of the Geological Society, London*, **158**, 461–474.

MORLEY, C. K. 2002. A tectonic model for the Tertiary evolution of strike-slip faults and rift basins in SE Asia. *Tectonophysics*, **347**, 189–215.

MORLEY, C. K. 2004. Nested strike-slip duplexes, and other evidence for Late Cretaceous–Palaeogene transpressional tectonics before and during India–Eurasia collision, in Thailand, Myanmar and Malaysia. *Journal of the Geological Society, London*, **161**, 799–812.

MORLEY, C. K. 2007. Variations in Late Cenozoic–Recent strike-slip and oblique-extensional geometries, within Indochina: The influence of pre-existing fabrics. *Journal of Structural Geology*, **29**, 36–58.

MORLEY, C. K. & WESTAWAY, R. 2006. Subsidence in the super-deep Pattani and Malay basins of Southeast Asia: A coupled model incorporating lower-crustal flow in response to post-rift sediment loading. *Basin Research*, **18**, 51–84.

MORLEY, C. K., SANGKUMARN, N., HOON, T. B., CHONGLAKMANI, C. & LAMBIASE, J. 2000. Structural evolution of the Li Basin, northern Thailand. *Journal of the Geological Society, London*, **157**, 483–492.

MORLEY, C. K., WOGANAN, N., SANKUMARN, N., HOON, T. B., ALIEF, A. & SIMMONS, M. 2001. Late Oligocene–Recent stress evolution in rift basins of Northern and Central Thailand: Implications for escape tectonics. *Tectonophysics*, **334**, 115–150.

MORLEY, C. K., WONGANAN, N., KORNASAWAN, A., PHOOSONGSEE, W., HARANYA, C. & PONGWAPEE, S. 2004. Activation of rift oblique and rift parallel pre-existing fabrics during extension and their effect on deformation style: Examples from the rifts of Thailand. *Journal of Structural Geology*, **26**, 1803–1829.

MORLEY, C. K., GABDI, S. & SEUSUTTHIYA, K. 2007a. Fault superimposition and linkage resulting from stress changes during rifting: Examples from 3D seismic data, Phitsanulok Basin, Thailand. *Journal of Structural Geology*, **29**, 646–663.

MORLEY, C. K., SMITH, M., CARTER, A., CHARUSIRI, P. & CHANTRAPRASERT, S. 2007b. Evolution of deformation styles at a major restraining bend, constraints from cooling histories, Mae Ping Fault zone, Western Thailand. *In*: CUNNINGHAM, W. D. & MANN, P. (eds) *Tectonics of Strike-slip Restraining and*

Releasing Bends. Geological Society, London, Special Publications, **290**, 235–249.

NIELSEN, C., CHAMOTE-ROOKE, N., RANGIN, C. & ANDAMAN CRUISE TEAM 2004. From partial to full strain partitioning along the Indo-Burmese hyper-oblique subduction. *Marine Geology*, **209**, 303–327.

O'LEARY, H. & HILL, G. S. 1989. Tertiary basin development in the Southern Central Plains, Thailand. *Proceeding of the International Conference on Geology and Mineral Resources of Thailand*. Department of Mineral Resources, Bangkok, 1–8.

PERTAMINA BPPKA, (ed.) 1996. *Petroleum Geology of Indonesia Basins. Volume 1: North Sumatra Basin*. Pertamina BPPKA, Jakarta.

PIGOTT, J. D. & SATTAYARAK, N. 1993. Aspects of sedimentary basin evolution assessed through tectonic subsidence analysis; Example: Northern Gulf of Thailand. *Journal of Southeast Asian Earth Sciences*, **8**, 407–420.

PIVNIK, D. A., NAHM, J., TUCKER, R. S., SMITH, G. O., NYEIN, K., NYUNT, M. & MAUNG, P. H. 1998. Polyphase deformation in a fore-arc/back-arc basin, Salin subbasin, Myanmar (Burma). *AAPG Bulletin*, **82**, 1837–1856.

POLACHAN, S. & RACEY, S. 1994. Stratigraphy of the Mergui Basin, Andaman Sea—implications for petroleum exploration. *Journal of Petroleum Geology*, **17**, 374–406.

POLACHAN, S., PRADIDTAN, S., TONGTAOW, C., JANMAHA, S., INTARAWIJITR, K. & SANGSUWAN, C. 1994. Development of Cenozoic basins in Thailand. *Marine and Petroleum Geology*, **8**, 84–97.

PRADIDTAN, S. & DOOK, R. 1992. Petroleum geology of the northern part of the Gulf of Thailand. *In*: PIANCHAROEN, C. (ed.) *Proceedings of a National Conference on Geological Resources of Thailand: Potential for Future Development*. Department of Mineral Resources, Bangkok, 235–246.

PRADIDTAN, S. & TONGTAOW, C. 1984. Cenozoic basins with petroleum potential in Thailand. *Conference on Applications of Geology and the National Development*. Chulalongkorn University, Bangkok, 34–43 [in Thai].

PUTTHAPIBAN, P. 1992. The Cretaceous–Tertiary granite magmatism in the west coast of peninsular Thailand and the Mergui Archipelago of Myanmar/Burma. *In*: PIANCHAROEN, C. (ed.) *Proceedings of a National Conference on Geological Resources of Thailand: Potential for Future Development*. Department of Mineral Resources, Bangkok, 75–88.

PUTTHAPIBAN, P. 2002. Geology and geochronology of the igneous rocks of Thailand. *In*: MANTAJIT, N. & POTISAT, S. (eds) *Proceedings of the Symposium on the Geology of Thailand*. Department of Mineral Resources, Bangkok, 261–283.

RACEY, A., LOVE, M. A., CANHAM, A. C., GOODALL, J. G. S., POLACHAN, S. & JONES, P. D. 1996. Stratigraphy and reservoir potential of the Mesozoic Khorat Group, NE Thailand. *Journal of Petroleum Geology*, **19**, 5–40.

RADHA KRISHNA, M. & SANU, T. D. 2000. Seismotectonics and rates of active crustal deformation in the Burmese arc and adjacent regions. *Journal of Geodynamics*, **30**, 401–421.

RAO, N. P. & KALPNA, 2005. Deformation of the subducted Indian lithospheric slab in the Burmese arc. *Geophysical Research Letters*, **32**, L05301, doi: 10.1029/2004GL022034.

RATANASTHIEN, B. 1989. Depositional environment of Mae Lamao Basin as indicated by palynology and coal petrology. *In*: THANASUTHIPITAK, T. & OUNCHANUM, P. (eds) *Proceedings of the International Symposium on Intermontane Basins: Geology and Resources*. Chiang Mai University, Chiang Mai, 205–215.

RATANASTHIEN, B. 2002. Problems of Neogene biostratigraphic correlation in Thailand and surrounding areas. *Revista Mexicana de Ciencias Geologicas*, **19**, 235–241.

REMUS, D., WEBSTER, M. & KEAWKAN, K. 1993. Rift architecture and sedimentology of the Phetchabun Intermontane Basin, Central Thailand. *Journal of Southeast Asian Earth Sciences*, **8**, 421–432.

REPLUMAZ, A. & TAPPONNIER, P. 2003. Reconstruction of the deformed collision zone between India and Asia by backward motion of lithospheric blocks. *Journal of Geophysical Research*, **108**, 101029–101053.

RHODES, B. P., BLUM, J. & DEVINE, T. 1997. Geology of the Doi Suthep metamorphic core complex and adjacent Chiang Mai Basin. *In*: DHEERADILOK, P., HINTHONG, C. ET AL. (eds) *Proceedings of the International Conference on the Stratigraphy and Tectonic Evolution of Southeast Asia and the South Pacific*. Bangkok, 305–313.

RHODES, B. P., BLUM, J. & DEVINE, T. 2000. Structural development of the mid-Tertiary Doi Suthep metamorphic core complex and western Chiang Mai basin, northern Thailand. *Journal of Asian Earth Sciences*, **18**, 97–108.

RHODES, B. P., CONEJO, R., BENCHAWAN, T., TITUS, S. & LAWSON, R. 2005. Palaeocurrents and provenance of the Mae Rim Formation, Northern Thailand: Implications for tectonic evolution of the Chiang Mai basin. *Journal of the Geological Society, London*, **162**, 51–63.

SAHU, V. K., GAHALAUT, V. K., RAJPUT, S., CHADHA, R. K., LAISHRAM, S. S. & KUMAAR, A. 2006. Crustal deformation in the Indo-Burmese arc region: Implications from the Myanmar and Southeast Asia GPS measurements. *Current Science*, **90**, 1688–1692.

SCHOONMAKER, A. & KIDD, W. S. F. 2006. Evidence for a ridge subduction event in the Ordovician rocks of north–central Maine. *Geological Society of America Bulletin*, **118**, 897–912, doi:10.1130/B25867.1.

SCHWARTZ, M. O., RAJAH, S. S., ASKURY, A. K., PUTTHAPIBAN, P. & DJASWADI, S. 1995. The Southeast Asian Tin Belt. *Earth-Science Reviews*, **38**, 95–293.

SEARLE, M. P., NOBLE, S. R., HURFORD, A. J. & REX, D. C. 1999. Age of crustal melting, emplacement andexhumation history of the Shivling leucogranite, Garhwal Himalaya. *Geological Magazine*, **136**, 513–525.

SEARLE, M. P., NOBLE, S. R., COTTLE, J. M., WATERS, D. J., MITCHELL, A. H. G., HLAING, T. & HORSTWOOD, M. S. A. 2007. Tectonic evolution of the Mogok metamorphic belt, Burma (Myanmar) constrained by U–Th–Pb dating of metamorphic and magmatic rocks. *Tectonics*, **26**, TC3014, doi:10.1029/2006TC002083.

SHAPIRO, N. M., RITZWOLLER, M. H. & ENGDAHL, R. E. 2008. Structural context of the Great Sumatra–Andaman Islands earthquake. *Geophysical Research Letters*, **35**, L05301, doi:10.1029/2008GL033381.

STAMPFLI, G. M. & BOREL, R. 2004. The TRANSMED transects in time and space. *In*: CAVAZZA, W., ROURE, F., SPAKMAN, W., STAMPFLI, G. M. & ZIEGLER, P. A. (eds) *The TRANSMED Atlas: The Mediterranean Region from Crust to Mantle*. Springer, Heidelberg, 53–80.

TAPPONNIER, P., PELTZER, G. & ARMIJO, R. 1986. On the mechanism of collision between India and Asia. *In*: COWARD, M. P. & RIES, A. C. (eds) *Collision Tectonics*. Geological Society, London, Special Publications, **19**, 115–157.

TEYSSIER, C. & WHITNEY, D. L. 2002. Gneiss domes and orogeny. *Geology*, **30**, 1139–1142.

THORKELSON, D. J. & TAYLOR, R. P. 1989. Cordilleran slab windows. *Geology*, **17**, 833–836, doi:10.1130/0091-7613(1989)017.

TORRES, C., SWAUGER, D. A., BERGMAN, S., TAPPONNIER, P., LACASSIN, R. & REPLUMAZ, A. 1997. The Sagaing Fault in Myanmar: Preliminary field observations and relevance to the Gulf of Martaban Cenozoic Petroleum System. *In*: HOWES, J. V. C. & NOBLE, R. A. (eds) *Proceedings of the Petroleum Systems of SE Asia and Australasia Conference 1997*. Indonesian Petroleum Association, Jakarta, 335.

UPTON, D. R. 1999. *A regional fission track study of Thailand: Implications for thermal history and denudation*. PhD thesis, University of London.

UPTON, D. R., BRISTOW, C. S., HURFORD, C. S. & CARTER, A. 1997. Tertiary denudation in Northwestern Thailand. Provisional results from apatite fission-track analysis. *In*: DHEERADILOK, P., HINTHONG, C. ET AL. (eds) *Proceedings of the International Conference on the Stratigraphy and Tectonic Evolution in Southeast Asia and the South Pacific*. Department of Mineral Resources, Bangkok, 421–431.

UTTAMO, W., ELDERS, C. F. & NICOLS, G. J. 2003. Relationships between Cenozoic strike-slip faulting and basin opening in northern Thailand. *In*: STORTI, F., HOLDSWORTH, R. E. & SALVINI, F. (eds) *Intraplate Strike-slip Deformation Belts*. Geological Society, London, Special Publications, **210**, 89–108.

VIGNY, C., SOCQUET, A. ET AL. 2003. Present-day crustal deformation around Sagaing fault, Myanmar. *Journal of Geophysical Research*, **108**, doi:10.1029/2002JB001999.

VIGNY, C., SIMONS, W. J. F. ET AL. 2005. Insight into the 2004 Sumatra–Andaman earthquake from GPS measurements in southeast Asia. *Nature*, **436**, 201–206.

WATANASAK, M. 1990. Mid Tertiary palynostratigraphy of Thailand. *Journal of Southeast Asian Earth Sciences*, **4**, 203–218.

WATCHARANANTAKUL, R. & MORLEY, C. K. 2000. Syn-rift and post-rift modelling of the Pattani Basin, Thailand, evidence for a ramp-flat detachment. *Marine and Petroleum Geology*, **17**, 937–958.

WATKINSON, I., ELDERS, C. & HALL, R. 2009. The kinematic history of the Khlong Marui and Ranong Faults, Southern Thailand. *Journal of Structural Geology*, **30**, 1554–1571.

WHITTAKER, J. M., MULLER, R. D., SDROLIAS, M. & HEINE, C. 2007. Sunda–Java trench kinematics, slab window formation and overriding plate deformation since the Cretaceous. *Earth and Planetary Science Letters*, **255**, 445–457.

ZAW, K. 1989. Comments and reply on 'Transcurrent movements in the Burma–Andaman Sea region'. *Geology*, **17**, 93–97.

Index

Page numbers in *italic* denote figures. Page numbers in **bold** denote tables.

Aasiaat domain 195, **199**, 206, 210, *218*, 223
Aasivik terrane **200**, 209
Abitibi–Wawa superterrane, Neoarchaean 170–176, **172–173**
Abloviak shear zone 220
Abner Formation 216
Abukuma Belt 18
ACCRETE project 262
accretionary margins, OMSZs 78, 113
accretionary orogens 1–4, *4*, **106**, 237
    characteristics 5
    crustal structure 5–12
    primitive 342, 345
    sediments 16–17
    types 12–14
accretionary prisms 7, 14–16, 17
    Canadian Cordillera 257, 259, 261–269
    Indo–Burma complex 375, 377, 382–383, 397
accretionary wedges, Fennoscandia 251–252
adakite, Archaean **156**, 162–163, **179**, 180, *181*, *182*
advancing orogens 12, 13–14, 42, 237
Aillik group **200**, 213, 221
Aishihik batholith 324, 325
Aishihik metamorphic suite 324
Akia terrane 60, *159*, 170, **200**, 209
Akilia association 133, **139–140**
Aksala Formation *312*
Åland, seismic profile *248*
Aleutian intra-oceanic arc subduction zone *107*, 111, 119
    crustal productivity 110, 112
    frontal prism 113
Aleutian Trench 7, *8*
Alexander terrane *258*, *274*, *276*
    Palaeozoic 284–285, 293–294, 299
Alice Springs orogeny 5
Alpine–Himalayan collisional orogen 1, *2*, 37, *38*, 157, **158**
Altaid–Cordilleran collisional orogen 157, **158**
Ambler arc *281*, 282
Ameralik dykes *130*, 131, *136*
Amîtsoq gneisses 131
amphibolite
    boninitic
        Isua supracrustal belt 138, **140**, 144, *145*
        Itsaq Gneiss Complex 138, *145*
Anap nunâ Group 195, **199**, 204
Andaman Sea region, tectonic evolution 380–382

Andean continental margin arc subduction zone *see* continental margin arc subduction zone
Andes
    crustal shortening 13–14
    metamorphism 50
    tectonic erosion 80–82, *84*
andesite, high-magnesian, Archaean **156**, 163–164, **179**, 180, *181*, *182*
Anshan, China
    Eoarchaean crustal growth 150–151
        geochronology *137*
    orthogneiss **143–144**
    trondhjemite **143–144**
Antelope Mountain Quartzite *276*, 289, 290, 294
Antler Orogeny 275, *298*
Appalachian orogen, metamorphism 59
Appalachian–Ouachita orogen 5
arc accretion
    sediment recycling 91–94
    steady-state 88–91
arc collision, Molucca Sea 364
arc imbrication, Canadian Cordillera 309–326
arc magmatism 5, 22, 105
    Alexander terrane 284
    CSSZs 108, 112
    extinction 111
    Lewes River Arc 324–325
    OMSZs 110–112
    Peri–Laurentia terranes 275, 277–280
    rates 87–88
    Southeast Asia accretionary orogen 364–365
arc–passive margin collision 88–91
Archaean, hot subduction 156–157, 158–164, 180
Archaean crustal growth
    geodynamic processes **157**
    West Greenland 155–170, *159*
Archaean–Proterozoic metamorphism 54–56
Arctic Alaska terrane, Palaeozoic 281–283, 294
Arfersiorfik intrusive suite **199**, 210–211, *212*
Astarte River Formation **199**, 202
aulacogens 4–5

BABEL seismic profiles, Fennoscandia 12, 240, 245, *246*, *248*, 249, *250*
Bacchus Formation 216
back-arc basins 12–13
    heat flow 18
    Lachlan orogen *343*
    ophiolites 17

back-arc mobile belt model, Thailand and Myanmar 394–396
back-arc volcanism, Yukon–Tanana terrane 279–280
Baffin Island 195, *196*
  Foxe fold belt 195, *196*, 217
  Narsajuaq island-arc terrane 214–215
  Rae craton 198, **199**, 201, 202
Baffin suture *196*, 217, *218*, 223
Ballantrae ophiolite 17
Baltican affinity terranes 276, 280–292
Banda suture 352, *354*, 362
Banda Volcanic Arc *353*, 362, 364, 365
banded iron formation
  Isua greenstone belt 165, *165*
  Itsaq Gneiss Complex 131, *132*, 133
Barbados, OMSZ *109*, 113
Barbados Ridge 17
Barberton greenstone belt 22, 61
Barentsia 281
Barnard Glacier pluton 285
Barrovian metamorphism 57, 59, 223
basalt
  augite-phenocrystic, Tally Ho shear zone 317
  Lachlan orogen 340
  Nb-enriched, Archaean **156**, 163–164, **179**
  see also island arc basalt (IAB); mid-ocean ridge basalt (MORB); ocean island basalt (OIB)
basalt-andesite-dacite-rhyolite (BADR) associations 156, 160–161, 163, **179**, *180*, *181*
basins, post-rift
  Gulf of Thailand 388, 390–391
  inversion, Thailand 375, 388, *389*
batholiths, andean-type 3
Belt–Purcell basin 261
Bendigo Zone *334*, 335
Bergeron suture *218*, 223
Bergslagen microcontinent 238, 239, 245, 247, 249, 251, 252
  seismic profile *250*
Bird's Head microcontinent 362
blueschist-facies 47, 51, 58–59
Bolivian orocline 14
Bone Gulf *353*, 361
boninite
  Archaean **156**, 158, 160–161, *181*, *182*
  Isua supracrustal belt 138, **140**, 144, *167*
  Lachlan orogen 340–341
Bonnet Plume River Intrusions 261
Border Ranges assemblage 7, *8*
Borneo
  sedimentation 360
  subduction 360
Borneo suture 352, *354*, 358, 360–361
Bothnia microcontinent 238, 239, 245
  BABEL seismic profiles *246*

Bothnian basin 245
  BABEL seismic profiles *246*, *248*
Bothnian belt, seismic profiles *243*
Bravo Lake Formation **199**, 202
Brooks Range 7, 281, 282
Browns Fork orogeny 284, 298
Buchan metamorphism 57
Burma block
  collision with India 382–383
  west, collision with Sundaland 378, 380
Burwell arc 220
Buttle Lake Group 286

Cache Creek terrane *258*, 267, *274*
Calderian orogeny 257
Caledonian affinity terranes 276, 280–292
Canada, northeastern
  crustal components 195, 201–216
  Precambrian correlation 193–225
Canadian Cordillera
  accretionary margin
    heat flow 268–269
    magnetic field 263–264, *265*
    origin 267–269
    Poisson's ratio 263, 266
  Proterozoic 257–269
    stratigraphy 259–261
    seismic profiles 259, 262–266
    suspect terranes *258*, 259, *263*, 267, 269
  arc imbrication 309–326
  Peri-Laurentian terranes 275
Canadian Shield 257, *258*
Cape Smith belt 221–223
Cassier Platform *263*
Cassier terrane *258*
Central Asian Orogenic Belt 3, 4
  evolution 39
Central Basin, Myanmar 18, *374*, 375, *382*, 394
Central Finland granitoid complex *238*, 239, 244
  seismic profiles *242*–*243*, *247*
Centralian Supergroup 5
Chapperon Group 287
Chase Formation 288
Chiang Mai basin *378*, 391, *393*
Chugach terrane 7, *8*, 17
Chukotat Group 215
Chukotka Peninsula 276, 281–283
Churchill domain *194*, 195, 214–215, 221, 223
CIA crust 105–120, **106**
Coast Range Ophiolite 7, *10*
Coates Lake Group 261
Coldfoot subterrane *281*, 283
collision, thick-skinned, Stikinia terrane 325–326
collisional orogens 3–4, *4*, 38, **106**, 237
Colville Hills *260*

INDEX

continental crust
 generation 75–76
  subduction recycling 75–98, **77**
   rates 84–88, **86**, **87**
  thickness 95–96
continental debris 96–97, 116
continental growth 22–24
 Neoarchaean 43
continental hinge line, Canadian Shield 257, 258, *258*, 266
continental-margin (Andean) arc subduction zone **106**, *107*, 110, 118
Coppermine River Group *260*, 261
Cordillera Deformation Front *258*
Cordilleran accretionary orogen 1–2, *2*
 terrane accretion 20, 50
 see also Canadian Cordillera; North American Cordillera
Cordilleran-type orogens 38
Corn Creek orogeny 261
correlation, northeastern Canada and western Greenland 193–225
Costa Rica, tephra geochemistry 83–84, *85*
coupling, plate 18–19, *19*
cratonization 18–21
Crescent terrane 7, *10*
Crocker Land *282*, 296
Crocker turbidite fan 360
crust, structure 5–12
crustal-suturing subduction zone (CSSZ) **106**, 108–110, *109*, *111*
 arc magmatism 108, 112
 CIA balance 115–120
 recycling losses 115, 118
Culbertson Lake allochthon 291
Cumberland batholith *196*, **199**, 202–203, 206, 217, 220
cyclomedusoids, Antelope Mountain 289

de Pas batholith 215
deformation
 brittle, Tally Ho shear zone 322, 324
 Isua supracrustal belt *132*, 133
 Sulawesi 361
 Thailand and Myanmar 397
Delamerian orogen *331*, 336, *344*
Denault Formation 216
Descon Formation 284
Devon Island *196*
 Rae craton 202
Dewar Lakes Formation **199**, 202
Dickson Hill *313*, *316*, *318*, 320, 322
Disko Bugt suture *197*, 206, 210, 211, *218*, 220, 223
Dismal Lakes Group 261
Doi Inthanon gneiss 378, 391, 394, 395, 397
Dookie segment 337, **338**, **340**
Doonerak fenster *282*, 282, 294

Dorset fold belt 195, *196*, 220–221
Duncan Peak allochthon 291
dunite, Itsaq Gneiss Complex 144–146

Eastern Klamath terrane *276*, *277*, 289–291, 294
eclogite–high-pressure granulite metamorphism (E–HPGM) 43–48, 51–52, *53*, 55, 56, 62
Ediacaran fauna 289
Ellesmere Island *196*, *218*
 Rae craton 202
Ellesmere–Devon terrane 195, *196*, **199**, 202, 217
Ellesmere–Inglefield belt 217, *218*
Ellesmerian orogeny 283, *298*
Endicott Group 283
Eoarchaean
 crustal growth
  Anshan, China 150–151
  West Greenland 127–151
 Isua greenstone belt 164–167, 176
 metamorphism 60–61
erosion
 control of crustal thickness 95–96
 orogenic topography 96–97
 tectonic
  geochemistry 82–84
  mechanisms 76–80
  non-steady state 80–82, *84*
 see also subduction, erosion
Etah Group **199**, 202, 217
Etah metaigneous complex **199**, 202, 217
Eurasian margin, Southeast Asia 351–366
exhumation rates 92–94

Færingehavn terrane *130*, 131, 133, 149, *159*, **200**, 209
Farewell terrane *276*, 283–284, 294, 299
faulting
 Lachlan orogen 333
 normal, Thailand 391–394
 strike-slip, Shan Plateau 383–386
fauna
 Baltican affinity 284, 289, 290
 Laurentian affinity 281, 283, 285, 289, 290, 291
 McCloud 280, 286, 292, 298
 Siberian affinity 283, 284, 289, 290
Fennia orogen **240**, *241*, 245, 252
Fennoscandia
 geology *238*, 239
 Palaeoproterozoic 237–253, *238*
 seismic data 240, *242–243*, *246*, *247*, *248*, *250*
Finlayson assemblage 279
FIRE seismic profiles, Fennoscandia 12, 240, *242–243*, 244, *247*
flat-slab subduction 5, 19–20

Flint Lake Formation **199**, 202, *203*
folding
    Kapisilik–Tre Brødre terranes *166*
    Tally Ho shear zone 322
    Thailand 377
Forest Mountain terrane *289*
Fort Simpson terrane 257, 258, 268
Foxe basin *196*
Foxe fold belt 195, *196*, 217, 223
Franciscan assemblage 7, *10*
frontal prism **106**, *108*, 112–113

gabbro, Itsaq Gneiss Complex **139**, 144
geochemistry, tectonic erosion 82–84, *85*
geochronology
    Isua supracrustal belt 133, 135
    Itsaq Gneiss Complex 136, *137*, 138
    Tally Ho shear zone 319–320, **321**, *323*
Georgia Strait 7, *10*
gneiss *see* Itsaq Gneiss Complex
gneiss domes, Thailand 393–394, 397
Golconda allochthon 292
Gold Hill *315*, 317
    rhyolite *313*, *314*, 319
Gondwana
    assembly 20–21, 39, 59
    subduction 39
Gorontalo Bay *353*, 361
Governor Fault Zone *334*
granite
    Itsaq Gneiss Complex 131, 133
    megacrystic
        Tally Ho shear zone *314*, 319
            geochronology 319–320, **321**, *323*
    Myanmar and Thailand *376*, 377–378, 397
granite magmatism, Sundaland 356–357, 360
granitoids, Lachlan orogen 333, *334*, 336
granodiorite, Tally Ho shear zone *313*, *314*, 319
granulite–ultrahigh-temperature metamorphism
        (G–UHTM) 43, 45, 46, 48–49, 51–52,
        53, 54–55, 62
Great Valley ophiolite 7, *10*
Greenland *197*
Greenland Caledonides *274*
greenschist-facies
    Kittilä arc 252
    Mount Wellington Belt 337
greenstone belts 21–22
    Archaean 60, 61, 155–184
    Knob Hill Complex 287
    Phanerozoic analogues 181–183
Grenville orogenic system 268
    evolution 39
Gulf of Thailand *374*, 377
    rift basins
        initiation 391
        thickness 388, 390–391

Halmahera Arc *353*, 362–363, 364
Hälsingland belt *238*, 245, 249
Häme belt *238*, 239, 245, 249, 252
    seismic profiles *247*, *248*
Hammond Subterrane *281*, 282
harzburgite, Itsaq Gneiss Complex 144–146
Havallah squence 292
Hearn Block *194*, 195
Heathcote Belt 336, **339**, **340**
Heathcote Fault Zone *334*
high-pressure–ultrahigh-pressure metamorphism
        (HPM–UHPM) 43, 44, 45, 46, 47,
        51–52, *53*
    Wilson cycles 58, 59
Himalayan orogenic system, evolution 39
Himalayan–Karakoram–Tibetan orogen 225, 373
Himalayan-type orogens 38
Himalayas, erosion rate 96, **97**
Hoare Bay Group 195
Hoffman-type breakup 58, 59–60
Hopedale block **199**, 208, 209
hotspots
    subduction 19
    see also mantle plumes
Hottah arc terrane 12, 257
Howqua Accretionary Complex 333, *344*, 345
Howqua segment *334*, 337, **339**, **340**
Hushe gneiss 225
Hyland Group *260*

Iapetus Ocean
    magmatic arcs 3
    opening 39
Iggavik mafic dykes **200**, 213
Ikertôq thrust zone 210, *212*
Imandra–Varzuga suture *238*, 241
imbrication 15, 129
    Canadian Cordillera 309–326
    Lachlan orogen 333
Inaluk dyke *130*
Inari island arc *238*, 239
Inari terrane 241
India
    collision with Burma block 382–383
    collision with Myanmar 380–394
        deformation 397
Indian Craton 224
Indo-Burma ranges *374*
    accretionary prism complex 375, 377, 382–383,
        397
Indonesia, Eastern *354*, 362–363
Indonesian island arc 3
Indosinian orogeny 377
Indus-Yarlung suture *379*
Inglefield mobile belt 195, *197*, **199**, 202, 217, 223
inner prism **106**, *108*, *109*, 113
Insular Superterrane 284–286

intra-oceanic arc crust *see* CIA
intra-oceanic arc subduction zone **106**, 110
intracratonic orogens 3, *4*, 5
inversion cycles 42, 58
island arc basalt, Itsaq Gneiss Complex *132*, 138, **139–140**, 148
island arcs 3
    Fennoscandia 239, 244–249
    see also arc magmatism
Island Harbour Bay plutonic suite 221
Isua supracrustal belt 131, *132*, 133, *134*, 135
    amphibolite 138
    boninite **140**, *145*, *167*
    geochronology 133, 135
    greenstone belt 22
        Eoarchaean 164–167, 176
    island arc basalt 138, **139–140**
    metamorphism 60
    orthogneiss 135
    picrite 160, 161–162, *167*
Isukasia terrane 131, *159*, **200**, 209
Itsaq Gneiss Complex 60
    amphibolite 138, *145*
    Eoarchaean crustal growth 127–151, *128*, *132*
        cycles 149
        geochronology 136, *137*, 138, 150
        thermal characteristics 147
    geological setting 129, 131, 133
    tonalite *130*, 131, **141–142**, 146–147, 150
Ivisaartoq greenstone belt *159*, *166*
    Mesoarchaean 167, 169–170, *171*, 177
Izu–Bonin–Mariana subduction zone **106**, 110, 119
    crustal productivity 110–111, *112*

jadeite belt 378
Jamieson–Licola Arc 342, *343*, *344*
Jamieson–Licola assemblage *334*, 337, **338–339**, **340**, 341
Java Trench 352, *354*
Jormua ophiolite 245
    seismic profiles *242*
Juan de Fuca Plate 7, *10*
Julianehåb batholith **200**, 221, *222*

Kabaw Fault 383
Kanairiktok plutonic suite **200**, 208
Kangâmiut dyke swarm **199**, 210, *211*, *212*
Kangdese batholith 3
Kap York metaigneous complex **199**, 201
Kapisilik terrane *159*, *166*, **200**, 209
Karelian craton *238*, 239, *241*, 245, 251
    seismic profiles *242–243*
Karheen Formation 284
Karrat Group 195, **199**, 203, 204–206, *219*
Kechika fault *263*

Keitele microcontinent *238*, 239, 244, 245, 247, 251
    BABEL seismic profiles *246*
Ketilidian orogen 195, *197*, **200**, 211, 213–214, 221, 223
Khorat Group 385–386
Kigarsima nappe *219*
Kikkertavak mafic dykes **200**, 213
Kittilä allochthon *238*, 239, 244, 251, 252
Klakas orogeny 284, 285
Klinkit assemblage 279–280
Klondike assemblage 279–280
Knaften island arc *238*, 239, 245, 252
    BABEL seismic profiles 245, *246*
Knob Hill complex 287
Kobberminebugt shear zone 221
Kohistan island arc 3
Kohistan terrane 225
Kola craton *238*, 239
Kootenay terrane *258*, *278*
    arc magmatism 275, 277–278, 298
Kosciusko batholith 333

Laberge Group 267, 311, *312*, 324, 325, 326
Labrador 195, *198*
    North Atlantic (Nain) craton 208, 209
Labradorian orogeny 221
Lachlan orogen
    Central *331*, 333, 336, 345
    Eastern 330, *331*, 333, 336, 345
    metamorphism 50, 333
    metavolcanics *334*, 336–342, **340**
    Palaeozoic 329–345
        geology 329–336, *331*
    subduction 336, *344*
    tectonic evolution 13, *332*, 336, *343*, *344*, 345
    turbidite fan 330, 333, 336, *343*, *344*
    Western 3, 332–333, *334*, *335*, 336, 345
        isotopic data 337, **338–339**
Ladakh–Gangdese granite 225
Lake Harbour Group **199**, 206–207, *208*
Lang sequence 291
Lapland granulite belt 241
Lapland–Kola orogen 239, **240**, 252
    accretion 241, 251
Lapland–Kola suture 241
Lapland–Savo orogen **240**
    accretion 241, *241*, 244–245, 251, 252
Lardeau Group 277, 279
large igneous province 105, **106**, 119
Lauge Koch Kyst supracrustal complex **199**, 201
Laurentia
    Palaeoproterozoic 195, 216–223, *224*, 225
    Palaeozoic 273
    Precambrian correlation 193–225
    Proterozoic 261, 268
    subduction 296–300
    suture with Amazonia 39

lawsonite 47, 51
leucogabbro, Tally Ho shear zone 312, *313*, 314–315, *316*, 317, 324
  geochronology 319–320, **321**, *323*
Lewes River Arc 310, 324, 325
Lewes River Group 310, *312*, 324, 326
Lewisian complex, metamorphism 55–56
Lhasa block *224*
LILE abundance 17, 148
Lima Basin, tectonic erosion *81*, 82, *83*
LITHOPROBE SNorCLE seismic profiles *258*, 259, 262–263
lithosphere, bouyant, accretion 20
Little Dal basalts 261
Llewellyn fault zone 311–312, 324
Loch Maree Group, metamorphism 56
Longstaff Bluff Formation **199**, 202
LOTI **156**, 158, 160
low velocity structures 58, 59
Luzon Arc 88–91, **92**, 111

McCloud fauna 280, 286, 292, 297
McCloud Limestone 291
Mackenzie Mountains, seismic profiles *262*
Mackenzie Mountains Supergroup *258*, *260*, 261, *266*, 268
Macquarie Volcanic Arc 333, 342, *343*, *344*, 345
Mae Cham basin, coal mine 391, 395
Mae Ping fault zone *374*, *379*, 384, *385*, *386*
Maggo gneiss **200**, 208
magmatism *see* arc magmatism
Makassar Straits *353*, 361
Makkovik orogen 195, *198*, **200**, 211, 213, 214, 221, 223
Maligiaq supracrustal suite **199**, 210, *212*
Manila Accretionary Complex 90
Manitoba promontory *194*, 195
Manitouwadge greenstone belt *160*, 176, *177*, *178*, 180, *181*
mantle convection, effect of continental lithosphere 57–58
mantle plumes 19, 22, 116–117
Marble Bar greenstone belt 22
Mariana Ridge 110, *see also* Izu–Bonin–Mariana subduction zone
Marmorilik Formation **199**, 203, 204
Mary River Group *196*, 201
mélange 15, 16
  Lachlan orogen 333, 337
Melbourne Zone *334*, *335*, *344*
Melville Bugt orthogneiss complex **199**, 201
Menihek Formation 216
Mesoarchaean
  Ivisaartoq greenstone belt 167, 169–170, *171*, 177
  metamorphism 60–61
  Qussuk greenstone belt 170

Meta Incognita microcontinent 195, *196*, **199**, 206–207, *207*, *208*, *218*, 220, 223
metamorphism
  Archaean–Proterozoic 54–56
  Canadian Cordillera 259, 261, 263
  classification 43–47
    Miyashiro 43, 56
  Fennoscandia 252
  Lachlan orogen 50, 333
  Myanmar 383, 397
  paired belts 50, 56–57, 62
  patterns 18, 37–63
  post-Cretaceous 47–49
  post-Neoarchaean 51–56
  pre-Cretaceous Phanerozoic 49–51
  pre-Neoarchaean 60–61
  subduction-to-collision systems 49, 50–51, 57
  supercontinent cycle 57–60
metavolcanics, Cambrian
  Lachlan orogen *334*, 336–342, **340**
    geochemistry 340–342
Mid-Lithuanian suture zone 249
mid-ocean ridge basalt (MORB) 17
  Knaften arc 245
  Tampere belt 245
middle prism **106**, 113
Miyashiro-type orogens 38
Mogok metamorphic belt 383, 386–387, 388, 397
Molucca Sea
  arc collision 364
  subduction 363
Molucca suture 352, *354*, 362–363
Mongolia–North China craton *224*
Moran Lake group **200**, 213
Mount Hodnett *313*, *315*, *318*, 320
Mount Kinabalu 360
Mount Reed Belt 342
Mount Wellington Belt *334*, 336–337
  geochemistry 340–342
  isotopic data 337, **338–339**, 340–342
Mount Wellington Fault Zone *334*, 337
Mount Wheaton *313*, *314*, 322
Mugford Group **199**, 209
Murmansk craton *238*, 239, *241*
Muskwa assemblage *258*, 261, *266*, 268
  seismic profiles 262–263, *264*
Myanmar 373, *374*
  back-arc mobile belt model 394–396
  Cenozoic tectonic provinces 375, 377
  Central Basin *374*, 375, *382*, 394
  collision with India 380–394
    deformation 397
  Mesozoic–early Cenozoic evolution 377–380
mylonite
  Isua supracrustal belt 135
  Tally Ho shear zone 317, *318*, 319, 320, 322
  Thailand 393

INDEX

Nagssugtoqidian orogen 195, *197*, 203, 205, 206, 209–210, *211*, 217, 220
Nahanni terrane 257
Nankai subduction zone 6–7
Nanok plutonic remnants **200**, 208
Narooma Accretionary Complex 333, *343*, *344*
Narsajuaq island-arc terrane 195, *196*, 214, 214–215, *218*, 223
Naternaq supracrustal belt **199**, 205
Nazca Plate 14
NEB *see* basalt, Nb-enriched
Neoarchaean
    Abitibi–Wawa superterrane 170–176, **172–173**
    continental growth 43
New Britain arc 111, 112, 119
New England orogen *331*, *344*, 345
New Quebec Orogen *198*, 215, 221
Nias forearc 363
Nicola horst 287
Nisling Assemblage 310, 312, 324–326
non-uniformitarian models 183–184
nonaccreting margins, OMSZs 113
Nordic orogen **240**, *241*, 251
Nordre Isortoq steep belt *218*, 220
Nordre Strømfjord supracrustal suite **199**, 205, 211, *212*
Norrbotton craton *238*, 239
North America, Western margin, crustal structure 7, *8*, *10*
North American Cordillera
    palaeogeography 292–300
    subduction 297
    terrane accretion 20
        Palaeozoic–Mesozoic 273–300, *274*
North American Plate 7, *8*, *10*
North Atlantic (Nain) Craton *194*, 195, *197*, *198*, **199**, 207–209, *218*
    south margin 211, 213–214
    western and northern margins 209–211
North Kara terrane 293
Northern Sierra terrane *276*, *277*, 291–292
Northwest Passage 295–297, *296*, 299
Northwest Territories, seismic profiles 262–263
Nuka Formation 283
Nûkavsak Formation **199**, 203, 204, *205*, *219*
Nulliak plutonic remnants **200**, 208
Nunatarsuaq supracrustal rock **199**, 205
Nuttio serpentinite and dunite 244

oblique accretion 12
ocean island basalt (OIB) 17
ocean-margin subduction zone (OMSZ) **106**, *107*, *111*
    accretionary margins *78*, 113
    arc magmatic productivity 110–112, 117

CIA balance 115–120
CIA recycling 107–108, *108*, *109*
    losses 112–115, **114**, 117–118
    nonaccretionary margins 113
Okanagan terrane *277*, *278*
    Palaeozoic 286–288, 294
Okinawa Trough 88, *89*, 91, **92**
Oklahoma aulacogen 5
Old Tom–Shoemaker assemblage 287
Olyutorsky arc complex 111
Ontong–Java Plateau 20, 111, 112
ophiolites 17, 21
    Bergeron suture 223
    Borneo 360
    Eastern Klamath terranes 289–290
    Lapland–Savo orogen 244
        Jormua 245
    Mount Wellington Belt 337
    Sulawesi 361
orogenesis 18–21
orogens, classification 3–5
orthogneiss
    Anshan, China **143–144**
    Isua supracrustal belt 135
    Itsaq Gneiss Complex *130*, 131, 133
    Nagssugtoqidian orogen 210
    Trail gneiss complex 287
    West Greenland 164–170
Oskarshamn–Jönköping belt *238*, 239, 249, 251, 252
Outokumpu area, seismic profiles *243*

Pacific Plate 7, *8*
    rollback 13
Pacific Rim
    back-arc basins 18
    metamorphism 50, 56
    orogenic belts 37, *38*
Pacific Rim terrane 7, *10*
Pacific-type orogens 38
palaeogeography, North American Cordillera 292–300
Palaeoproterozoic
    correlation Canada and West Greenland 195–225
    Fennoscandia 237–253
    metamorphism 55–56
Palau–Kyushu ridge 110
Pangaea, assembly 20–21, 57
Pangaean convection cell 57
Panthalassa 275, *296*
Panthalassan convection cell 57
Parafusulina 280
Parent Group 214
passive margin collision, sediment recycling 76, **77**, 88–91
Pearya terrane *274*, *277*, *281*, 282, 294–295

Pechenga–Imandra–Varzuga suture *238*, 241
Peninsula batholith, California 3
Peninsular terrane 7, *8*
Peri-Laurentian terranes 275–280
Peterman orogeny 5
Phanerozoic
    sediment recycling 75–98
    supercontinent cycle 57–59
Philip Island **339**, **340**
picrite, Archaean **156**, 161–162, 166–167, 177, *182*
Piling/Hoare Bay group 195, *196*, **199**, 202, 204, 217
pillow lava
    Isua supracrustal belt 131, *132*, 133, 165, *165*
    Ivisaartoq greenstone belt *166*
    Mount Wellington Belt 337
Pinguicula Group *260*, 261
Pirkanmaa belt 245, 247, 252
    seismic profiles *247*
Pistolet Subgroup 216
plate margins
    active *78*
    erosive *78*
    passive 76, **77**
plate tectonics
    establishment 21–22, 42–43
    orogenic systems 40–43
    pre-Neoarchaean 61–62
    reorganization 19, 20–21, 40
plumes *see* mantle plumes
Pontiac superterrane 172
Post Hill group **200**, 213
Povoas Formation *312*
Povungnituk Group 215, *216*
Prince Albert Group 201, 210
Prince William terrane 7, *8*
Proterozoic
    Canadian Cordillera accretionary margin 257–269
    metamorphism 55–56
Prøven igneous complex **199**, 206, 217
pull–push inversion cycles 42
Purcell Mountains 275
push–pull inversion cycles 42, 58
pyroxenite, Tally Ho shear zone *313*, *316*, 317, 324

Qeqertarssuaq Formation **199**, 203–204
Qiangtang block *224*
quartz diorite **142**, *313*, *314*, 319
Quebec promontory *194*, 195
Quesnellia terrane *258*, 267, 273, *274*, *278*
    Palaeozoic 286
Quidam terrane *224*
Qussuk greenstone belt, Mesoarchaean 170

Racklan Orogeny 259, 268
Rae Craton *194*, 195, *196*, *197*, **199**, 217
    Archaean, correlation 198, 201
    Palaeoproterozoic 202–206, 223, *224*, 225
Rajang Group 360
Ramah Group **199**, 209
Redding subterrane 290
retreating orogens 12–13, 42, 237, 252
Revsund granitoids 251
rhyolite, Tally Ho shear zone *313*, 319
ridge subduction 19, 50
ridge–trench interaction 40–41, 42, 59
rifting, Thailand 388–394, 398
Rinkian fold belt 195, *197*, 203, 217, *219*, 220
Roberts Mountains allochthon 275, 292
Rodinia 57, 59
rollback, trench 12–13, 14, 362
Romanzof orogeny *282*, 282–283, 296
Roo Rise 363, 364
Ruth Formation 216
Ryoke Belt 18
Ryukyu Arc 88, *89*, 90, *91*, **92**

Sabah orogeny 360
Sagaing Fault *376*, 377, 378
Sagaing Fault–Sumatran Fault strike-slipe system *374*, 375, 377
Saglek block **200**, 208, 209
Sambagawa Belt 18, 50
San Francisco Bay 7, *10*
Sangihe Arc *353*, 362–363, 364
Sarong suture 352
Savo island arc belt *238*, 239, 244, 252
    seismic profiles *242–243*
Schreiber–Hemlo greenstone belt *160*, 173–176, *177*, *178*, *181*
Schwaner Mountains 356, *357*, 364
seamounts, subduction 16, 364
sediments 16–17
    Proterozoic, Canadian Cordillera 259, *260*, 261
    recycling 23, 75–98
        arc accretion 91–94
        rate 84–88
        subduction 23, **107**, 108, 113–115, **114**, 116
Seward Peninsula 281–283, 294
Seward Subgroup 216
Shan Plateau *374*, 377, 380, 383–386, *383*, *393*, 395, 396, 397
Shan Scarp 377
Shan-Tai block 18, 397
shear 42
Shoofly Complex *277*, 291–292
Siberian affinity terranes *276*, 280–292
Sicker Group 286
Sierra City mélange 291–292, 294
Sierran foothills 7, *10*
Sikuleh terrane 378, 380

Silver Creek Formation 287, 288
Sisimiut charnockite suite **199**, 211
Skellefte district *238*, 239, 245, 249
slab 'graveyards' 57–58
Slave Craton *194*, 195, 257, *258*
Slide Mountain ocean 298–299
Slide Mountain terrane *258*, *274*, 275, *278*, 279–280, 298
Snake River Fault 261
SNorCLE seismic profiles *258*, 259, 262–263, *262*, 263, *264*
Snowcap assemblage 279, 298
Snyder Group **199**, 209
Sokoman Formation 216
Solomon arc massif 111, 112
Songpan–Ganzi terrane *224*
   crustal structure 7, 11
Sonoma orogeny 298
Soper River suture *196*, 214, *218*, 220–221
Sorong fault zone 362–363
Sortis group **200**, 214, 221
South American Cordillera, crustal shortening 13–14
South China Block *224*
Southeast Asia accretionary orogen 351–352, *354*, 363–366
   arc magmatism 364–365
   continental fragments 365
   Eurasian margin 351–366
   future growth 365–366
Spartan Group 214
Stawell Zone *335*
Stikinia terrane *258*, 266, 267, 273, *274*
   geology 310–312
   lithology 312–319
   subduction 325
   Tally Ho shear zone 309–326
   thick-skinned collision 325–326
Strel'na terrane 241
subduction *4*
   accretionary prisms 14–16
   Borneo 360
   channels 14, *15*, **107**, *108*, 109, *109*
   continental crust recycling 23, 76
      rates 84–88, **86**, **87**
   continental margin 5
   erosion 23, **107**, 108, 114–115, 116, 364
      mechanisms 76–80
   flat-slab 19–20
   hot 156–157, 158–164, 180–181
   Lachlan orogen 336, *344*
   Laurentia 296–300
   Molucca Sea 363
   North American Cordillera 296–300
   one-sided 39, *40*, 42–43
   retreating orogens 13
   sediment recycling 23, 75–98, **107**, 108, 113–115, **114**, 116

Stikinia terrane 325
Sulawesi 361
Sumatra–Andaman Trench 375, 395, 398
Sunda Arc 358, 362
Sundaland 356, 358, 363–364
thermal environments 39, 43, 49
two-sided *40*
western Laurentia 275
subduction zones
   choking 20, 38, 51
   crustal recycling 105–120, **106**
subduction-to-collision systems 49, 50–51, 57
Sula Spur 361
Sulawesi suture 352, *354*, 361, 365
   ophiolites 361
Sumatra–Andaman Trench *374*, 375, 395, 398
Sunda Arc 352, *353*, 361
   subduction 358, 362
Sunda suture 352, *354*, 358, 363
Sunda Trench 352, *354*, 358, 365, 380
Sundaland 351, *354*, 355
   accretion 363–364, 395
   continental core evolution 356–358
   granite magmatism 356
   structural elements 352, *353*, *354*
   subduction 356, 358, 363, 395
Sunrise Pynt shear zone 202, 217
supercontinent cycles 57–60, 116
Superior Craton *194*, 195, *196*, 215–216, *218*, 221–223
   western 55
Superior Province 155, *160*
   Archaean, adakite 163
   Neoarchaean 170–180
suspect terranes, Canadian Cordillera accretionary margin *258*, 259, *263*, 267
Svecobaltic orogen **240**, *241*, 249, 251, 252
Svecofennian orogen 237, **240**, *241*
   seismic data 12
Swampy Bay Subgroup 216

Taiwan
   arc accretion 88–91
      sediment recycling 91–94
Takhini deformation zone 310–311, 324
Talkeetna–Bonanza arc system 111, 112
Tally Ho mountain *313*, *314*, *316*, 320, 322
Tally Ho shear zone 309–326
   brittle deformation 322, 324
   folding 322
   geochronology 319–320, **321**, *323*
   geology *313*, *314*, *315*
   lithology 317, 319
   mylonite 317, *318*, 319, 320, 322
   structure 317, 319, 320–324
Taltson–Thelon magmatic zone 217

Tampere belt *238*, 239, 245, 247, 252
    seismic profiles *247*
Tasiusarsuaq terrane *159*, **200**, 209
Tasiussaq Bugt 206
Tasiuyak gneiss 206–207, 209–210, *218*, 220
Tasmanides 329, *330*, *331*
    metamorphism 49–50
tectonic mode switching 14, 40, 42
tectonics *see* plate tectonics
Terra Australis orogen 5
    evolution 39
    rollback 13
terrane accretion 20
terrane export 58
Tersk terrane *238*, 239, 241, 252
Tethysides, metamorphism 42
Thailand 373, *374*
    back-arc mobile belt model 394–396
    Cenozoic rift basins 377
    Cenozoic tectonic provinces 375, 377
    central extensional province 388–394
    extension 397–398
    low-angle normal faults 391–394
    Mesozoic–early Cenozoic evolution 377–380
    Palaeogene folding 377
    thermal gradient 395
    volcanism 395–396
thermal environment 39, 43–47, 49, 56–57
tholeiite, low-Ti *see* LOTI
Three Pagodas fault zone *374*, *379*, 384
Thule mixed gneiss complex **199**, 201
Tibetan plateau 225
    erosion rate 96, **97**
Timanide orogen 293
Tintina fault zone *263*
Tintina–Northern Rocky Mountain trench fault 266
tonalite
    Isua supracrustal belt 135
    Itsaq Gneiss Complex *130*, 131, **141–142**, 146–147, 150
tonalite–trondhjemite–granite suites (TTGs)
    formation 61
    intrusion 129, 147, 148, 150, 164
Tonanki subduction zone 6–7
Tonga–Kermadec subduction zone 110, 112, 119
Top Crocker Unconformity 360
Torngat orogen 195, *198*, 209, 220, 223
torque balance 40–41, 42
Trail gneiss complex 287
Trans-European suture zone *238*
Trans-Himalayan batholith 225
Trans-Hudson orogen 3, *194*, 195, 223, 225, 268
    Wilson cycle 58
transpression 42, 295
Transscandinavian igneous belt *238*, 239, 251, 252
    seismic profile *250*

transtension, dextral 386–388
Tre Brødre terrane *159*, *166*, **199**, 209
trench migration 41–42
    retreat rates 85, **86**
Trinity terrane 288–290, 292
triple junction 20
trondhjemite, Anshan, China **143–144**
Tsherkidium 284
tuff, silicic 17
turbidites 16–17
    see also Lachlan orogen turbidite fan
Turkic-type orogens 38

ultrahigh-pressure metamorphism (UHPM) 43, 44, 46, 47, 50, 54, 59
ultrahigh-temperature metamorphism (UHTM) 43, 44, 46, 51
Umba granulite terrane 241
uniformitarian models 183–184
Upernavik supracrustal rocks **200**, 208
Upper Arrow Lake 287–288
Uummannaq Island *204*
Uusimaa belt *238*, 239, 245, 249
    seismic profiles *247*
Uviak gneiss **200**, 208

Vallen group **200**, 213–214
Vancouver Island 7, *10*
volcaniclastic unit, Tally Ho shear zone 317

Wabigoon Subprovince, adakite, Archaean 163
Wagga–Omeo Metamorphic Complex 333, *334*, 344
Wales orogeny 284
Wann River shear zone 311, 312, 324
Waratah Bay **339**, **340**
Wasatch Front 257
Watts Group 214, 223
Wawa Subprovince *160*
    Neoarchaean 170, 172–176, **172–173**, 177, *178*, **179**, 180
    see also Abitibi–Wawa superterrane
Wernecke Mountains *260*
Wernecke Supergroup *258*, 259, *260*, 261, 268
West Coast Plutonic Complex 7, *10*
West Greenland
    Archaean crustal growth 155–170, *159*
    Eoarchaean crustal growth 127–151, *128*
    greenstone belts 164–170
    metamorphism 60
    North Atlantic craton 207–209
    Precambrian correlation 193–225
    Rae craton **199**, 201, 203–206
Wheaton Valley granodiorite *313*, *314*, 319
White Sea Assemblage 289

Whitehorse Trough 267, 310–311, *312*, 324, 325, 326
Wilson cycles 1, 3, *4*, 58–59
Windermere Supergroup *258*, *260*, 261, *266*
Winston Lake greenstone belt *160*, 176, *178*, 180, *181*
Wishart Formation 216
Wopmay collisional orogen 257, *258*, 268
  crustal loss 115
Wrangel Island 281–283
Wrangellia terrane 7, *8*, *10*, *258*, *274*, *277*
  Palaeozoic 285–286
Wyoming Craton *194*, 195

Yakutat terrane 7, *8*
Yreka terrane 289–290, 293
Yukon
  Proterozoic stratigraphy 259, 261
  seismic profiles 262–263
Yukon–Tanana terrane *258*, 267, 273, *274*
  arc magmatism 275, 280, 298–299